A FIELD GUIDE TO THE PLANTS OF
ALBERTA, MANITOBA AND SASKATCHEWAN

# BUDD'S FLORA
## OF THE
## CANADIAN PRAIRIE
## PROVINCES

Revised and enlarged by
J. Looman
Research Station
Swift Current, Saskatchewan
and
K.F. Best
Research Station
Regina, Saskatchewan

Research Branch
Agriculture Canada
Publication 1662   1987

**Budd's Flora of the Canadian Prairie Provinces**
*A Field Guide to the Plants of Alberta, Manitoba and Saskatchewan*

*by* Archibald C. Budd

ISBN: 978-1951682583

**AN ORCHARD INNOVATIONS REPRINT EDITION**
Printed in the United States of America
Contains information licensed under the Open Government License – Canada.

*Below:* Original publication data from *Budd's Flora of the Canadian Prairie Provinces* first published in 1979 and updated in 1987.

Catalogue No. A53-1662/1979          Canada: $38.50
ISBN 0-660-10233-1          Other Countries: $46.20

Price subject to change without notice
First printed 1979
Reprinted 1981
Revised 1987

## A tribute to A. C. Budd

Archibald Charles Budd was born in London, England, on 28 April 1889. He attended Bellenden Road Higher Grade School, where as top boy he earned a merit certificate. He took commercial training at Camberwell Grammar School, and attended Choumert Road Evening School and Kings College. From 1905 until 1909 he clerked in the Civil Service at the Customs House and the General Post Office.

A prize of 250 pounds for winning a limerick contest made it possible for Mr. Budd to emigrate to Canada in 1910 as a land seeker. He first worked as a farmhand at Waldeck, Saskatchewan, and later homesteaded and farmed at Burnham until 1922.

A lifelong interest in nature study helped him when he began working at the Dominion Experimental Station at Swift Current, Saskatchewan, in 1926. As an assistant to the late Dr. Sidney Barnes of the Soils Research Laboratory, Mr. Budd assisted in studies on the taxonomy of weeds and the physiology of weed seeds until 1944. Although he had no formal training in botany, Mr. Budd was a keen student of taxonomy. An avid reader of botanical works and an astute observer, he soon became an authority on the prairie flora.

3

From 1944 until his retirement in 1957, Mr. Budd was Range Botanist and he made an extensive study of plant life in the prairie region. As curator of the herbarium, he built up one of the finest plant collections in Western Canada. Mr. Budd also helped train many professional agrologists, who are managing and appraising rangeland and are carrying out ecological studies requiring a knowledge of applied botany.

For many years Mr. Budd was a member and director of the Saskatchewan Natural History Society. He often contributed articles to *The Blue Jay*, the magazine published by the Society. He assisted in the establishment of a herbarium in the Museum of Natural History in Regina, Saskatchewan, and he built up a large personal collection of plants, which he later donated to the Swift Current Museum.

In 1949 he prepared a preliminary draft of a key to prairie plants, which he called *Flora of the Farming and Ranching Areas of the Canadian Prairies*. Mimeographed copies were enthusiastically received by amateurs and professional botanists throughout Western Canada. In 1952 a revised and expanded version entitled *Plants of the Farming and Ranching Areas of the Canadian Prairies* was published by the Experimental Farms Service of the Canada Department of Agriculture. After many revisions and the addition of more drawings, *Wild Plants of the Canadian Prairies* was published in 1957, the year of Mr. Budd's retirement. Until his death on 30 December 1960, he cooperated on revisions that were incorporated in the editions printed in 1964 and 1969.

As a tribute to Archie and his faithful and devoted work, which was the basis for this edition, the title of this first major revision in which he did not have a direct input has been changed to:

*Budd's Flora of the Canadian Prairie Provinces*

J. Looman and K. F. Best

4

# Contents

# Preface

This publication, like its predecessor, *Wild Plants of the Canadian Prairies*, is intended for amateur botanists, agricultural representatives, farmers, and ranchers. For this reason, emphasis has been placed on keys for identification of species. However, the book will also be useful to professional botanists.

This publication is not just a revised edition of *Wild Plants of the Canadian Prairies*; it has also been very much expanded. The scope of the work has been greatly increased to include the areas of the Boreal forests and the Rocky Mountains. These additions were considered necessary because, since the last revision of *Wild Plants*, many parts of the Prairie Provinces have become accessible to travelers. The sections on grasses, sedges, and willows have been completely rewritten and expanded.

An attempt has been made to include all native species presently known to occur in the Prairie Provinces, as well as species, native or introduced, which are likely to be found along roads, railroads, rivers, and lakeshores; but undoubtedly some species have been missed. Introduced species, which occur occasionally as escapes from cultivation but do not persist outside the garden environment, are mentioned but not described.

Further changes include measurements given in metric units (SI, International System of Units) and photographs for illustrative purposes. In addition, rare plants have been identified as such.

The basis for rarity is the number of occurrences in the Prairie Provinces, as well as the number of plants per occurrence. Thus, a plant indicated as rare is known either from only a few locations in an area, or from several locations, but each with few plants. Very rare plants are known from very few places—perhaps only one or two—and are not abundant even there.

Please do not collect very rare and rare species. If identification of a species is uncertain, send a set of good photographs to a professional botanist for positive identification.

The 1987 revision was undertaken by Dr. J. Waddington of the Swift Current Research Station. It includes changes to the key for the chloripetalous dicotyledons (p. 26), the sympetalous dicotyledons (p. 29), the Leguminosae (p. 463), and group 3 of the Compositae (p. 704). A list of additions and corrections to the 1979 edition is provided on p. 804, including cross-references to the increasingly common European–Russian classification of perennial Triticeae. Finally, the index has been modified to reflect these changes.

*Metric conversion:*
*0.9 m   = 1 yard*
*15 cm   = 6 inches*
*2.5 cm  = 1 inch*
*25 mm   = 1 inch*

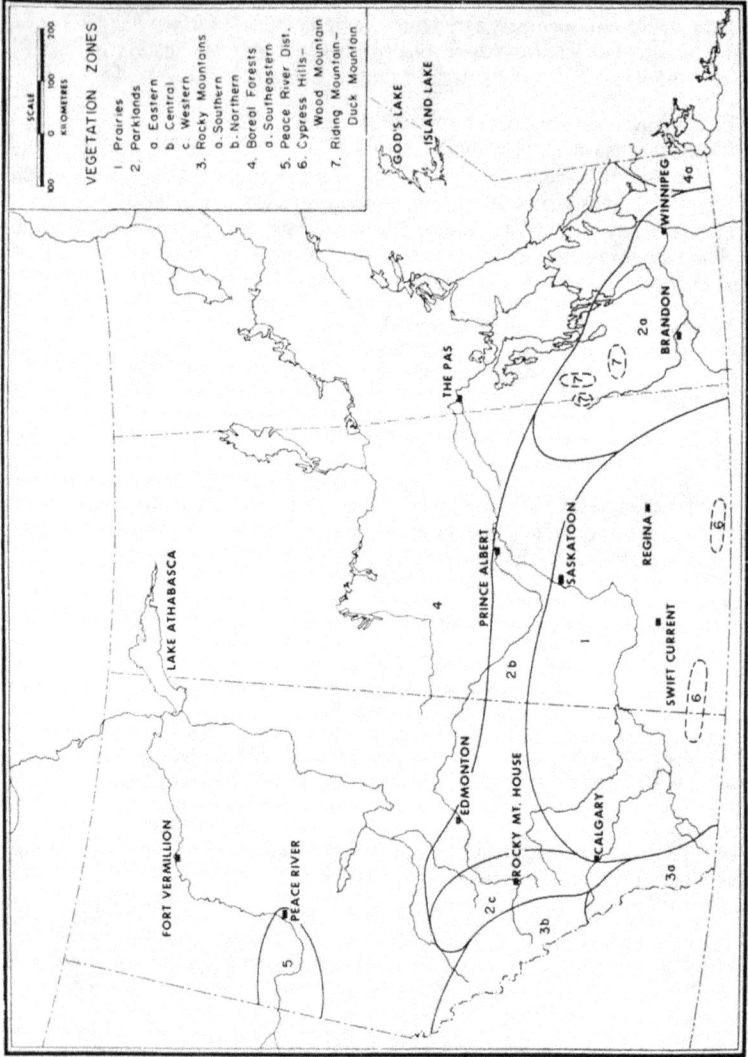

VEGETATION ZONES

1 Prairies
2 Parklands
  a Eastern
  b Central
  c Western
3. Rocky Mountains
  a. Southern
  b. Northern
4. Boreal Forests
  a. Southeastern
5. Peace River Dist.
6. Cypress Hills
7. Riding Mountain–
  Wood Mountain
  Duck Mountain

SCALE
100   0   100   200
KILOMETRES

GOD'S LAKE

ISLAND LAKE

WINNIPEG

BRANDON

THE PAS

LAKE ATHABASCA

PRINCE ALBERT

SASKATOON

REGINA

SWIFT CURRENT

FORT VERMILLION

PEACE RIVER

EDMONTON

ROCKY MT. HOUSE

CALGARY

Fig. 1.  Vegetation zones.

8

# Zones of vegetation

Manitoba, Saskatchewan, and Alberta occupy an area of almost 2 000 000 km². As a result of differences in climate and soil type, several vegetation types can be distinguished, each with some species occurring mainly there. Therefore, these vegetation types can be used to designate species distribution. The major types outlined in Fig. 1 follow.

## The Prairies

The Prairies include the almost treeless grasslands in the southern parts of Saskatchewan and Alberta. Rainfall in this area is low, summers are warm, and the vegetation is composed mainly of drought-tolerant species. Most of the grassland is short-grass prairie, with spear grass, blue grama, June grass, and other low-growing plants predominating. On eroded areas several uncommon, specially adapted cushion plants can be found, and the three native cacti belong to this vegetation type. Transitions to the midgrass prairie occur at the boundaries of the Prairies and on slopes with north to east exposures. Shrubs and trees occur almost entirely in deep coulees and ravines.

## The Parklands

The Parklands form a broad belt around the Prairies. They consist of open grassland alternating with tree groves. This belt can be divided into three sections: eastern, central, and western Parklands. In the eastern Parklands, the grassland is tall-grass prairie, and big bluestem, porcupine grass, prairie cord grass, and switch grass predominate. Tree groves contain bur oak, ash, Manitoba maple, and balsam poplar. The central Parklands have midgrasses in the grassland openings, which consist of rough fescue, western porcupine grass, and Hooker's oat grass, and on the light high-lime soils, little bluestem. Tree groves consist predominantly of aspen poplar and often include some willows. In the western Parklands, grassland is midgrass prairie, containing fescues, spear grass, and poverty oat grass. Tree groves are predominantly aspen poplar, often interspersed with willows and balsam poplar. Drought is not as prevalent in the Parklands as in the Prairies; soils are Dark Brown in the southern parts and Black in the more northern parts.

## The Boreal forest

The Boreal forest is the vast area covered with mostly coniferous forest situated north of the Parklands. Its southern margin intergrades with the Parklands, but farther north the Boreal forest is dominated by white spruce on

9

uplands, black spruce on lowlands, and jack pine on light sandy soils. Areas that have been logged in recent years are covered with deciduous forest, in which aspen poplar, balsam poplar, and birch are the main species. Wetlands are sedge–reed swamps, grading into muskeg, where large tussocks of peat moss, Labrador tea, willows, swamp birch, larch, and black spruce occur. An area in southeastern Manitoba varies from the general area in that it supports white pine, white cedar, ground hemlock, and other eastern or southeastern species. In Manitoba and northeastern Saskatchewan the Boreal forest includes outcrops of Precambrian rock, on which several ferns and saxifrages with numerous species of lichens can be found.

## Rocky Mountains

The western boundary of the Parklands and part of the Boreal forest grade into the Rocky Mountains. In the mountains the climate varies sharply with differences in elevation. In the lower montane zone to about 1500 m, fescue prairie occupies exposed slopes. Coniferous forest of spruce, lodgepole pine, Douglas-fir, and western red cedar, aspen poplar woods, or mixed forest are found on less exposed slopes and in ravines. At higher altitudes the open grasslands gradually disappear, to make place for vegetation dominated by shrubs, mainly willows and ground birch, and for the coniferous forest. At altitudes of about 2200–2400 m, the upper tree limit is reached, and alpine meadows, with many low sedges and grasses, mosses, and lichens predominating. The Rocky Mountains can be divided into the southern section, reaching north to approximately the Calgary–Banff line, and the northern section above this line. Several plant species, such as common camas, bear-grass, Pursh's silky lupine, and balsamroot, occur only in the southern section.

## Exceptional areas

Besides the major vegetation types, there are several types that occur in areas where the climate or soil differs from that prevailing in the area dominated by a major vegetation type. The most important ones follow.

## Cypress Hills — Wood Mountains

This ridge of hills rises above the surrounding Prairies with elevations ranging from about 870 m to 1350 m. The ridge is interrupted repeatedly by gaps of lowland, but in the lower elevations the vegetation is similar throughout. Midgrass prairie intergrades with fescue prairie and aspen poplar groves in coulees and ravines. At elevations above 1200 m the grasslands become pure fescue prairie, and above 1300 m, coniferous forest, similar to that of the southern Rocky Mountains, occurs.

## Riding Mountain — Duck Mountain

These areas of hills and plateaus rise 200–300 m above the surrounding plains in western Manitoba and eastern Saskatchewan. The vegetation is mainly coniferous forest, similar to that of the Boreal zone to the north. However, exposed areas and slopes support Parkland vegetation interspersed with small areas of fescue prairie vegetation.

## Peace River district

With a climate warmer and drier than what is customary for this latitude, the Peace River district offers a Parkland landscape, with southern exposed slopes and other exposed areas covered with midgrass prairie, dominated by spear grass. Somewhat eroded areas contain brittle prickly-pear, as well as several other species characteristic of the southern prairies. Tree groves are dominated by aspen poplar, with willows, chokecherries, pin cherries, and particularly Saskatoon berries often plentiful.

## Sand dune areas

Several areas of sandy soils and mobile sand dunes are characterized by a vegetation type that is quite different from that of the heavier soils surrounding them. The vegetation of sand dune areas depends on the zone in which they occur. The Middle, Great, and Little Sandhills in southeastern Alberta and southwestern Saskatchewan are in the Prairies, and are characterized by short-grass prairie on stabilized dry sand; sand grass, sand dropseed, Indian rice grass, and sand dock in open sand; little bluestem on moist calcareous sand; and large areas of western snowberry, silverberry, choke cherries, and aspen poplar in moist dune valleys. Several other species of shrubs occur in small groups.

In the Parklands, sand areas are mostly in midgrass prairie, with rough fescue and western porcupine grass dominating the dry locations and little bluestem the moister locations. Aspen poplar mixed with some balsam poplar and willows occupy the sites that have good moisture conditions. In the eastern Parklands, tall grasses dominate the grasslands in the sand dunes. These include big bluestem, porcupine grass, and switch grass. Willows, aspen poplar, and bur oak form the tree groves, with white spruce gradually taking over if the tree cover remains undisturbed.

## Saline areas

Many large and small areas with poor drainage and areas around lakes and sloughs have a high to very high salt content, often visible by a white salt

incrustation when they are dry. The vegetation on these areas is made up o
species with special adaptations for the extreme saline conditions. Most of th
species may also be found along the seacoasts, or they have affinities with
species found there: sea-blite, red samphire, seaside arrow-grass, and variou
salt grasses, which are indigenous in these areas.

## How to use the plant key

Plants are classified into various groups according to their structure an
method of reproduction. The first of these are divisions, which are separate
into subdivisions. The subdivisions are divided into classes, and the classe
into families. Families are split into genera, and each genus contains one o
more species. The species is sometimes further divided into forms, subspecies
and varieties.

The keys that follow contain paired contrasting statements of particula
characteristics, and a plant must agree with one of them. Find out which of th
pair it agrees with and then continue to the name or number following tha
statement until you find the family name. Go through the key to the family
then the genus, and finally the species. Continue to follow the statements unt
you find the species.

Imagine that the plant in the sketch (Fig. 2) is actually in front of you an
that you have no idea what it is and you want to identify it.

Turn to p. 20. Begin at the first key, and do not worry about the scientifi
names. If there are any terms or expressions that you do not understand, tur
to the Glossary and the diagrams.

Clause 1.   Because it obviously has flowers and will undoubtedly have seed
            it is in the SPERMATOPHYTA division. Go to the next entry
            numbered 2.
Clause 2.   It is certainly not an evergreen tree bearing cones, therefore it mus
            be an ANGIOSPERM. Go to Clause 3.
Clause 3.   Because the leaves are not parallel-veined, the plant must belong t
            the DICOTYLEDONEAE.

Now look through the section "KEY TO THE FAMILIES" and find th
portion headed "Class: DICOTYLEDONEAE" (p. 23). If you look carefull
at the flower, you will find that the petals are not entirely separate but ar
joined near the base, therefore the plant is a SYMPETALOUS dicotyledor
section C. Turn to the heading of that section on p. 29.

Clause 1.    It is obviously a herb and not a shrub or tree; go to Clause 8.
Clause 8.    It is a green plant; turn to Clause 11.
Clause 11.   The flowers are neither in heads nor in spikes; go on to Clause 19
Clause 19.   The ovary is superior, as you will find by referring to the Glossar
             and the diagram of flower parts; go to Clause 25.
Clause 25.   The petal portion (corolla) is regular (*see* Glossary); go to Claus
             28.

Fig. 2.    A specimen (*Lysimachia ciliata* L.) illustrating the use of the key.

Clause 28. It does not have a milky juice; go to Clause 30.
Clause 30. It is certainly not a twining plant; go to Clause 31.
Clause 31. The stamens are directly in front of the petals, therefore it should belong to the PRIMULACEAE, or primrose family, which is described on p. 580.

Turning to p. 580, under PRIMULACEAE, start again at Clause 1. Note that the leaves are not all basal, thus go to Clause 5.

Clause 5. The leaves are opposite; go to Clause 6.
Clause 6. Petals are present; go to Clause 7.
Clause 7. The stem leaves are normal; go to Clause 8.
Clause 8. The flowers are yellow, therefore it appears to belong to the genus *Lysimachia*, or loosestrife.

The genera are in alphabetical order; you will find *Lysimachia* on p. 583. Again begin with Clause 1 of the key to the species, and note that the plant is flowering, and does not have bulblets in the upper axils of the leaves, so go to Clause 2.

Clause 2. The flowers are borne on separate stalks in the leaf axils; go to Clause 3.
Clause 3. The leaves are lanceolate (*see* diagram of leaf shapes, p. 15) and rounded at the base; go to Clause 4.
Clause 4. The petioles of the lower leaves are about 1 cm long and ciliate, therefore it is apparently a specimen of *Lysimachia ciliata* L., or fringed loosestrife. Now read the description of that species and you will find that it agrees in every way with the specimen. Apparently the keying has been correct.

The *L.* after the scientific name means that the Swedish botanist Linnaeus gave that name to this plant. The other name is a synonym, or another name, given to it by another botanist. In this case, because Rafinesque thought that the plant should be in a separate genus, *Steironema*, he took the specific name from Linnaeus and placed it after *Steironema*, but divided the credit for the name with Linnaeus, and therefore the authority is (L.) Raf. Because later botanists agreed with Linnaeus, Rafinesque's name for this plant became a synonym.

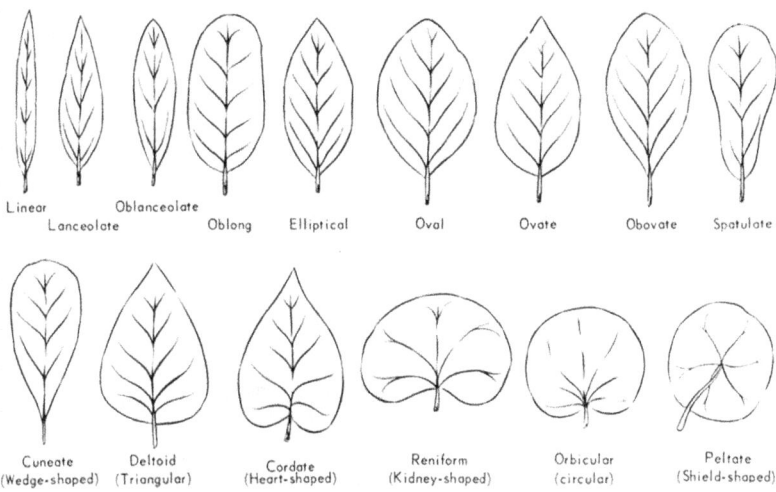

| Linear | Oblanceolate | | Elliptical | Oval | Ovate | Obovate | Spatulate |
| Lanceolate | | Oblong | | | | | |

A.C. Budd

Fig. 3.   Shapes of simple leaves.

| Cuneate (Wedge-shaped) | Deltoid (Triangular) | Cordate (Heart-shaped) | Reniform (Kidney-shaped) | Orbicular (circular) | Peltate (Shield-shaped) |

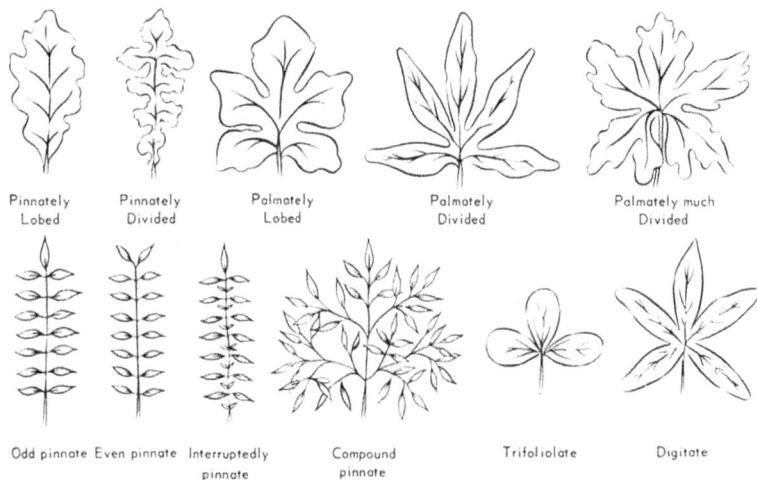

| Pinnately Lobed | Pinnately Divided | Palmately Lobed | Palmately Divided | Palmately much Divided |

| Odd pinnate | Even pinnate | Interruptedly pinnate | Compound pinnate | Trifoliolate | Digitate |

A.C. Budd

Fig. 4.   Types of divided leaves.

15

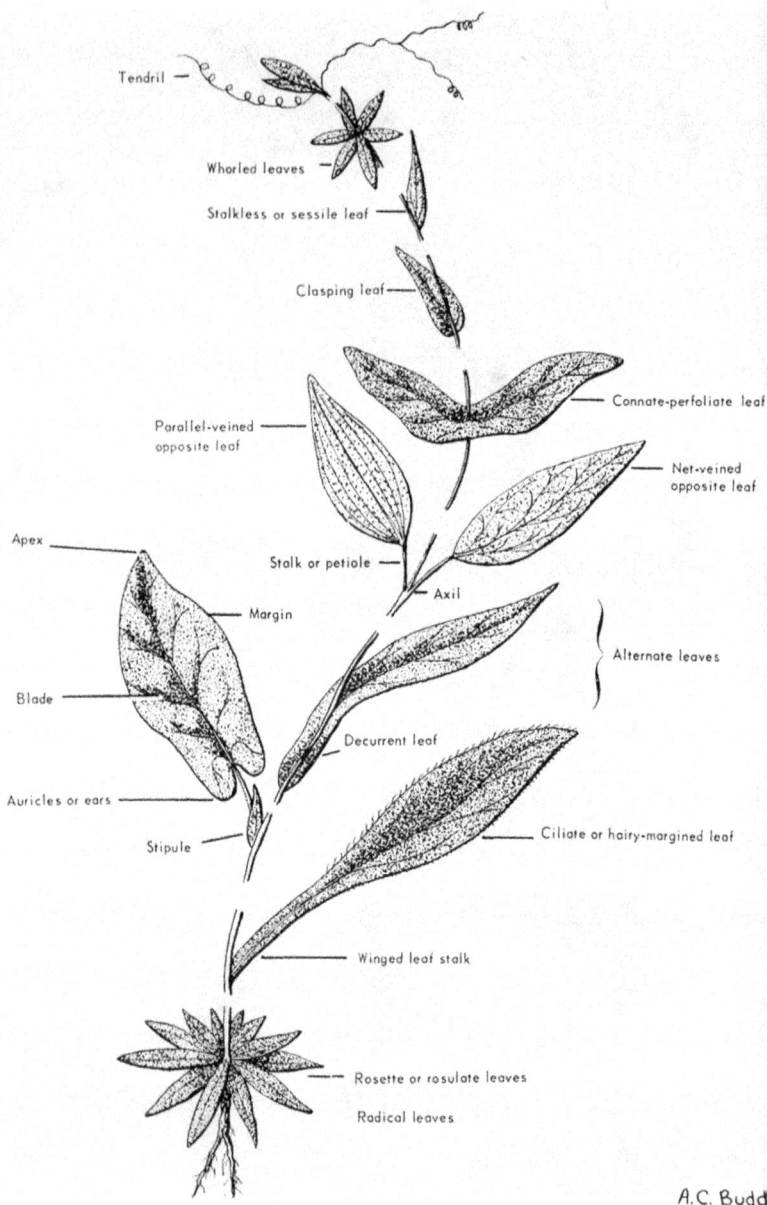

Tendril

Whorled leaves

Stalkless or sessile leaf

Clasping leaf

Parallel-veined opposite leaf

Connate-perfoliate leaf

Net-veined opposite leaf

Apex

Margin

Stalk or petiole

Axil

Blade

Alternate leaves

Decurrent leaf

Auricles or ears

Ciliate or hairy-margined leaf

Stipule

Winged leaf stalk

Rosette or rosulate leaves

Radical leaves

A.C. Budd

Fig. 5.   Leaf variations.

16

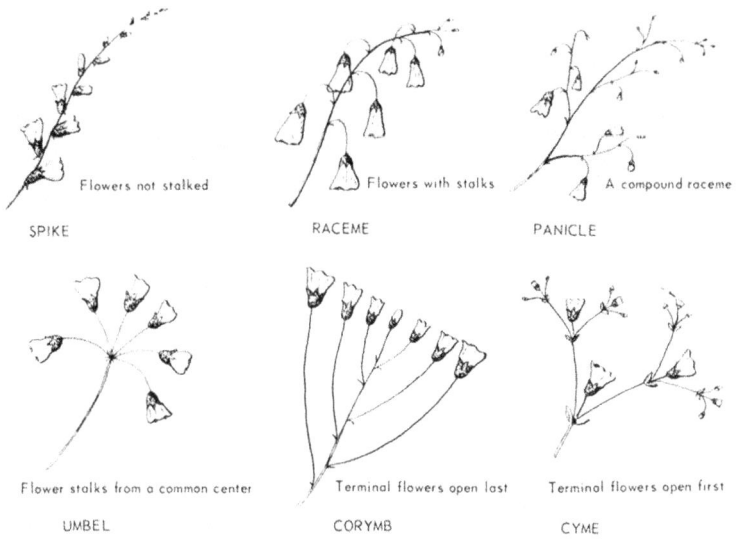

SPIKE — Flowers not stalked

RACEME — Flowers with stalks

PANICLE — A compound raceme

UMBEL — Flower stalks from a common center

CORYMB — Terminal flowers open last

CYME — Terminal flowers open first

A.C. Budd

Fig. 6.   Types of inflorescences.

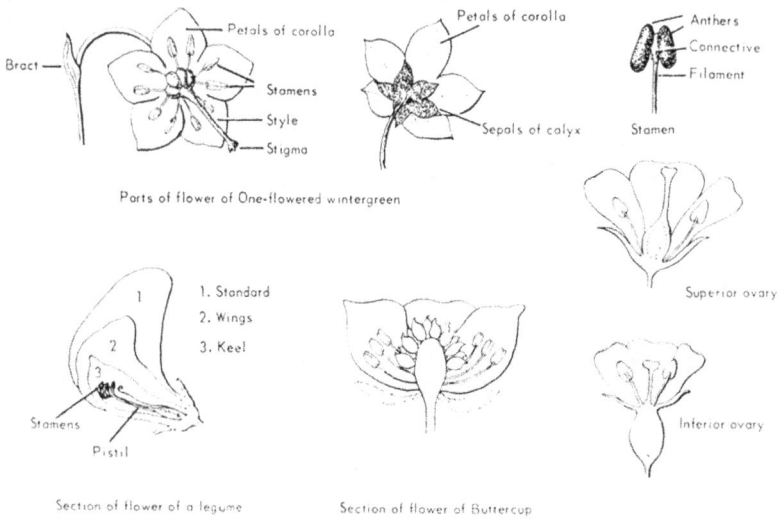

Parts of flower of One-flowered wintergreen

Bract — Petals of corolla — Stamens — Style — Stigma

Petals of corolla — Sepals of calyx

Anthers — Connective — Filament

Stamen

Superior ovary

Interior ovary

Section of flower of a legume

1. Standard
2. Wings
3. Keel

Stamens — Pistil

Section of flower of Buttercup

A.C. Budd

Fig. 7.   Flower parts.

17

**THREE-PETALED**
(Arrowhead)

**FOUR-PETALED**
(Mustard)

**FIVE-PETALED**
(Chickweed)

**MANY-PETALED**
(Purple cactus)

**URN-SHAPED**
(Bearberry)

**CYLINDRICAL**
(Gentian)

**CAMPANULATE**
(Harebell)

**FUNNELFORM**
(Morning-glory)

**SALVER-FORM**
(Collomia)

**ROTATE**
(Wild tomato)    (Bittersweet)

**REFLEXED PETALS**
(Shootingstar)

**PAPILIONACEOUS**
(Vetchling)

**BILABIATE**
(Marsh hedge-nettle)    (Monkeyflower)

**SPURRED**
(Toadflax)

(Violet)

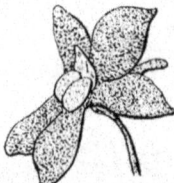
(Low larkspur)

(Leafy spurge)

(Lady's-slipper)

**IRREGULAR**

A.C. Bud

Fig. 8.  Types of flowers.

18

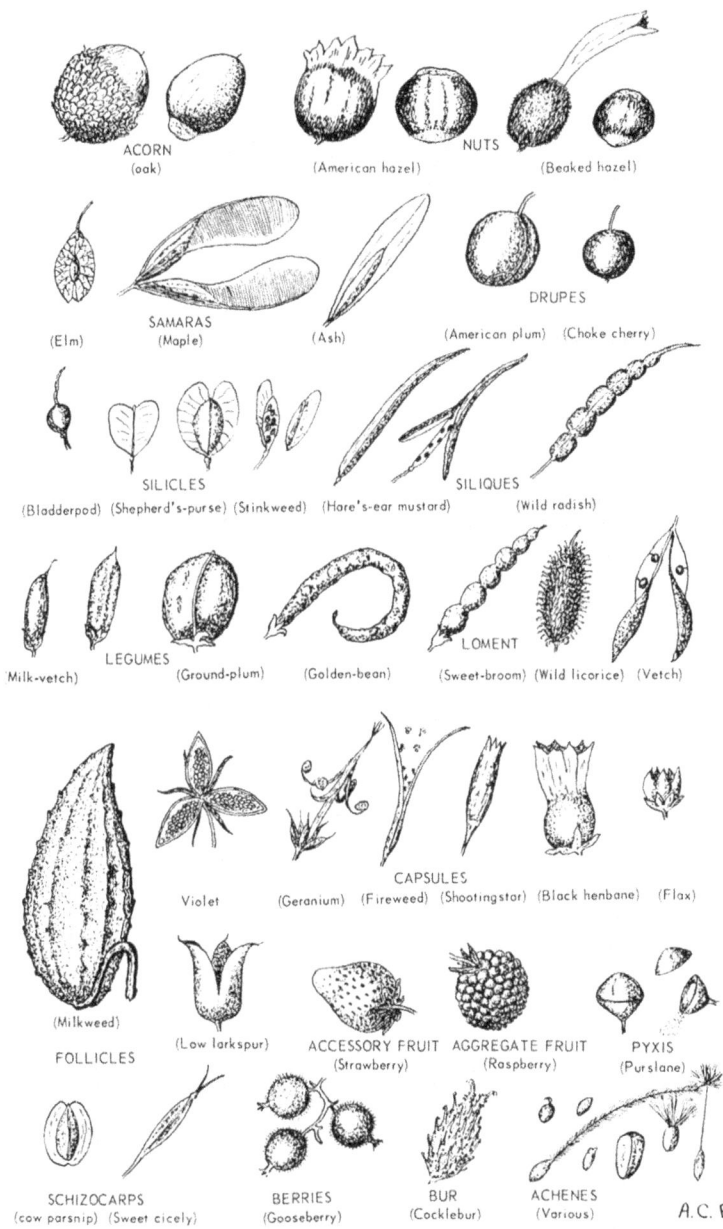

ACORN
(oak)

NUTS
(American hazel)      (Beaked hazel)

SAMARAS
(Maple)

(Elm)              (Ash)        DRUPES
(American plum)  (Choke cherry)

SILICLES
(Bladderpod) (Shepherd's-purse) (Stinkweed)    SILIQUES
(Hare's-ear mustard)      (Wild radish)

LEGUMES
(Milk-vetch)        (Ground-plum)      (Golden-bean)    LOMENT
(Sweet-broom) (Wild licorice) (Vetch)

CAPSULES
Violet   (Geranium)  (Fireweed)  (Shootingstar)  (Black henbane)   (Flax)

(Milkweed)
FOLLICLES
(Low larkspur)    ACCESSORY FRUIT    AGGREGATE FRUIT     PYXIS
(Strawberry)       (Raspberry)      (Purslane)

SCHIZOCARPS
(cow parsnip) (Sweet cicely)    BERRIES
(Gooseberry)        BUR
(Cocklebur)     ACHENES
(Various)      A.C. Budd

Fig. 9.   Types of fruits.

19

# Keys to main groups

## Key to Divisions, Subdivisions, and Classes

1. Plants without true flowers, reproducing by spores. ............................................ **Division: PTERIDOPHYTA, p. 32**

   Plants with flowers, reproducing by seeds. ............................................ **Division: SPERMATOPHYTA, p. 56** ....... 2

2. Plants bearing flowers having neither styles nor stigmas, the naked ovules (seeds) on the upper side of a scale. Mostly evergreen trees and shrubs bearing cones or cone-like fruit. ...................... **Subdivision: GYMNOSPERMAE, p. 56**

   Plants bearing flowers with the ovules enclosed in an ovary with styles or stigmas, and bearing seeds in a closed ovary. ................................................ **Subdivision: ANGIOSPERMAE, p. 61** .......... 3

3. Stems, when present, without central pith or annular layers; leaves usually parallel-veined; flower parts generally in threes or multiples of three; seeds with only one cotyledon or seed leaf. .................. **Class: MONOCOTYLEDONEAE, p. 61**

   Stems with a central pith, or if woody, the wood generally arranged in annular layers; leaves net-veined; seeds with two cotyledons or seed leaves. .................. **Class: DICOTYLEDONEAE, p. 279**

## Key to the Families

### Division: PTERIDOPHYTA—ferns and fern allies

1. Leaves reduced to scales, whorled at the internodes of strongly jointed stems, forming a sheath. ............................... **EQUISETACEAE, p. 47**

   Leaves reduced to green scales, awl-shaped, or larger, not forming a sheath; stems not strongly jointed. ............................... 2

2. Leaves linear, grass-like, sheathing at base. ................................ **ISOETACEAE, p. 50**

   Leaves not linear or grass-like. ................................ 3

3. Leaves scale-like or awl-shaped. ................................ 4

   Leaves larger and broader. ................................ 5

4. Sporangia borne in the upper axils of ordinary leaves, or in terminal terete strobiles; spores uniform in size. ..................... **LYCOPODIACEAE, p. 51**

   Sporangia borne in terminal four-sided strobiles; spores of two different sizes. .......... **SELAGINELLACEAE, p. 55**

5. Fronds (leaves) long-petioled, palmately
divided into 4 leaflets. ...................................... MARSILEACEAE, p. 47
Fronds not divided into 4 leaflets. .......................................................... 6

6. Plants with a single stalk, bearing a sterile
frond below, a fertile one above; spo-
rangia arranged in two rows. ......................... OPHIOGLOSSACEAE, p. 32
Plants not as above. ................................................................................. 7

7. Sporangia naked, borne on much
modified divisions at the tip or middle
of fronds. ......................................................... OSMUNDACEAE, p. 35
Sporangia covered, borne on fertile
fronds similar to the sterile ones or on
modified fronds. ............................................. POLYPODIACEAE, p. 35

## Division: SPERMATOPHYTA—seed-bearing plants

## Subdivision: GYMNOSPERMAE—plants with naked seeds

1. Cones reduced to a single ovule, fruit a
one-seeded berry. ........................................... TAXACEAE, p. 56
Cones many-seeded, fleshy or woody. ...................................................... 2

2. Leaves scale-like or awl-shaped, often
overlapping; opposite or whorled. .................. CUPRESSACEAE, p. 60
Leaves linear or needle-shaped, not over-
lapping; alternate or clustered. ........................ PINACEAE, p. 56

## Subdivision: ANGIOSPERMAE—plants with covered seeds

## Class: MONOCOTYLEDONEAE—monocotyledons

Stems, when present, without central pith or annual layers; leaves usually
parallel-veined; flower parts mostly in threes or multiples of three; seeds with
only one cotyledon (seed leaf).

1. Plants aquatic, floating or submerged, with
floating leaves or emersed inflorescence. ............................................... 2
Plants not aquatic, or if growing in water,
most of the plant emersed. ..................................................................... 4

2. Plants small, floating, leaf-like; without
differentiation between stems and
leaves. ............................................................. LEMNACEAE, p. 245
Plants larger, generally rooting; clear
differentiation between stem and leaves. ................................................ 3

3. Perianth absent or single and inconspicu-
ous; plants monoecious. ................................. ZOSTERACEAE, p. 65

# Class: DICOTYLEDONEAE—dicotyledons

Stems with a central pith, or, if woody, the wood generally arranged in annual layers; leaves net-veined; seeds with two cotyledons (seed leaves).

A. APETALOUS      Flowers with only one floral ring, with sepals but not petals (p. 23).

B. CHORIPETALOUS      Flowers with two floral rings, and with each petal distinct from the others (p. 25).

C. SYMPETALOUS      Flowers with two floral rings, but with the petals wholly or partly united forming a tube or bell (p. 29).


A. Apetalous dicotyledons
(flowers with only one floral ring, with sepals but not petals)

Stigmas 1; flowers with calyx; fruit
enclosed by 2–4 perianth segments;
leaves often bearing stinging hairs. ................ URTICACEAE, p. 304

11. Plants stemless with a single pair of large
reniform leaves; calyx 3-lobed. ...................... ARISTOLOCHIACEAE, p. 308

Plants with stems or stemless, but with
more than one pair of leaves. .......................................................................... 1

12. Leaves with stipules free or sheathing the
stem. ............................................................................................................................. 1

Leaves without stipules. ...................................................................................... 1

13. Leaves opposite; stipules membranous;
plant cushion-like. ............................................. CARYOPHYLLACEAE, p. 354
(*Paronychia*)

Leaves alternate; stipules forming a
sheath above nodes. ......................................... POLYGONACEAE, p. 308

14. Leaves opposite. ............................................................................................................. 1

Leaves alternate; lower ones sometimes
opposite. ..................................................................................................................... 18

15. Leaves scale-like; stem fleshy; flowers
imbedded in stem; plants of saline
areas. ................................................................. CHENOPODIACEAE, p. 338
(*Salicornia*)

Leaves not scale-like; calyx colored. ................................................................. 1

16. Plants climbing. ................................................... RANUNCULACEAE, p. 369
(*Clematis*)

Plants not climbing. ................................................................................................ 1

17. Erect herbs; inflorescence in panicles or
clusters. .............................................................. NYCTAGINACEAE, p. 344

Low plants; flowers solitary in leaf axils. ........... PRIMULACEAE, p. 583
(*Glaux*)

18. Fruit fleshy when ripe. ......................................... SANTALACEAE, p. 305
Fruit dry when ripe. ................................................................................................ 19

19. Leaves orbicular or reniform; sepals 4. ............... SAXIFRAGACEAE, p. 421
(*Chrysoplenium*)

Leaves not orbicular or reniform. ......................................................................... 2

20. Leaves compound or simple, but very
deeply divided. .................................................. RANUNCULACEAE, p. 364
(*Anemone*)

Leaves not compound and not very
deeply divided. ........................................................................................................ 2

21. Perianth segments 6, in two series; flowers
colored. ................................................................ POLYGONACEAE, p. 309
(*Eriogonum*)

Perianth segments 2–5; flowers small,
numerous, greenish. ............................................................................................. 2

22. Bracts and perianth segments dry and
membranous. .................................................... AMARANTHACEAE, p. 342

Membranous bracts absent; perianth seg-
ments greenish. ........................................ CHENOPODIACEAE, p. 324

23. Male flowers, at least, in catkins or
aments. ........................................................................................................ 24

Flowers not in catkins or aments. ........................................................ 27

24. Seeds each with a tuft of hairs. ........................... SALICACEAE, p. 279

Seeds without a tuft of hairs. ........................................................... 25

25. Styles 3 or more; fruit an acorn. ....................... FAGACEAE, p. 303

Styles 2; fruit not an acorn. ............................................................ 26

26. Three flowers in axil of each bract of male
catkin. ...................................................... BETULACEAE, p. 299

One flower in axil of each bract of male
catkin; fruit a nut. ........................................ BETULACEAE, p. 301
(*Corylus*)

27. Climbing plants; flowers with colored
sepals; fruit with a persistent feathery
style. ................................................................. RANUNCULACEAE, p. 369
(*Clematis*)

Plants not climbing; fruit without a feath-
ery style. ............................................................................................... 28

28. Leaves pinnately compound. ................................................................. 29

Leaves simple. ............................................................................................ 30

29. Fruit a double samara; leaflets mostly
3–5. ........................................................................ ACERACEAE, p. 515

Fruit a single samara; leaflets mostly
5–11. ...................................................................... OLEACEAE, p. 587

30. Leaves linear, evergreen, 2.5–7 mm long;
plants with decumbent stems. ......................... EMPETRACEAE, p. 514

Leaves wider, deciduous; plants with
erect stems. ........................................................................................... 31

31. Fruit fleshy when ripe. ................................................................................ 32

Fruit dry when ripe. ................................................................................... 33

32. Leaves silvery or brownish scurfy. ...................... ELAEAGNACEAE, p. 533

Leaves aromatic, not silvery or brownish
scurfy. .................................................................... MYRICACEAE, p. 299

33. Shrubs without stipules; branches spiny. ........... CHENOPODIACEAE, p. 339
(*Sarcobatus*)

Trees with stipules; flowers appearing
before leaves. ....................................................... ULMACEAE, p. 303


B.   Choripetalous dicotyledons
(flowers with two floral rings, and with each petal distinct from the others)

1. Succulent, spiny plants with leaves absent
or scale-like, inconspicuous. ........................... CACTACEAE, p. 531

Plants with leaves. ...................................................................................... 2

2. Leaves all basal, and tubular or with ten-
 tacles for catching insects. ................................................................. 3
 Leaves normal. ................................................................................... 4

3. Plants with solitary flowers; leaves tubu-
 lar with lid-like lobe on top. ................. SARRACENIACEAE, p. 418
 Plants with flowers in racemes; leaves flat
 with glandular tentacles. ........................... DROSERACEAE, p. 418

4. Herbs. ................................................................................................ 5
 Shrubs or trees. ............................................................................... 41

5. Plants aquatic; leaves submerged or
 floating. ........................................................................................... 6
 Plants terrestrial or semiaquatic. .................................................. 8

6. Leaves mostly floating, large, suborbicu-
 lar or reniform. ..................................... NYMPHAEACEAE, p. 361
 Leaves mostly submerged, small, coarsely
 to finely dissected. ......................................................................... 7

7. Flowers pedicellate, mostly white or yel-
 low, perfect. ......................................... RANUNCULACEAE, p. 362
 Flowers sessile, axillary, monoecious. .............. HALORAGACEAE, p. 543

8. Plants with a single ternately compound
 stem leaf, appearing as 3 long-petioled
 leaves. ................................................. BERBERIDACEAE, p. 384
                                                        (*Caulophyllum*)
 Plants with more than one leaf. .................................................. 9

9. Plants with colored milky juice. .................... PAPAVERACEAE, p. 384
 Plants not with colored milky juice. .......................................... 10

10. Leaves opposite, whorled, or basal. ............................................ 11
 Some or all leaves alternate. ........................................................ 18

11. Low plants with an involucre of 4 petal-
 like bracts; fruit a red drupe. ...................... CORNACEAE, p. 563
 Plants without large petal-like bracts. ........................................ 12

12. Mud plants with small axillary flowers. ............ ELATINACEAE, p. 525
 Terrestrial plants. .......................................................................... 13

13. Styles single. ................................................................................. 14
 Styles 2 or more. ........................................................................... 16

14. Ovary inferior, flowers 2- or 4-merous
 (parts). ................................................. ONAGRACEAE, p. 534
 Ovary superior. .............................................................................. 15

15. Petals 5, stamens 10. ................................... PYROLACEAE, p. 567
 Petals 4 or 6, stamens 12. ............................. LYTHRACEAE, p. 534

16. Sepals 2. ................................................... PORTULACACEAE, p. 345
 Sepals more than 2. ....................................................................... 17

17. Leaves glandular-dotted, stamens
 united at base into 3–5 bundles. ................. HYPERICACEAE, p. 522
 Leaves not glandular-dotted,
 stamens not united at base. ...................... CARYOPHYLLACEAE, p. 347

18. Calyx irregular, some sepals
    smaller than others. ....................................................................... 19
    Calyx regular. ............................................................................... 20

19. Leaves lobed or divided. ........................... RANUNCULACEAE, p. 362
    Leaves entire. ........................................ CISTACEAE, p. 525

20. Leaves with stipules. ...................................................................... 21
    Leaves without stipules, or having glands. ................................................ 26

21. Stamens numerous, united into a column;
    leaves palmately veined.                     MALVACEAE, p. 520
    Stamens usually separate or partly so, not
    in a column. .............................................................................. 22

22. Flowers irregular in shape. ................................................................ 23
    Flowers regular in shape. ................................................................... 24

23. Corolla pea-like; fruit a legume. ................... LEGUMINOSAE, p. 462
    Corolla with one petal spurred or sac-like;
    fruit a 3-valved capsule. ......................... VIOLACEAE, p. 526

24. Stamens usually numerous; ovary of one
    or more carpels, either separate or
    enclosed by a fleshy receptacle. ................. ROSACEAE, p. 436
    Stamens 5 or 10; ovary of 5 united
    carpels. .................................................................................. 25

25. Leaves palmately divided; fruit with a
    long beak. ........................................... GERANIACEAE, p. 504
    Leaves of 3 leaflets; fruit not beaked. ............. OXALIDACEAE, p. 505

26. Stamens usually more than 10. ............................................................ 27
    Stamens 10 or fewer. ...................................................................... 30

27. Carpels separate. ...................................... RANUNCULACEAE, p. 362
    Carpels united. ............................................................................ 28

28. Sepals 4; fruit a pod; annuals. ....................... CAPPARIDACEAE, p. 386
    Sepals 4–8; fruit a capsule opening at top. ............................................... 29

29. Flowers few, large, solitary, and terminal;
    petals over 2.5 cm long. .......................... LOASACEAE, p. 531
    Flowers many, small, in terminal spikes. .......... RESEDACEAE, p. 417

30. Ovary inferior. ........................................................................... 31
    Ovary superior. ........................................................................... 33

31. Parts of flowers in twos or fours. ................... ONAGRACEAE, p. 534
    Flowers in umbels; parts of flowers
    mostly in fives. .......................................................................... 32

32. Styles 5; fruit fleshy. ............................... ARALIACEAE, p. 544
    Styles 2; stems hollow. ............................... UMBELLIFERAE, p. 545

33. Carpels 3–5. ............................................................................. 34
    Carpels 1 or 2. ........................................................................... 37

34. Sepals 3; one petal-like and spurred; fruit
    an explosive capsule. .............................. BALSAMINACEAE, p. 508
    None of petals or sepals spurred; fruit not
    explosive. .................................................................... 35

C. Sympetalous dicotyledons
(flowers with two floral rings, but with the petals wholly or partly united forming a tube or bell)

9. Plants twining and attached to stems of
   other plants. ................................... CONVOLVULACEAE, p. 599
   (*Cuscuta*)

   Plants growing out of soil or attached to
   roots of other plants. ......................................................................... 1

10. Corolla regular; stamens 6–10. .................... MONOTROPACEAE, p. 570
    Corolla 2-lipped; stamens 4. ................... OROBANCHACEAE, p. 661

11. Flowers in heads or in form resembling a
    head. ......................................................................................................... 1
    Flowers in long or short spikes. ........................................................... 1
    Flowers not in heads or spikes. ............................................................. 2

12. Flowers in true heads with an involucre
    (bracts). ................................................................................................... 1
    Flowers in form resembling a head, open-
    ing in irregular order. ......................................................................... 1

13. Male and female flowers in separate
    heads. ........................................................ COMPOSITAE, p. 683
    (Group 1)

    Flowers perfect, or male and female
    flowers in same head. ......................................................................... 1

14. Stamens united by their anthers. .................... COMPOSITAE, p. 683
    (Groups 2 and 3)
    Stamens separate. ...................................... DIPSACACEAE, p. 681

15. Leaves mostly basal, 1–3 times divided
    into 3 leaflets; flowers very small,
    greenish. ..................................................... ADOXACEAE, p. 677
    Leaves not with 3 leaflets; flowers usually
    colored. ................................................................................................... 16

16. Corolla 2-lipped; leaves opposite, stems
    square. ........................................................ LABIATAE, p. 620
    Corolla regular; stems round. ....................... POLEMONIACEAE, p. 600

17. Flowers inconspicuous, on long spikes;
    leaves all basal. ......................................... PLANTAGINACEAE, p. 663
    Flowers brightly colored, irregular. ......................................................... 18

18. Fruit a many-seeded capsule. ....................... SCROPHULARIACEAE, p. 638
    Fruit not a many-seeded capsule. ........................................................... 19

19. Flowers alternate or very crowded; fruit
    of 4 nutlets. ............................................... VERBENACEAE, p. 619
    Flowers opposite, not crowded; fruit an
    achene, sharply reflexed. ............................ PHRYMACEAE, p. 663

20. Ovary inferior. ....................................................................................... 21
    Ovary superior. ..................................................................................... 26

21. Climbing plants with tendrils. ....................... CUCURBITACEAE, p. 677
    Not climbing plants. ............................................................................. 22

22. Leaves basal or alternate; plants often
    with milky juice. ................................................................................. 23
    Leaves opposite; plants without milky
    juice. ..................................................................................................... 24

# Genera and species

## Division: PTERIDOPHYTA——ferns and fern allies

## OPHIOGLOSSACEAE——adder's-tongue family

***Botrychium***    grape fern

Plants from fleshy roots with sheath at base of frond (leaf) stalk, containing the following year's bud. Fertile frond bearing the spore-bearing bodies spike-like or raceme-like. Sterile frond leaf-like and lobed.

1. Sterile frond oblong, ovate, or somewhat triangular-ovate, longer than wide, usually glabrous ........................................................................................... 2

   Sterile frond triangular, wider than long, pubescent, at least when young. ..................................................... 6

2. Sterile fronds simple, trilobed, or pinnate; pinnae without midrib. ........................................................................ 3

   Sterile fronds bipinnate; pinnae with conspicuous midrib. ............................................................................. 4

3. Sterile frond petiolate, simple, trilobed, or pinnate with 2 or 3 pairs of pinnae. ................................... *B. simplex*

   Sterile frond sessile, pinnate with 3–9 pairs of pinnae. ......................................................................... *B. lunaria*

4. Pinnae triangular to rhomboid-ovate, about as long as wide. ........................................................ *B. boreale*

   Pinnae lanceolate to oblong, longer than wide. ........................................................................................ 5

5. Pinnae lanceolate to linear-lanceolate, acute at apex; midrib pronounced. ..................... *B. lanceolatum*

   Pinnae ovate to oblong, obtuse at apex; midrib not pronounced. ................................... *B. matricariifolium*

6. Sterile frond petiolate, 2- or 3-pinnate, fleshy; pinnae obtuse at apex. ..................................... *B. multifidum*

   Sterile frond sessile, 3- or 4-pinnate, not fleshy; pinnae acute at apex. ........................... *B. virginianum*

*Botrychium boreale* Milde var. *obtusilobum* (Rupr.) Brown

Plants to 30 cm high. Leaf inserted near middle of stem, 10–20 cm long, subsessile, triangular-ovate, bipinnate; pinnae ovate, acute, pinnatifid. Rare; in grassland and open slopes near tree line; southern Rocky Mountains.

*Botrychium lanceolatum* (Gmel.) Angst.

Plants to 30 cm high. Leaf inserted at base of fertile frond, to 25 cm long, subsessile, triangular-ovate, bipinnate; pinnae lanceolate to linear-lanceolate, serrate or pinnatifid, with midrib pronounced. Rare; in moist grassland; southern Rocky Mountains, Cypress Hills.

*Botrychium lunaria* (L.) Swartz.                                    MOON FERN

Plants to 30 cm high. Leaf inserted near middle of stem, 5–20 cm long, sessile, oblong, pinnate, with 3–9 pairs of fan-shaped pinnae without midrib. Rare; in grassland and open areas; throughout the Prairie Provinces.

*Botrychium matricariifolium* Braun.          CHAMOMILE-LEAVED GRAPE FERN

Plants to 30 cm high. Leaf inserted above middle of stem, 5–20 cm long, subsessile, oblong to ovate, bipinnate; pinnae oblong to ovate, with midrib. Rare; in moist grassland, muskeg, wooded shores; throughout the Prairie Provinces.

*Botrychium multifidum* (Gmel.) Rupr.            THICK-LEAVED GRAPE FERN

Plants to 30 cm high. Leaf inserted near base of stipe, 5–20 cm long, triangular, wider than long, bi- or tri-pinnate; pinnae lanceolate, with midrib pronounced. Rare; in moist meadows, margins of woods; Boreal forest.

*Botrychium simplex* E. Hitchc.                                    GRAPE FERN

Plants 10–15 cm high. Leaf inserted near middle of stem, 5–8 cm long; stipe 2–4 cm long; pinnae subovate, cuneate at base. Rare; in meadows, margins of woods; Parklands and Boreal forest.

*Botrychium virginianum* (L.) Swartz. (Fig. 10)          VIRGINIA GRAPE FERN

Plants to 80 cm high. Leaf inserted near middle of stem, 20–50 cm long, triangular, wider than long, sessile, 3- or 4-pinnate; pinnae and segments oblong to ovate, dentate, with midrib pronounced. Rare; in moist woods; Boreal forest, Cypress Hills, Riding Mountain.

(Fig. 10 overleaf)

Fig. 10.  Virginia grape fern, *Botrychium virginianum* (L.) Swartz.

# OSMUNDACEAE—royal fern family

*Osmunda*        royal fern

*Osmunda claytoniana* L.                                    INTERRUPTED FERN

Plants with stout rhizomes. Fronds 50–100 cm high; stipes shorter than blade, 2–5 mm thick, tomentose when young; blades pinnate-pinnatifid; pinnae in 15–20 pairs, alternate or subopposite. Fertile pinnae 1–5 pairs, 3–6 cm long, 2 cm wide, in middle of blade, bipinnate, with segments oblong, soon withering; sterile pinnae oblong-lanceolate, to 12 cm long, 3 cm wide, acute, deeply pinnatifid, with segments in 10–15 pairs. Sporangia about 0.5 mm wide, dark brown. Southeastern boreal forest.


# POLYPODIACEAE—fern family

Ferns with fronds spirally coiled in the bud, and usually bearing green leaf-like fronds. Spores usually borne in little clusters (sori) and usually under a membranous cover (indusium).

1. Sori enclosed in modified globular seg-
    ments of fertile fronds; fronds becom-
    ing blackish in age. .................................................................. 2
   Sori exposed or covered by the reflexed
    margin of the fertile fronds. ...................................................... 3

2. Fronds scattered along rhizome; sterile
    fronds once pinnatifid; fertile fronds
    bipinnate. ............................................................... *Onoclea*
   Fronds tufted on a caudex; sterile fronds
    surrounding fertile ones. ................................... *Matteuccia*

3. Sori marginal or appearing marginal, cov-
    ered by margin or marginal lobes. ........................................... 4
   Sori not marginal. ....................................................................... 8

4. Sterile and fertile fronds dissimilar; fertile
    ones longer than sterile ones. ........................... *Cryptogramma*
   Sterile and fertile fronds not dissimilar. ................................... 5

5. Sori appearing marginal, but borne on
    veins of reflexed marginal lobes. ............................. *Adiantum*
   Sori marginal or submarginal, but not
    borne on reflexed lobes. ............................................................ 6

6. Rhizome without scales, hairy only;
    fronds and stipes (stalks) to 60 cm high. ..................... *Pteridium*
   Rhizomes scaly; fronds and stipes to 25
    cm high. ..................................................................................... 7

7. Stipes wiry, purplish brown; scales of rhi-
    zome rusty brown; blades with 6–12
    pairs of pinnae. ............................................................. *Pellaea*

Stipes soft and hairy or glabrate; scales of
   rhizome orange; blades 3-pinnate. ...................................................... *Cheilanthe*

8. Indusium absent; blades pinnatisect, with
   segments linear-oblong, denticulate. ...................................................... *Polypodiun*

   Indusium present or absent; blades bipin-
   natifid to tripinnate. ...................................................................................... •

9. Indusium attached by its base under sporangia; veins of fronds not reaching
   margins. ................................................................................................................ *Woodsi*

   Indusium laterally attached or absent;
   veins reaching margins. ................................................................................ 10

10. Indusium linearly attached. ...................................................................... 1

    Indusium attached at a point or absent. ...................................................... 1.

11. Indusia not crossing veins; scales of rhi-
    zomes with short, broad cells; plants
    small, up to 20 cm high. ...................................................................... *Asplenium*

    Indusia usually crossing veins; scales of
    rhizomes with long, narrow cells;
    plants large, up to 100 cm high. ...................................................... *Athyrium*

12. Indusium ovate or oblong; fronds deli-
    cate, light green. .............................................................................................. *Cystopteri*

    Indusium orbicular, reniform, or absent. ................................................ 1:

13. Indusium attached at its center; fronds
    leathery, with pinnae spinulose-
    toothed. ................................................................................................................, *Polystichun*

    Indusium attached laterally by a deep
    notch or absent; fronds not leathery. ...................................................... *Dryopteri*

## *Adiantum*   maidenhair

### *Adiantum pedatum* L.

   Plants with rhizomes 2–5 mm thick. Stipes reddish brown, forked above
into two branches, each bearing 3–5 pinnae on upper side. Pinnae 15–25 cm
long, 3–4 cm wide; pinnules 15–20 pairs, thin, glaucous, oblong, with outer
margins incised. Sori oblong-lunate, covered by the reflexed margin of pinnule
lobe. Rare; southern Rocky Mountains.

## *Asplenium*   spleenwort

### *Asplenium viride* Huds.   GREEN SPLEENWOR"

   Plants with a short, thin rhizome 1 mm thick. Stipes short, brownish a
base. Fronds delicate, 5–15 cm long; pinnae 9–16 pairs, opposite to suboppo
site at base, alternate above, 3–8 mm long, 2–4 mm wide, with apex obtuse
base cuneate, and margins toothed. Sori elongate; indusia hyaline, attached a
one side. Rare; on limestone cliffs; southern Rocky Mountains.

## *Athyrium*   spleenwort

Indusium absent; blades tripinnate-pinnatifid. ...................................................... *A. alpestr*

Indusium present; blades mostly subbipin-
nate. ................................................................................................ *A. filix-femina*

*Athyrium alpestre* (Hoppe) Rylands          ALPINE SPLEENWORT

Plants with a stout short-creeping rhizome, covered with old stipe bases.
Fronds 30–50 cm long, tufted, with stipe erect, shorter than blade; blade lanceolate, bi- or tri-pinnate; pinnae 20–25 pairs, ascending, oblong-lanceolate;
pinnules in 15 or more pairs, deeply toothed or lobed. Sori round; indusium
lacking. Rare; in alpine meadows; southern Rocky Mountains.

*Athyrium filix-femina* (L.) Roth          LADY FERN

Plants with thick creeping to suberect rhizomes, covered with scales.
Fronds 40–100 cm long, tufted, with stipe stout, scaly at base, shorter than, to
as long as, blade; blade lanceolate to ovate-lanceolate in outline, sparingly
scaly, glabrous or somewhat glandular, bipinnate or tripinnate; pinnae 20–30
pairs, ascending, linear-lanceolate; pinnules 1–2 cm long, serrate to deeply
pinnate-lobed. Sori elongate; indusia ciliate. Two varieties distinguished: var.
*filix-femina*, with stipe about as long as blade; basal scales to 6 mm long,
blackish; indusium hooked at apex, short ciliate; southeastern Boreal forest;
and var. *cyclosorum* (Ledeb.) Moore, with stipe about one-third as long as
blade; basal scales 10–12 mm long, pale brown; indusium often horseshoe-
shaped, long ciliate; Rocky Mountains.

**Cheilanthes**          lip fern

*Cheilanthes feei* Moore          LIP FERN

Plants from short branching rhizomes with abundant scales. Fronds 4–15
cm long; stipe as long as, or shorter than, blade, purplish, long pilose above;
blade linear-oblong, 1–3 cm wide, tripinnate; pinnae 7–15 pairs, subopposite,
distant below, ascending, ovate to oblong; pinnules 3–6 pairs, diminishing in
size toward tip, opposite to subopposite; segments 2 or 3 pairs, ovate to suborbicular, sessile, lobate or incised, sparsely pubescent above, densely brownish
pubescent below. Sori marginal, covered by the unmodified margin. Rare; on
limestone rocks and cliffs; southern Rocky Mountains.

**Cryptogramma**          rock brake

Rhizome short; fronds densely tufted. ................................................ *C. crispa*
Rhizome slender, long-creeping; fronds soli-
tary or few together, distant. ........................................................ *C. stelleri*

*Cryptogramma crispa* (L.) R. Br.          PARSLEY FERN

Plants with short rhizomes and numerous fibrous roots and scales. Fronds
densely tufted, glabrous. Sterile fronds erect to spreading; stipe straw-colored,
to 7 cm long; blade 5–7 cm long, ovate; pinnae 4–6 pairs, alternate or the lowest ones subopposite; pinnules 2–4 pairs; segments elliptic, with bases cuneate
and margins dentate. Fertile frond erect, to 20 cm high; stipe 1.5–2.0 times as
long as blade; pinnation similar to that of sterile fronds, but segments linear,
with margins revolute. Sori on vein tips. On outcrops of Precambrian rocks in

Boreal forest; Rocky Mountains. In Canada these plants are usually distinguishable from the Eurasian var. *crispa* as var. *acrostichoides* (R. Br.) Clarke (Fig. 11) with thicker fronds and paler scales.

### *Cryptogramma stelleri* (Gmel.) Prantl ROCK BRAKE

Plants with scaly, pilose, creeping rhizomes. Glabrous fronds scattered along the rhizome. Sterile fronds erect; stipe brownish at base, 6–9 cm long; blade ovate, 3–6 cm long; pinnae thin, 5–6 pairs, alternate or sometimes opposite, with 1–3 segments. Fertile fronds to 20 cm high; stipes 12–15 cm; blades bipinnate or sometimes tripinnate; segments linear-lanceolate, entire, margins revolute. Sori on vein tips. Rare; on shaded, usually limestone rock; southern Rocky Mountains.

### *Cystopteris* bladder fern

Stipes much shorter than blade; blades much
   narrower at base than long. ............................................................................... *C. fragilis*
Stipes often more than twice as long as blade;
   blades as wide at base as long. ........................................................................ *C. montana*

### *Cystopteris fragilis* (L.) Bernh. (Fig. 12) FRAGILE FERN

Plants with creeping unbranched rhizomes. Fronds solitary or in small tufts, 10–30 cm long; stipes 3–12 cm long; blade 7–18 cm long, light green, ovate or ovate-lanceolate, bipinnate to tripinnate, glabrous, or with a few hairs at base of pinnae; pinnae in 9–15 pairs, ascending, subopposite to alternate; segments ovate, margins dentate. Sori on veins; indusium hood-like. Fairly common; in deep, shaded, wooded ravines, on moist slopes, and rock ledges; throughout the Prairie Provinces.

### *Cystopteris montana* (Lam.) Bernh. MOUNTAIN BLADDER FERN

Plants with slender, black, creeping and branching rhizomes; scales conspicuous, pale brown. Fronds scattered, few, 20–35 cm high; stipes 12–20 cm; blades 8–15 cm long, tripinnate; pinnae in 7–13 pairs, the lower ones large, opposite or subopposite, bi- or tri-pinnatifid, the higher ones alternate, pinnules 12–15 mm long; segments dentate. Sori on veins; indusium inconspicuous. Rare; on wet calcareous rock or slopes; southern Rocky Mountains.

### *Dryopteris* shield fern

1. Blades short-pubescent above; segments
     ciliate; stipe bundles 2, united at base
     of blade. .................................................................................................................. 2
   Blades glabrous above; segments not ciliate; stipe bundles free above base of
     blade. ...................................................................................................................... 3
2. Blades triangular in outline; stipes scaly
     and pilose. ................................................................................................ *D. phegopteris*
   Blades lanceolate in outline; stipes
     glabrous. .................................................................................................. *D. thelypteris*
3. Rhizomes thin, blackish; stipe bundles 2;
     indusium absent. ........................................................................................ *D. disjuncta*

Fig. 11.    Parsley fern, *Cryptogramma crispa* (L.) R. Br. var. *acrostichoides* (R. Br.) Clarke.

Fig. 12.   Fragile fern, *Cystopteris fragilis* (L.) Bernh.

*Dryopteris austriaca* (Jacq.) Woynar          SPINULOSE SHIELD FERN

Plants with thick rhizomes. Fronds tufted, to 1 m high; stipes shorter than blades; blades 40–60 cm long, 20–30 cm wide at base, bipinnate-pinnatifid or tripinnate-pinnatifid; pinnae in 10–15 pairs; pinnules in 12–20 pairs; segments oblong, spinulose-toothed. Sori on veins; indusium kidney-shaped. A rather variable species, of which many varieties have been described, or treated as distinct species. In Canada the plant is usually named var. *spinulosa* (Muell.) Fiori (*D. spinulosa* (Muell.) Watt.). In moist woods and on slopes; throughout Boreal forest, Rocky Mountains. Rare in Parklands.

*Dryopteris cristata* (L.) Gray          CRESTED SHIELD FERN

Plants with short-creeping thick rhizomes. Fronds 35–80 cm high; stipes shorter than blades, 15–30 cm; sterile fronds spreading, shorter and wider than erect fertile fronds; blades lanceolate, pinnate-pinnatifid; pinnae in 10–20 pairs, to 8 cm long, with the lower ones spaced; segments ovate oblong, serrate or biserrate, with teeth acute. Sori on the veins; indusia smaller than sori. In damp woods and marshes; Boreal forests, Rocky Mountains.

*Dryopteris disjuncta* (Rupr.) Morton          OAKFERN

Plants with dark brown to blackish, scaly, slender rhizomes. Fronds to 50 cm high; stipes slender, longer than blades, shiny; blades deltoid, bipinnate-pinnatifid to tripinnate-pinnatifid, delicate; pinnae in 4–7 pairs, with basal ones up to half as long as blades; pinnules oblong, obtuse. Sori on the veins; indusium lacking. Rather rare; moist woods, talus slopes, damp; Precambrian rock outcrop, Boreal forests.

*Dryopteris filix-mas* (L.) Schott          MALE FERN

Plants with thick, erect rhizomes. Fronds 40–100 cm high, tufted; stipes shorter than blades, with brown, denticulate scales; blades 30–90 cm long, lanceolate, pinnate-pinnatifid to subbipinnate; pinnae in 20–25 pairs, to 15 cm long, 4 cm wide, lanceolate, acuminate; segments in 15–20 pairs, to 2.5 cm long, 1 cm wide, serrate. Sori on veins near midrib; indusia large, horseshoe-shaped. On slopes and cliffs; southern Rocky Mountains.

*Dryopteris fragrans* (L.) Schott                                        FRAGRANT SHIELD FERN

Plants with short, erect rhizomes. Fronds 25–40 cm high; stipes 6–10 cm
long, densely scaly, glandular; blades to 35 cm long, linear-lanceolate, pin
nate-pinnatifid; pinnae in 20–25 pairs, to 2.5–3.0 cm long, 1 cm wide, linea
oblong; segments in 6–10 pairs, oblong, crenate; veins scaly and glandular
Sori on veins; indusia large, rounded. Rare; on dry cliffs and slopes; Borea
forest.

*Dryopteris phegopteris* (L.) Christ                                      BEECH-FERN

Plants with slender, creeping, densely hairy and scaly rhizomes. Frond
40–50 cm long, solitary; stipes longer than blades, pilose and scaly through
out; blades to 25 cm long, almost as wide at base, pinnate-pinnatifid; pinna
in 8–12 pairs, with the lower ones subopposite, often reflexed; segment
oblong, obtuse, with the margins entire or somewhat crenate, pubescent. Sor
on veins, submarginal; indusia lacking. Rare; on moist slopes; Boreal forest.

*Dryopteris thelypteris* (L.) Gray                                        MARSH FERN

Plants with extensively creeping black, somewhat scaly rhizomes. Frond
solitary; sterile fronds to 40 cm long, with stipes about as long as blades; fertil
fronds to 75 cm long, with stipes longer than blades. Stipes black below, glab
rous; blades lanceolate, pinnate-pinnatifid; pinnae in 10–15 pairs; sterile pin
nae to 10 cm long, 2 cm wide, with segments oval, and margin entire; fertil
pinnae with the segments oblong, to 4 mm wide, and margins revolute. Sori on
veins; indusia small, ciliate. Rare; in the margins of marshes and wet woods
Manitoba.

**Matteuccia**          ostrich fern

*Matteuccia struthiopteris* (L.) Tod.                          OSTRICH FERN, FIDDLE HEADS

Plants large, with fronds typically densely tufted around crown of erect
scaly rhizome. Fronds of two kinds. Stipe of sterile fronds shorter than blade
to 40 cm long; blade up to 100 cm long, oblong-lanceolate, pinnate-pinnatifid
pinnae in 20–30 pairs; lower pinnae short, reflexed; middle pinnae long
ascending, with segments oblong and margins revolute. Fertile fronds 40–60
cm long; stipe thick, rigid, dark brown; pinnae to 5 cm long, linear, with mar
gins strongly revolute. Sori on the veins, covered by margin. In Canada the
plants are distinguishable as var. *pensylvanica* (Willd.) Mort. (Fig. 13) by hav
ing light brown scales on rhizome and lower stipe; in the Eurasian type these
scales have a black central band. Syn.: *Onoclea struthiopteris* (L.) Hoffm.
*Pteretis nodulosa* (Michx.) Nieuwl., *P. pensylvanica* (Willd.) Fern. Often plenti
ful; moist woods; Boreal forest, Rocky Mountains. Rare in Parklands.

**Onoclea**          sensitive fern

*Onoclea sensibilis* L.                                                   SENSITIVE FERN

Plants with thick, scaly rhizomes 5–7 mm in diam. Fronds to 100 cm long
solitary or few together; stipes longer than blades, glabrous. Sterile fronds pin
natifid; blades 20–40 cm long and as wide at base; rachis winged above; lowe
segments opposite, lobate to somewhat pinnatifid. Fertile fronds shorter than

Fig. 13.   Ostrich fern, *Matteuccia struthiopteris* (L.) Tod. var. *pensylvanica* (Willd.) Mort.

sterile ones; blade 12–15 cm long, bipinnate; pinnae in bead-like segments with revolute margins. Sori globose; indusium hood-like and covering sori. Southeastern Boreal forest.

**Pellaea**    cliff brake

*Pellaea glabella* Mett.    PURPLE CLIFF BRAKE

Plants with short, erect rhizomes with yellow brown scales. Fronds to 25 cm long; stipes shorter than blades, subdimorphic; blades pinnate or bipinnate at base; pinnae in 5–10 pairs, usually ascending; sterile segments ovate to ovate-oblong, to 10 mm wide at base; fertile segments linear-oblong, to 5 mm wide at base, with margins revolute. Sori submarginal, at tip of veins; indusia absent. Two varieties are recognized: var. *simplex* Butters, having cells of scales long-linear, 10–15 times as long as wide, occurring in southern Rocky Mountains; var. *nana* (Rich.) Cody, having plants usually smaller than in var. *simplex*, with cells of scales oblong-lanceolate, 3–5 times as long as wide. Both varieties occur on rock cliffs, usually limestone; Boreal forests.

**Polypodium**    rock tripe

*Polypodium vulgare* L.    COMMON ROCK TRIPE

Plants with stout rhizomes, 1.5–3 mm thick, covered with fibrous scales. Fronds mostly distant, 10–30 cm long; stipes shorter than blades, scaly at base. Blades pinnatisect, with segments in 12–20 pairs, alternate to subopposite, spreading. Sori round, submarginal; indusia absent. In Canada plants are distinguishable from the European var. *vulgare* by having paraphyses (thin sterile filaments) mixed in the sporangia, and smaller rhizome scales, and have been named var. *virginianum* (L.) D.C. Eaton (Fig. 14); occur in Precambrian rock outcrops in Boreal forest, northern Rocky Mountains. A var. *columbianum* Gilbert, usually smaller, with fewer segments in fronds, occurs in southern Rocky Mountains.

**Polystichum**    rock tripe

*Polystichum lonchitis* (L.) Roth    HOLLY FERN

Plants with erect or ascending, thick rhizomes, to 1 cm in diam. Fronds 25–50 cm long; stipes short or almost none, persistently scaly; blades linear-lanceolate in outline, pinnate; pinnae in 25–45 pairs, spreading at right angles, with margin spinulose-toothed, scaly below. Sori round; indusia arising from center of sori, lacerate. Southern Rocky Mountains.

**Pteridium**    brake fern

*Pteridium aquilinum* (L.) Kuhn    BRACKEN FERN

Plants with pubescent, creeping rhizomes. Fronds large; stipes about as long as blades; vascular bundles in cross section resembling "an eagle with spreading wings." Blades bipinnate- or tripinnate-pinnatifid; pinnae in 6–9 pairs, opposite or nearly so. Pinnulae alternate, with segments oblong or linear-oblong, and margins revolute. Sori linear, marginal, continuous, covered

Fig. 14.   Common rock tripe, *Polypodium vulgare* L. var. *virginianum* (L.)
D. C. Eaton.

by the revolute margin; indusium, if present, delicate, continuous along the margin. A variable cosmopolite, in which several species have been recognized, but now generally considered a single species with two subspecies and several more or less well-defined varieties. In Canada these plants are usually described as ssp. *aquilinum* var. *pubescens* Underw., with the fronds commonly pubescent below, especially along the midribs and margins. Boreal forest, Rocky Mountains.

## *Woodsia*

1. Blades stipitate glandular; stipes not articulate. ................................................................................................ 2

   Blades not glandular; stipes articulate below middle. ................................................................................................ 3

2. Fronds with long hairs on veins and midribs of pinnulae and segments. ....................................... *W. scopulina*

   Fronds lacking hairs. ............................................................................... *W. oregana*

3. Stipe brown, scaly, and pubescent. ................................................................................................ 4

   Stipe straw-colored, glabrous. ................................................................................... *W. glabella*

4. Stipe slender, less than 1 mm thick; midribs of pinnae not or scarcely scaly. ................................... *W. alpina*

   Stipe stout; midribs of pinnae scaly. ................................................................................... *W. ilvensis*

### *Woodsia alpina* (Bolton) S. F. Gray

Plants with scaly rhizomes; scales 4–6 mm long, 1–2 mm wide, brown, toothed. Fronds 6–15 cm long; stipes shorter than blades, brown, somewhat pilose and scaly, articulate below middle; blades 1–2 cm wide, linear, pinnate-pinnatifid; rachis somewhat hairy and scaly; pinnae in 8–15 pairs, deeply pinnatifid, hairy below, usually without scales; segments in 2 or 3 pairs, oblong to suborbicular. Sori at the apex of secondary veins; indusium with few long septate hair-like lobes. Shaded cliffs; southeastern Boreal forest.

### *Woodsia glabella* R. Br.

Plants with slender scaly rhizomes; scales 3–4 mm long, 1–1.5 mm wide, brown, toothed. Fronds 5–15 cm long; stipes shorter than blades, straw-colored, glabrous, scaly at base only, articulate below middle; blades 7–15 mm wide, linear, glabrous, without scales, pinnate; pinnae in 8–15 pairs, deltoid or rounded in outline, trilobate or pinnatifid, with 2 or 3 pairs of segments. Sori at apex of veins; indusia with numerous septate hairs. Shaded moist cliffs of calcareous or dolomitic rock; Boreal forest, Rocky Mountains.

### *Woodsia ilvensis* (L.) R. Br.

Plants with stout scaly rhizomes; scales 4–6 mm long, to 1 mm wide, brown and toothed. Fronds to 20 cm long; stipes shorter than blades, shiny brown, stout, to 1 mm thick, scaly and pubescent, articulate below middle; blades 2–3.5 cm wide, pinnate-pinnatifid; rachis scaly and hirsute; pinnae in 10–15 pairs, the lower ones subopposite, the higher ones alternate, deeply pinnatifid; segments in 2–7 pairs, oblong, with the apex rounded, scaly, and hairy below. Sori submarginal, confluent; indusia with a fringe of long hairs. Crevices in Precambrian rock outcrops; Boreal forest.

*Woodsia oregana* D. C. Eaton

Plants with stout scaly rhizomes; scales 3–5 mm long, to 0.5 mm wide, subentire. Fronds to 25 cm long; stipes shorter than blades, dark brown, stout to 1.5 mm thick, scaly at base, not articulate; blades lanceolate, to 5 cm wide, pinnate-pinnatifid to pinnate-bipinnatifid, glabrous or glandular pubescent; pinnae in 10–15 pairs, subopposite, deltoid to lanceolate, mostly bipinnatifid; segments in 5–7 pairs, with margins often revolute. Sori marginal, in part confluent; indusia divided into few narrow segments. Occur in rock crevices. The typical var. *oregana* has blades sparingly glandular to glabrous; rare in southern Rocky Mountains, Lake Athabasca. The var. *cathcartiana* (Robins.) Morton has blades copiously, finely glandular; rare in Cypress Hills.

*Woodsia scopulina* D. C. Eaton

Plants with stout scaly rhizomes; scales 4–5 mm long, 1.0–1.5 mm wide. Fronds 15–35 cm long; stipes shorter than blades, stout, brown, pubescent, scaly at base, not articulate; blades linear lanceolate, 3–7 cm wide, bipinnate; rachis somewhat pubescent, not scaly; pinnae in 9–17 pairs, subopposite, lanceolate, ovate or deltoid-ovate; pinnulae in 7–10 pairs, oblong, blunt, crenate to subpinnatifid, pubescent and somewhat glandular below. Sori submarginal at tips of veins; indusia with 3–6 filamentous scales. Occur in rock crevices; Boreal forest, Rocky Mountains.

## MARSILEACEAE—marsilea family

**Marsilea**      pepperwort

*Marsilea mucronata* A. Br.                              HAIRY PEPPERWORT

Low growing from slender creeping rootstocks. Leaves borne singly on thin stalks 5–15 cm long, divided into 4 triangular leaflets, each 3–10 mm long. Spores borne in ovoid bean-like containers (sporocarps) on short stalks near base of plant and covered with hair-like scales. Not common; in slough bottoms at several locations in the south central prairies. Syn.: *Marsilea vestita* Hook & Grev.

## EQUISETACEAE—horsetail family

**Equisetum**      horsetail

Perennial rush-like plants, with stems fluted or grooved, and joints or nodes solid but surrounded by a toothed sheath. Fertile stems with a terminal cone in which spores are borne.

1. Branches compound; sheaths bright reddish brown. ................................................................................ *E. sylvaticum*

Branches not compound, sometimes absent; sheaths not bright reddish brown. ....................................................................... 2

2. Stems usually much-branched with whorls of branches. ................................................... 3

   Stems usually not branched above the ground. ................................................................... 6

3. Plants with hollow branches. ............................................ 4

   Plants with solid branches. ............................................ 5

4. Center cavity of stem very small. ........................... *E. palustre*

   Center cavity of stem large. ................................... *E. fluviatile*

5. Fertile stems not branched, soon withering. ........................................................ *E. arvense*

   Fertile stems branching toward top, only the tip withering. ................................. *E. pratense*

6. Stems low, slender, tufted, 5- to 10-grooved. .................................................................... 7

   Stems tall, stouter, many-grooved. ............................ 8

7. Stems solid; sheaths 3-toothed. ......................... *E. scirpoides*

   Stems with small central cavity; sheaths 5- to 10-toothed. ...................................... *E. variegatum*

8. Sheaths almost as broad as high, cylindrical, turning black or gray with black bands above and below. ............................ *E. hyemale*

   Sheaths higher than broad, somewhat funnel-shaped, usually with a narrow black band. ................................................. *E. laevigatum*

### *Equisetum arvense* L. (Fig. 15)   COMMON HORSETAIL

A plant with annual stems withering at the end of the season; stems of two kinds: fertile, unbranched stems bearing the spore-containing cone at the summit; and sterile, much-branched stems. Plants 8–25 cm high; sheaths having 8–12 brownish teeth. Sterile stems with whorls of branches. Common; in wet places; throughout the Prairie Provinces, especially on sandy soils.

### *Equisetum fluviatile* L.   SWAMP HORSETAIL

An annual-stemmed species 10–100 cm high with whorls of hollow branches. Sheaths flaring and bearing about 18 dark brown teeth. Common; in marshes and shallow water; Boreal forest. Syn.: *E. limosum* L.

### *Equisetum hyemale* L. var. *affine* (Engelm.) A. A. Eaton
COMMON SCOURING-RUSH

A perennial-stemmed, unbranched species 30–100 cm high, with broad, conspicuous sheaths, usually grayish with black bands above and below. Common; where subsoil is moist, in sandhill areas, creek flats, and lake margins; throughout the Prairie Provinces. Syn.: *E. prealtum* Raf.

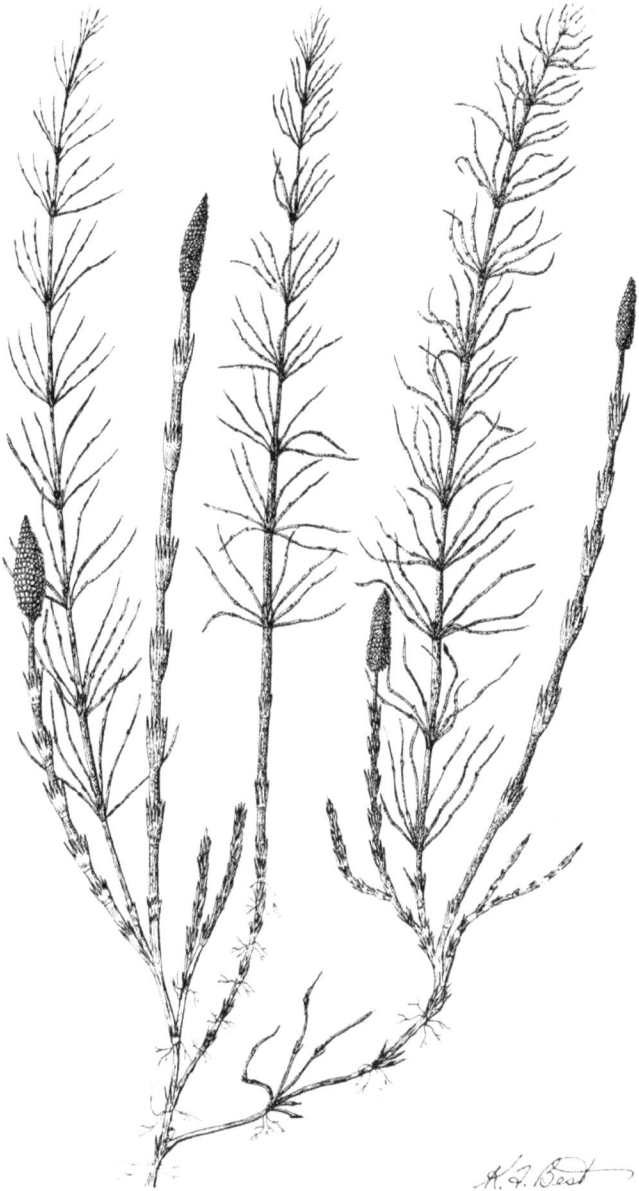

Fig. 15.   Common horsetail, *Equisetum arvense* L.

*Equisetum laevigatum* A. Br. SMOOTH SCOURING-RUSH

A species 30–100 cm high, usually with annual stems, but with some stems perennial. Sheaths green and flaring at the summit, often with a black band, or a gray or black band at the base of the sheath. Cone rounded or with a tip. On light soils, and in moist areas; throughout Prairies and Parklands. Includes *E. kansanum* Schaffn.

*Equisetum palustre* L. MARSH HORSETAIL

An annual-stemmed plant 15–50 cm high, with a very small cavity in the center of stem. Usually whorls of hollow branches. Sheaths usually somewhat flaring at the top and bearing 8 brownish but white-margined teeth. Fairly common; in wet soil; throughout Parklands and Boreal forest.

*Equisetum pratense* Ehrh. MEADOW HORSETAIL

An annual-stemmed species 15–50 cm high, with fertile stems appearing before sterile ones. Sterile stems much-branched, but fertile ones only branched when old. Fairly common; in moist, sandy soils; Parklands, Boreal forest. Cypress Hills, and other favorable locations.

*Equisetum scirpoides* Michx. DWARF SCOURING-RUSH

A tufted species with thread-like, unbranched, solid stems 8–15 cm high. Sheaths 3-toothed; cones very small. In swamps, wet spruce woods; Boreal forest, Cypress Hills, Riding Mountain – Duck Mountain.

*Equisetum sylvaticum* L. WOODLAND HORSETAIL

A pretty species with conspicuous sheaths bearing large, loose, reddish brown teeth. Both fertile and sterile stems, annual, and branched with solid, compound (divided) branches. Fairly common; in moist woodlands; Boreal forest, Cypress Hills, Riding Mountain – Duck Mountain.

*Equisetum variegatum* Schleich. VARIEGATED HORSETAIL

A low, tufted perennial-stemmed species 15–50 cm high, with 5–10 teeth on each sheath; teeth black with a white border and a bristle-like tip. Occur occasionally; in wet places; Boreal forest.

# ISOETACEAE—quillwort family

**Isoetes** quillwort

Submersed aquatic plants with a 2-lobed corm. All leaves with a basal sporangium, and a triangular ligule above the sporangium. Spores of two kinds: megaspores, 0.25 mm in diam or larger; and microspores, visible only through a microscope.

Megaspores 0.3–0.5 mm in diam, white or
 bluish; surface with small tubercles or
 wrinkled. ......................................................................................................... *I. bolanderi*

Megaspores 0.4–0.6 mm in diam; surface dis-
tinctly spinulose. ................................................................................ *I. echinospora*

*Isoetes bolanderi* Engelm.

Plants submersed. Leaves 6–25, bright green, soft, 6–15 or 20 cm long;
stomata few. Sporangia 3–4 mm long, covered for one-quarter to one-third of
their length by the ligule. Megaspores 0.3–0.5 mm, microspores 25–30 μm.
Rare; in lakes and ponds; southern Rocky Mountains.

*Isoetes echinospora* Dur. var. *braunii* (Dur.) Engelm.

Plants submersed or emersed. Leaves 10–45, coarse, pale green, 6–15 cm
long; stomata few toward the leaf tip. Sporangia 4–5 mm long, covered for
one-half to three-quarters of their length by the ligule. Megaspores 0.4–0.6
mm, microspores 25–30 μm. Rare; in lakes and ponds; Boreal forests.


# LYCOPODIACEAE—club-moss family

***Lycopodium***     club-moss

Perennial, low, usually trailing plants with short, stiff, single-nerved, overlapping leaves. Spores sulfur-colored, borne in spore cases (sporangia) on the
upper surfaces or on the axils of the leaves, on ascending or aerial branches,
often aggregated in strobili. Spores containing much oil, inflammable.

1. Sporangia borne in zones along stem, not
   in strobili. ........................................................................................ *L. selago*
   Sporangia borne in strobili. ........................................................................... 2
2. Strobili green, not much different from
   stems; plants without rhizomes. ............................................... *L. inundatum*
   Strobili yellow, much different from
   stems; plants with rhizomes. ......................................................................... 3
3. Strobili borne on a long peduncle. .............................................................. 4
   Strobili sessile on stem. ................................................................................. 5
4. Leaves to 7 mm long, awl-shaped, with a
   hair-like tip; branches not flattened. ...................................... *L. clavatum*
   Leaves to 3 mm long, scale-like, with apex
   acuminate; branches strongly flat-
   tened. ..................................................................................... *L. complanatum*
5. Erect stems much-branched, resembling a
   small shrub. ............................................................................. *L. obscurum*
   Erect stems not much-branched. ................................................................... 6
6. Leaves in 6–10 ranks; plants coarse, stems
   often very long, prickly. .......................................................... *L. annotinum*
   Leaves in 4–5 ranks; plants not coarse. ....................................................... 7
7. Branchlets much-flattened; leaves strong-
   ly appressed; trailing stems about 2
   mm thick. ................................................................................. *L. alpinum*

Branchlets not much-flattened; leaves
spreading; trailing stems about 1–2 mm
thick. ...................................................................................... *L. sabinifolium*

### *Lycopodium alpinum* L.

The horizontal stems 30–60 cm long, rooting throughout, 2–2.5 mm thick, with few bract-like yellow leaves. Erect stems repeatedly forked, 4–10 cm long, with branchlets flattened. Strobili sessile, solitary at the end of a leafy pedun-cle, 1–2 cm long. Rocky Mountains.

### *Lycopodium annotinum* L. STIFF CLUB-MOSS

A trailing, prostrate plant often 100–200 cm long, with stiff, linear-lanceo-late, sharp-tipped leaves about 3–7 mm long, crowded along stems and branches. Aerial or upright branches 10–20 cm high, tipped with yellowish fruiting spike 15–25 mm long. Often found in moist woodlands; Boreal forest, Riding Mountain – Duck Mountain, Cypress Hills.

### *Lycopodium clavatum* L. RUNNING-PINE

A prostrate, trailing plant, with the main stem up to 100–200 cm long. Leaves very similar to the previous species but tipped with a fine bristle. Fruit-ing spikes borne on a bract-covered stalk 3.5–12 cm long. Not common; Boreal forest.

### *Lycopodium complanatum* L. (Fig. 16) TRAILING CLUB-MOSS

A plant with the main stem trailing on or slightly under the ground sur-face. Leaves small and tightly clasping the stem, giving the stem a very narrow smooth appearance, quite different from other club-mosses. Fruiting spikes 15–25 mm long, cylindrical, and borne on a chaffy stalk 5–15 cm high. Fairly common in pine woods and damp forests; Boreal forest.

### *Lycopodium inundatum* L.

A prostrate plant with arching stems frequently rooting. Strobili few, 1.5–4 cm long, to 1 cm thick, at the tip of a leafy erect stem. Rare; in bogs; Boreal forest. Known only from Saskatchewan.

### *Lycopodium obscurum* L. (Fig. 17) GROUND-PINE

A species with the main stem creeping horizontally below the surface of the ground. Aerial upright branches appearing like miniature evergreen trees growing to 10–25 cm high, with tightly packed, spreading leaves about 3–4 mm long, linear-lanceolate, and sharp-tipped. Fruiting spikes almost stalkless, 1.5–3 cm long. Occasionally found in moist woods; Boreal forest, Cypress Hills.

### *Lycopodium sabinifolium* Willd. GROUND-FIR

A species with the creeping stem bearing few scaly leaves. Erect stems forking repeatedly; branchlets somewhat flattened at the tips, 5–20 cm long. Strobili sessile, solitary or in pairs on leafy stems, 1.5–2.5 cm long. Rare in open woods; Boreal forest. In Canada the plants are distinguishable as var.

Fig. 16.   Trailing club-moss, *Lycopodium complanatum* L.

Fig. 17. Ground-pine, *Lycopodium obscurum* L.

*sitchense* (Rupr.) Fern., having the fertile branches much longer than the sterile ones; in the typical form, fertile and sterile branches are about equal in length, with only the strobili elevated.

*Lycopodium selago* L.

Plants with short horizontal stems and tufted erect stems, forking repeatedly, to 30 cm high. Leaves 8-ranked, about 5 mm long, erect, and appressed. Sporangia borne in axils of leaves in alternating zones; each season's growth having a sterile basal zone, with a fertile zone at the summit. Peat bogs, muskeg; Boreal forest, Rocky Mountains.

## SELAGINELLACEAE—spike-moss family

**Selaginella**       little club-moss

1. Leaves acute, not bristle-tipped, rootlets
   few. ........................................................................................ *S. selaginoides*
   Leaves bristle-tipped, rootlets many. .................................................................. 2
2. Plants blue green, glaucous, loosely
   tufted. .................................................................................................. *S. wallacei*
   Plants green, densely tufted. ................................................................................ 3
3. Bristle 1.0–2.0 mm, conspicuous. .................................................................. *S. densa*
   Bristle 0.3–0.5 mm, inconspicuous. ......................................................... *S. rupestris*

*Selaginella densa* Rydb.                      PRAIRIE SELAGINELLA

A low, densely matted plant, with stems rooting almost their whole length and densely branched. Plants covered thickly with tiny leaves up to 3 mm long, each tipped by a minute bristle, and varying from green to yellowish according to age and condition. Strobili 10–25 mm long, covered with somewhat triangular, green, much overlapping bracts; spore containers occurring singly in the axils of these bracts. Although not usually noticed, this inconspicuous plant is probably one of the commonest and most dominant plants of the drier and more open prairies. It helps stop erosion and perhaps builds up soil by decaying organic matter, but it has no forage value and it increases with overgrazing and abuse of prairie pastures. Very plentiful on drier light soils, and eroded spots in grassland; throughout Prairies and Parklands.

*Selaginella rupestris* (L.) Spring

Very similar to the preceding species, but with bristles short and plants less conspicuously bristly. On rock outcrops, and in open pine forests; Boreal forest.

*Selaginella selaginoides* (L.) Link

Small, mostly inconspicuous plants. Prostrate stems 2–5 cm long, with few rootlets. Fertile stems erect, 6–10 cm high, or rarely to 20 cm; strobili 1.5–3 cm long, almost cylindric, open. Not common; in boggy areas, wet meadows; Boreal forest, Rocky Mountains.

*Selaginella wallacei* Hieron.

Very similar to *S. densa* and *S. rupestris*, but much more loosely spreading, and glaucous blue green. Apical leaves usually with a short bristle; stem leaves often not bristle-tipped. Over rocks and on dry slopes; southern Rocky Mountains.

## Division: SPERMATOPHYTA—seed-bearing plants

## Subdivision: GYMNOSPERMAE—naked-seeded plants

## TAXACEAE—yew family

**Taxus**   yew

Shrub, usually straggling, with ascending or
  rarely erect stems up to 2 m high. ................................................................ *T. canadensis*
Shrub or small tree with straight trunk, up to
  10 m high. ............................................................................................... *T. brevifolia*

*Taxus brevifolia* Nutt.                                                                WESTERN YEW

A tree 10–15 m high, erect, with a straight trunk. Leaves 2-ranked along branches, giving branches a flat appearance. Tip of leaves reflexed. Rare; southern Rocky Mountains.

*Taxus canadensis* Marsh.                                                      GROUND HEMLOCK

A shrub, usually with straggling to ascending branches, rarely with erect stems. Branches appearing flat. Tip of leaves not reflexed. Rare; in moist woods and muskeg; southeastern Boreal forest.

## PINACEAE—pine family

1. Leaves borne in clusters of 2 to many. ........................................................ 2
   Leaves borne singly. ...................................................................................... 3

2. Leaves in clusters of 2–5, evergreen; base
     of clusters enclosed in a chaffy sheath. ........................................... *Pinus*
   Leaves in clusters of 10–40, deciduous;
     base of clusters without sheath. ......................................................... *Larix*

3. Leaves 4-sided, not appearing 2-ranked;
     leaf scars prominent on twigs. ............................................................ *Picea*
   Leaves flat, appearing 2-ranked; leaf scars
     not prominent on twigs. ..................................................................... 4

4. Cones drooping, with scales persistent;
     bracts longer than scales, 3-lobed; leaf
     scars oval. ................................................................................. *Pseudotsuga*
   Cones erect, with scales deciduous; bracts
     shorter than scales, rounded; leaf scars
     round. ................................................................................................ *Abies*

## *Abies*    fir

*Abies balsamea* (L.) Mill.                                          BALSAM FIR

A tall tree, with gray fairly smooth bark having numerous resinous blisters. Leaves needle-like, 2–3 cm long, flat and stalkless, shiny dark green above
and whitish below, with 8–10 lines of stomata. Male and female flowers borne
on the same tree, with males yellowish to red and females purple. Cones dark
purple, somewhat oblong, 5–10 cm long. Common; in Boreal forest, Rocky
Mountains.

The var. *fallax* (Eng.) Boiv. ( = *A. lasiocarpa* (Hook.) Endl.) has needles
more glaucous, narrower, and with 10–12 lines of stomata below. Southern
Rocky Mountains.

## *Larix*    larch

1. Twigs tomentose pubescent, even when
   old. ........................................................................................ *L. lyallii*
   Twigs glabrous, or pubescent only when
   young. ........................................................................................ 2
2. Needles 1–2 cm long; scales of cones
   glabrous. ........................................................................................ *L. laricina*
   Needles 3 cm long or more; scales of
   cones puberulent. ........................................................................................ *L. occidentalis*

*Larix laricina* (DuRoi) K. Koch                                    TAMARACK

A rather slender tree 6–15 m high; bark reddish brown, with small, flaky
scales. Leaves needle-like, 1–2 cm long, very pale green, in clusters of 10–20
along the twigs, turning yellow in autumn and dropping off. Fruits small
cones, developing the first year and soon shed. Common; in swamps and marshy woods; Boreal forest, Rocky Mountains, Riding Mountain – Duck Mountain.

*Larix lyallii* Parl.                                              ALPINE LARCH

A tree, rarely over 10 m high; bark thin, furrowed, with reddish brown
loose scales. Leaves 30–40 in a cluster, bluish green, 4-sided. Seed cones ellip-
soid-oblong, with scales hairy below, fringed. At high altitudes, at timberline
in Rocky Mountains.

*Larix occidentalis* Nutt.                                        WESTERN TAMARACK

A large tree reaching to 50 m high; bark reddish brown, deeply furrowed
at base, forming large flutes. Leaves 15–30 in a cluster, triangular. Seed cones
oblong, with scales tomentose below when young. Rare; southern Rocky
Mountains.

## *Picea*    spruce

Shapely evergreen trees with 4-sided needle-like leaves scattered around
the twigs. Cones pendulous, maturing the first year.

Branchlets not hairy; cones falling the first winter; cones oblong-cylindric, often over 3 cm long. ............................................................................... *P. glauca*

Branchlets somewhat hairy; cones remaining on tree for several years; cones oval or ovoid, not over 3 cm long. ................................................. *P. mariana*

### Picea glauca (Moench) Voss        WHITE SPRUCE

A shapely tree 7–20 m high, with scaly, brown bark. Leaves bluish green 1–2.5 cm long. Female inflorescence crimson; cones cylindric 2.5–5 cm long, with smooth-margined scales. Very plentiful throughout the Prairie Provinces, except Prairies and Parklands. A variety, the western white spruce, *Picea glauca* (Moench) Voss var. *albertiana* (S. Brown) Sarg., in which the cone scales are erose or somewhat toothed at the margins, is the common variety of southern Rocky Mountains.

### Picea mariana (Mill.) BSP.        BLACK SPRUCE

A less shapely tree than the previous species 7–15 m high. Bark grayish to reddish brown, scaly. Needles 1–2 cm long, bluish green. Male inflorescence dark red, and female purplish; cones 1–4 cm long, ovoid, remaining on trees for several seasons. Fairly common; in wet or swampy woodlands; throughout Boreal forest.

### *Pinus*     pine

Tall evergreen trees with leaves borne in clusters; cones maturing in second season.

1. Leaves 2 in a cluster. ............................................................................... 2

   Leaves 3–5 in a cluster. ........................................................................... 4

2. Leaves 7–17 cm long; cones ovoid-conical, terminal or almost so. ..................................... *P. resinosa*

   Leaves less than 7 cm long; cones conical, lateral at maturity. ........................................ 3

3. Leaves 2–4 cm long, thick and rigid, twisted and spreading; cones curved toward tips of branches; scales without prickles. ............................................................. *P. banksiana*

   Leaves 3–6 cm long, not usually twisted and spreading; cones spreading at right angles to branches; scales with a prickle. ......................................................... *P. contorta*

4. Leaves 3 in a cluster, 8–20 cm long; cones to 18 cm long, scales with a recurved prickle. ..................................................................... *P. ponderosa*

   Leaves 5 in a cluster. ............................................................................. 5

5. Cones ovoid, with thick scales. ................................................................... 6

   Cones cylindric or subcylindric, with thin scales. .................................................. 7

6. Cones to 7 cm long. ............................................................................ *P. albicaulis*
   Cones to 20 cm long. ............................................................................ *P. flexilis*
7. Leaves 8–13 cm long, pale green and
   glaucous; cones to 15 cm long; eastern
   species. ................................................................................................ *P. strobus*
   Leaves 5–10 cm long, bluish green and
   glaucous; white bands of stomata;
   cones 10–20 cm long; western species. ................................................ *P. monticola*

*Pinus albicaulis* Engelm.                                    WHITEBARK PINE

Small alpine tree, with crooked and twisted trunk, or reduced to a shrub. Bark smooth, whitish; twigs yellowish, hairy. Leaves 4–8 cm long, dark green, stiff. Cones oval to subglobose, 3–7 cm long, purplish, with scales forming a stout protuberance, not prickly. Rare; at timberline; Rocky Mountains.

*Pinus banksiana* Lamb.                                        JACK PINE

A tree up to 20 m high, with thin, reddish brown bark. Needle-like leaves generally somewhat twisted, 2–5 cm long, borne in twos, yellowish green. Male inflorescence yellow, female dark purple; cones 2–5 cm long, curved, generally in pairs, with unarmed scales and usually pointing toward apex of branches. Abundant; on sandy soils; Boreal forest.

*Pinus contorta* Dougl. var. *latifolia* Engelm.              LODGEPOLE PINE

A tree similar to *P. banksiana*. Difficult to distinguish between the two species, but lodgepole pine usually having darker and less twisted needles, darker and thinner bark, and less curved cones often bending backward and pointing downward. Cone scales bearing a small, recurved prickle at tip. Very common; in southern Rocky Mountains and Cypress Hills, but apparently intergrading with jack pine in northwestern Boreal forest. Syn.: *P. murrayana* Balf. The species *P. contorta* and *P. banksiana* are sometimes considered as varieties of a single species: *P. divaricata* (Ait.) Dumont, var. *divaricata* (= *P. banksiana*) and var. *latifolia* (Engelm.) Boiv. (= *P. contorta* var. *latifolia*).

*Pinus flexilis* James                                         LIMBER PINE

A small tree with a short, stout trunk, and thick branches; bark light gray becoming dark in age. Leaves 3–7 cm long, rigid with 1–4 rows of stomata on all sides. Seed cones 8–20 cm long, light brown to purplish, scales thickened and curved inward at apex. Open rocky slopes and hilltops; at altitudes between 1300 and 2000 m in southern Rocky Mountains.

*Pinus monticola* Dougl.                                    WESTERN WHITE PINE

Trees to 50 m high, with a short-branched symmetrical crown. Leaves 5–10 cm long, in 5-leaved clusters, bluish green. Cones cylindric, 10–20 cm long, thin-scaled. Only found as young trees in southern Rocky Mountains.

*Pinus ponderosa* Dougl.                                   WESTERN YELLOW PINE

Trees to 75 m high, with a spire-like crown, or flat-topped in poor sites. Branchlets orange when young. Leaves in clusters of 3, to 25 cm long. Cones

7–15 cm long, with thin scales, thickened at apex, and armed with a slender prickle. Only found as young trees in southern Rocky Mountains.

*Pinus resinosa* Ait.                                                    NORWAY PINE

Trees to 40 m high, with thick, brown, furrowed bark. Leaves in pairs 10–15 cm long, dark green. Cones ovoid, 4–8 cm long, spreading, with apex thickened and a smooth protuberance. Southeastern Boreal forest.

*Pinus strobus* L.                                                   EASTERN WHITE PINE

Trees to 50 m high, with thick, furrowed bark. Leaves slender, in cluster of five, 8–13 cm long, pale green, glaucous. Cones cylindric, 10–15 cm long with scales having a protuberance at tip. Rare; southeastern Boreal forest.

**Pseudotsuga**        Douglas-fir

*Pseudotsuga menziesii* (Mirb.) Franco                              DOUGLAS-FIR

Trees to 50 m high, with dark brown, thick, furrowed bark. Lower branches often drooping. Leaves 2–3 cm long, flat, narrowed to a short stalk. Cones pendent, 5–10 cm long; bracts with 3 teeth, projecting beyond scales. Not common; Rocky Mountains.


# CUPRESSACEAE—cypress family

Seeds in a small dry cone; branchlets
flattened; medium-sized trees. ................................................................. *Thuja*
Seeds in a small, bluish, berry-like cone;
branchlets not flattened; low or trailing
shrubs. ............................................................................................. *Juniperus*

**Juniperus**        juniper

Low shrub with short, awl-shaped or scale-like leaves, opposite or in whorls. Fruit composed of 3–6 fleshy scales, each containing a seed and joined to form a fleshy berry-like cone.

1. Leaves in whorls of 3, linear, awl-shaped,
    sharp-pointed. ................................................................ *J. communis*
   Leaves opposite, scale-like. ............................................................... 2
2. Shrubs, prostrate or ascending. ............................................ *J. horizontalis*
   Shrubs or small trees, mostly with a well-
    developed trunk. .............................................................. *J. scopulorum*

*Juniperus communis* L.                                               LOW JUNIPER

A shrub about 1–1.5 m high in some places but usually very low. Leaves 5–12 mm long, narrowly awl-shaped, pointed and dark green below, whitish and grooved above. Cones berry-like in leaf axils, bluish with a bloom, 6–10 mm in diam. The var. *depressa* Pursh is fairly common on light, rocky soil throughout most of the Prairie Provinces. Syn.: *J. sibirica* Burgsd.


60 — CUPRESSACEAE                                                    *Conifers*

*Juniperus horizontalis* Moench (Fig. 18)    CREEPING JUNIPER

A prostrate shrub, with long, gnarled, woody stems often 3–5 m long. Leaves scale-like, overlapping, each about 1.5 mm long and forming narrow branchlets up to 12 mm long. Cones bluish and berry-like about 6 mm long, at the ends of branchlets. Very common throughout drier parts of area, forming large mats on dry banks and sandy hillsides. Syn.: *Sabina horizontalis* (Moench) Rydb.

*Juniperus scopulorum* Sarg.    ROCKY MOUNTAIN JUNIPER

Usually a small tree 2–3 m high, or a large upright shrub with several stems; otherwise hardly distinguishable from *J. horizontalis.* Southern Rocky Mountains.

**Thuja**    arbor-vitae

Leaves of branchlets and twigs with a conspic-
uous resin gland. ................................................................................ *T. occidentalis*
Leaves of branchlets and twigs not with a con-
spicuous resin gland. .................................................................................. *T. plicata*

*Thuja occidentalis* L.    WHITE CEDAR

A tree to 20 m high, with widely spreading branches. Branchlets forming a flattened spray, with soft twigs. Leaves strongly appressed, keeled, with a light green or yellow resin gland below apex. Cones about 10 mm long, oblong-ovoid. Moist woods; southeastern Boreal forest.

*Thuja plicata* D. Don    WESTERN RED CEDAR

Similar to the preceding species, but with leaves not distinctly keeled, and the resin gland inconspicuous. Moist woods; southern Rocky Mountains.

## Subdivision: ANGIOSPERMAE

## Class: MONOCOTYLEDONEAE

## TYPHACEAE—cattail family

**Typha**    cattail

Staminate part of spike contiguous with pistil-
late part. ................................................................................................ *T. latifolia*
Staminate part of spike separated from pistil-
late part by 3–5 cm. ......................................................................... *T. angustifolia*

*Typha angustifolia* L.    NARROW-LEAVED CATTAIL

Marsh or aquatic plants to 3 m high, with extensive creeping roots. Leaves 4–10 mm wide, to 30 cm long. Spikes 25–40 cm long, with staminate and pistillate parts about equal, separated by 3–5 cm of bare stem. Rare; southeastern Boreal forest.

Fig. 18. Creeping juniper, *Juniperus horizontalis* Moench.

*Typha latifolia* L. (Fig. 19) COMMON CATTAIL

Similar to the preceding species, but with leaves to 30 mm wide. Spikes with staminate and pistillate parts contiguous. Very common; in slough margins, marshes, lakeshores, and riverbanks; throughout the Prairie Provinces.

## SPARGANIACEAE—bur-reed family

**Sparganium** bur-reed

1. Stigmas 2; achenes broadly obpyramidal, truncate at summit. ................................................................ *S. eurycarpum*

   Stigma 1; achenes tapering at both ends, often stipitate. ........................................................................................ 2

2. Beak of achene less than 1.5 mm long or lacking; stipe of achene less than 1 mm long or lacking; staminate head solitary. .................................................................................................... 3

   Beak of achene 1.5–6 mm long; stipe of achene 1–5 mm long; staminate heads 2 or more. ..................................................................................... 4

3. Achenes short-beaked; pistillate heads all borne in leaf axils; staminate head separated from the uppermost pistillate one. ...................................................................................... *S. minimum*

   Achenes beakless; some pistillate heads borne above leaf axils; staminate head contiguous with the uppermost pistillate one. ............................................................................. *S. hyperboreum*

4. Beak of achene flattened, strongly curved; leaves all floating, not keeled. ................................. *S. fluctuans*

   Beak of achene not flattened or strongly curved; leaves erect or floating, keeled. ................................. 5

5. Beak of achene about as long as body. ............................. *S. chlorocarpum*

   Beak of achene about half as long as body. .................................................................................................... 6

6. Leaves 2–5 mm wide, rounded on back; fruiting heads 1–2 cm in diam. ........................................ *S. angustifolium*

   Leaves 5–10 mm wide, flat on back; fruiting heads 2–2.5 cm in diam. ................................. *S. multipedunculatum*

*Sparganium angustifolium* Michx. NARROW-LEAVED BUR-REED

Stems floating and elongated, or erect to 30–50 cm high; leaves 4–10 mm wide, flat. Inflorescence with staminate heads 2–4, distant; pistillate heads 2 or 3; fruiting heads 7–10 mm in diam. Not common; in slow running water.

*Sparganium chlorocarpum* Rydb. var *acaule* (Beeby) Fern. STEMLESS BUR-REED

Stems slender, erect, to 75 cm high; leaves 2–10 mm wide, overtopping inflorescence. Staminate heads 4–9, mostly distant; pistillate heads 1–4; fruiting heads 1.5–2.5 cm in diam.

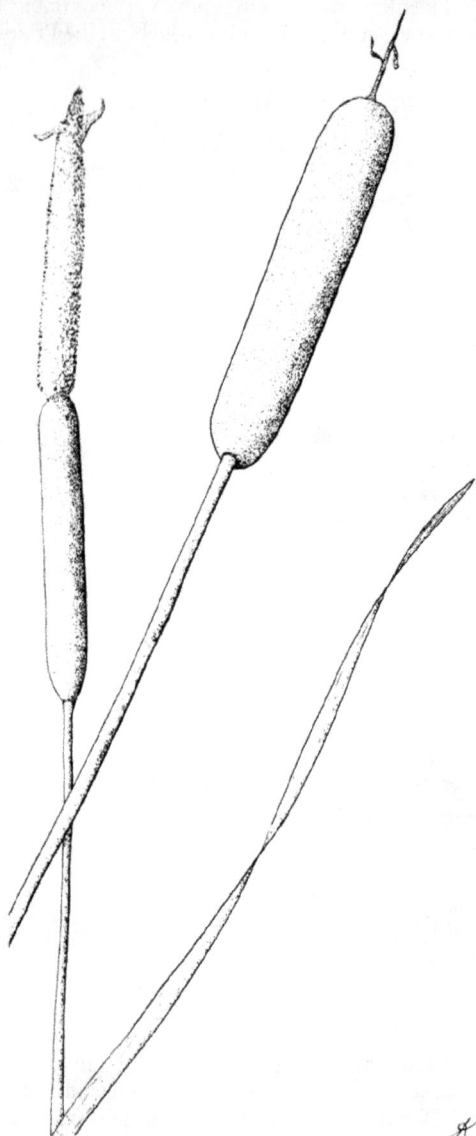

Fig. 19. Common cattail, *Typha latifolia* L.

*Sparganium eurycarpum* Engelm.                    BROAD-FRUITED BUR-REED

Stems stout, 50–150 cm high; leaves 7–10 mm wide, flat. Inflorescence branched; staminate heads 2–20; pistillate heads 1–3; fruiting heads 20–25 mm in diam.

*Sparganium fluctuans* (Morong) Robins.

Stems floating, to 1.5 m long; leaves flat, 3–8 mm wide, thin. Inflorescence branched; staminate heads 4–6; pistillate heads 3–6, mostly on the branches; fruiting heads 10–20 mm in diam.

*Sparganium hyperboreum* Laest.                    NORTHERN BUR-REED

Stems slender, 10–30 cm long; leaves 1–5 mm wide, thick. Staminate head one; pistillate heads 1–3, with at least 1 head above the axils; fruiting heads 5–12 mm in diam. Rare; in shallow water; Boreal forests.

*Sparganium minimum* (Hartm.) Fries                SMALL BUR-REED

Stems slender, floating, to 50 cm long; leaves 3–7 mm wide. Staminate head one; pistillate heads 1–3, all axillary; fruiting heads 5–7 mm in diam. Rare; in shallow water; Boreal forest.

*Sparganium multipedunculatum* (Morong) Rydb.     MANY STALKED BUR-REED

Stems stout, to 80 cm high; leaves 5–12 mm wide, flat. Staminate heads 1–4, crowded; pistillate heads 1–5, above the axils, with lower ones often on peduncles; fruiting heads 20–25 mm in diam. Not common; in shallow water; throughout the Prairie Provinces.


## ZOSTERACEAE—pondweed family

Annual or perennial aquatic or marsh plants growing entirely in water. Roots usually fibrous, often growing from the lower nodes of the stem. Leaves varying in shape from thread-like to broad, and all either floating or submersed. Flowers inconspicuous, with neither sepals nor petals.

1. Flowers bisexual, appearing above the surface of the water; stamens 2–4; leaves alternate. ......................................................................................................... 2

   Flowers unisexual, usually developed below the surface of the water; stamens 1; leaves opposite or all basal from the crown of the root. ................................................................................................ 3

2. Stamens 4; flowers in spikes; fruit without stem. .................................................................................... *Potamogeton*

   Stamens 2; flowers not in spikes; fruit long-stemmed. ............................................................................................ *Ruppia*

3. Annual plants; stemless; inflorescence either solitary in leaf axils or in spikes on summit of scape. ............................................................................................ *Lilaea*

   Plants with stems; leaves opposite. ........................................................................................ 4

4. Inflorescence of solitary flowers in axils of
    leaves. ............................................................................................................................ *Naja*
   Inflorescence clusters of flowers in axils of
    leaves. ............................................................................................................................ *Zannichelli*

## *Lilaea*     flowering quillwort

*Lilaea scillioides* (Poir.) Hauman                    FLOWERING QUILLWORT

An annual marsh or mud plant with narrow leaves, circular in cross sec
tion, and clustered. Plants 8–15 cm high, bearing spikes up to 1 cm long o
mixed male and female flowers and also solitary female flowers enclosed in a
sheath at the base of the leaves. Fruits small achenes, with those of the flower
of spikes winged and ridged, and those of the basal flowers larger and no
winged. A plant of the Pacific Coast, with very few records known from the
Canadian Prairie Provinces, these being reported in saline sloughs in the vicin
ity of Cypress Hills and the southeastern part of Alberta. Some authorities
place this species in a separate family, the Lilaeaceae, or quillwort, family
Syn.: *L. subulata* Humb. & Bonpl.

## *Najas*     naiad

*Najas flexilis* (Willd.) Rostk. & Schmidt                    SLENDER NAIAD

An annual aquatic plant. Leaves opposite, 1.0–2.5 cm long, with widened
bases and conspicuous sheaths. Flowers inconspicuous and borne in leaf axils
with male and female inflorescence borne on the same plant. Rare; in shallow
lakes and slow-moving water; near Edmonton and north of Winnipeg.

## *Potamogeton*     pondweed

Perennial aquatic plants with fibrous roots from the lower nodes of the
stems. Leaves generally submersed but in one species some floating; varying
from thread-like to broad. Flowers with neither sepals nor petals, sometimes
borne on spikes projecting from water and sometimes in axils of leaves.

1. Plants with all leaves similar and
    submersed. ............................................................................................................................ 2
   Plants with broader leaves floating and
    narrower leaves submersed. ............................................................................ *P. gramineus*

2. Leaves broad, their bases clasping stem. ............................................................ *P. richardsoni*
   Leaves linear and thread-like. ............................................................................................ 3

3. Stipules free from base of leaf. ............................................................................ *P. foliosus*
   Stipules fused with base of leaf, forming a
    sheath at least 1 cm long. ............................................................................................ 4

4. Stigmas raised on evident style; nutlets
    with 2 keels. ............................................................................................................ *P. pectinatus*
   Stigmas inconspicuous; nutlets not
    keeled. ............................................................................................................................ 5

5. Plant slender; sheaths close around stem. ............................................................ *P. interior*
   Plant coarse; sheaths 2–5 times diameter
    of stem. ............................................................................................................ *P. vaginatus*

*Potamogeton foliosus* Raf.                                    LEAFY PONDWEED

More likely to be found toward the eastern part of the Prairie Provinces, but not common.

*Potamogeton gramineus* L.                        VARIOUS-LEAVED PONDWEED

Easily distinguishable by broad floating leaves and narrower submersed ones. Not common; but may be expected in lakes; Boreal forest.

*Potamogeton interior* Rydb.                              INLAND PONDWEED

Fairly common in alkaline ponds in eastern portion of the Prairie Provinces.

*Potamogeton pectinatus* L.          SAGO PONDWEED, FENNEL-LEAVED PONDWEED

Our commonest thread-leaved pondweed; found in ponds, lakes, and streams; throughout the Prairie Provinces.

*Potamogeton richardsonii* (Benn.) Rydb.            RICHARDSON'S PONDWEED

Easily recognizable by broad clasping leaves. One of the commonest species; found in ponds and streams; throughout the Prairie Provinces.

*Potamogeton vaginatus* Turcz.                         SHEATHED PONDWEED

Distinguishable from sago pondweed by the broadened stipular sheath at the base of the narrow leaves. Fairly common; in larger sloughs and lakes; throughout the Prairie Provinces.

**Ruppia**       ditch-grass

Submersed perennial plants having hair-like stems and thread-like single-nerved leaves with a membranous sheath at base. Flowers perfect and clustered on a slender stem, the stem elongating into a spiral coil and curling up after fertilization of the flowers. Not common; in brackish and saline sloughs; throughout the entire Prairie Provinces.

Leaf sheaths 6–15 mm long; fruit 2 mm long
  or less. ................................................................................................ *R. maritima*
Leaf sheaths 18–30 mm long; fruit 3 mm or
  longer. .............................................................................................. *R. occidentalis*

*Ruppia maritima* L.                                          DITCH-GRASS

Widespread, but uncommon in the Prairie Provinces.

*Ruppia occidentalis* S. Wats.                       WESTERN DITCH-GRASS

Commoner than *R. maritima*; in sloughs over the entire Prairie Provinces.

**Zannichellia**      horned-pondweed

*Zannichellia palustris* L.                            HORNED-PONDWEED

A submersed, branching aquatic plant having thread-like, opposite, single-nerved leaves with membranous sheaths at the base. Flowers of both sexes borne in axils of leaves; fruits curved, nut-like, 3–5 mm long, with a short beak from which this plant derives its common name. Very common; in brackish ponds; throughout Prairies and Parklands.

# JUNCAGINACEAE—arrow-grass family

Flowers in a bracted, few-flowered raceme;
  stems leafy. ....................................................................................................... *Scheuchzeria*
Flowers in a bractless, many-flowered spike-
  like raceme; stems leafless. ........................................................... *Triglochin*

## *Scheuchzeria*

*Scheuchzeria palustris* L.

A rush-like bog plant with stems 10–40 cm high. Leaves 10–40 cm long, with the upper ones reduced to bracts; sheaths of basal leaves often 10 cm long, with ligule to 12 mm long. Inflorescence a few-flowered raceme; flowers white; perianth segments 3 mm long, 1-nerved, membranous. Follicles 4–8 mm long, with lower ones on pedicels to 25 mm long. Rare; in peat bogs and lake-shores; Boreal forest.

## *Triglochin*      arrow-grass

Perennial marsh or semiaquatic herbs with short rootstocks. Leaves linear or rush-like, semicylindric, all basal and clustered, and bearing membranous sheaths. Flowers perfect, borne in tall, slender, spike-like racemes. Fruit capsules splitting open at maturity. **Poisonous** to cattle and sheep.

Plant stout; carpels 6; fruit oblong or ovoid,
  obtuse at base. ........................................................................................... *T. maritima*
Plant slender; carpels 3; fruit linear or club-
  shaped, tapering at base. .................................................................... *T. palustris*

*Triglochin maritima* L. (Fig. 20)          SEASIDE ARROW-GRASS

A stout plant, with rootstock but no stolons. Leaves long, narrow, and half cylindric, up to 30 cm long and 3 mm wide, with bases usually covered with old leaf sheaths. Flowering stem 40–80 cm high, with flowers in a raceme at summit; raceme up to 50 cm long. Fruit about twice as long as thick, 6 mm long and 2–3 mm in diam, on short stalks. Common; found over the whole area in marshy and wet places. **Poisonous** to cattle and sheep.

*Triglochin palustris* L.          MARSH ARROW-GRASS

A slender plant, with rootstock and slender stolons. Leaves similar to the preceding species, but usually about 10–20 cm long. Flowers on slender stalks. Fruit about 3–5 times as long as thick, usually 6–8 mm long and 1 mm in diam, on fine stems paralleling the stalk. Found in marshy places throughout the Prairie Provinces, but not nearly so common as *T. maritima*. **Poisonous** to cattle and sheep.

Fig. 20.  Seaside arrow-grass, *Triglochin maritima* L.

## ALISMACEAE—water-plantain family

Flowers all perfect; fruit a single ring of
carpels. .......................................................................................................... *Alisma*

Flowers of one sex, with lower ones female
and upper ones male; fruit in dense globular
heads. .......................................................................................................... *Sagittaria*

### *Alisma*  water-plantain

Perennial aquatic or marsh plants growing from stout corm-like rootstocks producing offshoots. Leaves generally ovate or oblong, with several parallel veins; borne on long stems from the crown of the root. Flowers perfect,
with 3 green sepals and 3 white petals. Fruits flat-sided short-beaked achenes.

Flowering stems usually not extending above
the leaves, its branches generally curved
downward; achenes as wide as long; leaves
generally ovate-lanceolate. ........................................................... *A. gramineum*
Flowering stems extending above the leaves,
its branches ascending; achenes longer than
wide; leaves usually ovate. ............................................... *A. plantago-aquatica*

*Alisma gramineum* K. C. Gmel.          NARROW-LEAVED WATER-PLANTAIN

Leaves usually long and narrow or ovate-lanceolate, 30–80 mm long.
Flowering stems usually shorter than the leafage. Common in Prairies. Syn.:
*A. geyeri* Torr.

*Alisma plantago-aquatica* L. (Fig. 21)          COMMON WATER-PLANTAIN

Leaves oblong to ovate, 5–18 cm long. Inflorescence diffuse, to 30 cm or
higher, much overtopping the leafage. Common throughout the Prairie Provinces.

**Sagittaria**          arrowhead

Perennial water or marsh plants from fleshy rootstocks. Leaves all basal,
usually broadly arrow-shaped, borne on long stems; occasionally reduced to
mere thickened stems (called phyllodia) replacing blades. Flowers having 3
sepals and 3 waxy white petals; female flowers, borne lowest on the stem,
developing before male flowers. Fruits achenes, crowded into globular heads.

Beak of achenes erect and very minute. ................................................. *S. cuneata*
Beak of achene horizontal and long. .................................................. *S. latifolia*

*Sagittaria cuneata* Sheld. (Fig. 22)          ARUM-LEAVED ARROWHEAD

Readily identified by its broad arrow-shaped leaves and waxy white
flowers, borne in whorls of 3 on the long stem. Leaves up to 10–15 cm long,
but submersed stems often lacking blade, the thickened stem (phyllodia)
replacing the blade. Flowers 6–12 mm in diam, later forming globular seed
heads up to 15 mm in diam, green at first turning black later. Very common; in
water and marshy habitats; in all parts of the Prairie Provinces.

*Sagittaria latifolia* Willd.          BROAD-LEAVED ARROWHEAD

Resembling *S. cuneata*, but differing in having the beak of achene horizontal and fairly long. Leaves very variable, being either broadly or quite narrowly arrow-shaped. Not common over most of the Prairie Provinces, but
found in central and eastern Parklands.

Fig. 21.   Common water-plantain, *Alisma plantago-aquatica* L.

Fig. 22.  Arum-leaved arrowhead, *Sagittaria cuneata* Sheld.

# HYDROCHARITACEAE—frog's-bit family

*Elodea*        waterweed

*Elodea canadensis* Michx.                    CANADA WATERWEED

Submersed perennial aquatic plants with fibrous roots springing from the nodes of the stems. Leaves not stalked, borne in whorls of 2–4, oblong-ovate, to 5 mm long. Flowers rarely found, but borne on the end of an apparent stalk, with male flowers on one plant and female on another. Not common but probably widespread; in still or slow-moving water; throughout the Prairies and Parklands.

# GRAMINEAE—grass family

Annual or perennial herbs (Fig. 23), with stems usually hollow except at the nodes. Leaves borne on two sides of stem, one growing from each node, and consisting of a sheath, usually split, enveloping the main stem and a blade, which is a continuation of the sheath, growing at an angle to the stem. Blades flat or rolled, narrow, and without stalks. Inflorescence (Fig. 24) in spikes, racemes, or panicles, each composed of spikelets borne on a rachis. Each spikelet consisting of a series of bracts alternating on either side of a rachilla. The spikelet breaking off below or above the empty glumes; this point of articulation is important in identifying grasses. These bracts are called glumes. The lowermost glumes empty or sometimes one or both missing or replaced by bristles; other glumes containing the floret, each having an enveloping palea and lemma; flowering glumes often having bristles called awns. Flower usually perfect, but sometimes unisexual. In the perfect flower, feather-like stigmas arising from the style, usually having three stamens, and the ovary developing into a caryopsis (or grain). A magnifying glass may be needed for identifying the various parts of a grass plant.

Gramineae is such a large family that subdivisions have been made. The family has been divided into two subfamilies, Panicoideae and Festucoideae. These subfamilies are further divided into tribes. Species in 10 of these, Paniceae, Andropogoneae, Phalarideae, Hordeae, Chlorideae, Agrostideae, Aveneae, Festuceae, Zizanieae, and Oryzeae (Fig. 25) may be found in the Prairie Provinces.

The following characters distinguish grasses from sedges and rushes:

| Character | Gramineae—grasses (p. 73) | Cyperaceae—sedges (p. 181) | Juncaceae—rushes (p. 246) |
|---|---|---|---|
| Culm | Usually hollow; cylindrical or flattened | Filled with pith, rarely hollow; usually 3-sided | Filled with sponge-like pith, cylindrical |
| Nodes | Conspicuous | Indistinct | Indistinct |
| Leaf arrangement | Two-ranked | Three-ranked | Principal leaves basal or nearly so |
| Leaf blade | Usually flat, often folded, involute or bristle-like; glabrous or pubescent | Flat, plicate, or bristle-like; rarely pubescent | Channeled or round; usually glabrous |
| Leaf margins | Smooth, scabrous, or ciliate | Usually scabrous | Smooth |

*Monocots*

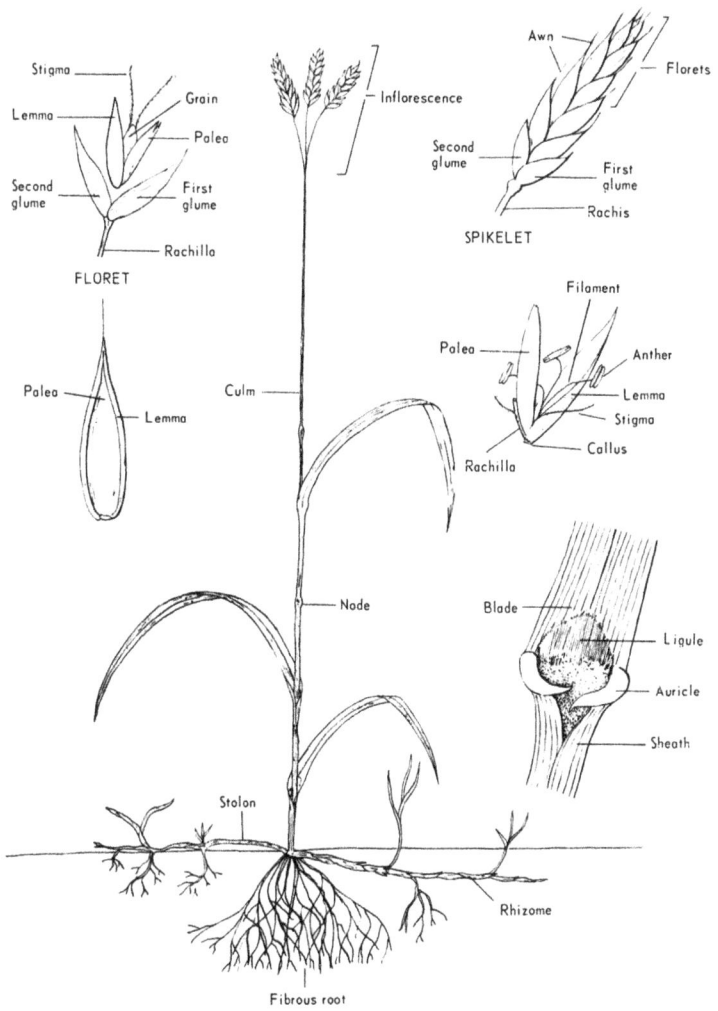

Fig. 23.   Structure of a typical grass.

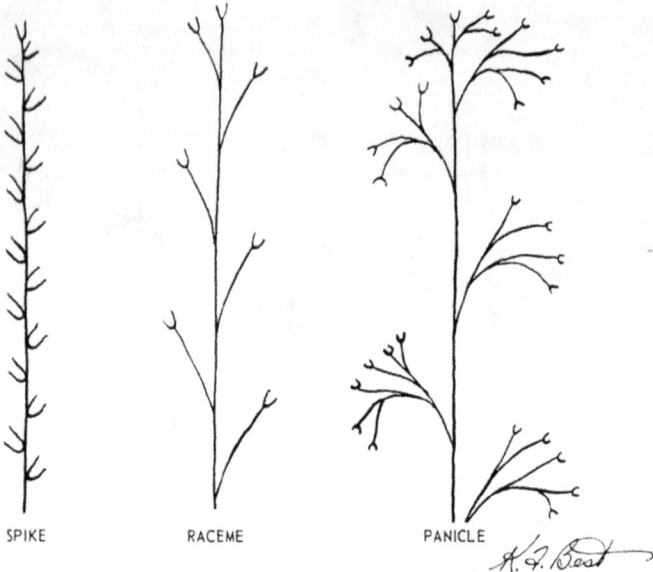

SPIKE  RACEME  PANICLE

Fig. 24.  Types of inflorescences of grasses.

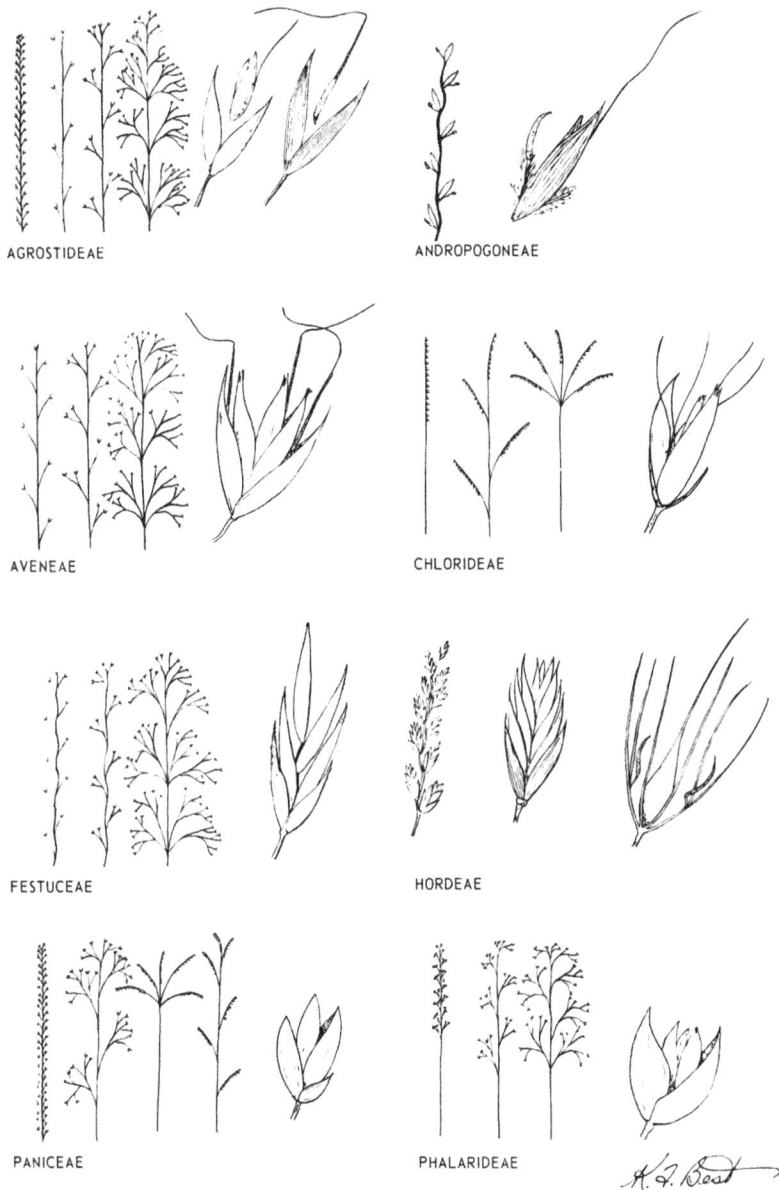

AGROSTIDEAE

ANDROPOGONEAE

AVENEAE

CHLORIDEAE

FESTUCEAE

HORDEAE

PANICEAE

PHALARIDEAE

Fig. 25.    Floral characteristics of grass tribes.

# Keys to the Grasses

1. Spikelets one- to many-flowered; sterile florets, if present, above the fertile ones (except in Phalarideae, having two sterile florets below the fertile one); articulation usually above the glumes; spikelets usually laterally compressed. ........... Subfamily 1. FESTUCOIDEAE 2

   Spikelets with one fertile floret, and a sterile floret, if present, below the fertile one; articulation below the glumes; spikelets compressed dorsally. .......................... Subfamily 2. PANICOIDEAE 9

2. Spikelets with one fertile floret above and two (or rarely one) sterile florets below; no sterile florets above the fertile florets. ................................................................... Tribe 6. PHALARIDEAE

   Spikelets with one to mànÿ fertile florets; sterile florets above the fertile florets. ..................... 3

3. Spikelets unisexual, one-flowered; the rachis disarticulating below the glumes. ............................. Tribe 8. ZIZANIEAE

   Spikelets perfect or if unisexual not as above; the rachis disarticulating above the glumes. ..................... 4

4. Spikelets disarticulating below the glumes, one-flowered; strongly compressed laterally; glumes small or absent. ................................................................ Tribe 7. ORYZEAE

   Spikelets disarticulating above or below the glumes, at least one of these well-developed. ..................... 5

5. Spikelets sessile on the rachis in spikes. ..................... 6

   Spikelets pedicellate in open or contracted, sometimes spike-like, panicles. ..................... 7

6. Spikes terminal, solitary; spikelets arranged singly or in twos or threes on opposite sides of the rachis; the rachis disarticulating in some species. ................................................ Tribe 2. HORDEAE

   Spikes usually digitate or racemose; spikelets arranged on one side of the rachis in two rows. ................................................ Tribe 5. CHLORIDEAE

7. Spikelets one-flowered, with only occasionally two florets, in some spikelets in *Muhlenbergia*. ................................................ Tribe 4. AGROSTIDEAE

   Spikelets two- to many-flowered. ..................... 8

8. Glumes shorter than the first floret; lemmas awnless or awned from the tip or a bifid apex. ................................................ Tribe 1. FESTUCEAE

   Glumes as long as the lowest floret, sometimes enclosing the whole spikelet; lemmas awnless or awned from the back. ................................................ Tribe 3. AVENEAE

*Monocots*

9. Glumes membranous; fertile lemma and palea indurate, sterile lemma like the glumes. ................................................... Tribe 9. PANICEAE

Glumes indurate; fertile and sterile lemmas equal in texture. ................................. Tribe 10. ANDROPOGONEAE

# FESTUCOIDEAE subfamily

## Tribe 1.  FESTUCEAE

1. Rachilla with long silky hairs; inflorescence a large, plume-like panicle; tall reeds. ................................................ *Phragmites*

   Rachilla without silky hairs; inflorescence not plume-like; not tall reeds. .......................... 2

2. Glumes exceeding the uppermost floret in the spikelet, these shiny. ............................. *Dupontia*

   Glumes shorter than the lowest floret. ................ 3

3. Plants dioecious with long rhizomes or stolons; lemmas without a tuft of hairs at the base. ....................................................... *Distichlis*

   Plants monoecious, or if dioecious lemmas with a tuft of hairs at the base, or plants annual. ................................................. 4

4. Lemmas prominently 3-nerved. ............................ 5

   Lemmas 5- to many-nerved, often obscurely so. ........................................................ 7

5. Inflorescence hidden among sharp-pointed leaves; plants annual (Chlorideae). ............................................ *Munroa*

   Inflorescence not as above. ................................ 6

6. Spikelets many-flowered; glumes and lemmas keeled; plants annual. ........................ *Eragrostis*

   Spikelets 2-flowered; perennial. ....................... *Catabrosa*

7. Callus of florets bearded. .................................. 8

   Callus of florets not bearded. ........................... 9

8. Plants with rhizomes; lemmas erose at tip, awnless. .......................................................... *Scolochloa*

   Plants with fibrous roots; lemmas bifid at tip, awned. ......................................................... *Schizachne*

9. Spikelets strongly compressed, arranged in one-sided clusters on stiff panicle branches. ........................................................ *Dactylis*

   Spikelets and inflorescence not as above. ........ 10

10. Lemmas obscurely nerved; panicle loose, with drooping branches; plants large. ............. *Arctophila*

    Lemmas distinctly nerved, or if not, plants not as above. ..................................... 11

## Tribe 2. HORDEAE

## Tribe 3. AVENEAE

## Tribe 4. AGROSTIDEAE

Lemmas not thicker and harder than the glumes. ............................................................................... 5

2. Spikelets dorsally compressed; glumes shiny; lemmas awnless; rachis disarticulating below the glumes. .............................. *Milium*

Spikelets not dorsally compressed; glumes dull; lemmas awned; rachis disarticulating above the glumes. ........................ 3

3. Awn weak, short, and deciduous; fruit plump. ...................................................... *Oryzopsis*
Awn firm, persistent; fruit slender. ........................................... 4

4. Awn of lemma simple. ............................................................ *Stipa*
Awn of lemma trifid. ............................................................... *Aristida*

5. Rachis disarticulating below the glumes. ....................................... 6
Rachis disarticulating above the glumes. ....................................... 8

6. Glumes long-awned; panicle dense, spike-like, appearing silky. ........................................ *Polypogon*
Glumes awnless. ................................................................................. 7

7. Panicle open, with spreading branches; florets stipitate; glumes not united. ........................... *Cinna*
Panicle dense, spike-like; florets not stipitate; glumes united at base. ....................... *Alopecurus*

8. Lemmas awned at the apex or short-pointed. ................................... *Muhlenbergia*
Lemmas awnless, or the awn inserted below the apex. ................................................... 9

9. Inflorescence dense and spike-like. .............................. *Phleum*
Inflorescence an open, branched, panicle. ....................... 10

10. Florets bearing a tuft of hairs at the base. ....................... 11
Florets without hairs at the base. ................................... 12

11. Lemmas awned from the middle or near the base; 3- to 5-nerved; the glumes longer than the lemma. ........................ *Calamagrostis*
Lemmas awnless, 1-nerved; the first glume shorter than the lemma. .................... *Calamovilfa*

12. Glumes longer than the lemma; palea small or obsolete; callus often somewhat hairy. ...................................... *Agrostis*
Glumes shorter than the lemma; palea well-developed; callus smooth. ........................ 13

13. Lemma 1-nerved; mature grain plump, free from the lemma and palea. ............... *Sporobolus*
Lemma 3- to 5-nerved; mature grain not plump, not readily freed from the lemma and palea. ................................ *Arctagrostis*

# Tribe 5.   CHLORIDEAE

1. Plants with imperfect flowers; monoe-
   cious or dioecious. .................................................................. *Buchloe*
   Plants with perfect flowers. ..................................................... 2

2. Spikelets with more than one perfect
   floret; inflorescence a few-flowered
   head hidden among sharp-pointed
   leaves. ................................................................................... *Munroa*
   Spikelets with only one perfect floret;
   inflorescence not as above. ..................................................... 3

3. Spikelets with one or more sterile florets
   above the fertile ones. ........................................................ *Bouteloua*
   Spikelets without sterile florets. ............................................. 4

4. Spikes very slender; leaves short and
   narrow. ......................................................................... *Schedonnardus*
   Spikes not very slender; leaves long and
   wide. ...................................................................................... 5

5. Glumes unequal, narrow; spikelets 6–15
   mm long; plants with scaly rhizomes. ................................... *Spartina*
   Glumes equal, boat-shaped; spikelets 3
   mm long; plants with fibrous roots,
   annual. ............................................................................. *Beckmannia*

# Tribe 6.   PHALARIDEAE

Lower florets staminate; spikelets brownish,
   shiny; glumes rounded on the back. ............................... *Hierochloe*
Lower florets neutral; spikelets greenish, not
   shiny; glumes boat-shaped, keeled. ................................... *Phalaris*

# Tribe 7.   ORYZEAE

One genus; glumes lacking; lemmas awnless. ....................... *Leersia*

# Tribe 8.   ZIZANIEAE

One genus; spikelets unisexual. ............................................. *Zizania*


# PANICOIDEAE subfamily

# Tribe 9.   PANICEAE

1. Spikelets with one to many bristles form-
   ing an involucre. ...................................................................... *Setaria*
   Spikelets without bristles. ....................................................... 2

2. Glumes or sterile lemmas awned. ................................... *Echinochloa*
   Glumes and sterile lemmas awnless. ....................................... 3

3. Spikelets in digitate slender racemes at tip
of culms. ............................................................................ *Digitaria*

Spikelets in open panicles with slender
branches. ............................................................................ *Panicum*

## Tribe 10.  ANDROPOGONEAE

Racemes reduced to one or a few joints; these
racemes numerous in large open panicles. .................................... *Sorghastrum*

Racemes several- to many-jointed; these
racemes solitary or digitately clustered. ........................................ *Andropogon*

### *Agrohordeum*     wild rye

*Agrohordeum macounii* (Vasey) Lepage     MACOUN'S WILD RYE

Plants densely tufted, erect, 50–100 cm high. Sheath glabrous or some
times pubescent; blades to 5 mm wide, scabrous. Spike slender, erect to some
what nodding, 4–12 cm long; rachis disarticulating at maturity; spikelets over
lapping, 8–10 mm long, mostly 2-flowered; glumes 5–8 mm long, very narrow
lemmas 7–10 mm long; awns 1–2 cm long. Moist, often alkali flats; Prairie
and Parklands.

### *Agropyron*     wheatgrass

1. Plants with rhizomes. ............................................................................ 2
   Plants with fibrous roots. ...................................................................... 7

2. Stomata forming fine white lines on
   underside of blades; rhizomes long,
   yellow white. ................................................................................ *A. repens*
   Stomata not forming white lines. ...................................................... 3

3. Glumes rigid, 10–12 mm long; plants
   glaucous, bluish green; auricles often
   purplish. ........................................................................................ *A. smithii*
   Glumes not rigid, mostly 6–9 mm long;
   plants green or gray green; auricles yel-
   lowish green. ................................................................................ 4

4. Awns of lemmas recurved or divergent. ........................................ 5
   Awns of lemmas straight. .................................................................... 6

5. Lemmas pubescent. ........................................................................ *A. albicans*
   Lemmas glabrous or scabrous. ...................................................... *A. albicans*

6. Lemmas pubescent. .................................................... *A. dasystachyum*
   Lemmas glabrous or scabrous. .................................................... *A. riparium*

7. Plants annual, introduced. ............................................................ *A. triticeum*
   Plants perennial ................................................................................ 8

8. Spikes short, with spikelets very closely
   spaced on the rachis; introduced
   species. ...................................................................................... *A. cristatum*
   Spikes elongated, with spikelets not
   closely spaced on the rachis; native
   species. ...................................................................................... 9

9. Spikelets awnless or awn-tipped. ............................................................................ 10

Spikelets long-awned, with awns 1–4 cm
long. ........................................................................................................................ 12

10. Glumes narrow; rachilla scaberulous;
blades involute. ....................................................................................... *A. spicatum*

Glumes wide, 2–2.5 mm; rachilla villous;
blades flat or nearly so. ............................................................................................ 11

11. Glumes with thin margins, awn-tipped;
lemmas usually pubescent; spike 3–8
cm long. ........................................................................................... *A. latiglume*

Glumes not with thin margins, awnless;
lemmas glabrous; spike 10–25 cm long. ............................... *A. trachycaulum*

12. Awn straight or nearly so; blades glab-
rous, lax; spikes 5–20 cm long. ...................................................... *A. subsecundum*

Awns divergent or bent. ............................................................................................ 13

13. Spikelets not closely spaced on the rachis,
barely overlapping. ........................................................................... *A. spicatum*

Spikelets closely spaced on the rachis,
overlapping. ............................................................................................................... 14

14. Culms tufted, decumbent to ascending,
often flexuous; spikes often nodding. .................................................... *A. scribneri*

Culms erect; spikes erect. ......................................................................... *A. bakeri*

*Agropyron albicans* Scribn. & Smith          AWNED NORTHERN WHEATGRASS

Plants with tufted culms 40–70 cm high arising from slender rhizomes.
Blades flat to involute, 1–3 mm wide, glabrous. Spike 6–15 cm long; spikelets
loosely overlapping, 4- to 8-flowered, 1–1.5 cm long; glumes 6–9 mm long,
sparsely pubescent; lemmas 7–10 mm long, densely or sparsely pubescent;
awn 1–1.5 cm long, divergent at maturity. Prairies.

A variety with glabrous lemmas has been named var. *griffithsii* (Scribn. &
Smith) Beetle.

*Agropyron bakeri* E. Nels.          BAKER'S WHEATGRASS

Plants loosely tufted, with erect culms 50–100 cm high. Sheaths and
blades glabrous, scabrous, or sparsely pubescent. Blades flat, to 8 mm wide.
Spikes 5–10 cm long; spikelets 10–15 mm long, 3- to 5-flowered, loosely imbricate; glumes 8–10 mm long, awn-tipped; lemmas 10–12 mm long, with the
awn to 4 cm long, divergent and recurved when dry. Prairies, southern Rocky
Mountains.

*Agropyron cristatum* (L.) Gaertn. (Fig. 26)          CRESTED WHEATGRASS

Plants densely tufted, with culms 30–50 cm high. Sheaths scabrous or the
lowest ones pubescent; blades to 8 mm wide, scabrous to pubescent above.
Spikes 2–7 cm long, flat; spikelets 8–15 mm long, 3- to 5-flowered, densely
crowded, spreading to ascending; glumes 4–6 mm long, awn-tipped; lemmas
6–8 mm long, awnless or awn-tipped. Introduced forage grass; widely sown for
pasture and hay, and escaped from cultivation in many areas. Usually, crested
wheatgrass includes several taxa besides *A. cristatum*: *A. desertorum* (Fisch.)

Fig. 26.  Crested wheatgrass, *Agropyron cristatum* (L.) Gaertn.

*Monocots*

Schult., with more rounded spikes and smaller, less spreading spikelets; *A. cristatiforme* Sarkar, with the spikelets very densely crowded and the culm villose below the spike; and *A. pectiniforme* R. & S., with the spikelets spaced in the spike.

*Agropyron dasystachyum* (Hook.) Scribn.          NORTHERN WHEATGRASS

Plants with tufted culms 40–70 cm high arising from slender creeping rhizomes. Sheath glabrous to slightly scabrous; blades to 6 mm wide, often involute, strongly veined and scabrous above. Spike 6–15 cm long, often involute, strongly veined, and scabrous above; spikelets 10–15 mm long, 4- to 8-flowered, loosely to closely imbricate; glumes 6–9 mm long, acute or awn-tipped; lemmas 8–10 mm long, more or less densely pubescent, awnless or with a short awn. Grasslands; throughout Prairies and Parklands.

*Agropyron latiglume* (Scribn. & Sm.) Rydb.      BROAD-GLUMED WHEATGRASS

Plants loosely tufted, with culms 20–50 cm high, ascending to geniculate. Blades 3–5 mm wide, flat, short-pubescent on both sides. Spikes 3–7 cm long; spikelets 10–15 mm long, 3- to 5-flowered, closely imbricated; glumes 7–10 mm long, very broad, thin-margined, awn-tipped; lemmas 10–12 mm long, pubescent, awn-tipped or awnless. Alpine meadows; Rocky Mountains, southwest Alberta.

*Agropyron repens* (L.) Beauv. (Fig. 27)      QUACK GRASS, COUCH GRASS

Plants with culms 50–100 cm high, arising in tufts from long, creeping, thick yellowish white rhizomes. Sheaths at first often soft pubescent, later glabrous; blades 6–10 mm wide, flat, usually sparsely pubescent above, and the underside glabrous, with stomata visible as fine white lines. Spikes 5–15 cm long; spikelets 10–15 mm long, 4- to 7-flowered; glumes 6–8 mm long, awn-tipped; lemmas 8–10 mm long, with the awn 2–8 mm long. An introduced species, often becoming weedy in waste places and gardens, rarely in cultivated fields.

*Agropyron riparium* Scribn. & Sm.      STREAMBANK WHEATGRASS

Plants with culms 30–80 cm high, arising in tufts from creeping slender rhizomes. Sheaths glabrous; blades 1–3 mm wide, often involute. Spikes 5–10 cm long; spikelets 10–15 mm long, 5- to 7-flowered, closely imbricated. Glumes 6–10 mm long, awnless; lemmas 8–10 mm long, glabrous or sparsely pubescent along margins, awnless. Ravines; Prairies.

*Agropyron scribneri* Vasey

Plants densely tufted, with culms 10–30 cm high, ascending-spreading. Sheaths usually more or less pubescent; blades 1–3 mm wide, flat, more or less pubescent, mostly basal; culm leaves very short. Spikes 3–7 cm long, often nodding or flexuous; spikelets 8–12 mm long, 3- to 5-flowered, densely crowded; glumes 6–8 mm long, tapering into an awn; lemmas 8–10 mm long, tapering into an awn; awn 15–20 mm long, strongly divergent. Alpine slopes; Rocky Mountains, southwest Alberta.

Fig. 27.  Quack grass, *Agropyron repens* (L.) Beauv.

*Agropyron smithii* Rydb. WESTERN WHEATGRASS

Plants with culms 30–60 cm high, arising singly or in small tufts from long creeping rhizomes. Sheaths glabrous, with auricles often purplish; blades 3–6 mm wide, stiff, prominently veined, usually very glaucous. Spikes 7–15 mm long, erect; spikelets 10–20 mm long, 6- to 10-flowered; glumes 10–12 mm long, rigid, tapering to a short sharp awn; lemmas 10–12 mm long, glabrous to somewhat pubescent. Moist areas, moderately alkaline river flats; Prairies and Parklands. A form with both glumes and lemmas pubescent has been described as var. *molle* (Scribn. & Sm.) Jones, and has the same distribution as the species.

*Agropyron spicatum* (Pursh) Scribn. & Smith BLUEBUNCH WHEATGRASS

Plants densely tufted, often forming large bunches, with erect culms 60–100 cm high. Sheaths glabrous; blades 3–5 mm wide, flat to loosely involute, glabrous to minutely pubescent below, finely pubescent above, usually glaucous. Spikes 7–17 cm long, sometimes longer; spikelets 10–15 mm long, 6- to 8-flowered, mostly shorter than the internodes; glumes 6–8 mm long, acute or awn-tipped; lemmas 8–10 mm long, with the awn 10–20 mm long, strongly divergent. Prairies, southwest Alberta. A form with awnless lemmas has been described as var. *inerme* Heller.

*Agropyron subsecundum* (Link) Hitchc. (Fig. 28) AWNED WHEATGRASS

Plants loosely tufted, with erect culms 50–100 cm high. Sheaths densely pubescent in young plants, becoming glabrous; blades 6–10 mm wide, flat, pubescent when young. Spikes 6–15 or 20 cm long, erect to slightly nodding; spikelets 12–20 mm long, imbricated, 5- to 7-flowered, often twisted to one side of the rachis; glumes 12–15 mm long, broad, tapering into a short awn; lemmas 10–12 mm long, with the awn 10–30 mm long. Moist areas, fescue prairie, woodlands; Parklands and Boreal forest. Often considered to be a variety of *A. trachycaulum* (Link) Malte var. *unilaterale* (Cass.) Malte.

*Agropyron trachycaulum* (Link) Malte (Fig. 29) SLENDER WHEATGRASS

Plants tufted, with culms 50–100 cm high, erect or somewhat decumbent at base. Sheaths glabrous, sometimes purplish at base; blades 4–6 mm wide, flat, somewhat scabrous on both sides. Spikes 10–25 cm long, usually erect to slightly nodding; spikelets 15–20 mm long, 5- to 8-flowered, somewhat imbricated or remote; glumes 10–12 mm long, awnless; lemmas 12–15 mm long, awnless or awn-tipped. Moist areas, around sloughs and lakes, open woods and meadows; throughout the Prairie Provinces.

*Agropyron triticeum* Gaertn.

Plants loosely tufted, annual, with culms 10–30 cm high, decumbent at base or erect. Blades 2–3 mm wide, flat. Spikes 10–15 mm long, oval to ovate, thick; spikelets 5–7 mm long, densely crowded; glumes 4–6 mm long, awnless; lemmas 6–8 mm long, awnless. Introduced; southeast Alberta.

Fig. 28.   Awned wheatgrass, *Agropyron subsecundum* (Link) Hitchc.

*Monocots*

Fig. 29.   Slender wheatgrass, *Agropyron trachycaulum* (Link) Malte.

***Agrostis***    bent grass

1. Palea present, at least half as long as the
   lemma. ................................................................................................ 2

   Palea absent or minute. ........................................................................ 4

2. Plants tufted; rachilla prolonged behind
   the palea as a bristle. .................................................... *A. thurberiana*

   Plants rhizomatous or stoloniferous;
   rachilla not prolonged behind the
   palea. .................................................................................................. 3

3. Plants often with long stolons; panicle
   contracted even when in flower; blades
   to 3 mm wide. ........................................... *A. stolonifera* var. *major*

   Plants often rhizomatous, sometimes
   rooting at the nodes of decumbent
   stems; panicle open when in flower;
   blades usually more than 3 mm wide. .............................. *A. stolonifera*

4. Panicle narrow, with the branches
   appressed or ascending. ...................................................................... 5

   Panicle open, with the branches spreading
   or reflexed. .......................................................................................... 6

5. Culms usually less than 3 dm high; pani-
   cle to 6 cm long, purple. ................................................. *A. variabilis*

   Culms usually more than 3 dm high;
   panicle to 30 cm long, greenish. .................................... *A. exarata*

6. Panicle very diffuse; spikelets about 2 mm
   long; lemmas awnless. ....................................................... *A. scabra*

   Panicle not very diffuse; spikelets about 3
   mm long; lemmas awned. ................................................. *A. borealis*

*Agrostis borealis* Hartm.                              NORTHERN BENT GRASS

Plants tufted, with erect culms 5–40 cm high. Sheaths glabrous; leaves 1–3 mm wide, mostly basal. Panicle 1–10 cm long, with lower branches whorled, spreading; glumes 2–3 mm long; lemma awned, slightly shorter than glumes, with the awn usually twisted, bent, and exserted. Slopes and moist areas; Rocky Mountains, Boreal forest.

*Agrostis exarata* Trin.                                      SPIKE REDTOP

Plants tufted, with culms 20–100 cm high, slender to somewhat stout. Sheaths glabrous to somewhat scabrous; blades 2–10 mm wide, scabrous. Panicle 5–30 cm long, narrow, dense or slightly open; glumes 2.5–4 mm long, acuminate to awn-tipped, scabrous in the keel; lemma 1.5–2 mm long, with the midrib excurrent as a short awn. Moist areas; Prairies and Parklands.

*Agrostis scabra* Willd. (Fig. 30)                        ROUGH HAIR GRASS

Plants tufted, sometimes densely so, with erect culms 30–70 cm high. Sheaths glabrous; blades 1–3 mm wide, scabrous. Panicle 15–25 cm long, diffuse, with the branches spreading to ascending, branching above the middle; glumes 2–2.5 mm long, acuminate; lemma 1.5–1.7 mm long, sparsely pubescent at base. Meadows, open woods, waste places; throughout the Prairie Provinces.

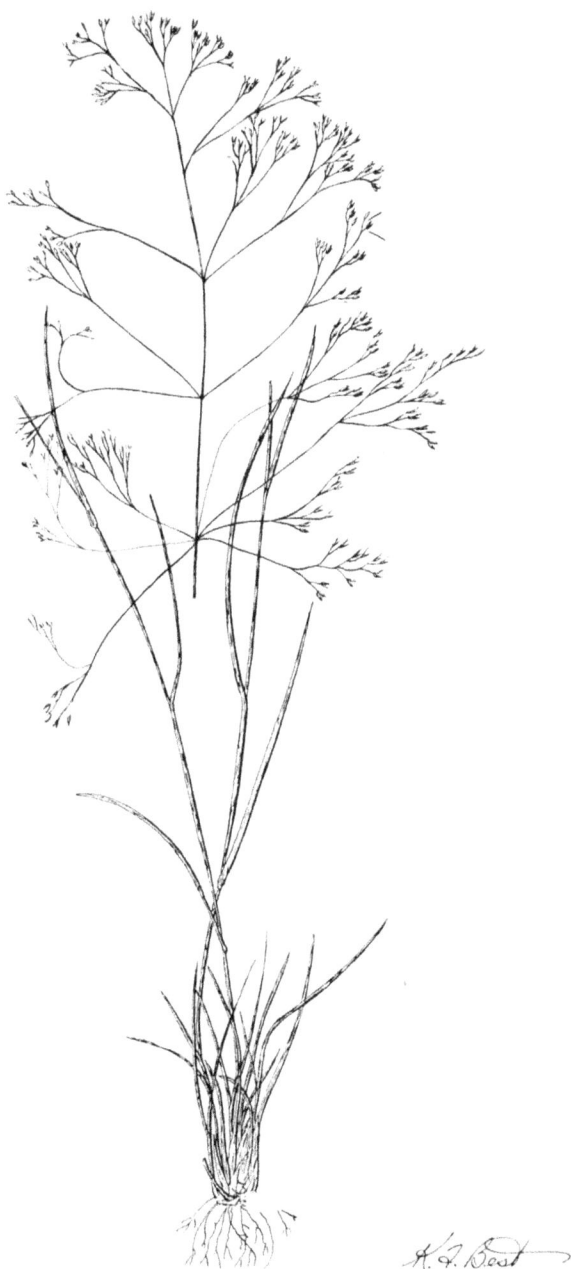

Fig. 30.   Rough hair grass, *Agrostis scabra* Willd.

*Agrostis stolonifera* L. REDTOP

Plants tufted, often with rhizomes, with culms erect or decumbent, or forming stolons, which root at the internodes. Sheaths glabrous; blades 4–8 mm wide, flat or somewhat involute to folded, scabrous on both sides. Panicle 5–30 cm high, usually pyramidal; glumes 2–3.5 mm long, acuminate, somewhat pubescent on the keel; lemma 1.5–2.3 mm long, sometimes short-awned; palea about half as long as lemma. Moist areas in Boreal forest, or as escape from cultivation. Often listed as *A. alba* L., but this name appears to be based on a rather depauperate specimen of *Poa nemoralis* L. and is therefore invalid. Our plants are in part var. *genuina* (Schur) A. & G. (Fig. 31), with rather numerous and creeping rhizomes, and var. *major* Farw. forma *palustris* (Huds.) J. & W., lacking or with few short rhizomes.

*Agrostis thurberiana* Hitchc.

Plants in small tufts, with erect culms 20–40 cm high. Sheaths glabrous; blades 1–2 mm wide, mostly basal, crowded. Panicle 5–7 cm long, somewhat drooping; glumes 1.5–2 mm long, often purplish; lemmas about as long as glumes; palea about 1.5 mm. Bogs and meadows; Rocky Mountains, southwest Alberta.

*Agrostis variabilis* Rydb. ALPINE REDTOP

Plants densely tufted, with erect culms 10–25 cm high. Sheaths glabrous; blades about 1 mm wide, flat. Panicle 2–5 cm long, narrow, with branches ascending; glumes 2–2.5 mm long, purple; lemma 1.5 mm long. Rocky slopes and creek banks; Rocky Mountains, southwest Alberta.

**Alopecurus**     water foxtail

1. Panicles thick, about 10 mm wide. ............................................................................ 2
   Panicles slender, about 5 mm wide. ......................................................................... 3
2. Panicles very woolly, ovoid, to 4 cm long;
     plants with rhizomes. ........................................................................ *A. alpinus*
   Panicles not very woolly, cylindric, to 10
     cm long; plants without rhizomes. ......................................... *A. pratensis*
3. Culms erect or spreading; awn of lemma
     barely exserted; anthers orange yellow. ................................... *A. aequalis*
   Culms usually decumbent, geniculate;
     awn of lemma much exserted; anthers
     pale yellow or purplish. .................................................... *A. geniculatus*

*Alopecurus aequalis* Sobol. SHORT-AWNED FOXTAIL

Plants forming large tufts or sods, with culms 15–60 cm high, erect or spreading. Blades 1–4 mm wide, gray green. Panicle slender, cylindric, 2–7 cm long, 4 mm wide; glumes 2.5–3 mm long; lemmas 2.5 mm long, with the awn inserted at the middle, hardly exserted. Anthers orange. Moist to wet soils, around sloughs, lakeshores, river flats; common throughout the Prairie Provinces.

*Alopecurus alpinus* J. E. Smith ALPINE FOXTAIL

Plants from creeping rhizomes, with stiff culms 10–60 cm high, erect or decumbent at base. Sheaths glabrous, often inflated. Blades 3–5 mm wide.

Fig. 31.   Redtop, *Agrostis stolonifera* L. var. *genuina* (Schur.) A. & G.

Panicle thick, ovoid or oblong, 1–4 cm long, 1 cm wide, woolly; glumes 3–4 mm long, woolly; lemmas 3–4 mm long, with the awn inserted near the base exserted. Rare; southwest Saskatchewan, southern Rocky Mountains.

*Alopecurus geniculatus* L. WATER FOXTAIL

Plants forming tufts or sods, with culms to 60 cm high, at first spreading later geniculate ascending, often rooting at the lower nodes. Blades to 6 mm wide, gray green, strongly nerved. Panicle cylindric, to 5 cm long, 7 mm wide glumes 2.5–3 mm long, short pubescent toward the apex; lemmas 2.5 mm long short pubescent, with the awn inserted at the base, well-exserted. Anthers light yellow. Rare; moist and wet soils; throughout the Prairie Provinces.

*Alopecurus pratensis* L. MEADOW FOXTAIL

Plants from short rhizomes, with culms to 100 cm high, erect or decum bent at base. Sheaths glabrous; blades to 6 mm wide or wider, flat. Panicle cylindric, 3–10 cm long, often wider than 1 cm; glumes about 5 mm long, whit ish with green veins; keel and margins ciliate; lemmas 4 mm long, with the awn inserted at the base, 1 cm long, exserted. Introduced forage grass, occa sionally seeded in moist meadows; throughout the Prairie Provinces.

**Andropogon**     bluestem

1. Racemes solitary on peduncles. ................................................................. *A. scoparius.*
   Racemes 2–5 on peduncles, digitate. .......................................................... 2

2. Plants with long rhizomes; spikes villous. ........................................ *A. hallii*
   Plants with short rhizomes; spikes not
   villous. .............................................................................................. *A. gerardi*

*Andropogon gerardi* Vitman (Fig. 32) BIG BLUESTEM

Plants in large tufts, with offshoot on short rhizomes; culms 100–150 cm high. Sheaths pubescent; blades flat, 6–10 mm wide, somewhat pubescent at base, otherwise glabrous, blue green to glaucous. Racemes 3–6, 5–10 cm long usually purplish; sessile spikelet 7–10 mm long, with first glume having awn 10–20 mm long, geniculate, twisted below; pedicellate spikelet 7–10 mm long awnless, staminate. Prairies, southeastern Parklands.

*Andropogon hallii* Hack. SAND BLUESTEM

Plants in tufts with extensive creeping rhizomes; culms 100–150 cm high; sheaths somewhat pubescent to glabrous; blades 6–10 mm wide, glabrous blue green to purplish. Racemes 3–6, 5–10 cm long, purplish, densely villous with gray or yellowish hairs; sessile spikelet 7–10 mm long, with first glume having awn 5 mm long; pedicellate spikelet 7–10 mm long, awnless, staminate Sand hills; southeastern Parklands.

*Andropogon scoparius* Michx. (Fig. 33) LITTLE BLUESTEM

Plants densely tufted, with short scaly rhizomes; erect culms 30–70 cm high. Sheaths keeled, glabrous; blades 5–8 mm wide, flat to folded, light green to blue green, glaucous. Racemes 3–6 cm long, several to many on long ascending peduncles, with rachis pilose; sessile spikelet 6–8 mm long, with the

Fig. 32.   Big bluestem, *Andropogon gerardi* Vitman.

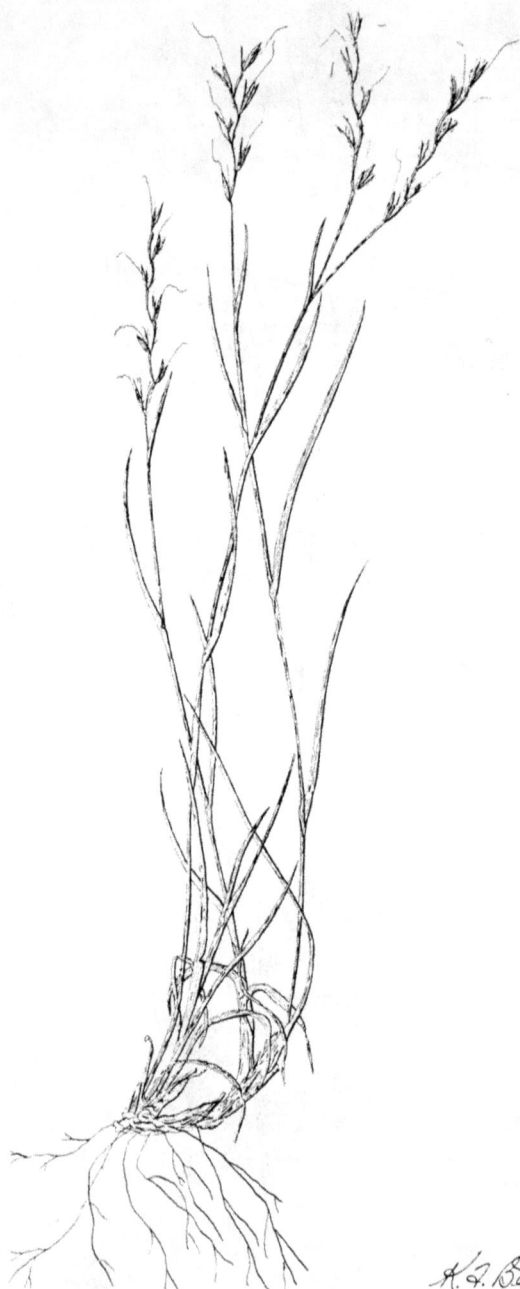

Fig. 33. Little bluestem, *Andropogon scoparius* Michx.

awn 8–15 mm long; pedicellate spikelet reduced, short-awned. Usually calcareous soils; Prairies, southern Manitoba, southern Saskatchewan, southwest Alberta.

### Arctagrostis

*Arctagrostis latifolia* (R. Br.) Griseb.

Plants with creeping rhizomes; tufted culms 30–100 cm high, or occasionally to 150 cm. Sheath smooth, glabrous; blades 4–10 mm wide, flat, rather lax. Panicle 3–25 or to 30 cm long, somewhat open to narrow, with the branches 1–12 cm long, erect and appressed to ascending, usually 3 or 4 at one side of the stem, often interrupted or lobed; spikelets 2.5–5 mm long, with first glume 2–3.5 mm long, 1- to 3-nerved, and second glume 3–5 mm long, 3-nerved; lemmas 3.5–5 mm long, 5-nerved, pubescent, with the palea as long as the lemma, obscurely to strongly 3-nerved, pubescent. A species of variable size, appearance, and color; panicle ranging from yellowish green through purple-tinged to deep purple; anthers yellow or purple. Rare; in marshes and tundra; Boreal forest.

### Arctophila

*Arctophila fulva* (Trin.) Anderson

Plants with creeping rhizomes; culms 30–90 cm high, erect or decumbent. Sheaths strongly ribbed; blades 4–10 mm wide, flat. Panicle 6–20 cm long, usually open, with branches 5–8 cm long, ascending to spreading, and the lower ones often reflexed; spikelets 3–5 mm long, usually purplish, 2- to 6-flowered, with the first glume 2–3 mm long, and the second glume 3–4 mm long; lemmas 2.5–4 mm long, glabrous, obtuse. Rare; wet shores and marshes; Boreal forest.

### Aristida          three-awn

Plants easily distinguished by the 3 lemmas.

*Aristida longiseta* Steud.                                    RED THREE-AWN

Plants tufted, with erect culms 20–30 cm high. Sheaths scabrous with minute prickles. Blades 1–2 mm wide, often involute or bristle-like, very scabrous on both sides. Panicle 3–4 cm long without the awns, narrow, erect, few-flowered; branches ascending or appressed; first glume 8–10 mm long; second glume 15–20 mm long; lemmas 12–15 mm long, terete, with awns 6–8 cm long, divergent. Rare; in grassland, on slopes; Prairies.

### Avena          oat

| | |
|---|---|
| Spikelets mostly 2-flowered; awn usually straight or absent; lemmas glabrous. | *A. sativa* |
| Spikelets mostly 3-flowered; awn stout, twisted, geniculate; lemmas pubescent. | *A. fatua* |

*Avena fatua* L. (Fig. 34) WILD OAT

Plants tufted, with culms 30–75 cm high or sometimes to 100 cm, erect, stout. Sheaths smooth; blades 4–8 mm wide, flat. Panicle open, with branches ascending, spreading; spikelets mostly 3-flowered; glumes 2–2.5 cm long; the lower lemmas 15–20 mm long, the upper ones shorter, pubescent with stiff brown hairs, with the awn to 4 cm long, twisted and geniculate. A noxious weed in cultivated fields. Has become very common in many areas; throughout Prairies and Parklands.

*Avena sativa* L. OAT

The cultivated form, derived from *A. fatua*, differing from it in having mostly 2-flowered spikelets, glabrous lemmas, and the awn small and straight or lacking. Widely cultivated throughout the Prairie Provinces.

**Beckmannia**        slough grass

*Beckmannia syzigachne* (Steud.) Fern. (Fig. 35) SLOUGH GRASS

Plants tufted, with culms 30–70 cm high, erect or decumbent, and sometimes spreading, geniculate ascending. Sheaths glabrous; blades to 12 mm wide, light green. Panicle 10–25 cm long; branches 1–5 cm long, appressed to ascending; spikes crowded, 1–2 cm long; spikelets 1-flowered, 3 mm long, with glumes 3 mm long, wrinkled, deeply keeled; lemmas 7.5 mm long. Wet areas, slough margins, lakeshores, valuable as hay, and readily eaten by cattle; throughout the Prairie Provinces.

**Bouteloua**        grama

Spikes 10–50, pendulous. ............................................................................. *B. curtipendula*
Spikes 1–3, spreading. ...................................................................................... *B. gracilis*

*Bouteloua curtipendula* (Michx.) Torr. SIDE-OATS GRAMA

Plants tufted, from short scaly rhizomes, with erect culms 50–75 cm high. Sheaths usually pubescent; blades to 7 mm wide, flat to convolute, with few scattered hairs. Spikes 35–50, usually pendulous, secund; spikelets 5–8 in a spike, 6–10 mm long; fertile lemma 6 mm, mucronate; sterile lemma with 3 awns. Rare; in grassland; southeastern Parklands.

*Bouteloua gracilis* (HBK.) Lag. (Fig. 36) BLUE GRAMA

Plants densely tufted, from very short scaly rhizomes, with erect culms 20–50 cm high. Sheaths glabrous or sparsely pubescent; blades to 3 mm wide, light green to dark or reddish green, twisted. Spikes 1–3, usually 2, 2.5–5 cm long, falcate, spreading at maturity; spikelets numerous, 70–80, about 5 mm long; first glume 2–3 mm, second glume 3–4.5 mm; fertile lemma 3–4 mm long, pubescent; sterile lemma densely bearded. Common; in grassland; Prairie and Parkland.

Fig. 34. Wild oat, *Avena fatua* L.

Fig. 35. Slough grass, *Beckmannia syzigachne* (Steud.) Fern.

Fig. 36.   Blue grama, *Bouteloua gracilis* (HBK.) Lag.

## *Bromus*    brome

1. Plants with rhizomes. ................................................................................................. 2
   Plants with fibrous roots. ........................................................................................... 3

2. Lemmas glabrous; blades and culms glabrous or somewhat scabrous. ................................................... *B. inermis*
   Lemmas pubescent, at least along the margins; blades pilose above; culms pubescent at nodes. ................................. *B. pumpellianus*

3. First glume 1-nerved. ............................................................................................ 4
   First glume 3- to 5-nerved. ...................................................................................... 7

4. Plants annual; sheath and blade softly pubescent; spikelets drooping; culms to 5 dm high. ................................... *B. tectorum*
   Plants perennial; sheath and blades pubescent or glabrous; culms to 1 m high. ............................................... 5

5. Lemmas rather evenly pubescent on the back, but more densely so on lower margins. ......................................... *B. purgans*
   Lemmas pubescent along margins and lower part only, with the upper part glabrous. ........................................... 6

6. Sheath and blades soft pubescent; blades 5–8 mm wide; lemmas narrow; ligules prominent; awns 6–8 mm long. ..................... *B. vulgaris*
   Sheath and blades glabrous or short pubescent; blades 4–6 mm wide; lemmas broad; ligures inconspicuous. ...................... *B. ciliatus*

7. Plants annual; sheath and blade softly villose. ................................................................ 8
   Plants perennial; sheath and blade glabrous or sparingly pubescent. ............................................. 9

8. Spikelets 5–8 mm wide, inflated; first glume 5-nerved; awns strongly divergent. ............................................... *B. squarrosus*
   Spikelets less than 5 mm wide; first glume 3-nerved; awns flexuous. .......................................... *B. japonicus*

9. Spikelets strongly flattened, with lemmas keeled; panicle with ascending branches. ........................................ *B. marginatus*
   Spikelets not strongly flattened, with lemmas rounded on the back; panicle drooping. ....................................... 10

10. Second glume 3-nerved. ....................................................................................... *B. porteri*
    Second glume 5-nerved. ....................................................................................... *B. kalmii*

*Bromus ciliatus* L.  FRINGED BROME

Plants loosely cespitose, with 2–4 culms together 60–100 cm high. Culms and sheaths glabrous or short pubescent at the nodes; blades 4–10 mm wide, glabrous to sparsely villose. Panicle 10–20 cm long, with branches slender, often flexuous, spreading or drooping, up to 15 cm long. Spikelets drooping, 1–2 cm long, 4- to 10-flowered, with first glume lance-subulate, 5–7 mm long, 1-nerved, and second glume lanceolate, 7–10 mm long, 3-nerved; lemmas 8–11 mm long, pubescent along the lower half to three-quarters margin, glabrous or nearly so on the back, with the awn to 5 mm long. Woods, and forest openings; throughout the Prairie Provinces.

*Bromus inermis* Leyss.  SMOOTH BROME

Erect culms 50–100 cm high, from long creeping rhizomes. Culms, sheaths, and blades usually glabrous or nearly so; blades 5–12 mm wide. Panicle 6–20 cm long, usually contracted and secund at maturity, with 1–4 branches at each node. Spikelets 1.5–2.5 cm long, 7- to 10-flowered, with first glume 6–9 mm long, 1-nerved, and second glume 8–10 mm long, 3-nerved; lemmas 10–12 mm long, 5- to 7-nerved, somewhat pubescent at base, awnless or short-awned. Introduced for forage, and now acclimatized in many areas, particularly in low-lying areas; throughout Prairies and Parklands.

*Bromus japonicus* Thunb.  JAPANESE CHESS

Plants cespitose, with culms erect or ascending 40–70 cm high. Culms long pubescent on the nodes, with sheaths densely villous; blades 1–3 mm wide, long villous. Panicle 10–20 cm long, open, with branches slender, spreading or drooping, somewhat flexuous. Spikelets 1–2 cm long, 5- to 9-flowered, with first glume 4–6 mm long, broad, 3-nerved, and second glume 6–8 mm long, 5-nerved; lemmas 7–9 mm long, 9-nerved, with awn 8–10 mm long, twisted and flexuous at maturity. Weed along roadsides, waste places, and fields; southwest Alberta.

*Bromus kalmii* Gray

Plants loosely cespitose, with few culms together 50–100 cm high. Culms usually pubescent at the nodes, with lower sheaths pubescent, and upper ones glabrous or nearly so; blades 5–10 mm wide, pubescent on both sides. Panicle 5–10 cm long, with branches slender, flexuous, drooping. Spikelets few, 1.5–2.5 cm long, 5- to 10-flowered, with first glume 6–7 mm long, 3-nerved, and second glume 7–9 mm long, 5-nerved; lemmas 8–10 mm long, 7-nerved, pubescent on the back, densely so along margins, with the awn 2–3 mm long. Rare; margins of woods, openings; southeastern Parklands.

*Bromus marginatus* Nees

Plants cespitose, with few culms together 50–100 cm high. Sheaths usually retrorsely pubescent; blades to 12 mm wide, scabrous to sparsely pubescent. Panicles 10–20 cm long, with branches ascending to spreading; spikelets 15–25 mm long, 6- to 10-flowered, with first glume 6–9 mm long, 3-nerved, and second glume 10–15 mm long, 5-nerved; lemmas 10–15 mm long, keeled, minutely pubescent, with the awn 5–7 mm long. Open woods; southern Rocky Mountains.

*Bromus porteri* (Coult.) Nash

Plants loosely cespitose, with culms 30–60 cm high, slender, pubescent on the nodes. Sheaths pilose to glabrous; blades 2–4 mm wide, scabrous. Panicle nodding, 5–10 cm long, with branches slender flexuous, spreading or drooping. Spikelets 2–2.5 cm long, 5- to 7-flowered, with first glume 5–7 mm long, 3-nerved, and second glume 7–9 mm long, 3-nerved; lemmas 9–12 mm long, densely and evenly pubescent on the back, with the awn 2–4 mm long. Margins of woods and openings; Alberta.

*Bromus pumpellianus* Scribn.                    NORTHERN AWNLESS BROME

Culms erect from long creeping rhizomes. Sheaths usually pubescent, purplish at base; blades to 12 mm wide, scabrous below, often pubescent above. Panicle 6–20 cm long, with branches ascending or the lower ones divergent, usually contracted at maturity, 1–4 branches at each node; spikelets 1.5–3.5 cm long, 7- to 10-flowered, with first glume 6–9 mm long, 1-nerved, and second glume 8–11 mm long, 3-nerved; lemmas 9–12 mm long, 5- to 7-nerved, pubescent at base and along margins, awnless or the awn 1–2 mm long. Open woods, margins, and shrubbery; Cypress Hills, Rocky Mountains.

*Bromus purgans* L.                              CANADA BROME

Plants loosely cespitose, with few culms together 60–100 cm high. Culms pubescent at nodes; sheaths usually retrorsely pubescent; blades 5–10 mm wide, sparsely pubescent to glabrous on both surfaces. Panicle 5–20 cm long, nodding, with branches elongate, spreading or drooping; spikelets 2–3 cm long, 7- to 10-flowered, with first glume 5–8 mm long, 1-nerved, and second glume 6–10 mm long, 3-nerved; lemmas 9–12 mm long, 7-nerved, thinly to densely and evenly pubescent on whole back, with the awn 2–8 mm long. Woods and forest margins; Parklands and Boreal forest.

*Bromus squarrosus* L.                            FIELD BROME

Plants cespitose, with culms 20–30 cm high. Sheaths and blades densely pubescent. Panicles 10–20 cm long, nodding, with branches slender, spreading or drooping; spikelets 1.5–2.0 cm long, 7- to 10-flowered, with first glume 4–6 mm long, 5-nerved, and second glume 5–8 mm long, 7-nerved; lemmas 6–9 mm long, 7-nerved, glabrous, with the awn about 10 mm long, spreading or recurved. Waste places and roadsides.

*Bromus tectorum* L. (Fig. 37)                    DOWNY CHESS

Plants cespitose, with erect culms 30–60 cm high. Sheaths and blades densely long pubescent. Panicle drooping, 10–20 cm long, with many rather short flexuous branches drooping; spikelets 2–3 cm long, 6- to 10-flowered, pubescent throughout, with first glume 5–7 mm long, 1-nerved, and second glume 8–10 mm long, 3-nerved; lemmas 10–12 mm long, 5- to 7-nerved, with the awn 12–17 mm long, straight. Waste places and roadsides.

*Bromus vulgaris* (Hook.) Shear

Plants cespitose, with culms 70–100 cm high, erect, slender, pubescent on the nodes. Sheaths and blades more or less pubescent; blades 7–12 mm wide.

Fig. 37.   Downy chess, *Bromus tectorum* L.

Panicle 10–15 cm long, with branches slender, ascending to divergent, some-what drooping; spikelets 1.5–2.5 cm long, 5- to 9-flowered, with first glume 5–8 mm long, 1-nerved, and second glume 7–9 mm long, 3-nerved; lemmas 8–10 mm long, sparsely pubescent on the back, with the awn 6–8 mm long. Woods and openings; southwest Alberta.

### *Buchloe*   buffalo grass

*Buchloe dactyloides* (Nutt.) Engelm.   BUFFALO GRASS

Plants stoloniferous, forming a dense sod, with culms of staminate plants to 20 cm high and culms of pistillate plants shorter than the leaves. Sheaths somewhat inflated; blades 1–2 mm wide, sparsely pubescent. Staminate spikes 5–15 mm long, 2 or 3 on a peduncle; spikelets 2-flowered, sessile, closely imbri-cated, with first glume 1–1.5 mm long, and second glume 1.5–2 mm long; both glumes 1-nerved; lemmas to 4 mm long, 3-nerved. Pistillate spikes with 4 or 5 spikelets, usually 2 spikes on a peduncle; the peduncle short, included in the sheaths of the upper leaves; first glume of pistillate floret 4 mm long, and sec-ond glume 5 mm long, firm, thick, rigid, rounded and expanded in the middle, enveloping the floret, and having 3 awn-like lobes at the summit; lemma mem-branous, 3-nerved, with the summit 3-lobed. Very rare; Prairies.

### *Calamagrostis*   reed grass

Medium- to tall-growing creeping-rooted grasses, with open or narrow panicles of small-flowered spikelets. Few or no basal leaves. The tall species usually occur in low, moist localities; found in all types of vegetation across the Prairie Provinces.

1. Awn geniculate, longer than or about as
   long as the glumes, usually protruding
   from glumes. ............................................................................................. 2
   Awn straight, as long as or shorter than
   the glumes, usually not protruding
   from glumes. ............................................................................................. 5
2. Awn longer than the glumes. ................................................................. 3
   Awn not longer than the glumes. ........................................................... 4
3. Panicle open, with the branches spread-
   ing; glumes acuminate, glabrous, 4–6
   mm long; callus hairs half as long as
   lemma; culms to 40 cm high. ...................................... *C. deschampsioides*
   Panicle narrow, with the branches
   appressed to erect; glumes acute, minutely scabrous, 6–8 mm long; callus
   hairs one-third as long as lemma;
   culms 60–100 cm high. ..................................................... *C. purpurascens*
4. Plants strongly rhizomatous, with culms
   often solitary; blades about 2 mm
   wide, stiff, involute; collar glabrous. ............................. *C. montanensis*
   Plants rhizomatous, with culms tufted;
   blades about 4 mm wide, rather lax,
   flat; collar pubescent. .......................................................... *C. rubescens*

5. Panicle open, with the branches spreading
   and often drooping. ............................................................ *C. canadensis*

   Panicle narrow and more or less contracted, with the branches appressed or
   ascending. ................................................................................ 6

6. Blades and upper part of culm scabrous;
   ligules of upper leaves 4–8 mm long,
   lacerate; panicle dense, to 25 mm
   wide. ...................................................................................... *C. inexpansa*

   Blades smooth or scabrous only at tip;
   culms smooth except under the panicle; ligules of upper leaves 1–3 mm
   long, entire; panicle dense or loose,
   about 10 mm wide. ................................................................ 7

7. Panicle lax, purplish; spikelets 4.5–5.5
   mm long; lemmas 3.5–4 mm long;
   leaves short, about 2 mm wide. ........................................ *C. lapponica*

   Panicle stiff, usually brownish; spikelets
   2.5–4.5 mm long; lemmas 2–3.5 mm
   long; leaves long, about 4 mm wide. .................................. *C. neglecta*

*Calamogrostis canadensis* (Michx.) Beauv.          MARSH REED GRASS

Plants tufted, with creeping rhizomes; culms 60–120 to 150 cm high.
Sheaths glabrous; blades 6–10 mm wide, flat, rather lax. Panicle 10–20 cm
long, mostly nodding, open to dense and somewhat contracted; glumes 3–4
mm long, acute, somewhat scabrous; lemma 2–3 mm long, thin, smooth, with
callus hairs abundant, as long as the lemma; awn delicate, straight, inserted
just below the middle. In marshes, lakeshores, meadows, and moist woods;
throughout the Prairie Provinces.

In the northern part of the boreal zone var. *scabra* (Presl) Hitchc., with
larger spikelets, glumes 4.5–6 mm long, and lemmas 3.5–4 mm, and var.
*robusta* Vasey, with glumes 3–5.6 mm and lemmas 3–3.5 mm, have been found.

*Calamagrostis deschampsioides* Trin.

Plants small, tufted; culms 15–40 cm high. Sheaths smooth; blades 2–4
mm wide, glabrous. Panicle 4–8 cm long, open, pyramidal, with branches
spreading, bearing spikelets toward the tips; spikelets purplish; glumes 4–6
mm long, acuminate; lemma 3–5 mm long, with callus hairs half as long as the
lemma; awn 8–10 mm long, bent, exserted. Boreal forests, Hudson Bay, Manitoba.

*Calamagrostis inexpansa* A. Gray          NORTHERN REED GRASS

Plants tufted, with slender creeping rhizomes; culms 40–100 cm high.
Sheaths glabrous, often purplish at base; blades 2–4 mm wide, flat or somewhat involute, firm. Panicle 5–15 cm long, contracted, dense, with branches
ascending-appressed at maturity; glumes 3–4 mm long, acuminate; lemmas
2.5–3.5 mm long, with callus hairs half to three-quarters as long as the lemma;
awn inserted above the middle, a little longer than the lemma, straight. One of
the common slough grasses; around sloughs, lakeshores, marshes; throughout
the Prairie Provinces.

*Calamagrostis lapponica* (Wahl.) Hartm.

Plants with creeping rhizomes; culms 30–60 cm high, mostly solitary. Sheaths scabrous; leaves 1–3 mm wide, scabrous toward the tip. Panicle to 10 cm, narrow, contracted to somewhat open, with branches ascending; glumes 4.5–5.5 mm long; lemmas 3.5–4 mm long, with callus hairs as long as the lemma; awn delicate, curved or weakly bent. Boreal forest. Generally considered to be var. *nearctica* Porsild.

*Calamagrostis montanensis* Scribn. (Fig. 38)          PLAINS REED GRASS

Plants with extensively creeping rhizomes; culms 20–40 cm high, solitary or in small tufts, stiffly erect. Sheaths rather chartaceous; leaves 2–3 mm wide, bluish green, stiff, flat to involute. Panicle 5–10 cm long, erect, interrupted, dense, usually whitish; spikelets 4–5 mm long; glumes acuminate; lemmas 4–4.5 mm long, with callus hairs about half as long as the lemma; awn inserted above the base, 5 mm long, geniculate, exserted. In moist or moderately dry grassland; throughout Prairies and Parklands.

*Calamagrostis neglecta* (Ehrh.) Gaertn., Mey. & Schreb.   NARROW REED GRASS

Plants with long slender rhizomes; culms to 100 cm high, solitary or few in a small tuft. Sheaths glabrous; blades 2–5 mm wide, flat to convolute, usually scabrous on both sides, gray green. Panicle 5–10 cm long, very narrow, usually contracted, greenish to brownish; spikelets 3–3.5 mm; glumes acuminate; lemmas about as long as glumes, with callus hairs about three-quarters as long as the lemma; awn as long as the lemma, straight. Mostly associated with *C. canadensis* and *C. inexpansa* in somewhat acid wet soils; throughout the Prairie Provinces.

*Calamagrostis purpurascens* R. Br.          PURPLE REED GRASS

Plants tufted; rhizomes, if present, very short; erect culms 40–60 cm high. Sheaths scabrous; blades 2–4 mm wide, flat or somewhat involute, scabrous. Panicle 5–15 cm long, dense, reddish to purplish; spikelets 6–8 mm long; glumes acuminate, scabrous; lemmas as long as the glumes, with callus hairs less than half as long as the lemma; awn exserted near the base, about 1 cm long, geniculate, exserted. Rocky Mountains, Cypress Hills.

*Calamagrostis rubescens* Buckl.          PINE REED GRASS

Plants with creeping rhizomes; culms closely spaced, often sod-forming or solitary. Sheaths smooth, pubescent at the collar; blades 2–5 mm wide, flat, somewhat lax, scabrous. Panicle 6–15 cm long, often somewhat loose, with ascending branches, mostly contracted at maturity, usually rather dense, often purplish; spikelets 4–5 mm long; glumes narrow, acuminate; lemmas 4.5 mm long; callus hairs less than half as long as the lemma; awn inserted at the base, geniculate, exserted. Open pine woods; Boreal forest, Cypress Hills, southern Rocky Mountains.

Fig. 38. Plains reed grass, *Calamagrostis montanensis* Scribn.

*Calamovilfa*        sand grass

*Calamovilfa longifolia* (Hook.) Scribn. (Fig. 39)

Plants with long, yellowish, thick and tough, scaly rhizomes; culms 50–150 cm high, usually solitary. Sheaths usually smooth, rarely somewhat pubescent; blades to 12 mm wide, flat to involute, firm, dark green. Panicle 15–35 cm long, with branches ascending, and often appressed, contracted at maturity; spikelets 6–7 mm long; first glume 4 mm; lemma 6–6.5 mm, with callus hairs copious, 4–5 mm long. In sandy areas of Prairies and Parklands; throughout the Prairie Provinces.

*Catabrosa*        brook grass

*Catabrosa aquatica* (L.) Beauv.        BROOK GRASS

Plants with long rhizomes; culms 10–40 cm long, decumbent, rooting at the nodes. Sheaths often somewhat compressed, glabrous, often purplish at base; blades 4–8 mm wide, glabrous, light green, flat, lax. Panicle 10–30 cm long, pyramidal, with branches slender, spreading or sometimes reflexed; spikelets about 3 mm long, usually purplish, with first glume 1 mm long, and second glume 1.5 mm long; lemma 2.5–3 mm long, broad, strongly 3-nerved, with the apex scarious, irregular; palea as long as the lemma. In streams, ponds, swampy areas; throughout the Prairie Provinces in Prairies and Parklands.

*Cinna*        wood grass

*Cinna latifolia* (Trev.) Griseb.        SLENDER WOOD GRASS

Plants loosely tufted, with culms to 150 cm high. Sheaths smooth; blades 8–15 mm wide, flat, lax. Panicle 15–30 cm long, greenish or yellowish, with branches slender, spreading or drooping; spikelets 3.5–4.0 mm long; glumes 3–3.5 mm long; lemma 3 mm long, 3-nerved, with awn to 1 mm long. Throughout Boreal forests.

*Dactylis*        orchard grass

*Dactylis glomerata* L. (Fig. 40)        ORCHARD GRASS

Plants forming large tussocks; culms to 100 cm high. Sheaths strongly compressed, keeled; blades 4–10 mm wide, flat or folded, pale green or somewhat glaucous. Panicles 5–20 cm long, with branches solitary or 2 together, secund, ascending, appressed at maturity; spikelets 5–7 mm long, mostly 3- to 4-flowered, clustered at the tips of the branches, with first glume 5–6 mm long, and second glume 6–7 mm long, often ciliate on the keel; lemmas 6–8 mm long, mucronate, ciliate on the keel and margins. Introduced; seeded in orchards and irrigation projects; occasionally escaped.

Fig. 39. Sand grass, *Calamovilfa longifolia* (Hook.) Scribn.

Fig. 40.  Orchard grass, *Dactylis glomerata* L.

### *Danthonia*     oat grass

1. Lemmas glabrous except on the margins. ...................................................................... 2

   Lemmas pubescent on whole back. .......................................................................... 4

2. Panicle usually with a single spikelet;
   culms seldom more than 25 cm high. .............................................. *D. unispicata*

   Panicle usually with several spikelets;
   culms often more than 25 cm high. ........................................................................ 3

3. Panicle narrow, with the branches
   appressed; glumes 10–15 mm long;
   blades pilose. ........................................................................ *D. intermedia*

   Panicle open, with the branches spreading; glumes 15–20 mm long; blades
   glabrous. ........................................................................ *D. californica*

4. Glumes about 10 mm long; lemmas about
   4 mm long; blades often curled. ................................................ *D. spicata*

   Glumes about 20 mm long; lemmas about
   10 mm long; blades not curled. ............................................... *D. parryi*

*Danthonia californica* Boland.        CALIFORNIA OAT GRASS

     Plants tufted, with culms 30–80 cm high, glabrous, tending to break at the nodes. Sheaths glabrous; blades 1–3 mm wide, glabrous, pilose at the collar. Panicle 5–8 cm long, usually with 2–5 spikelets on slender, flexuous branches; spikelets 20–25 mm long, with glumes 15–20 mm long; lemma 8–10 mm long, with apical teeth long aristate, pilose at base and callus; awn twisted, geniculate, 15–20 mm long. Rare; Prairies, southern Rocky Mountains. Generally considered to be var. *americana* (Scribn.) Hitchc., usually having foliage pilose.

*Danthonia intermedia* Vasey (Fig. 41)        TIMBER OAT GRASS

     Plants tufted, with culms to 50 cm high. Sheaths pilose-pubescent; blades to 3 mm wide, flat to involute, short pubescent above, long scattered pubescent below. Panicle 2–5 cm long, with branches appressed ascending, each bearing a single spikelet; spikelets 12–15 mm long, with glumes 10–15 mm long; lemmas 7–8 mm long, with apical teeth acuminate-aristate, appressed pilose along lower margins and callus; awn twisted, geniculate, 10–15 mm long. In fescue grasslands; in Parklands, Boreal zone, Rocky Mountains, Cypress Hills, Wood Mountain.

*Danthonia parryi* Scribn.        PARRY OAT GRASS

     Plants forming large tough tussocks, with stout culms 30–60 cm high. Sheaths glabrous; blades to 3 mm wide, flat to involute, pilose at the collar, otherwise glabrous. Panicle 3–7 cm long, with branches appressed-ascending, 1–2 cm long, each bearing a single spikelet; spikelets 20–25 mm long, with glumes about 20 mm long; lemma about 10 mm long, with apical teeth acuminate, pilose on the back and margins; awn twisted and geniculate, 15–20 mm long. Prairies, southern Rocky Mountains.

*Danthonia spicata* (L.) Beauv.        POVERTY OAT GRASS

     Plants usually in rather small tufts, with culms 20–50 cm high. Sheaths usually pilose-pubescent; blades 1–3 mm wide, flat to involute, often flexuous,

Fig. 41.   Timber oat grass, *Danthonia intermedia* Vasey.

usually pilose-pubescent, especially at the collar. Panicle 2–5 cm long, with stiff branches 2–3 cm long, the lower ones with 2 or 3 spikelets, the upper ones bearing a single spikelet; spikelets to 15 mm long, with glumes 10–12 mm long; lemmas 3–5 mm long, with apical teeth acuminate, sparsely pubescent across the back. Rock outcrops, dry prairie; Parklands, Boreal forest, Rocky Mountains.

*Danthonia unispicata* (Thurb.) Munro ONE-SPIKE OAT GRASS

Plants tufted, sod-forming when abundant, with culms 15–25 cm high. Sheaths pilose, or the lower ones glabrous; blades to 3 mm wide, flat, light green, pilose. Panicle a single spikelet, or occasionally 2 or 3 spikelets, 10–15 mm long, with glumes 10–12 mm long; lemmas 3–5 mm long, glabrous, with the callus pubescent. Cypress Hills, southern Rocky Mountains.

**Deschampsia**   hair grass

1. Plants annual; very few, short leaves. ............................................... *D. danthonioides*
   Plants perennial; several to many leaves. ........................................................ 2

2. Panicle spike-like, with the branches appressed; leaves filiform. .................................................... *D. elongata*
   Panicle open, with the branches spreading; leaves not filiform. ........................................................ 3

3. Glumes longer than the florets; blades flat, rather soft. ........................................................ *D. atropurpurea*
   Glumes shorter than the florets; blades rigid, translucent between the ribs. ................................... *D. caespitosa*

*Deschampsia atropurpurea* (Wahl.) Scheele MOUNTAIN HAIR GRASS

Plants loosely tufted, with culms 40–80 cm high, purplish at base. Sheaths glabrous; blades 4–6 mm wide, flat, thin. Panicle 5–10 cm long, loose, open, with few slender branches; spikelets purplish, with glumes about 5 mm long, the second glume 3-nerved; lemmas 2.5 mm long, with callus hairs 0.8–1.2 mm long; awn of the first floret straight, that of the second one geniculate, exserted. Southern Rocky Mountains.

*Deschampsia caespitosa* (L.) Beauv. (Fig. 42) TUFTED HAIR GRASS

Plants densely tufted, with culms 60–120 cm high. Sheaths compressed, keeled, glabrous; blades to 5 mm wide, flat to folded, prominently veined, translucent between the veins. Panicle 10–25 cm long, loose, open, often nodding, with branches slender, bearing spikelets toward the tips, spreading, the lower ones to 10 cm long; spikelets 4–5 mm long, mostly 2-flowered, with the first glume 2 mm long, 1-nerved, and the second glume 3 mm long, 3-nerved; florets distant, with the lemma of the first floret about 2 mm long, and that of the second floret 1.5 mm long; awns about 1 mm long, fragile. On poorly drained, rather fertile soils; throughout the Prairie Provinces.

*Deschampsia danthonioides* (Trin.) Munro ANNUAL HAIR GRASS

Plants tufted, with culms 15–60 cm high, slender, erect. Sheaths glabrous; blades 2 mm wide, glabrous, very few. Panicle 7–25 cm long, open, with branches very slender, mostly in twos, ascending; spikelets borne toward the

Fig. 42.   Tufted hair grass, *Deschampsia caespitosa* (L.) Beauv.

tips, with glumes 4–8 mm long, 3-nerved; lemmas 2–3 mm long, pilose at base; awn 4–6 mm long, geniculate. Not yet reported, but probably in southern Rocky Mountains.

## *Deschampsia elongata* (Hook.) Munro          SLENDER HAIR GRASS

Plants densely tufted, with culms 30–100 cm high, slender, erect. Sheaths glabrous to somewhat scabrous; blades 1–1.5 mm wide, flat or folded, soft. Panicle to 30 cm long, narrow, with slender branches appressed-ascending; spikelets short appressed, pedicellate, with glumes 4–6 mm long, 3-nerved; lemmas 2–3 mm long; awns about 4 mm long, straight. Meadows and slopes; southern Rocky Mountains.

## *Digitaria*          crabgrass

First glume evident, with sheaths pubescent. ............................................. *D. sanguinalis*
First glume rudimentary or lacking, with
   sheaths glabrous. ................................................................................... *D. ischaemum*

## *Digitaria ischaemum* (Schreb.) Schreb.          SMOOTH CRABGRASS

Plants sod-forming, with culms to 60 cm high, at first erect, but soon decumbent and spreading, and branching. Sheaths glabrous, smooth; blades to 10 mm wide, narrowed at the base. Racemes 4–10 cm long, usually 2–6, reddish purple; spikelets 2–2.5 mm long, with the first glume rudimentary or lacking, and the second glume 1.5 mm long, 3-nerved; sterile lemma about 1.5 mm, 5-nerved; fertile lemma 1.5 mm, pubescent. Rare; in waste areas; Manitoba.

## *Digitaria sanguinalis* (L.) Scop.          CRABGRASS

Plants sod-forming, with culms to 60 cm high, decumbent and spreading, often rooting at the nodes. Sheaths, at least the lower ones, papillose-pilose; blades 5–10 mm wide, pubescent. Racemes to 15 cm long, usually having 3–6 spikelets 3 mm long, with the first glume scale-like, and the second glume 1.5–2 mm long, 3-nerved; sterile lemma to 2 mm long, 7-nerved; fertile lemma to 2 mm long, dark reddish purple. Rare; in waste areas; Manitoba, Saskatchewan, Alberta.

## *Distichlis*          alkali grass

## *Distichlis stricta* (Torr.) Rydb. (Fig. 43)          ALKALI GRASS, SALT GRASS

Plants sod-forming, from scaly creeping rhizomes, with tufted culms 10–40 cm high. Sheaths glabrous; blades 2–4 mm wide, flat to involute, yellowish green. Panicles 2–6 cm long, with branches erect to ascending; staminate spikelets 10–15 mm long, 8- to 15-flowered; pistillate spikelets about 10 mm long, 7- or 9-flowered; glumes 3–6 mm long; lemma 2.5–4 mm long, firm. Salt marshes, around saline sloughs, and dry saline areas; throughout Prairies and Parklands.

Fig. 43.  Alkali grass, *Distichlis stricta* (Torr.) Rydb.

## Dupontia

*Dupontia fisheri* R. Br.

Plants from creeping rhizomes, with culms 15–60 cm high, erect to decumbent at the base. Sheaths smooth, glabrous; blades 2–4 mm, flat or involute, with pronounced midrib, yellowish green. Panicle 3–10 cm long, narrow, with short branches ascending to appressed; spikelets 4–8 mm long, shiny bronze to purplish, having 3–5 florets, with the first glume 1.5–5 mm long, and the second glume 2–6 mm long, enclosing the spikelet; lemmas 3–5 mm long, pubescent, especially on the midrib toward the base, occasionally at the base only. Rare; saline marshes; along Hudson Bay.

## Echinochloa      barnyard grass

*Echinochloa crusgalli* (L.) Beauv. (Fig. 44)      BARNYARD GRASS

Plants tufted, with culms to 100 cm high, erect to decumbent, stout. Sheaths flat, compressed, keeled, glabrous; leaves 6–15 mm wide, flat or V-shaped, keeled below. Panicle 10–20 cm long; racemes erect to spreading; spikelets crowded, ovate, about 3 mm long, green- or reddish-tinged, with the first glume 1 mm long, broadly ovate, acute, 3-nerved, and the second glume 3 mm long, 5-nerved, acuminate; sterile lemma about 3 mm long, with awn 1–5 cm long; glumes and sterile lemma pubescent; fertile lemma 2.5 mm long, glabrous. Considered a weed in gardens, fields, and waste places; Prairies and Parklands. *Echinochloa pungens* (Poir.) Rydb. has been distinguished by the pubescence of glumes and sterile lemma, having hairs with a papillate base. However, *E. crusgalli* in Europe has, usually, as much papillosity as *E. pungens*. The var. *frumentacea* (Roxb.) Wight, with thick racemes and inflated spikelets, has been grown for forage, and is occasionally found. In Europe, plants with erect culms have been differentiated as var. *erecta* Soest, and those with the culms all prostrate and only the panicle ascending, as var. *depressa* Soest. Further differentiation can be made on the basis of length of awn: those with long awns, 2–5 cm, are f. *longiseta* (Doll) J. & W.; those with most of the awns shorter than 1 cm, or reduced to a mere tip, are f. *breviseta* (Doll) J. & W.

(Fig. 44 overleaf)

Fig. 44. Barnyard grass, *Echinochloa crusgalli* (L.) Beauv.

*Monocots*

*Elymus*    wild rye

Tufted, usually fairly tall grasses with spike-like awned heads. Closely related to the wheat grasses, but the double floret at each node distinguishing the rye grasses from the wheat grasses, having only one floret at the node of the rachis. Fairly palatable to livestock.

> Glumes not indurated or bent; lemmas
> pubescent, at least on the margins;
> blades stiff. ................................................................................ *E. angustus*

## *Elymus angustus* Trin.
ALTAI WILD RYE

Plants densely tufted, with culms 60–100 cm high. Sheaths glabrous; blades to 15 mm wide, flat, stiff, prominently veined. Spike 10–20 cm long; spikelets 20–25 mm long; glumes 10–18 mm long; lemmas 10–20 mm long. Introduced; forage grass; Saskatchewan.

## *Elymus arenarius* L.
SEA LIME GRASS

Plants from creeping rhizomes, with culms to 100 cm high, stout, erect. Sheaths glabrous; blades to 15 mm wide, hard and stiff, involute in drying. Spike to 30 cm long, 2.5 cm wide, stiffly erect; spikelets to 2.5 cm long, usually 2 at the upper and lower nodes, 3 at the middle nodes, mostly 3-flowered; glumes 15–25 mm long, lanceolate, pubescent; lemmas 20–25 mm long, acuminate, short pubescent. Boreal zone, along coast of Hudson Bay and Lake Athabasca. Plants with the culm pubescent below the spike and the glumes and lemmas more densely pubescent are *E. arenarius* ssp. *mollis* (Trin.) Hult.

## *Elymus canadensis* L. (Fig. 45)
CANADA WILD RYE

Plants tufted, with short rhizomes at least when young; culms 100–150 cm high. Sheaths glaucous; blades 10–20 mm wide, flat, sometimes convolute, dark green to glaucous. Spike 10–25 cm long, nodding; spikelets 20–30 mm long without awns, mostly 3 or 4 at a node; glumes 15–20 mm long, pubescent, with the awn as long as the body; lemmas 10–15 mm long, pubescent, with the awns 2–4 cm long, divergent. Beaches, riverbanks, sandy areas, and sand dunes; throughout the Prairies and Parklands.

## *Elymus cinereus* Scribn. & Merr.
GIANT WILD RYE

Plants densely tufted, forming large tussocks, with very short rhizomes; erect culms 60–120 cm high. Sheaths glabrous; blades 10–15 mm wide, flat to convolute, thick and stiff. Spike 10–25 cm long, thick, dense; spikelets 15–20 mm long, 3–5 at a node; glumes awl-shaped, with the first one 8–10 mm long, and the second one 10–15 mm long; lemmas 10–15 mm long, somewhat pubescent. Rare; riverbanks, slopes, and ravines; southern Saskatchewan, southern Alberta.

## *Elymus glaucus* Buckl.
SMOOTH WILD RYE

Plants tufted, with culms 60–120 cm high, erect or ascending. Sheaths glabrous or scabrous; blades 10–15 mm wide, flat, becoming involute on drying. Spike 5–20 cm long, erect to somewhat nodding; spikelets 20–25 mm long; glumes 8–15 mm long, lanceolate, acuminate or awn-tipped; lemmas about 10 mm long, with the awn 10–20 mm long, erect to spreading. Not common; open woods and meadows; Parklands.

## *Elymus hirtiflorus* Hitchc.
BLUE WILD RYE

Plants tufted, with slender creeping rhizomes; erect culms 40–90 cm high. Sheaths glabrous; blades 1–4 mm long, flat or involute, firm. Spike 5–15 cm

Fig. 45.   Canada wild rye, *Elymus canadensis* L.

long, erect; spikelets about 10 mm long; glumes 8–10 mm long, hirsute, tapering into an awn; lemmas 8–9 mm long, hirsute, with an awn 5–10 mm long. Woods, riverbanks; southern Rocky Mountains.

*Elymus innovatus* Beal                              HAIRY WILD RYE

Plants in small tufts from long creeping rhizomes; erect culms 50–80 cm high. Sheaths glabrous, or the lower ones scabrous; blades 6–12 mm wide, flat to convolute, scabrous on both sides. Spike 5–12 cm long, usually dense, villous; spikelets 10–15 mm long; glumes 10–12 mm long, subulate, purplish or grayish villous; lemmas about 10 mm long, awn-tipped, villous. Common throughout Boreal forest; in woods, clearings, and openings; rare in Parklands and Prairies.

*Elymus interruptus* Buckl.                    VARIABLE-GLUMED WILD RYE

Plants tufted, with erect culms 70–100 cm high. Sheaths glabrous; blades 5–12 mm wide. Spike 8–15 cm long, flexuous or somewhat nodding; spikelets 20–30 mm long; glumes about 10 mm long, awl-shaped; lemmas 10–12 mm long, scabrous or hirsute, with the awn 1–3 cm long, flexuous or divergent. Rare; woods and openings; Boreal forest, Saskatchewan, Riding Mountain.

*Elymus junceus* Fisch.                              RUSSIAN WILD RYE

Plants densely tufted, with erect culms 30–60 cm high. Sheaths glabrous, smooth; blades 3–5 mm wide, flat to convolute, scabrous on both sides, grayish green. Spike 5–12 cm long, erect; spikelets 12–15 mm long, 2 or 3 at a node; glumes 4–5 mm long; lemmas 5–7 mm long. Introduced forage grass; becoming established in coulees and roadsides in various locations in the Prairies, rarely in Parklands.

*Elymus virginicus* L.                              VIRGINIA WILD RYE

Plants in loose, small tufts; with erect culms 60–120 cm high. Sheaths glabrous; blades 5–12 mm wide, flat, scabrous on both sides. Spike 5–15 cm long, usually erect or somewhat flexuous; spikelets 10–15 mm long; glumes 13–17 mm long, 5-nerved, much widened at base; lemmas 10–12 mm long, tapering into an awn, about 1 cm long. In woods and openings, along rivers in Parklands; rare in Prairies.


***Eragrostis***      love grass

1. Plants stoloniferous, rooting at the nodes
    of prostrate culms. ................................................................... *E. hypnoides*
    Plants not stoloniferous, with culms erect
    or decumbent. ...................................................................................... 2

2. Palea long ciliate on the keels; spikelets
        mostly 1.5–2 cm long, 15- to
        40-flowered. ................................................................... *E. cilianensis*
    Palea scabrous; spikelets shorter, 8- to
        15-flowered. ................................................................... *E. poaeoides*

*Eragrostis cilianensis* (All.) Lutati                    STINKGRASS

Plants mat-forming, with culms 10–50 cm high, ascending to spreading. Sheaths pilose at the collar; blades 2–7 mm wide, flat; the nerves and margins with depressed glands, causing the unpleasant odor of the plant. Panicles 5–20 cm long, erect; the branches ascending with glands; glumes about 2 mm long, deciduous; lemmas 2–2.5 mm long. Rare in southern Parklands of Manitoba and southeast Saskatchewan; a weed in waste areas and gardens.

*Eragrostis hypnoides* (Lam.) BSP.                    CREEPING LOVE GRASS

Plants mat-forming, with culms stoloniferous, creeping and flowering culms 10–20 cm high. Sheaths glabrous; blades 1–2 mm wide, scabrous to pubescent above. Panicle 1–6 cm long, few-flowered; spikelets 5–15 mm long, 8- to 12-flowered; glumes 1.5–2 mm long; lemmas 1.5–2.0 mm long. Rare; on sandy shores in southern Parklands of Manitoba, southeast Saskatchewan.

*Eragrostis poaeoides* Beauv.                    STINKGRASS

Plants mat-forming, with culms 10–50 cm high, ascending to spreading, branching. Sheaths long pubescent toward the collar; blades glabrous. Panicle 5–10 cm long; branches with depressed glands; glumes 1.5–2 mm; lemmas 1.5–2 mm, with the keel glandular. Very rare weedy species; at Saskatoon, Sask.

*Festuca*        fescue

Small- to medium-growing grasses with paniculate heads. Useful forage grasses and very palatable to stock.

1. Leaf blades flat, up to 10 mm wide; plants
     not densely tufted. ........................................................................................... 2
   Leaf blades narrow, to 4 mm wide, usu-
     ally involute; plants mostly densely
     tufted. ........................................................................................................... 4

2. Lemmas tipped with an awn 5–20 mm
     long; panicle open, drooping. ................................................ *F. subulata*
   Lemmas awnless. ........................................................................................... 3

3. Panicle erect; spikelets 8–12 mm long;
     lemmas thin, with scarious margins. ........................................ *F. elatior*
   Panicle diffuse, with the branches spread-
     ing to reflexed; lemmas firm. ................................................ *F. obtusa*

4. Plants annual; culms solitary or in small
     tufts. ........................................................................................ *F. octoflora*
   Plants perennial. ........................................................................................... 5

5. Glumes firm, much shorter than the
     spikelet. ....................................................................................................... 6
   Glumes thin, dry, not much shorter than
     the spikelet. ............................................................................................. 10

6. Panicle open, with the branches mostly in
     pairs, ascending or spreading; lemmas
     awnless or nearly so. ................................................................ *F. viridula*

Panicle narrow, with the branches mostly appressed; lemmas awned. .................................................... 7

7. Plants loosely tufted, often with short matted rhizomes; culms often decumbent, red at the base; underside of leaf glossy. ...................................................................... *F. rubra*

   Plants densely tufted; roots fibrous. ............................ 8

8. Awns as long as, or longer than, the lemmas; blades filiform, soft; panicle secund. .......................................................... *F. occidentalis*

   Awns shorter than the lemmas; blades firm, involute, rigid. .................................................... 9

9. Culms usually more than 40 cm high; panicle 10–20 cm long, with branches somewhat spreading; blades glaucous blue green. ............................................... *F. idahoensis*

   Culms usually less than 40 cm high; panicle to 10 cm long, narrow; blades gray green. ...................................................................... *F. ovina*

10. Spikelets with 2 or 3 florets, the third floret sterile; glumes subequal, as long as the first lemma. ........................................ *F. hallii*

    Spikelets with 3–6 florets, mostly all fertile; glumes unequal, shorter than the first lemma. .................................................. 11

11. Plants grayish or bluish green; shoots coarse, with the culms 40–100 cm high. ..................... *F. campestris*

    Plants yellowish or dark green; shoots small, with the culms 30–60 cm high. .......................... *F. altaica*

### *Festuca altaica* Trin. ex Ledeb.                    NORTHERN ROUGH FESCUE

Plants tufted, with fibrous root system; culms 30–60 cm high. Sheaths glabrous to scabrous; blades 1–2.5 mm wide, 7-nerved, somewhat pubescent, flat to involute. Panicle 8–15 cm long, open to somewhat contracted; nodes with 1–3 flexuous branches, the lower ones spreading to ascending; spikelets 10–13 mm long, 3- to 5-flowered, light green often diffused with purple; first glume 4–5 mm long, 1-nerved; second glume 6–7 mm long, faintly 3-nerved; lemmas 6–8 mm long, 5-nerved, scabrous to more or less pubescent; palea equaling the lemma. In the typical form the culm and sheath are glabrous and usually lustrous. The form with scabrous to softly pubescent culms and scabrous sheaths is f. *scabrella* (Torr.) Loom. (*F. scabrella* Torr. ex Hook.). Grassland and open woods; northern Rocky mountains.

### *Festuca campestris* Rydb.                            ROUGH FESCUE

Plants tufted, with fibrous root system, often forming large tussocks; shoots coarse, with culms 40–120 cm high, erect, stout. Sheaths and culms glabrous, often lustrous; blades 2–4 mm wide, 7-nerved, rough, usually sparsely short pubescent, gray green or bluish green, flat or involute, often breaking off at the collar. Panicle 10–20 cm long, open to somewhat contracted at anthesis; nodes with 1–3 flexuous branches, the lower ones spreading to reflexed; spikelets 11–16 mm long, 3- to 6-flowered; first glume 3–6 mm long, 1-nerved; second glume 4–8 mm long, 3-nerved; lemmas 7–9 mm long, 5-nerved, glabrous to scabrous or slightly pubescent on the margins, occasionally diffused with purple. One of the important grasses in the prairies in the southern Rocky Mountains and foothills. Syn.: *F. scabrella* Torr. var. *major* Vasey, *F. doreana* Loom.

*Festuca elatior* L.                                                MEADOW FESCUE

Plants sod-forming; culms 50–100 cm high, the decumbent culms sometimes rooting at the nodes. Sheaths glabrous, reddish or purplish at the base; blades 5–8 mm wide, prominently veined, dull green above, and slightly keeled, glossy green below. Panicle 10–20 cm long, erect or nodding at the top, often somewhat secund, contracted before and after flowering; lower branches mostly in twos; spikelets linear-cylindric, 8–12 mm long, 6- to 10-flowered, yellow green or violet-tinged; first glume 3 mm long, 1-nerved; second glume 4 mm long, 3-nerved; lemmas 6–7 mm long, obscurely 5-nerved, with the apex scarious. Introduced forage grass, for seeding in moist meadows; Parklands. Tall fescue, *F. elatior* var. *arundinacea* (Schreb.) Wimm., is used for seeding in wet areas.

*Festuca hallii* (Vasey) Piper                            PLAINS ROUGH FESCUE

Plants tufted, with fibrous roots and often more or less well-developed rhizomes; culms 20–60 cm high, glabrous, often lustrous. Sheaths glabrous, lustrous, often diffused with purple; blades 1–1.5 mm wide, sparsely short pubescent, mostly gray green, 5-nerved, always involute. Panicle 6–15 cm long, open to contracted at flowering; the lowest node with 1 or 2 branches, these ascending to contracted; spikelets 7–8 mm long, with 2 or 3 florets, the third floret usually infertile; glumes membranous, often diffused with purple, lustrous; first glume 6–7 mm long, 1-nerved; second glume 6–8 mm long, 1- to 3-nerved; lemmas 7–8 mm long, 5-nerved, scabrous to short pubescent, especially on the margins, often diffused with purple. An important grass in the Parkland prairie, Cypress Hills, and Wood Mountain; sparse on sheltered slopes in Prairies.

*Festuca idahoensis* Elmer. (Fig. 46)                        IDAHO FESCUE

Plants densely tufted, with culms to 100 cm high; whole plant blue green and glaucous. Sheaths flattened, keeled, smooth; blades to 2 mm wide, folded, filiform, erect. Panicle 10–20 cm long; branches ascending to appressed; spikelets 4–7 mm long, 5- to 7-flowered; glumes 2.5 mm and 3 mm long; lemmas almost terete, 6–7 mm, with the awn 2–4 mm long. Fescue prairie, Cypress Hills, southern Rocky Mountains.

*Festuca obtusa* Biehler

Plants loosely tufted, with culms 50–100 cm high, solitary or few together. Sheaths glabrous; blades 4–7 mm wide, flat, lax, glossy. Panicle 10–20 cm long, loose, somewhat nodding; spikelet 4–8 mm long, 3- to 5-flowered; glumes about 3 and 4 mm long; lemmas about 4 mm long, coriaceous. Open woods; southeast Boreal forest and Parklands.

Fig. 46.  Idaho fescue, *Festuca idahoensis* Elmer.

*Monocots*

*Festuca occidentalis* Hook.

Plants densely tufted, with erect culms 40–100 cm high. Sheaths glabrous; blades to 2 mm wide, involute, filiform, soft and smooth. Panicle 7–20 cm long, loose, often drooping at the top; branches solitary or in twos; spikelets 6–10 mm long, 3- to 5-flowered, mostly on slender pedicels; glumes 3 and 4 mm long; lemmas 5–6 mm long, thin, with awn 6–10 mm long, straight. Woods and shrubbery, southern Rocky Mountains.

*Festuca octoflora* Walt. SIX-WEEKS FESCUE

Plants tufted, with erect culms 15–30 cm high. Sheaths flattened, keeled, scabrous or minutely pubescent; blades to 2 mm wide, filiform, twisted, strongly veined, dark green. Panicle 4–8 cm long; spikelets 6–8 mm long, 5- to 10-flowered, crowded; first glume 3–3.5 mm long, 1-nerved; second glume 3.5–4.5 mm long, 3-nerved; lemmas 4–5 mm long, with the awn 3–7 mm long. Not common; moist open ground in Prairies.

*Festuca ovina* L. SHEEP FESCUE

Plants densely tufted, with culms 10–30 cm high; whole plant gray green. Sheaths glabrous, smooth; blades about 1 mm wide, tightly rolled, filiform. Panicle 5–7 cm long, narrow, contracted before and after flowering; spikelets 4–7 mm long, green, sometimes violet-tinged, 3- to 8-flowered; first glume 1.5 mm long; second glume 2 mm long; lemmas 3–4 mm long, obscurely 5-nerved. In open prairie and forest margins throughout Parklands; in openings and open woods in Boreal forest; rare in moist areas in Prairies. A variable species, which has been divided into several varieties and forms. Plants from the Prairie Provinces are usually distinguished as *F. ovina* var. *saximontana* (Rydb.) Gl. (Syn.: *F. saximontana* Rydb.). However, these plants conform quite well to *F. ovina* ssp. *eu-ovina* Hack. var. *vulgaris* Koch, with very narrow filiform leaves, and the lemma with an awn more than 1 mm long.

*Festuca rubra* L. RED FESCUE

Plants loosely tufted, with short rhizomes; culms mostly 30–80 cm high, erect to somewhat decumbent at the base. Sheaths finely pubescent, reddish to purplish at the base; blades to 3 mm wide, thick, V-shaped to tightly folded, dark green. Panicle 5–20 cm long, erect or somewhat nodding; branches spreading during and after flowering; spikelets 7–10 mm long, 4- to 6-flowered; first glume 2.5–3 mm long; second glume 3–3.5 mm long; lemmas 5–7 mm long, often somewhat pubescent, awn-tipped or with a short awn, to about 3 mm long. Rocky Mountains, Boreal forest in Alberta. Introduced and cultivated for pasture, hay, and lawns throughout the Prairie Provinces. Very variable; most of our plants belong to ssp. *eu-rubra* Hack. var. *genuina* Hack., the loosely tufted rhizomatous form. Plants with a small panicle and densely pubescent, rather small spikelets are subvar. *arenaria* (Osb.) Hack.; Lake Athabasca, York Factory.

*Festuca subulata* Trin.

Plants tufted, with erect culms 50–100 cm high. Sheaths glabrous; blades 3–10 mm wide, flat, thin, lax. Panicle 15–40 cm long, open, drooping; branches in twos and threes, spreading to reflexed, the lower ones to 15 cm long; spikelets 7–10 mm long, 3- to 5-flowered, very open; first glume 3 mm long; second glume 5 mm long; lemmas about 4 mm long, scabrous toward the apex, tapering into an awn 5–20 mm long. Rare; moist woods; Rocky Mountains.

## *Festuca viridula* Vasey

Plants loosely tufted, with erect culms 50–100 cm high. Sheaths glabrous or scabrous; blades 4–6 mm wide, flat or involute, soft, erect. Panicle 10–15 cm long, open; branches mostly in pairs, ascending to spreading; spikelets 6–12 mm long, 3- to 6-flowered; first glume 4–6 mm long; second glume 5–7 mm long; lemma 6–8 mm long, membranous. Montane meadows; southern Rocky Mountains.

## *Glyceria*    manna grass

Mostly medium to tall semiaquatic grasses, found in sloughs, shallow water, and marshes. Quite palatable to livestock.

1. Spikelets linear, almost terete; sheaths strongly flattened. ...................................................................... 2
   Spikelets ovate or oblong; sheaths not strongly flattened. ...................................................................... 3

2. Spikelets less than 2 cm long, slender pediceled; the upper glume 2–3 mm long; lemmas 3–4 mm long; blades 2–5 mm wide. ...................................................................... *G. borealis*
   Spikelets 2–4 cm long, subsessile; the upper glume 3–4 mm long; lemmas 6–8 mm long; blades 4–8 mm wide. ...................................................................... *G. fluitans*

3. Upper glume 3-nerved; lemmas with 5 prominent nerves; ligules 5 mm or longer. ...................................................................... *G. pauciflora*
   Upper glume 1-nerved; lemmas with 7 prominent nerves; ligules less than 5 mm long. ...................................................................... 4

4. Upper and lower glumes of about equal length, both 2–2.5 mm long. ...................................................................... 5
   Upper glumes about 1 mm long, lower glumes about 0.5 mm long. ...................................................................... 6

5. Glumes acute, greenish white; lemmas purple with very narrow scarious margins; blades to 15 mm wide, with culms to 1.5 m high. ...................................................................... *G. grandis*
   Glumes obtuse, bronze purple; lemmas purple with broad scarious margins; blades 3–6 mm wide, with culms to 0.6 m high. ...................................................................... *G. pulchella*

6. Plants strongly rhizomatous, with culms to 0.8 m high; blades usually less than 5 mm wide; panicle to 2 dm long. ...................................................................... *G. striata*

Plants not strongly rhizomatous tufted, with culms to 1.5 m high; blades 5–10 mm wide; panicle 2–3 dm long. .................................................................. *G. elata*

*Glyceria borealis* (Nash) Batch.                    NORTHERN MANNA GRASS

Plants with creeping rhizomes; culms to 100 cm high, solitary or in tufts. Sheaths compressed, keeled; blades 2–4 mm wide, flat or folded. Panicle 20–40 cm long, with branches to 10 cm long, mostly drooping; spikelets 10–15 mm long, 6- to 12-flowered, often appressed, linear; glumes about 1.5 and 3 mm long; lemmas 3–4 mm long, thin, strongly 7-nerved. Wet areas, slough margins, lakeshores; common in Boreal forest; rare in Parkland and Prairie.

*Glyceria elata* (Nash) Hitchc.                       TALL MANNA GRASS

Plants with creeping rhizomes; culms 100–150 cm high, often in large tufts or tussocks, dark green. Sheaths compressed, keeled; blades 6–12 mm wide, flat, lax. Panicle 15–30 cm long, oblong, with branches spreading, drooping, the lower ones often reflexed; spikelets 4–6 mm long, ovate, 6- to 8-flowered; glumes about 0.6 and 1.3 mm long; lemmas 3 mm long, strongly 7-nerved. Moist woods, riverbanks, wet meadows; southern Rocky Mountains.

*Glyceria fluitans* (L.) R. Br.                       MANNA GRASS

Plants with creeping rhizomes, loosely sod-forming; culms to 100 cm high, lax, ascending. Sheaths compressed, keeled; blades 4–7 mm wide, dull green, flat, with margins and midrib scabrous. Panicle 10–50 cm long, narrow; spikelets 20–25 mm long, loosely 5- to 12-flowered; first glume about 3 mm long; second glume 4 mm long; lemmas 6–7 mm long, strongly 7-nerved, the 5 middle ones almost parallel. Probably found in Manitoba.

*Glyceria grandis* S. Wats. (Fig. 47)                TALL MANNA GRASS

Plants with creeping rhizomes, loosely sod-forming; culms to 200 cm high, solitary or in small tufts. Sheaths to 1 cm thick, round or compressed above, somewhat scabrous; blades 10–15 mm wide, flat, scabrous on the margins. Panicle 20–40 cm long, usually nodding, rather dense; spikelets 6–8 mm long, 4- to 7-flowered; first glume 1.5–2 mm long; second glume 2–2.5 mm long; lemmas about 2.5 mm long, mostly purplish, strongly 7-nerved. Slough margins, lakeshores, riverbanks; throughout the Prairie Provinces. Often considered a variety or subspecies of *G. maxima* (Hartm.) Holmb., namely var. *americana* (Torr.) Boiv., or ssp. *grandis* (Wats.) Hult.

*Glyceria pauciflora* Presl                          SMALL-FLOWERED MANNA GRASS

Plants with long creeping rhizomes; culms 50–100 cm high, in small tufts or solitary. Sheaths compressed, keeled, smooth; blades 5–15 mm wide, thin, flat, lax. Panicle 10–20 cm long, open to somewhat dense, flexuous; spikelets 4–5 mm long, 5- or 6-flowered, mostly purplish; glumes 1 and 1.5 mm long; lemmas 2–2.5 mm long, prominently 5-nerved. Wet areas; Rocky Mountains.

*Glyceria pulchella* (Nash) K. Schum.                GRACEFUL MANNA GRASS

Plants with rhizomes; culms 40–60 cm high, loosely tufted. Sheaths compressed, keeled; blades 2–6 mm wide, flat, yellow green. Panicle 10–15 cm

A.C. Budd

Fig. 47.   Tall manna grass, *Glyceria grandis* S. Wats.

long, with branches spreading to drooping; spikelets 5–6 mm long, mostly purplish; glumes 1 and 1.5 mm long; lemmas 2–2.5 mm long, purplish with a broad scarious margin. Uncommon; wet areas, meadows; Boreal forest, Peace River district.

*Glyceria striata* (Lam.) Hitchc.                                    FOWL MANNA GRASS

Plants with long creeping rhizomes; culms 30–80 cm high, often in large clumps. Sheaths compressed, keeled; blades 2–5 mm wide, flat or folded, light green. Panicle 10–20 cm long, erect or nodding at the tip; spikelets 3–4 mm long; first glume 0.5–0.8 mm long; second glume 1–1.5 mm long; lemmas 1.5–2.2 mm long, strongly 7-nerved, somewhat scarious at the apex. Common in shallow water; throughout the Prairie Provinces.

### *Helictotrichon*    oat grass

*Helictotrichon hookeri* (Scribn.) Henr.                         HOOKER'S OAT GRASS

Plants densely tufted, with erect culms 20–40 cm high. Sheaths compressed, keeled; blades 2–5 mm wide, flat or folded, blue green, glaucous. Panicle 5–10 cm long, narrow, with branches erect or ascending, mostly bearing a single spikelet; spikelets 12–18 mm long, 3- to 6-flowered; glumes 12–15 mm long; lemmas 10–12 mm long, firm, brown; rachilla white villous; awn 10–15 mm long, twisted and geniculate. Fairly common; in moist to moderately dry prairie; in Prairies and Parklands, Rocky Mountains, Cypress Hills, Riding Mountain, Duck Mountain.

### *Hierochloe*    sweet grass

Staminate lemmas bearing awns 5–8 mm long. ...................................................... *H. alpina*
Staminate lemmas, awnless. .............................................................. *H. odorata*

*Hierochloe alpina* (Swartz) R. & S.                                  HOLY GRASS

Plants with short rhizomes; culms 10–40 cm high, erect, tufted, leafy shoots at base. Sheath glabrous; blades 1–2 mm wide, with the basal ones elongate; culm leaves short. Panicle 3–4 cm long, narrow, with branches short, ascending; spikelets 6–8 mm long, broad; glumes 4–5 mm long, ovate; staminate lemmas about 5 mm long, ciliate on the margins, with awns 5–8 mm long, inserted below the tip; fertile lemma 5 mm long, appressed pubescent. Alpine and Arctic meadows, Boreal forest, Hudson Bay, possibly Rocky Mountains.

*Hierochloe odorata* (L.) Beauv. (Fig. 48)                         SWEET GRASS

Plants with long rhizomes; culms 30–60 cm high, erect, often solitary or with a few leafy shoots. Sheaths smooth; blades 2–6 mm wide, those of the shoots elongate; culm leaves seldom more than 5 cm long. Panicle 10–15 cm long, pyramidal, to 7 cm wide at the base; branches spreading, somewhat flexuous; spikelets about 6 mm long, lustrous golden yellow; glumes 5–6 mm long, ovate; staminate lemmas about 4 mm long, awnless; fertile lemma 4 mm long, brown, appressed pubescent. Wet areas; throughout the Prairie Provinces.

Fig. 48.  Sweet grass, *Hierochloe odorata* (L.) Beauv.

## *Hordeum*    barley

1. Plants perennial; awns slender; auricles
   missing. ............................................................................ *H. jubatum*
   Plants annual; awns stout; auricles pres-
   ent or missing. .......................................................................................... 3

2. Blades with prominent auricles. .................................................. *H. vulgare*
   Blades without auricles. ............................................................... *H. pusillum*

*Hordeum jubatum* L. (Fig. 49)                                     WILD BARLEY

Plants densely tufted; culms 30–60 cm high, erect or decumbent at the base. Sheaths pubescent, glaucous, often purplish; blades 2–6 mm wide, flat, often twisted, usually villose above, densely puberulent below, bluish green. Spike 5–10 cm long, often nodding; lateral spikelets reduced to 1–3 spreading awns; glumes of perfect floret awn-like, spreading; lemma 6–8 mm long, with the awn as long as the glumes. Several varieties are distinguishable by the length of the awns. In var. *jubatum* the awns 2.5–6 cm long, the spike about as wide as long; var. *caespitosum* (Scribn.) Hitchc. having awns 1.5–3 cm long, spikes about twice as long as wide; var. *boreale* (Hitchc.) Boiv. (= *H. brachyantherum* Nevski) having awns less than 1.5 cm long, and the spike several times longer than wide. Varieties *jubatum* and *caespitosum* very common in wetlands, brackish marshes, roadsides; throughout the Prairie Provinces. Variety *boreale* not common; occurring occasionally in saline areas; in Prairies.

*Hordeum pusillum* Nutt.                                          LITTLE BARLEY

Plants annual, with tufted culms 10–30 cm high. Sheaths soft pubescent; blades 1–3 mm wide, flat, gray green, pubescent. Spike 2–7 cm long, erect; spikelets 8–15 mm long without the awns; glumes of fertile spikelets 8–10 mm i ːg, much widened above the base, narrowing into an awn 8–15 mm long; lemmas 6–8 mm long, that of the central spikelet awned, those of the later spikelets awn-tipped. Open ground, alkali flats; southern Rocky Mountains.

*Hordeum vulgare* L.                                                 BARLEY

Plants annual, with erect tufted culms 30–100 cm high. Sheaths somewhat pubescent; blades 5–15 mm wide, strongly auricled, flat, usually pubescent. Spike 2–10 cm long without the awns; spikelets 8–15 mm long, sessile; glumes narrow, divergent at base, with a stout awn; lemmas with an awn 10–15 cm long. Cultivated in two main forms: 2-rowed barley (*H. distichon* L.), having sterile lateral spikelets; and 6- or 4-rowed barleys (*H. hexastichon* L.), having all fertile florets. In 4-rowed barley the lateral spikelets overlapping. Both types consisting of several cultivated varieties (cultivars) occurring along roadsides and railways, and in waste areas, but lacking persistence.

## *Koeleria*    june grass

*Koeleria gracilis* Pers. (Fig. 50)                                JUNE GRASS

Plants densely tufted, with erect culms 10–50 cm high. Sheaths glabrous to more or less densely pubescent; blades 1–4 mm wide, flat or involute, bluish

Fig. 49.  Wild barley, *Hordeum jubatum* L.

*Monocots*

Fig. 50.   June grass, *Koeleria gracilis* Pers.

green, more or less pubescent on one or both sides or glabrous. Panicle 3–10 cm long, spike-like, often interrupted below, contracted at maturity; spikelets 4–4.5 mm long, 2- or 3-flowered; glumes 3–4 mm long; lemmas 3–4 mm long, lustrous. Occurring in four forms: ssp. *nitida* (Nutt.) Domin is glabrous throughout, or has at most the lower sheaths somewhat pubescent; in ssp. *eugracilis* Domin at least the sheaths are pubescent; in var. *typica* Domin the blades are pubescent with long spreading hairs; in var. *glabra* Domin the blades are short pubescent or glabrate. Very common throughout grasslands and occasionally found in forest openings or open forest, especially on light calcareous soils throughout the Prairie Provinces.

### *Leersia*    cut grass

*Leersia oryzoides* (L.) Sw.                                                 RICE CUT GRASS

Plants with slender rhizomes; culms 70–100 cm high, erect or decumbent at the base. Sheaths retrorse scabrous; blades 6–12 mm wide, with the margins very rough, spinulose. Panicle 10–20 cm long, with branches slender, spreading or ascending, bearing spikelets along the upper half or two-thirds, the lower branches often included in the sheath; spikelets 4–6 mm long, 3–8 forming a spike-like raceme; lemmas about 4 mm long, stiffly ciliate on the keel and nerves; glumes absent. Wet meadows, riverbanks, and lakeshores; southeastern Parklands, Boreal forest.

### *Lolium*    rye grass

1. Glumes as long as, or longer than, the
   spikelets. ................................................................................................. 2
   Glumes shorter than the spikelets. ................................................................. 3

2. Florets compressed, 9–10 mm long. ............................................. *L. persicum*
   Florets rounded, 6–8 mm long. ................................................. *L. temulentum*

3. Lemmas awnless or nearly so. ......................................................... *L. perenne*
   Lemmas, at least the upper ones, awned;
   awns 3–5 mm long. ...................................................................... *L. multiflorum*

*Lolium multiflorum* Lam.                                            ITALIAN RYE GRASS

Plants annual or short-lived perennial, with culms to 100 cm high, erect or ascending. Sheaths glabrous, or somewhat scabrous; blades 3–5 mm wide, prominently nerved, scabrous above, with auricles usually well-developed. Spike 10–30 cm long; spikelets 1–4 cm long, numerous, mostly 10- to 15-flowered; glumes 8–10 mm long, mostly 7-nerved; lemmas 7–8 mm long, usually awned. Introduced; often in mixtures for lawn grass and occasionally persisting.

*Lolium perenne* L.                                                PERENNIAL RYE GRASS

Plants sod-forming; culms 30–60 cm high, erect or geniculate ascending. Sheaths smooth and glabrous, usually compressed, not keeled; blades to 6 mm wide, keeled, prominently veined above, glossy bright green below, with small auricles. Spike to 25 cm long, often somewhat nodding; spikelets 15–20 mm long, acute, awnless. Introduced; seeded in short-term pasture and hayland, and in mixtures for lawn grass.

*Lolium persicum* Boiss. & Hohen. (Fig. 51)          PERSIAN DARNEL

Plants annual, tufted; culms 30–60 cm high, branching at the lower nodes. Sheaths glabrous, round or slightly compressed; blades 2–6 mm wide, flat to convolute, twisted. Spike 8–12 cm long; spikelets 15–20 mm long, distant to somewhat overlapping; glumes 10–15 mm long; lemmas about 10 mm long, with the awn 5–12 mm long. Introduced; a troublesome weed in grainfields, gardens, and waste areas; has become widespread throughout the Prairies and Parklands.

*Lolium temulentum* L.          DARNEL

Plants annual, loosely tufted; culms 30–80 cm high, mostly unbranched or with little branching. Sheaths somewhat scabrous; blades 4–8 mm wide, flat, scabrous above. Spike to 25 cm; spikelets 10–25 mm long, numerous, barely imbricate; 6- to 15-flowered; glumes 12–20 mm long; lemmas 6–8 mm long, with the awn 3–5 mm long, or rarely awnless. Introduced; weedy in fields and gardens; Prairies and Parklands. The seed of this species contains a **poisonous** narcotic.

## *Melica*          melic grass

1. Lemmas awned; plants not bulbous at base. ................................................................................ *M. smithii*

   Lemmas awnless; plants mostly bulbous at base. ................................................................................ 2

2. Glumes narrow; lemmas narrow, long acuminate, pubescent. ................................................................ *M. subulata*

   Glumes broad; lemmas broad, acute or obtuse, glabrous. ................................................................ 3

3. Spikelets ascending on stout pedicels; first glume more than half as long as the spikelet. ................................................................ *M. bulbosa*

   Spikelets spreading on slender flexuous pedicels; first glume less than half as long as the spikelet. ................................................................ *M. spectabilis*

*Melica bulbosa* Geyer          ONION GRASS

Plants with short rhizomes; culms 30–60 cm high, bulbous at base. Sheaths flat; blades 2–4 mm wide, flat to involute, glabrous to somewhat pubescent. Panicle 10–15 cm long, narrow, with the short branches appressed; spikelets 7–15 mm long, papery; glumes 6–8 mm long; lemmas 5–8 mm long. Open woods and meadows; southern Rocky Mountains.

*Melica smithii* (Porter) Vasey          MELIC GRASS

Plants with elongated rhizomes; culms 50–100 cm high, not bulbous at the base. Sheaths retrorsely scabrous; blades 6–12 mm wide, flat, lax. Panicle 10–25 cm long, erect to nodding at the tip; branches solitary, distant, spreading to reflexed; spikelets 15–20 mm long, 3- to 6-flowered, often purplish; first glume 3–6 mm long; second glume 4–8 mm long; lemmas 8–10 mm long, with the awn 3–5 mm long, inserted at the bifid apex. Moist woods, meadows; southern Rocky Mountains.

Fig. 51.   Persian darnel, *Lolium persicum* Boiss. & Hohen.

*Melica spectabilis* Scribn.                                    PURPLE ONION GRASS

Plants with short rhizomes; culms 30–60 cm high, bulbous at base. Sheaths pubescent; blades 2–4 mm wide, flat to involute; panicle 10–15 cm long, with the flexuous branches spreading-ascending; spikelets 10–15 mm long, purplish; first glume 4–6 mm long; second glume 5–7 mm long; lemmas 6–10 mm long, strongly 7-nerved. Open woods and meadows; southern Rocky Mountains.

*Melica subulata* (Griseb.) Scribn.                             ALASKA ONION GRASS

Plants with short rhizomes; culms 60–100 cm high, mostly bulbous at base. Sheaths retrorsely scabrous to long pubescent; blades 2–5 mm wide, flat, thin, scabrous above. Panicle 10–20 cm long, narrow, with the branches appressed-ascending; spikelets 15–20 mm long, narrow; first glume 5–7 mm long; second glume 6–10 mm long; lemmas about 12 mm long, 7-nerved, pubescent. Moist woods; Rocky Mountains.


**Milium**      millet grass

*Milium effusum* L.                                             MILLET GRASS

Plants with short, rather stout rhizomes; culms 30–70 cm high, erect from a bent base. Sheaths smooth; blades 7–12 mm wide, flat, lax. Panicles 10–20 cm long, open, pyramidal, with the slender branches spreading, and the lower ones often reflexed; spikelets 3–3.5 mm long; glumes 3–3.5 mm long, rounded, scaberulous; lemmas 2.5–3 mm long. Very rare; moist woods and clearings; Boreal forest.


**Muhlenbergia**      muhly

A very variable genus with panicles usually narrow.

1. Panicle open with divergent capillary branches; spikelets long pedicellate. ................................................. *M. asperifolia*

   Panicle contracted, with the branches appressed or ascending; spikelets short pedicellate or sessile. ................................................................................ 2

2. Panicle very narrowly linear, usually not more than 2 mm wide; blades 1–2 mm wide. ......................................................................................... 3

   Panicle not narrowly linear, usually about 5 mm wide; blades 2–8 mm wide. ........................................................ 4

3. Plant with rhizomes; glumes ovate, 1–1.5 mm long, less than half as long as the spikelet. ................................................................. *M. richardsonis*

   Plant with fibrous roots and hard, scaly, bulb-like base; glumes acuminate-cuspidate, 2–2.5 mm long, more than half as long as the spikelet. ........................................ *M. cuspidata*

4. Lemmas awned, the awn to 10 mm long; hairs at base of lemma copious, as long as the lemma; glumes awnless or awn-tipped. ........................................................................................................ M. andina

   Lemmas not awned; hairs at base of lemma not conspicuous, usually less than half as long as lemma; glumes awnless, awn-tipped, or awned. ........................................................................................ 5

5. Glumes awnless or awn-tipped, about as long as the lemma. ............................................................ M. mexicana

   Glumes awned, much longer than the lemmas. ................................................................................................ 6

6. Sheath keeled; ligule 1–1.5 mm long; culms usually branching from the middle nodes; internodes smooth. .......................................... M. racemosa

   Sheath not keeled; ligule minute; culms simple or branching from the base; internodes puberulent. .......................................... M. glomerata

*Muhlenbergia andina* (Nutt.) Hitchc. (Fig. 52)  FOXTAIL MUHLY

Plants with elongated, wiry, and scaly rhizomes; culms 20–60 cm high, erect, puberulent below the nodes. Sheaths somewhat scabrous; blades 2–6 mm wide, scabrous. Panicles 7–15 cm long, spike-like; spikelets 3–4 mm long; glumes 3–4 mm long, scabrous on the keel; lemmas 2.5–3.5 mm long, tapering into an awn 4–8 mm long, with copious hairs at base. Very rare; wet mud soils; Duck Mountain, Boreal forest.

*Muhlenbergia asperifolia* (Nees & Mey.) Parodi  SCRATCH GRASS

Plants with thin, scaly rhizomes; culms 10–30 cm high, compressed, branching at the base. Sheaths flattened, keeled; blades to 2 mm wide, flat, very scabrous above, smooth below. Panicles 5–15 cm long, diffuse, with branches very slender, at first erect or ascending, later spreading; spikelets 1.5–2 mm long; glumes 1–2 mm long; lemmas 2 mm long. Not common; damp or marshy calcareous or moderately alkaline soils; Prairies.

*Muhlenbergia cuspidata* (Torr.) Rydb.  PRAIRIE MUHLY

Plants with hard, bulb-like scaly bases; culms 10–30 cm high, slender, wiry, densely tufted. Sheaths somewhat flattened, glabrous; blades 3 mm wide, flat to folded, prominently veined, hard. Panicles 5–10 cm long, very narrow, with branches short, appressed; spikelets 2–3 mm long; glumes 1–1.5 mm long; lemmas about 2.5–3 mm long. Slopes and crests of moderately to strongly eroded calcareous slopes; Prairies.

*Muhlenbergia glomerata* (Willd.) Trin.  BOG MUHLY

Plants with long, branching scaly rhizomes; culms 20–50 cm high, occasionally branching at the base. Sheaths scabrous; blades 2–5 mm wide, ascending, flat. Panicles 3–7 cm long, usually interrupted, narrow; spikelets 5–6 mm long; glumes about 2 mm long, with a stiff awn 3–5 mm long; lemmas about 3 mm long, awnless, long pilose at the base. Bogs and swamps; Boreal forest, where var. *cinnoides* (Link) Hermann is found.

Fig. 52.   Foxtail muhly, *Muhlenbergia andina* (Nutt.) Hitchc.

*Muhlenbergia mexicana* (L.) Trin.                                    WOOD MUHLY

Plants with creeping scaly rhizomes; culms 30–60 cm high, erect or ascending, somewhat branching below. Sheaths scabrous; blades 2–4 mm wide, flat, lax. Panicles 10–15 cm long, with short densely flowered ascending branches; spikelets 2–3 mm long; glumes 1.5–2 mm long, awn-tipped; lemmas about 2 mm long, awn-tipped, long pilose at the base. Not common; margins of woods, moist grassland; eastern Parklands.

*Muhlenbergia racemosa* (Michx.) BSP.                          MARSH MUHLY

Plants with creeping scaly rhizomes; stout culms 20–50 cm high. Sheaths flattened, keeled, scabrous; blades 3–6 mm wide, flat to folded, scabrous on both sides. Panicles 5–15 cm long, narrow, often interrupted; spikelets 5–7 mm long; glumes 4–4.5 mm long, awn-tipped; lemmas 2.5–3.5 mm long, occasionally awn-tipped, long pilose at base. Not common; meadows, margins of woods, coulees and ravines; Prairies and Parklands.

*Muhlenbergia richardsonis* (Trin.) Rydb. (Fig. 53)            MAT MUHLY

Perennial with numerous thin, hard, scaly rhizomes; culms 5–40 cm high, densely tufted, wiry, erect or decumbent. Sheaths round, smooth; blades 1–2 mm wide, flat or involute, scabrous above, smooth below. Panicles 3–10 cm long, very narrow; spikelets 2–3 mm long; first glume 1 mm long; second glume 1.5 mm long; lemmas 2.5–3 mm long. Common; in moist, often alkaline, grasslands; Prairies, Parklands, parts of Boreal forest, Peace River district.


**Munroa**      false buffalo grass

*Munroa squarrosa* (Nutt.) Torr.                        FALSE BUFFALO GRASS

Plants annual, forming mats to 50 cm across; culms prostrate, with internodes to 10 cm long. Sheaths round, pilose or ciliate at the throat, inflated; blades to 3 mm wide, stiff, in bundles at nodes and tips of branches. Spikelets 8–12 mm long, 2- to 4-flowered, in groups of two or three at the tips of culms; first glume 4–6 mm long; second glume 6–8 mm long; both glumes 1-nerved; lemmas 3–5 mm long, 3-nerved, with a conspicuous tuft of hairs halfway along the margin. Very rare; in dry grassland; Prairies.


**Oryzopsis**      rice grass

Long-leaved, tufted grasses of medium height with rather large, rice-like seeds, found in various locations.

1. Panicles diffuse, with regularly dichotomous branches; glumes with a long
   firm tip, much exceeding the long silky
   lemma. ......................................................................... *O. hymenoides*

   Panicles not diffuse, with branches erect
   or somewhat spreading; glumes not
   sharp-pointed; lemmas short pubescent
   or glabrous. .......................................................................... 2

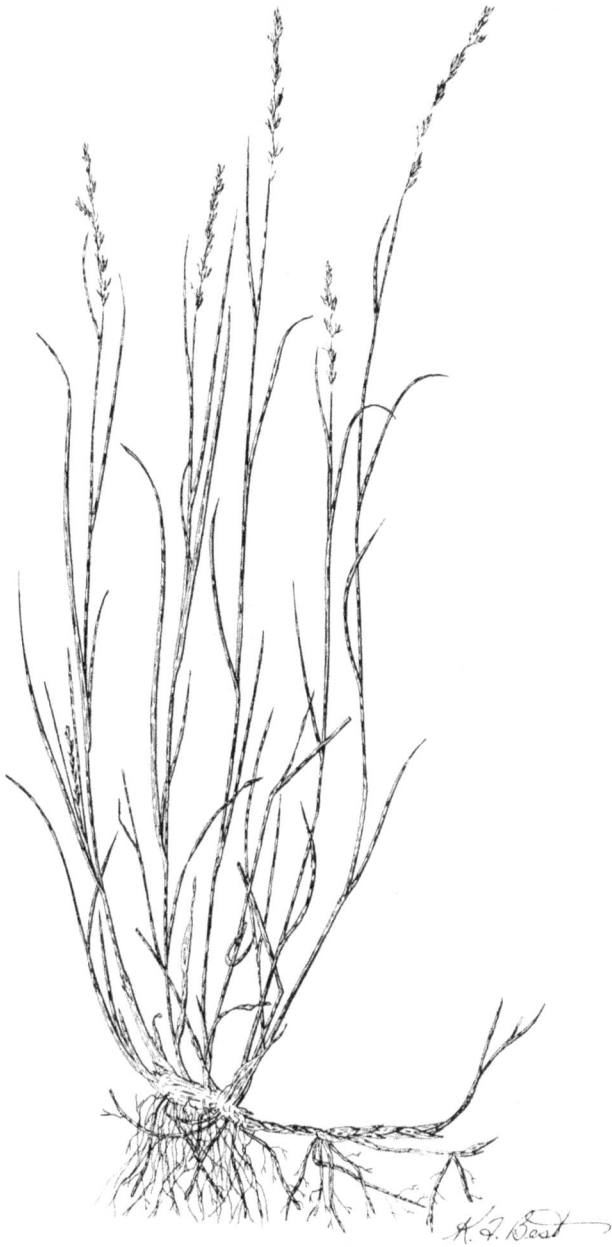

Fig. 53.    Mat muhly, *Muhlenbergia richardsonis* (Trin.) Rydb.

2. Lemmas glabrous, with awns 5–10 mm long; blades flat or involute, 0.5–2 mm wide. ........................................................ *O. micrantha*

   Lemmas pubescent. ........................................................ 3

3. Blades flat, evergreen, 4–10 mm wide; spikelets 6–8 mm long not including the awns; awns 5–10 mm long. ........................ *O. asperifolia*

   Blades mostly involute or filiform; spikelets 3–5 mm long. ........................................ 4

4. Blades flat to involute; panicle open, lax, with flexuous, ascending or spreading branches. ........................................................ *O. canadensis*

   Blades filiform; panicle narrow, with ascending or appressed branches. ........................................ 5

5. Panicle branches erect or appressed; awns 4–6 mm long, geniculate; glumes acute. ........................................................ *O. exigua*

   Panicle branches loosely ascending; awns 0.5–2 mm long, straight; glumes obtuse. ........................................................ *O. pungens*

*Oryzopsis asperifolia* Michx.          WHITE-GRAINED MOUNTAIN RICE GRASS

   Plants tufted, with culms 20–70 cm high, at first erect, later spreading to prostrate. Sheaths smooth or somewhat scabrous, dark purple at base; blades of two types: some 1–5 cm long, others 20–40 cm long, all 3–10 mm wide, flat to convolute, dark green. Panicles 5–10 cm long, few-flowered, narrow, with branches ascending-appressed; spikelets 6–9 mm long; glumes 6–8 mm long, mostly 5- to 7-nerved; lemmas 7–9 mm long, sparsely pubescent on the back, densely so at the base, with the awn 5–10 mm long. Rather common in woods, Boreal forest; less common in Parklands.

*Oryzopsis canadensis* (Poir.) Torr.          CANADIAN RICE GRASS

   Plants tufted, with erect culms 30–60 cm high. Sheaths scabrous; blades 2–4 mm wide, flat to involute, scabrous. Panicles 5–10 cm long, open, with branches flexuous, ascending to spreading; spikelets about 5 mm long, on long slender pedicels; glumes 4–5 mm long; lemmas about 3 mm long, appressed-pubescent, with the awn 1–2 cm long, weakly twice geniculate. Not common; in woods; Boreal forest, Parklands.

*Oryzopsis exigua* Thurb.          LITTLE RICE GRASS

   Plants densely tufted, with culms 15–30 cm high, stiffly erect. Sheaths smooth or somewhat scabrous; blades 2–4 mm wide, filiform to involute, stiffly erect. Panicles 3–6 cm long, narrow, with branches ascending-appressed; spikelets about 4 mm long, short-pediceled; glumes 3–4 mm long; lemmas 3–4 mm long, appressed pubescent, with the awn 4–8 mm long, geniculate. Not common; woods and clearings; southern Rocky Mountains.

*Oryzopsis hymenoides* (Roem. & Schult.) Ricker (Fig. 54)          INDIAN RICE GRASS

   Plants densely tufted, with culms 30–60 cm high. Sheaths smooth or somewhat scabrous, prominently veined; blades 2–5 mm wide, to 50 cm long,

Fig. 54.  Indian rice grass, *Oryzopsis hymenoides* (Roem. & Schult.) Ricker.

mostly involute, coarsely veined. Panicles 10–20 cm long, diffuse, with branches slender, in pairs, branching forked, and flexuous pedicels; spikelets 6–7 mm long, solitary; glumes 6–7 mm long, papery; lemmas about 3 mm long, almost black at maturity, densely pilose with white hairs as long as the lemma; the awn about 4 mm long, straight. On sandy soils and slopes, sand dunes; Prairies and Parklands. An important grass in the sand hills, very resistant to wind action, and a good sand-binder. Palatable to livestock, and fairly resistant to grazing.

*Oryzopsis micrantha* (Trin. & Rupr.) Thurb.     LITTLE-SEED RICE GRASS

Plants rather densely tufted, with culms 30–70 cm high, erect, slender. Sheaths glabrous; blades 0.5–2 mm wide, flat or involute, scabrous. Panicles 10–15 cm long, open, with branches spreading to reflexed, single or in pairs; spikelets about 4 mm long; glumes 3–4 mm long, thin; lemmas 2–2.5 mm long, glabrous or appressed pubescent, with the awn 5–10 mm long, straight. Rare; in shrubbery on sandy soils; Prairies, eastern Parklands.

*Oryzopsis pungens* (Torr.) Hitchc.     NORTHERN RICE GRASS

Plants densely tufted, with culms 20–40 cm high, erect, slender. Sheaths smooth or somewhat scabrous; blades 1–2 mm wide, flat or involute, strongly nerved, erect. Panicles 3–6 cm long, narrow, with branches ascending-appressed or spreading; spikelets 3–4 mm long, few; glumes 3–4 mm long, often bronze-colored; lemmas 3–4 mm long, densely pubescent, with the awn 1–3 mm long. Rather common; open woods, clearings, mostly on light soils; Boreal forest.

*Panicum*     millet

Annual or perennial grasses of various habits and habitats, with glumes unequal, the first often being very minute. Mostly found in the moister, eastern part of the Prairie Provinces.

1. Plants annual. ................................................................................................ 2
   Plants perennial. ........................................................................................... 3
2. Panicles erect, diffuse; spikelets 2–4 mm
      long. .......................................................................................... *P. capillare*
   Panicles arching, not diffuse; spikelets
      4.5–5.5 mm long. ..................................................................... *P. miliaceum*
3. Plants with hard, scaly creeping rhizomes. ............................................. *P. virgatum*
   Plants without rhizomes, often with spe-
      cialized form in autumn. .......................................................................... 4
4. Spikelets less than 2 mm long. ............................................................. *P. lanuginosum*
   Spikelets more than 2 mm long. ......................................................................... 5
5. Spikelets 3.5–4 mm long, soft villous;
      sheaths papillose-hispid. ............................................................. *P. leibergii*
   Spikelets glabrous or pilose, not long
      villous. ...................................................................................................... 6
6. Spikelets of autumn plants often hidden
      in lower sheaths. ...................................................................................... 7

Spikelets of autumn plants on branches of
the culms. ............................................................................................................................ 9

7. Spikelets 3–4.5 mm long, beaked, exceed-
ing the fruit. ........................................................................................ *P. depauperatum*

Spikelets 2–3.5 mm long, beakless, as long
as the fruit. ........................................................................................................................ 8

8. Spikelets 2.5–3.5 mm long, sheaths
densely pilose. ............................................................................................ *P. perlongum*

Spikelets 2.2–2.7 mm long, sheaths glab-
rous or nearly so. ..................................................................................... *P. linearifolium*

9. Spikelets 2.5–3 mm long, pilose; culms of
autumn plants branching at base,
forming bushy tufts. ..................................................................................... *P. wilcoxianum*

Spikelets 3–4 mm long, glabrous or minutely pubescent; autumn plants
branching from the nodes, not forming
bushy tufts. ....................................................................................................................... 10

10. Spikelets 3–3.4 mm long; panicle
branches spreading; culms of autumn
plants branching at upper nodes. ...................................................... *P. oligosanthes*

Spikelets 3.5–4 mm long; panicle
branches stiffly erect; culms of autumn
plants branching at lower nodes. ...................................................... *P. xanthophysum*

*Panicum capillare* L.                                                        WITCH GRASS

Plants annual, tufted, with culms 20–80 cm high, erect or spreading, papillose-hispid. Sheaths dull green, conspicuously papillose-hispid; leaves 5–15
mm wide, hispid on both sides, papillose-ciliate at the base. Panicles to more
than half the height of plants; the plant very diffuse, densely flowered; spikelets 2–2.5 mm long; the entire panicle breaking off at maturity. Not common;
waste places, sandy prairie. The var. *occidentale* Rydb. having the panicle to
two-thirds the height of the plants, long exserted; rare in west; Parklands.

*Panicum depauperatum* Muhl.                                        PANIC GRASS

Plants tufted; the summer plants with several to many erect culms 20–30
cm high. Sheaths glabrous or papillose-pilose; blades 2–5 mm wide, flat to
involute when drying. Panicles 5–10 cm long, exserted; spikelets 3–4 mm long,
pointed, glabrous or sparsely pubescent. Autumn plants similar, but with panicles reduced, partly hidden in the basal leaves. Rare; sandy areas, open
woods; southeastern Boreal forest, Parklands.

*Panicum lanuginosum* Ell.                                            SOFT MILLET

Summer plants in large clumps; culms 40–70 cm high, lax, spreading, villous throughout, except under the nodes. Sheaths densely velvety pubescent;
blades 5–10 mm wide, flat, pubescent on both sides. Panicles 5–15 cm long;
spikelets about 1–2 mm long, pubescent. Autumn plants with culms decumbent or spreading, repeatedly branching from the middle nodes, with branches
again repeatedly branching and forming leafy inflorescences. Rare; open pine
woods, sandy areas; southeastern Boreal forests, Parklands.

*Panicum leibergii* (Vasey) Scribn.

Summer plants tufted; culms 30–70 cm high, erect or geniculate at the base, pilose to scabrous. Sheaths papillose-hispid, with hairs spreading; blades 6–15 mm wide, erect or ascending, thin, papillose-hispid on both sides. Panicles 5–15 cm long; spikelets 3.5–4 mm, strongly papillose-hispid. Autumn plants spreading, branching from the middle and lower nodes. Rare; dry prairie and clearings; southeastern Boreal forest, Parklands.

*Panicum linearifolium* Scribn.

Summer plants densely tufted; culms 20–40 cm high, slender, erect. Sheaths papillose-pilose; blades 2–4 mm wide, erect, usually exceeding the panicles. Panicles 5–10 cm long, with branches flexuous, ascending; spikelets 2.2–2.7 mm long, sparsely pilose. Autumn plants with reduced panicles hidden among the basal leaves. Rare; rock outcrops in southeastern Boreal forest.

*Panicum miliaceum* L.                                              BROOMCORN MILLET

Plants annual, tufted; culms 20–80 cm high, stout, erect or decumbent at the base. Sheaths pilose; blades 10–20 mm wide, to 30 cm long, pilose to glabrate on both sides. Panicles 10–30 cm long, included at the base, nodding, branches ascending, scabrous; spikelets 4.5–5.5 mm long; fruit 3 mm long, straw-colored to reddish brown. Introduced; escaped in various places.

*Panicum oligosanthes* Schult. var. *scribnerianum* (Nash) Fern.

Summer plants tufted; culms 40–80 cm high, appressed-pubescent. Sheaths papillose-pubescent; blades 5–8 mm wide, to 15 cm long, glabrous or nearly so above, coarsely puberulent below. Panicles 5–15 cm long; spikelets 3–3.5 mm long, sparsely pubescent. Autumn plants freely branching from the upper nodes, erect to spreading. Rare; sandy prairie, southeastern Parklands.

*Panicum perlongum* Nash                                    LONG-STALKED PANIC GRASS

Summer plants in small tufts, with the whole plant pilose; culms 20–40 cm high. Sheaths pilose; blades 2–4 mm wide, pubescent on both sides. Panicles 5–10 cm long, with branches appressed-ascending; spikelets 2.5–3.5 mm long, sparingly pubescent. Autumn plants with numerous reduced panicles. Rare; open woods on sandy soil; southeastern Parklands.

*Panicum virgatum* L. (Fig. 55)                                       SWITCH GRASS

Plants with long, scaly rhizomes; culms 80–150 cm high, tufted, erect, tough. Sheaths round, glabrous, white to purplish-tinged below; blades 6–12 mm wide, distinctly veined, long pubescent above at base, otherwise glabrous. Panicles 15–50 cm long, open to diffuse; spikelets 3.5–5 mm long; fruit narrowly ovate. Very rare; prairies, open woods; southeastern Parklands.

*Panicum wilcoxianum* Vasey                                         SAND MILLET

Summer plants tufted, papillose-hirsute throughout; culms 10–25 cm high; blades 3–6 mm wide, to 10 cm long, involute-acuminate. Panicles 2–5 cm long; spikelets 2.5–3 mm long, papillose-pubescent. Autumn plants branching from the nodes, forming bushy tufts with rigid, erect blades. Rare; sand hill prairie, clearings; Parklands.

Fig. 55. Switch grass, *Panicum virgatum* L.

*Panicum xanthophysum* A. Gray

Summer plants tufted, yellowish green; culms 20–60 cm high, scabrous. Sheaths somewhat papillose-pilose; blades 10–20 mm wide, to 15 cm long, erect, with base ciliate, otherwise glabrous. Panicles 5–10 cm long, very narrow, with branches stiffly erect, few-flowered; spikelets 3.5–4 mm long, pubescent. Autumn plants branching from the lower nodes, erect or ascending. Open pine woods; southeastern Boreal forest.

### *Phalaris*     canary grass

Perennial with rhizomes; panicles narrow;
outer glumes not winged. ............................................................. *P. arundinacea*
Annual; panicles spike-like, ovate; outer
glumes winged. ........................................................................... *P. canariensis*

*Phalaris arundinacea* L. (Fig. 56)                    REED CANARY GRASS

Plants with thick rhizomes, to 4 mm in diam, scaly, dark brown; culms to 2 m high, stiffly erect, smooth, with up to 10 leaves. Sheaths glabrous, distinctly veined; blades to 20 mm wide, to 20 cm long, flat, somewhat scabrous. Panicles 10–20 cm long, with the lower branches 3–5 cm long, spreading during flowering, later appressed; glumes 3–5 mm long, narrowly winged; lemmas 3–4 mm long, appressed-pubescent; sterile lemmas 1 mm long, villous. Native; in wet areas, marshes, riverbanks, and lakeshores; Boreal forest. Introduced for forage in irrigated haylands.

*Phalaris canariensis* L.                                CANARY GRASS

A medium-growing annual with rather pretty ovoid head, 20–35 mm long, pale with green stripe on glumes. Found where seed for a caged bird has been scattered.

### *Phleum*     timothy

Panicles elongate-cylindric; awns less than
half the length of empty glumes. ........................................................ *P. pratense*
Panicles short, ovoid or oblong; awns about
half the length of empty glumes. ......................................................... *P. alpinum*

*Phleum alpinum* L. (Fig. 57)                          ALPINE TIMOTHY

Plants with short rhizomes; culms 20–50 cm high, densely tufted, decumbent at the base. Sheath round, glabrous, inflated near the middle; blades 4–8 mm wide, 2–15 cm long, scabrous above and on the margins. Panicles 1–4 cm long, to 1 cm wide, bristly; glumes about 5 mm long, ciliate on the keel, with the awn 2–3 mm long. Mountain meadows, bogs, and wet areas; Rocky Mountains, Cypress Hills; occasionally in Boreal forest.

*Phleum pratense* L. (Fig. 58)                         TIMOTHY

Plants with a swollen, bulb-like base; culms 50–80 cm high, often forming large clumps. Sheaths glabrous, green, often purplish at base; blades 6–12 mm wide, to 30 cm long, flat, often twisted, light green or grayish green. Panicles, usually 5–10, occasionally to 20 cm long, to 1 cm thick; glumes 3–5 mm long,

Fig. 56. Reed canary grass, *Phalaris arundinacea* L.

Fig. 57.   Alpine timothy, *Phleum alpinum* L.

*Monocots*

Fig. 58.  Timothy, *Phleum pratense* L.

with a green midrib, ciliate on the keel, with the awn 1–2 mm long, stout. Introduced forage grass; escaped and established in many parts of Parklands, Boreal forest, and in moist areas in Rocky Mountains.

### *Phragmites*        giant reed grass

*Phragmites communis* Trin.                    COMMON REED GRASS

Plants with long, extensively creeping rhizomes, to 3 cm thick; culms 1–4 m high, erect, occasionally decumbent and stoloniferous. Sheaths glabrous, purplish at the base; blades 20–40 mm wide, to 40 cm long, acuminate, glabrous. Panicles 15–40 cm long, with branches spreading in flower, often drooping, densely flowered; spikelets 6–15 mm long; first glume 5–7 mm long; second glume 10–12 mm long; lemmas about 12 mm long; hairs of the rachilla exceeding the spikelets. Wet places in Parklands, Boreal forest; rarely in springy places in the Prairies.

### *Poa*        blue grass

A very large and difficult genus, mainly of low- to medium-growing species and found in all types of habitat. The leaves have boat-shaped tips and usually have a double line down the midrib.

Valuable forage grasses. Many species grow very early in the season.

1. Plants with rhizomes. ............................................................................ 2
   Plants with fibrous roots. ..................................................................... 10
2. Culms strongly compressed, flat; panicle narrow, with the branchlets bearing spikelets to near the base. ........................................ *P. compressa*
   Culms terete or somewhat flattened. ...................................................... 3
3. Lemmas pubescent, at least on the nerves. ........................................... 5
   Lemmas not pubescent. ......................................................................... 4
4. Panicle open, with branches spreading; lemmas keeled; rhizomes long. ............................................ *P. nervosa*
   Panicle narrow, dense; lemmas rounded on the back; rhizomes short. ................................................. *P. ampla*
5. Lemmas less than 4 mm long. ............................................................. 6
   Lemmas more than 4 mm long. ............................................................ 7
6. Lemmas cobwebby at base; keel and marginal nerves sparsely pubescent toward the base. ................................................... *P. pratensis*
   Lemmas not cobwebby at base; densely pubescent on lower keel and marginal nerves. ............................................................. *P. arida*
7. Lemmas cobwebby at base. ............................................................... *P. arctica*
   Lemmas not cobwebby at base. ............................................................ 8
8. Panicle narrow, with branches ascending; spikelets 7–10 mm long; leaves folded or involute, stiff. ................................... *P. fendleriana*

Panicle open, with branches spreading; leaves flat or folded, not stiff. .................................................................. 9

9. Panicle usually small, less than 10 cm long; lemmas purplish, suffused with orange at the tip. ....................................................... *P. arctica*

Panicle usually large, 15–20 cm long; lemmas not purplish; foliage blue green. ................................. *P. glaucifolia*

10. Plants annual; blades soft and flat, often rugose; ligule white. ........................................................... *P. annua*

Plants perennial. ................................................................................................ 11

11. Plants densely tufted; blades narrow. ........................................................ 12

Plants not densely tufted; blades flat. ........................................................ 20

12. Lemmas cobwebby at base. ........................................................................ 13

Lemmas not cobwebby at base. ................................................................ 14

13. Spikelets less than 4.5 mm long; lemmas less than 3 mm long; blades 2–3 mm wide, flat. ....................................................... *P. nemoralis*

Spikelets 5–6 mm long; lemmas about 4 mm long; blades 1 mm wide, folded, lax. ............................................................ *P. pattersonii*

14. Lemmas not pubescent. .............................................................................. 15

Lemmas crisp pubescent, silky or villose, at least at base. ................................................................................ 16

15. Blades of culm leaves filiform; spikelets 7–9 mm long. ........................................................... *P. cusickii*

Blades of culm leaves flat; spikelets 5–6 mm long. ........................................................... *P. epilis*

16. Lemmas rounded on the back; spikelets not flattened. ......................................................................... 17

Lemmas keeled; spikelets flattened. ...................................................... 18

17. Blades short, often curled; culms usually less than 30 cm high; panicles less than 10 cm long. ........................................................... *P. secunda*

Blades 10 cm or longer, not curled; culms 30–60 cm high; panicles to 15 cm long. ................................. *P. canbyi*

18. Plants usually blue green; spikelets 2- and 3-flowered; panicle rather lax. ................................. *P. glauca*

Plants not blue green; spikelets 4- and 5-flowered; panicle stiff, erect. ...................................................... 19

19. Blades 5–10 cm long; culms 5–20 cm high; blades about 1 mm wide, lax. ................................. *P. pattersonii*

Blades 1–5 cm long; culms 10–20 cm high; blades 1–1.5 mm wide, stiff. ................................. *P. rupicola*

20. Lemmas cobwebby at base. ........................................................................ 21

Lemmas not cobwebby at base. ................................................................ 23

21. Spikelets 5–6 mm long; panicle usually 5–10 cm long; culms solitary or few together. ........................................................... *P. leptocoma*

*Poa alpina* L.                                        ALPINE BLUE GRASS

   Plants tufted, sod-forming, with erect culms 10–30 cm high. Sheaths glabrous, keeled; blades 2–5 mm wide, flat, short. Panicles 1–8 cm long, compact
short pyramidal or ovoid, with branches spreading, or the lower ones reflexed;
spikelets 4–6 mm long, purplish; glumes 1.5–2 mm long; lemmas 3–4 mm long,
pubescent to villous on keel and marginal nerves. Rocky Mountains, Boreal
forest.

*Poa ampla* Merr.                                        BIG BLUE GRASS

   Plants densely tufted, with short rhizomes; erect culms 60–80 cm high.
Sheaths smooth or somewhat scabrous; blades 1–3 mm wide, green or glaucous. Panicles 10–15 cm long, narrow, mostly dense; spikelets 8–10 mm long,
4- to 7-flowered; glumes 2.5–5 mm long; lemmas 4–6 mm long. Meadows and
slopes; Rocky Mountains.

*Poa annua* L.                                        ANNUAL BLUE GRASS

   Plants annual or biennial, sod-forming; culms 5–20 cm high, ascending or
spreading, sometimes rooting at the internodes. Sheaths somewhat compressed, glabrous; blades 1–4 mm wide, flat or somewhat folded, thin, light
green. Panicles 3–10 cm long, pyramidal, often secund, with branches spreading, the lower ones often reflexed; spikelets about 3 mm long, 4- and 5-
flowered, green- or purple-tinged; first glume 1.5–2 mm long, 1-nerved; second
glume 2–2.5 mm long, 3-nerved; glumes 2.5–3 mm long, 5-nerved, subglabrous

to short pubescent. Introduced; weedy in gardens in the Prairies; increasingly common in Parklands and Boreal forest.

*Poa arctica* R. Br.                                                    ARCTIC BLUE GRASS

Plants with creeping rhizomes; culms 10–30 cm high, erect, with the base decumbent, tufted. Sheaths glabrous; blades 2–3 mm wide, flat to folded. Panicles 5–10 cm long, open, pyramidal, with the lower branches spreading to reflexed; spikelets 5–8 mm long, 3- and 4-flowered; first glume 1.5–2 mm long; second glume 2–2.5 mm long; lemmas 3.5–4 mm long, densely villous-pubescent on the keel and marginal nerves. Alpine and high boreal meadows; Rocky Mountains, Boreal forests.

*Poa arida* Vasey                                                     PLAINS BLUE GRASS

Plants with creeping rhizomes; culms 20–50 cm high, erect, solitary or few together. Sheaths glabrous; blades 2–3 mm wide, folded, stiff. Panicles 2–10 cm long, narrow, with branches appressed-ascending; spikelets 5–7 mm long, 4- to 8-flowered; first glume 2–4 mm long; second glume 2.5–4.5 mm long; lemmas 3–4 mm long, densely pubescent on the back. Dry to moist, often somewhat alkali prairies; Prairies, Parklands.

*Poa canbyi* (Scribn.) Piper (Fig. 59)                                 CANBY BLUE GRASS

Plants densely tufted, with erect culms 30–80 cm high. Sheaths somewhat compressed, scabrous; blades to 4 mm wide, flat to folded, green and glaucous. Panicles 10–15 cm long, narrow, compact or somewhat loose, with branches short, appressed or ascending; spikelets 5–6 mm long, 3- to 5-flowered; first glume 2.5–3.5 mm long; second glume 3–4.5 mm long; lemmas 3–4.5 mm long, obscurely 5-nerved, more or less pubescent on the nerves, at least toward the base. Moist, often alkali meadows; Prairies, Parklands.

*Poa compressa* L.                                                    CANADA BLUE GRASS

Plants with creeping rhizomes; culms 15–50 cm, solitary or few together, flat, wiry. Sheaths strongly compressed and sharply keeled, glabrous; blades 2–5 mm wide, flat to folded, bluish green. Panicles 3–10 cm long, with branches usually short, in pairs; spikelets 4–6 mm long, 3- to 6-flowered; first glume 1.5–2.5 mm long; second glume 2–3 mm long; lemmas 2–3 mm long, somewhat pubescent. Introduced; dry, poor, often stony soils; throughout the Prairie Provinces.

*Poa cusickii* Vasey                                                  EARLY BLUE GRASS

Plants densely tufted, with erect culms 20–40 cm high. Sheaths compressed, sharply keeled, scabrous; blades 1–3 mm wide, flat to folded and bristle-like, erect. Panicles 3–8 cm long, dense, usually obovoid, pale; spikelets 7–9 mm long; glumes 3–4 mm long; lemmas 4–6 mm long, smooth or somewhat scabrous. Dry to moist prairie; throughout the Prairies and Parklands.

*Poa epilis* Scribn.                                                  SKYLINE BLUE GRASS

Plants densely tufted, with erect culms 20–40 cm high. Sheaths compressed; blades 2–3 mm wide, folded or involute. Panicles 2–6 cm long, dense, oblong, usually purplish; spikelets about 5 mm long, 3-flowered; glumes about 5 mm long; lemmas 6 mm long, glabrous. Mountain meadows; Rocky Mountains.

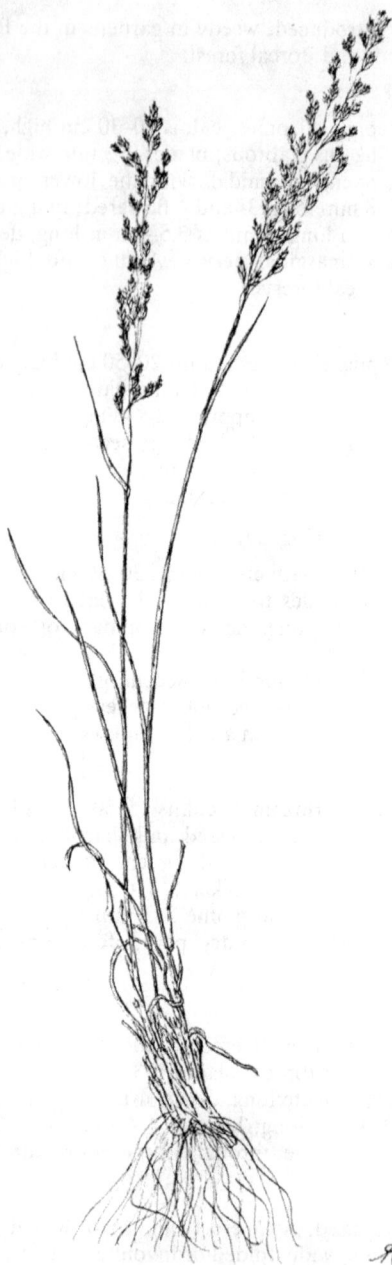

Fig. 59.   Canby blue grass, *Poa canbyi* (Scribn.) Piper.

*Poa fendleriana* (Steud.) Vasey                                    MUTTON GRASS

Plants tufted, with culms 30–50 cm high, erect, scabrous below the panicle. Sheaths scabrous; blades 1–2 mm wide, folded or involute, stiff, erect.
Panicles 2–7 cm long, contracted; spikelets 6–8 mm long, 4- to 6-flowered;
glumes about 3 and 4 mm long; lemmas 4–5 mm long, long pilose on lower
keel and marginal nerves. Dry slopes, prairie; Prairies.

*Poa glauca* Vahl

Plants tufted, sometimes densely so, with erect culms 20–50 cm high.
Sheaths glabrous; blades 1–2 mm wide, flat, blue green, glaucous. Panicles
3–10 cm long, narrow, often compact, with branches ascending, spreading in
flower; spikelets 5–6 mm long, 2- or 3-flowered; glumes 2.5–4 mm long, subequal; lemmas 2.5–4 mm long, densely pubescent on lower half of keel and
marginal nerves. Stony areas, sandy soils; Boreal forest, Rocky Mountains.

*Poa glaucifolia* Scribn. & Will.                        GLAUCOUS BLUE GRASS

Plants with slender rhizomes; culms 30–60 cm high, loosely tufted, erect.
Sheaths somewhat compressed, glabrous; blades 2–4 mm wide, flat to folded,
glabrous, glaucous on both sides. Panicles 10–20 cm long, open, with branches
ascending to spreading; spikelets 5–7 mm long, 2- to 4-flowered; glumes about
3.5 and 4.5 mm long; lemmas about 4 mm long, villous on lower half of keel
and marginal nerves. Not common; moist, often somewhat alkali areas and
meadows; Prairies, Parklands.

*Poa gracillima* Vasey                                   PACIFIC BLUE GRASS

Plants loosely tufted, with culms 30–60 cm high, mostly decumbent at the
base. Sheaths somewhat scabrous; blades 0.5–1.5 mm wide, flat or folded, with
the basal ones filiform. Panicles 3–10 cm long, pyramidal, with branches
spreading, the lower ones often reflexed; spikelets 4–6 mm long; first glume
2.5–3 mm long; second glume 3–4 mm long; lemmas about 4 mm long, pubescent on the lower part of back. Slopes, riverbanks, lakeshores; Rocky Mountains.

*Poa juncifolia* Scribn.                                 ALKALI BLUE GRASS

Plants tufted, with erect culms 30–60 cm high. Sheaths somewhat
flattened; blades 1–3 mm wide, involute, rather stiff. Panicles 10–20 cm long,
narrow, with branches appressed; spikelets 7–10 mm long, 3- to 6-flowered;
glumes about 4 mm long; lemmas about 4 mm long, glabrous or nearly so.
Alkaline meadows; Prairies.

*Poa leptocoma* Trin.                                    BOG BLUE GRASS

Plants loosely tufted, with culms 20–50 cm high, often decumbent at the
base. Sheaths slightly scabrous; blades 2–4 mm wide, short, flat, lax. Panicles
7–15 cm long, open, nodding at the tip, with branches very slender, ascending
or spreading; spikelets 4–6 mm long, 2- to 4-flowered; glumes about 2.5 and 3
mm long; lemmas 3.5–4.5 mm long, pubescent on the back. Not common;
bogs and wet meadows; Rocky Mountains.

*Monocots*                                    GRAMINEAE (POACEAE) — 163

*Poa nemoralis* L.                                                WOOD BLUE GRASS

Plants sod-forming, with culms 30–70 cm high, loosely tufted, erect, often decumbent at the base. Sheaths glabrous, smooth; blades to 2 mm wide, flat, lax. Panicles 5–10 cm long, loose, often nodding, with branches spreading in flower, later appressed, scabrous; spikelets 3–5 mm long, 1- to 6-flowered, light green; glumes about 2.5 and 3.5 mm long; lemmas 3–4 mm long, pubescent on the lower back. Meadows and open woods; Parklands, Boreal forests. Var. *interior* (Rydb.) Abbe & Butters (*P. interior* Rydb.), more densely tufted, stiffer, with the branches of the panicles more contracted, is very similar to var. *firmula* Gaud. of Europe and may be identical. Plants of this variety with reduced panicles are forma *rariflora* (Desf.) A. & G. In dry areas, sandy open forest, slopes; Boreal forest, Rocky Mountains.

*Poa nervosa* (Hook.) Vasey                                       WHEELER'S BLUE GRASS

Plants with long rhizomes; culms 30–60 cm high, erect, somewhat tufted. Sheaths often purplish below, the lower ones retrorsely pubescent; blades 2–4 mm wide, flat or folded. Panicles 5–10 cm long, open, nodding at the tip, with branches spreading, the lower ones often reflexed; spikelets 4–6 or sometimes 8 mm long; glumes subequal, about 2 mm long; lemmas 3–4 mm long, strongly nerved, pubescent to glabrous on lower back. Slopes, open woods; Rocky Mountains, Cypress Hills.

*Poa palustris* L.                                                FOWL BLUE GRASS

Plants loosely tufted, with culms 30–100 cm high, decumbent, purplish at base. Sheaths somewhat flattened, keeled, often somewhat scabrous; blades 2–4 mm wide, flat or loosely folded, lax, scabrous on both sides. Panicles 10–30 cm high, pyramidal or oblong, with branches spreading; spikelets 3–4 mm long, 3- or 4-flowered; glumes subequal, 2.5–3 mm long; lemmas 2.5–3 mm long, often bronzed at the tip, pubescent on lower back. Meadows, moist areas, lakeshores, and riverbanks; throughout Prairie Provinces.

*Poa pattersonii* Vasey

Plants densely tufted, with erect culms 5–20 cm high. Sheaths smooth; blades about 1 mm wide, folded, lax. Panicles 1–4 cm long, dense, purplish; spikelets 5–6 mm long, 4- or 5-flowered; glumes about 3 and 3.5 mm long; lemmas about 4 mm long, strongly pubescent on keel and marginal nerves. Rare; in alpine meadows and slopes; southern Rocky Mountains.

*Poa pratensis* L.                                                KENTUCKY BLUE GRASS

Plants with long rhizomes; culms to 100 cm high. Culms compressed and slightly keeled, glabrous, dark green; blades to 5 mm wide, often to 40 cm long, linear, dark green, the lower side glossy, soft. Panicles 5–15 cm long, contracted before flowering, with branches spreading during flowering; spikelets 3–6 mm long, 3- to 5-flowered, green- or purplish-tinged; glumes subequal, 3–3.5 mm long, the first 1-nerved, the second 3-nerved; lemmas 3.5–4 mm long, distinctly 5-nerved, densely short pubescent on the nerves, and strongly webbed at the base. Common throughout the Prairie Provinces. Sown in lawn mixtures, pastures; also found in meadows, moist prairies, and forest openings. Probably native, as well as introduced. Very variable, consisting of many races. Low plants, 15–20 cm high, with small, few-flowered panicles, and often rather grayish green, are var. *humilis* (Ehrh.) Griseb.; on dry ground. Plants 15–30 cm high, culms stiff, usually with a single culm leaf, and panicles small, stiff, with branches single or in pairs, are var. *arenaria* J. & W.; in dry pastures, sand dunes. The form described as *P. agassizensis* Boivin & Love seems to fit in both these varieties.

*Poa rupicola* Nash                    TIMBERLINE BLUE GRASS

Plants densely tufted, with culms 10–20 cm high, erect, stiff. Sheaths smooth; blades 1–1.5 mm wide, erect. Panicles 2–5 cm long, narrow, with branches short, ascending to appressed; spikelets 4–5 mm long, mostly 3-flowered, purple; glumes about 3 mm long; lemmas about 3.5 mm long, villous on lower keel and marginal nerves. Slopes, openings, and meadows; Rocky Mountains.

*Poa secunda* Presl                    SANDBERG'S BLUE GRASS

Plants densely tufted, with erect culms 10–30 cm high. Sheaths compressed, often somewhat scabrous; blades 1–2 mm wide, flat or folded, twisted to erect. Panicles 2–10 cm long, narrow, with branches short ascending to appressed; spikelets 4–6 mm long, pale green; glumes about 3–3.5 mm long; lemmas 3.5–4 mm long, pubescent on lower back. Apparently rare or lacking in Manitoba; dry grasslands; Prairies, Parklands.

*Poa stenantha* Trin.

Plants tufted, with erect culms 30–50 cm high. Sheaths smooth; blades 1–2 mm wide, flat or somewhat involute, lax. Panicles 5–15 cm long, nodding, with branches drooping; spikelets 6–8 mm long, 3- to 5-flowered; glumes about 2.5 mm and 3.5 mm long; lemmas about 5 mm long, pubescent on lower back. Moist meadows and openings; Rocky Mountains.

*Poa trivialis* L.                    ROUGH-STALKED BLUE GRASS

Plants sod-forming, with culms erect or ascending, sometimes stoloniferous below. Sheaths compressed, sharply keeled; blades 2–5 mm wide, flat, lax, glossy green below. Panicles 10–20 cm long, oblong, often somewhat contracted, with branches scabrous, ascending-spreading; spikelets about 4 mm long, 2- to 5-flowered, usually green, sometimes bronze- or purplish-tinged; glumes about 2 and 3 mm long; lemmas 2.5–3 mm long, finely but distinctly 5-nerved, strongly webbed at the base. Introduced; occasionally sown in pastures; escaped in various locations.

**Polypogon**        beard grass

*Polypogon monspeliensis* (L.) Desf.                    RABBITFOOT GRASS

Plants annual, with culms 15–50 cm high, erect or decumbent. Sheaths scabrous; blades 4–6 mm wide, scabrous. Panicles 1–15 cm long, spike-like, dense, silky, yellowish when mature; glumes 2 mm long, with the apex somewhat lobed and the awn 6–8 mm long, pubescent on the back; lemmas about 1 mm long, short-awned. Introduced; established in various locations as a weedy plant in waste areas, dry banks, coulees, along streams.

*Puccinellia*    salt-meadow grass

Low to medium, tufted, feathery panicled grasses of moist alkaline soils.

1. Plants stoloniferous, with the stolons
   bearing bulblets; lemmas entire,
   3.5–4.5 mm long. ................................................................................ *P. phryganodes*
   Plants not stoloniferous, with the culms
   ascending or decumbent; lemmas 1.5–3
   mm long. ...................................................................................................... 2

2. Lemmas rounded or truncate at the tip;
   lower panicle branches reflexed at
   maturity. ........................................................................................... *P. distans*
   Lemmas blunt, narrowed to a triangular
   tip; panicle branches not reflexed. .................................................. *P. nuttalliana*

*Puccinellia distans* (L.) Parl.                    SLENDER SALT-MEADOW GRASS

Plants sod-forming, bluish green, with culms 15–60 cm high, mostly genic-ulate-ascending. Sheaths broad, smooth; blades 1–2 mm wide, linear, flat. Panicles 5–15 cm long, pyramidal, with branches scabrous, reflexed at maturi-ty; spikelets 4–6 mm long, 4- to 7-flowered; glumes 1 and 2 mm long; lemmas 1.5–2 mm long, 5-nerved, often reddish-tinged with a hyaline or yellowish membranous margin. Introduced from Europe; not common; on alkaline soils around lakes.

*Puccinellia nuttalliana* (Schultes) Hitchc. (Fig. 60)

NUTTALL'S SALT-MEADOW GRASS

Plants sod-forming or tufted, with culms 30–60 cm high, usually erect or ascending, slender. Sheaths smooth; blades 1–3 mm wide, flat or involute, glaucous. Panicles 10–20 cm long, open, pyramidal, with branches scabrous, spreading; spikelets 4–7 mm long, 3- to 6-flowered; first glume 1–1.5 mm long; second glume 1.5–2 mm long; lemmas 2–3 mm long, narrowed to an obtuse apex. Common; on moist to rather dry alkaline soils; throughout the Prairie Provinces.

*Puccinellia phryganodes* (Trin.) Scribn. & Merr.

Plants sod-forming, with sterile culms stoloniferous, bearing bulblets at the nodes; fertile culms 20–40 cm high, erect or decumbent. Sheaths smooth; blades 1–3 mm wide, mostly folded, thick. Panicles 5–15 cm long, open; spike-lets 8–12 mm long, 4- to 7-flowered; first glume 2–3 mm long; second glume 3–4 mm long; lemmas 3.5–4.5 mm long. Around sloughs and wet calcareous gravelly soils; Churchill.

*Schedonnardus*    tumble grass

*Schedonnardus paniculatus* (Nutt.) Trel.                    TUMBLE GRASS

Plants spreading, tufted, with culms 20–40 cm high, erect or ascending. Sheaths compressed, sharply keeled; blades 1–2 mm wide, folded, twisted, and wavy. Spikes 2–5 cm long; spikelets about 4 mm long, narrow, distant. Inflorescence elongating at maturity into a spiral that breaks off and rolls with the wind. Dry, sandy, or infertile soils; Prairies.

Fig. 60.   Nuttall's salt-meadow grass, *Puccinellia nuttalliana* (Schultes)
Hitchc.

**Schizachne**     purple oat grass

*Schizachne purpurascens* (Torr.) Swallen (Fig. 61)     PURPLE OAT GRASS

Plants with rhizomes; culms 50–100 cm high, loosely tufted, usually decumbent at the base. Sheaths round or slightly flattened; blades 2–6 mm wide, flat to loosely folded. Panicles 6–15 cm long, often secund, with branches more or less drooping, single or in pairs; spikelets 20–25 mm long; first glume 4–5 mm long; second glume 6–8 mm long, purplish; lemmas 8–10 mm long, with the awn 10–15 mm long. In woods throughout the Prairie Provinces; rare in Prairies.

**Scolochloa**     spangletop

*Scolochloa festucacea* (Willd.) Link (Fig. 62)     SPANGLETOP

Plants with thick, long rhizomes; culms to 150 cm high, stout. Sheaths glabrous, prominently veined; blades 5–10 mm wide, flat to convolute. Panicles 15–20 cm long, loose, with branches fascicled, ascending; spikelets 6–9 mm long, 4- to 7-flowered; first glume 5–7 mm long; second glume 6–8 mm long; lemmas 5–7 mm long, villous at the base. Wet areas throughout the Prairie Provinces.

**Secale**     rye

*Secale cereale* L.     RYE

Annual or winter annual; culms to 2 m high, usually densely pubescent above. Leaves flat, often glaucous, to 15 mm wide. Spikes to 15 cm long; spikelets usually 2-flowered; glumes awl-shaped, to 1 cm long; lemmas 3- to 5-nerved, to 18 mm long, with the awn 4–8 cm long. A cereal grain grown on light soils; occasionally established in roadsides and waste areas.

**Setaria**     foxtail

Weedy annuals with cylindric, dense, spike-like, bristly heads, found in waste places and as a weed in cultivated fields. Introduced from Europe but rapidly spreading.

Spikelets with more than 5 bristles, these
mostly less than 3 times as long as the
spikelet. ................................................................................................ *S. glauca*
Spikelets with fewer than 5 bristles, these 3–5
times as long as the spikelet. ................................................................ *S. viridis*

*Setaria glauca* (L.) Beauv.     YELLOW FOXTAIL

Plants annual, tufted, with culms 50 cm or more high, erect, occasionally branching above. Sheaths compressed, keeled, pale green; blades 5–12 mm wide, flat or loosely folded, twisted, soft, gray green. Panicles 5–10 cm long, spike-like, cylindrical; spikelets 3–3.5 mm long; first glume about 1.2 mm long, 3-nerved; second glume about 1.7 mm long, 5-nerved; lemmas about 3 mm long, rugose; bristles usually 6–8 or up to 20, 6–10 mm long. Not common; introduced; weedy in gardens and waste places.

Fig. 61.   Purple oat grass, *Schizachne purpurascens* (Torr.) Swallen.

Fig. 62.  Spangletop, *Scolochloa festucacea* (Willd.) Link.

*Setaria viridis* (L.) Beauv. (Fig. 63)

Plants annual, tufted, with culms 20–50 cm high, erect or geniculate-ascending. Sheaths slightly compressed, glabrous or appressed-pubescent; blades 5–12 mm wide, flat, light green. Panicles 3–10 cm long; spikelets 2–2.5 mm long; first glume less than 1 mm long, 1-nerved; second glume about 2 mm long, 5-nerved; lemmas 2–2.5 mm long, smooth; bristles 1–3, below each spikelet, 8–10 mm long. Introduced; weedy in gardens, cultivated land, roadsides; throughout the Prairies and Parklands.

**Sitanion**        squirreltail

*Sitanion hystrix* (Nutt.) J. G. Smith                    SQUIRRELTAIL

Plants densely tufted, with culms 10–50 cm high, erect to spreading, stiff. Sheaths somewhat keeled above, softly pubescent or glabrous; blades 1–3 mm wide, flat to involute, finely pubescent on both sides. Spikes 2–7 cm long, the lower parts often included in the sheaths; spikelets about 15 mm long excluding the awns; glumes about 6 mm long excluding the awn, very narrow; lemmas about 10 mm long excluding the awn; awns of glumes and lemmas 3–10 cm long. Very rare; slopes, dry grassland, open areas; Prairies.

**Sorghastrum**        Indian grass

*Sorghastrum nutans* (L.) Nash                    INDIAN GRASS

Plants with long, scaly rhizomes; culms 60–150 cm high, loosely tufted, somewhat pubescent on the nodes. Sheaths round, distinctly veined; blades 5–10 mm wide, flat, to 50 cm long, dull green to glaucous. Panicles 10–25 cm long, narrow, yellowish, brownish at maturity; spikelets 6–8 mm long, with the awn 10–15 mm long; summit of branchlets, rachis joints, pedicels, and spikelets long grayish hirsute. Rare; in grassland; southeastern Parklands.

**Spartina**        cord grass

Erect coarse grasses with scaly rhizomes found in moist areas throughout the Prairie Provinces. Spikelets crowded on one side of panicle branches, which are borne alternately on the main stem and closely parallel it. Leaves very coarse.

Lemmas awned, found only in eastern prairies. ............................................. *S. pectinata*
Lemmas not awned. ................................................................................. *S. gracilis*

*Spartina gracilis* Trin. (Fig. 64)                    ALKALI CORD GRASS

Plants with tough rhizomes; culms 60–100 cm high, solitary or in small tufts. Sheaths glabrous, yellowish green, to 5 mm wide, very scabrous above. Inflorescence 6–15 cm long, with 4–8 closely appressed spikes 2–4 cm long; spikelets 6–8 mm long; first glume about 5 mm long; second glume about 10 mm long; lemmas about 10 mm long; glumes and lemmas ciliate on the keel. Common; in dry to wet alkali areas, and sandy soils; Prairies and Parklands.

Fig. 63.   Green foxtail, *Setaria viridis* (L.) Beauv.

*Monocots*

Fig. 64. Alkali cord grass, *Spartina gracilis* Trin.

*Spartina pectinata* Link

Plants with long, tough rhizomes; culms to 2 m high, solitary or a few together. Sheaths glabrous, distinctly veined; blades 5–15 mm wide, to 60 cm long, flat to involute. Inflorescence 10–30 cm long, with 10–20 spikes 4–8 cm long, ascending to appressed; first glume 5–6 mm long, short-awned; second glume 8–12 mm long, with the awn 4–10 mm long; lemmas 7–9 mm long. Moist prairie, riverbanks, and lakeshores; southeastern Prairies and Parklands.

**Sphenopholis**        wedge grass

Tall tufted grasses with narrow panicles and many shining spikelets.

Second glume much wider than lemma,
  wedge-shaped; panicle dense, spike-like,
  erect. ................................................................................................. *S. obtusata*
Second glume not much wider than lemma,
  obtuse or acute; panicle not dense or spike-
  like, drooping. .............................................................................. *S. intermedia*

*Sphenopholis intermedia* Rydb.

Plants in small tufts, with erect culms 30–60 cm high. Sheaths glabrous or somewhat pubescent; blades 2–5 mm wide, flat, soft. Panicles 8–15 cm long, to 3 cm in diam, more or less lobed, lustrous; first glume 1.5–2.5 mm long; second glume 2–3 mm long; lemmas 2.5–3 mm long. Moist soil, ravines, margins of woods, and meadows; throughout the Prairie Provinces.

*Sphenopholis obtusata* (Michx.) Scribn.

Plants tufted, with erect culms 20–60 cm high. Sheaths glabrous or slightly scabrous, distinctly veined; blades 2–5 mm wide, flat, scabrous on both sides. Panicles 5–15 cm long, narrow, dense, mostly spike-like, sometimes lobed below, lustrous; first glume 1–2 mm long; second glume 1.5–2.5 mm long; lemmas 1.5–2.5 mm long. Moist areas, shrubbery, coulee bottoms, creek banks, and lakeshores; throughout the Prairie Provinces.

**Sporobolus**        dropseed

Tufted plants with paniculate inflorescence, the seeds readily falling at maturity, giving the common name to the genus. Fairly palatable forage plants.

1. Plants annual; panicle contracted or
     spike-like. .......................................................................................... *S. neglectus*
   Plants perennial; panicle open and
     branches more or less spreading. ........................................................... 2
2. Spikelets less than 3.2 mm long. ............................................... *S. cryptandrus*
   Spikelets more than 3.2 mm long. .............................................. *S. heterolepis*

*Sporobolus cryptandrus* (Torr.) A. Gray (Fig. 65)     SAND DROPSEED

Plants tufted, with culms 30–100 cm high, solitary or few together, erect or spreading to prostrate. Sheaths prominently veined, often purplish at the base, with the margin ciliate; blades 2–5 mm wide, flat to involute, with conspicuous tufts of hairs at the base. Panicles 10–25 cm long, the base often included in the sheath, with branches spreading or reflexed; spikelets 2–2.5 mm long, grayish; first glume about 1 mm long; second glume 2–2.5 mm long; lemmas about 2.5 mm long. Prairies and open woods on sandy soils; Prairies and Parklands.

*Sporobolus heterolepis* A. Gray     PRAIRIE DROPSEED

Plants densely tufted, with culms 30–70 cm high, erect, slender. Sheaths somewhat flattened, glabrous or pubescent; blades 1–3 mm wide, to 45 cm long, flat to involute, erect or slightly drooping. Panicles 5–20 cm long, oblong to narrowly ovoid, with branches 3–6 cm long, ascending to appressed; spikelets 3–6 mm long, grayish; first glume 2–4 mm long; second glume 4–6 mm long; lemmas 5–5.5 mm long. Moist grasslands; southeastern Prairies and Parklands.

*Sporobolus neglectus* Nash     ANNUAL DROPSEED

Plants annual, tufted, with culms 20–40 cm high, erect or spreading. Sheaths glabrous; blades 1–2 mm wide, with the upper culm leaves 1–2 cm long. Panicles 2–5 cm long, rarely exserted, often exceeded by the uppermost blade; spikelets 2.5–3 mm long; first glume 1.5–2.5 mm long; second glume 2–3 mm long; lemmas 2–3 mm long. Rare; dry open areas; Prairies.

**Stipa**     needle grass

Medium- to tall-growing tufted grasses with narrow panicles of fairly large spikelets bearing twisted awns. Spikelets breaking off at maturity above glumes. Lemmas remaining attached to the seed characterized by a hard, sharp-pointed, hairy callus, and a long twisted awn. The needle-like lemmas injurious to the mouthparts and skin of grazing animals, particularly sheep. Except when the seeds are mature and still attached, the needle grasses are excellent forage plants, both for hay and pasture, and are one of the most important groups of forage plants on the western range.

1. Glumes 15–40 mm long; lemmas 8–20 mm long; awns 10–25 cm long. ........................................................ 2

   Glumes 8–10 mm long; lemmas 5–7 mm long; awns 2–3 cm long. ........................................................ 4

2. Awns indistinctly geniculate, with the upper part strongly curled and flexuous; glumes 15–30 mm long; lemmas 8–15 mm long. ........................................ *S. comata*

   Awns distinctly twice geniculate. ........................................ 3

3. Glumes 30–40 mm long; lemmas 15–25 mm long; awns to 25 cm long. ........................ *S. spartea*

   Glumes 20–30 mm long; lemmas 12–15 mm long; awns to 12 cm long. ........................ *S. spartea* var. *curtiseta*

Fig. 65.  Sand dropseed, *Sporobolus cryptandrus* (Torr.) A. Gray.

4. Panicle open, with branches spreading and drooping; spikelets few at ends of branches. ............................................................................... *S. richardsonii*

   Panicle narrow with appressed branches. .................................................................... 5

5. Callus blunt; collar and lower nodes of panicle villose. .......................................................................... *S. viridula*

   Callus sharp-pointed; collar and lower nodes of panicle glabrous. ................................................. *S. columbiana*

*Stipa columbiana* Macoun                                    COLUMBIA NEEDLE GRASS

Plants tufted, with culms 30–60 cm high, erect or ascending. Sheaths glabrous; blades 1–3 mm wide, to 25 cm long, usually involute. Panicles 10–25 cm long, narrow, with branches appressed-ascending; glumes about 1 cm long; lemmas 6–7 mm long, pubescent, with the callus sharp-pointed; awns 2–2.5 cm, twisted below, twice geniculate. Prairies and openings; Rocky Mountains, Peace River, Cypress Hills.

*Stipa comata* Trin. & Rupr. (Fig. 66)                              SPEAR GRASS

Plants in small, dense tufts, with culms 30–60 cm high, erect. Sheaths round or somewhat compressed; blades 1–3 mm wide, flat to involute, prominently ridged above. Panicles 10–20 cm long, commonly included at the base; glumes 15–25 mm long; lemmas 8–12 mm long, at first pale, shiny brown at maturity, sparsely pubescent; awns 10–15 cm long, twisted below, curled toward the tip, flexuous. Very common; in grasslands of the Prairies and Parklands, and in the Peace River region. A very important forage species in the drier parts of the grasslands.

*Stipa richardsonii* Link                                    RICHARDSON NEEDLE GRASS

Plants tufted, with culms 50–80 cm high, usually erect. Sheaths slightly flattened, smooth or scabrous; blades 1–3 mm wide, involute, scabrous, indistinctly veined. Panicles 10–20 cm long, open, with branches slender, flexuous, spreading or drooping, bearing spikelets at the tips; glumes 8–10 mm long; lemmas about 5 mm long, pubescent, brown when mature; awns 25–35 mm long, weakly twice geniculate. Moist grasslands; Rocky Mountains, Cypress Hills, Parklands, Riding Mountain, Duck Mountain.

*Stipa spartea* Trin.                                           PORCUPINE GRASS

Plants in large tufts or tussocks, with erect culms to 1 m high. Sheaths round, with the outer margin usually ciliate; blades 3–5 mm wide, flat to convolute, ridged above. Panicles 15–20 cm long, narrow, nodding, with branches slender, each bearing one or two spikelets; glumes 3–4 cm long, tapering to a slender point; lemmas 15–25 mm long, brown, with the callus 7 mm long, villous; awns stout, 15–25 cm long, twice geniculate. Becoming rare; grassland and openings; southeastern Prairies and Parklands.

*Stipa spartea* Trin. var. *curtiseta* Hitchc.                WESTERN PORCUPINE GRASS

Very similar in stature to spear grass, but with wider, usually flat leaves; glumes 2–3 cm long; lemmas 12–15 mm long, brown; awns to 10 cm long, twice geniculate. Common; in moist prairie; throughout Prairies, Parklands, and Rocky Mountains; replacing spear grass as dominant in the moister sites.

Fig. 66. Spear grass, *Stipa comata* Trin. & Rupr.

*Stipa viridula* Trin. (Fig. 67)                              GREEN NEEDLE GRASS

Plants loosely tufted, with erect culms 50–100 cm high. Sheaths round, prominently veined, villous at the throat; blades 2–5 mm wide, to 25 cm long, flat. Panicles 10–20 cm long, narrow, with branches appressed-ascending; glumes 7–10 mm long; lemmas 5–6 mm long, plump, with the callus blunt, appressed-pubescent. Moderately dry to moist areas, in shrubbery, forest margins; throughout Prairies and Parklands. A variety of this species has been improved, and named green stipa grass, occasionally seeded for forage, mostly in comparative trials.

**Trisetum**          trisetum

Tufted perennials, with open or contracted panicles; spikelets usually 2- or 3-flowered, shiny. Not common, of almost no forage value.

1. Lemmas awnless, or the awn less than 2
   mm long, hidden by the glumes. ............................................................ *T. wolfii*
   Lemmas awned, with the awn bent and
   exserted. .................................................................................................. 2

2. Panicle open, nodding, with branches
   flexuous. ........................................................................................... *T. cernuum*
   Panicle contracted, with branches short,
   ascending. ............................................................................................... 3

3. Plants densely tufted; panicles 5–15 cm
   long, spike-like. ............................................................................. *T. spicatum*
   Plants loosely tufted, with culms often
   solitary; panicles 10–25 cm long, nar-
   row, loose. ................................................................................... *T. canescens*

*Trisetum canescens* Buckl.                              TALL TRISETUM

Plants loosely tufted, with culms 60–100 cm high, erect or decumbent at the base, often solitary. Sheaths, at least the lower one, sparsely to densely retrorse pubescent; blades 2–7 mm wide, flat, scabrous to pubescent. Panicles 10–25 cm long, narrow, usually loose, occasionally interrupted and spike-like, with branches appressed-ascending; spikelets about 8 mm long, 2- or 3-flowered; glumes 5–7 mm long; lemmas 5–6 mm long; rachilla hairs copious; awns to 12 mm long, geniculate. Moist meadows, woods, and openings; southern Rocky Mountains.

*Trisetum cernuum* Trin.                              NODDING TRISETUM

Plants loosely tufted, with culms 60–100 cm high, rather lax. Sheaths glabrous or sparsely pubescent; blades 6–12 mm wide, flat, lax, scabrous. Panicles 15–30 cm long, open, with branches flexuous, spreading or ascending; spikelets 6–12 mm long, usually with 3 florets; first glume 0.5–2 mm long, 1-nerved; second glume 3–4 mm long, 3-nerved; lemmas 5–6 mm long; callus hairs to 1 mm long, those of the rachilla to 2 mm long; awns 5–10 mm long, flexuous. Moist woods and openings; southern Rocky Mountains.

Fig. 67. Green needle grass, *Stipa viridula* Trin.

*Monocots*

*Trisetum spicatum* (L.) Richt.                                          SPIKE TRISETUM

Plants tufted, with culms 10–50 cm high, erect, pubescent below the pani-cle. Sheaths retrorse-pubescent; blades 1–3 mm wide, flat or involute, pubescent to subglabrous. Panicles 3–10 cm long, dense; spikelets about 6 mm long, shiny; first glume 3–5 mm long, 1-nerved; second glume 3.5–5.5 mm long, 3-nerved; lemmas 3.5–5 mm long; awns to 6 mm long, inserted below the apex, flexuous. Not common; grasslands, open woods; Rocky Mountains, Cypress Hills.

*Trisetum wolfii* Vasey                                                  AWNLESS TRISETUM

Plants loosely tufted, with culms 50–100 cm high, erect. Sheaths scabrous, or the lower sparsely pubescent; blades 2–4 mm wide, flat, scabrous or pilose above. Panicles 8–15 cm long, rather dense, with short branches appressed-ascending; spikelets 5–7 mm long, usually 2-flowered; glumes about 5 mm long, subequal; lemmas 4–5 mm long, awnless or with a short awn below the tip. Meadows and moist woods; southern Rocky Mountains, Cypress Hills.

**Triticum**          wheat

*Triticum aestivum* L.                                                   WHEAT

Annual plants, with stems to 1 m high. Leaves flat, to 2 cm wide. Spike to 12 cm long, dense; spikelets 2- to 5-flowered; glumes ovate, the upper part keeled, mucronate; lemmas glabrous to pubescent, awnless or awned, depending on the variety. The most important cereal grain, with many varieties, in the Prairie Provinces; occasionally established in roadsides and waste areas.

**Zizania**          wild-rice

*Zizania aquatica* L.                                                    ANNUAL WILD-RICE

Plants aquatic, with stout culms 1–2 m high. Sheaths long, with the ligule 10–15 mm long; blades 5–15 mm wide, flat. Panicles 30–50 cm long, with branches 10–20 cm long; spikelets unisexual; upper half of panicle pistillate, with branches appressed-ascending; lower half of panicle staminate, with branches spreading, spikelets pendulous, 6–11 mm long; pistillate spikelets with awns 1–6 cm long. Borders of streams and lakes; native in southeastern Boreal forest; introduced and established in several lakes in Boreal forests of Saskatchewan and Alberta.


# CYPERACEAE—sedge family

Grass-like or rush-like annual or perennial plants, usually with solid stems. Root fibrous or of long running rootstocks. The long, narrow leaves with closed sheaths on three sides of the stem, or 3-ranked. Flowers either perfect or unisexual with no sepals or petals, their places being taken by bristles or scales (perigynia). Although usually associated with moist or marshy areas, many species occur in very arid localities.

1. Flowers perfect, with spikelets uniform; achenes not enclosed in a perigynium or bract. ...................................................................... 2

   Flowers imperfect, with staminate and pistillate flowers in same or different spikelets; achenes enclosed in a perigynium or bract. ...................................................... 6

2. Scales of the spikelets 2-ranked, keeled; perianth lacking; spikelets in simple or compound terminal umbels. ........................... *Cyperus*

   Scales of the spikelets several-ranked, spirally arranged, not keeled; perianth present as bristles; spikelets solitary and terminal or partly lateral. ..................... 3

3. Base of the style persistent as a tubercle on top of the achene. ................................... 4

   Base of the style not persistent, the achene without a tubercle. ............................... 5

4. Culms leafless, with sheaths usually bladeless; spikelet solitary, terminal. .......... *Eleocharis*

   Culms with bristle-like leaves; spikelets several. ................................................. *Rhynchospora*

5. Perianth bristles 1–8, occasionally lacking, usually little longer or shorter than the achene. ................................................. *Scirpus*

   Perianth bristles numerous, often 2–3 cm long, silky. ...................................... *Eriophorum*

6. Achenes partly enclosed by a spathe-like scale. .............................................. *Kobresia*

   Achenes entirely enclosed by a perigynium. ................................................. *Carex*

## *Carex*     sedge

The genus *Carex* is very large, and its taxonomy is difficult. To distinguish between species, characters of the mature inflorescence are needed. The species can be divided into several groups of species that share common characters. However, determination within groups is often not as easy, because the characters are variable, hybridization between species is known to occur, and the opinions of experts differ on the validity of species and varieties.

Sedges are important in the Prairie Provinces. Some species, such as the awned sedge, beaked sedge, and water sedge, form a large part of the "slough hay," harvested in many areas. In dry prairie, several species of sedges increase in abundance as a result of overgrazing, and thereby they help prevent soil erosion. In Boreal forest, sedges are often dominant in swamp vegetation, and are helpful in building peat soils. Also, in alpine tundra and meadows in the Rocky Mountains, sedges are often the dominant constituents of the vegetation.

Because of their importance, the sedges have been treated in detail in this publication. A total of 125 species and many varieties are described. The treatment followed is conservative, in that several closely related species are grouped into one large species. Keys are provided to separate the small species within these groups.

Well-developed mature fruits (perigynia) and inflorescences are needed for successful determinations. The various types of perigynia and inflorescences are shown in Figs. 68 and 69.

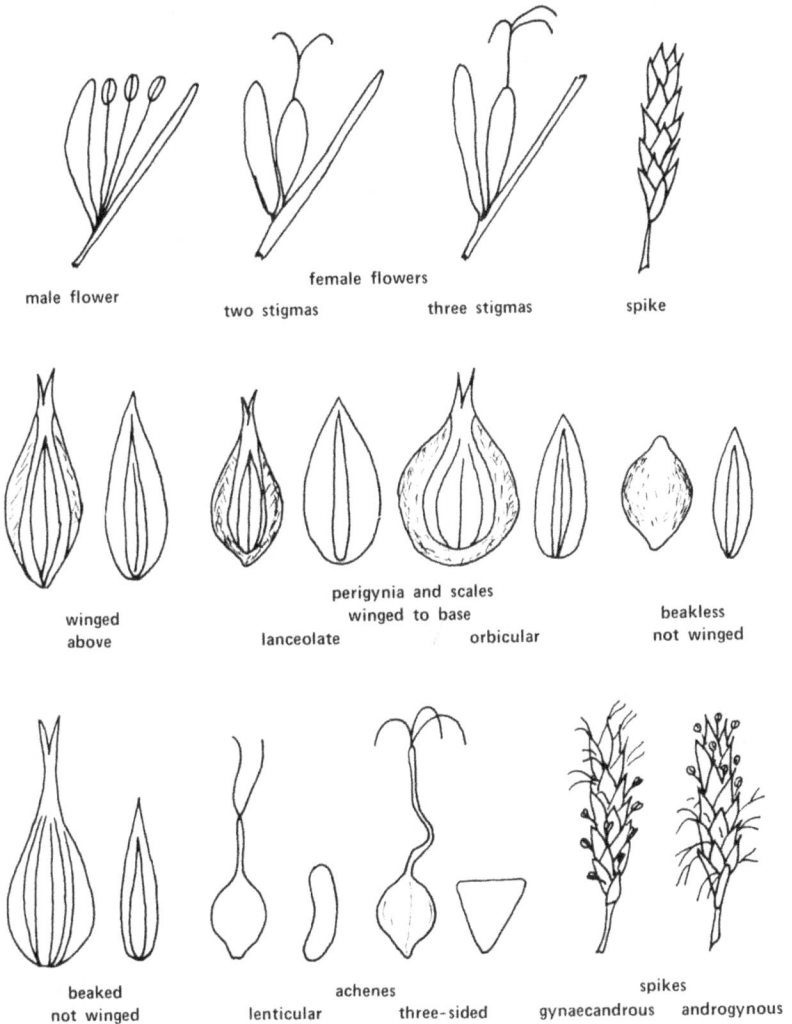

Fig. 68.    Flowering and fruiting characters of sedges.

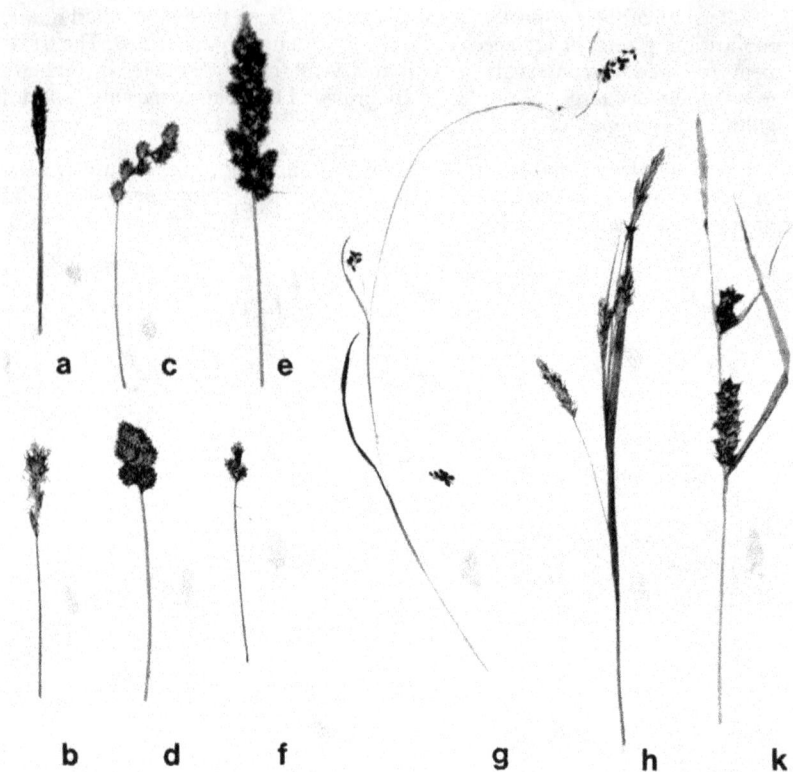

Fig. 69. Inflorescences of sedges: *a*, single spike; *b–k*, two or more spikes; *b–f*, mostly bisexual spikes; *b*, *c*, spikes approximate; *d*, spikes crowded; *e*, compound spikes; *f*, gynaecandrous spikes; *g–k*, terminal spikes staminate, lateral spikes pistillate; *g, h*, pistillate spikes more or less long-peduncled; *k*, pistillate spikes sessile.

# Key to the Groups of *Carex*

1. Spike one, usually bractless; leaves usually all basal.     Group 1

   Spikes two or more, with bracts often present; stem leaves often present.     2

2. Spikes mostly bisexual, sessile; achenes lenticular; stigmas 2.     Group 2

   Spikes of two kinds, the terminal spikes staminate, the lower spikes pistillate.     3

3. Achenes lenticular; stigmas 2.     Group 3

   Achenes triangular; stigmas 3.     4

4. Perigynia with a conspicuously bidentate beak.     Group 4

   Perigynia beakless, or the beak not conspicuously bidentate.     Group 5

# Key to Sections and Species in Group 1

1. Achenes lenticular; stigmas 2.     2

   Achenes triangular; stigmas 3.     4

2. Spikes usually completely staminate or completely pistillate; perigynia dark, lustrous, spreading to reflexed.     Section 3. Dioicae *(C. gynocrates)*

   Spikes androgynous; perigynia ascending to spreading.     3

3. Perigynia stipitate, striate.     Section 1. Nardinae *(C. nardina)*

   Perigynia sessile, not striate.     Section 2. Capitatae *(C. capitata)*

4. Pistillate scales deciduous; at least the lower perigynia reflexed at maturity.     5

   Pistillate scales not deciduous; perigynia spreading to ascending at maturity.     6

5. Perigynia dark brown; style jointed to the achene.     Section 4. Callistachys

   Perigynia yellowish to light brown; style continuous with the achene.     Section 44. Orthocerates

6. Plants dioecious, rhizomatous; perigynia pubescent.     Section 23. Scirpinae *(C. scirpoidea)*

   Plants monoecious; spikes androgynous.     7

7. Margins of the staminate scales united in the lower part.     8

   Margins of the staminate scales free to the base.     9

8. Pistillate scales leaf-like; perigynia clearly
beaked.

Section 19. Phyllostachyae
(*C. backii*)

    Pistillate scales not leaf-like; perigynia
rounded at the tip, beakless.

Section 18. Politrichoideae
(*C. leptalea*)

9. Perigynia pubescent or puberulent; plants
densely cespitose; leaves stiffly filiform.

Section 20. Filifoliae
(*C. filifolia*)

    Perigynia not pubescent.

10

10. Perigynia stipitate, striate.

Section 1. Nardinae
(*C. nardina*)

    Perigynia sessile, not striate.

11

11. Pistillate scales without hyaline margins,
shorter than the perigynia; prairie
species.

Section 21. Obtusatae
(*C. obtusata*)

    Pistillate scales with hyaline margins, con-
cealing the perigynia; mountain species.

12

12. Perigynia 3–4 mm long.

Section 25. Rupestres

    Perigynia 5–7 mm long.

Section 26. Firmiculmes
(*C. geyeri*)

## Key to Sections and Species in Group 2

1. Culms single or few together from long
creeping rhizomes or stolons.

2

    Culms more or less densely tufted; rhi-
zomes, if any, short.

6

2. Culms stoloniferous, producing shoots
from the nodes of old decumbent
stems.

Section 9. Chordorrhizae
(*C. chordorrhiza*)

    Culms not stoloniferous; new shoots
arising from rhizomes.

3

3. Spikes closely aggregated into a globose
or ovoid head; bracts scarious; spikes
androgynous.

Section 5. Foetidae
(*C. maritima*)

    Spikes not closely aggregated; bracts not
scarious.

4

4. Perigynia 4–6 mm long, wing-margined at
the tip, with bidentate beak.

Section 8. Arenariae
(*C. siccata*)

    Perigynia to 4 mm long, not wing-mar-
gined, not conspicuously bidentate
beaked.

5

5. Ventral strip of sheath green, nerved to the top; collar with two dark warts; spikes 10–20 or more in a head.

Section 7. Intermediae
*(C. sartwellii)*

Ventral strip of sheath hyaline or brown, unveined; collar without warts; spikes usually less than 10 in a head.

Section 6. Divisae

6. Spikes androgynous. 7
Spikes gynaecandrous. 11

7. Spikes with only 2 or 3 perigynia; perigynia beakless, not flattened on the inside.

Section 14. Heleonastes
*(C. disperma)*

Spikes with more than 3 perigynia; perigynia beaked, flattened on the inside. 8

8. Spikes 4–10 in the inflorescence; all rachis nodes bearing a single spike.

Section 10. Bracteosa

Spikes usually more than 10 in the inflorescence; at least the lowest rachis node bearing 2 or more spikes. 9

9. Culms soft and weak, the angles sharply winged; perigynia widest at base, the beak about as long as the body; sheaths loose.

Section 13. Vulpinae

Culms stiff, not winged; perigynia widest above the middle, the beak about half as long as the body; sheaths tight. 10

10. Perigynia straw-colored; pistillate scales awned; ventral strip of sheaths cross-wrinkled.

Section 11. Multiflorae
*(C. vulpinoidea)*

Perigynia brown to black; pistillate scales not awned; ventral strip of sheaths smooth, with reddish spots.

Section 12. Paniculatae

11. Perigynia winged along part or all of the margin; always with a more or less distinctly bidentate beak.

Section 17. Ovales

Perigynia not winged along the margin; the beak, if present, not distinctly bidentate. 12

12. Achenes filling the perigynia almost fully; perigynia white puncticulate under magnification.

Section 14. Heleonastes

Achenes filling only the upper half or two-thirds of the perigynia; perigynia not puncticulate. 13

13. Achene visible through the wall of the perigynium; perigynia more than 4 mm long; awn-like bracts present.

Section 16. Deweyanae
*(C. deweyana)*

Achene not visible through the wall of the
perigynium; perigynia less than 4 mm
long; bracts only slightly developed.                    Section 15. Stellulatae
                                                          *(C. muricata)*

## Key to Sections and Species in Group 3

1. Lower pistillate spikes long-peduncled,
   pendulous.                                             Section 43. Cryptocarpae

   Lower pistillate spikes sessile or short-pe-
   duncled, erect or ascending.                                                      2

2. Perigynia shiny, inflated, clearly ribbed,
   the beak pronounced; style continuous
   with the achene.                                       Section 48. Vesicariae
                                                          *(C. saxatilis)*

   Perigynia not shiny or inflated, the ribs
   obscure or absent, the beak not pro-
   nounced; style jointed to the achene.                                            3

3. Perigynia rounded, not beaked; bracts
   with a sheath.                                                                    4

   Perigynia flattened, with a short beak;
   bracts sheathless.                                     Section 42. Acutae

4. Sheaths of bracts long; perigynia serrulate
   toward the tip.                                        Section 27. Albae
                                                          *(C. rufina)*

   Sheaths of bracts short; perigynia not
   serrulate.                                             Section 28. Bicolores

## Key to Sections and Species in Group 4

1. Perigynia pubescent or puberulent.                                               2
   Perigynia glabrous.                                                              3

2. Perigynia rounded in cross section, not
   ribbed; lower bracts shorter than or
   equaling the inflorescence.                            Section 22. Montanae

   Perigynia triangular in cross section,
   ribbed; lower bracts longer than the
   inflorescence.                                         Section 39. Hirtae

3. Style continuous with and usually
   strongly curved above the achene, persistent; leaves often strongly and
   clearly cross-veined.                                                            4

   Style jointed to the achene, straight, with-
   ering; leaves slightly or not cross-veined.                                      8

4. Perigynia about 6 times as long as wide;
   lower bracts sheathing.                                Section 45. Folliculatae
                                                          *(C. michauxiana)*

   Perigynia about 3–4 times as long as
   wide; lower bracts sheathless.                                                   5

5. Perigynia shiny, conspicuously 7- to 9-ribbed, usually contracted at base of beak.                                      Section 48. Vesicariae

   Perigynia not shiny, 12- to 25-ribbed, though often obscurely, tapering to the beak.                                                                        6

6. Perigynia 9–18 mm long; pistillate spikes globose.                                       Section 49. Lupulinae
                                                     *(C. intumescens)*

   Perigynia to 10 mm long; pistillate spikes cylindric.                                                                       7

7. Lower pistillate spikes long-peduncled, pendulous.                                  Section 46. Pseudo-Cyperae
   Lower pistillate spikes short-peduncled, ascending or erect.                         Section 47. Paludosae

8. Pistillate spikes long-peduncled, loosely flowered.                                 Section 35. Longirostres
                                                     *(C. sprengelii)*

   Pistillate spikes sessile or short-peduncled, densely and closely flowered.         Section 36. Extensae

# Key to Sections and Species in Group 5

1. Margins of staminate scales united toward the base; pistillate scales leaf-like.      Section 19. Phyllostachyae
                                                     *(C. backii)*

   Margins of staminate scales free to the base; pistillate scales not leaf-like.                                              2

2. Perigynia pubescent or puberulent.                                                                                         3
   Perigynia glabrous.                                                                                                         6

3. Pistillate bracts developed, sheathed; perigynia 5–6 mm long, lanceolate.                                                  4

   Pistillate bracts lacking either sheath or blade; perigynia less than 5 mm long, ovoid.                                     5

4. Spikes 3–8, the upper 3 or 4 closely aggregated; perigynia with a minute beak.      Section 37. Ferrugineae
   Spikes about 3, widely separated; perigynia with a beak about as long as the body.  Section 33. Sylvaticae
                                                     *(C. assiniboinensis)*

5. Pistillate bracts sheathless, blades more or less well developed; perigynia rounded in cross section.   Section 22. Montanae
   Pistillate bracts long-sheathing, bladeless; perigynia triangular in cross section.  Section 24. Digitatae

6. Leaves and sheaths pubescent or puberulent.                                         Section 38. Virescentes
                                                     *(C. torreyi)*

Leaves and sheaths glabrous. — 7

7. Lower pistillate bracts long-sheathing. — 8
   Lower pistillate bracts sheathless or almost sheathless. — 13

8. Bracts bladeless; perigynia less than 2 mm long; leaves involute, to 0.5 mm wide. — Section 27. Albae *(C. eburnea)*

   Bracts with blades; perigynia longer than 2 mm; leaves wider than 0.5 mm. — 9

9. Pistillate spikes long-peduncled, pendulous. — 10
   Pistillate spikes short-peduncled, erect to ascending. — 11

10. Pistillate spikes 1–6 cm long; culms to 60 cm high; leaves 3–9 mm wide, deep green. — Section 32. Gracillimae *(C. gracillima)*

    Pistillate spikes usually less than 1 cm long; culms 10–20 cm high; leaves less than 2 mm wide, yellowish brown. — Section 34. Capillares *(C. capillaris)*

11. Culms soft, somewhat winged; sheaths loose, dilated at summit; perigynia abruptly constricted and bent into a beak. — Section 30. Laxiflorae *(C. laxiflora)*

    Culms firm, not winged; sheaths tight; perigynia not contracted and bent into the beak. — 12

12. Perigynia obscurely veined; veins less than 10; perigynia tapered at base, somewhat triangular in cross section. — Section 29. Paniceae

    Perigynia clearly veined; veins more than 10; perigynia rounded at base, rounded in cross section. — Section 31. Granulares

13. Pistillate spikes long-peduncled, spreading or pendulous. — Section 33. Sylvaticae *(C. castanea)*

    Pistillate spike short-peduncled or sessile, ascending or erect. — 14

14. Bracts of the pistillate spikes small or absent; pistillate spikes close together, about 15 mm long. — 15

    Bracts of the pistillate spikes well-developed; pistillate spikes not close together, longer than 15 mm. — 16

15. Pistillate spikes oblong; culms longer than the leaves; plants cespitose. — Section 25. Rupestres *(C. glacialis)*

Pistillate spikes spherical; culms about as long as the leaves; plants rhizomatous.

Section 21. Obtusatae
*(C. supina)*

16. Perigynia 7–9 mm long, shiny, inflated, strongly 7- to 9-veined; style continuous with the achene.

Section 48. Vesicariae
*(C. oligosperma)*

Perigynia less than 7 mm long, not shiny or inflated, obscurely veined; style jointed to the achene.

17

17. Perigynia with a beak about one-quarter as long as the body, not flattened, spreading at maturity.

Section 36. Extensae

Perigynia beakless or almost beakless, clearly flattened, ascending to erect.

18

18. Root system covered with a yellowish felt; terminal spike staminate; perigynia yellowish to greenish.

Section 40. Limosae

Root system without felt cover; terminal spike gynaecandrous or staminate; perigynia greenish to brown.

Section 41. Atratae

## Section 4. Callistachys

Perigynia erect or spreading at maturity; leaves involute, 1 mm wide; staminate flowers few; plants cespitose. ............................................................................ *C. pyrenaica*

Perigynia deflexed at maturity; leaves flat, 1.5–2.5 mm wide; staminate flowers numerous; plants rhizomatous. ................................................................................ *C. nigricans*

## Section 6. Divisae

1. Inflorescence 1–2 times as long as broad; culms about 20 cm high; rhizomes brownish, 1–2 mm in diam. ...................................................................... 2

Inflorescence 2–5 times as long as broad; culms usually more than 20 cm high; rhizomes brown or black, 2–3 mm thick. .......................................................... *C. praegracilis*

2. Plants monoecious; perigynia 2.5–3 mm long, not concealed by the scales; the beak about 0.5 mm long. ................................................................ *C. stenophylla*

Plants dioecious; perigynia 3.5–4 mm long, fully concealed by the scales; the beak about 1.5 mm long. .......................................................... *C. douglasii*

## Section 10. Bracteosae

1. Spikes barely overlapping in the head, erect with a bract; beak of the perigynium about as long as the body. ........................................................ *C. hookerana*

Spikes densely crowded in the head, only
the lowest spike with a bract; beak of
the perigynium shorter than the body. ......................................................... 2

2. Heads ovoid, about as long as broad;
perigynia and scales brown. ....................................................... *C. hoodii*

Heads elongate, more than twice as long
as broad; perigynia and scales
yellowish. ............................................................................................... 3

3. Culms firm, sharply triangular, not
winged; perigynia about 4 mm long,
the beak about 1 mm long. ......................................................... *C. gravida*

Culms soft, the angles winged; perigynia
about 3 mm long, the beak about 1.5
mm long. ........................................................................ Section 13. Vulpinae
*(C. alopecoidea)*

## Section 12. Paniculatae

Perigynia dark brown, not concealed by the
scales; sheaths with red spots. .................................................. *C. diandra*
Perigynia light brown to brown, concealed by
the scales; sheaths without red spots. ....................................... *C. prairea*

## Section 13. Vulpinae

Perigynia lanceolate, the beak about as long as
the body, the base spongy. .......................................................... *C. stipata*
Perigynia ovoid, the beak about half as long as
the body, the base not spongy. ............................................... *C. alopecoidea*

## Section 14. Heleonastes

1. Spikes few, widely spaced in the
inflorescence; perigynia 1–5 in a spike. ........................................ 2
Spikes usually several, at least the upper
ones close together; perigynia usually
more than 5 in a spike. ..................................................................... 3

2. Spikes androgynous; bracts obsolete or
that of the lowest spike about 1 cm
long. ............................................................................................ *C. disperma*
Spikes gynaecandrous; bract of the lowest
spike as long as or longer than the
inflorescence. ............................................................................. *C. trisperma*

3. Beak of the perigynia merely indicated;
scales silvery, more or less transparent. ....................................... 4
Beak of the perigynia at least 0.5 mm
long; scales whitish to brown, not
transparent. .......................................................................................... 5

4. Spikes closely crowded into a head at the
summit of the stem; scales about as
long as the perigynia. .............................................................. *C. tenuiflora*

Spikes not crowded into a head, spaced along the stem; scales about half as long as the perigynia. ...................................................... *C. loliacea*

5. Spikes more or less crowded into a head at the summit of the stems. ............................................................ 6

Spikes spaced in an elongated inflorescence. .................................................................................. 7

6. Perigynia broadest at the base; dorsal veins pronounced and raised. ........................................ *C. arcta*

Perigynia broadest at the middle, tapering to the base; dorsal veins not pronounced and raised. ......................................... *C. heleonastes*

7. Perigynia stipitate; the terminal spike with a long staminate base; scales reddish orange brown, as long as the perigynia. ........................................................... *C. mackenziei*

Perigynia not stipitate; the terminal spike with few staminate flowers; scales whitish to brown, shorter than the perigynia. ...................................................................... 8

8. Perigynia 5–10 in a spike; the beak about 1.5 mm long. .................................................... *C. brunnescens*

Perigynia 10–30 in a spike; the beak about 0.25 mm long. ..................................................... *C. curta*


## Section 17. Ovales

1. Bracts leaf-like or considerably longer than the inflorescence or both. .......................................... 2

Bracts not leaf-like or longer than the inflorescence. ............................................................... 3

2. Bracts leaf-like and much longer than the inflorescence; perigynia about 5 times as long as broad. ......................................... *C. sychnocephala*

Bracts not leaf-like but much longer than the inflorescence; perigynia about 2–3 times as long as broad. .......................................... *C. athrostachya*

3. Perigynia subulate, about 4–5 times as long as broad; spikes closely crowded into a head. ........................................................... *C. crawfordii*

Perigynia lanceolate to ovate, 2–3 times as long as broad; spikes crowded to distant in the head. ............................................................... 4

4. The pistillate scales about as long as the perigynia, completely or almost completely concealing the perigynia. ........................ 5

The pistillate scales shorter and narrower than the perigynia, exposing the beak and margins of the perigynia. .................................... 9

5. Beak of the perigynium flat, the teeth spreading, usually serrulate to the tip. ............................................ 6

Beak of the perigynium terete, the teeth, if developed, parallel, not serrulate to the tip. ............................................................................ 8

6. Inflorescence flexuous, the spikes hardly or not at all overlapping. ................................ *C. argyrantha*

Inflorescence stiffly erect, the spikes crowded or at least well overlapping. ............................ 7

7. Pistillate scales and perigynia closely appressed; bracts poorly developed. ................................ *C. xerantica*

Pistillate scales and perigynia ascending, not appressed; lower bracts prominent. ........................... *C. adusta*

8. Perigynia 6–8 mm long, prominently veined on both sides. .................................... *C. petasata*

Perigynia 4.5–6 mm long, obscurely veined or veinless. .................................... *C. phaeocephala*

9. Perigynia narrowly lanceolate, about 3 times as long as wide. .................................... 10

Perigynia lanceolate to ovoid, 2–2.5 times as long as wide. .................................... 11

10. Spikes crowded into a head at the tip of the culm; perigynia 4–7 mm long, veined on both sides; leaves 1–3 mm wide. .................................... *C. scoparia*

Spikes not crowded into a head, at least the lower spikes separated; perigynia 3–5 mm long, prominently veined ventrally, veinless or obscurely veined dorsally; leaves 3–7 mm wide. .................................... *C. tribuloides*

11. Perigynia broadly ovate, the body almost orbicular, rather abruptly beaked. .................................... *C. straminea*

Perigynia elliptic to ovoid, not almost orbicular, tapering to the beak. .................................... 12

12. Beak of the perigynia terete, the teeth, if developed, parallel. .................................... 13

Beak of the perigynia flattened, the teeth usually well-developed, spreading. .................................... 14

13. Spikes, at least the lower ones, well-separated in the head. .................................... *C. preslii*

Spikes crowded in the head, hardly separated from each other. .................................... *C. macloviana*

14. Perigynia prominently 3- to 5-veined dorsally; pistillate scales pale green to light brown; inflorescence elongate, flexuous. .................................... *C. festucacea*

Perigynia not prominently veined dorsally; pistillate scales reddish to deep brown; inflorescence dense, stiff. .................................... *C. tincta*

## Section 21. Obtusatae

Head a solitary spike, staminate at the tip. ........................................................... *C. obtusata*

Head with 2 or more spikes, the upper one entirely staminate or with a few pistillate flowers at the base. ................................................................................................. *C. supina*

## Section 22. Montanae

1. Pistillate spikes all on well-developed culms, 1–4 dm high. ................................................................................. 2

   Pistillate spikes in part on very short culms, more or less hidden among the leaf bases. ................................................................................................................ 4

2. Staminate spike to 20 mm long; pistillate scales about as long as the perigynia; perigynia subglobose. ........................................................ *C. pensylvanica*

   Staminate spike to 10 mm long; pistillate scales much shorter than the perigynia; perigynia ellipsoid. ................................................................................. 3

3. Staminate spikes 5–10 mm long; pistillate spikes 3 or 4, the lower ones pedun-cled; perigynia 3–4 mm long; culms 1.5–3 dm high. ................................................................ *C. nigromarginata*

   Staminate spikes 2–5 mm long; pistillate spikes usually 2 or 3, sessile; perigynia 2.5–3 mm long; culms 0.5–1.5 dm high. ........................ *C. deflexa*

4. Culms bearing a staminate spike, occa-sionally with one pistillate spike; bracts poorly developed, shorter than the inflorescence. ........................................................................................................... 5

   Culms bearing a staminate spike, usually with 2 or 3 staminate spikes; bracts well-developed, surpassing the inflorescence. ........................................................................................................... 6

5. Perigynia pubescent; blades 1–3 mm wide. ........................................................................................ *C. umbellata*

   Perigynia glabrous; blades 2–5 mm wide. ........................................ *C. tonsa*

6. Staminate spike 2–5 mm long; perigynia 2.5–3 mm long, with a beak about 0.5 mm long. ................................................................................................ *C. deflexa*

   Staminate spike 3–15 mm long; perigynia 3–3.5 mm long, with a beak about 1 mm long. ................................................................................................ *C. rossii*

## Section 24. Digitatae

1. Lower pistillate spikes long-peduncled; bracts usually with a short blade; pistil-late scales short-awned. ........................................................ *C. pedunculata*

Lower pistillate spikes short-peduncled or
sessile; bracts bladeless or almost so;
pistillate scales not awned. ............................................................................. 2

2. Staminate spike 3–6 mm long, crowded
by the pistillate spikes; pistillate scales
light brownish. ......................................................................... *C. concinna*

Staminate spike 8–25 mm long, well overtopping the pistillate spikes; pistillate
scales reddish or dark brown. ............................................................... 3

3. Lower pistillate spikes short-peduncled,
often somewhat remote. ......................................................... *C. richardsonii*

Lower pistillate spikes almost sessile,
approximate and usually overlapping. ......................................... *C. concinnoides*

## Section 25.  Rupestres

Inflorescence a solitary androgynous spike. ........................................... *C. rupestris*
Inflorescence consisting of a sessile staminate
spike, and one or more pistillate spikes, the
lower ones peduncled. ......................................................................... *C. glacialis*

## Section 27.  Albae

Stigmas 3; achenes triangular in cross section. ....................................... *C. eburnea*
Stigmas 2; achenes lenticular in cross section. ........................................ *C. rufina*

## Section 28.  Bicolores

Terminal spike staminate; pistillate scales pale
brown to whitish. ................................................................................... *C. aurea*
Terminal spike gynaecandrous; pistillate
scales purplish brown to black. ............................................................... *C. bicolor*

## Section 29.  Paniceae

1. Perigynia beaked, with the beak bidentate, about one-quarter the length of
the body; pistillate spikes loosely
flowered. .......................................................................................... *C. vaginata*

Perigynia beakless, or the beak very short,
not bidentate; pistillate spikes rather
closely flowered. .................................................................................. 2

2. Pistillate spikes sessile or short-peduncled, ascending to erect; leaves whitish
glaucous, usually folded or involute. ...................................................... *C. livida*

Pistillate spikes, at least the lower ones,
long-peduncled, ascending to spread-
ing; leaves not very glaucous, flat. ....................................................... *C. tetanica*

*Monocots*

## Section 31.  Granulares

Staminate spike sessile or short-peduncled, not
overtopping the pistillate spikes. .................................................................... *C. granularis*

Staminate spike long-peduncled, overtopping
the pistillate spikes. .................................................................................... *C. crawei*

## Section 33.  Sylvaticae

Perigynia pubescent;  pistillate  spikes  very
loosely flowered. ...................................................................................... *C. assiniboinensis*

Perigynia glabrous;  pistillate  spikes  rather
closely flowered. ...................................................................................... *C. castanea*

## Section 36.  Extensae

Perigynia 3.5–6 mm long, spreading to reflexed
in the spike, with the beak about half to
three-quarters the length of the body. .................................................................... *C. flava*

Perigynia 2–3.5 mm long, spreading to ascending in the spike, with the beak about one-
quarter to half the length of the body. .................................................................... *C. viridula*

## Section 33.  Ferrugineae

1. Perigynia beakless or almost beakless;
    terminal spike entirely staminate. ........................................................ *C. atrofusca*

   Perigynia tapering into a beak; terminal
    spike pistillate above. ........................................................................................ 2

2. Perigynia 3.5–5 mm long, 1 mm wide;
    blackish above, greenish or yellowish
    below. ........................................................................................................ *C. misandra*

   Perigynia 4.5–6 mm long, 1.5–2 mm wide;
    yellowish brown. ................................................................................ *C. petricosa*

## Section 39.  Hirtae

1. Perigynia 4–7 mm long, prominently
    veined; beak half as long as the body,
    its teeth spreading; rhizomes long. .................................................... *C. houghtonii*

   Perigynia 2.5–4 mm long, the veins not
    prominent, hidden under the pubes-
    cence; beak less than half as long as the
    body, its teeth slightly spreading; rhi-
    zomes short. ........................................................................................................ 2

2. Leaves to 2 mm wide, involute; perigynia
    4–5 mm long; the body oval, somewhat
    flattened. ................................................................................................ *C. lasiocarpa*

   Leaves to 4 mm wide, flat; perigynia
    2.5–3.5 mm long; the body almost orbi-
    cular, terete. .......................................................................................... *C. lanuginosa*

## Section 40. Limosae

1. Pistillate scales long acuminate, to twice
   as long as but much narrower than the
   perigynia. ................................................................ *C. magellanica*

   Pistillate scales ovate, about as long and
   wide as the perigynia. ............................................................ 2

2. Staminate spike short-peduncled; pistillate scales partly clasping and enclos-
   ing the perigynia. ...................................................... *C. rariflora*

   Staminate spike long-peduncled; pistillate
   scales not clasping or enclosing the
   perigynia. ...................................................................... *C. limosa*

## Section 41. Atratae

1. Culms arising from long rhizomes, the
   base without the previous year's dry
   leaves. ........................................................................ *C. buxbaumii*

   Culms arising from a tuft of the previous
   year's dry leaves, or the lower sheaths
   not filamentose. .......................................................................... 2

2. Lower pistillate spikes long-peduncled,
   the peduncles as long as or longer than
   the spikes. ................................................................................... 3

   Lower pistillate spikes sessile or short-peduncled, the peduncles shorter than the
   spikes. ........................................................................................ 5

3. The terminal spike gynaecandrous. ............................ *C. mertensii*
   The terminal spike entirely staminate. ...................................... 4

4. Perigynia round in cross section, not
   flattened; achenes not stipitate. ................................ *C. raynoldsii*

   Perigynia strongly flattened; achenes sti-
   pitate. ........................................................................ *C. podocarpa*

5. Perigynia 2–3.5 mm long; pistillate scales
   1.5–2.5 mm long. ........................................................................ 6

   Perigynia 3.5–4 mm long; pistillate scales
   about 2.5 mm long. ...................................................... *C. atrata*

6. The terminal spike usually completely pistillate; scales much shorter than the
   perigynia; rhizomes short. ........................................ *C. norwegica*

   The terminal spike staminate or gynaecandrous; scales about equaling the
   perigynia; rhizomes long. ........................................ *C. parryana*

## Section 42. Acutae

1. The terminal spike gynaecandrous; leaf
   blades 0.5–1.5 mm wide; pistillate
   scales shorter and narrower than the
   perigynia. .................................................................. *C. eleusinoides*

The terminal spikes staminate; leaf blades
  2–8 mm wide; pistillate scales about as
  long and as wide as the perigynia. ........................................................................................ 2

2. Perigynia unveined or the veins obscure. ................................................................. 3
   Perigynia clearly veined. ................................................................................................. 6

3. Culms in large, dense clumps; lowest new
   leaf sheaths bladeless. .............................................................................. *C. stricta*
   Culms singly or in small tufts; lowest new
   leaf sheaths with blades. ............................................................................. 4

4. Staminate spikes usually several; culms in
   tufts, together with sterile shoots. ........................................................ *C. aquatilis*
   Staminate spike solitary; culms arising
   singly or a few together from rhizomes. .................................................... 5

5. Plants low, 1–4 dm high; pistillate scales
   as wide as the perigynia. .......................................................................... *C. bigelowii*
   Plants taller, 3–10 dm high; pistillate
   scales about half as wide as the
   perigynia. ................................................................................................. *C. aperta*

6. Perigynia thick, leathery; beak 0.5–1 mm
   long, bidentate. ........................................................................ *C. nebraskensis*
   Perigynia not thick and leathery; beak 0.1
   mm long or less, blunt. .............................................................................. 7

7. Perigynia prominently veined, stipitate;
   pistillate scales with a green center
   about one-fifth the width of the scale. .................................................... *C. kelloggii*
   Perigynia finely veined, hardly stipitate;
   pistillate scales with a green center one-
   third the width of the scale. ...................................................... *C. lenticularis*


## Section 43.   Cryptocarpae

1. Pistillate spikes short-peduncled, ascend-
   ing to erect. ......................................................................................... *C. salina*
   Pistillate spikes long-peduncled, spread-
   ing to pendulous. ....................................................................................... 2

2. Staminate spikes spreading to pendulous;
   pistillate scales tapering to a long awn. ................................................. *C. paleacea*
   Staminate spikes erect; pistillate scales
   abruptly contracted to a long awn. ........................................................ *C. crinita*


## Section 44.   Orthocerates

Perigynia 6–8 mm long; rachilla rudimentary;
  culms with 2 or 3 leaves. ........................................................................ *C. pauciflora*
Perigynia 4–5 mm long; rachilla well-devel-
  oped; culms with 4–8 leaves. ................................................................ *C. microglochin*

## Section 46. Pseudo-Cyperae

Perigynia spreading or ascending at maturity,
thin-walled, scarcely stipitate. ................................................................ *C. hystricina*

Perigynia spreading to reflexed at maturity,
thick-walled, clearly stipitate. .......................................................... *C. pseudo-cyperus*


## Section 47. Paludosae

1. Sheaths and base of leaf blades
   pubescent. ................................................................................................ *C. atherodes*

   Sheaths and blades glabrous. ................................................................................... 2

2. Teeth of the perigynium 0.5 mm long. ........................................................ *C. lacustris*

   Teeth of the perigynium 1–2 mm long. ...................................................... *C. laeviconica*


## Section 48. Vesicariae

1. Stigmas 2; achenes lenticular. ..................................................................... *C. saxatilis*

   Stigmas 3; achenes triangular. ............................................................................... 2

2. Pistillate spikes globose to very short
   oblong with 3–15 perigynia; beak of
   the perigynium short-bidentate or
   edentate. ................................................................................................. *C. oligosperma*

   Pistillate spikes oblong to cylindric with
   15 to many perigynia; beak of the peri-
   gynium usually prominently bidentate. ................................................................ 3

3. The lower perigynia reflexed at maturity;
   lower bracts two or more times as long
   as the inflorescence. ..................................................................................... *C. retrorsa*

   The lower perigynia not reflexed at maturity, ascending to spreading; lower
   bracts slightly exceeding the
   inflorescence. ......................................................................................................... 4

4. Culms thick, spongy at base, and bluntly
   angled; leaves and sheaths conspicu-
   ously cross-veined. ...................................................................................... *C. rostrata*

   Culms slender, not spongy at base, and
   sharply angled; leaves and sheaths
   cross-veined but not conspicuously so. ................................................................ 5

5. Perigynia 3–3.5 mm long, obscurely
   veined; pistillate scales purplish black;
   leaves 1–3 mm wide, involute. ............................................................... *C. rotundata*

   Perigynia 4–10 mm long, clearly veined;
   pistillate scales yellowish to purplish
   brown; leaves 3–7 mm wide, flat. ........................................................... *C. vesicaria*

*Carex adusta* Boott                                                    BROWNED SEDGE

Plants tufted, with culms 20–60 cm high, erect, much taller than the leaves; blades 3–4 mm wide. Inflorescence 2–3 cm long; spikes crowded, sub-globose, 8–12 mm long; pistillate scales ovate, hiding the perigynia, brown, with the margin hyaline; perigynia 4–5 mm long, brown, ovate, sharp-edged below the middle. Not common; dry, usually sandy, soil; Prairies and Parklands.

*Carex alopecoidea* Tuck.                                               FOXTAIL SEDGE

Plants tufted, with culms 30–70 cm high, erect, soft, mostly as long as or shorter than the leaves; blades 3–4 mm wide. Inflorescence 2–5 cm long, simple or compound, with the upper spikes crowded, the lower ones distant; scales shorter than the perigynia, brown; perigynia 3–4 mm long, tapering into a beak nearly as long as the body, brown. Wet meadows, slough margins, banks of creeks and rivers; Prairies, Parklands, Boreal forest.

*Carex aperta* Boott                                                    OPEN SEDGE

Plants with stout, woody rhizomes; culms 30–70 cm high, tufted, stiff, and sharply 3-sided; leaves 2–5 mm wide, flat to channeled. Inflorescence 15–20 cm long; terminal spike staminate, 2–3.5 cm long; pistillate spikes 1–5 cm long, usually 2 or 3, erect, sessile or on short peduncles, distant to approximate; scales mostly longer but narrower than the perigynia, reddish black with light midrib; perigynia 2.5–3.5 mm long, greenish to straw-colored. Moist areas; southern Rocky Mountains.

*Carex aquatilis* Wahl.                                                 WATER SEDGE

Plants with long, scaly rhizomes; culms 10–80 cm high, densely tufted, often in large clumps; blades 3–8 mm wide, flat to channeled, light green or bluish green. Inflorescence 10–20 cm long; terminal spikes staminate, 1.5–3 cm long, usually 1–3; pistillate spikes 2–4 cm long, usually 2–6; scales usually narrower than the perigynia, brown with light green center; perigynia 2–4 mm long, purplish green or green, minutely beaked. Slough margins, marshes, wet meadows; throughout the Prairie Provinces. A rather variable species var. *altior* (Rydb.) Fern. is a larger and coarser plant; var. *stans* (Drej.) Boott, a smaller form. Both grade into the typical form. Water sedge is eaten by cattle, and forms part of slough hay.

*Carex arcta* Boott                                                     NARROW SEDGE

Plants with short rhizomes; culms 20–40 cm high, densely tufted; blades 2–4 cm wide. Inflorescence 2–4 cm long, with spikes crowded, 5–7 or more; scales somewhat shorter and as wide as perigynia, pale brown; perigynia 2–2.5 mm long, dull brown or greenish brown, white puncticulate, with the beak about one-quarter as long as the body. Not common; swamps, wet woods, muskeg; Boreal forest.

*Carex argyrantha* Tuck.                                         SILVERY-FLOWERED SEDGE

Plants densely tufted, with culms 30–70 cm high, equaling or exceeding the leaves; blades 3–6 mm wide, flat to somewhat involute. Inflorescence 5–10 cm long, flexuous, often nodding at the tip; spikes 6–12 mm long, with the upper ones close, the lower distant; scales as long as, but narrower than, the perigynia, shiny greenish or light brown; perigynia 3.5–4.5 mm long, distinctly nerved. Moist grassland, open woods, clearings; throughout the Prairie Provinces. Includes *C. aenea* Fern.

*Carex assiniboinensis* Boott                                    ASSINIBOIA SEDGE

Plants loosely tufted, with culms 30–60 cm high, slender, weak; blades 1–3 mm wide. Inflorescence 10–20 cm long; terminal spike staminate; pistillate spikes 2–4 cm long, loosely flowered, spreading or drooping; scales equaling the perigynia in size, long acuminate; perigynia 5–8 mm long, narrowly lanceolate, densely short pubescent, with the beak slender, about as long as the body. Moist open woods; eastern Boreal forest, Parklands.

*Carex atherodes* Spreng. (Fig. 70)                              AWNED SEDGE

Plants with creeping rhizomes; culms 50–120 cm high, loosely tufted, stout, erect; sheaths, especially the lower ones, pubescent; blades 4–12 mm wide, mostly pubescent toward the base; sheaths and blades rarely glabrous. Inflorescence to 25 cm long; terminal 2–6 spikes staminate, 2–4 cm long; pistillate spikes 4–10 cm long, usually 2–4, distant, erect, sessile or short-peduncled; scales shorter than the perigynia, pale brown; the green midrib prolonged into an awn; perigynia 6–10 mm long, prominently nerved, with the beak bearing long, often divergent or recurving teeth. Common; in slough margins, marshes, and wet places; throughout the Prairie Provinces. An important species, palatable to livestock, and a major part of slough hay.

*Carex athrostachya* Olney                                       LONG-BRACTED SEDGE

Plants with short rhizomes; erect culms 10–60 cm high, slender, about as tall as, or a little taller than, the leaves; blades 2–5 mm wide, mostly basal; inflorescence 1–2 cm long, and about the same width, with bracts to 6 cm long; spikes 4–10, crowded, gynaecandrous; scales a little shorter than the perigynia, reddish brown with green center; perigynia 3–5 mm long, straw-colored, nerved, the upper half serrulate. Not common; wet meadows; Prairies, Cypress Hills, Rocky Mountains.

*Carex atrata* L.

Plants with short rhizomes; culms 20–50 cm high, tufted; blades 2–5 mm wide, soft, much shorter than the culm. Inflorescence 3–10 cm long, with bracts leaf-like, the lowest one about equaling the inflorescence; terminal spike gynaecandrous, with the lower 2 or 3 pistillate, 1–3 cm long; at first erect, later spreading or drooping; scales about equaling the perigynia, dark brown to purplish black; perigynia 2.5–4 mm long, greenish to blackish. A large species, in which the following taxa have been recognized:

1. Lower pistillate spikes long-peduncled;
    perigynia greenish brown. ............................................ *C. raymondii* Calder
    Lower pistillate spikes sessile or short-pe-
    duncled; perigynia greenish to purplish
    black. ...................................................................................... 2

2. Perigynia flattened. ...................................................................... 3

A.C. Budd

Fig. 70. Awned sedge, *Carex atherodes* Spreng.

>  Perigynia somewhat inflated, not
>  flattened. ................................................................ *C. atrosquama* Mack.
>
> 3. Perigynia granular on the surface, pur-
>  plish black. ............................................................ *C. albo-nigra* Mack.
>
>  Perigynia smooth, yellowish green to
>  brown. .................................................................... *C. epapillosa* Mack.

Of these, *C. raymondii* occurs in meadows throughout Boreal forest; the others, in meadows in the Rocky Mountains.

## Carex atrofusca Schk.

Plants with rather stout rhizomes; culms 20–40 cm high, erect; blades 2–4 mm wide, flat. Inflorescence 3–6 cm long, with bracts 1–3 cm long; terminal spikelets entirely staminate, usually 1 or 2, 6–12 mm long, slender-peduncled; pistillate spikes 10–20 mm long, ascending to spreading, or later drooping on slender peduncles; scales shorter and narrower than the perigynia, blackish; perigynia 3–4 mm long, blackish, often with a green margin. Very rare; Churchill.

## Carex aurea Nutt.        GOLDEN SEDGE

Plants with slender creeping rhizomes; culms 10–30 cm high, usually spreading or ascending; blades 2–4 mm wide. Inflorescence 5–10 cm long; blades of the lowest bracts exceeding the inflorescence; terminal spikelet gynaecandrous or staminate throughout, 5–15 mm long; pistillate spikes 5–15 mm long, usually 3–5, ascending to spreading on slender peduncles; scales as long as, but narrower than, the perigynia, straw-colored to brownish with green midrib; perigynia 2–3 mm long, elliptic to obovoid, at maturity golden orange. Wet meadows, springy places; throughout Parklands, Cypress Hills, Riding Mountain, Duck Mountain, Rocky Mountains, Boreal forest. Includes *C. garberi* Fern. var. *bifaria* Fern., having deep brown scales with conspicuous green midrib, and prominently veined perigynia.

## Carex backii Boott        BACK'S SEDGE

Plants densely cespitose, with culms 5–35 cm high; blades 3–6 mm wide, often exceeding the culms. Inflorescence, without the scales, 1–3 cm long, the staminate portion of the spike 2–4 mm long, inconspicuous; scales foliaceous, to 35 mm long, prominently veined, broadened at the base, long acuminate; perigynia 5–6 mm long, tapering into a stout beak. Dry, open, sandy, or gravelly areas; Parklands, Rocky Mountains.

## Carex bicolor Bell.        TWO-COLORED SEDGE

Plants with slender rhizomes; culms 1–12 cm high, very slender, flexuous, spreading; blades 1–3 mm wide, flat to channeled. Inflorescence 2–5 cm long, with the terminal spike gynaecandrous; pistillate spikes 5–10 mm long, usually 1–4, crowded; scales purplish black with green midrib; perigynia 1.5–3 mm long, whitish, beakless. Very rare; in muskeg and marl areas; Churchill.

## Carex bigelowii Torr.        STIFF SEDGE

Plants with stout, scaly rhizomes; culms 10–40 cm high, solitary, or few tufted; blades 2–6 mm wide, flat, smooth. Inflorescence 5–15 cm long; the terminal spike staminate, 5–20 mm long; peduncle 5–15 mm long; pistillate spikes 5–35 mm long, usually 2 or 3, erect, the lower ones with peduncles 5–20 mm long; the lowest bract shorter than the inflorescence; scales hiding the perigynia, dark brown, or with a narrow green midrib; perigynia 2–4 mm long, dull green or purple-tinged, occasionally completely purplish. Bogs, marshes, tundra; Boreal forest.

*Carex brunnescens* (Pers.) Poir.                 BROWNISH SEDGE

Plants densely tufted, with culms 20–70 cm high, erect or spreading, much exceeding the leaves; blades 1–3 mm wide, flat or channeled. Inflorescence 3–6 cm long, with 5–8 spikes 4–6 mm long, short ovoid or subglobose, with 5–10 perigynia, the lower ones distant, the upper ones approximate; scales shorter than the perigynia, yellowish or tinged with brown; perigynia 2–2.5 mm long, lightly nerved, puncticulate. Bogs, muskeg, and wet woods; Boreal forest.

*Carex buxbaumii* Wahl.                       BROWN SEDGE

Plants with long, slender, brown rhizomes; culms 30–70 cm high, loosely tufted, erect, sharply 3-sided; sheaths reddish brown, becoming filamentose at the base; blades 2–3 mm wide, flat or channeled, gray green or blue green. Inflorescence 3–10 cm long, terminal spike gynaecandrous, 2–3 cm long; pistillate spikes 1–2 cm long, short-peduncled or subsessile; bract of the lowest spike leaf-like, the same length as, or somewhat longer than, the inflorescence; scales 3.5–4 mm long, awn-tipped, reddish brown, with green midrib; perigynia 3–4 mm long, light green, beakless or with two minute teeth. Swamps, wet meadows, and riverbanks; Boreal forest, Rocky Mountains.

*Carex capillaris* L.                         HAIR-LIKE SEDGE

Plants densely tufted, with culms 5–40 cm high; blades 2–4 mm wide, flat, light green. Inflorescence 5–15 cm long; the terminal spike staminate, 4–8 mm long; pistillate spikes 5–15 mm long, usually 2 or 3, drooping on slender peduncles 1–4 cm long; scales 1–2 mm long, hyaline with a green midrib; perigynia 2.5–4 mm long, brown, with two strong veins. Springy areas, marshes, and bogs; Boreal forest. Large plants, 20–40 cm high, with leaves to 20 cm long have been named var. *elongata* Olney; plants 5–20 cm high with leaves 3–10 cm long are var. *capillaris*. However, intermediate sizes also occur.

*Carex capitata* L.                         CAPITATE SEDGE

Plants with thin, hard, ascending rhizomes; culms 10–40 cm high, wiry, stiff; blades mostly less than 1 mm wide, channeled, stiffly erect, often as long as the culm. Inflorescence a single, androgynous, ovoid spike; scales much smaller than the perigynia, center brown with a strong midrib, and a wide hyaline margin; perigynia 2–3 mm long, straw-colored to brown, spreading. Meadows, bogs, open woods; Boreal forest, northern Rocky Mountains.

*Carex castanea* Wahl.                      CHESTNUT SEDGE

Plants tufted, with culms 30–70 cm high, purplish at base, erect; blades 3–6 mm wide, softly pubescent. Inflorescence 4–10 cm long; terminal spike staminate, 1–2 cm long; pistillate spikes 1–2 cm long, usually 3, spreading or drooping on slender peduncles 1–3 cm long; scales about equaling the perigynium, brownish; perigynia 4–6 mm long, with the beak slender, bidentate, light brown. Very rare; in damp woods; southeastern Boreal forest.

*Carex chordorrhiza* L. f.                          PROSTRATE SEDGE

Plants sod-forming, with culms elongate, prostrate; fertile culms 10–30 cm high, arising from nodes of the old culms; blades 1–2 mm wide, 1–3 on fertile culms, the sterile shoots with several long leaves. Inflorescence with 3–8 spikes crowded in a head 5–15 mm long; spikes androgynous; scales equaling perigynia, straw-colored; perigynia 2.5–3.5 mm long, oblong-ovoid, plump, strongly veined. Bogs, muskeg, and lakeshores; Boreal forest.

*Carex concinna* R. Br.                             BEAUTIFUL SEDGE

Plants with slender rhizomes; culms 10–25 cm high, slender, often recurved or curved ascending, loosely tufted; blades 2–4 mm wide, flat. Inflorescence 1–3 cm long, with the terminal spike staminate, 2–5 mm long, sessile; pistillate spikes 5–10 mm long, approximate, usually 2 or 3; scales about half as long as the perigynia, light to dark reddish brown, with the margins hyaline; perigynia 2.5–3 mm long, thinly pubescent, with a short beak. Damp woods, meadows, and riverbanks; Boreal forest, Rocky Mountains.

*Carex concinnoides* Mack.                     LOW NORTHERN SEDGE

Plants with slender rhizomes; culms 10–35 cm high, slender, erect to curved; blades 3–5 mm wide, flat, stiff. Inflorescence 2–3 cm long, with the terminal spike 15–20 mm long, sessile; pistillate spikes 5–10 mm long, approximate, sessile; scales shorter than the perigynia, the center reddish brown with wide hyaline margins; perigynia 2.5–3 mm long, pubescent, green, with a short beak. In woods, on stony riverbanks; southern Rocky Mountains.

*Carex crawei* Dewey                              CRAWE'S SEDGE

Plants with slender, elongate rhizomes; culms 5–30 cm high, one to several tufted, slender, stiffly erect; blades 2–5 mm wide, stiff. Inflorescence 5–20 cm long, with the terminal spike staminate, 5–20 mm long, on a peduncle 1–7 cm long; pistillate spikes 5–30 mm long, usually 1–4, ascending on peduncles 1–3 cm long, the upper ones often subsessile; scales much shorter than the perigynia, light brown, with a green center; perigynia 2.0–3.5 mm long, light green. Wet meadows, lakeshores, and moist woods; Parklands, Boreal forest.

*Carex crawfordii* Fern.                           CRAWFORD'S SEDGE

Plants densely tufted, with culms 10–70 cm high, mostly barely exceeding the leaves, stiffly erect; blades 1–3 mm wide, erect, flat. Inflorescence 1–3 cm long, with 3–12 spikes, crowded in an oblong or ovoid head; spikes 6–10 mm long, gynaecandrous; scales lanceolate, much shorter than the perigynia, dull brown; perigynia lanceolate, 3–4.5 mm long, 0.7–1 mm wide, long-tapering into the beak. Meadows, swamps, margins of woods; Boreal forest, Rocky Mountains.

*Carex crinita* Lam.                              LONG-HAIRED SEDGE

Plants densely cespitose, with culms 50–80 cm high, exceeding the leaves, 3-sided; blades 6–12 mm wide, flat. Inflorescence 10–20 cm long, with terminal

1–3 spikes staminate, to 5 cm long; pistillate spikes 3–10 cm long, usually 2–5, approximate or somewhat distant, drooping and nodding on slender peduncles; scales shorter and narrower than the perigynia, brownish with green midrib; perigynia 2–4.5 mm long, inflated, green, abruptly short-beaked. Very rare; southeastern Boreal forest.

*Carex curta* Good.                                                    SHORT SEDGE

Plants with slender rhizomes; culms 20–50 cm high, densely tufted, sharply 3-sided; blades 2–3 mm wide, about as long as the culms, flat, rather stiff, gray green. Inflorescence 3–5 cm long, usually with 4–6 spikes, the upper ones approximate, the lower ones distant; spikes about 5 mm long, ovoid, gynaecandrous, with 10–20 or more flowers; scales shorter than the perigynia, straw-colored, with the midrib light green; perigynia 2–2.5 mm, yellowish green. Muskeg, bogs, and marshes; throughout Boreal forest. Syn.: *C. canescens* L.

*Carex deflexa* Hornem.                                               BENT SEDGE

Plants with rather stout horizontal or ascending rhizomes; culms 5–20 cm high, in leafy tufts, purplish at the base; blades 1–3 mm wide, soft, flat. Inflorescence 5–20 mm long, with the terminal spike staminate, 2–5 mm long, often hidden by the 2–4 pistillate spikes, these 5–10 mm long, approximate, the lower ones with a leaf-like bract 5–20 mm long; scales shorter than the perigynia, light brown, with midrib green; perigynia 2–3 mm long, stipitate, green, pubescent. Not common; open forest, clearings; Boreal forest.

*Carex deweyana* Schw.                                                DEWEY'S SEDGE

Plants densely tufted, with culms 30–80 cm high, erect to spreading, weak. Sheaths glabrous; blades 2–4 mm wide. Inflorescence 3–8 cm long, with 3–5 spikes, the lower one with a bract to 3 cm long; spikes sessile, the upper ones approximate, the lower ones remote; scales exceeding the perigynia, pale yellowish brown; perigynia 4–5.5 mm long, tapering to a beak more than half as long as the body, rough-margined. Moist or shady woods, margins of woods, meadows; Boreal forest.

*Carex diandra* Schrank                                               TWO-STAMENED SEDGE

Plants with short rhizomes; culms 40–80 cm high, in dense tufts, erect. Sheaths scabrous; leaves 2–4 mm wide, erect. Inflorescence 3–5 cm long, with 6–10 spikes, the lower ones somewhat distant, the upper ones approximate to crowded; scales equaling the perigynia, brownish; perigynia 2–3 mm long, shiny dark brown at maturity, tapering into a rough-margined beak, more than half as long as the body. Swamps, wet meadows, and lakeshores; Boreal forest, Cypress Hills, Riding Mountain.

*Carex disperma* Dewey                                                TWO-SEEDED SEDGE

Plants with slender, creeping rhizomes, sod-forming, culms 15–50 cm high, very slender, weak, spreading. Sheath smooth; blades 1–2 mm wide, flat, soft. Inflorescence 3–5 cm long, with 2–5 androgynous spikes, 3–6 mm long, distant; scales shorter than the perigynia, white hyaline with green midrib; perigynia 2–3 mm long, minutely beaked. Bogs, wet woods, springy places; throughout the Prairie Provinces.

*Carex douglasii* Boott

Plants with slender, long-creeping rhizomes; culms 10–30 cm high, solitary or in small tufts. Sheaths smooth; blades 1–2 mm wide, erect. Inflorescence 2–3 cm long, ovoid, usually dioecious, but occasionally monoecious heads are found; spikes 8–12 mm long; scales as long as or slightly longer than the perigynia, light to dark brown, with green midrib; perigynia 3.5–4 mm long, with the beak more than half as long as the body. Dry grassland, sandhills; Prairies, Parklands.

*Carex eburnea* Boott                                          BRISTLE-LEAVED SEDGE

Plants with long, slender rhizomes; culms 10–30 cm high, densely tufted. Sheaths smooth; blades to 0.5 mm wide, flat or involute, erect. Inflorescence 2–6 cm long, with the staminate spike sessile, exceeded by the 2 or 3 long-peduncled pistillate spikes, all 3–6 mm long; scales much shorter than the perigynia, whitish to light brown with green midrib; perigynia 1.5–2 mm long, dark brown when ripe. Open woods and riverbanks; Boreal forest, Parklands.

*Carex eleusinoides* Turcz.                                     WIRE-GRASS SEDGE

Plants with slender rhizomes; culms 15–30 cm high. Sheaths smooth; blades 0.5–1.5 mm wide, tufted at the base. Inflorescence 3–6 cm long, crowded, with the terminal spike gynaecandrous; the 2–3 pistillate spikes 7–10 mm long, sessile or subsessile, with the lowermost bract exceeding the inflorescence; scales much shorter and narrower than the perigynia, purplish black; perigynia 3.–3.5 mm long, stipitate, distinctly many nerved, white to light green. Very rare; northern Rocky Mountains.

*Carex festucacea* Schk.                                        BROAD-FRUITED SEDGE

Plants densely tufted, with culms 30–80 cm high, exceeding the leaves. Sheaths smooth; blades 2–7 mm wide. Inflorescence 1–6 cm long, with 5–10 spikes, 6–10 mm long, crowded or more or less distant below; scales shorter and narrower than the perigynia, tinged with brown, the margins hyaline; perigynia 2.5–4.5 mm long, ovate to obovate. Meadows, open woods, clearing, lakeshores; in Boreal forest, Parklands. A large species, including the following small species:

1. Perigynia distinctly veined; inflorescence
   elongate, flexuous. ............................................................... *C. tenera* Dewey
   Perigynia finely veined; inflorescence
   compact. ................................................................................................ 2

2. Perigynia 2.5–3.5 mm long; scales pale to
   dull brown; sheaths tight; blades 2–4
   mm wide. ...................................................... *C. bebbii* Olney
   Perigynia 3–4.5 mm long; scales hyaline
   to light brown; sheaths loose; blades to
   7 mm wide. ...................................................... *C. normalis* Mack.

*Carex filifolia* Nutt. (Fig. 71)                               THREAD-LEAVED SEDGE

Plants densely tufted, with hard bases; culms 8–30 cm high, stiff, wiry. Sheaths smooth, reddish; blades 0.25 mm wide, needle-like, stiff. Inflorescence

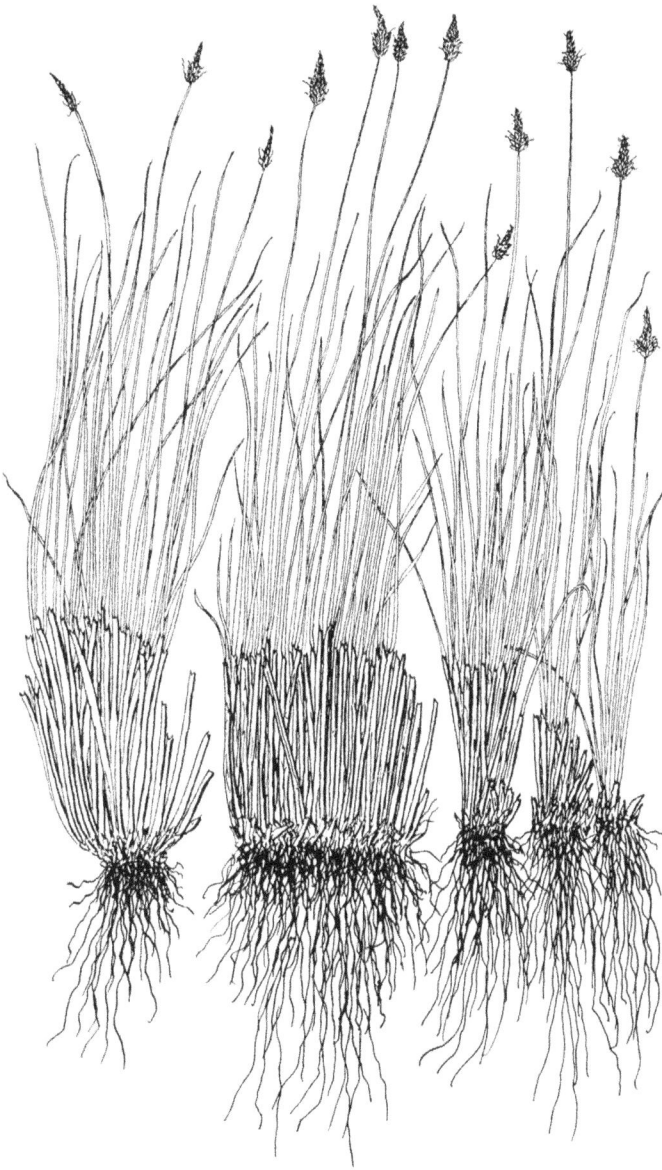

A.C. Budd

Fig. 71.  Thread-leaved sedge, *Carex filifolia* Nutt.

a solitary, androgynous spike, 1–3 cm long, erect; scales hiding the perigynia, the center dark brown, the broad margins hyaline; perigynia 3–4 mm long, straw-colored to light brown, puberulent. Dry grassland, eroded slopes, and hills; Prairies, Parklands. Palatable to livestock, and of moderate forage value.

*Carex flava* L.                                                                YELLOW SEDGE

Plants with short rhizomes, sod-forming; culms 30–70 cm high, sharply 3-sided, stiff, as long as or slightly exceeding the leaves. Sheaths smooth, straw-colored to light brown; blades 4–5 mm wide, flat, lax. Inflorescence 3–6 cm long, with the terminal spike staminate, 5–20 mm long; pistillate spikes 2–4, 5–15 mm long, distant or approximate; bracts leaf-like, the lowest one exceeding the inflorescence; scales shorter than the perigynia, reddish brown with green midrib; perigynia 4.5–7 mm long, with the beak about as long as the body, yellow. Not common; wet meadows, riverbanks, and swamps; Boreal forest.

*Carex geyeri* Boott                                                        GEYER'S SEDGE

Plants with thick, short, scaly rhizomes, to 3 mm thick; culms 10–40 cm high, rough. Sheaths loose; blades 1–3 mm wide, the lower ones short. Inflorescence a solitary androgynous spike; scales hiding the perigynia, brownish with hyaline margins; perigynia 4–6 mm long, light brown, beakless or nearly so. Dry woods, slopes; Rocky Mountains.

*Carex glacialis* Mack.                                                    GLACIER SEDGE

Plants densely tufted, with erect culms 5–15 cm high. Sheaths scabrous; blades 1–1.5 mm wide, channeled, recurved. Inflorescence 1–1.5 cm long, with the terminal spike staminate, 2–6 mm long; 1–3 pistillate spikes, 2–6 mm long, the lowest short-peduncled, the upper sessile; scales slightly shorter than the perigynia, brownish to purplish black, with hyaline margins; perigynia 2–2.5 mm long, light to dark brown, with a hyaline beak. Very rare; barren tundra; Boreal forest.

*Carex gracillima* Schw. (Fig. 72)                              SLENDER SEDGE

Plants tufted, with culms 40–80 cm high, slender, erect, purplish brown at the base. Sheaths smooth, tight; blades 3–10 mm wide, flat, thin. Inflorescence to 30 cm long, with the terminal spike gynaecandrous, 2–3 cm long; pistillate spikes 3–5 cm long, usually 3–5, long-peduncled, remote, the lower ones separated by as much as 15 cm, spreading or drooping; scales half to nearly as long as the perigynia, hyaline with green midrib; perigynia 2.5–4 mm long, distinctly nerved, green. Uncommon; moist woods; southeastern Parklands, Boreal forest.

*Carex granularis* Muhl.                                              GRANULAR SEDGE

Plants with short rhizomes; culms 30–80 cm high, tufted. Sheaths loose, glabrous; blades 4–10 mm wide, flat. Inflorescence 5–15 cm long, with the terminal spike staminate, often exceeded by the pistillate spikes, 1–2 cm long; pistillate spikes 15–20 mm long, usually 2–4, the upper ones subsessile, the lower ones progressively longer peduncled; bracts leaf-like, the upper ones exceeding the spikes; scales shorter than the perigynia, hyaline with a green

Fig. 72.   Slender sedge, *Carex gracillima* Schw.

midrib; perigynia 2.5–4 mm long, green, strongly ribbed, with a short bent beak. Damp meadows, springy areas; southeastern Parklands, Boreal forest.

*Carex gravida* Bailey                                                 HEAVY SEDGE

Plants densely tufted, with culms 30–60 cm high, scabrous at the summit. Sheaths scabrous; blades 4–8 mm wide. Inflorescence 1–3 cm long, dense, oblong-ovoid, with 4–7 spikes; bracts shorter than the head; scales slightly shorter than the perigynia, brown; perigynia 4–5 mm long, yellowish brown, with the beak about one-third as long as the body. Rare; dry grasslands; Prairies.

*Carex gynocrates* Wormsk. (Fig. 73)                    NORTHERN BOG SEDGE

Plants with slender, creeping rhizomes; culms 10–30 cm high, erect, solitary or a few tufted. Sheaths smooth; blades 0.5 mm wide, involute. Inflorescence a single spike, usually either staminate or pistillate, occasionally androgynous; scales shorter than the perigynia, brown, with hyaline margins and midrib; perigynia 3–3.5 mm long. Bogs, marshes, and muskeg; Boreal forest.

*Carex heleonastes* L. f.                                          HUDSON BAY SEDGE

Plants with short rhizomes; culms 10–30 cm high, tufted, slender. Sheaths smooth; blades 2 mm wide, flat, with the margins revolute. Inflorescence 3–5 cm long, with 1–4 spikes, approximate; spikes gynaecandrous, 3–5 mm long; scales shorter than the perigynia, with the center yellowish brown, the margins hyaline; perigynia 2–3.5 mm long, greenish, white punctate. Muskeg, meadows, and swamps; Boreal forest.

*Carex hoodii* Boott                                                   HOOD'S SEDGE

Plants with short, stout rhizomes; culms 20–80 cm high, slender, sharply 3-sided. Sheaths glabrous, tight; blades 1.5–4 mm wide, flat or channeled. Inflorescence 1–2 cm long, ovoid to orbicular, with 4–8 spikes, crowded; spikes androgynous; scales hiding the perigynia or nearly so, shiny brown, with green midrib and hyaline margins; perigynia 3.5–5 mm long, broadly winged, with the beak about one-third as long as the body. Meadows and slopes; Rocky Mountains.

*Carex hookerana* Dewey                                          HOOKER'S SEDGE

Plants with short rhizomes; culms 10–40 cm high, erect, slender. Sheaths tight, glabrous; leaves 1–2.5 mm wide, flat, thin. Inflorescence 2–5 cm long, with 5–8 spikes, distant to approximate; spikes androgynous, each with a short bract; scales hiding the perigynia, pale brown, distinctly awned; perigynia 2.5–3.5 mm long, brownish. Rare; dry grasslands, openings, and clearings; Prairies, Parklands.

*Carex houghtonii* Torr.                                             SAND SEDGE

Plants with long-creeping, thick rhizomes; culms 30–60 cm high, loosely tufted. Sheaths glabrous, loose; blades 4–10 mm wide. Inflorescence 5–15 cm long; staminate spike terminal, usually solitary, occasionally with 1 or 2 small staminate spikes or 1 or 2 perigynia at the base; pistillate spikes 1–3, 1–4 cm

Fig. 73.  Northern bog sedge, *Carex gynocrates* Wormsk.

long, remote, sessile or short-peduncled, erect; perigynia 5–7 mm long, strongly many nerved, finely pubescent, with the beak about 2 mm long, bidentate. Sandy or rocky soils and open forests; Boreal forest.

*Carex hystricina* Muhl.                          PORCUPINE SEDGE

Plants densely tufted, with erect culms 20–70 cm high. Sheaths smooth, loose; blades 2–10 mm wide; sheaths and blades often conspicuously cross-veined. Inflorescence 5–15 cm long, with the terminal spike staminate, 2–4 cm long, occasionally gynaecandrous; 1–4 pistillate spikes, 10–40 mm long, the upper ones approximate, the lowest one often 6 or 7 cm distant, peduncled, erect to spreading; perigynia 5–8 mm long, substipitate, ascending to spreading above, somewhat reflexed below, with the beak 2–2.5 mm long, slender, bidentate. Not common; swamps and wet meadows; Boreal forest.

*Carex intumescens* Rudge                          SWOLLEN SEDGE

Plants in large tufts, with culms 30–70 cm high, erect to spreading, sharply triangular. Sheaths smooth; blades 3–8 mm wide. Inflorescence 3–5 cm long, with the terminal spike staminate, 15–25 mm long; pistillate spikes 1–3, globose, 10–20 mm in diam, crowded; perigynia 10–15 mm long, tapering into the bidentate beak. Rare; damp woods; southeastern Boreal forest.

*Carex kelloggii* Boott                          KELLOGG'S SEDGE

Plants with slender, short rhizomes; culms 10–60 cm high, more or less densely tufted. Sheaths smooth, brownish at the base; blades 1–3 mm wide, equaling to exceeding the culms. Inflorescence 5–10 cm long, with the terminal spike staminate, 20–30 mm long, occasionally gynaecandrous; 3–5 pistillate spikes, 1–7 cm long, the upper ones sessile or subsessile, the lower ones short-peduncled; perigynia 1.5–3 mm long, flattened, light green; scales smaller than the perigynia, with the center dark reddish brown. Moist, springy areas; Rocky Mountains.

*Carex lacustris* Willd.                          LAKESHORE SEDGE

Plants with long rhizomes; culms tufted, 50–125 cm high, stout, sharply triangular, rough on the edges. Sheaths reddish below, with the inner membrane disintegrating to the appearance of a ladder; blades 6–15 mm wide; sheaths and blades usually cross-veined. Inflorescence 15–35 cm long, with the terminal 2–4 spikes staminate, 1–8 cm long; pistillate spikes 2–4, usually distant, 3–10 cm long, erect, sessile or short-peduncled; perigynia 5–8 mm long, dull green, tapering into the beak. Occasionally staminate and pistillate spikes are compound, with 3–5 flowers on the rachillae. Marshes, swamps, and lakeshores; Boreal forest.

*Carex laeviconica* Dewey                          SMOOTH-FRUITED SEDGE

Plants with rhizomes; culms 50–70 cm high, stout, loosely tufted. Sheaths glabrous, with the inner membrane disintegrating and laddering; blades 3–6 mm wide, flat. Inflorescence to 25 cm long, with the upper 2–6 spikes staminate, 2–10 cm long; pistillate spikes 2–4, distant, 3–7 cm long, erect, sessile or short-peduncled. Perigynia 5–8 mm long, strongly many nerved, tapering into the beak, with the teeth 1–2.5 mm long, straight. Swampy areas, slough margins; Prairies, Parklands.

*Carex lanuginosa* Michx.                                    WOOLLY SEDGE

Plants with slender, long-creeping rhizomes; culms 30–70 cm high, usually in small tufts, sharply triangular. Sheaths glabrous, reddish at the base, with the inner membrane laddering; blades 2–5 mm wide, flat or involute, often exceeding the culms. Inflorescence 5–20 cm long, with the terminal 1 or 2 spikes staminate, 1–5 cm long; pistillate spikes 1–3, distant, 15–40 mm long, erect, sessile or the lower spikes short-peduncled. Perigynia 2.5–3.5 mm long, densely pubescent, abruptly beaked, with the teeth about 0.5 mm long, erect. Slough margins, marshes, and wet places; throughout the Prairie Provinces. Syn.: *C. lasiocarpa* Ehrh. var. *latifolia* (Böck.) Gl.

*Carex lasiocarpa* Ehrh.                                 HAIRY-FRUITED SEDGE

Plants with stout long-creeping rhizomes, 2–3 mm thick; culms 50–120 cm high, tufted, bluntly triangular, scabrous above. Sheaths glabrous, loose, cross-veined, with the inner membrane laddering; blades 1–2 mm wide, folded or convolute, those of sterile shoots exceeding the flowering culms. Inflorescence 6–35 cm long, with the terminal 1–3 spikes staminate, 1–7 cm long, occasionally with a few perigynia at the base; pistillate spikes 2 or 3, distant, 1–3 cm long, often with a few staminate flowers at the tip, sessile or short-peduncled; perigynia 4–5 mm long, densely pubescent, tapering into the beak, with the teeth about 1 mm long. Bogs, lakeshores, and riverbanks; Boreal forest.

*Carex laxiflora* Lam. var. *varians* Bailey                 PLEASING SEDGE

Plants tufted, with culms 20–50 cm high, somewhat winged. Sheaths glabrous, loose, expanded at the throat; blades 3–7 mm wide, flat. Inflorescence 5–15 cm long, with the terminal spike staminate, 1–2 cm long; pistillate spikes 2–4, the upper ones approximate to the pistillate spike, the lower distant by up to 10 cm; perigynia 2.5–4 mm long, nerveless to obscurely or distinctly 10- to 12-nerved, with the body abruptly contracted into an oblique short beak; scales hyaline with a green midrib, obtuse or with a short awn. Apparently very rare; moist woods; Boreal forest.

*Carex lenticularis* Michx.                              LENS-FRUITED SEDGE

Plants with short rhizomes; culms 20–60 cm high, densely tufted. Sheaths glabrous, brown at the base; blades 2–3 mm wide, as long as or exceeding the culms. Inflorescence 5–12 cm long, with the terminal spike staminate, 1–3 cm long, occasionally gynaecandrous; pistillate spikes 3–5, crowded, 15–40 mm long; perigynia 2–3 mm long, glaucous green; scales smaller than the perigynia, with the midrib green, the margins purple brown. Rare; riverbanks, lakeshores, and marshes; Boreal forest, Rocky Mountains.

*Carex leptalea* Wahl.                                 BRISTLE-STALKED SEDGE

Plants with slender creeping rhizomes; culms 10–40 cm high, very slender, densely tufted. Sheaths smooth, tight; blades about 0.5 mm wide, shorter than to as long as the culms. Inflorescence a single spike, 5–15 cm long, androgynous; staminate scales blunt, with the margins overlapping around the rachis; perigynia 2.5–6 mm long, green, finely many nerved. Bogs and marshes; Boreal forests, Parklands.

*Carex limosa* L. MUD SEDGE

Plants with stout creeping rhizomes, to 3 mm thick; rootlets clothed with a yellow to gray or brownish felt-like layer; culms 30–50 cm high, sharply 3-angled. Sheaths glabrous, fibrillose; blades 1–1.5 mm wide, stiff, channeled or folded. Inflorescence 3–6 cm long, with the terminal spikelet staminate, 1–2 cm long, erect; pistillate spikes 1 or 2, 10–15 mm long, on filiform pedicels, 10–25 mm long, spreading to drooping; perigynia 2.5–4 mm long, grayish green, distinctly 8- to 10-nerved; scales about as long as the perigynia, light or dark brown, with green midrib. Not common; in bogs, marshes, and muskeg; Boreal forest.

*Carex livida* (Wahl.) Willd. LIVID SEDGE

Plants with slender rhizomes; culms 20–50 cm high, solitary. Sheaths glabrous; blades 1–3 mm wide, equaling or surpassing the leaves, strongly glaucous. Inflorescence 3–6 cm long, with the terminal spike staminate, rarely gynaecandrous 10–30 mm long; pistillate spikes 1–3, 5–15 mm long, sessile or short-peduncled; perigynium 3–5 mm long, glaucous green, finely many nerved; scales purplish with green midrib. Rare; bogs, marshes, and muskeg; Boreal forest.

*Carex loliacea* L. RYE-GRASS SEDGE

Plants with slender rhizomes; culms 20–60 cm high, weak, tufted, bluntly 3-angled. Sheaths glabrous, tight; blades 1–2 mm wide, flat or channeled, shorter than to as long as the culms. Inflorescence 2–3 cm long, with 2–6 gynaecandrous spikelets, 3–5 mm long, at least the lower ones distant; perigynia 2–3 mm long, distinctly many veined; scales white hyaline, with midrib green, smaller than the perigynia. Bogs, wet places; Boreal forests.

*Carex mackenziei* Krecz. MACKENZIE SEDGE

Plants with slender rhizomes; erect culms 15–40 cm high. Sheaths scabrous; blades 1–3 mm wide. Inflorescence 2–5 cm long; spikes 3–6, about 1 cm long, distant, with the terminal spike gynaecandrous, about one-half staminate; perigynia 2.5–3.5 mm long, stipitate; scales as long as the perigynia, reddish brown. Brackish marshes; Hudson Bay.

*Carex macloviana* d'Urv. THICK-SPIKE SEDGE

Plants more or less densely tufted from short rhizomes, with erect culms 10–40 cm high. Sheaths glabrous, open; blades 2–6 mm wide. Inflorescence 5–20 mm long, with 3–8 spikes crowded into a head; perigynia 3.5–4.5 mm long, brown, finely nerved; scales shorter than the perigynia. Moist areas; throughout the Prairie Provinces. A large species in which several small species have been distinguished.

1. Perigynia much flattened, thin, scale-like
   except over the achene. ............................................................................................................ 2
   Perigynia not very thin and scale-like,
   with the front convex, the back flat. ................................................................................. 6

2. Scales and perigynia copper brown; peri-
   gynia obscurely veined. .......................................................................... *C. macloviana*

*Carex magellanica* Lam. (Fig. 74)                    BOG SEDGE

Plants with long slender rhizomes; the rootlets covered with a yellow, grayish, or brownish felt-like layer; culms 30–60 cm high, very slender. Sheaths glabrous, mostly brownish at the base; blades 2–4 mm wide, flat. Inflorescence 5–12 cm long, with the terminal spike staminate, 5–15 mm long; pistillate spikes 2–4, the lower ones distant, the upper approximate, 8–20 mm long, on slender peduncles 1–4 cm long, spreading to drooping; perigynia 3–3.5 mm long, glaucous green to dull brown, veinless to finely veined; scales equaling or exceeding, but narrower than, the perigynia, brown with green midrib, or all brown. Bogs and muskeg; Boreal forests. Syn.: *C. paupercula* Michx.

*Carex maritima* Gunn.                    SEASIDE SEDGE

Plants with extensively creeping, tough, slender rhizomes; culms 5–15 cm high, mostly solitary, often curved at the summit. Sheaths glabrous, loose; blades 1–2 mm wide, flat or involute, often exceeding the culms, curved or curled. Inflorescence 5–20 mm long, with spikes 3–5, densely packed in a head, androgynous; perigynia 3.5–5 mm long, greenish to golden brown, finely veined or veinless; scales smaller than the perigynia, brown. Rare; gravelly or rocky areas; Boreal forest, Rocky Mountains.

*Carex mertensii* Prescott                    PURPLE SEDGE

Plants with short, stout rhizomes; culms 30–100 cm high, densely tufted, sharply 3-angled, winged, erect. Sheaths glabrous, loose, purplish red below; blades 4–8 mm wide, flat or the margins revolute. Inflorescence 5–12 cm long, with the terminal spike staminate or gynaecandrous, lateral spikes gynaecand-

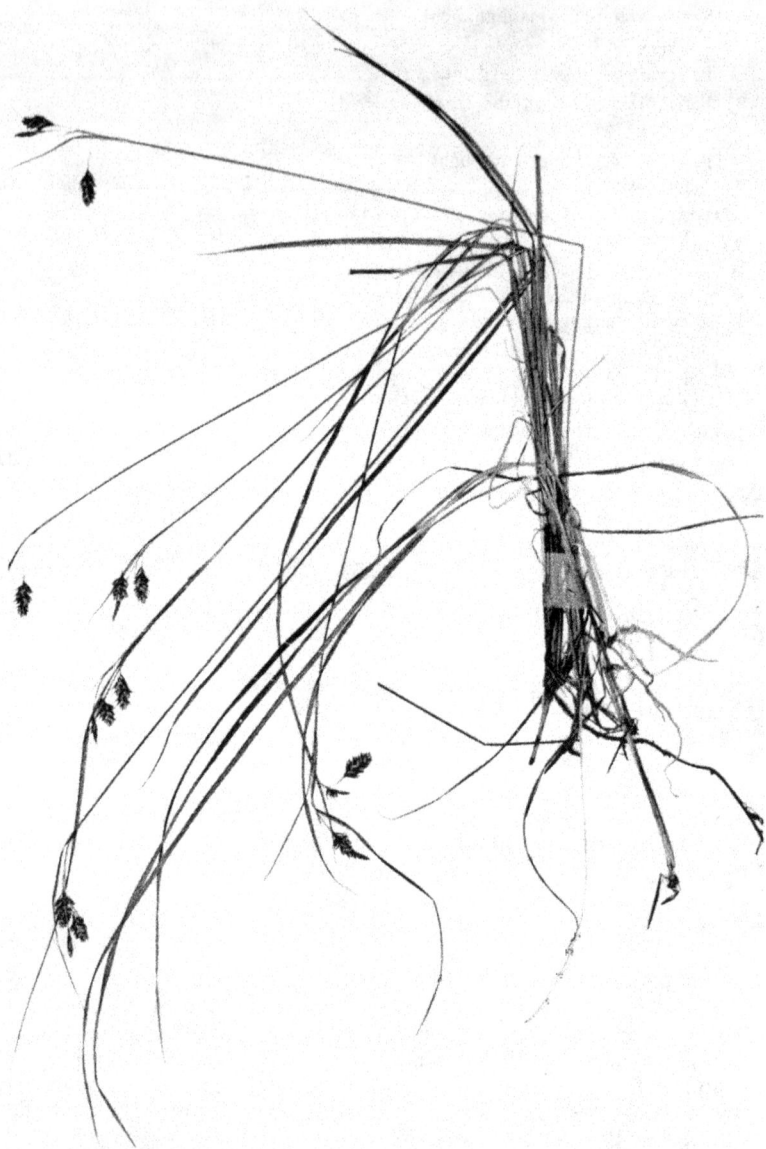

Fig. 74. Bog sedge, *Carex magellanica* Lam.

rous, with a few staminate flowers, 1–4 cm long, drooping on slender peduncles; perigynia 4.5–5 mm long, light green to pale brown, often purplish spotted; scales purplish brown, with light midrib and hyaline margins, smaller than the perigynia. Open slopes and forests; Rocky Mountains.

*Carex michauxiana* Böck.                                    LONG-FRUITED SEDGE

Plants rather densely tufted, with erect culms 20–60 cm high. Sheaths glabrous, loose; blades 2–4 mm wide. Inflorescence 2–10 cm long, with the terminal spike staminate, 5–15 mm long, often hidden by the upper pistillate spike; pistillate spikes 2–4, ovoid to suborbicular, 15–25 mm long, the lower ones distant; bracts exceeding the culms; perigynia 8–12 mm long, very slender, sharply veined; scales much smaller than the perigynia, brown. Bogs and wet meadows; southeastern Boreal forest.

*Carex microglochin* Wahl.                                   SHORT-AWNED SEDGE

Plants with slender creeping rhizomes; culms 8–25 cm high, erect, tufted. Sheaths glabrous, tight, brown below; blades about 0.5 mm wide, involute, stiffly erect. Inflorescence a solitary spike, 5–15 mm long, androgynous; perigynia 3–5 mm long, linear-lanceolate, light green to brownish, at first ascending, later spreading to reflexed; scales wider but shorter than the perigynia, early deciduous. Rare; alpine meadows; Rocky Mountains.

*Carex misandra* R. Br.                                       NODDING SEDGE

Plants with short rhizomes; culms 10–30 cm high, densely tufted, slender, erect. Sheaths glabrous, brownish below; blades 1–3 mm wide, densely clustered at base of culms. Inflorescence 3–6 cm long, nodding, with the terminal spike gynaecandrous, 5–10 mm long; pistillate spikes 2 or 3, approximate, 5–20 mm long, erect to nodding on slender peduncles; perigynia 3.5–5 mm long, purplish black above, straw-colored below; narrowly lanceolate; scales wider but shorter than the perigynia, blackish with hyaline margin and apex. Rare; tundra and alpine meadows; Boreal forest, Rocky Mountains.

*Carex muricata* L.                                           INLAND SEDGE

Plants densely tufted, with culms 20–70 cm high, slender, erect. Sheaths glabrous, tight; blades 1–4 mm wide. Inflorescence 3–5 cm long, with 3–7 spikes in an interrupted oblong head; the terminal spike gynaecandrous, lateral spikes pistillate or gynaecandrous; perigynia 2.5–5.0 mm long, tapering into a more or less distinctly bidentate beak; scales shorter than to about as long as the perigynium, pale brown with green midrib. Moist woods, meadows, and bogs; Boreal forest, Rocky Mountains. Includes *C. interior* Bailey and *C. angustior* Mack.

*Carex nardina* Fries                                         FRAGRANT SEDGE

Plants densely tufted, with culms 30–50 cm high, slender, erect. Sheaths smooth, tight, with the old sheaths persisting; blades 0.5–1 mm wide, stiff, shorter to longer than the culms. Inflorescence a single androgynous spike, 5–15 mm long; perigynia 3–4.5 mm long, striate, short-stipitate; scales about equaling the perigynia, brown, with greenish or straw-colored center. Not common; dry slopes, rocky areas; Rocky Mountains. Plants with culms mostly 15–30 cm high have been distinguished as var. *hepburnii* (Boott) Kük.

*Carex nebraskensis* Dewey                                      NEBRASKA SEDGE

Plants with long creeping rhizomes; culms 20–100 cm high, tufted, coarse. Sheaths smooth; blades 3–8 mm wide, flat, usually cross-veined, glaucous green. Inflorescence 5–10 cm long, the upper 1 or 2 spikes staminate, 15–40 mm long; pistillate spikes 2–5, sessile or nearly so, 1–6 cm long, approximate; lowest bract commonly exceeding the inflorescence; perigynia 3–3.5 mm long, straw-colored, red-dotted, strongly veined; scales much smaller than the perigynia. Very rare; marshy ground; Prairies (Alberta).

*Carex nigricans* C. A. Mey.                                    BLACKENING SEDGE

Plants with stout, long-creeping rhizomes; culms 5–20 cm high, stiff. Sheaths smooth, the old ones persisting; blades 1–2.5 mm wide, flat or channeled. Inflorescence a single androgynous spike, 8–15 mm long; perigynia 3.5–4.5 mm long, at first appressed, later spreading to reflexed; stipitate, veinless; scales much shorter than the perigynia, dark brown. Exposed alpine meadows; Rocky Mountains.

*Carex nigromarginata* Schw. var. *elliptica* (Boott) Gl.   BLACK-MARGINED SEDGE

Plants with short rhizomes; culms 5–50 cm high, tufted, very slender. Sheaths glabrous, tight; blades 1.5–3 mm wide, flat, soft, mostly shorter than the culms. Inflorescence 1–3 cm long, with the spikes approximate or crowded; terminal spike staminate, 5–10 mm long; pistillate spikes 1–4, with the lowest bract often exceeding the inflorescence; perigynia 3–4 mm long, pubescent, with the beak 0.5–1 mm, distinctly bidentate; scales much smaller than the perigynia, reddish brown with hyaline margins. Not common; open woods, margins of woods, and riverbanks; Parklands, Boreal forest.

*Carex norvegica* Retz.                                         NORWAY SEDGE

Plants with short, stout rhizomes; culms 20–60 cm high, erect or arched, loosely tufted. Sheaths glabrous; blades 1–3 mm wide, soft, flat. Inflorescence 10–35 mm long, usually with 3 spikes, crowded or the lower ones approximate, with the terminal spike pistillate or gynaecandrous, the lateral ones pistillate, 1–3 cm long, sessile or nearly so; perigynia 2–3.5 mm long, glaucous green, finely veined; scales deep purplish brown with white hyaline margins. Moist woods, lakeshores, and swamps; Boreal forest, Rocky Mountains.

*Carex obtusata* Lilj.                                          BLUNT SEDGE

Plants with slender, long-creeping, purplish black rhizomes; culms 6–20 cm high, slender, solitary or few together. Sheaths glabrous, usually reddish brown below; blades 1–1.5 mm wide, involute or flat. Inflorescence a single androgynous spike, 5–12 mm long, with 1–6 perigynia, 3–4 mm long, dark chestnut or blackish brown at maturity; scales smaller than the perigynia, pale brown. Locally common; dry to moist grassland; Prairies, Parklands.

*Carex oligosperma* Michx.                                      FEW-FRUITED SEDGE

Plants with long creeping rhizomes; culms 40–100 cm high, slender, stiffly erect. Sheaths glabrous, loose; blades 1–2 mm wide, involute, stiff, to 80 cm long. Inflorescence 6–10 cm long; the lowest bract exceeding the inflorescence; the terminal spike staminate, 1–5 cm long; the lateral spikes 1–3, pistillate, 1–2 cm long, ovoid or subcylindric, with 2–15 perigynia, 4–7 mm long, with the beak 1–2 mm long, bearing short teeth; scales smaller than the perigynia, pale brown to hyaline. Not common; bogs and sedge meadows; Boreal forest.

*Carex paleacea* Wahl. CHAFFY SEDGE

Plants with long stolons; culms 10–60 cm high, forming small tussocks. Sheaths glabrous; blades 3–8 mm wide, flat. Inflorescence 6–15 cm long, the terminal spikes 1–3 staminate, 2–5 cm long; the lateral spikes 2–4, pistillate or gynaecandrous, 2–5 cm long, drooping on slender peduncles; perigynia 2.5–3 mm long, glaucous green, short-beaked; scales shorter than the perigynia, brown, with the pale midrib prolonged into an awn, up to 1 cm long. Very rare; salt marshes; Boreal forest.

*Carex parryana* Dewey PARRY'S SEDGE

Plants with slender, long creeping rhizomes; culms 15–40 cm high, erect. Sheaths glabrous, brown to reddish below; blades 2–3 mm wide, flat or the margins revolute, erect. Inflorescence 4–8 cm long; the terminal spike gynae-candrous or staminate; the lateral spikes 2–4, pistillate, 1–3 cm long, approximate to crowded; perigynia 1.5–2.5 mm long, straw-colored or greenish, punctulate; scales equaling the perigynia, dark reddish brown, with green midrib. Rare; moist grasslands; Prairies, Parklands.

*Carex pauciflora* Lightf. FEW-FLOWERED SEDGE

Plants with long, slender rhizomes; culms 10–40 cm high, few together or solitary. Sheaths glabrous; blades 1–2 mm wide, stiff, involute, shorter than the culm. Inflorescence a single androgynous spike, to 1 cm long; perigynia 6–7 mm long, narrowly lanceolate, long-tapering into a beak, spreading or reflexed; scales smaller than the perigynia, pale brown. Bogs and muskeg; Boreal forest.

*Carex pedunculata* Muhl. STALKED SEDGE

Plants with short, thick rhizomes; culms 10–30 cm high, often barely exceeding the leaves. Sheaths glabrous, the lowest ones with a very short, stiff leaf, brownish; blades 2–4 mm wide, flat, thick. Inflorescence 4–10 cm long, with the terminal spike androgynous or staminate, 5–15 mm long; the lateral spikes 2 or 3, pistillate or sometimes androgynous, 1–3 cm long, with the upper ones ascending on short pedicels, the lower ones long-pediceled, spreading or ascending; perigynia 3.5–5 mm long, thinly pubescent, sharply angled, pale green; scales smaller than the perigynia, purplish brown, with the green midrib prolonged into a short awn. Rich woods; southeastern Boreal forest.

*Carex pensylvanica* Lam. SUN-LOVING SEDGE

Plants with extensively creeping slender rhizomes; culms 10–30 cm high, usually exceeding the leaves, tufted. Sheaths glabrous, disintegrating; blades 1–3 mm wide, flat, erect. Inflorescence 15–50 mm long, with the terminal spike staminate, the lateral spikes 2 or 3, pistillate, 5–10 mm long, sessile or short-pe-duncled; perigynia 2–4 mm long, finely pubescent, abruptly narrowed to a bidentate beak, stipitate. Dry to moist grassland, open woods, and thickets;

throughout Prairies and Parkland. A variable species, the following varieties are present:

1. Perigynia 3–3.5 mm long, round; leaves
   stiff and scabrous. ..................................................................... var. *digyna* Böck.
   Perigynia 2–4 mm long, angled; leaves
   soft, not scabrous. ........................................................................................ 2

2. Perigynia 2–3 mm long, with the beak
   about 0.5 mm long. .............................................................. var. *pensylvanica*
   Perigynia 3–4 long, with the beak more
   than 1 mm long. ............................................................................ var. *distans* Peck

The var. *pensylvanica* occurs in southeastern Parklands and in a few locations in Boreal forest; var. *digyna* Böck. (*C. heliophila* Mack.) is the common form in grassland of Prairies and Parklands; var. *distans* Peck occurs occasionally in openings in Boreal forest.

## *Carex petasata* Dewey                                    PASTURE SEDGE

Plants with short rhizomes; culms 20–80 cm high, densely tufted, smooth, mostly nodding at the top. Sheaths glabrous, tight; blades 2–5 mm wide, flat. Inflorescence 2–5 cm long, with 3–6 gynaecandrous spikes, approximate; spikes about 1 cm long; perigynia 4–8 mm long, oblong to lanceolate; scales equaling the perigynia, reddish brown with white hyaline margins. Meadows, open woods, and clearings; Parklands, Boreal forest. A large species, including the following:

1. Perigynia 4–4.5 mm long; blades 3–5 mm
   wide. ......................................................................................... *C. platylepis* Mack.
   Perigynia 4.5–8 mm long; blades 2–3 mm
   wide. ........................................................................................................ 2

2. Perigynia 6–8 mm long, prominently
   veined. ....................................................................................................... *C. petasata*
   Perigynia 4.5–6 mm long, obscurely
   veined. ................................................................................................. *C. praticola* Rydb.

## *Carex petricosa* Dewey                                    STONE SEDGE

Plants with slender, long rhizomes; erect culms 10–30 cm high. Sheaths glabrous; blades 1–3 mm wide, stiff. Inflorescence 4–8 cm long, with 3–8 spikes; uppermost spikes androgynous or staminate, 7–20 mm long; lateral spikes 2 or 3, pistillate or androgynous, 10–30 mm long; perigynia 4–6 mm long, ciliate on the margins, yellowish brown to dark brown; scales shorter than the perigynia, reddish brown. Riverbanks, slopes, and alpine tundra; Rocky Mountains. Plants of the lower altitudes are usually larger, to 90 cm high, and have broad yellowish brown perigynia, 2–2.5 mm wide. These have been distinguished as var. *franklinii* (Boott) Boiv.

## *Carex phaeocephala* Piper                                HEAD-LIKE SEDGE

Plants with short, matted rhizomes, forming large clumps; culms 10–30 cm high, slender, scabrous above, stiff. Sheaths smooth, tight; blades 3–6 mm wide, mostly basal. Inflorescence 10–25 mm long, with 2–5 gynaecandrous spikes, approximate to crowded; perigynia 4–6 mm long, oblong-ovate, straw-colored to dark brown; scales equaling the perigynia, dark brown with a light center and hyaline margins. Slopes and alpine meadows; Rocky Mountains.

*Carex podocarpa* R. Br.                                                    ALPINE SEDGE

Plants with slender creeping rhizomes; culms 10–50 cm high, loosely tuft-ed, bluntly triangular, more or less nodding above. Sheaths glabrous, reddish brown below; blades 2–4 mm wide, flat, deep green. Inflorescence 4–7 cm long, with the terminal spike staminate, 5–20 mm long; lateral spikes 2–4, pis-tillate or somewhat androgynous, 1–2 cm long, approximate, short peduncu-late, ascending or spreading; perigynia 3.5–4.5 mm long, blackish brown; scales much smaller than the perigynia, brownish black. Alpine meadows, riverbanks; Rocky Mountains.

*Carex praegracilis* W. Boott                                           GRACEFUL SEDGE

Plants with blackish, long-creeping rhizomes; culms 15–60 cm high, sharply triangular, scabrous above. Sheaths scabrous, loose; blades 1–3 mm wide, flat or channeled, mostly basal. Inflorescence 1–5 cm long, with 5–10 androgynous spikes crowded into a dense head, or the lower ones approxi-mate; spikes 5–10 mm long; perigynia 3–4 mm long, dull brownish black; scales larger than the perigynia, with the center chestnut brown, the margins hyaline. In marshes, around sloughs, wet meadows; throughout the Prairie Provinces.

*Carex prairea* Dewey                                                    PRAIRIE SEDGE

Plants with short stout rhizomes; culms 50–100 cm high, erect. Sheaths smooth, brown below, the lower ones with short blades, the membrane contin-ued 2–3 mm beyond base of blades, copper-colored; principal blades 2–3 mm wide, to 40 cm long. Inflorescence 3–8 cm long, compound, with the branches, at least the lower ones, bearing 2–4 androgynous spikes; perigynia 2.5–3 mm long, dull dark brown, tapering to a beak with serrulate margins; scales equal-ing the perigynia, reddish brown. Swamps and wet meadows; Parklands, Boreal forest.

*Carex preslii* Steud.                                                    PRESL SEDGE

Plants with thick, corm-like, short rhizomes; culms 20–70 cm high, slen-der, scabrous, densely tufted. Sheaths smooth, very thin, with the membrane white hyaline, loose; blades 1–4 mm wide, flat. Inflorescence 10–20 mm long, with 2–4 gynaecandrous spikes aggregated into a head; perigynia 3–4.5 mm long, yellowish to greenish brown; scales smaller than the perigynia, olive brown to chestnut. Alpine slopes; Rocky Mountains.

*Carex pseudo-cyperus* L.                                            CYPERUS-LIKE SEDGE

Plants with short rhizomes, densely sod-forming; culms to 1 m high, often nodding at the top, sharply 3-angled, very scabrous. Sheaths scabrous, loose; blades 5–15 mm wide, light green, flat, scabrous. Inflorescence 10–20 cm long; the terminal spike staminate, 2–5 cm long; pistillate spikes 3–5, the upper ones approximate, the lower distant, 3–7 cm long, spreading or drooping on slender peduncles, 3–10 cm long; perigynia 5–6 mm long, spreading to reflexed at maturity, with the beak 0.5–2 mm long, bidentate; scales 2–3 mm long, ovoid, with the midrib prolonged into an awn, 3–4 mm long. Swamps and bogs; Boreal forests.

*Carex pyrenaica* Wahl.                                                    SPIKED SEDGE

Plants densely tufted, with culms 5–25 cm high, equaling or exceeding the leaves. Sheaths smooth, whitish, tight; blades 0.3–1.5 mm wide, folded. Inflorescence a single androgynous spike, 5–20 mm long; perigynia 3–4.5 mm long, glossy dark brown, stipitate, tapering into a hyaline beak; scales shorter but wider than the perigynia, brown to dark brown. Rocky or grassy alpine slopes; Rocky Mountains.

*Carex rariflora* (Wahl.) Smith                                          SCANT SEDGE

Plants with slender rhizomes, the rootlets covered with a yellow or grayish felt-like layer; culms 20–30 cm high, bluntly 3-angled. Sheaths glabrous; blades 1–3 mm wide. Inflorescence 4–8 cm long; the terminal spike staminate or somewhat gynaecandrous, 6–15 mm long; pistillate spikes 1–3, approximate, 10–15 mm long, short-peduncled; perigynia 3–4 mm long, glaucous green, beakless; scales about equaling the perigynia, dark purple. Bogs and muskeg; southeastern Boreal forest.

*Carex raynoldsii* Dewey (Fig. 75)                                      RAYNOLD'S SEDGE

Plants with short, stout rhizomes; culms 20–80 cm high, erect, sharply 3-angled. Sheaths glabrous; blades 2–5 mm wide, flat, to 50 cm long. Inflorescence 4–8 cm long; the terminal spike staminate, 1–2 cm long; pistillate spikes 2–4, approximate or the lower one somewhat distant, 1–2 cm long; perigynia 3.5–4.5 mm long, yellowish green or brown, distinctly several-veined; scales shorter than the perigynia, dark reddish brown or black. Meadows and open woods; Rocky Mountains, Cypress Hills.

*Carex retrorsa* Schw. (Fig. 76)                                         TURNED SEDGE

Plants with short, stout rhizomes; culms 40–100 cm high, densely tufted, stout, obtusely 3-angled. Sheaths glabrous, often cross-veined; blades 3–12 mm wide, to 40 cm long, flat, soft. Inflorescence 5–15 cm long; the upper 1 or 2 spikes staminate, 10–25 mm long, often hidden among the 3–8 pistillate spikes, 15–60 mm long, all aggregated, or the lower ones approximate; perigynia 7–10 mm long, tapering into a long bidentate beak, shiny, with many distinct veins, light green or straw-colored; scales smaller than the perigynia. Swampy areas, lakeshores, and wet meadows; Boreal forest.

*Carex richardsonii* R. Br.                                              RICHARDSON'S SEDGE

Plants with long rhizomes; culms 10–25 cm high, tufted, short pubescent. Sheaths glabrous, open; blades 1.5–2.5 mm wide, stiff, short, mostly basal. Terminal spike staminate, 10–25 mm long, peduncled; lateral 2 or 3 spikes pistillate, 1–2 cm long, short-peduncled, erect, approximate or somewhat distant; perigynia 2.5–3.5 mm long, pubescent, green; scales larger than the perigynia, purplish brown, with the margins hyaline. Dry woods, open areas; Boreal forest, Rocky Mountains.

Fig. 75.   Raynold's sedge, *Carex raynoldsii* Dewey.

Fig. 76.  Turned sedge, *Carex retrorsa* Schw.

*Monocots*

*Carex rossii* Boott ROSS' SEDGE

Plants with slender rhizomes; culms 10–30 cm high, erect, often exceeded by the leaves. Sheaths smooth, purplish brown below; blades 1–3 mm wide, stiffly erect or curly. Inflorescence about 2 cm long; the terminal spike staminate, to 15 mm long; lateral spikes pistillate, 5–10 mm long; basal spikes abundant; perigynia 3–4 mm long, green to light brown, pubescent, with the beak 1–2 mm long; scales equaling the perigynia, keeled. Dry slopes, rocky areas, and clearings; throughout Prairies, Parklands, Rocky Mountains, Boreal forest.

*Carex rostrata* Stokes BEAKED SEDGE

Plants with short rhizomes, and long stolons; culms 50–100 cm high, stout, erect, usually exceeded by the leaves. Sheaths smooth, strongly cross-veined; blades 5–10 mm wide, usually cross-veined, flat. Inflorescence 10–30 cm long; the upper spikes 2–4, staminate, 15–50 mm long, well separated from the 2–5 pistillate spikes, 4–10 cm long, short-peduncled to sessile; perigynia 4–8 mm long, narrowed to a bidentate beak; scales narrower than the perigynia, light brown. Swamps, marshes, lakeshores, riverbanks; throughout Parklands, Boreal forest.

*Carex rotundata* Wahl. ROUND SEDGE

Plants with short rhizomes; culms 20–60 cm high, sharply 3-angled, erect. Sheaths smooth, reddish brown below; blades 1–3 mm wide, involute. Inflorescence 5–15 cm long; the terminal spikes 2 or 3, staminate, 10–20 mm long; the lateral spikes 3 or 4, pistillate, 20–40 mm long, sessile or short pedunculate; perigynia 3–3.5 mm long, tapering to the bidentate beak, obscurely veined; scales smaller than the perigynia, purplish black. Peat marsh and tundra; eastern Boreal forest.

*Carex rufina* Drejer REDDISH SEDGE

Plants with slender rhizomes; culms 10–25 cm high, loosely tufted. Sheaths glabrous, short; leaves 1–3 mm wide, often surpassing the culms. Inflorescence 5–10 cm long, with bracts long-sheathing; the terminal spike gynaecandrous, 5–10 mm long; the lateral spikes 2 or 3, pistillate, 5–15 mm long, ascending on elongate peduncles; perigynia 2–3 mm long, brownish green, serrulate above, with a beak 0.1–0.2 mm long; scales smaller than the perigynia, purplish brown. Very rare; arctic and Boreal tundra; Nueltin Lake, Manitoba.

*Carex rupestris* All. ROCK SEDGE

Plants with slender rhizomes; culms 8–15 cm high, solitary or few together. Sheaths reddish brown below; blades 1–2 mm wide, to 15 cm long, usually curved, exceeding the culms. Inflorescence a single androgynous spike, 1–2 cm long; perigynia about 3 mm long; scales dark chestnut brown with hyaline margins, somewhat shorter but broader than, and partly enveloping, the perigynia. Rare; open, usually calcareous soils; Boreal forest.

*Carex salina* Wahl. SALT SEDGE

Plants with slender rhizomes; culms 5–60 cm high, curving or erect. Sheaths glabrous; blades to 9 mm wide, flat or involute. Inflorescence 5–12 cm long; the upper spikes 1–3, staminate, 2–3 cm long; the 1–3 pistillate spikes 5–80 mm long, short-peduncled to subsessile; perigynia 2–3.5 mm long, pale green; scales somewhat longer than the perigynia, purple brown, with the green midrib prolonged into an awn. Salt marshes; Hudson Bay. Three forms are recognized:

1. Culms 5–15 cm high, slender, obtusely angled; blades 1–2.5 mm wide; pistillate spikes 5–15 mm long, 3–4 mm thick. ..................... var. *subspathacea* (Wormsk.) Tuck.

   Culms 10–60 cm high, stiff; blades 2–9 mm wide; pistillate spikes 1–8 cm long, 3–10 mm thick. .................................. 2

2. Culms 10–30 cm high, obtusely angled; blades 2–4 mm wide, with the margins revolute; pistillate spikes 1–3 cm long, 3–4 mm thick. .................................. var. *salina*

   Culms 15–60 cm high, rather sharply angled; blades 2–9 mm wide, flat; pistillate spikes 2–8 cm long, 4–10 mm thick. ..................... var. *kattegatensis* (Fries) Almq.

## *Carex sartwellii* Dewey                SARTWELL'S SEDGE

Plants with slender rhizomes; culms 40–80 cm high, sharply triangular, stiff. Sheaths with a green, clearly veined membrane; blades 2–4 mm wide. Inflorescence 2–6 cm long, with 6–20 androgynous spikes 5–10 mm long; perigynia 2.5–3.5 mm long, brown, finely veined; scales smaller than the perigynia, reddish brown, with the midrib green, margin hyaline. Wet meadows, bogs, and margins of woods; Boreal forest.

## *Carex saxatilis* L.                ROCKY-GROUND SEDGE

Plants with creeping purplish rhizomes; culms 20–80 cm high, decumbent at the base, leafy. Sheaths smooth; blades 2–5 mm wide, flat, scabrous toward the tip. Inflorescence 3–10 cm long; the upper 1 or 2 spikes staminate, 1–3 cm long; the 1–3 pistillate spikes 5–20 mm long; perigynia 3–5 mm long, purplish black; scales smaller than the perigynia, reddish black. Gravelly soil and muskeg; Boreal forest. Two varieties can be distinguished:

Culms 20–40 cm high; pistillate spikes subsessile or short-peduncled, approximate. ......................................... var. *saxatilis*

Culms to 80 cm high; pistillate spikes long-peduncled, drooping, distant. ................................... var. *major* Olney

## *Carex scirpoidea* Michx.                RUSH-LIKE SEDGE

Plants with long, creeping rhizomes; culms 20–50 cm high, solitary or few together. Sheaths glabrous, loose, the lower ones reddish brown; blades 1–3 mm wide. Inflorescence a single spike, staminate or pistillate, 1–3 cm long; perigynia 2.5–3 mm long, densely pubescent, green; scales smaller than the perigynia, deep brown. Marshy areas, slopes, and meadows; throughout the Prairie Provinces.

*Carex scoparia* Schk. BROOM SEDGE

Plants with short rhizomes; culms 20–100 cm high, loosely to densely tufted, sharply 3-angled. Sheaths smooth; blades 1–3 mm wide, 15–60 cm long. Inflorescence 25–40 mm long, with 3–12 gynaecandrous spikes 6–12 mm long, approximate or crowded into a head; perigynia 4–7 mm long, lanceolate, flat, greenish or straw-colored; scales smaller than the perigynia, light brown, with green midrib. Moist areas, swamp, muskeg; Boreal forest, a few locations in Parklands.

*Carex siccata* Dewey (Fig. 77) HAY SEDGE

Plants with long, tough rhizomes; culms 15–80 cm high, solitary or few together. Sheaths smooth; blades 1–3 mm wide, often almost as long as the culms. Inflorescence 2–4 cm long, with 4–12 spikes; the uppermost spike androgynous, the middle spikes staminate, the lower ones pistillate; perigynia 3–6 mm long, nerved on both sides, winged in the upper half, narrowing into a bidentate beak; scales shorter than the perigynia, light brown with hyaline margins. Common; sandy areas, open pine woods, sometimes covering large areas; Boreal forest. Syn.: *C. foenea* Willd.

*Carex sprengelii* Dewey SPRENGEL'S SEDGE

Plants with long, creeping rhizomes; culms 15–80 cm high, tufted. Sheaths smooth, loose; blades 2–4 mm wide, flat, soft. Inflorescence 10–20 cm long; the upper 1–3 spikes staminate, 1–2 cm long; the lateral 2–4 spikes pistillate, 2–4 cm long, distant, on long slender ascending or drooping peduncles; perigynia 5–6 mm long, with the body subglobose, contracted into a bidentate beak as long as, or longer than, the body; scales narrower than the perigynia, greenish white. Open woods, shrubbery, and moist semiopen areas; throughout the Prairie Provinces.

*Carex stenophylla* Wahl. ssp. *eleocharis* (Bailey) Hulten LOW SEDGE

Plants with slender, long creeping rhizomes; culms 3–20 cm high, solitary or tufted. Sheaths glabrous, loose; blades 1–2 mm wide, channeled to involute. Inflorescence 5–20 mm long, with 3–5 androgynous spikes, closely aggregated in a head; perigynia 2.5–3.5 mm long, dark brown at maturity; scales wider than the perigynia. Often very abundant; dry grasslands; Prairies.

*Carex stipata* Muhl. AWL-FRUITED SEDGE

Plants with short, thick rhizomes; culms 10–100 cm high, erect, soft, winged on the angles. Sheaths glabrous, loose; blades 2–8 mm wide, often equaling or surpassing the culms. Inflorescence 1.5–10 cm long, compound, with 5–15 spikes aggregated into a head, the lower spikes sometimes approximate, the upper ones crowded; perigynia 4–6 mm long, yellowish, sharply nerved on both sides; scales smaller than the perigynia, brownish. Moist woods, swamps, and bogs; Parklands, Boreal forest.

*Carex straminea* Willd. STRAW-COLORED SEDGE

Plants with short, stout rhizomes; culms 30–80 cm high, densely tufted. Sheaths glabrous, brownish; blades 2–6 mm wide. Inflorescence 1–6 cm long,

Fig. 77.  Hay sedge, *Carex siccata* Dewey.

with 3–10 gynaecandrous spikes 7–10 mm long, approximate to somewhat distant; perigynia 3.5–7 mm long, suborbicular; scales smaller than the perigynia. Meadows, grasslands, open forest; Parklands, Boreal forest. A large species, which has been divided as follows:

1. Perigynia broadest above the middle; leaves 3–6 mm wide; sheaths loose. ........................... *C. cumulata* (Bailey) Mack.
   Perigynia broadest below the middle; leaves 2–4 mm wide; sheaths tight. ........................................................................ 2

2. Perigynia thin, 5–7 mm long, 3–5 mm broad, distinctly veined on both sides. ..................................... *C. bicknellii* Britt.
   Perigynia thick, firm, 3.5–5 mm long, 2–3.5 mm broad, obscurely veined or veinless on the inside. ............................................................... 3

3. Perigynia 4–5 mm long, strongly veined on the back, obscurely veined inside; wings thick; scales almost reaching tip of beak. ...................................................... *C. brevior* (Dewey) Mack.
   Perigynia 3.5–4 mm long, obscurely veined on both sides; wings thin; scales reaching to base of beak. ............................................................... 4

4. Heads 1–3 cm long; spikes crowded; leaves 2–3 mm wide. ...................................................... *C. molesta* Mack.
   Heads to 8 cm long; spikes more or less separated; leaves 3–4 mm wide. ............................... *C. merritt-fernaldii* Mack.

*Carex stricta* Lam.                                                    ERECT SEDGE
   Plants with long, scaly, rather stout rhizomes; culms 40–100 cm high, loose-tufted into large clumps. Sheaths brown or reddish below, tight; blades 3–5 mm wide, flat. Inflorescence 5–15 cm long; the upper 1–3 spikes staminate, 1–4 cm long; the 1–4 lateral spikes pistillate, or sometimes androgynous, 2–6 cm long, erect, sessile or the lower ones short-peduncled, approximate or somewhat distant; perigynia 1.5–3 mm long, light green to straw-colored; scales variable, from narrower and shorter to broader and longer than the perigynia, reddish brown to purple brown. Marshy areas; southeastern Boreal forest.

*Carex supina* Wahl. ssp. *spaniocarpa* (Steud.) Hulten          WEAK SEDGE
   Plants with slender brown rhizomes; culms 8–30 cm high, sharply 3-angled, slender. Sheaths glabrous; blades 0.5–2 mm wide, channeled, crowded, often almost as high as the culms. Inflorescence 2–3 cm long; the terminal spike staminate, 10–15 mm long, pale; the 1–3 lateral spikes pistillate, 5 mm long, with 3–5 perigynia, approximate; perigynia 2.5–3 mm long, plump, lustrous, with a short beak; scales equaling the perigynia, light reddish brown. Rare; stony areas, beaches; southeastern Boreal forest, Parklands.

*Carex sychnocephala* Carey                              LONG-BEAKED SEDGE
   Plants densely tufted, with culms 10–50 cm high, erect or spreading. Sheaths smooth, tight; blades 2–5 mm wide, often exceeding the inflorescence,

flat. Inflorescence without the bracteal leaves 2–3 cm long; perigynia 5–6 mm long, lanceolate, green; scales smaller than the perigynia, hyaline with a green midrib. Wet meadows and lakeshores; Parklands, Boreal forest.

*Carex tenuiflora* Wahl.                                      THIN-FLOWERED SEDGE

Plants with very slender rhizomes; culms 20–60 cm high, slender. Sheaths smooth, loose; blades 1–2 mm wide, flat. Inflorescence 5–15 mm long, with 2–4 gynaecandrous spikes approximate to crowded; perigynia 3–3.5 mm long, greenish; scales equaling the perigynia, white hyaline with green midrib. Rare; wet wood, bogs, and muskeg; Boreal forest.

*Carex tetanica* Schk.                                      RIGID SEDGE

Plants with slender rhizomes; culms 30–60 cm high, tufted. Sheaths glabrous; blades 2–5 mm wide, flat. Inflorescence 5–10 cm long; the terminal spike staminate, 1–2 cm long, with the peduncle to 10 cm long; the 1–3 lateral spikes pistillate, 1–3 cm long, slender-peduncled, ascending; perigynia 3–4 mm long, light brown, finely veined; scales shorter than the perigynia, dark reddish brown with green midrib. Open grassland; southeastern Parklands, Boreal forest. Two varieties are represented:

Perigynia 2-ranked; the pistillate spikes 3–5
  mm thick. .............................................................. var. *woodii* (Dewey) Wood
Perigynia 6-ranked; the pistillate spikes 5–7
  mm thick. .............................................................. var. *meadii* (Dewey) Bailey

*Carex tincta* Fern.                                      TINGED SEDGE

Plants with short rhizomes; culms 40–80 cm high, slender, sharply triangular, tufted. Sheaths somewhat cross-veined, with the membrane wrinkled; blades 2–4 mm wide, flat, firm. Inflorescence 1–4 cm long, with 4–10 gynaecandrous spikes, approximate to closely crowded into a head; perigynia 3.5–4 mm long, brownish, sharply many nerved on the back, tapering into a serrate beak; scales about equaling the perigynia, reddish brown with green midrib and hyaline margins. Meadows and open woods; Parklands, Boreal forest. Presumably a hybrid of species in the *C. festucacea* group.

*Carex tonsa* (Fern.) Bickn.                                      BALD SEDGE

Plants with creeping rhizomes; culms poorly developed, hidden among the basal leaves. Sheaths reddish brown; blades 2–4 mm wide, spreading, curved. Inflorescence 10–15 mm long, hidden among the leaves; perigynia 1.5–3 mm long, glabrous or nearly so, yellowish brown, with the beak about 1.0–1.5 mm long, curved, bidentate; scales equaling the perigynia, pale brown with green midrib. Open woods, sandy areas; Boreal forest.

*Carex torreyi* Tuck.                                      TORREY'S SEDGE

Plants with short rhizomes; culms 25–50 cm high, slender, weak, sharply triangular, somewhat pubescent below the inflorescence. Sheaths soft pubescent, tight; blades 1.5–4 mm wide, flat, pilose. Inflorescence 3–5 cm long; the terminal spike staminate, 5–15 mm long; the 1–3 lateral spikes pistillate, 5–15 mm long, sessile or short-peduncled; perigynia 2.5–3.5 mm long, yellowish green, strongly nerved, abruptly short-beaked; scales smaller than the perigynia, light brown with broad hyaline margins. Rare; in meadows and moist woods; Prairies, Parklands.

*Carex tribuloides* Wahl. PRICKLY SEDGE

Plants tufted, with culms 30–70 cm high, stout. Sheaths glabrous, tight; blades 3–7 mm wide. Inflorescence 2–5 cm long, with 7–12 spikes 6–12 mm long, approximate or crowded; perigynia 3–5 mm long, straw-colored, obscurely veined on both sides; scales smaller than the perigynia, pale brown with hyaline margins. Wet areas, marshes, and shores; southeastern Boreal forest. Includes *C. projecta* Mack., *C. cristatella* Britt.

*Carex trisperma* Dewey THREE-SEEDED SEDGE

Plants with slender rhizomes; culms 20–70 cm high, very slender and weak, often straggling. Sheaths glabrous, tight; blades to 30 cm long, 1–3 mm wide, flat. Inflorescence 3–6 cm long, with 1–3, but commonly 2, spikes 2–4 cm apart; the lower bract 2–4 cm long; spikes 3–5 mm long, gynaecandrous, with 1–5 perigynia 2.5–4.0 mm long, greenish brown; scales shorter than to equaling the perigynia, with the center green and margins hyaline. Wet woods, bogs, and marshes; Boreal forest.

*Carex umbellata* Schk. UMBELLATE SEDGE

Plants with stout rhizomes; culms 3–20 cm high. Sheaths purplish or reddish brown below; blades 1–3 mm wide, stiff, usually exceeding the culms. Inflorescence of two kinds: 1 culm with 1 terminal spike staminate, 5–10 mm long, sometimes with a pistillate spike below it; and 1–3 very short culms bearing 1 pistillate spike 4–10 mm long; perigynia 1.5–3 mm long, pubescent, yellowish green, with a prominent, often curved, beak; scales equaling the perigynia, with the center green and the margins hyaline. Open woods on sandy or stony soils; Boreal forest, Rocky Mountains.

*Carex vaginata* Tausch SHEATHED SEDGE

Plants with long, slender, creeping rhizomes; culms 20–60 cm high, solitary or few together. Sheaths glabrous, tight; blades 2–5 mm wide, flat. Inflorescence 10–15 cm long; the terminal spike staminate, 1–2 cm long, peduncled; lateral spikes 1–3, often androgynous, 10–25 mm long, short-peduncled, loosely spreading, or the lower ones drooping; perigynia 3–4 mm long, light brown, beaked; scales smaller than the perigynia, purplish brown. Wet woods, bogs, and muskeg; Boreal forest.

*Carex vesicaria* L. BLISTER SEDGE

Plants with stout, creeping rhizomes; culms 50–100 cm high, tufted. Sheaths glabrous, cross-veined; blades 3–7 mm wide, equaling the culms, or those of sterile shoots exceeding them. Inflorescence 10–20 cm long; the 1–4 upper spikes staminate, 2–6 cm long; the lower spikes 1–3, but usually 2, pistillate, 3–5 cm long; perigynia 7–9 mm long, yellow green to light brown, inflated, strongly veined, with the beak about 2 mm long and having divergent teeth; scales smaller than the perigynia, light brown. Very rare; marshes and lakeshores; southeastern Boreal forest.

*Carex viridula* Michx. GREEN SEDGE

Plants more or less densely tufted, with erect culms 5–40 cm high. Sheaths glabrous, loose; blades 1–3 mm wide, usually yellowish green. Inflorescence 1–5 cm long; the terminal spike staminate, 5–15 mm long, peduncled, often barely exceeding the 2–6 pistillate spikes 5–10 mm long, crowded to approximate; the lowest bract much exceeding the inflorescence; perigynia 2–3.5 mm long, green or brownish green; scales about equaling the perigynia, yellowish brown. Wet areas, particularly calcareous or slightly alkali lakeshores; throughout the Prairie Provinces.

*Carex vulpinoidea* Michx. FOX SEDGE

Plants with short, stout rhizomes; culms 30–80 cm high, densely tufted. Sheaths glabrous, with the membrane greenish, clearly veined, and cross-wrinkled; blades 2–4 mm wide, commonly exceeding the leaves. Inflorescence 4–8 cm long, compound, with many densely crowded spikes; bracts up to 5 cm long, mostly well exserted; perigynia 2–3 mm long, yellowish green; scales about equaling the perigynia. Wet areas, meadows, and bogs; Boreal forest.

*Carex xerantica* Bailey WHITE-SCALED SEDGE

Plants with short, thick rhizomes; culms 30–60 cm high, densely tufted. Sheaths glabrous; blades 2–3 mm wide, flat, stiff. Inflorescence 3–8 cm long, with 3–8 spikes 6–12 mm long, crowded toward the tip, approximate to distant below; perigynia 4–6 mm long, straw-colored to light brown; scales about equaling the perigynia, light brown, with the midrib green and the margins hyaline. Open woods, meadows, disturbed areas; Parklands, southern Boreal forest, Cypress Hills.


**Cyperus**      nut-grass

Triangular-stemmed plants with long grass-like leaves, mostly basal. Flowers perfect and borne in spikelets or in an umbel-like inflorescence at the summit of the stem. Fruit a 3-angled achene.

1. Tip of the scales slender, recurved; spikelets greenish to pale brown, 2–10 mm
   long; annual with fibrous roots. ............................................... *C. inflexus*
   Tip of the scales not recurved; spikelets
   yellowish to brownish, to 20 mm long;
   perennials with corm-like thickened
   roots. .................................................................................................... 2

2. Joints of the rachilla with conspicuous
   hyaline wings; scales 3.5–6 mm long,
   acute or mucronate. ................................................................ *C. strigosus*
   Joints of the rachilla sharp-edged, not
   winged; scales 2–4 mm long, awned or
   mucronate. .......................................................................................... 3

3. Upper scales with an awn 0.5–1 mm long;
   achenes 2–3 mm long, about 1 mm
   thick. ............................................................................... *C. schweinitzii*

Upper scales obtuse or short mucronate; achenes 1.5–2 mm long, about 1.2 mm thick. ................................................................................................ *C. houghtonii*

### *Cyperus houghtonii* Torr.                    HOUGHTON'S NUT-GRASS

Plants with short rhizomes; culms 20–70 cm high, obtusely angled, smooth; leaves 2–4 mm wide. Heads loosely fan-shaped; bracts 2–4, exceeding the inflorescence; spikelets 5–20 mm long; scales almost orbicular, 2–3 mm long; achene 1–1.5 mm long, ellipsoid, dark brown. Rare; sandy areas; southeastern Parklands, Boreal forest.

### *Cyperus inflexus* Muhl.                    AWNED NUT-GRASS

Plants annual, low, tufted, fragrant when bruised or dried; culms 1–15 cm high, very slender; leaves 1–2 mm wide. Heads dense; bracts 2–4, much exceeding the inflorescence; spikelets 5–10 mm long; scales 1.5–3 mm long, oblong, strongly nerved, narrowed to a long, recurving, slender tip; achene 0.6–1 mm long, 3-sided obovoid, pale brown. Rare; slough margins, Prairies.

### *Cyperus schweinitzii* Torr.                    SAND NUT-GRASS

Plants with hard, short, scaly rhizomes, with corm-like branches; culms 10–70 cm high, sharply angled; leaves 2–6 mm wide, firm. Heads mostly obovoid; bracts 3–6, exceeding the inflorescence; spikelets 6–15 mm long; scales 3.5–4.5 mm long, ovate, strongly nerved, with the midrib prolonged into a short awn; achene 2.5–3.5 mm long, 3-sided ellipsoid, light brown. Rare; sandhills; Parklands.

### *Cyperus strigosus* L.                    STRAW-COLORED NUT-GRASS

Plants with a hard, short, corm-like rhizome; culms tufted, 10–60 cm high, stout; leaves 2–12 mm wide, flat, soft. Heads compound, condensed to somewhat open; bracts 2–6 or 7, exceeding the inflorescence; spikelets compressed, 5–10 mm long, yellowish; scales 3–4.5 mm long, oblong, strongly nerved; achene 1.5–2 mm, 3-sided, linear. Rare; moist meadows and swamps; eastern Parklands, southeastern Boreal forest.

### *Eleocharis*        spike-rush

Perennial, sedge-like, slough margin plants, from coarse, creeping roots, or annuals with fibrous roots. Leaves generally reduced to sheaths without blades and all basal. Flowers perfect and borne in a spike at summit of stem.

1. Achenes lenticular or biconvex; stigmas
   2. ........................................................................................... 2

   Achenes triangular; stigmas 3. ..................................................... 6

2. Plants annual; the root system fibrous. ......................... *E. engelmannii*

   Plants perennial; the root system rhizomatous. ..................................... 3

3. Sterile basal scale 1 encircling the culm at base of spikelet. ..................................... 4

   Sterile basal scales 2 or 3 at base of spikelet. ..................................... 5

4. Tubercle narrowly conical, the base less
   than half as wide as the achene; spike-
   lets closely many-flowered. ........................................................ *E. calva*

   Tubercle deltoid-conical, the base more
   than half as wide as the achene; spike-
   lets loosely few-flowered. ........................................................ *E. uniglumis*

5. Tubercle as broad as long; scales firm,
   stiffly acuminate. ........................................................ *E. smallii*

   Tubercle much longer than broad; scales
   soft, scarious at the tip. ........................................................ *E. palustris*

6. Tubercle confluent with the achene, form-
   ing a beak. ........................................................ 7

   Tubercle distinctly jointed to the achene,
   forming a cap. ........................................................ 8

7. Spikelets to 20 mm long, many-flowered,
   fusiform. ........................................................ *E. rostellata*

   Spikelets to 7 mm long, 2- to 8-flowered,
   ovoid. ........................................................ *E. pauciflora*

8. Culms to 2 dm high, loosely tufted, form-
   ing mats from slender rhizomes. ........................................................ *E. acicularis*

   Culms 2–10 dm high, not matted, arising
   from stout rhizomes. ........................................................ 9

9. Culms 4- to 8-angled; achene 0.6–1.1 mm
   long, including the tubercle. ........................................................ *E. elliptica*

   Culms flattened; achene 1–1.5 mm long,
   including the tubercle. ........................................................ *E. compressa*

*Eleocharis acicularis* (L.) R. & S.                NEEDLE SPIKE-RUSH

Plants with very slender rhizomes, tufted, forming dense mats; culms slender, 2–15 cm high; sheaths loose. Spikelets flattened, 2–7 mm long, 3- to 15-flowered; scales 1–2.5 mm long, membranous, with a green midrib; achenes 0.7–1.2 mm long, whitish, ellipsoid. Common; around sloughs, in mud flats, along shores, and other wet places; throughout the Prairie Provinces.

*Eleocharis calva* Torr.                BALD SPIKE-RUSH

Plants with slender, reddish rhizomes; culms 10–50 cm high, solitary or loosely tufted; sheaths reddish, tight. Spikelets 10–20 mm long, lanceolate, densely flowered; scales 2–3 mm long, oblong-ovate, membranous, pale brown; achenes 1–1.5 mm long, narrowly obovoid, light brown. Not common; marshes, lakeshores, mud flats; southeastern Parklands.

*Eleocharis compressa* Sulliv.                FLATTENED SPIKE-RUSH

Plants with thick, hard rhizomes, 2–4 mm thick; culms 15–50 cm high, flat, wiry; sheaths tight. Spikelets 5–15 mm long, ovoid to ellipsoid; scales 2–3 mm long, ovate, brown or purplish; achenes 1–1.5 mm long, bluntly 3-sided, golden brown. Marshy areas, sandy shores; Parklands.

*Eleocharis elliptica* Kunth                               SLENDER SPIKE-RUSH

Plants with slender purplish rhizomes, 1–2 mm thick; culms 5–50 cm high, scattered or in small tufts; sheaths tight, reddish below. Spikelets 3–10 mm long, loosely flowered; scales 2–3.5 long, ovate, reddish brown; achenes about 1 mm long, yellow or orange, prominently cross-ridged. Wet areas, sand dunes, shores, meadows; southeastern Parklands.

*Eleocharis engelmannii* Steud.                         ENGELMANN'S SPIKE-RUSH

Plants annual, with culms 10–50 cm high, densely tufted; sheaths loose. Spikelets 4–15 mm long, ovoid-cylindric; scales 1–1.5 mm long, elliptic-obovate, brownish; achenes 1–2 mm long, lustrous brown when ripe. Wet areas; Prairies, Parklands.

*Eleocharis palustris* (L.) R. & S. (Fig. 78)           CREEPING SPIKE-RUSH

Plants with slender, long, creeping, reddish rhizomes; culms 10–60 cm high, firm; sheaths loose, red or brownish. Spikelets 5–20 mm long; scales 3–4 mm long, oblong-ovate, reddish brown; achenes 1–1.5 mm long, obovoid, light brown. Common; in shallow water, mud flats, and lakeshores; throughout the Prairie Provinces.

*Eleocharis pauciflora* (Lightf.) Link var. *fernaldii* Svenson
                                                        FEW-FLOWERED SPIKE-RUSH

Plants with long, creeping rhizomes, often with scaly tubers; culms 10–30 cm high, in small tufts. Spikelets 3–6 mm long; scales 3–5 mm long, ovate, lustrous brown; achenes 2–2.5 mm long, net-veined, gray brown. Calcareous wet soils; Boreal forest.

*Eleocharis rostellata* Torr.                              BEAKED SPIKE-RUSH

Plants with short caudex; culms 10–60 cm high, tufted; sheaths loose. Spikelets 6–12 mm long; scales 2–3 mm long, ovate, green with brown midrib; achenes 2–3 mm long, oblong-obovoid, olive brown. Wet soils; northwestern Boreal forest.

*Eleocharis smallii* Britt.                                SMALL'S SPIKE-RUSH

Plants with long-creeping rhizomes; culms 10–60 cm high; sheaths loose, reddish. Spikelets 5–20 mm long, loosely flowered; scales 3–5 mm long, lanceolate to narrowly ovate; achenes 1–1.5 mm long, obovoid, brown. Mud flats, peaty swamps, and shores; throughout the Prairie Provinces.

*Eleocharis uniglumis* (Link) Schultes                  ONE-GLUMED SPIKE-RUSH

Plants with slender reddish rhizomes; culms 10–50 cm high, loosely tufted, very slender; sheaths tight, reddish. Spikelets 1–1.5 cm long, loosely flowered; scales 2–3 mm long, thin; achenes 1–1.5 mm long, ellipsoid to obovoid. Wet areas, lakeshores; throughout the Prairie Provinces.

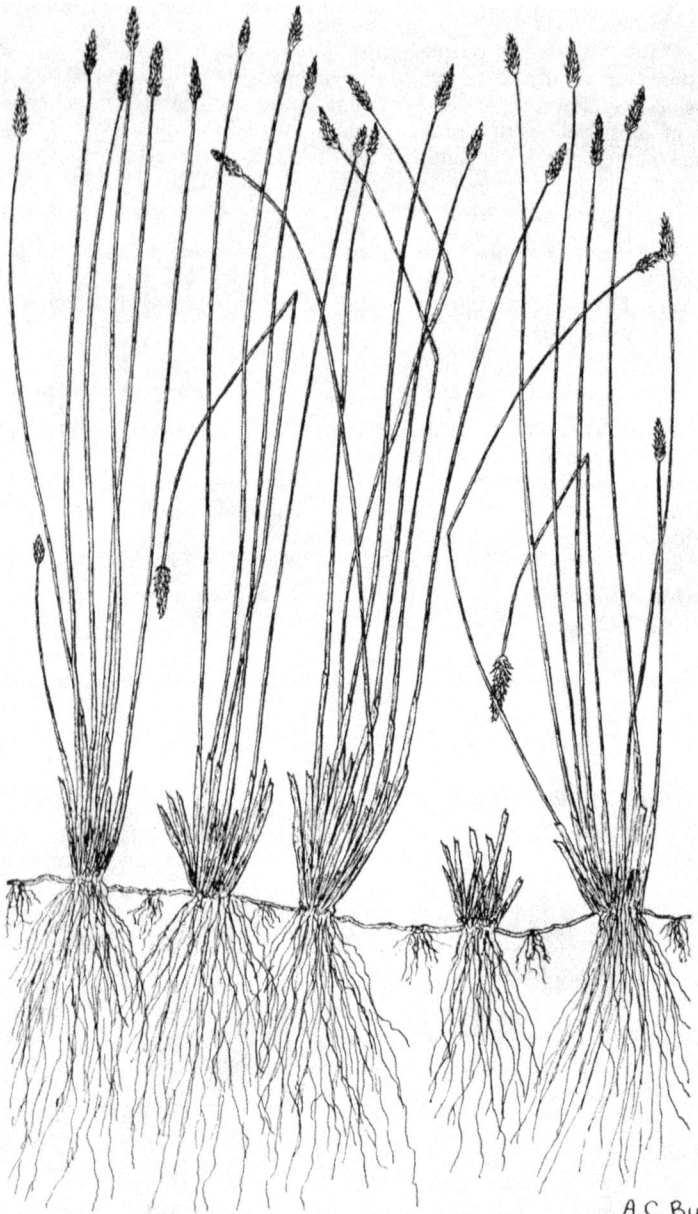

A.C. Budd

Fig. 78.  Creeping spike-rush, *Eleocharis palustris* (L.) R. & S.

*Monocots*

## *Eriophorum*  cotton-grass

### *Eriophorum angustifolium* Honck.  TALL COTTON-GRASS

Plants with short, stout rhizomes; culms 20–60 cm high, mostly solitary, soft blunt-angled. Sheaths smooth, reddish below, with blades 3–6 mm wide, flat to convolute, scabrous. Inflorescence with 2–10 spikelets; bracts 2–3, dark purple at the base; spikelets 1–2 cm long, divergent or drooping, with bristles 2–5 cm long, white; scales lead gray to brownish; achenes 2.5–3.5 mm long. Muskeg, swamps, wet meadows; Boreal forest.

### *Eriophorum brachyantherum* Trautv.  CLOSE-SHEATHED COTTON-GRASS

Plants tufted, with culms 20–60 cm high, terete, slender. Sheaths smooth, with blades 1–2 mm wide, filiform, the upper sheaths without blades.

Inflorescence a solitary spikelet 1–1.5 cm long, with bristles 2–3 cm long, cream-colored; scales lead gray; achenes 2–2.5 mm long. Bogs and swamps; Boreal forest.

*Eriophorum callitrix* Cham.                          BEAUTIFUL COTTON-GRASS

Plants tufted, with culms 10–25 cm high, stiff, stout. Sheaths smooth, with blades 2–4 mm wide, coarse; cauline sheaths with a short blade, the upper ones inflated. Inflorescence a single spikelet about 1 cm long, with bristles 2–3 cm long, bright white; scales grayish; achenes 2–2.5 mm long. Bogs and swamps; southern Boreal forest.

*Eriophorum chamissonis* C. A. Mey.                    RUSSET COTTON-GRASS

Plants with short rhizomes and somewhat stoloniferous; culms 10–35 cm high, solitary or few together. Sheaths loose, the lower ones with blades 1–3 mm wide, channeled or involute; the upper sheath bladeless, tight. Inflorescence a single spikelet 1–1.5 cm long, with bristles 2.5–3 cm long, reddish brown; scales brownish to blackish, with white margins; achenes 2–2.5 mm long. A form with white bristles is f. *albidum* (Nyl.) Fern. Muskeg and swampy areas; Boreal forest.

*Eriophorum gracile* Koch                              SLENDER COTTON-GRASS

Plants in small tufts, with culms 20–40 cm high, very slender, weak, mostly without basal leaves. Sheath of upper culm leaf 3–6 cm long, with the blade 1–3 cm long, 1–1.5 mm wide. Inflorescence with 2–5 spikelets, with a single bract 1–2 cm long, lead gray or blackish; spikelets spreading or nodding, 7–10 mm long, with bristles 1–1.5 cm long, bright white; scales lead gray to blackish; achenes 1.5–2 mm long. Muskeg, swamps, and bogs; Boreal forest.

*Eriophorum scheuchzeri* Hoppe                         ONE-SPIKE COTTON-GRASS

Plants with rhizomes; culms 10–40 cm high, solitary, somewhat leafy at the base. Sheaths loose, the upper ones bladeless, black-tipped; the lower sheaths with blades 2–5 mm wide, channeled or involute. Inflorescence a solitary spikelet about 1 cm long, with bristles 1–1.5 cm long, bright white; scales lead gray to blackish, with narrow pale margins; achenes 1.5–2 mm long. Muskeg areas, sedge swamps; Boreal forest.

*Eriophorum vaginatum* L.                              SHEATHED COTTON-GRASS

Plants densely tufted, forming large clumps, with culms 20–60 cm high, stiff, 3-sided. Sheaths glabrous, with blades stiff, filiform; culm leaves consisting of one or two inflated bladeless sheaths. Inflorescence a solitary spikelet 1–1.5 cm long, with bristles 2–2.5 cm long, bright white; scales lead gray to blackish, divergent or reflexed; achenes 2.5–3.5 mm long. In the Prairie Provinces, our plants are ssp. *spissum* (Fern.) Hulten, differing from the typical Eurasian plants in having longer and more inflated upper sheaths, larger anthers, and narrower achenes. Bogs and marshes; Boreal forest.

*Eriophorum viridi-carinatum* (Engelm.) Fern.          THIN-LEAVED COTTON-GRASS

Plants in small tufts, with culms 20–70 cm high, 3-sided, slender. Sheaths smooth; basal leaves numerous, with blades 2–6 mm wide, flat to channeled;

cauline leaves 2 or 3, short. Inflorescence with 3–15 spikelets; bracts 2–4, green or with a brownish base; spikelets 5–10 mm long, with bristles 1–2 cm long, cream-colored to grayish; scales olive green to lead gray, with the midrib prominent, reaching the tip, sometimes protruding; achenes 1.5–2 mm long. Muskeg, bogs, and marshes; Boreal forest.

## *Kobresia*   bog-sedge

Spikes linear, undivided; glumes 3–3.5 mm
    long; achenes about 2.5 mm long. ................................................. *K. myosuroides*
Spikes ovoid, compound; glumes 2.5–3 mm
    long; achenes about 3 mm long. ................................................ *K. simpliciuscula*

### *Kobresia myosuroides* (Vill.) F. & P.                    BOG-SEDGE

Plants densely tufted, with culms 10–30 cm high, erect. Sheaths brown, becoming fibrillose and very conspicuous, with blades about 1 mm wide, filiform, erect, about equaling the culms. Inflorescence 1–3 cm long, linear, simple; upper spikelets staminate, the lower ones androgynous; scales 2–3 mm long; glumes 3–3.5 mm long, light brown, shiny, tightly enclosing the achenes, these 2.5 mm long, short stipitate, light brown. Rare; mountain slopes; Rocky Mountains.

### *Kobresia simpliciuscula* (Wahl.) Mack.             SIMPLE BOG-SEDGE

Plants densely tufted, with culms 10–30 cm high, erect. Sheaths cinnamon brown, with blades about 1 mm wide, filiform, involute; old sheaths and blades very conspicuous. Inflorescence 1–4 cm long, compound; the terminal spike staminate, the lateral ones androgynous or pistillate, 3–8 mm long; scales small, reddish brown; glumes 2.5–3 mm long, reddish brown, shiny, enclosing the achene. Rare; slopes and calcareous bogs; Rocky Mountains, Hudson Bay.

## *Rhynchospora*   beak-rush

Glomerules turbinate; uppermost bract about
    as long as the glomerules; bristles 10–12. ............................................. *R. alba*
Glomerules ovoid or oblong; uppermost bract
    much longer than the glomerules; bristles 6. ............................ *R. capillacea*

### *Rhynchospora alba* (L.) Vahl                    WHITE BEAK-RUSH

Plants densely tufted, with stems 15–30 cm high, slender. Leaves 0.5–2.5 mm wide, short, mostly basal. Inflorescence 8–15 mm long, at first white, later brownish; spikelets 3.5–5 mm long, usually 2-flowered; bristles 10–12, retrorsely barbed. Rare; bogs, wet places, muskeg; Boreal forest.

### *Rhynchospora capillacea* Torr.                   SLENDER BEAK-RUSH

Plants tufted, with stems 10–40 cm high, very slender. Leaves about 1 mm wide, the upper ones commonly exceeding the inflorescence. Inflorescence 8–10 mm long, with up to 10 spikelets 5–7 mm long; bristles 6, retrorsely barbed. Rare; bogs, swamps, and wet sands; Boreal forest.

***Scirpus***     bulrush

Spikelets all sessile or subsessile, or in
part sessile, in part on peduncles to 5
cm long; stigmas 2. ................................................................ *S. paludosus*

14. Bristles much longer than the scales,
protruding. ................................................................ *S. cyperinus*
Bristles shorter than the scales. ................................................................ 15

15. Achene triangular; stigmas 3; sheaths
green. ................................................................ *S. atrovirens*
Achene lenticular; stigmas 2; sheaths usu-
ally reddish purple. ................................................................ *S. microcarpus*

*Scirpus acutus* Muhl.                    VISCID GREAT BULRUSH

Plants with extensive rhizomes; culms 50–200 cm high, round, olive green,
hard, and firm; sheaths firm, often with a short blade. Panicle 5–15 cm long,
with branches ascending or spreading, rarely pendulous; spikelets 1–2 cm
long; scales red-dotted, viscid-pubescent above; achene shiny black at maturity. Sloughs, riverbanks, and lakeshores; Prairies, Parklands.

*Scirpus americanus* Pers.                    THREE-SQUARE BULRUSH

Plants with long creeping, stout rhizomes; culms 20–100 cm high, solitary
or few together, sharply 3-angled; leaves 2–8 mm wide, linear, channeled.
Inflorescence 1–5 cm long, with 1–8 spikelets; bract 2–15 cm long; spikelets
5–20 mm long, ovoid to almost cylindric, reddish brown. Wet, often brackish
to moderately saline areas; Prairies, Parklands.

*Scirpus atrovirens* Willd.                    GREEN BULRUSH

Plants tufted, with culms 30–100 cm high, bluntly 3-angled; leaves 5–15
mm wide. Inflorescence 5–20 cm long, umbelliform, with forked branches;
spikelets 2–8 mm long, greenish brown; bracts 3–5, leafy, with the longest one
much longer than the inflorescence. Bogs and marshes; Prairies, Parklands.

*Scirpus caespitosus* L. var. *callosus* Bigel.                    TUFTED BULRUSH

Plants densely tufted, forming hard tussocks, with culms 10–30 cm high,
wiry, filiform, round; basal sheaths bearing a blade to 15 mm long.
Inflorescence a single spikelet 3.5–6 mm long, ovoid to lanceolate. Bogs,
marshes, and tundra; Boreal forest.

*Scirpus clintonii* Gray                    CLINTON'S BULRUSH

Plants densely tufted, with culms 10–30 cm high, very slender, sharply 3-
angled; leaves 0.5–1 mm wide, short. Inflorescence a single spikelet 3–6 mm
long, lanceolate to ovoid, pale brown; the outer scales with the midrib prolonged into an awn. Rocky Mountains.

*Scirpus cyperinus* (L.) Kunth                    WOOL-GRASS

Plants tufted, with culms 50–100 cm high, obscurely 3-angled; leaves 2–5
mm wide, to 30 cm long. Inflorescence 5–20 cm long, with branches 3–10 cm
long, ascending or spreading to drooping; spikelets 3–6 mm long, very numerous, ovoid, with the bristles exceeding the scale, elongating at maturity; bracts
of the involucre 5–30 cm long. Wet meadows, swamps, and lakeshores; Boreal
forests, rarely in Parklands.

*Scirpus fluviatilis* (Torr.) Gray          RIVER BULRUSH

Plants with long rhizomes having corm-like thickenings; culms 50–150 cm high, sharply 3-angled, leafy to near the inflorescence; leaves 5–15 mm wide, pale green, flat. Inflorescence 6–10 cm long, with 1–5 spikelets; bracts to 15 cm long; spikelets 15–40 mm long, brown. Along streams, lakeshores; Boreal forest.

*Scirpus hudsonianus* (Michx.) Fern.          ALPINE COTTON-GRASS

Plants with creeping rhizomes; culms 10–45 cm high, in small tufts, sharply angled. Inflorescence a single spikelet 4–7 mm long; the lowest scale with an awn 2–4 mm long, the other scales awnless; bristles 6, exserted in flowering, elongating to 1–3 cm at maturity, barbless. Marshes and boggy places; Boreal forest, Rocky Mountains.

*Scirpus microcarpus* Pers.          SMALL-FRUITED BULRUSH

Plants with a thick rhizome; culms 30–80 cm high, stout; leaves 4–15 mm wide, firm, flat. Inflorescence 5–20 cm long, with branches 3–15 cm long, the shorter ones ascending, the longer ones drooping; spikelets 3–6 mm long, numerous, dark green; bracts usually 3, the lower one exceeding the inflorescence. Damp areas, swamps, and bogs; Boreal forest, Parklands.

*Scirpus nevadensis* Wats.          NEVADA BULRUSH

Plants with creeping rhizomes; culms 30–50 cm high, solitary or few tufted, blunt-angled or almost terete. Inflorescence 1–5 cm long, with 2–10 spikelets; the bract 3–12 cm long; spikelets 5–20 mm long, ovoid to lanceolate, reddish brown. Not common; lakeshores and riverflats, often brackish; western Prairies, Parklands.

*Scirpus paludosus* Nels.          PRAIRIE BULRUSH

Plants with stout rhizomes, corm-like thickened internodes; culms 30–120 cm high, 5–20 mm thick at the base; leaves 5–15 mm wide, channeled. Inflorescence 5–10 cm long, with spikelets 10–25 mm long, sessile or more or less long-peduncled, ovoid, usually pale brown; the bracts 5–20 cm long. Common; in alkaline marshes and lakeshores; throughout the Prairie Provinces.

*Scirpus pumilus* Vahl          DWARF BULRUSH

Plants densely tufted on short rhizomes, with culms 10–20 cm high, sharply angled, wiry. Inflorescence a solitary spikelet, 3–5 mm long, ovoid, light brown. Bogs and marshes; Boreal forest, Parklands.

*Scirpus rufus* (Huds.) Schrad.          RED BULRUSH

Plants with extensively creeping rhizomes; culms 10–40 cm high, tufted, bluntly angled; leaves 1–3 mm wide, erect. Inflorescence a spike, 10–20 mm long, with two rows of spikelets, involucre a single bract, to 5 cm long; spikelets 5–10 mm long, 2- to 5-flowered, lustrous reddish brown; bristles lacking or small. Saline or brackish marshes; Boreal forest, Parklands. Syn.: *Blysmus rufus* (Huds.) Link.

*Scirpus torreyi* Olney                                        TORREY'S BULRUSH

Plants with a slender, lax, creeping rhizome; culms 40–100 cm high, 3-angled, solitary; leaves 5–10 mm wide, channeled, firm or lax. Inflorescence 3–5 cm long, spikelets 10–15 mm long, ovoid to cylindric; bracts 3–15 cm long. Lakeshores and ponds; southeastern Boreal forest, Parklands.

*Scirpus validus* Vahl                                         GREAT BULRUSH

Plants with stout, extensively creeping, reddish rhizomes; culms 50–300 cm high, to 20 mm thick at the base, round, soft, easily compressed; sheaths usually bladeless. Inflorescence 5–10 cm long, with branches 1–7 cm long, the longer ones repeatedly branching, drooping; spikelets 3–7 mm long, brown. Very common; often forming extensive borders along sloughs and lakeshores; throughout the Prairie Provinces.

# LEMNACEAE—duckweed family

Aquatic perennials with small disk-like, leafy fronds floating on or in the water, with one or several rootlets hanging from the lower surface of the frond. Flowers unisexual, minute, and borne on the surface of the fronds; fruit a minute achene with a thin-walled covering. Duckweeds are found in still water throughout the Prairie Provinces.

Rootlets solitary, one to a frond. ............................................................... *Lemna*
Rootlets several to a frond. ............................................................. *Spirodela*

## *Lemna*          duckweed

Fronds short-stalked or stalkless, floating. .......................................... *L. minor*
Fronds long-stalked, usually submerged. .......................................... *L. trisulca*

*Lemna minor* L.                                               LESSER DUCKWEED

Small, floating fronds 3–5 mm across, found floating in large quantities in still water. Common; throughout the Prairie Provinces in suitable locations.

*Lemna trisulca* L.                                            IVY-LEAVED DUCKWEED

Small oblong or lanceolate fronds tapering to a narrow stalk, and found floating at various depths in the water. Fairly common; throughout the Prairie Provinces.

## *Spirodela*          larger duckweed

*Spirodela polyrhiza* (L.) Schleid.                            LARGER DUCKWEED

Found floating solitary or in small groups on the surface of water. Has 5–11 rootlets hanging from beneath fronds. Fairly common; in Parklands, Boreal forest.

# ARACEAE—arum family

Plants with flowers closely borne in a dense spike called a spadix and with a large leaf-like bract called a spathe at the back of, or sometimes partly enclosing, the spadix.

Leaf blades linear; spathe green and leaf-like. ......................................................... *Acorus*
Leaf blades ovate and cordate; spathe white
   and petal-like. ................................................................................................ *Calla*

## *Acorus*       calamus

*Acorus calamus* L.                                                    SWEET FLAG

Perennial herbs growing from stout rhizomes to 40–100 cm high with long, narrow sword-like leaves. Inflorescence a spike-like spadix 4–10 cm long, borne at an angle with the stem and with a leaf-like spathe projecting 3–10 cm beyond the inflorescence. Plant has a peculiar taste and odor. Found in swamps and along water courses; Boreal forest.

## *Calla*       water-arum

*Calla palustris* L.                                                    WATER CALLA

Low-growing perennial herb from long rhizomes, rooting at the nodes. The long-stalked leaves are usually broadly ovate, cordate-based, 5–10 cm long on stalks 7–20 cm long. Inflorescence, borne on a long stalk, a spike-like spadix 1.5–2.5 cm long, backed by an oval white spathe 2.5–7.0 cm long. Fruits red berries, each containing a few seeds. Found in swamps and shallow water; Boreal forest.

# COMMELINACEAE—spiderwort family

## *Tradescantia*       spiderwort

*Tradescantia occidentalis* (Britt.) Smyth.                    WESTERN SPIDERWORT

Plants with slender stems, 20–60 cm high; the leaves linear, less than 1 cm wide, involute. Cymes solitary, terminal; sepals 6–10 mm long; petals 10–15 mm long, rose to blue; pedicels and sepals glandular pubescent. Very rare; dry grassland; southeastern Parkland.

# JUNCACEAE—rush family

Grass-like plants growing from either fibrous roots or rhizomes, and sometimes bearing round or stem-like leaves. Flowers perfect, regular, with 3 sepals and 3 petals, both very similar and scale-like. Flowers borne on top of the stem, but the projecting stem-like bract often makes the flowers appear to be on the side of the stem. The fruit, a capsule, splitting at maturity.

Capsule containing many small seeds; plants
glabrous. ............................................................................................. *Juncus*

Capsule containing 3 large seeds; plants with
pubescent leaves. ............................................................................... *Luzula*

### *Juncus*     rush

1. Bract of the inflorescence terete, appearing as a continuation of the stem;
   inflorescence appearing lateral. .................................................. Group 1

   Bract of the inflorescence flat or channeled; inflorescence appearing terminal
   or lateral. ......................................................................................... 2

2. Leaves not septate, terete, or flat with the
   flat side toward the stem. ............................................................. Group 2

   Leaves septate, terete, or flat with the
   edge of the leaf toward the stem. ................................................. Group 3

### Group 1

1. Inflorescence few-flowered, usually only 2
   or 3 flowers; seeds with tail-like
   appendages. ....................................................................................... 2

   Inflorescence usually with several to
   many flowers; seeds without tail-like
   appendages. ....................................................................................... 3

2. Blade of the uppermost leaf sheath well-
   developed; capsule acute. ........................................................... *J. parryi*

   Blade of the uppermost leaf sheath greatly
   reduced; capsule retuse. ...................................................... *J. drummondii*

3. Seeds about 0.5 mm long; flowers green-
   ish; rhizomes slender. .......................................................... *J. filiformis*

   Seeds about 1 mm long; flowers brown to
   blackish; rhizomes stout. ................................................................ 4

4. Inflorescence many-flowered; anthers 2–4
   times as long as the filament. ............................................... *J. balticus*

   Inflorescence few-flowered; anthers about
   as long as the filaments. ........................................................ *J. arcticus*

### Group 2

1. Flowers borne singly in the inflorescence,
   each one with a pair of bractlets. ..................................................... 2

   Flowers borne in heads with several
   flowers. .............................................................................................. 7

2. Plants annual; inflorescence diffuse,
   about half the height of the plant. ........................................ *J. bufonius*

   Plants perennial; inflorescence much less
   than half the height of the plant. ...................................................... 3

3. Sepals obtuse, the tips curved inward; leaf sheaths extending halfway up the culm; rhizomes spreading. ..................... *J. compressus*

Sepals acute, the tips not curved inward; leaf sheaths confined to the lower one-third of the culm; rhizomes erect. ........................ 4

4. Seeds with a tail-like appendage at each end; leaves somewhat channeled. ...................... *J. vaseyi*

Seeds without appendages; leaves flat or involute. ........................ 5

5. Capsule completely 3-celled. ........................ *J. confusus*

Capsule not completely 3-celled, with the partitions reaching less than halfway. ........................ 6

6. Perianth 4–6 mm long; leaf blades mostly less than half as long as the culm; auricles short, yellowish. ........................ *J. dudleyi*

Perianth 3–4.5 mm long; leaf blades mostly more than half as long as the culm; auricles long, whitish. ........................ *J. tenuis*

7. Seeds without tail-like appendages. ........................ *J. longistylis*

Seeds with tail-like appendages. ........................ 8

8. Plants rhizomatous; heads 1–3, each with 6–8 flowers. ........................ *J. castaneus*

Plants tufted; heads solitary, with 1–6 flowers. ........................ 9

9. Perianth dark brown to blackish; the bract leaf-like, as long as or slightly longer than the inflorescence. ........................ *J. biglumis*

Perianth whitish to pinkish or light brown; the bract scale-like. ........................ *J. albescens*

## Group 3

1. Leaves flat, with the edge toward the stem; the septa incomplete. ........................ 2

Leaves terete or occasionally somewhat flattened; the septa complete. ........................ 4

2. Seeds with tail-like appendages. ........................ *J. tracyi*

Seeds without tail-like appendages. ........................ 3

3. Stamens 3; stems strongly flattened; sheaths not auricled. ........................ *J. ensifolius*

Stamens 6; stems not strongly flattened; sheaths mostly auricled. ........................ *J. saximontanus*

4. Heads spherical, short-peduncled; stamens 6; capsules lanceolate. ........................ 5

Heads at most hemispherical, long-peduncled; stamens 3 or 6; capsules oblong. ........................ 6

5. Sepals longer than the petals; heads 10–15
   mm in diam with 25–80 flowers. ............................................................... *J. torreyi*

   Sepals as long as the petals; heads 7–10
   mm in diam with 10–20 flowers. ............................................................. *J. nodosus*

6. Seeds with tail-like appendages. ...................................................................... 7
   Seeds without tail-like appendages. ............................................................... 9

7. Heads usually solitary, densely many-
   flowered; perianth about 4 mm long,
   dark brown. ................................................................. *J. mertensianus*

   Heads few to many in a more or less
   diffuse inflorescence; perianth 2.5–5
   mm long, color various. ............................................................................. 8

8. Perianth about 2.5 mm long; seeds short-
   tailed, 0.7–1.2 mm long, the body
   longer than the tails. ................................................... *J. brevicaudatus*

   Perianth 4–5 mm long; seeds long-tailed,
   1.2–2 mm long, the body shorter than
   the tails. ............................................................................. *J. canadensis*

9. Branches of inflorescence erect or stiffly
   ascending; plants in small tufts from
   slender rhizomes; capsule straw-col-
   ored or light brown. .............................................................. *J. alpinus*

   Branches of inflorescence loosely ascending or spreading; plants densely tufted
   from short rhizomes; capsule dark
   shiny brown. .................................................................... *J. articulatus*

*Juncus albescens* (Lange) Fern.                                     WHITE RUSH

   Plants loosely tufted, with stems 5–15 cm high, very slender; leaves 1–5
cm long, narrow. Inflorescence a single head, 1- to 5-flowered; lower bract
long-acuminate or awned as long as or longer than the first flower; perianth
whitish or pinkish; sepals 3–4 mm long; capsule about 3.5 mm long; seeds
1.5–2 mm long, smooth, brown, with tails shorter than the body. Moist areas;
Rocky Mountains.

*Juncus alpinus* Vill.                                               ALPINE RUSH

   Plants with creeping rhizomes; stems 15–50 cm high, in small tufts; stem
leaves 1 or 2, seldom reaching the inflorescence, narrow. Inflorescence a cyme
5–15 cm long; glomerules 3- to 12-flowered, obpyramidal; perianth segments
2–3 mm long; capsule 2.5–3.5 mm long. Wet meadows, bogs, and lakeshores;
Parklands, Boreal forest.

*Juncus arcticus* Willd.                                             ARCTIC RUSH

   Plants with stout rhizomes; stems 10–50 cm high, stout. Inflorescence a
cyme, few-flowered, with branches mostly simple; the bract appearing like a
continuation of the stem; perianth segments 2.5–3.5 mm long, dark brown;
capsule about 3 mm long; seeds about 1 mm long. Sedge tundra; Boreal forest.

*Juncus articulatus* L. JOINTED RUSH

Plants tufted, with slender stems 20–60 cm high; stem leaves usually 2–4, mostly 3–8 cm long, strongly septate. Inflorescence a cyme 3–10 cm long, with branches ascending to divergent; glomerules 3- to 11-flowered; perianth segments 2–3 mm long, brown; capsule 2.5–4 mm long, chestnut or purple brown. Bogs, wet meadows, lakeshores; Boreal forest.

*Juncus balticus* Willd. BALTIC RUSH

Plants with long-creeping, thick rhizomes; stems 20–60 cm high, in more or less regular rows from the rhizomes, or singly; basal sheath 8–15 cm long, bladeless. Inflorescence appearing lateral, cymose or densely head-like, mostly 2–8 cm long, occasionally to 12 cm long; perianth 4–5 mm long, purplish brown; capsule about 5 mm long, chestnut brown; seeds 0.8–1 mm long. Wet meadows, slough margins, sandhills, lakeshores, often in great abundance. A variable species in which several varieties have been distinguished. Slender plants with a small dense inflorescence have been named var. *montanus* Engelm. (Fig. 79); stout plants with a more open inflorescence, 4–8 or 12 cm long, are var. *littoralis* Engelm. Both varieties occur throughout the Prairie Provinces.

*Juncus biglumis* L. TWO-GLUMED RUSH

Plants tufted, with stems 5–20 cm high; leaves basal, filiform. Inflorescence a single head; the lowest bract mostly well-developed; perianth 3–3.5 mm long; capsule 3.5 mm long, purplish black; seeds about 1.5 mm long, with white tails. Tundra; Hudson Bay, Boreal forest.

*Juncus brevicaudatus* (Engelm.) Fern. SHORT-TAIL RUSH

Plants densely tufted, with stems 10–50 cm high, slender; leaves 1–2 mm in diam, erect. Inflorescence 3–12 cm long, narrow, with branches ascending to erect; glomerules few to many, each 2- to 7-flowered; perianth segments 2.5–3 mm long, 3-nerved; capsule 3.5–4.5 mm long; seeds about 1 mm long, with short tails. Marshes and lakeshores; Boreal forest, Parklands.

*Juncus bufonius* L. TOAD RUSH

Plants annual, with stems 3–20 cm high, tufted, erect or spreading; leaves about 1 mm wide. Inflorescence 2–10 cm long, spreading, freely branching; flowers singly or in twos or threes; perianth segments unequal, with sepals 3.5–6.5 mm long, petals 3–5.5 mm long. Common in low, wet ground; throughout the Prairie Provinces.

*Juncus canadensis* J. Gay CANADA RUSH

Plants tufted, with stems 40–100 cm high, stout and rigid; leaves 1.5–2.5 mm in diam, erect. Inflorescence 2–20 cm long, varying from compact to loosely branching; glomerules few to many, 2- to 10-flowered; perianth segments 2.5–4 mm long; capsule 3.5–5 mm long, brown; seeds about 1 mm long, with white tails. Rare; marshes and wet meadows; northeastern Parkland and Boreal forest.

Fig. 79.   Baltic rush, *Juncus balticus* Willd. var. *montanus* Engelm.

*Juncus castaneus* Sm. CHESTNUT RUSH

Plants with slender rhizomes; stems 20–40 cm high, erect; leaves 1–2 mm wide, basal. Inflorescence 2–5 cm long; the involucral leaf 2–8 cm long; glomerules 1–3, the lowest ones sessile or nearly so, the upper one peduncled, 10–18 mm in diam; perianth segments 5–7 mm long, chestnut brown; capsule 6–9 mm long, with a conspicuous beak; seeds 2.5–4 mm long, with long slender tails. Arctic and alpine meadows and tundra; Boreal forest, Rocky Mountains.

*Juncus compressus* Jacq. FLATTENED RUSH

Plants with slender rhizomes; stems 20–60 cm high, erect, tufted; leaves 1 mm wide, erect. Inflorescence 2–10 cm long, cymose, with branches ascending, bearing many-flowered glomerules; perianth segments 2–2.5 mm long, purple brown; capsule 2–3 mm long; seeds about 1 mm long. Introduced; wet ground, salt marshes; Manitoba.

*Juncus confusus* Cov. FEW-FLOWERED RUSH

Plants tufted, with stems 30–50 cm high, slender; leaves filiform. Inflorescence 5–20 mm long, compact, pale; perianth segments 3.5–4 mm long, brown with a yellowish midrib and narrow margins; capsule 3–3.5 mm long; seeds about 1 mm long. Moist grassland, open woods, and meadows; southern Rocky Mountains.

*Juncus drummondii* E. Mey. DRUMMOND'S RUSH

Plants densely tufted, with the base matted, persistent; stems 10–30 cm high, erect; sheaths short, bladeless, or the blade bristle-like. Inflorescence 1- to 5-flowered; the bract appearing like a continuation of the stem; perianth segments 5–7 mm long, brown with a light midrib; capsule 5–7 mm long, triangular; seeds about 2 mm long, tailed at both ends. Moist places; Rocky Mountains.

*Juncus dudleyi* Wieg. DUDLEY'S RUSH

Plants densely or loosely tufted, with stems 30–60 cm high; leaves basal, flat, 10–30 cm long; auricles of sheaths rounded, papery, yellowish. Inflorescence 2–5 cm long, with the few branches loosely ascending; perianth segments 3.5–5.5 mm long; capsule 3–4 mm long; seeds about 0.4 mm long. Moist or dry soil, waste areas; throughout the Prairie Provinces. Syn.: *J. tenuis* Willd. var. *dudleyi* (Wieg.) Hermann.

*Juncus ensifolius* Wikstr. EQUITANT-LEAVED RUSH

Plants tufted, with stems 30–60 cm high, flattened, sharply 2-keeled; leaves 2–6 mm wide, flat. Inflorescence 15–70 mm long, with 2–6 heads, about 1 cm thick, many-flowered, brownish; perianth segments about 3 mm long, dark brown; capsule about 3 mm long; often somewhat exceeding the perianth, dark brown; seeds about 0.3 mm long. Riverbanks, lakeshores; Rocky Mountains, Cypress Hills.

*Juncus filiformis* L. THREAD RUSH

Plants with slender rhizomes; stems 10–40 cm high, arising from the rhizomes in rows, or in small tufts; leaves short, thick. Inflorescence 2–4 cm long, sparingly branched, few-flowered; involucral leaf erect, to 20 cm long; perianth segments 2.5–4.5 mm long, greenish, with hyaline margins; capsules 2.5–4 mm long; seeds about 0.3 mm long. Lakeshores, bogs, and alpine meadows; Boreal forest, rare in northeastern Parkland.

*Juncus longistylis* Torr.                                LONG-STYLED RUSH

Plants with slender rhizomes; stems solitary, 20–60 cm high; leaves mostly basal, 1–4 mm wide, flat; stem leaves short, flat or involute. Inflorescence 2–3 cm long, with 1–5 heads 8–15 mm across, 2- to 8-flowered; perianth segments 4.5–6 mm long, greenish brown; capsule 4–5 mm long, brownish; seed about 0.5 mm long, tailed at both ends. Fairly common; wet meadows, lakeshores, and riverbanks; Prairies, Parklands, Boreal forest.

*Juncus mertensianus* Bong.                              SLENDER-STEMMED RUSH

Plants with matted rhizomes; stems 10–40 cm high, densely tufted; leaves 1–2 mm wide, compressed, septate. Inflorescence usually a solitary head, occasionally 2 or 3 heads, dark brown, almost spherical; perianth segments 3.5–4.5 mm long, dark brown; capsule 3.5–4.5 mm long, dark brown; seeds 1 mm long, reticulate. Rare; wet ground, meadows, and slopes; Rocky Mountains, Cypress Hills.

*Juncus nodosus* L.                                       KNOTTED RUSH

Plants with slender rhizomes; stems 20–60 cm high, erect; leaves to 10 cm long, and 1.5 mm thick, septate. Inflorescence a cyme 2–5 cm long, contracted or more or less open with 2–20 globose heads, 7–12 mm in diam, 10- to 25-flowered; perianth segments 2.5–4 mm long, reddish brown; capsule 3.5–4.5 mm long; seeds 1.5 mm long. Bogs, marshes, and lakeshores; Prairies, Parklands, Boreal forest.

*Juncus parryi* Engelm.                                   PARRY'S RUSH

Plants densely tufted, with stems 10–30 cm high, slender; leaves only from the upper sheaths. Inflorescence a single head, 1- to 3-flowered; perianth segments 6–7 mm long; capsule about 7 mm long; seeds about 0.5 mm long, tailed at both ends. Mountain slopes and meadows; Rocky Mountains.

*Juncus saximontanus* A. Nels.                            ROCKY MOUNTAIN RUSH

Plants with stout creeping rhizomes; stems 20–50 cm high, compressed; leaves 2–4 mm wide, flat, with the edge toward the stem, 10–20 cm long. Inflorescence 6–10 cm long, with 2–12 heads, 4–6 mm across, few- to many-flowered; perianth segments 2.5–3.5 mm long, dark brown; capsule about 3 mm long; seeds 0.5 mm long, reticulate. Marshes and wet areas; Rocky Mountains, Cypress Hills.

*Juncus tenuis* Willd.                                    SLENDER RUSH

Plants loosely or densely tufted, with stems 10–60 cm high, mostly erect; leaves mostly basal, 10–25 cm long, flat or involute; the sheaths conspicuously long auricled. Inflorescence a cyme 2–8 cm long, commonly exceeded by the involucral leaf, compact to open; perianth segments 3–5 mm long; capsule 3–4 mm long, straw-colored; seeds about 0.5 mm long. Moist to wet areas; Prairies, Parklands, Boreal forest.

*Juncus torreyi* Cov.                                    TORREY'S RUSH

Plants with slender tuberous-thickened rhizomes; stems 40–80 cm high, stout, rigid, erect; leaves 1–3 mm thick, 10–30 cm long; sheaths with auricles 2.5–3.5 mm long. Inflorescence 2–4 cm long, with 2 to many heads, usually crowded; heads 10–15 mm in diam, globose; 25- to 100-flowered; perianth segments unequal, with the petals 4.5–5 mm long, the sepals 3.5–5 mm, linear-lanceolate; capsule 4.5–5.5 mm long, narrowly lanceolate; seeds about 0.3 mm long. Wet areas; Prairies, Parklands.

*Juncus tracyi* Rydb.                                    MUD RUSH

Plants with stout rhizomes; stems 30–60 cm high, stout, compressed; leaves 5–20 cm long, 2–4 mm wide. Inflorescence 4–8 cm long, with 5–9 heads, 8–10 mm in diam; perianth segments 3–4 mm long, light brown; capsule about 3–3.5 mm long, brown; seeds 0.5 mm long, reticulate. Rare; wet areas; northern Rocky Mountains.

*Juncus vaseyi* Engelm.                                  BIG-HEAD RUSH

Plants with short rhizomes; stems 30–80 cm high, erect, tufted; basal leaves to 30 cm long, terete or nearly so. Inflorescence 1–4 cm long, compact; perianth segments 3.5–4.5 mm long, brown; capsule 4–5.5 mm long, truncate, straw-colored; seeds 0.5–0.8 mm long, conspicuously tailed at both ends. Moist soil, meadows, and lakeshores; Prairies, Parklands, Boreal forest.


**Luzula**          wood-rush

Similar to *Juncus* but capsule 1-celled with 3 seeds; foliage more or less pilose.

Appendage of seeds small; perianth seg-
ments 2–2.5 mm long. ................................................................................. 7
7. Sepals ciliate on the margins; leaves 1–4
mm wide, subulate at the tip. ................................................. *L. spicata*
Sepals with entire or lacerate margins;
leaves 2–6 mm wide, blunt,
callus-tipped. ........................................................................... *L. groenlandica*

## *Luzula campestris* (L.) DC.          FIELD WOOD-RUSH

Plants tufted, with stems 20–40 cm high; leaves 2–6 mm wide, flat; cauline leaves 2 or 3, short. Inflorescence 3–5 cm long, with slender ascending peduncles, and one or more sessile glomerules; perianth segments 2.5–3.5 mm long; capsule 3.5 mm long; seeds 1–1.3 mm long, with the appendage 0.4–0.5 mm. Dry to moist woods; Boreal forest, Cypress Hills.

## *Luzula confusa* Lindeb.          NORTHERN WOOD-RUSH

Plants densely tufted, with stems 10–30 cm high; leaves 1–3 mm wide; stem leaves 2 or 3, short. Inflorescence 3–5 cm long, with 2–5 glomerules on slender peduncles, or occasionally the glomerules crowded into a short spike; perianth segments 2.5–3 mm long, brown; capsule 3 mm long, red brown; seeds 1–1.2 mm long, with a tuft of minute hairs. Gravel ridges and barrens; Hudson Bay, Boreal forest, also reported from Rocky Mountains.

## *Luzula glabrata* (Hoppe) Desv.          SMOOTH WOOD-RUSH

Plants with scaly rhizomes; stems 20–50 cm high; leaves 4–10 mm wide, with stem leaves well-developed. Inflorescence 6–10 cm long, with the branches ascending or spreading; perianth segments 3–3.5 mm long, dark brown; capsule about 3 mm long, almost black; seeds 1–1.3 mm long, with a tuft of hairs. Meadows and slopes; southern Rocky Mountains.

## *Luzula groenlandica* Boecher          GREENLAND WOOD-RUSH

Plants tufted; stems 20–50 cm high; leaves 3–7 mm wide, flat, with stem leaves 2 or 3, short. Inflorescence 3–5 cm long; perianth segments 2–2.5 mm long; capsule 2.5–3 mm long, dark brown; seeds about 1 mm long, with the appendage short. Tundra; Hudson Bay.

## *Luzula parviflora* (Ehrh.) Desv.          SMALL-FLOWERED WOOD-RUSH

Plants with creeping rhizomes; stems 20–60 cm high, erect; leaves 5–10 mm wide. Inflorescence 5–12 cm long, with branches loosely spreading to drooping; perianth segments 1.5–2.5 mm long, pale brown; capsule 2–2.5 mm long, dark brown; seeds 1–1.4 mm long, with a tuft of long hairs. Infertile soil, open forests; Boreal forest, Rocky Mountains.

## *Luzula pilosa* (L.) Willd. var. *saltuensis* (Fern.) Boiv.          HAIRY WOOD-RUSH

Plants tufted from short rhizomes; stems 10–40 cm high; leaves 3–10 mm wide, and up to 30 cm long, with stem leaves 2–4, shorter and narrower than the basal leaves. Inflorescence 3–6 cm long, with branches loosely spreading; perianth segments 2.5–4.5 mm long, the center chestnut brown, with narrow margins; capsule 3–4.5 mm long, brown; seeds about 1 mm long, with the appendage about equaling the body. Open woods; Boreal forest.

*Luzula spicata* (L.) DC.                                    SPIKED WOOD-RUSH

Plants densely tufted, with stems 10–30 cm high, slender; leaves 1–4 mm wide; stem leaves 1–3, short. Inflorescence 1–3 cm long, spike-like, often nodding, with the glomerules dense, sessile; perianth segments 2–2.5 mm, bristle-pointed, brown; capsule 2.5 mm long, red brown; seeds 1–1.2 mm long, with a short appendage. Arctic and alpine meadows; Boreal forest, Rocky Mountains.

*Luzula wahlenbergii* Rupr.                              MOUNTAIN WOOD-RUSH

Plants with rhizomes; stems 20–60 cm high, erect; leaves 3–8 mm wide; stem leaves well-developed, thick, dull. Inflorescence 5–10 cm long, with the slender branches spreading and drooping; perianth segments 1.5–2.5 mm long, pale; capsule 2–2.5 mm long, dark brown; seeds 1–1.3 mm long, gray brown to yellow with a tuft of long hairs. Meadows and open forests; southern Rocky Mountains.


# LILIACEAE—lily family

Perennial herbs growing from bulbs, corms, or rootstocks. With a few exceptions, the leaves are parallel-veined. Except in the *Maianthemum* genus, the flowers have 3 sepals and 3 petals, all colored, and 6 stamens in 2 whorls. The fruit is a capsule or a berry.

1. Leaves reduced to thin, dry scales; branchlets filiform; flowers small, greenish. ..................................................... *Asparagus*
   Leaves normal, not reduced to scales. ........................................ 2

2. Plants climbing with tendrils; leaves roundish, net-veined, flowers in axillary umbels. ........................................................ *Smilax*
   Plants not climbing. ...................................................... 3

3. Leaves linear, many times as long as wide. ................................. 4
   Leaves variously shaped, not more than five times as long as wide. .......... 12

4. Flowers orange, large, 5–7 cm long. ................................. *Lilium*
   Flowers not orange, smaller. .............................................. 5

5. Inflorescence an umbel, head-like cluster, or few-flowered. .............................................. 6
   Inflorescence a many-flowered raceme. ...................................... 7

6. Inflorescence few-flowered; flowers yellowish white, large, to 25 mm long. ................. *Calochortus*
   Inflorescence an umbel or head-like cluster; flowers 5–15 mm long. ........................ *Allium*

7. Flowers purplish blue, to 2 cm long. ........................... *Camassia*
   Flowers not purplish blue. ................................................. 8

8. Flowers greenish or reddish brown, 10–15 mm long, nodding on slender pedicels. ........................................... *Stenanthium*

Flowers some shade of white, in erect racemes. .................................................................................................... 9

9. Flowers large, 3–5 cm long; leaves with a rigid point at the tip. .................................................................... *Yucca*

Flowers smaller, to 1.5 cm long; leaves not with a rigid point. ............................................................................. 10

10. Racemes large, densely flowered; leaves rough; stems to 1.2 m high. ................................................ *Xerophyllum*

Racemes slender, loosely flowered; leaves smooth; plants not more than about 0.6 m high. .......................................................................................... 11

11. Leaves keeled, V-shaped in cross section; flowers 5–15 mm long. ...................................................... *Zygadenus*

Leaves flat; flowers to 5 mm long. ...................................................................................................... *Tofieldia*

12. Flowers solitary or in umbels, terminal. .................................................................................................. 13

Flowers axillary, in panicles or in racemes. ................................................................................................ 17

13. Stems leafy. .......................................................................................................................................... 14

Stems leafless. ...................................................................................................................................... 16

14. Flowers white or pinkish. ................................................................................................................ *Trillium*

Flowers orange yellow or straw yellow. .................................................................................................... 15

15. Flowers nodding; plants with a deep-seated bulb. ................................................................................... *Fritillaria*

Flowers erect or nodding; plants with rhizomes. .................................................................................. *Uvularia*

16. Flowers bright yellow, nodding; fruit a capsule; leaves glabrous. ............................................................. *Erythronium*

Flowers white or greenish white; fruit a bluish berry; leaves pilose beneath. .......................................... *Clintonia*

17. Flowers in a large, diffuse panicle. ................................................................................................. *Veratrum*

Flowers axillary or in racemes, or at tips of branches. ................................................................................ 18

18. Flowers 1–4 at the tip of the stem or branches; fruit a red or orange berry. ......................................... *Disporum*

Flowers axillary or in racemes. ................................................................................................................ 19

19. Flowers axillary. .................................................................................................................................... 20

Flowers in racemes. ............................................................................................................................... 21

20. Perianth segments free at the tip only. ........................................................................................ *Polygonatum*

Perianth segments free to the base. ................................................................................................ *Streptopus*

21. Plants small, usually with only 2 or 3 leaves; raceme small. .............................................................. *Maianthemum*

Plants large, with leafy stems; racemes simple or paniculate. ............................................................ *Smilacina*

*Allium*    onion

Strong-scented herbs growing from a bulbous root, which is generally covered with a fibrous coat. Leaves linear and narrow, flowers brightly colored and borne in umbels. Fruit a capsule containing dark seeds.

1. Flower stalks shorter than individual
   flowers; leaves circular in cross section
   and hollow. ............................................................................. *A. schoenoprasum*
   Flower stalks longer than individual
   flowers; leaves not hollow. ....................................................................... 2

2. Umbel nodding or drooping. .................................................... *A. cernuum*
   Umbel erect. ............................................................................................... 3

3. Flowers rose-colored. ................................................................ *A. stellatum*
   Flowers white or pale pink. ........................................................ *A. textile*

*Allium cernuum* Roth                                           NODDING ONION

Growing from coarse-necked bulbs, usually on a short rhizome; flower heads rose or rarely white. Found occasionally; in grassy openings; Parklands.

*Allium schoenoprasum* L. var. *sibiricum* (L.) Hartm.          WILD CHIVES

Perennial from small bulbs, 20–50 cm high, with many hollow leaves up to 3 mm thick and circular in cross section. Inflorescence a dense, compact umbel 2–5 cm across; flower stalks very short and flowers bright rose pink. Not common; moist areas, meadows, and coulees; Rocky Mountains, Boreal forests.

*Allium stellatum* Fraser                                  PINK-FLOWERED ONION

Erect umbels of pink or rose flowers from a membranous-coated bulb. Fairly common; in wooded lands; throughout Parklands.

*Allium textile* Nels. & Macbr. (Fig. 80)                      PRAIRIE ONION

A low plant, from an onion-like bulb having a net-like coating and a very pungent odor. From 8–25 cm high, with umbels of white or pale pink flowers, which blossom early in season. Common; on dry prairie; throughout Prairies, Parklands.

*Asparagus*    asparagus

*Asparagus officinalis* L.                                         ASPARAGUS

Plants with succulent rhizomes; stems 60–150 cm high; branches filiform; leaves scaly. Flowers small, yellowish green, axillary, with the pedicels jointed; fruit a red berry. Introduced, occasionally escaped from cultivation.

*Calochortus*    mariposa lily

*Calochortus apiculatus* Baker                                  MARIPOSA LILY

Plants with ovoid bulbs and membranous scales; stems 10–30 cm high; leaf solitary, basal, 5–15 mm wide. Flowers 1–4, yellowish white, 3–5 cm across; the sepals shorter than the petals; stamens 6; stigmas 3; fruit a 3-angled capsule. Dry slopes, open woods; southern Rocky Mountains.

Fig. 80. Prairie onion, *Allium textile* Nels. & Macbr.

***Camassia***     camas

*Camassia quamash* (Pursh) Greene                    COMMON CAMAS

   Herbaceous perennial plants from an ovoid bulb 1–3 cm broad, the leaves mostly basal, narrow, and grass-like, 20–40 cm long. Blue flowers with 6 petal-like segments 10–25 mm long, borne in a raceme at the summit of scapose stems 30–60 cm high. Moist meadows; in southern Rocky Mountains.

***Clintonia***

Flowers white, usually solitary. ........................................................................... *C. uniflora*
Flowers yellowish green, in 3- to 8-flowered
   umbel. ................................................................................................................ *C. borealis*

*Clintonia borealis* (Ait.) Raf.                          BLUEBEAD LILY

   Plants with slender rootstocks; scapes 15–40 cm high; leaves 2–5, to 30 cm long, obovate to oblong-elliptic, glossy green. Flowers on pedicels 1–3 cm long; perianth segments 15–18 mm long, greenish yellow, nodding; fruit a blue berry, 8 mm in diam. Moist woods; southeastern Boreal forest.

*Clintonia uniflora* (Schult.) Kunth                        CORN LILY

   Plants with slender rootstocks; scapes 6–20 cm high; leaves 2–5, to 20 cm long, oblanceolate, pubescent below. Flowers showy; perianth segments about 1 cm long, white, villose; fruit a deep blue berry 8–10 mm in diam. Moist woods; southern Rocky Mountains.

***Disporum***     fairybells

Stigma 3-lobed; flowers solitary or in pairs. ............................................. *D. trachycarpum*
Stigma entire; flowers in clusters of 1–4. ................................................... *D. hookeri*

*Disporum hookeri* (Torr.) Britt. var. *oreganum* (Wats.) Q. Jones
                                                   OREGON FAIRYBELLS

   Plants 30–80 cm high, with leaves ovate to oblong-lanceolate, 3–9 cm long, 2–6 cm wide, cordate-based. Flowers 1–4 in a cluster, 10–12 mm long, creamy white or tinged with green; stigma entire; berry ovoid, 8–16 mm in diam, 6-seeded. Moist woods; southern Rocky Mountains.

*Disporum trachycarpum* (S. Wats.) B. & H. (Fig. 81*A*)        FAIRYBELLS

   Plants 30–60 cm high, with leaves ovate to oblong-lanceolate, 3–8 cm long, 2–5 cm wide, short pubescent, rounded or subcordate at the base. Flowers solitary or in pairs, occasionally 3 together, 8–14 mm long, creamy white; stigma 3-lobed; berry depressed globose, 8–10 mm in diam, 4- to 18-seeded. Moist woods, ravines, and coulees; throughout Prairies, Parklands, southern parts of Boreal forest.

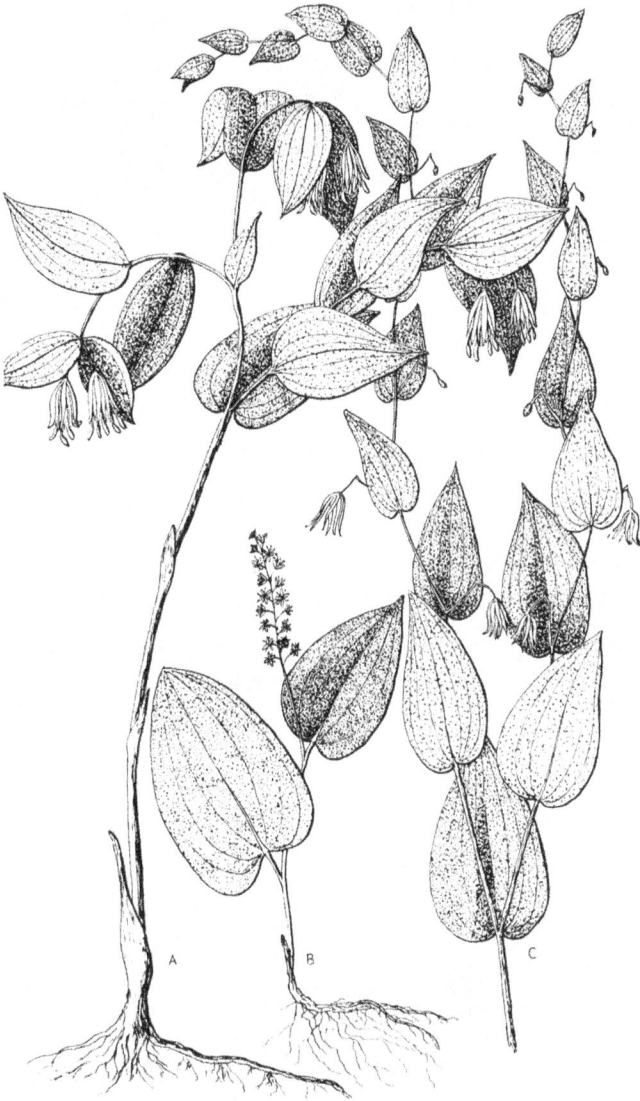

A.C. Budd

Fig. 81.   *A*, Fairybells, *Disporum trachycarpum* (S. Wats.) B. & H.; *B*, two-leaved Solomon's-seal, *Maianthemum canadense* Desf. var. *interius* Fern.; *C*, clasping-leaved twistedstalk, *Streptopus amplexifolius* (L.) DC.

*Erythronium*        fawn lily

*Erythronium grandiflorum* Pursh (Fig. 82)        GLACIER LILY

Plants with deep-seated corm, arising from a rootstock; scape 20–30 cm high. Leaves 7–15 cm long, oblong-lanceolate. Flowers solitary or in a 2- to 6-flowered raceme; perianth segments 3–5 cm long, bright yellow, recurved; capsule 3–5 cm long. Open forests, hillsides, and grassy slopes; southern Rocky Mountains.

*Fritillaria*        fritillary

*Fritillaria pudica* (Pursh) Spreng.        YELLOWBELL

A short plant, 10–30 cm high, from a very scaly bulb. Leaves about 8 cm long and up to 6 mm wide, somewhat whorled about halfway up the stem. Flower, nodding and single, bell-shaped, yellow or orange, on the summit of the stem, about 15–20 mm high. Very rare; on moist banks; southern Rocky Mountains. Syn.: *Ochrocodon pudicus* (Pursh) Rydb.

*Lilium*        lily

*Lilium philadelphicum* L.        WOOD LILY

Very showy, erect plants 20–60 cm high, from whitish, scaly bulblets. Leaves linear, in whorls. Flowers showy, 8 cm long, sepals and petals red or orange with black spots, 1–5 flowers on a plant. Found in the eastern part of the Prairie Provinces, but westward giving place to the western red lily, *L. philadelphicum* L. var. *andinum* (Nutt.) Ker (Fig. 83), having the lower leaves alternate and the upper ones whorled. The variety is found in moist meadows throughout the entire area, but is becoming scarcer with the advance of settlement and overpicking. Plants found in the Foothills region have wider, lanceolate leaves and were formerly considered a separate species.

*Maianthemum*        wild lily-of-the-valley

*Maianthemum canadense* Desf. var. *interius* Fern. (Fig. 81*B*)
                TWO-LEAVED SOLOMON'S-SEAL

Low, erect plant, 5–15 cm high, with 2, sometimes 3, ovate leaves, cordate-based, borne alternately on the stem. Flowers small, white, in a rather dense raceme. Berries pale red, speckled, about 5 mm in diam. Fairly common; in rich, moist woods; throughout the Prairie Provinces.

*Polygonatum*        Solomon's-seal

*Polygonatum canaliculatum* (Muhl.) Pursh        COMMON SOLOMON'S-SEAL

Coarse plants, from jointed rootstocks, 30–100 cm high. Leaves large, alternate, up to 15 cm long and to 10 cm wide. White, cylindric flowers suspended in little bunches below the stem, about 15 mm long. Berries dark blue, 8–12 mm in diam. Fairly common; in open woodlands in the eastern part. Syn.: *P. commutatum* (R. & S.) Dietr.

Fig. 82.   Glacier lily, *Erythronium grandiflorum* Pursh.

A.C. Budd

Fig. 83.  Western red lily, *Lilium philadelphicum* L. var. *andinum* (Nutt.) Ker.

*Monocots*

*Smilacina*      false spikenard

Plants from rootstocks, with alternate, ovate to lanceolate leaves, and small, perfect, greenish white flowers borne in racemes or panicles. Fruits small berries.

1. Inflorescence a branched, dense panicle. ................................................. *S. racemosa*
   Inflorescence an unbranched raceme. ........................................................................ 2

2. Leaves 6–12; berries green. ........................................................................ *S. stellata*
   Leaves 2–4; berries dark red. ........................................................................ *S. trifolia*

*Smilacina racemosa* (L.) Desf.

Large-leaved plants, 30–90 cm high. Leaves elliptic or oval, up to 15 cm long and to 7 cm wide, parallel-veined, without stalks, except lower ones, which may have very short stalks. White flowers in a densely branched panicle at the end of the stem. Berries red, purple-specked, up to 6 mm across. Not common; found in moist woods; throughout southeastern Boreal forest, Riding Mountain. Replaced in western Boreal forest, Parklands, Rocky Mountains, and Cypress Hills by *S. racemosa* (L.) Desf. var. *amplexicaulis* (Nutt.) S. Wats., having all leaves sessile and clasping.

*Smilacina stellata* (L.) Desf.          STAR-FLOWERED SOLOMON'S-SEAL

Erect, low plants, 15–50 cm high. Leaves arranged on opposite sides of the stem, folded at the base, overlapping each other in early stages, but spreading and flattening out later. Flowers small, white, in a spike-like raceme on the stem. Fruits green berries, each with 6 black stripes. Very common; in moist soil, meadows, woods, and low areas; throughout the Prairie Provinces. Readily eaten by livestock.

*Smilacina trifolia* (L.) Desf.          THREE-LEAVED SOLOMON'S-SEAL

Short, slender plants from a thin rootstock, growing to 5–30 cm high and usually bearing 3 alternate, oblong-lanceolate leaves without stalks, but often sheathing the stem at their base. Flowers few, in a terminal raceme. Berries dark red, up to 6 mm in diam. Wet woods and bogs; Boreal forest.


*Smilax*      carrionflower

Unlike most monocotyledons, this genus has netted-veined leaves. The parallel veins, however, are much more distinct and conspicuous than the others.

*Smilax herbacea* L. var. *lasioneura* (Hook.) DC.          CARRIONFLOWER

Plants climbing by tendrils, up to 1.5 m long. Leaves alternate, long-stalked, oval to ovate, cordate-based, up to 8 cm long and to 6 cm across. Flowers small, greenish, in many-flowered umbels. Fruits purplish berries. These plants derive their common name from the carrion-like odor of the flowers, which attracts flies that aid in fertilization. Fairly common; in shady, moist woodlands; throughout Prairies, Parklands.

***Stenanthium***     stenanthium

*Stenanthium occidentale* A. Gray     BRONZEBELLS

Bulbous-rooted herbs, with grass-like, linear, mostly basal leaves, 15–30 cm long. Flowers greenish purple, drooping, 10–15 mm long, in a raceme on a stem 30–50 cm high. Moist woods and slopes; Rocky Mountains.


***Streptopus***     twistedstalk

Leaves clasping the stem; flowers greenish white. ................................................................................. *S. amplexifolius*

Leaves sessile, not clasping the stem; flowers pink or rose. ................................................................. *S. roseus* var. *perspectus*


*Streptopus amplexifolius* (L.) DC. (Fig. 81*C*)     CLASPING-LEAVED TWISTEDSTALK

Herbs, from rootstock, 30–100 cm high. Leaves ovate to lanceolate, 5–10 cm long and 3–6 cm wide, clasping the stem at their bases. Greenish white flowers, 8–12 mm long, on long twisted stalks, usually in pairs beneath the axils of the leaves. Fruits red berries, 10–15 mm long, on a long twisted stalk. Represented by var. *americanus* Schultes, as described. Moist woods, ravines, and thickets; Parklands, Boreal forest. Locally, var. *denticulatus* Fassett, with the leaf margins finely denticulate, has been found.

*Streptopus roseus* Michx. var. *perspectus* Fassett     ROSE MANDARIN

Herbs with matted rootstocks; stems 30–80 cm high, simple or branched, finely pubescent. Leaves lanceolate, broadly rounded at the base, sessile, 5–9 cm long, 2–4 cm wide. Flowers about 1 cm long, rose or purplish, nodding. Not common; moist woods and thickets; southeastern Parklands, Boreal forest.


***Tofieldia***     asphodel

Stem not hairy; flowers single in a short raceme. ................................................................................................ *T. pusilla*

Stem slightly hairy near top and with sticky glands; flowers in bunches of 3 in a short raceme. ............................................................................................. *T. glutinosa*


*Tofieldia glutinosa* (Michx.) Pers.     STICKY ASPHODEL

Plants with slender rootstocks; stems 10–50 cm high, viscid above; basal leaves 5–20 cm long, 3–8 mm wide. Inflorescence 2–5 cm long; flowers white, with segments about 4 mm long; capsules 5–6 mm long, ovoid. Marshes, bogs, and riverbanks; Boreal forest, Rocky Mountains.

*Tofieldia pusilla* (Michx.) Pers.     BOG ASPHODEL

Plants with short rootstocks; stems 5–20 cm high, slender, smooth; basal leaves 2–6 cm long, 1–3 mm wide. Inflorescence 5–20 mm long; flowers greenish white 1.5–2.5 mm long; capsules 2.5–3 mm long. Marshes, bogs, and riverbanks; Boreal forest, Rocky Mountains.

**_Trillium_**   wakerobin

Flowers on recurved peduncles, to below the
leaves. ................................................................................ _T. cernuum_ var. _marcranthum_
Flowers on erect peduncles, above the leaves. ...................................................... _T. ovatum_

_Trillium cernuum_ L. var. _macranthum_ Eam. & Wieg.   NODDING WAKEROBIN

Plants with short rootstocks; stems 20–40 cm high, slender; leaves rhombic-ovate, 6–10 cm long, 5–12 cm broad; flowers on peduncles 1–4 cm long, reflexed to below the leaves; perianth segment 15–25 mm long; petals white. Rare; moist woods; eastern Boreal forest.

_Trillium ovatum_ Pursh   WESTERN WAKEROBIN

Plants with short, stout rootstocks; stems 30–50 cm high; leaves rhombic-ovate, 6–10 cm long, 5–9 cm wide. Flowers on an erect peduncle 25–75 mm long; perianth segments 25–50 mm long; petals white, turning deep rose. Very rare; moist woods; southern Rocky Mountains.

**_Uvularia_**   bellwort

_Uvularia sessilifolia_ L.   SMALL BELLWORT

Slender, smooth-stemmed herbs, 15–30 cm high, with stalkless, lanceolate-oblong leaves, 3–8 cm long. Flowers borne singly 12–30 mm long, pale greenish yellow. Fruit sharply 3-angled, a capsule about 25 mm long. Very rare; in shady woodlands; southeastern Boreal forest.

**_Veratrum_**   false hellebore

_Veratrum eschscholtzii_ Gray   GREEN FALSE HELLEBORE

Plants with a short, thick rootstock; stems stout, very leafy, somewhat pubescent, 1–2 m high. Leaves oval to elliptic, 10–30 cm long, acute, short-petioled to sessile or clasping, strongly veined, and often plaited. Flowers numerous in a panicle 20–50 cm long; perianth greenish or yellowish green, 15–20 mm across; capsule ovoid, 20–25 mm long, to 12 mm thick, with many large flat seeds. Moist forests and wet places; southern Rocky Mountains.

**_Xerophyllum_**   turkey-beard

_Xerophyllum tenax_ (Pursh) Nutt.   BEAR-GRASS

Plants in dense clumps; stems slender 20–150 cm high, with scattered ascending leaves; basal leaves 50–80 cm long, 3–6 mm wide, rigid, flat, with rough margins. Inflorescence a dense raceme 30 cm or more long; flowers creamy white, with the segments 6–9 mm long; fruit a small capsule 5–7 mm long. Rare; dry mountain slopes; southern Rocky Mountains.

**_Yucca_**   yucca

_Yucca glauca_ Nutt.   YUCCA, SOAPWEED

Coarse plants with short, woody stems and stiff, narrow basal leaves with sharp, hard tips. Flowers cream-colored or greenish white, with petals and sepals 25–50 mm long, in a raceme surmounting a stem 20–60 cm high. Fruits capsules containing numerous black seeds in layers. This plant is fertilized only by the yucca moth, which feeds on it, thereby making the insect and the plant interdependent. Very rare; on dry slopes; Prairies.

***Zygadenus***       camas

Plants with an onion-like bulbous root and long, linear, grass-like leaves, somewhat flattened in cross section. Flowers perfect, with sepals and petals very similar, in racemes or panicles. Fruit an ovoid capsule containing many seeds.

Flowers pale yellow, about 5 mm long. ............................................................ *Z. gramineus*
Flowers greenish yellow or straw-colored,
   about 10 mm long. ............................................................................... *Z. elegans*

*Zygadenus elegans* Pursh                    SMOOTH CAMAS

Grows 30–60 cm high; leaves pale green; flowers grayish white to greenish, in an open raceme. Very common; in moist places, saline meadows; throughout the Prairie Provinces. Syn.: *Anticlea elegans* (Pursh) Rydb.

*Zygadenus gramineus* Rydb. (Fig. 84)              DEATH CAMAS

A low-growing early herb from an onion-like bulb, which is usually 6–15 cm below the soil surface. Leaves linear and grass-like, 10–20 cm high. **Very poisonous** to sheep and somewhat poisonous to cattle. Common; in draws, around grassy sloughs and uplands; throughout the south central and southwestern Prairies. Because of its very early spring growth, it is sought after by livestock, and, wherever dense stands occur, poisoning may result. Syn.: *Toxicoscordion gramineum* Rydb.


# AMARYLLIDACEAE—amaryllis family

***Hypoxis***        star-grass

*Hypoxis hirsuta* (L.) Coville                 YELLOW STAR-GRASS

Plants 10–20 cm high, from a bulbous root. Leaves very narrow, grasslike, slightly hairy. Flowers perfect, 3 or 4 to an umbel, with 3 sepals and 3 petals, greenish outside and yellow inside. Flowers 5–10 mm long; fruit a narrow capsule. Not common; found growing among grass in eastern Parklands.


# IRIDACEAE—iris family

Flowers large, irregular; leaves to 3 cm wide;
   plants with stout rootstocks. .................................................................. *Iris*
Flowers small, regular or nearly so; leaves to 5
   mm wide; plants with fibrous roots. ............................................. *Sisyrinchium*

Fig. 84.   Death camas,
*Zygadenus gramineus*
Rydb.

***Iris***      flag

Flowers yellow. ....................................................................................................... *I. pseudacorus*
Flowers blue violet. ................................................................................................. *I. versicolor*

*Iris pseudacorus* L.                                                        WATER FLAG

Stems 50–60 cm high, with leaves stiff, erect, 1–2 cm broad, equaling or exceeding the stems. Perianth 7–9 cm across, bright yellow to cream. Capsule 5–8 cm long, 6-angled. Introduced, occasionally escaped; southeastern Parklands, Boreal forest.

*Iris versicolor* L.                                                          BLUE FLAG

Stems 20–80 cm high, arising from a thick, creeping rhizome; leaves erect, arching, 1–2 cm broad, shorter than to equaling the stems. Perianth 6–8 cm across, pale to dark blue or violet. Capsule 3–6 cm long, 3-angled. Marshes, swamps, and lakeshores; southeastern Parklands, Boreal forest.

*Sisyrinchium*        blue-eyed grass

Perennial plants from rootstocks, with 2-edged stems. Leaves narrowly linear or grass-like and borne on two sides of the stem. The flowers perfect, with 3 sepals and 3 petals, all alike, blue, and with sharp-pointed tips. Fruit a round capsule.

*Sisyrinchium montanum* Greene        COMMON BLUE-EYED GRASS

Stems 10–30 cm high with narrow-winged edges. Flowers bright blue, with petals 3 mm long. Common, often found in large colonies; in meadows and moist places; throughout Prairies and Parklands. This, the commonest of the species, was formerly called *Sisyrinchium angustifolium* Miller.

## ORCHIDACEAE—orchis family

Perennial herbs, with roots often fleshy. Leaves entire, usually sheathing the stems. Flowers irregular, usually with a large lip. *Cypripedium* has 2 stamens but the other genera have a single stamen, with the pollen grains united into 4–8 masses called pollinia. A difficult, widely distributed family with pretty flowers.

1. Flowers large, to 8 cm long; lip inflated and mocassin-shaped; fertile anthers 2. ........................................................................ *Cypripedium*
   Flowers mostly smaller; lip concave or flat, not mocassin-shaped; fertile anther 1. ............................................................................................ 2

2. Flower solitary; plants with a single basal leaf; scape with sheathing bracts. ................................................ 3
   Flowers 2 to many, in racemes or spikes. ................................................ 4

3. Flower purplish, 15–20 mm long; leaf oval to round-ovate, 2–6 cm long. ............................ *Calypso*
   Flower rose purplish, 20–30 mm long; leaf linear, 10–20 cm long. ................................ *Arethusa*

4. Plants without chlorophyll; leaves reduced to scales. .............................. *Corallorhiza*
   Plants with chlorophyll; leaves normal. ................................................ 5

5. Flowers distinctly spurred. ................................................ 6
   Flowers not spurred. ................................................ 7

6. Leaves 2, oval, basal; lip white, spotted with purple; sepals and petals roseate. ...................... *Orchis*
   Leaves 1 to several; flowers uniformly greenish or white. ............................ *Habenaria*

7. Leaves 2, sessile near the middle of the stem. ................................................ *Listera*
   Leaves 1 to several, basal or scattered. ................................................ 8

8. Plants with a subbasal, solitary leaf. ................................................ 9
   Plants with 2 to several leaves. ................................................ 10

9. Leaf linear to narrowly oblong; flowers
25–50 mm long, pink purple. ................................................... *Calopogon*

Leaf oval to elliptic; flowers small, 2–3
mm long, greenish. ................................................... *Malaxis*

10. Leaves somewhat fleshy, strongly reticu-
late, ovate to obovate, basal. ................................................... *Goodyera*

Leaves not fleshy, not strongly reticulate,
linear or lanceolate. ................................................... 11

11. Leaves 2, strongly keeled, sheathing the
stem at the base; flowers greenish. ................................................... *Liparis*

Leaves several, not strongly keeled, basal
or scattered on the stem; flowers
creamy white. ................................................... *Spiranthes*

### *Arethusa*    arethusa

*Arethusa bulbosa* L.                                 DRAGON'S-MOUTH

Plants with solid bulbs, 10–30 cm high; the leaf 2–4 mm wide, nearly equaling the scape. Flower with a pair of small bracts at the base; sepals and petals lanceolate, magenta; lip pinkish white with purple and yellow spots and streaks. Very rare; bogs and wet meadows; southeastern Boreal forest.

### *Calopogon*    grass-pink

*Calopogon pulchellus* (Salisb.) R. Br.              PURPLE GRASS-PINK

Plants with solid bulbs, 30–90 cm high; the leaf to 40 cm long, grass-like. Inflorescence a loose raceme with 3–15 rose purple flowers, 15–20 mm long. Very rare; bogs and swamps; southeastern Boreal forest.

### *Calypso*    Venus-slipper

*Calypso bulbosa* (L.) Oakes (Fig. 85A)              VENUS-SLIPPER

Growing from bulbous corm, with stem 8–15 cm high. One round-ovate leaf at base and a single pink flower 15–20 mm long with large pink sac having purple lines and inner tuft of yellow hairs. An exceedingly delicate and beautiful flower. In cool woods; Boreal forest, Cypress Hills. Syn.: *Cytherea bulbosa* (L.) House.

### *Corallorhiza*    coralroot

Plants growing on dead and decaying organic matter; therefore, found only in wooded areas. Lacking green color; stems scaly, pinkish yellow; roots coral-like.

1. Flowers small, not longer than 10 mm. ................................................... *C. trifida*
Flowers longer than 10 mm. ................................................... 2

2. Flowers conspicuously striped; lip entire. ................................................... *C. striata*
Flowers more spotted than striped; lip
3-lobed. ................................................... *C. maculata*

A.C. Budd

Fig. 85. Orchids: *A*, Venus-slipper, *Calypso bulbosa* (L.) Oakes; *B*, small yellow lady's-slipper, *Cypripedium calceolus* L. var. *parviflorum* (Salisb.) Fern.; *C*, green-flowered bog orchid, *Habenaria hyperborea* (L.) R. Br.; *D*, northern twayblade, *Listera borealis* Morong; *E*, round-leaved orchid, *Orchis rotundifolia* Banks.

*Corallorhiza maculata* Raf. LARGE CORALROOT

Plant with stem 20–50 cm high, purplish, having scaly leaves. Flowers 12–20 mm long, reddish purple; lip white, spotted with red. Not common; in shady woods; throughout the Prairie Provinces.

*Corallorhiza striata* Lindl. STRIPED CORALROOT

Plant with coarse, stout stem 20–50 cm high. Flowers dark purple, with darker purple stripes. Very rare; in shady woodlands; southeastern Boreal forest, Cypress Hills.

*Corallorhiza trifida* Chat. EARLY CORALROOT

Small species with slender stem 10–30 cm high, and small greenish flowers. Not common; in shaded woods; throughout the Prairie Provinces.

***Cypripedium*** lady's-slipper

Perennial plants with thick fibrous roots and erect stems, bearing two to several large leaves and 1–3 large flowers. All are rare or very rare; the most colorful orchids in the Prairie Provinces.

1. Lower sepals separate and spreading. ...................................................... *C. arietinum*
   Lower sepals united, usually 2-veined or 2-toothed. ........................................................................................................... 2
2. Flowering stem a naked scape, with 2 leaves basal, subopposite. ........................................................ *C. acaule*
   Flowering stem leafy, with 2–6 leaves along the stem. ....................................................................................................... 3
3. Sepals and petals white; 25–40 mm long; lip white with rose purple streaks. ............................................ *C. reginae*
   Sepals greenish, brownish, or yellowish. ...................................................................... 4
4. Sepals yellowish; petals greenish yellow to brownish; lip yellow. ............................................................ *C. calceolus*
   Sepals greenish or brownish; lip not yellow. ....................................................................................................... 5
5. Flower solitary; sepals and petals greenish, crimson-striped. ............................................................ *C. candidum*
   Flowers 1–3 on stem; petals brownish or white. ....................................................................................................... 6
6. Sepals and petals greenish or purplish brown, 40–60 mm long. ........................................................ *C. montanum*
   Sepals green, with the lateral petals white, 15–20 mm long. ........................................................ *C. passerinum*

*Cypripedium acaule* Ait. (Fig. 86) STEMLESS LADY'S-SLIPPER

Scape leafless, 20–40 cm high, with 2 basal leaves, 10–20 cm long, elliptic, thinly pubescent. Flower solitary; sepals and lateral petals yellowish green to greenish brown, 30–50 mm long, lip 30–60 mm long, pink with reddish veins. Rare; dry to moist woods, bogs, and swamps; Boreal forest.

Fig. 86. Stemless lady's-slipper, *Cypripedium acaule* Ait.

*Cypripedium arietinum* R. Br.            RAM'S-HEAD LADY'S-SLIPPER

Stem with 3–5 sessile leaves above and 2 or 3 sheathing scales below, 10–40 cm high; leaves 5–10 cm long, lanceolate, finely ciliate. Flower solitary; sepals and lateral petals greenish brown, 15–25 mm long; lip whitish, veined with red, 10–20 mm long. Very rare; moist coniferous woods; southeastern Boreal forest.

*Cypripedium calceolus* L.            YELLOW LADY'S-SLIPPER

Plants 20–40 cm high, with leafy stems; the leaves more or less sheathing 6–20 cm long, elliptic. Flowers 1–2, with an erect leaf-like bract at the base; sepals and petals greenish yellow to purplish brown, the lip yellow, usually more or less purple-veined. Small yellow lady's-slipper, *C. calceolus* var. *parviflorum* (Salisb.) Fern. (Fig. 85B), has sepals 3–5 cm long, the lip 20–35 mm long; moist woods; Boreal forest. *C. calceolus* var. *pubescens* (Willd.) Correll has sepals 5–8 cm long, and the lip 35–60 mm long. Both varieties are becoming increasingly rare; moist woods; southeastern Boreal forest.

*Cypripedium candidum* Muhl.            SMALL WHITE LADY'S-SLIPPER

A somewhat glandular or sticky-haired plant 10–25 cm high. The ovate to lanceolate leaves 6–12 cm long. The single flower has a white lip with purplish stripes inside. Scarce; but has been found in moist spots; southeastern Boreal forest.

*Cypripedium montanum* Dougl.            MOUNTAIN LADY'S-SLIPPER

Plants 20–50 cm high, more or less glandular pubescent throughout; leaves 4–6, ovate to lanceolate 5–16 cm long. Flowers with brownish green sepals and petals, 4–6 cm long, the lip 2–3 cm long, white, purple-veined. Moist woods; southern Rocky Mountains.

*Cypripedium passerinum* Richardson            NORTHERN LADY'S-SLIPPER

Differs from other lady's-slippers by having rounded sepals and smaller flowers, with the lip 15–20 mm long, white or pale lilac with purple spots. Rare; moist coniferous woods; Cypress Hills, Boreal forest.

*Cypripedium reginae* Walt.            SHOWY LADY'S-SLIPPER

A very showy species, 30–60 cm high; leaves oval to elliptic, 8–20 cm long, covered with soft hairs; flowers white, with a large, inflated sac-like lip 30–40 mm long, with reddish purple stripes. Rare; swampy woodlands; southeastern Boreal forest, Riding Mountain.

**Goodyera**       rattlesnake-plantain

Plants with fleshy rootstocks and a basal clump of dark green mottled leaves and greenish white flowers borne on long stem.

Stem 10–20 cm high; lip of flower decidedly
     inflated; flowers to 5 mm long. ......................................................................... *G. repens*
Stem 30–45 cm high; lip of flower scarcely
     inflated, margins turned inward; flowers to
     10 mm long. ................................................................................. *G. oblongifolia*

*Goodyera oblongifolia* Raf. RATTLESNAKE-PLANTAIN

Plants 30–45 cm high, with ovate-lanceolate to narrowly elliptic leaves to 8 cm long, 20–25 mm wide. Raceme 6–12 cm long; flowers 8–10 mm long, with the lip 6–7 mm long, the body semiglobose. Very rare; dry to moist woods; Boreal forest, Rocky Mountains, Cypress Hills.

*Goodyera repens* (L.) Br. LESSER RATTLESNAKE-PLANTAIN

Plants 10–30 cm high, with ovate to oblong leaves 15–30 mm long, 8–12 mm wide. Raceme 3–6 cm long; flowers 3.5–5 mm long, with the lip 3–3.5 mm long, the body deeply pouch-like. Rare; dry to moist woods; Boreal forest, Rocky Mountains, Cypress Hills, Riding Mountain. The var. *repens* has the leaves reticulate green-veined; plants with the leaves white reticulate distinguished as var. *ophioides* Fern.

### *Habenaria*    bog orchid

1. Leaves basal. ............................................................................................... 2
   Leaves cauline. ............................................................................................. 5
2. Leaf solitary, 5–15 cm long, 1–5 cm wide;
   scape usually bractless. ........................................................... *H. obtusata*
   More than one basal leaf. ............................................................................ 3
3. Leaves lanceolate or oblanceolate, 6–12
   cm long, 1–3 cm wide; spur 5 mm long. ........................................ *H. unalascensis*
   Leaves broadly ovate or rotund, to 15 cm
   long, and often as wide; spur longer
   than 10 mm. ............................................................................................... 4
4. Stem bractless; spur 10–25 mm long. ........................................... *H. hookeri*
   Stem with several small bracts; spur
   25–45 mm long. ...................................................................... *H. orbiculata*
5. Lip 3-toothed, with the central tooth
   short; spur 2–3 mm long. ........................................ *H. viridis* var. *bracteata*
   Lip entire; spur 4–8 mm long. ..................................................................... 6
6. Spur 4–5 mm long, pouch-like, often
   purplish. ................................................................................... *H. saccata*
   Spur 5–8 mm long, cylindrical. .................................................................... 7
7. Flowers white, with the lip abruptly wid-
   ened at base. ............................................................................. *H. dilatata*
   Flowers greenish, with the lip gradually
   widened toward base. ............................................................. *H. hyperborea*

*Habenaria dilatata* (Pursh) Hook. WHITE BOG ORCHID

Plant with slender, leafy stem 30–60 cm high; flowers small, white, about 10–15 mm long, in a spike-like raceme. Rare; in bogs; Boreal forest. Syn.: *Limnorchis dilatata* (Pursh) Rydb.

*Habenaria hookeri* Torr. HOOKER'S BOG ORCHID

Plant with 2 basal leaves, 20–40 cm high; the leaves broadly elliptic to rotund; flowers yellowish green. Very rare; moist woods; southeastern Boreal forest.

*Habenaria hyperborea* (L.) R. Br. (Fig. 85*C*)     GREEN-FLOWERED BOG ORCHID

Fairly stout, leafy-stemmed plant 20–60 cm high; flowers greenish, in spike-like raceme. In moist woodlands, meadows, and stream banks; Boreal forest, Riding Mountain, Cypress Hills. Syn.: *Limnorchis viridiflora* (Cham.) Rydb.

*Habenaria obtusata* (Pursh) Richardson     SMALL NORTHERN BOG ORCHID

Slender plant without stem leaves, 10–25 cm high. Solitary, obovate basal leaf and loose raceme of greenish yellow flowers. Moist woods, marshes, and bogs; Boreal forest. Syn.: *Lysiella obtusata* (Pursh) Rydb.

*Habenaria orbiculata* (Pursh) Torr.     ROUND-LEAVED BOG ORCHID

Plants with 2 basal leaves; scape 30–60 cm high; leaves broadly elliptic to rotund; flowers greenish white, about 2 cm long, with the spike to 20 cm long. Rare; moist woods and bogs; Boreal forest.

*Habenaria saccata* Greene     SLENDER BOG ORCHID

Plants with leafy stems 20–50 cm high; leaves 4–12 cm long, lanceolate or oblanceolate; flowers green, tinged with purple or brown, with the lip 4–7 mm long, the spur 3–5 mm long, pouch-shaped, green, often purplish-tinged. Rare; wet meadows, bogs, and forests; southern Rocky Mountains.

*Habenaria unalascensis* (Spreng.) Wats.     ALASKA BOG ORCHID

Plants with basal leaves; scape 20–60 cm high, with a few scale-like leaves; basal leaves 1–4, narrowly oblanceolate; flowers green, in a spike 10–30 cm long, with the spur 3–5 mm long. Rare; moist or dry woods and meadows; Rocky Mountains, Boreal forest.

*Habenaria viridis* (L.) R. Br. var. *bracteata* (Muhl.) Gray
LONG-BRACTED ORCHID

Stout, leafy-stemmed plant 15–60 cm high, with very conspicuous green bracts, the lower ones at least twice as long as the greenish flowers. In moist meadows and open woods; throughout the Prairie Provinces. Syn.: *H. bracteata* (Muhl.) R. Br.

**Liparis**     twayblade

*Liparis loeselii* (L.) Rich.     TWAYBLADE

Stem strongly ribbed, 5–20 cm high, with 2 lanceolate leaves at the base; flowers few, small, greenish, in a raceme. Rare; in bogs and moist woods; Boreal forest.

**Listera**     twayblade

Low-bog or moist-soil plants with 2 almost opposite leaves about halfway up the stem.

1. Lip cleft to below middle; lobes linear;
   leaves cordate to deltoid. .................................................................... *L. cordata*

Lip shallowly cleft; lobes oblong to ovate;
leaves ovate to elliptic. ..................................................................................... 2

2. Lip nearly oblong, narrowed at the middle, auriculate at base. .................................................................. *L. borealis*

Lip widening toward the tip, not auriculate at base. ............................................................................................ 3

3. Lip 8–10 mm long, minutely toothed on the margins near the base. ......................................... *L. convallarioides*

Lip 4–6 mm long, prominently toothed on the margins near the base. ............................................. *L. caurina*

### *Listera borealis* Morong (Fig. 85*D*)   NORTHERN TWAYBLADE

Plants 5–25 cm high, with the leaves ovate to elliptic, 1–5 cm long. Flowers pale to yellow green, with deep green veins in sepals and petals. Rare; mossy woods, moist meadows and slopes; Rocky Mountains, Cypress Hills.

### *Listera caurina* Piper   WESTERN TWAYBLADE

Plants 10–20 cm high, with the leaves broadly ovate to orbicular, 3–6 cm long; stems glandular puberulent above. Flowers greenish or yellowish. Rare; moist woods; southern Rocky Mountains.

### *Listera convallarioides* (Sw.) Torr.   BROAD-LIPPED TWAYBLADE

Plants 10–20 cm high, with the leaves broadly oval, 3–5 cm long; stems and inflorescence glandular pubescent. Flowers green or yellow green. Rare; meadows, bogs, and moist woods; Boreal forest.

### *Listera cordata* (L.) R. Br.   HEART-LEAVED TWAYBLADE

Plants 10–20 cm high, with the leaves broadly round-ovate, 15–30 mm long, truncate to cordate at the base. Flowers green, tinged with purple. Very rare; moist woods and swamps; Boreal forest.

## *Malaxis*   adder's-mouth

1. Stem leaves 2 or more; lip erect. ................................................................ *M. paludosa*
Stem leaf solitary; lip deflected. .................................................................................... 2

2. Lip not lobed, pointed. ............................................ *M. monophylla* var. *brachypoda*
Lip deeply lobed. ...................................................................................................... *M. unifolia*

### *Malaxis monophylla* (L.) Sw. var. *brachypoda* (Gray) Morris & Eames   ADDER'S-MOUTH

A slender bog plant 5–20 cm high, from an ovoid corm or solid bulb-like base. The single leaf clasping, oval to elliptic, to 8 cm long; greenish yellow flowers in a raceme, very small, about 3 mm long. Rare; Boreal forest.

### *Malaxis paludosa* (L.) Sw.   BOG ADDER'S-MOUTH

Plants 5–15 cm high, with the 2–5 leaves ovate-lanceolate, 1–3 cm long; flowers yellowish green. Rare; bogs; Boreal forest.

*Malaxis unifolia* Michx.    GREEN ADDER'S-MOUTH

Scape 10–30 cm high, with the single leaf sessile, oval or elliptic, 3–6 cm long, 1–3 cm wide; flowers greenish. Rare; damp woods and bogs; southeastern Boreal forest.

**Orchis**    orchid

*Orchis rotundifolia* Banks (Fig. 85*E*)    ROUND-LEAVED ORCHID

An orchid 20–25 cm high, with 1 oval to almost round leaf near the base, and often with 1 or 2 sheathing scales below it. Leaf 3–8 cm long and 2–5 cm wide. Flowers on stem 2–5 cm long, rose, 10–15 mm long, with white lips, spotted with purple. In moist woodlands; especially in Boreal forests and in Cypress Hills.

**Spiranthes**    lady's-tresses

Flowers in 1 spiral. ............................................................................................ *S. gracilis*
Flowers in 2 or 3 spirals. ........................................................................ *S. romanzoffiana*

*Spiranthes gracilis* (Bigel.) Beck    SLENDER LADY'S-TRESSES

An orchid 15–40 cm high. Leaves mostly basal, short-stalked, ovate or elliptic, and usually withering before the flowers appear. Rare; open woods and bogs; Boreal forest.

*Spiranthes romanzoffiana* Cham.    HOODED LADY'S-TRESSES

An orchid 15–40 cm high, with the lower leaves linear to linear-lanceolate; flowers in 3 spirals, delicately fragrant. In swampy places; throughout the Prairie Provinces.


## Class: DICOTYLEDONEAE

Seedings with 2 seedling leaves (cotyledons); stems with a central pith or, if woody, the wood arranged in annual layers; the leaves netted-veined; and the flower parts usually in fours and fives, or multiples of four and five.


## SALICACEAE—willow family

Deciduous (that is, shedding their leaves in fall, not evergreen) trees or shrubs. Leaves alternate on stem. Flowers in aments (catkins) without sepals and petals, these being replaced by glands or a cup-like disk. Male and female flowers produced on separate plants. The numerous seeds bear a tuft of hairs at the apex, which aids in dissemination.

Winter buds covered by several scales; bracts
 below flowers fringed; stamens usually more
 than 10; a cup-shaped disk below each
 flower. ............................................................................................................. *Populus*

Winter buds covered by one scale; bracts below flowers entire, not fringed; stamens 2–10; one or more glands below each flower. ........................................................................ *Salix*

## *Populus*     poplar

Fairly tall trees with either smooth or furrowed bark and light-colored, soft, straight-grained wood. Leaves usually broad and pointed, with petioles sometimes as long as the leaf itself. Flowers unisexual, those of each sex in long catkins on separate trees. Fruit capsule containing small seeds, each bearing a tuft of white hairs.

The genus *Populus* consists of species that appear to hybridize freely, which has resulted in several forms, intermediate between the parents. Because of the considerable variability in the size and form of the leaves in the species, many species have been described that are now considered to be varieties or hybrids.

1. Petioles distinctly flattened. ........................................................................ 2
   Petioles terete. ........................................................................................... 4
2. Leaves coarsely toothed. ........................................ *P. grandidentata*
   Leaves not coarsely toothed. ......................................................... 3
3. Leaves roundish in outline, with the margins finely crenate. ....................................... *P. tremuloides*
   Leaves deltoid in outline, with the margins coarsely dentate. ........................................ *P. deltoides*
4. Leaves lanceolate to linear-lanceolate. ......................... *P. angustifolia*
   Leaves ovate or roundish. ........................................................... 5
5. Leaves roundish, coarsely toothed, white below. ...................................................... *P. alba*
   Leaves ovate, dark green above, light green or glaucous below. ............................... *P. balsamifera*

### *Populus alba* L.                                WHITE POPLAR

Trees up to 30 m high, with bark grayish, furrowed; branches brownish; branchlets coarse, glabrous, grayish brown; leaves 4–6 cm long, 3–4 cm broad, oval to round in outline, with margins coarsely toothed, green above, white tomentose below. Aments appearing before the leaves, on leafless peduncles. Capsules 5–7 mm long, subsessile. Male flowers with 6–10 stamens. Introduced from Europe, and planted in shelterbelts.

### *Populus angustifolia* James            NARROW-LEAVED COTTONWOOD

Trees to 15 m high, with bark greenish; branches greenish brown; branchlets green, coarse, glabrous; leaves 5–12 cm long, lanceolate to long-ovate, acute to acuminate, base cuneate to subcordate, with margins finely crenate, dark green above, pale green below. Aments appearing with the leaves, drooping. Capsules 5–7 mm long, subsessile. Male flowers with 6–10 stamens. Along rivers in Prairies.

*Populus balsamifera* L. BALSAM POPLAR

Large trees, to 25 m high, with bark at first grayish white, becoming dark gray and furrowed in age; branches grayish brown; twigs coarse, light gray; leaves 5–15 cm long, ovate to ovate-lanceolate, acute to acuminate, base obtuse to cuneate or subcordate, with margins minutely crenulate to subentire, dark green and somewhat shiny above, light yellow green to rusty below. Aments appearing before the leaves, on drooping peduncles. Capsules 6–10 mm long. Stamens 12–30. Along rivers and lakes, in coulees and ravines, throughout the Prairie Provinces, but becoming more common in Boreal forest. This species is very variable in leaf shape and size; it includes several forms and varieties, sometimes described as separate species: *Populus candicans* Michx., *P. tacamahaca* Mill., *P. trichocarpa* T. & G., and *P. gileadensis* Rouleau.

*Populus deltoides* Marsh. COTTONWOOD

Large trees 15–25 m high, with bark greenish gray, furrowed; branches grayish brown; twigs greenish gray; leaves 5–10 cm long, broadly deltoid, acuminate, base often somewhat cordate, with margins coarsely serrate-crenate, shiny green above, lighter green below. Aments appearing before the leaves, on drooping peduncles. Capsules 6–10 mm long. Stamens about 60. Along rivers and lakeshores, sand dunes, Prairies. Western plants are classified as var. *occidentalis* Rydb. based on the coarser serration; in eastern specimens the number of teeth on one side of the leaves is about 50–60% higher than that in the western specimens.

*Populus grandidentata* Michx. LARGE-TOOTHED ASPEN

Medium-sized trees with bark greenish gray, becoming brown and furrowed in age; branches gray brown; twigs gray; leaves 6–10 cm long, broadly ovate; with margins usually having 5–10 large rounded teeth on each side; densely white tomentose when young, later thinly pubescent. Aments appearing before the leaves. Capsules 5–7 mm long. Male flowers with 5–12 stamens. Southeastern Boreal forest.

*Populus tremuloides* Michx. ASPEN POPLAR

Slender trees, to 30 m high, with bark grayish white, furrowed in lower part of stem in age; branches brown; twigs yellow green; leaves 3–10 cm long, broadly ovate to orbicular, abruptly pointed, base truncate to subcordate, with margins finely crenate or serrate to subentire, dark green or dark yellow green above, pale green below. Aments appearing before the leaves, on drooping peduncles. Capsules 4–6 mm long. Male flowers with 5–12 stamens. Very common; in depressions and other moist areas; in the Prairies, widespread in the Parklands and Boreal forest.

*Salix*    willow

The willows are very difficult to identify; the flowers or fruits of both sexes, and also in many instances the mature leaves, are needed for positive identification. Even then, several of the diagnostic characters are not constant as a result of hybridization between species, which may occur quite often.

Nevertheless, it is possible to determine typical species with a high degree of accuracy, and come close with most of the more or less intermediate forms. Determinations should be made on fertile material using the female plants. Matching of the male plants is then possible on the basis of the characters common to both sexes, for example, color of bark, leaf characters, and characters of the catkins. In the determination of the female plants, the following characters are used:

position of catkin: sessile or on a leafy branchlet (also for male plants);
capsules: pubescent or glabrous, size, sessile or pedicellate;
length of pedicel: longer or shorter than the bract;
bracts: light or dark in color.

With the use of well-developed material, that is, with mature capsules and, if possible, expanded leaves also, most specimens can be determined accurately based on the combinations of these four characters.

Sterile specimens can usually be determined with a fair degree of accuracy by anyone who has experience in working with willows. New shoots often bear leaves that are not typical in size, and therefore it is best always to use mature leaves in all determinations. In Fig. 87, the main characters of both fruiting and sterile material, used in the keys, are illustrated.

## Key to species of *Salix* by fruiting characters

1. Capsules pubescent. .................................................................................................. 2
   Capsules glabrous. ................................................................................................... 23

2. Bracts of the capsules light in color, yellowish to light brownish or purplish. ........................................................ 3
   Bracts of the capsules dark in color, brown to black. ................................................................................................ 8

3. Plants very small, with creeping stems mostly underground; underside of leaves reticulate-veined. ............................................... *S. reticulata*
   Plants not creeping, with stem at least partly upright. ................................................................................ 4

4. Leaves narrowly linear, the length-to-width ratio 10:1 or more; leaf margins remotely dentate. ........................................................... *S. interior*
   Leaves not narrowly linear, the length-to-width ratio seldom more than 5:1. ............................................. 5

5. Capsules on long pedicels; catkins loose. ........................................................... 6
   Capsules with short pedicels; catkins compact. ................................................................................................. 7

6. Shrubs with stems to 2 m high; leaves oblanceolate or ellipsoid, dull green, mostly somewhat pubescent, rugose-veiny below. ................................................................. *S. bebbiana*
   Shrubs with stems seldom higher than 1 m; leaves narrowly elliptic oblanceolate, green above, glaucous below, not rugose-veiny. ............................................................. *S. pedicellaris*

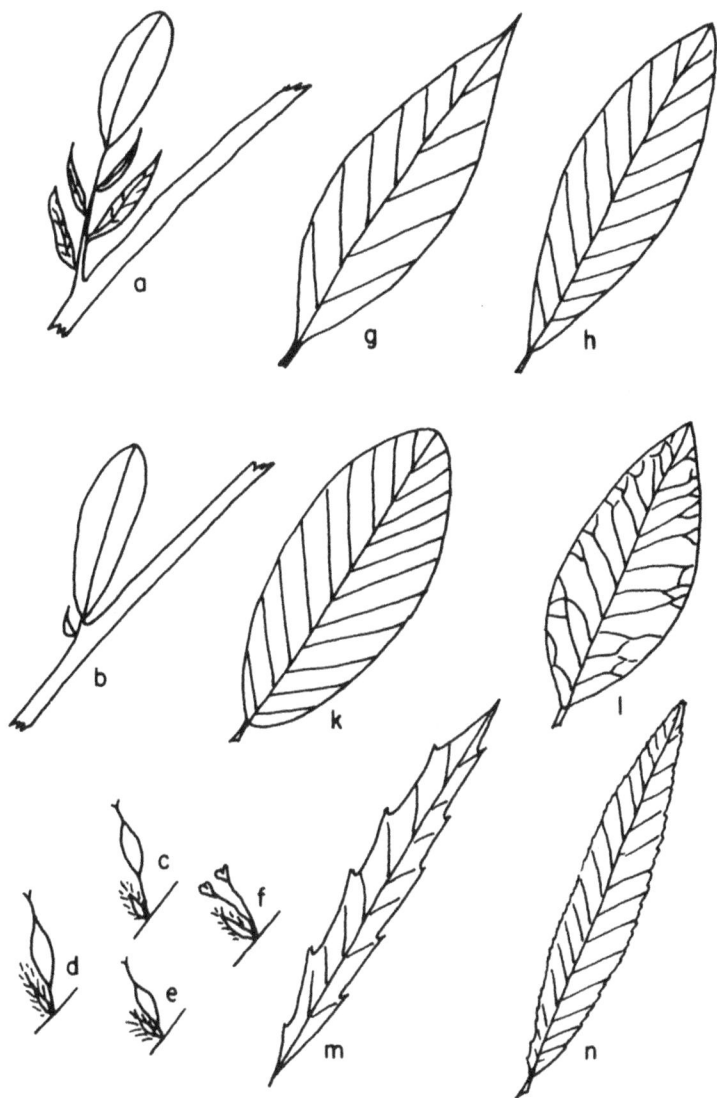

Fig. 87. Main characters of willows. Catkins: *a*, on leafy branchlet; *b*, sessile. Capsules: *c*, pedicel longer than bract; *d*, pedicel as long as bract; *e*, pedicel shorter than bract. Male flower: *f*. Leaves: *g*, acuminate; *h*, acute; *k*, rounded; *l*, irregularly veined; *m*, remotely dentate; *n*, finely serrate.

7. Undersurface of leaves densely silky-pubescent; leaves subsessile; catkins to 2 cm long. ............................................... *S. brachycarpa*

Undersurface of leaves glaucous, pubescent but becoming glabrous; leaves petioled; catkins to 7 cm long. ................... *S. glauca*

8. Leaf margins distinctly toothed. ...................9

Leaf margins entire or indistinctly toothed. .......................................................... 12

9. Leaves deep green and rugose above, silvery silky and strongly veined below. ...................... *S. vestita*

Leaves not as above. .............................................. 10

10. Leaves shiny green above, glaucous and silky pubescent below. ........................... *S. arbusculoides*

Leaves glabrous below. ...................................... 11

11. Leaves glabrous on both sides, with underside light green or glaucous, strongly reticulate-veined. ............................................ *S. maccalliana*

Leaves pubescent above, glabrous below, not noticeably reticulate-veined. ....................... *S. barclayi*

12. Leaves lanceolate, densely white tomentose below, dark green and pubescent above; leaf margins entire, involute. ......................... *S. candida*

Leaves not as above. .............................................. 13

13. Shrubs strongly depressed, with stems partly or entirely underground. ................................ 14

Shrubs not strongly depressed, with stems mostly upright. ................................................ 15

14. Capsules silky-villose, grayish; leaves green above, yellowish below. ........................ *S. arctica*

Capsules thinly pubescent, purplish; leaves shiny green above, glaucous below. ................................................................ *S. arctophila*

15. Catkins on leafy peduncles. ................................ 16

Catkins not on leafy peduncles. .......................... 17

16. Leaves densely pubescent below. ...................... *S. sitchensis*

Leaves glabrous below. ...................................... *S. petiolaris*

17. Leaves pubescent on both sides, more or less cordate at the base. ................................ *S. barrattiana*

Leaves glabrous or nearly so above, not cordate at base. .................................................. 18

18. Young twigs and branches densely tomentose. ...................................................... *S. alaxensis*

Young twigs and branches not densely tomentose. ...................................................... 19

19. Leaves pubescent below; branchlets yellowish to brown. ........................................... 20

Leaves not pubescent below, mostly glaucous; branchlets reddish or yellowish. ................. 21

20. Leaves narrowly lanceolate, the length-to-width ratio about 6:1; capsules 4–6 mm long. ................................................................. *S. pellita*

Leaves oblanceolate or obovate, the length-to-width ratio less than 5:1; capsules 7–9 mm long. .................................... *S. humilis*

21. Leaves lanceolate, the length-to-width ratio about 6:1; capsules 4–6 mm long; twigs yellow. ................................. *S. pellita* var. *psila*

Leaves elliptic or elliptic-oblanceolate, the length-to-width ratio about 3:1; capsules 6–12 mm long; twigs reddish. ................. 22

22. Lateral veins of leaves regular, parallel; capsules 6–7 mm long, subsessile. ................. *S. planifolia*

Lateral veins of leaves irregularly spaced; capsules 7–12 mm long, clearly pedicellate. ................................................. *S. discolor*

23. Bracts of the capsules dark brown to black. ............................................................... 24

Bracts of the capsules yellow to light brown or purplish. ...................................... 31

24. Catkins sessile or subsessile, leafless. ................. 25
Catkins on leafy peduncles. ................................. 26

25. Leaves ovate to orbicular, strongly veined below, entire to somewhat glandular dentate. .......................................................... *S. calcicola*

Leaves elliptic-ovate to narrowly ovate, reticulate-veined and glaucous below, distinctly serrate. ................................ *S. monticola*

26. Leaves pubescent. ............................................. 27
Leaves glabrous. ................................................ 28

27. Leaves green, pubescent on both sides. ................. *S. commutata*

Leaves green, pubescent above, glabrous and glaucous below. ........................... *S. barclayi*

28. Pedicels of capsules much longer than bracts. .............................................................. 29

Pedicels of capsules as long as or shorter than bracts. .......................................................... 30

29. Young twigs dark brown or yellow brown; leaves lanceolate to obovate, rounded to cordate at base. ................. *S. mackenzieana*

Young twigs yellow or reddish, shiny; leaves ovate to lanceolate, fragrant when crushed. ...................................... *S. pyrifolia*

30. Leaves green and reticulate-veined on both sides; stipules usually lacking. .............................. *S. myrtillifolia*

Leaves yellowish green above, glaucous below; stipules usually present. .............................. *S. lutea*

31. Shrubs depressed; stems underground with only leaves and catkins showing. .............................. *S. herbacea*

Shrubs not depressed; stems upright. .............................. 32

32. Leaves narrowly linear, length-to-width ratio about 10:1; leaf margins remotely dentate. .............................. *S. interior*

Leaves not narrowly linear, length-to-width ratio 5:1 or less. .............................. 33

33. Foliage fragrant when young or crushed. .............................. 34

Foliage not fragrant. .............................. 35

34. Leaves firm, dark glossy green above; petioles with glands at leaf base; pedicels of capsules shorter than bracts. .............................. *S. pentandra*

Leaves thin, dull or slightly shiny above; petioles without glands; pedicels of capsules longer than bracts. .............................. *S. pyrifolia*

35. Young twigs fragile, breaking off at the base in strong wind or on touch. .............................. *S. fragilis*

Young twigs not fragile. .............................. 36

36. Leaves pubescent, especially below. .............................. *S. alba*

Leaves glabrous. .............................. 37

37. Petioles with glands at leaf base. .............................. 38

Petioles glandless. .............................. 39

38. Catkins appearing at the same time as the leaves; capsules 4–7 mm long. .............................. *S. lucida*

Catkins appearing later than the leaves; capsules 7–12 mm long. .............................. *S. serissima*

39. Leaves usually large, to 12 cm long, closely and finely serrate. .............................. *S. amygdaloides*

Leaves usually small, to 6 cm long, often with revolute margins. .............................. *S. pedicellaris*

## Key to species of *Salix* by leaf characters

1. Leaves densely pubescent beneath. .............................. 2

Leaves glabrous or only sparsely pubescent beneath. .............................. 15

2. Leaf margins toothed or entire, revolute. .............................. 3

Leaf margins toothed or entire, not revolute. .............................. 6

3. Length-to-width ratio of leaves about 3:1. .................................................................... 4

   Length-to-width ratio of leaves about 6:1. .................................................................... 5

4. Young twigs densely felty tomentose; leaf margins entire or minutely glandular serrulate. .................................................. *S. alaxensis*

   Young twigs not densely tomentose; leaf margins entire or coarsely glandular serrate. ....................................................... *S. bebbiana*

5. Leaf margins entire; pubescence often interspersed with brown hairs. ................................ *S. candida*

   Leaf margins obscurely glandular serrulate; pubescence silvery. ........................................ *S. pellita*

6. Length-to-width ratio of leaves 10:1 or more. ............................................................. *S. interior*

   Length-to-width ratio of leaves usually not more than about 5:1. ....................................... 7

7. Leaves glabrous or glabrate above. ............................................... 8

   Leaves clearly pubescent above. ................................................. 10

8. Leaves obovate, often almost orbicular, length-to-width ratio about 1.5:1, deep green and rugose above. ................................... *S. vestita*

   Leaves elliptic to oblanceolate, length-to-width ratio about 4:1, not rugose above. ......................................................... 9

9. Length-to-width ratio of leaves usually about 3:1. .................................................... *S. discolor*

   Length-to-width ratio of leaves usually about 5:1. .................................................... *S. humilis*

10. Leaf margins finely glandular serrulate. ......................... *S. alba*

    Leaf margins entire or obscurely dentate. .................................. 11

11. Leaves subsessile, the petiole usually less than 1 mm long. ...................................... *S. brachycarpa*

    Leaves with well-developed petioles. ..................................... 12

12. Twigs and young branches grayish or light brown. ................................................... 13

    Twigs and young branches dark brown to black. ..................................................... 14

13. Leaf scars on older branches very prominent; young branches yellowish brown. .......................... *S. barrattiana*

    Leaf scars on older branches not prominent; young branches grayish brown. ......................... *S. glauca*

14. Leaves as pubescent above as below, pubescence tomentose. ...................................... *S. commutata*

    Leaves much less pubescent above than below, pubescence silky. ................................... *S. sitchensis*

15. Length-to-width ratio of leaves 10:1 or
    more. ................................................................ *S. interior*

    Length-to-width ratio of leaves usually
    not more than 6:1. .................................................. 16

16. Leaf margins clearly toothed. ................................. 17
    Leaf margins entire or obscurely toothed. ............... 31

17. Petioles glandular at leaf base. ............................ 18
    Petioles glandless. ................................................ 21

18. Twigs very fragile, readily breaking in
    strong wind or on touch. ....................................... *S. fragilis*

    Twigs not fragile. ................................................. 19

19. Leaves very dark glossy green above,
    lighter and bluish green below, acute to
    short acuminate; introduced small tree. ............... *S. pentandra*

    Leaves not very glossy above. ............................... 20

20. Leaves green below, long acuminate. ................... *S. lucida*
    Leaves glaucous below, short acuminate. ............. *S. serissima*

21. Twigs drooping or arching; leaves long
    acuminate. ........................................................ *S. amygdaloides*

    Twigs not drooping; leaves not long
    acuminate. ........................................................ 22

22. Leaves ovate to orbicular; branchlets
    stout, with conspicuous leaf scars. ..................... *S. calcicola*

    Leaves lanceolate to oblong; branchlets
    more slender, not with conspicuous
    leaf scars. ......................................................... 23

23. Length-to-width ratio of leaves about
    6:1. ................................................................. 24

    Length-to-width ratio of leaves less than
    5:1. ................................................................. 25

24. Leaf margins coarsely serrate; leaves
    glabrous below. ................................................ *S. petiolaris*

    Leaf margins closely and shallowly ser-
    rate; leaves somewhat silky pubescent
    below. .............................................................. *S. arbusculoides*

25. Leaves thin, aromatic when young or
    crushed. ........................................................... *S. pyrifolia*

    Leaves not thin, not aromatic. ............................ 26

26. Plants small shrubs, usually less than 1 m
    high; stems often decumbent. ............................. *S. myrtillifolia*

    Plants usually more than 1 m high; stems
    usually upright. ................................................. 27

27. Leaves pubescent above, especially on the
    veins, glabrous and glaucous below. ................... *S. barclayi*

    Leaves glabrous on both sides. ............................ 28

40. Leaves obscurely veined below. ................................................................ *S. pellita*
    Leaves clearly reticulate-veined below. ................................................. *S. pedicellaris*

41. Leaves yellowish green above, pale green
    or glaucous below; usually some leaves
    serrate or undulate. ......................................................................... *S. lutea*

    Leaves not yellowish green; leaf margins
    entire or obscurely and remotely
    crenate. ......................................................................................................... 42

42. Lateral veins regular, parallel, rather
    closely spaced. ............................................................................. *S. planifolia*
    Lateral veins irregular, not closely spaced. ................................................ 43

43. Stipules usually present, large, clasping
    most of the stem. ...................................................................... *S. monticola*

    Stipules usually absent, when present,
    small, hardly exceeding width of
    petiole base. ................................................................................ *S. discolor*

44. Leaves usually more or less pubescent on
    both sides. ..................................................................................................... 45

    Leaves glabrous above, more or less
    pubescent below, particularly on the
    veins. ............................................................................................. *S. discolor*

45. Leaves rugose below, glaucous. ......................................................... *S. bebbiana*
    Leaves glaucous below, clearly veined but
    not rugose. ...................................................................................... *S. glauca*

*Salix alaxensis* (Anderss.) Cov.                                    ALASKA WILLOW

A shrub to 4 m high; bark dark to chestnut brown, more or less persist-ently gray villous tomentose; branchlets densely white or yellow tomentose. Leaves 5–11 cm long, 1.5–3.5 cm broad, narrowly ovate to oblong, acute, with base cuneate; margins revolute, entire; bright green above, densely white pubescent below. Catkins appearing before the leaves; sessile. Capsules 4–5 mm long, densely puberulent; scales black; pedicels shorter than the scales. Male flowers with 2 stamens. Along streams and lakeshores; in the Rocky Mountains south to Jasper National Park.

*Salix alba* L.                                                      WHITE WILLOW

A tree up to 20 m high; bark grayish; branches ascending, appearing sil-very gray. Leaves 5–10 cm long, 1–1.5 cm broad, lanceolate, acuminate, with base cuneate; margins finely serrate; appressed white silky pubescent on both sides. Catkins appearing at the same time as the leaves; terminating short leafy peduncles. Capsules 3–5 mm long, glabrous; scales yellowish green; pedicels shorter than the scales. Male flowers with 2 stamens. A Eurasian species, often planted in shelterbelts.

*Salix amygdaloides* Anderss.                                PEACH-LEAVED WILLOW

A shrub or tree 3–15 m high; bark rough, dark reddish brown; branches gray brown; twigs yellowish to reddish brown, drooping. Leaves 5–12 cm long,

1.5–3 cm broad, ovate-lanceolate, acuminate, tapering to base; margins closely serrate; light to yellowish green above, glaucous below, glabrous on both sides. Catkins appearing at the same time as the leaves; terminating short leafy peduncles. Capsules 4–7 mm long, glabrous; scales yellow, deciduous; pedicels 2–3 mm long. Male flowers with 4–7 stamens. Not common; along rivers and streams; Prairies.

*Salix arbusculoides* Anderss.                                     SHRUBBY WILLOW

Shrubs or occasionally small trees 1–6 m high; bark reddish brown; branches reddish brown; bark peeling; branchlets velvety. Leaves 5–8 cm long, 1–2 cm broad, narrowly ovate to elliptic, acute, with base acute; margins glandular serrulate; glossy and glabrous above, sparsely pubescent to almost glabrous below. Catkins appearing shortly before or at the same time as the leaves; subsessile, with bract-like leaves at base. Capsules 4–5 mm long, sparsely pubescent; scales dark brown; pedicels shorter than the scales. Male flowers with 2 stamens. Along rivers and in muskeg areas; Boreal forest.

*Salix arctica* Pall.                                             ARCTIC WILLOW

Dwarf shrubs, usually prostrate, occasionally to 50 cm high; branches stout to long, rooting at some nodes, yellowish to chestnut brown; branchlets yellowish green to brown, often villous when young. Leaves 2–7 cm long, 1–2.5 cm broad, oblanceolate to elliptic, acute to obtuse, with base acute; margins entire; mostly glabrous above, glabrous or very sparsely pubescent, somewhat reticulately veined below. Catkins appearing at the same time as the leaves; terminating leafy peduncles. Capsules 6–9 mm long, sparsely to densely pubescent; scales black; pedicels shorter than the scales. Male flowers with 2 stamens. Alpine tundra and alpine meadows; Rocky Mountains.

*Salix arctophila* Cock.                                         TRAILING WILLOW

Dwarf shrubs, usually trailing along the ground; branches chestnut brown to greenish gray brown; branchlets yellowish green. Leaves 2–4 cm long, 1–2 cm broad, obovate to broadly elliptic, obtuse to acute, with base acute; margins entire to shallowly crenulate, glandular; glabrous, often glossy green above, very sparsely pubescent or glabrous, glaucous below. Catkins appearing at the same time as the leaves; terminating leafy peduncles. Capsules 5–6 mm long, thinly pubescent to subglabrous; scales black; pedicels shorter than the scales. Male flowers with 2 stamens. Tundra; Boreal forest.

*Salix barclayi* Anderss.                                        BARCLAY'S WILLOW

Shrubs 1–3 m high; branches dark brown; branchlets yellowish green at first villous, later subglabrous. Leaves 3–8 cm long, 1.5–4 cm broad, acuminate to acute, with base acute to subcordate; margins glandular serrulate to subentire; the upper side pubescent along the midrib, glabrous and glaucous below. Catkins appearing at the same time as the leaves; terminating leafy peduncles. Capsules 5–7 mm long, glabrous; scales dark brown or brown only in the upper half; pedicels shorter than to as long as the scales. Male flowers with 2 stamens. Along creeks, rivers, and lakeshores; Rocky Mountains.

*Salix barrattiana* Hook.                                    BARRATT'S WILLOW

Low shrub to 1 m high; branches reddish brown, usually pubescent with conspicuous leaf scars; branchlets short, with short internodes, villous. Leaves 4–8 cm long, 1–2 cm broad, elliptic, acute, with base acute; margins entire to obscurely serrulate; sparsely pubescent above, densely white pubescent below. Catkins appearing before the leaves; sessile. Capsules 4–6 mm long, densely pubescent; scales black; pedicels shorter than scales. Male flowers with 2 stamens. Alpine meadows and slopes; Rocky Mountains.

*Salix bebbiana* Sarg.                                        BEAKED WILLOW

Shrubs to small trees 1–10 m high; branches reddish brown; branchlets light brown, at first densely pubescent. Leaves 2–7 cm long, 1–2.5 cm broad, acute to obtuse, with base acute to obtuse; margins entire to crenate, somewhat revolute; thinly pubescent to subglabrous above; glaucous, rugose, subglabrous to pubescent below. Catkins appearing shortly before or at the same time as the leaves; terminating short leafy peduncles. Capsules 5–9 mm long, pubescent to subglabrous; scales yellow, in part deciduous; pedicels as long as or longer than the scales. Male flowers with 2 stamens. One of the most common willows; around sloughs, along rivers and lakeshores, and in woods; throughout the Prairie Provinces.

*Salix brachycarpa* Nutt. (Fig. 88)        SHORT-CAPSULED WILLOW

Small shrubs, usually 0.5–1.0 m high, occasionally to 3 m high; branches stout, grayish to reddish brown, the outer layers of bark peeling; branchlets pubescent. Leaves 1–3 cm long, 0.5–1.5 cm broad, obovate to elliptic, obtuse to acute, with base acute; petioles very short. Catkins appearing at the same time as the leaves; terminating short leafy peduncles; seldom more than 2 cm long. Capsules 3–5 mm long, pubescent; bracts yellowish; pedicels shorter than the bracts. Male flowers with 2 stamens. Sand dunes, gravel bars, and marsh areas; in various parts of the Prairie Provinces, and in the Rocky Mountains.

*Salix calcicola* Fern. & Wieg.                               LIME WILLOW

Shrubs 0.5–2 m high; branches pubescent, becoming glabrate with age; branchlets stout with inconspicuous leaf scars, pubescent. Leaves 2–5 cm long, 1.5–3.5 cm broad, ovate to almost orbicular, with base rounded to cordate; margins subentire to glandular denticulate, acute; rugose green above, glaucous below. Catkins appearing before the leaves; sessile. Capsules 7–9 mm long, glabrous; scales dark brown to black; pedicels shorter than the scales. Male flowers with 2 stamens. Tundra; Hudson Bay; rare in Rocky Mountains.

*Salix candida* Fluegge                                       HOARY WILLOW

Shrubs to 3 m high; branches dark brown; branchlets densely tomentose to sparsely pubescent. Leaves 5–8 cm long, 0.7–2 cm broad, narrowly elliptic; margins entire to subentire, revolute; dull green, somewhat tomentose, with sunken veins above; densely white tomentose below; acute; base acute. Catkins appearing at the same time as the leaves; terminating short leafy peduncles. Capsules 6–8 mm long, white tomentose; scales brown; pedicels shorter than the scales. Male flowers with 2 stamens. In wet, usually somewhat saline, swampy areas; throughout the Prairie Provinces.

A.C. Budd

Fig. 88. Short-capsuled willow, *Salix brachycarpa* Nutt.

*Salix commutata* Bebb                                      CHANGEABLE WILLOW

Low shrub 0.2–1.2 m high; branches dark brown, usually pubescent; branchlets densely white tomentose. Leaves 2.5–5.5 cm long, 1.5–3.5 cm broad, elliptic to ovate-elliptic; margin subentire or entire to serrulate; pubescent on both sides. Catkins appearing after the leaves; terminating leafy peduncles. Capsules 4–6 mm long, glabrous; scales brown to dark brown, or brown only in upper half; pedicel shorter than to as long as the scales. Male flowers with 2 stamens. Gravel beds along rivers and lakes; in Rocky Mountains.

*Salix discolor* Muhl.                            PUSSY WILLOW, DIAMOND WILLOW

Large shrubs or small trees up to 6–7 m high; bark grayish brown; branches reddish brown; branchlets rather stout yellowish to reddish brown, at first pubescent, later glabrous. Leaves 5–8 cm long, 2–3.5 cm broad, acute to short acuminate, with base acute to obtuse; margin entire to undulate-crenate; dark green, often somewhat glossy above, glaucous, with strongly raised veins below; somewhat pubescent to glabrous on both sides; veins branching before reaching margin. Catkins appearing long before the leaves; subsessile, often with 2 or 3 small bracts. Capsules 6–10 mm long, densely puberulent; scales black; pedicels shorter than to as long as the scales. Male flowers with 2 stamens. One of the most common and earliest flowering willows; around sloughs and lakes, along riverbanks, and in woods and swamps; throughout the Prairie Provinces.

*Salix fragilis* L.                                              BRITTLE WILLOW

Tree up to 20 m high; bark rough and fissured on old trees; branches grayish brown; branchlets yellow to reddish, easily breaking at base. Leaves 7–15 cm long, 2–4 cm broad, lanceolate; margin glandular serrate; dark green above, glaucous below. Catkins appearing at the same time as the leaves; terminating leafy peduncle. Capsules 4–6 mm long, glabrous; scales yellow, deciduous; pedicels shorter than the scales. Male flowers with 2 stamens. Introduced from Eurasia, and planted in shelterbelts.

*Salix glauca* L.                                               SMOOTH WILLOW

Shrubs to about 1 m or occasionally 2 m high; branches dull grayish; branchlets grayish tomentose. Leaves 2.5–10 cm long, 1–3.5 cm broad, obovate to elliptic; margins entire; dark green, pubescent to glabrous above, glaucous and usually villous below; acute to acuminate or obtuse, with base acute to obtuse. Catkins appearing at the same time as the leaves; terminating leafy peduncles. Capsules 3–5 mm long, pubescent; scales yellow to light brown; pedicels shorter than the scales. Male flowers with 2 stamens. Along rivers, in muskeg; Rocky Mountains, Boreal forest.

*Salix herbacea* L.

Small, creeping shrubs; stems and branches buried, glabrous; branchlets chestnut brown, glabrous. Leaves 1–3 cm long, 1–2.5 cm broad, oval to suborbicular; margins serrate or crenate-serrate; glabrous on both sides. Catkins appearing later than the leaves; terminating leafy peduncles. Capsules 3–6 mm long, glabrous; scales brown to dark brown; pedicels shorter than the scales. Male flowers with 2 stamens. Tundra; Hudson Bay.

*Salix humilis* Marsh. GRAY WILLOW

Shrubs 1–3 m high; branches grayish brown; branchlets pubescent to sub-glabrous, yellowish. Leaves 3–12 cm long, 1–3 cm broad, oblanceolate to ob-ovate; margins entire to sparsely dentate; acute to short acuminate; base acute; dark green above, glaucous; somewhat rugose and pubescent below. Catkins appearing before the leaves; sessile to subsessile. Capsules 7–9 mm long, pubescent; scales black; pedicels shorter than to as long as the scales. Male flowers with 2 stamens. Openings around sloughs and meadows; Boreal forest.

*Salix interior* Rowlee SANDBAR WILLOW

Shrubs 0.5–4 m high, with extensively creeping root systems; stems and branches grayish brown; branchlets reddish, at first sparsely pubescent, becoming glabrous. Leaves very variable, 3–15 cm long, 0.3–1 cm broad, narrowly oblong to linear; margins distantly denticulate to subentire, with teeth 0.5–1 cm apart; green and glabrous above, pale green, pubescent to subglabrous below. Catkins appearing at the same time as the leaves; terminating leafy peduncles. Capsules 5–8 mm long, glabrous to somewhat pubescent; scales yellowish, deciduous; pedicels shorter than the scales. Male flowers with 2 stamens. This species belongs to the group Longifoliae, which has several highly variable species. *S. interior* is also highly variable, and has been subdivided into several varieties and species on the basis of leaf length and width, pubescence, and other characters. The form with very narrow leaves, 3–4 mm broad, is var. *pedicellata* (Anderss.) Ball; the form with small, permanently pubescent leaves is var. *wheeleri* Rowlee (*S. exigua* Nutt.). The species is also considered to be *S. fluviatalis* Nutt. with the two previously mentioned varieties then named var. *sericans* (Nees) Boiv. with the forma *hindsiana* (Berth.) Boiv.

The "typical" plants occur around sloughs and lakes, along creeks, rivers, and canals throughout the Prairie Provinces; var. *pedicellata* and var. *wheeleri* appear to be more restricted to drier locations and sand dune areas.

*Salix lucida* Muhl. SHINING WILLOW

Shrubs or small trees 1–6 m high; bark brown; branches brown, sometimes sparsely pubescent; branchlets reddish brown, sparsely pubescent. Leaves 5–15 cm long, 1.5–4 cm broad, usually ovate, long acuminate; base acute to obtuse or cordate, with several glands at apex of petiole; margin glandular serrate; dark green and often glossy above, pale green to glaucous below. Catkins appearing at the same time as the leaves; terminating leafy peduncles. Capsules 5–7 mm long, glabrous; scales yellow, deciduous; pedicels shorter than the scales. Western specimens usually have the leaves glaucous below (*S. lasiandra* Berth.), eastern ones have the leaves light green. Along rivers and in meadows; throughout the Prairie Provinces.

*Salix lutea* Nutt. YELLOW WILLOW

Shrubs 2–5 m high; branches grayish brown; branchlets reddish brown. Leaves 4–10 cm long, 1.5–4 cm broad, lanceolate, acute to short acuminate, with base obtuse to somewhat cordate; margins serrulate to entire; yellowish green above, glaucous below. Catkins appearing at the same time as or shortly before the leaves; subsessile on a short leafy peduncle. Capsules 4–5 mm long, glabrous; scales brown; pedicels as long as or longer than the scales. Male flowers with 2 stamens. Along streams and lakeshores; throughout the Prairie Provinces.

*Salix maccalliana* Rowlee                               VELVET-FRUITED WILLOW

Shrubs 1–3 m high; branches reddish to dark brown; branchlets reddish brown. Leaves 4.5–7 cm long, 0.6–2 cm broad, elliptic to oblong; margins entire to glandular serrate or crenulate; acute to short acuminate; base acute to obtuse; dark green and glossy above, pale green, somewhat reticulate-veined below. Catkins appearing at the same time as the leaves; terminating leafy peduncles. Capsules 7–10 mm long, pubescent; scales brown, occasionally only in the upper half; pedicels shorter than to as long as the scales. Male flowers with 2 stamens. Lakeshores and muskeg; throughout the Prairie Provinces.

*Salix mackenzieana* (Hook.) Barratt                      MACKENZIE WILLOW

Shrubs or small trees 2–5 m high; branches grayish brown; branchlets chestnut brown. Leaves 6–10 cm long, 2–3.5 cm broad, ovate-lanceolate; margins glandular serrulate; acute to long acuminate; base obtuse to cordate; somewhat glossy green above, glaucous and reticulate below. Catkins appearing shortly before the leaves; terminating leafy peduncles. Capsules 4–6 mm long, glabrous; scales brown; pedicel longer than the scales. Male flowers with 2 stamens. Along creeks and rivers, on lakeshores; Boreal forest.

*Salix monticola* Bebb                                     MOUNTAIN WILLOW

Shrubs 1–3 m high; branches dark reddish brown; branchlets yellowish. Leaves 4–6 cm long, 1.5–3 cm broad, elliptic to ovate-elliptic; margins glandular serrulate to crenate; acute to short acuminate; base obtuse to subcordate; green above, glaucous below. Catkins appearing before the leaves; sessile to subsessile; sometimes with one or two small bracts at base. Capsules 6–8 mm long, glabrous; scales dark brown; pedicels smaller than to as long as the scales. Male flowers with 2 stamens. Lakeshores, muskeg, and marshy areas; throughout the Prairie Provinces.

*Salix myrtillifolia* Anderss.                           MYRTLE-LEAVED WILLOW

Low shrubs usually less than 1 m high; branches grayish brown; branchlets greenish to reddish brown. Leaves 2–6 cm long, 0.8–2.5 cm broad, elliptic to obovate; margins fine glandular serrulate to crenate; green, often glossy above, pale green, often somewhat reticulate below. Catkins appearing at the same time as the leaves; terminating leafy peduncles. Capsules 4.5–7 mm long, glabrous; scales black or black-tipped only; pedicels shorter than to as long as the scales. Male flowers with 2 stamens. In muskeg, along wet lakeshores, and riverbanks; throughout Boreal forest.

*Salix pedicellaris* Pursh                                    BOG WILLOW

Low shrubs 0.2–1.5 m high; branches grayish brown; branchlets reddish yellow to brownish. Leaves 2–6 cm long, 5–15 mm broad, elliptic to narrowly obovate; margins entire often revolute; acute to obtuse; base obtuse to acute; dull green and somewhat glaucous above, glaucous below. Catkins appearing at the same time as the leaves; terminating leafy peduncles. Capsules 5–6.5 mm long, glabrous; scales light brown; pedicels longer than scales. Male flowers with 2 stamens. In var. *athabascensis* (Raup) Boiv. the capsules and leaves are somewhat pubescent. Muskeg and marshy areas; in Boreal forest.

*Salix pellita* Anderss.                                                    SATIN WILLOW

Shrubs or small trees 3–5 m high; branches brown; branchlets yellowish or greenish brown, brittle. Leaves 4–12 cm long, 0.8–2.0 cm broad, linear-lanceolate; acuminate; base acute to obtuse; margins entire to obscurely serrate, revolute; glabrous above, densely silky pubescent below. Catkins appearing shortly before or at the same time as the leaves; subsessile, often with 2 or 3 bracts at base. Capsules 4–6 mm long, pubescent; scales black; pedicels shorter than the scales. Male flowers with 2 stamens. Lakeshores, riverbanks, swamps, and muskeg; in Boreal forest. In forma *psila* Schn., the leaves are glabrescent and glaucous below. Local in the area where the species occurs. In Alberta specimens occur in which the pubescence is shorter and more uniform; these have been named var. *angustifolia* (Bebb) Boiv., but are considered to be a distinct species, *S. drummondiana* Barr., by some taxonomists.

*Salix pentandra* L.                                                  BAY-LEAVED WILLOW

Usually small trees up to 7 m high; branches brown; branchlets reddish brown, shiny. Leaves 4–12 cm long, 2–4 cm broad, ovate to elliptic; long acuminate; base obtuse; dark green and usually very glossy above, pale green below; margins glandular serrulate. Catkins appearing at the same time as the leaves; terminating leafy peduncles. Capsules 5–7 mm long, glabrous; scales yellow, deciduous; pedicels shorter than the scales. Male flowers with 5 stamens. A Eurasian species, introduced and planted as an ornamental.

*Salix petiolaris* Sm.                                                   BASKET WILLOW

Shrubs or small trees 2–7 m high; branches reddish brown; branchlets at first yellowish, later reddish. Leaves 4–12 cm long, 0.8–2.5 cm broad, narrowly lanceolate; acute to acuminate; base acute; margins subentire to rather closely glandular serrulate; dark green above, glaucous below. Catkins appearing at the same time as the leaves; terminating leafy peduncles. Capsules 6–8 mm long, pubescent; scales brown; pedicels as long as or longer than the scales. Male flowers with 2 stamens. Meadows, lakeshores, and along streams; throughout the Prairie Provinces.

*Salix planifolia* Pursh                                            FLAT-LEAVED WILLOW

Shrubs 1–4 m high; branches dark grayish to reddish brown; branchlets greenish brown, often at first pubescent, later glabrous. Leaves 3–6 cm long, 1–2 cm broad, elliptic to lanceolate; acute; base acute; margins subentire to glandular serrulate; glossy green with somewhat sunken veins above; glaucous below. Catkins appearing before the leaves; sessile, sometimes with 2 or 3 small bracts at the base. Capsules 5–6 mm long, pubescent; bracts dark brown or black; pedicels shorter than the bracts. Male flowers with 2 stamens. One of the most common species; around sloughs and lakes, along streams, in swampy areas; throughout the Prairie Provinces. This species is considered to be *S. phylicifolia* L. or *S. phylicifolia* ssp. *planifolia* (Pursh) Hiitonen by some authors. *S. phylicifolia,* however, has catkins terminating leafy peduncles.

*Salix pyrifolia* Anderss.                                            BALSAM WILLOW

Shrubs usually to 4 m high; branches brown; branchlets greenish yellow when young, becoming reddish brown. Leaves 4–12 cm long, 2–4 cm broad; ovate to lanceolate-oblong; margins glandular serrulate to crenate; acute to acuminate, with the apex often asymmetric; base obtuse to cordate; dull green above, glaucous and finely reticulate below; leaves thin and somewhat translucent; often purplish when young, fragrant, especially when crushed. Catkins appearing at the same time as the leaves; terminating leafy peduncles. Capsules 5–9 mm long, glabrous; scales light brown or purplish; pedicels longer than the scales. Male flowers with 2 stamens. Sloughs and lakeshores, muskeg, swamps, and riverbanks; Boreal forest.

*Salix reticulata* L.                                               SNOW WILLOW

Dwarf shrubs, with stem buried; branches short, light brown; branchlets greenish yellow to light brown. Leaves 1–6 cm long, 0.8–5 cm broad; orbicular to suborbicular; obtuse to somewhat retuse; base obtuse to cordate; margins subentire to somewhat crenate, revolute; glossy dark green above, with the veins sunken; pale green, conspicuously reticulately veined, thinly pubescent to glabrous below. Catkins appearing at the same time as the leaves; terminating leafy peduncles. Capsules 4–5 mm long, pubescent; scales yellowish to reddish; pedicels shorter than the scales. Male flowers with 2 stamens. Tundra; in the Boreal forest of northern Manitoba and Saskatchewan. A somewhat smaller plant with only the branchlets clearly above ground occurs in alpine tundra in the Rocky Mountains; this plant is considered as var. *nivalis* (Hook.) Anderss. or *S. nivalis* Hook. by some authors.

*Salix serissima* (Bailey) Fern.                                   AUTUMN WILLOW

Shrubs to 4 or 5 m high; stems grayish brown; branches brown to olive brown; branchlets yellowish brown, shiny. Leaves 4–8 cm long, 1–3 cm broad, lanceolate to elliptic-lanceolate; acute to short acuminate; base acute; margins fine glandular serrulate; green and often glossy above, lighter green and subglaucous below. Catkins appearing after the leaves, maturing in late summer; terminating leafy peduncles. Capsules 7–12 mm long, glabrous; scales yellow, deciduous; pedicels shorter than to as long as the scales. Male flowers with 5 stamens. Swampy areas; Boreal forest.

*Salix sitchensis* Sanson                                           SITKA WILLOW

Shrubs 0.5–3 m high; stems grayish; branches dark to grayish brown; branchlets densely pubescent when young, later glabrescent. Leaves 3–10 cm long, 1.5–5 cm broad, narrowly elliptic to obovate; obtuse, often with an asymmetric acuminate tip; base acute; margins subentire, distantly glandular; dull green, sparsely pubescent above, densely silky pubescent below. Catkins appearing at the same time as the leaves; terminating leafy peduncles. Capsules 3–5.5 mm long, pubescent; scales dark brown, or with a dark brown apex only; pedicels shorter than the scales. Male flowers with 1 stamen. Banks of streams and lakeshores; Rocky Mountains.

*Salix vestita* Pursh                                        ROCK WILLOW

   Depressed shrubs 0.25–0.75 m high. Branches brown with conspicuous
leaf scars; branchlets brown, pubescent. Leaves 2–6 cm long, 1.5–5 cm broad,
oval to broadly oblong or suborbicular; obtuse to retuse; base obtuse to subcordate; margins subentire, often somewhat revolute; deep green, rugose
above, densely long pubescent and glaucous below. Catkins appearing after
the leaves; subsessile or on naked peduncles. Capsules 4–5 mm long, pubescent; scales dark brown; pedicels shorter than the scales. Wet, rocky, usually
shaded habitats on lakeshores in Boreal forest, and subalpine locations in the
Rocky Mountains.


## MYRICACEAE—bayberry family

*Myrica*          sweet gale

*Myrica gale* L.                                 SWEET GALE, BOG-MYRTLE

   A shrub to 100–120 cm high, with wedge-shaped leaves up to 5 cm long
and 1–2 cm wide, which have a pleasing odor when bruised. Flowers in catkins
before leaves appear, unisexual, the male and female inflorescence usually on
separate plants. Plants bearing flowers of one sex one year have been known to
produce flowers of the opposite sex the next year. Fruit in the form of a small
nutlet, coated with a resinous wax and having two wing-like scales, borne in
small heads. Found along stream banks and in swamps on acid soil; Boreal
forest.


## BETULACEAE—birch family

   Trees or shrubs with unisexual flowers, both sexes borne on the same
plant. Flowers in catkins, fruit a nutlet borne in a cone-like head or a nut
enclosed in bracts.

1. Nuts wingless, enclosed in a leafy
     involucre. ....................................................................................... 2
   Nuts mostly winged, not enclosed in a
     leafy involucre. .............................................................................. 3
2. Pistillate flowers 2–4; inflorescence capi-
     tate; involucre not inflated; nut large,
     acorn-like. ......................................................................... *Corylus*
   Pistillate flowers several; inflorescence an
     ament; involucre inflated; nut small,
     achene-like. ........................................................................ *Ostrya*
3. Fruiting bracts woody, persistent; nutlets
     narrow-winged or with leathery mar-
     gins; stamens 4. ................................................................ *Alnus*
   Fruiting bracts thin, deciduous; nutlets
     mostly broadly winged; stamens 2. ................................... *Betula*

***Alnus***   alder

Fruit thin-margined; flowering before the
leaves. ......................................................................................................... *A. incana*
Fruit distinctly winged; flowering after the
leaves. .................................................................................. *A. viridis* var. *sinuata*

## *Alnus incana* (L.) Moench   SPECKLED ALDER

A tall shrub or small tree with elliptic leaves, dark green above and paler below, usually with some hairiness on the veins on the underside. Flowers developing before leaves appear; fruit not winged. Common; along streams, river valleys, and other habitats; Boreal forest, less common in Parkland. Syn.: *Alnus rugosa* (Du Roi) Spreng. var. *americana* (Regel) Fern.; *A. tenuifolia* Nutt.

## *Alnus viridis* (Chaix) DC. var. *sinuata* Regel   GREEN ALDER

A shrub 2–3 m high with oval leaves, which are often sticky on the underside. Fruit with thin wings on both sides. Fairly abundant; in dry sandy, coniferous woods; Boreal forest. Syn.: *Alnus crispa* (Ait.) Pursh.

***Betula***   birch

1. Trees with white, papery bark. ............................................... *B. papyrifera*
   Shrubs or small trees with reddish or
   brown bark. ................................................................................................... 2
2. Wing of seed broader than body of seed;
   tall shrub or small tree; leaves sharply
   double-toothed. ........................................................................ *B. occidentalis*
   Wing of seed narrower than body of seed;
   low bushes or shrubs. ..................................................................................... 3
3. Young twigs and branches with scattered
   hairs, slightly glandular; leaves tapering at base; fruit with wing half as wide
   as body. ...................................................................................... *B. glandulifera*
   Twigs and branches not hairy, densely
   glandular; leaves rounded at base; fruit
   with very narrow wing. ............................................... *B. nana* var. *sibirica*

## *Betula glandulifera* (Regel) Butler   SWAMP BIRCH

A shrub, with leaves 20–30 mm long, dark green above and yellowish or reddish green beneath. Twigs and undersides of young leaves with very fine hairs. Seed with a distinct wing half the width of the seed. Around swamps and bogs; in Boreal forest and Rocky Mountains. Sometimes considered *B. pumila* L. var. *glandulifera* Regel.

## *Betula nana* L. var. *sibirica* Led.   SCRUB BIRCH

A shrub 30–150 cm high, with glandular twigs and small rounded leaves 5–20 mm long. Twigs and leaves without any hairiness, but with resinous glands. Seed with a very narrow wing, sometimes almost lacking. Fairly common; in marshes, sloughs, and boggy places; Boreal forest, Rocky Mountains. An arctic-alpine species. Syn.: *Betula glandulosa* Michx.

*Betula occidentalis* Hook.                                          RIVER BIRCH

A small tree or large shrub up to 10 m high, growing in many-stemmed clumps, with ovate or almost round leaves 2–5 cm long. Inflorescence a cylindric catkin; fruit borne in cylindric, cone-like heads; seeds with broad wings. Common in low places, stream banks, depressions in sandhills; throughout the Prairie Provinces. Syn.: *B. fontinalis* Sarg.

*Betula papyrifera* Marsh.                          WHITE BIRCH, CANOE BIRCH

A tree with white bark, up to 15 m high in favorable locations. Leaves ovate to rhomboid, serrate, dark green above. Along rivers, in openings, and somewhat moister sites; commonly in cutover areas. Three forms can be distinguished:

1. Leaves glabrous; twigs densely glandular. ........................................................................ var. *neoalaskana* (Sarg.) Raup
   Leaves pubescent along the nerves below; twigs not densely glandular. ............................................................................................... 2
2. Pubescence consisting of tufts of hairs in the axils of the veins; leaves rounded to truncate at the base. ......................................................... var. *papyrifera*
   Pubescence more or less velvety below, often also pubescent above; leaves cordate at the base. ........................................... var. *cordifolia* (Regel) Fern.


**Corylus**        hazelnut

Shrubs or small trees with much-branched stems and smooth bark. Leaves broadly oval, to 10 cm long. Flowers in catkins, which are produced before leaves develop, the slender red stigmas quite conspicuous. Fruit a true nut enclosed in two leaf-like bracts.

Bracts not much united, barely covering nut. ............................................. *C. americana*
Bracts united and produced into long beak, completely enclosing nut, and extending about 3 cm beyond nut. ................................................................................ *C. cornuta*

*Corylus americana* Walt. (Fig. 89*A*)                    AMERICAN HAZELNUT

Shrub 1–3 m high, with ovate leaves covered with pinkish hairs. Nuts rather compressed and enclosed in two distinct leafy bracts, which are slightly longer than the nut. Common throughout eastern Boreal forest and Parkland, but unusual farther west.

*Corylus cornuta* Marsh. (Fig. 89*B*)                     BEAKED HAZELNUT

Similar in growth to *C. americana*, but with leaves sometimes hairless or with flattened sparse hairs. Nuts, which are very little compressed, enclosed in the united bracts, extending about 3 cm beyond the nut, forming a kind of beak. Fairly plentiful in woodlands and moist hillsides; throughout Parklands and Boreal forest.

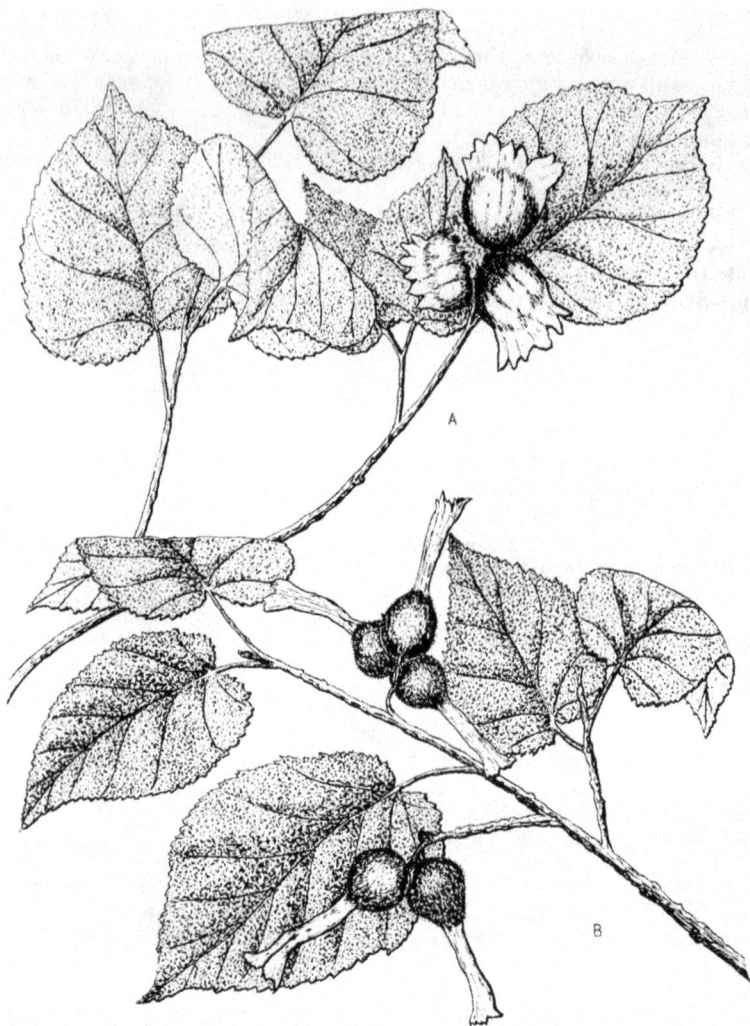

Fig. 89.   *A*, American hazelnut, *Corylus americana* Walt.; *B*, beaked hazelnut, *Corylus cornuta* Marsh.

A.C. Budd

***Ostrya***      ironwood

*Ostrya virginiana* (Mill.) K. Koch                       HORN HOPBEAM

    Trees to 20 m high; the leaves narrowly to broadly oblong or ovate, short acuminate, serrate. Catkins 3–5 cm long, at maturity resembling a hops strobile, the involucral bracts inflated, 1–3 cm long, loosely enclosing the fruit. Very rare; dry to moist woods; southeastern Boreal forest.

## FAGACEAE—beech family

***Quercus***      oak

*Quercus macrocarpa* Michx.                                BUR OAK

    Trees with gray, flaky bark and very hard wood, 15 m high in eastern locations, but a small scrubby tree farther west. Leaves bright, shiny green above and grayish white and slightly woolly beneath, deeply cut with rounded lobes. Flowers unisexual, in slender catkins. Fruit an acorn resting in a shallow fringed cup. Common in southeastern Parkland, but farther west only found in the valleys of the rivers tributary to the Assiniboine River in Saskatchewan.

## ULMACEAE—elm family

Leaves with 2 prominent veins beside the
    midrib; fruit a drupe with large stone. ................................................................... *Celtis*
Leaves with a single midrib; fruit a thin,
    broadly winged samara. ........................................................................ *Ulmus*

***Celtis***      nettletree

*Celtis occidentalis* L.                                 HACKBERRY

    Trees to 15 m high, with prominently ridged bark. Leaves 6–12 cm long, conspicuously serrate. Fruit ellipsoid, 7–13 cm long, dark red to nearly black. Very rare; moist woods; southeastern Boreal forest.

***Ulmus***      elm

*Ulmus americana* L.                             AMERICAN ELM

    A tall tree with reddish bark and smooth twigs. Leaves oval, with soft hairs beneath and rough above, veins very prominent. Flowers perfect, in little bunches early in spring before the leaves appear. Fruit a one-seeded, flat samara, or winged fruit. A common tree in southeastern Parkland, rare along rivers in Prairie, but in various locations planted and escaped.

# CANNABINACEAE—hemp family

***Humulus***          hop

*Humulus lupulus* L.                                      COMMON HOP

A perennial twining climber, often 3–6 m long; leaves usually large, palmately 3- to 7-lobed, with cordate bases, opposite, 2–8 cm across, and bearing many tiny whitish or yellowish glandular spots on the undersides. Upper leaves near the inflorescence often not divided, merely toothed. Stems rough to the touch, with fine, stiff, reflexed hairs. Male flowers in loose panicles, green, with 5 sepals. Female flowers on separate plants, in catkin-like or cone-like heads in the axils of the leaves, with broad greenish or yellowish imbricated bracts covering the fruits. These clustered fruits (hops) vary from 2 to 5 cm in length. Fairly common in moist places in southeastern Parklands, particularly in river valleys, and scattered throughout the Prairies. Contact with this plant causes a form of dermatitis in some individuals.

# URTICACEAE—nettle family

Plants often with stinging hairs and greenish flowers, which are borne in clusters at the junction of the stem and the leaf stalk.

1. Plants without stinging hairs. ........................................................................ *Parietaria*
   Plants with stinging hairs. ............................................................................................ 2
2. Leaves alternate; sepals 5. ............................................................... *Laportea*
   Leaves opposite; sepals 4. ................................................................... *Urtica*

***Laportea***          wood nettle

*Laportea canadensis* (L.) Gaud.                          WOOD NETTLE

A perennial with alternate and toothed leaves with stinging hairs. Fairly common; in rich woodlands; southeastern Parklands and Boreal forest.

***Parietaria***          pellitory

*Parietaria pensylvanica* Muhl.                       AMERICAN PELLITORY

A weak-stemmed, hairy, annual plant, 10–50 cm high, with opposite, lanceolate, 3-nerved leaves 2–8 cm long. Greenish flowers in dense clusters around the leaf axils. Very rare; shaded areas; southeastern Parklands.

***Urtica***          nettle

Plants perennial, with extensive rootstocks;
   stipules 5–15 mm long. ....................................................... *U. dioica* var. *procera*
Plants annual; stipules 1–3 mm long. ...................................................... *U. urens*

*Urtica dioica* L. var. *procera* (Muhl.) Wedd. (Fig. 90)          STINGING NETTLE

A perennial with very coarse rootstocks, which have pink offshoots. Plants have tall, straight stems, usually square in cross section, and with ovate to lanceolate, serrate (toothed), opposite leaves bearing stinging hairs. Greenish flowers borne in clusters at the junction of stem and leaf stalk. Very common around sloughs, moist places, and bushes; throughout the Prairie Provinces. Syn.: *U. gracilis* Ait.

*Urtica urens* L.                                              ENGLISH NETTLE

Annual plants with erect stems 10–50 cm high, simple or branched, with numerous stinging hairs. Leaves elliptic to ovate, deeply coarse serrate; flower clusters oblong. A rare weed; in gardens; Prairies and Parklands.

## SANTALACEAE—sandalwood family

Smooth, hairless perennials with rootstocks, partly parasitic on the roots of other plants. Flowers perfect, without petals, but with 5 sepals, which are united to form a bell-like tube. Fruit a 1-seeded drupe (dry or fleshy fruit with a hard nut-like seed in the center).

**Comandra**          comandra

Flowers axillary. ............................................................................... *C. livida*
Flowers in terminal or subterminal clusters. ................................... *C. umbellata*

*Comandra livida* Rich.                              NORTHERN COMANDRA

A slender, erect plant, 10–30 cm high, with oval leaves 1–3 cm long on short stalks. Flowers in the axils of the leaves. Fruit a spherical red drupe about 3 mm in diam, edible. Not very common; but found in moist woods; Boreal forest, Riding Mountain, Rocky Mountains. Syn.: *Geocaulon lividum* (Richards.) Fern.

*Comandra umbellata* (L.) Nutt. (Fig. 91)          BASTARD TOADFLAX

An erect plant, from a white creeping rootstock, 6–30 cm high, usually in bunches of several plants from the same rootstock. Leaves linear or linear-lanceolate, 10–25 mm long, without stalks, borne alternately on the stems. Flowers at the summit of stems, greenish white to pinkish, small, about 5 mm long. Fruit ovoid, 3–8 mm long. This plant is often attacked by a tiny insect that causes small round galls about 6 mm in diam that are sometimes mistaken for the fruit.

Three forms can be distinguished:

1. Leaves thin, distinctly veined; calyx lobes
   2–3 mm long; fruit 3–6 mm across. .................................................. var. *umbellata*
   Leaves thick, not distinctly veined; calyx
   lobes 3–4 mm long; fruit 5–8 mm
   across. ............................................................................................................... 2
2. Plants usually small, to 15 or 20 cm high;
   panicle branches short; the panicle
   ovoid. .......................................................... var. *angustifolia* (DC.) Torr.

Fig. 90.  Stinging nettle, *Urtica dioica* L. var. *procera* (Muhl.) Wedd.

Fig. 91.   Bastard toadflax, *Comandra umbellata* (L.) Nutt.

Plants usually 20–40 cm high; panicle
 branches long; the panicle more
 corymbiform. ............................................................ var. *pallida* (DC.) G. N. Jones

The var. *umbellata*, Richards comandra, is the common form in openings,
open forests, and disturbed areas throughout Boreal forest (syn.: *C.
richardsiana* Fern.); var. *angustifolia*, pale comandra, is very common in dry
grasslands of the Prairies (syn.: *C. pallida* A. A.); var. *pallida* occurs in grassland in southern Rocky Mountains (syn.: *C. pallida* A. DC.). Many plants,
such as those of the Peace River and Parkland areas, are intermediate between
these forms, but they approach one form more closely than the others.


## LORANTHACEAE—mistletoe family

***Arceuthobium***        dwarf mistletoe

Stems 5–10 cm high; staminate flowers
 terminal. ...................................................................................... *A. americanum*
Stems 0.5–2 cm high; all flowers axillary. .......................................... *A. pusillum*

*Arceuthobium americanum* Nutt.                AMERICAN MISTLETOE

A parasitic plant on conifers, mostly *Pinus*, often causing witches'-broom.
Stems fragile, greenish yellow, often much-branched; leaves opposite, scale-
like. Fruit 1-seeded, berry-like, 2–3 mm long, ejecting the seed explosively.
Boreal forest.

*Arceuthobium pusillum* Peck                DWARF MISTLETOE

A parasitic plant, mostly on black spruce, more rarely on white spruce or
tamarack. Stems fragile, greenish brown, simple or sparingly branched; leaves
scale-like. Fruit about 2 mm long. Eastern Boreal forest.


## ARISTOLOCHIACEAE—birthwort family

***Asarum***        wild ginger

*Asarum canadense* L.                WILD GINGER

Plants with slender, branched rootstocks and 2 large, rotund or reniform
leaves with a deeply cordate base, pubescent especially on the petioles. Flower
solitary, with the petals minute or absent. Fruit a capsule with large wrinkled
seeds. Rare; woods; southeastern Boreal forest.


## POLYGONACEAE—buckwheat family

Herbs with mostly alternate, entire leaves and, except in *Eriogonum*, with
sheathing stipules called ocrea. Flowers small, without petals, but with a perianth of 4–6 more or less united sepals. Achenes either triangular or lens-
shaped.

1. Leaves without stipules; a whorl of bracts below flower clusters; stamens 9. .......................................... *Eriogonum*

   Leaves with stipules; without whorl of bracts below flowers; stamens less than 9. ....................................................................................... 2

2. Sepals 4 or 6; stigmas tufted. ..................................................... 4

   Sepals 5, more or less equal; stigmas not tufted. ..................................................................................... 3

3. Seeds generally enclosed by somewhat enlarged calyx, but, if protruding from calyx, the leaves are long and narrow. ............................... *Polygonum*

   Seeds protruding from calyx; leaves broadly arrow-shaped. ....................................................... *Fagopyrum*

4. Sepals 6, the three inner ones large, 15–20 mm, enclosing the fruit; leaves lanceolate. ...................................................................... *Rumex*

   Sepals 4, the inner ones erect, 4–6 mm; leaves reniform. ............................................................... *Oxyria*

### *Eriogonum*      umbrellaplant

1. Plants annual; inflorescence a diffusely branched raceme. .................................................. *E. cernuum*

   Plants perennial; inflorescence a more or less dense umbel. ............................................................... 2

2. Inflorescence subtended by leaf-like bracts. ...................................................................................... 3

   Inflorescence subtended by small, reduced bracts. ..................................................................... 4

3. Involucral bracts linear, ascending. ............................................. *E. flavum*

   Involucral bracts lanceolate or oblong, spreading or reflexed. ............................................. *E. umbellatum*

4. Leaves oval, obovate or orbicular, about 1 cm long. ....................................................................... *E. ovalifolium*

   Leaves lanceolate or linear-lanceolate. ..................................... 5

5. Leaves in dense crowded rosettes; flowers yellow or reddish. ............................................ *E. androsaceum*

   Leaves in loose rosettes; flowers whitish or pinkish. ............................................................. *E. multiceps*

*Eriogonum androsaceum* Benth.          CUSHION UMBRELLAPLANT

Plant forming small, compact cushions; scapes 2–10 cm high. Leaves 1–2 cm long, oblanceolate to spatulate, densely villose, later becoming glabrate. Inflorescence often a single small umbel; flowers 4–5 mm long. Not common; exposed, often rocky, areas; southern Rocky Mountains.

*Eriogonum cernuum* Nutt.          NODDING UMBRELLAPLANT

Low annual plants up to 30 cm high, with a whorl of small leaves near base of stem. Leaves almost circular, 1–2 cm in diam, white woolly beneath.

Flowers white or rose, borne in small clusters about 5 mm wide on stems in a much-branched inflorescence, usually drooping. Not common but abundant locally; found on badlands, abandoned fields, dry and sandy soil; throughout southwestern Prairies.

*Eriogonum flavum* Nutt. (Fig. 92)      YELLOW UMBRELLAPLANT

A low perennial with a very coarse, scaly, dark brown, woody, tufted root. Leaves all basal, usually linear-oblong or spatulate, 2–5 cm long, green and slightly hairy above, densely white woolly beneath. Flowers pale yellow, in umbel-like clusters at head of stems. Common; on dry and eroded hillsides, badlands, and canyons; throughout Prairies.

*Eriogonum multiceps* Nees      BRANCHED ERIOGONUM

A low perennial, rarely over 20 cm high, with branched stems. Leaves very narrow, to 5 cm long; the whole plant, leaves, and stem densely white woolly. Whitish or pale pink flowers in small clusters at the summits of the stems. Uncommon; on eroded banks and badlands; Prairies.

*Eriogonum ovalifolium* Nutt.      SILVERPLANT

A low plant with the caudex closely branched; leaves densely white tomentose. Scapes 5–15 cm high; inflorescence usually a single dense umbel, with very small bracts. Not common; dry hillsides and plains; southern Rocky Mountains.

*Eriogonum umbellatum* Torr.      UMBRELLAPLANT

Plant with short woody branches forming loose mats; leaves clustered in numerous rosettes, 3–5 cm long, glabrous above, dense white tomentose below. Scapes 10–30 cm high, pubescent below; umbel with rays 1–6 cm long; bracts 10–25 mm long. Exposed sites, summits, and slopes; southern Rocky Mountains. Syn.: *E. subalpinum* Greene.

**Fagopyrum**      buckwheat

*Fagopyrum tataricum* (L.) Gaertn. (Fig. 93)      TARTARY BUCKWHEAT

An annual weed 50–100 cm high, with somewhat triangular leaves often as broad as long, 3–10 cm long and wide. Flowers small, white, in bunches on the flowering stems, arising from the junction of the leaf stalks and stems. Seeds about the same size as wheat kernels. This weed, introduced from Asia, is a serious pest in northern and central Alberta; rare in Boreal forest. Common cultivated buckwheat, *F. esculentum* Moench, has larger white or pinkish flowers, and smooth shiny seeds. Rarely found as a weed.

**Oxyria**      mountain sorrel

*Oxyria digyna* (L.) Hill      MOUNTAIN SORREL

Plant with rootstock and stout root; stems 5–30 cm high. Leaves mostly basal, alternate, long-petioled; stipule sheath loose. Flowers whorled in a dense raceme 4–6 mm long; achenes lens-shaped. Mountain meadows, rock slopes; Rocky Mountains.

Fig. 92. Yellow umbrellaplant, *Eriogonum flavum* Nutt.

Fig. 93. Tartary buckwheat, *Fagopyrum tataricum* (L.) Gaertn.

***Polygonum***     knotweed

1. Plants twining; leaves broad, hastate (ar-
   row-shaped) at base. ..................................... Section: Bilderdykia

   Plants not twining; leaves linear, lanceo-
   late, or oblong. ............................................................................. 2

2. Flowers solitary or in clusters in axils of
   leaves; leaves small. ............................................. Section: Avicularia

   Flowers in dense spikes, either terminal or
   in axils of leaves; leaves large. .................................................... 3

3. Basal leaves with long stalks, stem leaves
   smaller and stalkless; stem single and
   unbranched. .................................................................. Section: Bistorta

   Leaves all similar; stem much-branched. .................... Section: Persicaria

***Polygonum*** (Sect.: Avicularia)     doorweed

Annual low-growing plants, often prostrate on ground, with tough stems
and alternate leaves, usually narrow. Sheathing stipules usually fairly conspicuous and often white and translucent. Flowers small, borne in clusters in axils
of leaves, and having a perianth of five partly united sepals. Seeds 3-angled
achenes, black or brown. Plants appear to thrive when they are trampled.

1. Inflorescence crowded at summit of stem;
   bracts leaf-like. ............................................................................ 2

   Inflorescence not crowded at summit of
   stem; flower cluster axillary. ...................................................... 3

2. Bracts broadly white-margined; achenes
   jet black. ..................................................................... *P. confertiflorum*

   Bracts mostly not white-margined;
   achenes brown. ..................................................................... *P. kelloggii*

3. Fruit reflexed or pendulous; achenes
   black. ............................................................................................. 4

   Fruit erect or ascending; achenes
   brownish. ....................................................................................... 6

4. Calyx 3–4 mm long; plants 10–40 cm
   high. .......................................................................... *P. douglasii*

   Calyx 2–3 mm long; plants 5–15 cm high. ................................ 5

5. Leaves linear or linear-lanceolate; bracts
   awl-shaped. ............................................................. *P. engelmannii*

   Leaves ovate to oblanceolate; bracts
   oblong or lanceolate. ..................................................... *P. austiniae*

6. Leaves oval or blunt; length-to-width
   ratio 2.5:1. .................................................................................... 7

   Leaves linear to lanceolate; length-to-
   width ratio 4:1 or more. ............................................................ 8

7. Plants erect, 3–15 cm high; flower clusters
   crowded at the tips of branches. ...................................... *P. minimum*

   Plants prostrate or decumbent, much
   larger; flower clusters scattered along
   the stems. ..................................................................... *P. achoreum*

8. Plants erect or ascending, usually quite high. ...................................................................................... 9

   Plants usually prostrate, depressed. ...................................................................... 10

9. Leaves 10–25 mm wide, margins white. .................................................... *P. erectum*

   Leaves 5–15 mm wide, margins not white. ..................................................... *P. ramosissimum*

10. Petioles included in ocreae; margins of perianth segments white or pinkish; achene 2.5–3.5 mm. .............................................. *P. aviculare*

    Petioles exserted; margins of perianth segment bright pink; achene 3.5–4.5 mm. ...................................................................... *P. boreale*

## *Polygonum* (Sect.: Bilderdykia)    wild buckwheat

1. Plants annual. ................................................................. *P. convolvulus*

   Plants perennial. .......................................................................................... 2

2. Ocreae reflexed-bristly at base; stems sharply angled; fruit not winged. ........................................ *P. cilinode*

   Ocreae smooth; stems nearly terete; fruit broadly winged. ............................................................. *P. scandens*

## *Polygonum* (Sect.: Bistorta)    bistort

Basal leaves tapering gradually to stalk; inflorescence not bulblet-bearing. ............................................ *P. bistortoides*

Basal leaves cordate or blunt-based; inflorescence bearing bulblets at lower part. ................................... *P. viviparum*

## *Polygonum* (Sect.: Persicaria)    smartweed

The smartweeds, especially the perennial species, are a rather confusing group of plants: they can have three phases, each phase differing from the others by so much that in some instances they have been classed as different species instead of phases of the same species. The aquatic phase is found submersed in water, is usually without hairs, and has broad leaves. The paludose (or marshy) phase is found where water has recently receded, and the plants grow in wet mud. The terrestrial phase is found where the soil has dried somewhat; the leaves are usually narrow and the plant often more or less hairy. The flowers of smartweeds are borne in conspicuous racemes, usually white and pink or greenish, and the perennial species are usually the dominant plant in the location they occupy. Most species are palatable and nutritious. Heavy pasturing may cause bighead or yellows in sheep.

1. Plants perennial, with extensively creeping rootstocks. .......................................... *P. amphibium*

   Plants annual. ............................................................................................ 2

2. Leaves 3–6 cm long, oblong-sagittate, cordate at base. ................................................. *P. sagittatum*
   Leaves not sagittate, cuneate at base. ........................................................ 3
3. Perianth covered with brownish glands. ..................................................... 4
   Perianth not glandular. ...................................................................... 5
4. Cleistogamous flowers present in the leaf axils. ................................................. *P. hydropiper*
   Cleistogamous flowers not present. ........................................ *P. punctatum*
5. Ocreae ciliate; peduncles not glandular. ........................................ *P. persicaria*
   Ocreae not ciliate; peduncles and often perianth bearing yellow subsessile glands. ..................................................................... 6
6. Achene 1.5–2.0 mm wide; leaves glabrous below. ......................................................... *P. lapathifolium*
   Achene 2.0–3.0 mm wide; leaves sparingly pubescent below. ....................................... *P. scabrum*

### *Polygonum achoreum* Blake    STRIATE KNOTWEED

An annual growing to about 40 cm high, with oval to elliptic bluish green leaves 6–25 mm long. Common around yards and waste places, appearing to thrive on abuse. If kept mowed, it takes the place of a lawn, and helps to prevent soil erosion by wind and water. In waste places and farmyards; throughout the Prairie Provinces.

### *Polygonum amphibium* L.    SWAMP PERSICARIA

Perennial with far-creeping blackish, branching rootstocks. Leaves 5–15 cm long, oblong-ovate, truncate, cordate or cuneate at the base, glabrous or somewhat pubescent. Ocreae often with foliaceous collar when young. Spikes 1–3 cm long. Very common; in standing water; throughout the Prairie Provinces. Syn.: *P. coccineum* Muhl. forma *natans* (Wiegand) Stanford; *P. natans* A. Eaton forma *genuinum* Stanford. The terrestrial form, with larger inflorescence, spikes 5–10 cm long, leaves often densely pubescent, and ocreae with a broad foliaceous collar, is *P. amphibium* L. var. *emersum* Michx. Very common; in mud flats, slough margins, and lakeshores; throughout the Prairie Provinces. Syn.: *P. coccineum* Muhl. forma *terrestre* (Willd.) Stanford; *P. natans* A. Eaton forma *hartwrightii* (Gray) Stanford; *P. amphibium* L. var. *stipulaceum* (Coleman) Fern.

### *Polygonum austiniae* Greene

Plants erect, 4–12 cm high, branched from the base. Lower leaves oval to obovate or lanceolate, petioled; upper leaves reduced, sessile. Flowers in the leaf axils, solitary or in pairs, reflexed. Achenes 2.5–3 mm long, shiny black. Dry, usually disturbed ground; southern Rocky Mountains.

### *Polygonum aviculare* L. (Fig. 94)    DOORWEED

A semiprostrate annual plant, 10–50 cm long. Leaves fairly broad, oblong, 1.5–6 cm long, pale bluish green. The ocreae (or stipules) silvery and translucent. Very common; in waste places; throughout the Prairie Provinces.

Fig. 94. Doorweed, *Polygonum aviculare* L.

*Polygonum bistortoides* Pursh WESTERN BISTORT

A perennial plant growing from a horizontal fleshy rootstock to 10–40 cm high. Basal leaves 6–15 cm long, lanceolate to oblong, on a stalk about as long as the blade. Stem leaves much shorter and without stalks. Inflorescence a dense, oblong raceme of white or pinkish flowers, borne at the head of the stem. A plant of swamps, stream beds, and wet places; Rocky Mountains, Cypress Hills.

*Polygonum boreale* (Lange) Small NORTHERN DOORWEED

Stems prostrate or decumbent to erect, 10–50 cm long. Leaves 3–5 cm long, obovate; petioles well exserted from the ocreae. Perianth green with conspicuous pink margins. Gravel banks; Hudson Bay.

*Polygonum cilinode* Michx. BINDWEED

Stems to 2 m long, twining, trailing, sometimes erect; pubescent. Leaves triangular-ovate, deeply cordate. Racemes long-peduncled, 4–10 cm long, perianth white; achene glossy black, 1.5–2 mm, with the outer calyx rarely narrowly winged. Clearings and rock outcrops; eastern Boreal forest.

*Polygonum confertiflorum* Nutt. SMALL KNOTWEED

Plants annual, stems erect, 3–15 cm high, glabrous, simple or branched. Leaves 1–3 cm long, linear-lanceolate. Flower clusters crowded into short terminal spikes; perianth segments with a conspicuous white margin; achene jet black. Rare or overlooked; dry, disturbed areas; Cypress Hills.

*Polygonum convolvulus* L. WILD BUCKWHEAT

An annual, tap-rooted weed, with weak twining stems and pale green arrow-shaped leaves, 1–5 cm long, and pointed at the tip. Flowers greenish or pink, in small racemes in the axils of the leaves or at the end of the stem. Seeds black, granular, three-angled, and often enclosed in a green or brown coating, and about 3 mm long. Introduced from Europe, but very common in cultivated land and waste places; throughout the Prairie Provinces.

*Polygonum douglasii* Greene DOUGLAS KNOTWEED

An erect annual to 45 cm high; narrow leaves 1–5 cm long. Seeds oblong, narrow, reflexed or pendulous on the stem. Not very common; various locations; southwestern Prairies.

*Polygonum engelmannii* Greene SLENDER KNOTWEED

An erect annual 5–15 cm high; stems slender, often diffusely branched at base; linear to lanceolate leaves, and axillary clusters of 1–4 flowers. Moist banks and slopes; Rocky Mountains.

*Polygonum erectum* L. ERECT KNOTWEED

Plants with erect or decumbent stems 10–50 cm high; leaves elliptic to elliptic-lanceolate, finely crenulate and white-margined, 1–4 cm long, 0.5–2 cm broad. Achenes of two kinds: dark brown, shiny, punctate, about 2.5 mm long; or dull brown, 3–3.5 mm long. Waste areas and lakeshores; southeastern Parklands and Boreal forest.

*Polygonum hydropiper* L.                                    WATER-PEPPER

Stems 10–50 cm long, erect or spreading, often reddish. Leaves narrowly lanceolate to ovate-lanceolate, 2–5 cm long; ocreae short ciliate. Racemes 3–6 cm long; perianth greenish; achenes dull dark brown, 2–3 mm long. Rare; wet places; southeastern Boreal forest.

*Polygonum kelloggii* Greene                                DWARF KNOTWEED

A small annual 3–15 cm high, erect, simple or branched; leaves 1–3 cm long, linear to linear-lanceolate; flower clusters crowded at the ends of branches; achenes dark brown. Rare; moist areas; southwestern Prairies.

*Polygonum lapathifolium* L.                                PALE PERSICARIA

A variable annual plant, with lanceolate to ovate leaves, sometimes deep green and hairless, but more often pubescent beneath. The type with pubescent leaves was formerly listed as a separate species, *P. tomentosum* Schrank. Although hirsute and whitish underneath when young, mature plants appear glabrous. Flowers in erect spikes, usually pale pink to white; those growing under moist conditions usually paler than those of the drier locations. Very common along slough margins and low areas; occasionally in cultivated fields; throughout most of the Prairie Provinces.

*Polygonum minimum* S. Wats.                                LEAST KNOTWEED

Small annuals 3–15 cm high, erect or spreading, branched at the base, leafy. Leaves oval to obovate, 5–15 mm long. Flower clusters of 2 or 3, axillary, perianth white-margined; achenes 2–2.5 mm long. Dry ground; southern Rocky Mountains.

*Polygonum persicaria* L. (Fig. 95)                          LADY'S-THUMB

Large annuals, with stems 30–80 cm high, erect or ascending. Leaves narrowly lanceolate, 3–7 cm long; ocreae short-ciliate, minutely pubescent. Spikes dense, 1–5 cm long; perianth pink or reddish; achenes black, 2–2.5 mm long. Not common; wet areas; throughout the Prairie Provinces.

*Polygonum punctatum* Ell. var. *confertiflorum* (Meisn.) Fassett

                                                            WATER SMARTWEED

Large annuals, with stems to 60 cm high, simple or branched, erect or ascending. Leaves to 15 cm long, 2 cm wide, narrowly lanceolate. Racemes to 10 cm long, much interrupted; perianth greenish, glandular; achenes 2.5–3 mm long. Wet areas, lakeshores, waste places; eastern Parklands and Boreal forest.

*Polygonum ramosissimum* Michx.                             BUSHY KNOTWEED

A yellowish green erect plant 15–90 cm high, with a much-branched stem and lanceolate to linear-oblong leaves to 3 cm long. Found occasionally; in sandy soil, river bottoms, and lakeshores; throughout Prairies and Parklands.

*Polygonum sagittatum* L.                               ARROW-LEAVED TEAR-THUMB

Slender annuals, with stems 10–50 cm high, weak, reclining, with reflexed barbs. Leaves lanceolate-sagittate, deeply cordate at base. Inflorescence head-like, terminal on stems or branches; perianth pink; achene dark brown or black, 2–3 mm long. Rare; wet areas; southeastern Boreal forest.

Fig. 95.   Lady's-thumb, *Polygonum persicaria* L.

*Polygonum scabrum* Moench                    GREEN SMARTWEED

Large annuals, with stems 30–80 cm high, simple or branching, erect to spreading-ascending. Leaves lanceolate to oblong-lanceolate, 3–8 cm long, white pubescent below, especially the lower ones. The upper leaves glandular-dotted. Spikes erect, usually greenish, very glandular below; achenes 2.5–3.5 mm long. Waste areas, lakeshores; western Parklands and Boreal forest.

*Polygonum scandens* L.                    FALSE BUCKWHEAT

Twining perennials, with sharply angled stems to 2 m long. Leaves oblong-ovate, deeply cordate at base. Inflorescences from most upper leaf axils 5–10 cm long; perianth white; fruit broadly winged, the outer sepals expanding, to 15 mm long; achene glossy black, 3–5 mm long. Rare; shores, clearings, and margins of woods; southeastern Parklands and Boreal forest.

*Polygonum viviparum* L.                    ALPINE BISTORT

A perennial plant 5–25 cm high, from a scaly corm-like rootstock, with cordate-based basal leaves 5–10 cm long. Stem leaves narrow and tapering to the base, which is enclosed in a membranous sheath or ocreae. Inflorescence a dense raceme of whitish or pinkish flowers terminating the stem, but the lower flowers replaced by small bulbs. An alpine species; Rocky Mountains.

**Rumex**      dock

Mostly perennial plants with thick roots, leaves of varying shapes and sizes. Flowers, perfect or unisexual, consisting of 6 sepals, the 3 inner ones enlarging and forming wings or valves enclosing the fruit. These valves sometimes entire and sometimes fringed with bristle-like teeth. In some species one or all three of the valves may bear a protuberance called a tubercle, useful as a character for identification. Usually in wet or marshy places; throughout the Prairie Provinces, but at least one species is a plant of the sandhills.

1. Valves greatly enlarging in fruit, becoming bright pink. ...................................................... *R. venosus*
   Valves not greatly enlarging. ........................................................................... 2

2. Flowers unisexual. .......................................................................................... 3
   Flowers perfect. .............................................................................................. 5

3. Leaves cuneate at base, long-petioled. ................................. *R. paucifolius*
   Leaves hastate or sagittate at base. ...................................................... 4

4. Valves much larger than achene. ........................................... *R. acetosa*
   Valves equaling the achene, or a little larger. ................................... *R. acetosella*

5. Valves distinctly lobed or toothed on the margin. ............................................................................ 6
   Valves entire or nearly so. .................................................................... 7

6. Inflorescence continuous; valves 2.5–3 mm. ............................................. *R. maritimus* var. *fueginus*
   Inflorescence with distant whorls; valves 4–5 mm. ..................................................... *R. dentatus*

7. Pedicels not jointed near the middle. ........................................................................ 8
   Pedicels jointed. ................................................................................................................ 9

8. Valves all with a grain-like body. ............................................................ *R. orbiculatus*
   Valves without grains. ................................................................................ *R. occidentalis*

9. Fruit with 1–3 large grains. ............................................................................................. 10
   Fruit without grains, or with only 1 small
   grain. ...................................................................................................................................... 12

10. Plants with prostrate or ascending stems;
    most stem leaves with axillary shoots. ............................................... *R. salicifolius*
    Plants with erect stems; axillary shoots
    absent or few. .................................................................................................................. 11

11. Valves entire, 3–6 mm long; leaf margins
    strongly undulate. .................................................................................................. *R. crispus*
    Valves with small teeth, 3.5–4 mm long;
    leaf margins undulate, but not strongly
    so. ........................................................................................................... *R. stenophyllus*

12. Basal and lower stem leaves almost as
    wide as long, deeply cordate at base. ................................................... *R. confertus*
    Basal and lower stem leaves much longer
    than wide. ........................................................................................................................ 13

13. Lower stem leaves linear-lanceolate,
    length-to-width ratio about 8:1; valves
    3.5–5 mm long, less than 5 mm wide. ..................................................... *R. fennicus*
    Lower stem leaves broadly lanceolate,
    length-to-width ratio about 4:1; valves
    4.5–6 mm long, more than 5 mm wide. ............................................... *R. longifolius*

*Rumex acetosa* L.                                                                      SOUR DOCK

An erect plant sometimes over 60 cm high, with broadly lanceolate arrow-shaped leaves 2–12 cm long. Used as food; escaped from gardens in many parts of the Prairie Provinces.

*Rumex acetosella* L. (Fig. 96)                                           SHEEP SORREL

A low plant not over 30 cm high, with leaves narrowly hastate or spear-shaped, 2–10 cm long. In this species sepals do not enlarge in fruit. An introduced weed but only rarely found; in dry or sandy places and on acid soils.

*Rumex confertus* Willd.

Fall perennial, 60–100 cm high, with the basal leaves triangular cordate, to 20 cm wide; the petiole shorter than the blade. Valves about 6 mm long, 8 mm wide. Very rare; waste areas; eastern Parklands.

*Rumex crispus* L.                                                                   CURLED DOCK

An introduced weed, to over ' m high in favorable localities, with leaves crinkled at the edge, the lower ones 15–30 cm long, the upper ones smaller. Fairly common; in moist places; eastern Prairies and Parklands.

Fig. 96.   Sheep sorrel, *Rumex acetosella* L.

*Rumex dentatus* L.                                                    TOOTHED DOCK

Plants annual, 20–70 cm high, with the basal leaves small, 2–3 times as long as wide, truncate to subcordate at the base. Valves 4–6 mm long, 2–3 mm wide, with teeth 3–6 mm long. Rare; weed in cultivated land; southwestern Alberta.

*Rumex fennicus* Murb.                                                 FIELD DOCK

An introduced perennial growing to more than 1 m high, with lanceolate leaves 15–25 cm long, tapering toward the stalk. Erect, dense fruiting heads turn brown in late summer and are very conspicuous in fields and along road-sides. A tiny swelling on the stalk of fruiting valves about a third of the way up from the base distinguishes this species from western dock, which it superfi-cially resembles. Originally introduced from Europe but has spread rapidly across the south central part of the Prairies and Parklands. Now a common plant on almost every type of soil; in upland fields and moist areas. Syn.: *R. domesticus* Hartm. var. *pseudonatronatus* Borb.

*Rumex longifolius* DC.                                                LONG-LEAVED DOCK

Stout perennial plants 60–120 cm high, with leaves 20–40 cm long, 10–15 cm wide. Valves 4.5–5.5 mm long, 5.5–6.5 mm wide. Rare; weed in moist areas; eastern Parklands and Boreal forest.

*Rumex maritimus* L. var. *fueginus* (Phil.) Dusen              GOLDEN DOCK

An annual species to 60 cm high, from a very fleshy root, and usually diffusely branched, with leaves lanceolate, 6–25 cm long, pale green. Stems covered with very short hairs. Fruiting heads conspicuous, golden yellow, very dense. Valves of calyx have a spine-like bristly margin, which gives the heads a prickly appearance and makes this species easy to distinguish. Found in moist and saline places, lake flats, and lakeshores, and often the dominating plant in such sites, sometimes in masses covering 1000 m$^2$ or more. Common in suita-ble locations; throughout the Prairie Provinces.

*Rumex occidentalis* S. Wats.                                          WESTERN DOCK

A tall, coarse, erect plant to 1 m high, with leaves broadly oblong or lan-ceolate, up to 30 cm long, often cordate or heart-shaped at the base. Fruiting heads dense and very conspicuous. Common; in wet or moist places; through-out western part of the Prairie Provinces.

*Rumex orbiculatus* Gray                                               GREAT WATER DOCK

Stout, erect plants to 1.5 m high, with leaves lanceolate, acute or rounded at base. Valves 5–8 mm long and wide. Rare; swamps and shallow water; Boreal forest.

*Rumex paucifolius* Nutt.                                              ALPINE SHEEP SORREL

Perennial plants with simple or branched taproot; stems 20–70 cm high; leaves mostly basal, 4–10 cm long, cuneate at the base. Valves 2.5–3 mm long, suborbicular, becoming reddish. Mountain meadows and streams; southern Rocky Mountains.

*Rumex salicifolius* Weinm. NARROW-LEAVED DOCK

A narrow-leaved species, erect or sprawling, in low and moist places. Leaves pale green, 5–10 cm long and 10–25 mm wide. Fruiting heads narrow and usually not so dense as in most docks. Common in moist or saline places, roadside ditches, and similar sites; throughout the Prairie Provinces. Syn.: *R. triangulivalvis* (Danser) Rech. f.; *R. mexicanus* Meisn.

*Rumex stenophyllus* Ldb. NARROW-LEAVED FIELD DOCK

An introduced perennial, 1–2 m high, with narrowly lanceolate leaves to 15 cm long, narrowed at both ends and crinkly margined. Inflorescence tall, narrow, and somewhat leafy at the base. Fruiting valves toothed on the margins and bearing a tubercle on each valve; a tiny swollen joint on the lower part of each valve stalk. This plant resembles field dock in many ways. Becoming increasingly common; in southern part of Prairies.

*Rumex venosus* Pursh (Fig. 97) SAND DOCK

A perennial species with running, woody rootstocks and branched stems. Sheathing stipules whitish and papery, very conspicuous. Leaves pale green, 5–15 cm long, and nearly half as wide. Calyx valves conspicuous in fruit, large, sometimes to 3 cm broad, almost round, bright red to pink. Often picked and dried to form winter bouquets. Common in suitable localities, becoming dominant in some places; on sandy soil, sand dunes, and often on railway grades and roadsides; often an early invader of a shifting sand dune; throughout Prairies and Parklands.


# CHENOPODIACEAE—goosefoot family

A large and varied family consisting of both annual and perennial herbs and shrubs. Leaves usually alternate, without stipules, and often white mealy. Flowers have no petals but 2–5 sepals, and may be either perfect or unisexual. Many species are weedy and most produce vast quantities of seed. Some are edible, and nearly all are palatable to livestock, but one species is **poisonous.**

1. Leaves and branches opposite; fleshy-stemmed plants; leaves reduced to scales; flowers sunk into stem. ............................................... *Salicornia*
   Leaves alternate. ......................................................................................... 2

2. Plants with spiny branches or spine-tipped leaves; calyx forming wing around mature fruit. ........................................................................... 3
   Plants neither spiny nor with spine-tipped leaves. ................................................................................ 4

3. Shrubs with spiny branches. ........................................................ *Sarcobatus*
   Annual herbs with fleshy spine-tipped leaves. ........................................................................... *Salsola*

4. Flowers unisexual, female flowers enclosed in two bracts that enlarge after flowering. ............................................................................. 5

Fig. 97.   Sand dock, *Rumex venosus* Pursh.

Flowers perfect. ..................................................................................... 9

5. Fruiting bracts covered with silvery, silky
hairs; low wooly shrubs. ........................................................... *Eurotia*

Fruiting bracts not covered with silky
hairs. ............................................................................................ 6

6. Fruiting bracts not united; seed coat
often winged at apex. ................................................................ *Axyris*

Fruiting bract united, at least at base;
seed coat not winged at apex. ...................................................... 7

7. Fruit with 2 teeth at apex; seed coat
membranous. ............................................................................ *Suckleya*

Fruit not 2-toothed at apex; seed coat
leathery. ...................................................................................... 8

8. Female flowers with no calyx or sepals,
merely a pair of bracts. ............................................................ *Atriplex*

Female flowers with 2 or 3 translucent
sepals, which are much shorter than the
bracts. ...................................................................................... *Endolepis*

9. Calyx with only one sepal; 1–3 stamens. ..................................... 10

Calyx with 3–5 sepals; 1–5 stamens. ........................................... 11

10. Leaves spear-shaped; stems fleshy;
flowers in clusters in leaf axils; seeds
round and small. ..................................................................... *Monolepis*

Leaves narrow and linear; flowers single
in axils of leaf-like bracts; seeds flat
and large. ............................................................................... *Corispermum*

11. Mature calyx with wings, or minute
spines. ...................................................................................... 12

Mature calyx without wings or spines. ......................................... 14

12. Mature calyx with a tiny spine on each
lobe. ........................................................................................... *Bassia*

Mature calyx without spines, but winged. ..................................... 13

13. Leaves flat,wavy-edged, ovate. ............................................. *Cycloloma*

Leaves linear, usually turning red when
mature. ..................................................................................... *Kochia*

14. Leaves circular in cross section, fleshy. .................................. *Suaeda*

Leaves not circular in cross section. ..................................... *Chenopodium*


*Atriplex*    atriplex

Annual or perennial plants with unisexual flowers. Male flowers usually
with 3–5 sepals and the same number of stamens and no bracts; female flowers
usually without sepals but with a pair of bracts.

1. Plant a small shrub, with pale leaves covered with small silvery scales; leaves
without stalks. ........................................................................ *A. nuttallii*

*Atriplex argentea* Nutt.                  SILVERY ATRIPLEX

A bushy annual growing to 40–80 cm high, but usually much smaller; stem angled. Leaves usually deltoid or triangular, 6–25 mm long, grayish green and scurfy. Male flowers usually in terminal spikes, and female flowers in clusters in axils of leaves. Very common in saline flats and similar sites in southern Prairies, but apparently rare elsewhere.

*Atriplex hortensis* L. (Fig. 98)             GARDEN ATRIPLEX

A fairly tall, erect annual plant to 60–120 cm high. Leaves of various shapes, cordate to triangular and ovate. Peculiarly interesting, because it bears two kinds of female flowers on the same inflorescence. One kind has no calyx but two large bracts, becoming quite large at maturity, over 6 mm; these flowers produce large, flat, pale brown seeds, about 3–10 mm in diam, which germinate a few days after being sown. The other kind of female flower has a small 3- to 5-lobed calyx but no bracts, and produces small, black, shiny seeds about 2 mm in diam, which sometimes remain dormant in the soil for many months before germination. Inflorescence often large, sometimes up to 30 cm long. A garden escape, it has become very common around towns and cities; in many parts of the Prairie Provinces. A red-leaved variety, *atrosanguinea* Hort., is often found escaped from gardens.

*Atriplex nuttallii* S. Wats.                NUTTALL'S ATRIPLEX

A perennial shrub or subshrub with a very deep rooting system. May be nearly prostrate, or with branches to 75 cm high. Leaves and stem pale green, with a fine scurfiness; leaves linear-oblong to spatulate or obovate, 2–5 cm long. Almost certainly dioecious (male flowers on one plant and female flowers on another). Readily eaten by livestock and considered a useful pasture plant

Fig. 98.   Garden atriplex, *Atriplex hortensis* L.

because of its high mineral content and palatability. Found on badlands, eroded soils, and alluvial flats; throughout southern Prairies.

*Atriplex patula* L.                                                    ORACHE

An annual plant 30–90 cm high; stem coarse, erect, often marked with vertical grooves or channels. There are two varieties: the common orache, *A. patula* L.; and the halberd-leaved orache, *A. patula* L. var. *hastata* (L.) Gray. In the common orache the leaves taper to their stalks, but in the halberd-leaved orache the base of the leaf is flatter and abruptly narrowed to the stalk. Both varieties common; in saline meadows and waste places; particularly in southwestern Prairies. Often mistaken for lamb's-quarters.

*Atriplex powellii* S. Wats.

Annual plants 20–50 cm high, freely branched; leaves 1–2 cm long, silvery white. Flowers clustered in the axils; fruiting bracts about 5 mm broad, toothed below. Rare; alkali flats; southwestern Prairies.

*Axyris*        axyris

*Axyris amaranthoides* L. (Fig. 99)                        RUSSIAN PIGWEED

An erect, bushy annual weed 20–60 cm high. Stem much-branched, with ascending branches. Leaves lanceolate, pale green, to 8 cm long. Flowers borne in dense, leafy clusters; male flowers toward the ends of the stems. The whole plant turns a pale straw color at maturity. Seeds are of two types on the same plant: one type bearing a 2-lobed, long oval-shaped membranous wing at the top; and the other type almost round with no wings. The long winged seeds germinate readily, but the round wingless ones have a long period of dormancy. This introduced annual is very common in shelterbelts, on cultivated fields and abandoned lands, and in waste places; throughout the Prairie Provinces. It can be found often among bushes far from any cultivated land.

*Bassia*        bassia

*Bassia hyssopifolia* (Pall.) Kuntze                      FIVE-HOOK BASSIA

A slender-stemmed, weedy, branching annual 15–60 cm high. Leaves linear, 10–25 mm long, grayish green, usually closely appressed to the stem. Inflorescence in form of narrow interrupted spikes of tiny, inconspicuous flowers. Fruit not much larger than the head of a pin but enclosed with 5 tiny bracts, each bearing a small hooked spine. Very plentiful in some saline areas in southwestern Prairies, appears to be spreading eastward along railway grades.

*Chenopodium*        goosefoot

Annual plants, usually weedy, with alternate, often farinose or mealy leaves. Flowers perfect, with no bracts, calyx of 2–5 sepals enclosing the fruit.

1. Calyx fleshy, red, and globular, resembling small strawberry. ............................................................. *C. capitatum*
   Calyx not fleshy, red, or strawberry-like. .................................................................. 2

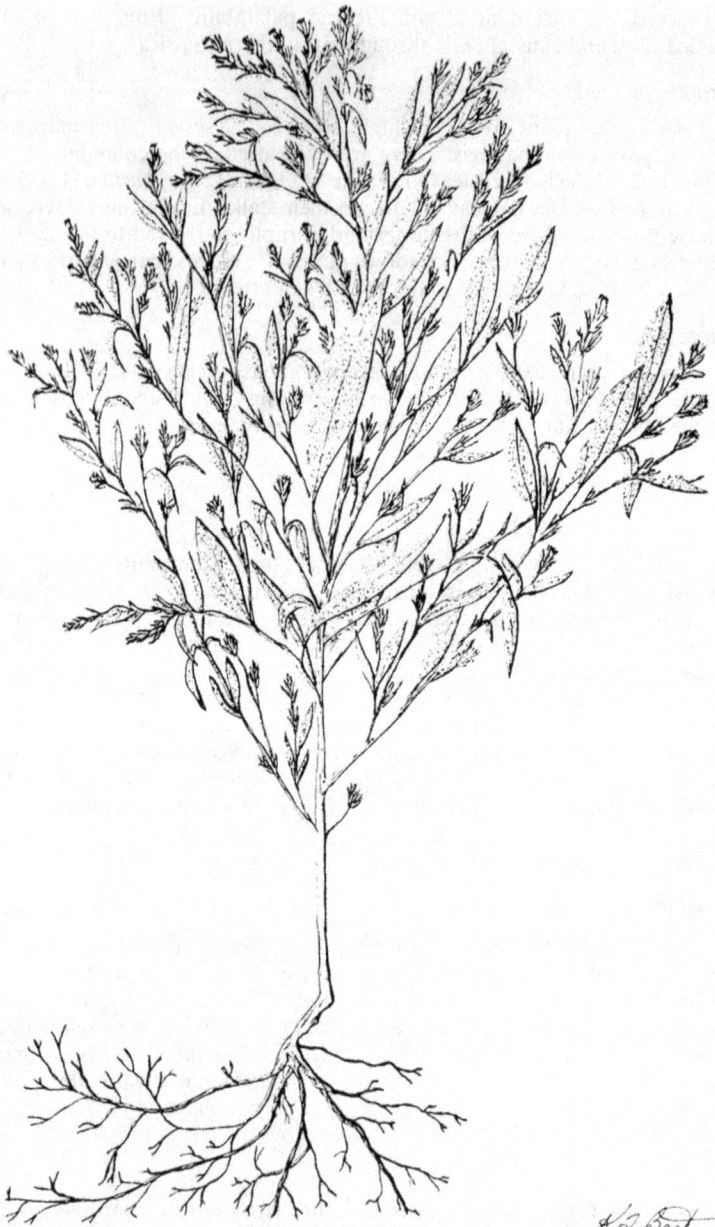

Fig. 99.  Russian pigweed, *Axyris amaranthoides* L.

2. Stamens 1 or 2; calyx only slightly fleshy
and reddish. ................................................................. *C. rubrum*

Stamens 5; calyx not fleshy. ................................................................. 3

3. Leaves with large, divaricate lobes, not
mealy, shiny green; seeds about 2 mm
in diam. ................................................ *C. hybridum* var. *gigantospermum*

Leaves entire or wavy-edged; seeds not
over 1 mm in diam. ................................................................. 4

4. Calyx lobes not keeled; panicles of
flowers in axils of leaves and shorter
than the leaves. ................................................ *C. glaucum* var. *pulchrum*

Calyx lobes keeled; upper panicles of
flowers longer than the leaves. ................................................................. 5

5. Seeds separating readily from their coat-
ing. ................................................................. 6

Coating rather firmly attached to the
seeds. ................................................................. *C. album*

6. Leaves ovate to triangular or hastate,
many-nerved. ................................................................. 7

Leaves linear, entire, 1- to 3-nerved. ................................................................. 8

7. Leaves triangular or hastate, to ovate,
almost as broad at the base as long. ................................................ *C. fremontii*

Leaves ovate to oblong-lanceolate,
rounded to cuneate at the base. ................................................ *C. polyspermum*

8. Leaves grayish mealy, especially below. ................................................ *C. leptophyllum*

Leaves pale green, almost glabrous, not
mealy. ................................................................. *C. subglabrum*


*Chenopodium album* L. (Fig. 100)                LAMB'S-QUARTERS

A rank annual weed 20–80 cm high. Leaves variable, usually wavy-margined and roughly ovate with a mealy coating, up to 7 cm long. Stems usually
longitudinally grooved, often with reddish lines and blotches. Inflorescence
usually of a bluish tinge, in dense panicles in leaf axils and at summit of stem.
Although this species has been introduced from Europe, it is probably also a
native. Young plants of this species are often eaten in place of spinach. One of
the commonest weeds; found in waste places, roadsides, and gardens;
throughout the Prairie Provinces. Syn.: *C. lanceolatum* Muhl.; *C. dacoticum*
Standl.


*Chenopodium capitatum* (L.) Aschers.                STRAWBERRY BLITE

An annual herb with stem either erect or spreading, to 40 cm high. Leaves
pale green, roughly triangular, 3–7 cm long. Inflorescence in form of small
round clusters at intervals on the stem, turning red, resembling small strawber-
ries. Found on rocky or stony soil and around bluffs and woodlands; throughout the Prairie Provinces, particularly in Boreal forest.

Fig. 100. Lamb's-quarters, *Chenopodium album* L.

*Chenopodium fremontii* S. Wats. (Fig. 101)          FREMONT'S GOOSEFOOT

An annual with rather slender, erect stems, usually with longitudinal dark green lines. Grows to about 60 cm high, and has broadly triangular leaves up to 5 cm long, pale green on both sides. Inflorescence open, interrupted, and not dense, usually appearing rather straggly. Common among bushes and bluffs; throughout the Prairies and Parklands.

*Chenopodium glaucum* L. var. *pulchrum* Aellen          SALINE GOOSEFOOT

A low prostrate plant 10–40 cm long, with rather fleshy, often reddish stem. Leaves usually small, varying from somewhat triangular to oval or oblong, sinuately dentate or lobed, resembling very small oak leaves, 10–25 mm long, and mealy on underside. Inflorescence in small spikes in axils of leaves. Common in moist, saline locations; throughout Prairies and Parklands. Syn.: *C. salinum* Standl.

*Chenopodium hybridum* L. var. *gigantospermum* (Aellen) Rouleau
MAPLE-LEAVED GOOSEFOOT

A tall annual 50–120 cm high. Leaves bright green, rather shiny, to 15 cm long, usually having 2–4 large sharp-pointed lobes on either margin and a long pointed apex. Inflorescence somewhat similar to that of *C. fremontii*, with which it is often associated, loose and interrupted, in long open panicles. Fairly common in shady wooded places; throughout the Prairie Provinces.

*Chenopodium leptophyllum* Nutt.          NARROW-LEAVED GOOSEFOOT

A tall erect annual with a somewhat mealy stem, striate or longitudinally grooved with alternate yellow and green lines. Leaves linear or linear-lanceolate, pale green and mealy above and densely mealy below, usually entire, and to 2–4 cm long and 6 mm wide. Inflorescence in small spikes in upper leaf axils and at summit of stem, densely mealy. Very plentiful on dry, sandy soil; in southwest Prairie Provinces, especially in the sandhills.

*Chenopodium polyspermum* L.          MANY-SEEDED GOOSEFOOT

An erect or procumbent annual to 1 m high, with glabrous stems and leaves. Leaves ovate-elliptic, entire or a slight tooth above the base. Inflorescence long, lax with many axillary cymes. Rare; an introduced weed; Parklands.

*Chenopodium rubrum* L.          RED GOOSEFOOT

A tall erect plant to 75 cm high, with ascending branches. Leaves not mealy, thick, and dark green, pointed at both ends, coarsely toothed, 3–10 cm long. Flower clusters in leafy spikes in leaf axils, fruit turning dull red when ripe. Very common in saline, moist soil; throughout the Prairie Provinces.

*Chenopodium subglabrum* (Wats.) Nelson

An erect annual 20–80 cm high, with lanceolate or oblong leaves, glabrous and green above, 3–6 mm long. Inflorescence diffuse; the stem leaves mealy. Rare or overlooked; on sandy soils; Prairies.

Fig. 101. Fremont's goosefoot, *Chenopodium fremontii* S. Wats.

*Corispermum*        bugseed

Annual plants with very branched stems and narrow, linear leaves; flowers solitary in the axils of the leaf-like bracts and the seeds flattened, giving rise to the common name.

Plants not hairy; fruit with a distinct wing
around it. .................................................................................... *C. hyssopifolium*
Plants slightly hairy; fruit scarcely winged,
merely acute-margined. ...................................................... *C. orientale* var. *emarginatum*


*Corispermum hyssopifolium* L.                                   BUGSEED

A much-branched annual 20–50 cm high, with narrow, linear, pale green leaves about 3–5 cm long. Inflorescence in small clusters at axils of leaves, each separate flower having a small leaf-like bract. Seeds medium brown, about 3 mm across, flat on one side and slightly convex on the other, with a decided wing around them. Not very common; but has been found on sandy soil in several locations; in southwest and south central Prairies. Syn.: *C. marginale* Rydb.

*Corispermum orientale* Lam. var. *emarginatum* (Rydb.) Macbr. (Fig. 102)
VILLOSE BUGSEED

An annual, very similar to *C. hyssopifolium,* but with stems and leaves having fine, white, stellate or star-shaped hairs. Seeds similar to the preceding but without a wing. Plant eaten by livestock; also a good sand binder. Fairly plentiful in sandy soil; especially the sandhills of southwest. Syn.: *C. villosum* Rybd.


*Cycloloma*        winged pigweed

*Cycloloma atriplicifolium* (Spreng.) Coulter          WINGED PIGWEED

A much-branched annual 15–50 cm high, sometimes with cobwebby hairiness on stem and leaves. Leaves up to 6 cm long, lanceolate, with a toothed margin. Inflorescence borne on open panicles; each fruit having a broad membranous wing below it. When mature, the plant becomes purplish and breaks off, blowing with the wind and scattering its seed. Rare; but has been found on sandy soils; in Prairies.


*Endolepis*        endolepis

*Endolepis suckleyi* Torr.                                      ENDOLEPIS

A low erect annual to 30 cm high, with short leaves 15–25 mm long, lanceolate and with 1 nerve. Stem and midrib often purplish. Flowers in clusters on short spikes in axils of leaves and at ends of stem. Not common, but plentiful locally; found on saline flats, eroded clay slopes, and badlands; throughout southern Prairies. Syn.: *Atriplex dioica* (Nutt.) Macbr.

Fig. 102.  Villose bugseed, *Corispermum orientale* Lam. var. *emarginatum* (Rydb.) Macbr.

A.C. Budd

Fig. 103.   Winterfat, *Eurotia lanata* (Pursh) Moq.

*Eurotia*        winterfat

*Eurotia lanata* (Pursh) Moq. (Fig. 103)                    WINTERFAT

A perennial subshrub or herb 15–50 cm high. Plant covered with fine, star-like white hairs, which become reddish as the plant ages. Leaves 1–5 cm long, linear, with rolled margins. Flowers unisexual, both male and female flowers being on the same plant, the male flower clusters above the female clusters. Female flowers and fruit enclosed in 2 bracts, which are united almost to the top and have 2 horns on the top. Whole female inflorescence clothed with long, silky, white hairs, making the plants very conspicuous. Much relished by livestock as an excellent forage plant. Very useful on winter range because its protein and mineral content remain high throughout the fall and winter. Common on dry prairies and on heavy soils; throughout Prairies.


*Kochia*        kochia

*Kochia scoparia* (L.) Schrad.                    SUMMER-CYPRESS

An annual plant, escaped from gardens, about 60 cm high, erect, and of regular pyramidal or ovoid shape. Leaves linear and closely compacted, pale green when young, but purplish red when mature. The plant produces many seeds, which germinate readily and form a mat of seedlings the following year. Used as an ornamental in gardens because of its symmetrical shape and its red coloring in the fall, but its prolific seeding habits have caused it to become a weed in the vicinity of most towns and cities in Prairies and Parklands.


*Monolepis*        monolepis

*Monolepis nuttalliana* (R. & S.) Greene        SPEAR-LEAVED GOOSEFOOT

A prostrate annual with fleshy reddish stems and hastate or spear-shaped leaves, which form a rosette in the plant's early stage. Spreading on the soil, 20–50 cm across. Inflorescence in small clusters in axils of leaves, produced early in season. One of the first weeds to start in the spring on fallow and in waste places, where its fleshy taproots and succulent stems cause a great drain on moisture reserves. Common around gopher holes; a native that has become weedy and, though originally found on saline soils, is now found on cultivated land; throughout Prairies and Parklands.


*Salicornia*        samphire

*Salicornia rubra* A. Nels.                    RED SAMPHIRE

A low annual with no true leaves, their places being taken by scales at the nodes of the stems. Stems circular in cross section (terete), the branches opposite. Flowers very minute, sunk into the tissue of the stems. Plant turning a bright crimson at maturity and giving a reddish color to dry sloughs in late summer and early autumn. Very common in strongly saline sloughs; throughout Prairies and Parklands. Syn.: *S. europaea* L. ssp. *rubra* (Nels.) Breitung; *S. europaea* L. var. *prona* (Lunell) Boiv.

**Salsola**     glasswort

*Salsola kali* L. var. *tenuifolia* Tausch.          RUSSIAN-THISTLE

An introduced annual weed 10–60 cm high. Early leaves dark green, thread-like, about 2 cm long; later leaves shorter and broader, coming to a sharp hard point. Flowers borne in axils of upper leaves, with a membranous wing on calyx around seed. As plants age they become dry and the spiny tips of the leaves and bracts harden, making the whole plant prickly. Plants usually become reddish and at maturity break off at the ground and drift across the country with the winds, scattering their seeds. Palatable to livestock, but highly laxative. Very common; in fields, waste places; throughout Prairies; less common in Parklands. Syn.: *S. pestifer* A. Nels.

**Sarcobatus**     greasewood

*Sarcobatus vermiculatus* (Hook.) Torr. (Fig. 104)          GREASEWOOD

A much-branched, shrubby perennial, with spiny branches, in some localities 1.5–2 m high. Stems usually almost white. Leaves pale yellowish green, linear, about 2–4 cm long. Male flowers borne in small cylindric spikes at ends of stems, female flowers borne singly in axils of leaves. A broad membranous wing forms on calyx around fruit. If eaten in large amounts, greasewood is rather **poisonous** to livestock, especially to lambs during the spring and summer. It contains sodium and potassium oxalates and during dry seasons these salts occur in a concentrated form. Around strongly saline sloughs and flats; throughout Prairies.

**Suaeda**     sea-blite

*Suaeda depressa* (Pursh) S. Wats.          WESTERN SEA-BLITE

A low-growing annual or perennial, with very dark green, rather fleshy, narrow leaves. There are two forms: the species *depressa* and its variety *erecta* or erect sea-blite. In the *depressa* form the plant is low and spreading and the leaves are 10–30 mm long. The var. *erecta* is upright, with longer leaves 15–40 mm long. Flowers dark greenish, clustered in axils of upper leaves. At maturity plants turn very dark, almost black. Common in saline areas around sloughs and saline flats; throughout Prairies and Parklands. The var. *erecta* S. Wats. is probably a little more plentiful than the species. Syn.: *S. maritima* L. var. *americana* (Pers.) Boiv.

**Suckleya**     suckleya

*Suckleya suckleyana* (Torr.) Rydb. (Fig. 105)          POISON SUCKLEYA

A somewhat prostrate, much-branched annual, with succulent stems up to 30 cm long. Leaves stalked, alternate, round to diamond-shaped, up to 3 cm long, irregularly blunt-toothed along the upper margins. Flowers unisexual, the male in the axils of the upper leaves, the female in the lower leaf axils; female flowers enclosed in a couple of stiff winged bracts. Fruit 5–6 mm long with an abruptly pointed end. This plant is **poisonous** to livestock because it contains hydrocyanic acid. Not plentiful; however, it has been found in moist and saline areas; in the south central Prairies.

A.C. Budd

Fig. 104.   Greasewood, *Sarcobatus vermiculatus* (Hook.) Torr.

Fig. 105. Poison suckleya, *Suckleya suckleyana* (Torr.) Rydb.

# AMARANTHACEAE—amaranth family

***Amaranthus***     amaranth

Coarse, annual, weedy plants, somewhat similar to the goosefoot family, but having flowers enclosed in three dry and persistent bracts.

1. Plants tall; upper flowers in dense terminal spikes. ........................................................................... *A. retroflexus*

   Plants low, much-branched; flowers in small clusters in axils of leaves. ........................................................... 2

2. Plants erect and bushy; leaves pale green and not shiny; sepals 3, much shorter than the somewhat spine-tipped bracts. ..................................... *A. albus*

   Plants prostrate, with succulent reddish stems; leaves dark green and rather shiny; sepals 4 or 5, almost as long as the blunt-tipped bracts. ................................................... *A. graecizans*

*Amaranthus albus* L.                                    TUMBLEWEED

A bushy, much-branched, whitish-stemmed annual 25–60 cm high, with pale green, dull, spatulate leaves 10–35 mm long. The end of the midrib of the leaf is usually projected as a tiny spine at the end of the leaf, a condition often found in other species of amaranth, but more pronounced in this species. Flowers in small clusters in leaf axils; seed black and shiny, enclosed in a small utricle, or envelope, the top of which falls off at maturity. A native plant and an early invader of newly broken ground, fireguards, and rough ground. When dry, this plant breaks off and blows across the prairie, dispersing its seeds. Now a common weed; on waste ground, in gardens, and along roadsides; throughout the Prairies and Parklands. This species has usually been called *A. graecizans*, but that name was intended by Linnaeus for the prostrate amaranth.

*Amaranthus graecizans* L. (Fig. 106)        PROSTRATE AMARANTH

A coarse, prostrate, weedy annual, with reddish, fleshy stems forming mats 15–60 cm in diam. Leaves usually spatulate, the broadest part beyond the middle, 5–25 mm long, dark, shiny green. Flowers borne in leaf axils. Not considered a native of the area, but has come in from the southwest, where it is a native of dry sites in the intermountain areas. A very common garden weed; throughout Prairies and Parklands. This species has been known as *A. blitoides* S. Wats.

*Amaranthus retroflexus* L.                          RED-ROOT PIGWEED

A coarse, rough, erect annual with reddish-colored roots. Often growing to 1 m high and having rough, angular stems somewhat hairy near the top. Leaves usually ovate, on fairly long stalks, rough to the touch, and 6–10 cm long. Inflorescence harsh and rough, borne in dense spikes in leaf axils and in a large terminal spike at summit of stem. Sepals dry and parchment-like with spiny tips, the 3 bracts around each flower also spine-tipped, giving the inflorescence a decidedly rough appearance. Seeds shiny, black, and in a small

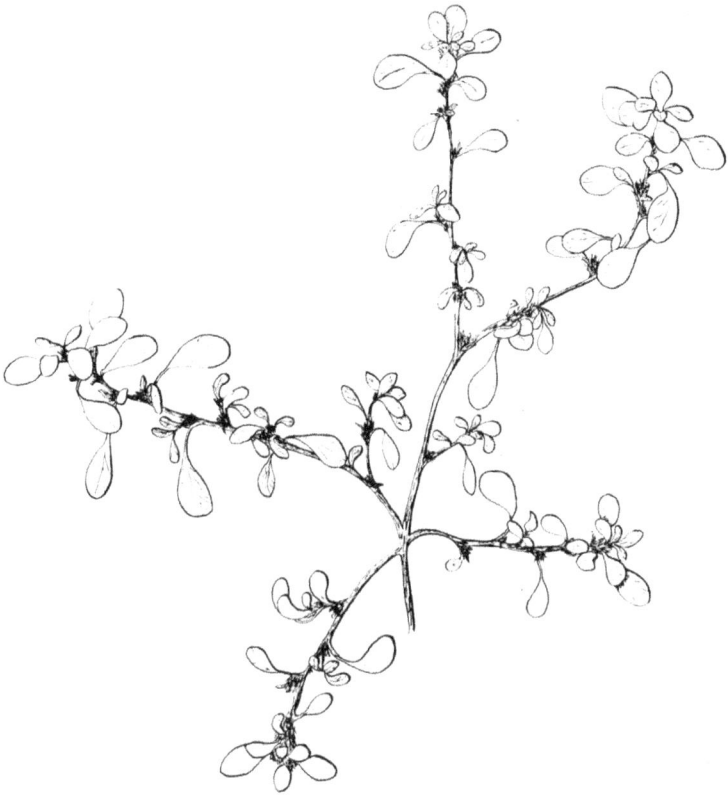

Fig. 106.   Prostrate amaranth, *Amaranthus graecizans* L.

utricle, similar to other amaranths. A native of the subtropical states of USA, but a common weed over all North America. Very common; in waste places, roadsides, gardens, and fields; throughout Prairies and Parklands.

## NYCTAGINACEAE—four-o'clock family

Plants annual; bracts free. ............................................................................................ *Abronia*

Plants perennial; bracts united into a tube. ................................................................ *Mirabilis*

### *Abronia*　　　sand verbena

*Abronia micrantha* Torr.　　　　　　　　　　　　　　　SAND PUFFS

　　Annual plants with decumbent stems, 10–30 cm high, branching at base, often somewhat pubescent. Leaves opposite, petioled, 2–4 cm long, prominently veined. Inflorescence a head supported by 4–6 bracts; perianth a corolla-like salverform tube with 5 petal-like lobes. Fruit with wings 10–15 mm, rarely 20 mm, broad. Rare; sand dunes in Prairies.

### *Mirabilis*　　　umbrellawort

　　Perennial herbs with opposite, entire leaves without stipules. Stems usually swollen at nodes. Flowers regular and perfect, with no petals, but sepals brightly colored and petal-like, and generally united into a bell-like or funnelform tube. Flowers usually in clusters of 3–5 with a saucer-like involucral bract beneath them.

1. Leaves cordate, heart-shaped, with distinct stalks. ............................................................ *M. nyctaginea*

   Leaves not cordate, but ovate-lanceolate or linear, mostly without stalks or the lower ones with short stalks. ........................................ 2

2. Stem more or less hairy, sticky. ............................................................ *M. hirsuta*

   Stem not at all hairy below, but fine, sticky-haired above. ........................................ *M. linearis*

### *Mirabilis hirsuta* (Pursh) MacM.　　　HAIRY UMBRELLAWORT

　　A slender erect plant 30–60 cm high, with glandular hairs on the stem. Leaves usually lanceolate or linear-lanceolate and hairy, 2–7 cm long. Another form equally common has the lower part of the stem and leaves almost devoid of hairs, except just under the nodes of the stem, and slightly narrower leaves. This form has sometimes been designated *Oxybaphus pilosus*, a separate species. Most authorities, however, do not separate the forms. Common; on sandy soils; throughout Prairies and Parklands. Syn.: *Oxybaphus hirsutus* (Pursh) Sweet.

### *Mirabilis linearis* (Pursh) Heimerl　　　LINEAR-LEAVED UMBRELLAWORT

　　Erect, with very narrow, single-nerved leaves, 15–65 mm long. It was reported from two locations in the southwestern area by the early botanist Macoun. However, its range is south of the Prairie Provinces and its appearance north of the International Boundary is unusual. Syn.: *Oxybaphus linearis* (Pursh) Robinson.

*Mirabilis nyctaginea* (Michx.) MacM.      HEART-LEAVED UMBRELLAWORT

A tall erect species 40–90 cm high, with an almost hairless stem and large cordate or heart-shaped leaves, 5–10 cm long and 2–7 cm wide. Inflorescence usually deep reddish, with a greenish involucral wing below the flowers. Fairly common; in rich soils in southeastern Parklands and Boreal forest, gradually spreading westward along the railway tracks. Several large clumps have been located on the main line of the Canadian Pacific Railway in southern Saskatchewan, and, where found, it appears to be the dominant plant on the cinder fill of the railway grade. Syn.: *Oxybaphus nyctagineus* (Michx.) Sweet.

# PORTULACACEAE—purslane family

Rather succulent plants with perfect flowers, 2 sepals, and 5 petals. The fruit is a capsule, which opens with 2 or 3 valves at the top or with the top falling off like a lid (circumscissile).

1. Capsule opening by valves. ........................................................... *Claytonia*
   Capsule circumscissile at maturity. ........................................................... 2

2. Ovary partly inferior; prostrate annual
   weed. ........................................................... *Portulaca*
   Ovary superior; perennial with taproot
   and bracteate scape. ........................................................... *Lewisia*

## *Claytonia*      springbeauty

1. Leaves alternate. ........................................................... 2
   Leaves opposite. ........................................................... 3

2. Plants annual; leaves linear. ........................................................... *C. linearis*
   Plants perennial; leaves ovate to lanceolate. ........................................................... *C. parvifolia*

3. Plants annual; stem leaves numerous. ........................................................... *C. fontana*
   Plants perennial. ........................................................... 4

4. Plants with a globose corm; stem leaves 2. ........................ *C. caroliniana* var. *lanceolata*
   Plants with a thick taproot; leaves basal. ........................................................... *C. megarrhiza*

*Claytonia caroliniana* Michx. var. *lanceolata* (Pursh) Wats.
     LANCE-LEAVED SPRINGBEAUTY

A low-growing early spring flowering plant, growing from a corm or tuber-like globose root 6–10 cm below the soil surface. In some plants several stems arise from a single corm. Occasionally a plant bears one or two basal, stalked leaves, but they are usually absent. Two stalkless (sessile) opposite leaves are borne on the stem, lanceolate in shape, 15–50 mm long, with three distinct veins. Flowers have petals about 1 cm long, white with pink lines or faintly pinkish. A western species, common along the margin between grass-land and wooded areas in the Foothills, but eastward has been collected only in the Cypress Hills, where it is found flowering in large masses soon after snow melts in spring. Found along edges of woodlands and around margins of clearings.

### *Claytonia fontana* (L.) Davis INDIAN LETTUCE

Low annual with weak branching stems, often rooting at the nodes. Leaves opposite, 5–15 mm long. Inflorescence leafy, the flowers nodding. Rare; wet places; Hudson Bay.

### *Claytonia linearis* Dougl. LINEAR-LEAVED SPRINGBEAUTY

A low-branching annual, growing from fibrous roots to 6–10 cm high. Leaves linear or almost thread-like, 10–30 mm long. Flowers nodding, 2–7 in a raceme, with pale pinkish petals about 3 mm long. A western species, found in very early spring in higher meadows in the Cypress Hills and southern Rocky Mountains. Syn.: *Montia linearis* (Dougl.) Greene.

### *Claytonia megarrhiza* (Gray) Parry ALPINE SPRINGBEAUTY

Plants usually not more than 10 cm high; basal leaves numerous, fleshy, spatulate or obovate, 1–8 cm long; flowering stalks numerous, flowers 10–15 mm across, white or pink. Rocky places, scree fields; southern Rocky Moun-tains.

### *Claytonia parvifolia* Moc. SMALL-LEAVED SPRINGBEAUTY

Perennial plants with branched rootstocks, often developing stolons. Basal leaves 10–35 mm long, obovate, long-petioled, more or less fleshy; stem leaves reduced, alternate. Inflorescence racemose, with 1–10 flowers, 10–15 mm across, white or pink. Moist areas, mountain meadows; southern Rocky Mountains.

### *Lewisia* bitter-root

### *Lewisia pygmaea* (Gray) Rob. DWARF BITTER-ROOT

Perennial with a thick fleshy taproot bearing several stems. Basal leaves linear, 3–6 cm long, 1.5–3 mm wide; stems 15–50 mm long, with a pair of opposite bracts. Flowers 1–3, about 15 mm across, white or pinkish. Rare; alpine slopes; southern Rocky Mountains.

### *Portulaca* purslane

### *Portulaca oleracea* L. PURSLANE

A succulent, prostrate annual, sometimes making a mat up to 40 cm across. Stems reddish, very thick, fleshy, juicy, and hairless. Leaves alternate, dark shiny green, spatulate or obovate, 5–25 mm long, thick, and fleshy. Flowers borne singly, without stalks, in the axils of the leaves, and usually open only in bright sunshine, bright yellow, about 6 mm in diam, with 2 sepals and 5 petals. Seeds numerous and minute, contained in a pointed capsule, or pyxis, the top of which breaks off, releasing the seeds. An introduced weed, common in gardens; throughout the Prairie Provinces.

## CARYOPHYLLACEAE—pink family

Herbs with stems swollen at the joints and entire, opposite leaves. Flowers usually with 4 or 5 petals, sometimes none, and 4 or 5 sepals. The fruit is a capsule opening by valves at the top. A difficult family to identify, often requiring the fruiting capsule for positive identification.

1. Plants with woody base; leaves with spiny tips; fruit with a single seed. ..................................................... *Paronychia*
   Plants not with woody base; fruit many-seeded. ........................................................................ 2

2. Sepals united part way, forming a tube. ................................................................ 3
   Sepals entirely separate. ........................................................................ 8

3. Calyx with 1–3 pairs of bracts below. ........................................................ *Dianthus*
   Calyx without bracts. ........................................................................ 4

4. Styles 2 only. ........................................................................ 5
   Styles more than 2. ........................................................................ 6

5. Calyx becoming inflated and wing-angled; flowers few, pink. ........................................................ *Saponaria*
   Calyx not inflated or angled; flowers very numerous. ........................................................ *Gypsophila*

6. Sepals with long, leaf-like lobes, over 20 mm long; styles 5, opposite the petals; flowers dark purple. ........................................................ *Agrostemma*
   Lobes of sepals not over 20 mm long; styles alternate with the petals. ........................................................ 7

7. Styles usually 3; capsule usually divided into partitions at base. ........................................................ *Silene*
   Styles 5; capsule 1-celled to base. ........................................................ *Lychnis*

8. Plants with small, ovate, papery stipules at junction of stem and leaf, and stem and branch. ........................................................ 9
   Plants without stipules. ........................................................................ 10

9. Styles 5; leaves in whorls. ........................................................ *Spergula*
   Styles 3; leaves opposite. ........................................................ *Spergularia*

10. Capsule opening with twice as many valves or teeth as there are styles; petals deeply 2-cleft. ........................................................ 11
    Capsule opening with as many entire (or later 2-cleft) valves as there are styles; petals entire or merely notched at apex. ........................................................ 12

11. Capsule short, ovate or oblong, usually opening with 6 valves; styles usually 3. ........................................................ *Stellaria*
    Capsule long, cylindric, often curved, usually opening with 10 valves; styles usually 5. ........................................................ *Cerastium*

12. Flowers 5-merous. ..................................................................... *Sagina*

Flowers with 5 sepals, 3 styles, and 3
valves. .................................................................................. *Arenaria*

## *Agrostemma*   corn cockle

*Agrostemma githago* L. (Fig. 107)                    PURPLE COCKLE

An erect hairy annual plant with a taproot, 30–75 cm high, with hairy linear leaves up to 10 cm long. Flowers borne singly at head of stems with hairy sepals united at base, but with long lobes often up to 2.5 cm, much longer than the petals. Petals purple and flowers 25–40 mm in diam. Seed about 3 mm in diam, somewhat flattened, black, and roughened with rows of minute protuberances. Seeds **poisonous** to chickens. A weed introduced from Europe, but not common; in grainfields throughout the Prairie Provinces.

## *Arenaria*   sandwort

Low tufted herbs with opposite leaves. Flowers white, with 4 or 5 sepals and 4 or 5 entire or slightly notched petals.

1. Leaves oval to oblong. ............................................................. 2
   Leaves linear or filiform. ......................................................... 6
2. Bracts of the inflorescence small, scarious. ........................... 3
   Bracts of the inflorescence large, leaf-like. ........................... 4
3. Leaves pubescent below along the
   midrib; sepals rounded. ................................... *A. lateriflora*
   Leaves glabrous below; sepals acuminate. ........ *A. macrophylla*
4. Plants fleshy, glabrous. ......................... *A. peploides* var. *diffusa*
   Plants not fleshy, at least the pedicels
   puberulent. .............................................................................. 5
5. Plants perennial; stems 3–10 cm high,
   puberulent; leaves glabrous. ................................... *A. humifusa*
   Plants annual; stems 5–20 cm high,
   puberulent; leaves puberulent. ......................... *A. serpyllifolia*
6. Sepals rounded or obtuse at the tip. ...................................... 7
   Sepals distinctly acute to acuminate. ..................................... 9
7. Leaves 3–8 mm long, about 1 mm wide. ........ *A. laricifolia* var. *occulta*
   Leaves 1–6 cm long, linear to filiform. ................................. 8
8. Plants glabrous. ................................... *A. congesta* var. *lithophila*
   Plants glandular puberulent, at least in
   the inflorescence. ............................. *A. capillaris* var. *americana*
9. Leaves clearly spine-tipped, recurved
   spreading. ................................................................ *A. nuttallii*
   Leaves not spine-tipped. ......................................................... 10
10. Flowers solitary, terminal. ...................................................... 11
    Flowers in cymose inflorescence, 2 or
    more together. ........................................................................ 12

Fig. 107. Purple cockle, *Agrostemma githago* L.

11. Inflorescence cymose, open. ................................. *A. stricta* ssp. *dawsonensis*
    Inflorescence with flowers solitary on
    leafy stems. ........................................ *A. rossii* var. *columbiana*
12. Plants usually glabrous; stem leaves often
    with axillary branches. ...................... *A. stricta* ssp. *dawsonensis*
    Plants usually glandular puberulent, at
    least in the inflorescence; axillary
    branches few or none. ............................................... *A. verna*

*Arenaria capillaris* Poir. var. *americana* (Maguire) Davis

Plants tufted, with a slender caudex; stems 5–15 cm high. Leaves 2–7 cm long, filiform, mostly erect. Inflorescence a few-flowered cyme, with the pedicels 5–20 mm long, sepals 3–4 mm long, petals 6–8 mm long. Alpine meadows; southern Rocky Mountains.

*Arenaria congesta* Nutt. var. *lithophila* (Rydb.) Maguire
ROCKY-GROUND SANDWORT

An erect perennial to 30 cm high, with tufted basal leaves and a few stem leaves. Leaves linear and thread-like, 1–5 cm long. Flowers in open clusters on tops of stems, with white petals 4–6 mm long, and straw-colored sepals. Very local, but found on rocky benchland and slopes; Cypress Hills, Wood Mountain. Syn.: *A. lithophila* Rydb.

*Arenaria humifusa* Wahl.

Plants tufted, mat-forming; stems 3–10 cm high, slender, few-flowered; leaves 2–10 mm long; sepals 3–5 mm long; petals 5–7 mm long. Sand and gravel bars; Boreal forest.

*Arenaria laricifolia* (L.) Rob. var. *occulta* (Ser.) Boiv.

Plants with more or less woody, branched caudex, mat-forming; leaves 3–7 mm long, 0.5–1 mm wide; flowering stems ascending, puberulent, often glandular; 1- to 6-flowered; sepals 3–5 mm long, glandular puberulent; petals 4–7 mm long. Stony slopes, scree fields; Rocky Mountains. Syn.: *Minuartia laricifolia* (L.) S. & T.; *A. obtusiloba* (Rydb.) Fern.

*Arenaria lateriflora* L.
BLUNT-LEAVED SANDWORT

An erect perennial with thin, weak, slightly hairy stems and oval to oblong, thin, pale green leaves. Flowers about 6–12 mm across, usually borne in pairs, and petals about twice the length of sepals; sometimes 4 sepals and petals, and sometimes 5 of each. Common; in moist woodlands; throughout Prairies and Parklands. Syn.: *Moehringia lateriflora* (L.) Fenzl.

*Arenaria macrophylla* Hook.
LARGE-LEAVED SANDWORT

Resembling the preceding species in habit and pubescence. Leaves mainly 2–5 cm long, 3–8 mm wide; sepals 3–6 mm long; petals 5–10 mm long. Rare; rocky areas; Boreal forest, Saskatchewan.

*Arenaria nuttallii* Pax

Plants with a deep taproot, many-stemmed, loosely matted stems 6–12 cm high, glandular pubescent; leaves 6–10 mm long; inflorescence a few- to many-flowered cyme, with the pedicels 5–15 mm long, sepals 4–5 mm long, petals shorter than sepals. Slopes; southern Rocky Mountains.

*Arenaria peploides* L. var. *diffusa* Hornem.                    SEA-PURSLANE

Plants glabrous; stems trailing, succulent, rooting at the nodes. Leaves 5–20 mm long, 5–10 mm wide; flowers solitary in leaf axils, and in few-flowered terminal cymes; capsule 6–10 mm wide, globose. Beaches; along Hudson Bay.

*Arenaria rossii* R. Br. var. *columbiana* Raup

Plants densely tufted, with the flowering stems spreading. Leaves 2–5 mm long, awl-shaped, with axils of upper leaves having short branches; flowers solitary, with the pedicels 5–15 mm long, sepals 1.5–2.5 mm long, petals slightly shorter than sepals. Rare; alpine slopes; southern Rocky Mountains.

*Arenaria serpyllifolia* L.                    THYME-LEAVED SANDWORT

A much-branched annual with a slightly downy stem 8–20 cm high. Leaves small and ovate, 3–8 mm long; flowers very small, usually 6–10 mm across, and borne on summit of stem. Introduced from Europe and very rare; a weed in cultivated fields.

*Arenaria stricta* Michx. ssp. *dawsonensis* (Britton) Maguire   DAWSON SANDWORT

An annual, branched from the base, 10–30 cm high, with thread-like leaves usually 6–12 mm long; petals no longer than the sepals, which are 3-nerved and 2.5–4 mm long. Lakeshores, river flats; Boreal forest, Rocky Mountains.

*Arenaria verna* L.                    EARLY SANDWORT

Plants loosely tufted; stems 2–15 cm high, usually glandular pubescent above. Leaves 5–20 mm long, linear-lanceolate to oval-shaped; cymes 1- to 6-flowered; pedicels 2–25 mm long; sepals 3 mm long; petals 2–4.5 mm long. Slopes, meadows, and rock slides; Rocky Mountains, rare in Cypress Hills.

***Cerastium***          chickweed

Low-growing annuals or perennials with opposite leaves; white flowers with cleft petals and long capsule, usually opening with 10 valves.

1. Annuals; capsule 2–3 times as long as
   sepals. ................................................................................ *C. nutans*
   Perennials; capsules once or twice as long
   as sepals. ................................................................................ 2

2. Axillary shoots numerous, at the axils of
   most leaves. ................................................................................ 3
   Axillary shoots few or lacking. ................................................................................ 4

3. Plants glandular pubescent throughout. ................................................... *C. arvense*
   Plants densely white tomentose, not glan-
   dular. ................................................................................ *C. tomentosum*

4. Stems usually glandular pubescent, with
   the leaves stiffly hirsute, ciliate. ................................................ *C. vulgatum*

Stems and leaves pilose or glandular
pilose. ................................................................................................... *C. alpinum*

*Cerastium alpinum* L.                                                         ALPINE CHICKWEED

Perennial plants 5–20 cm high; leaves about 10 mm long, 5 mm wide, soft long pubescent, sometimes also glandular. Flowers on peduncles 1–4 cm long, 20–25 mm across, with the petals about twice as long as the sepals. Alpine slopes and scree fields; southern Rocky Mountains, Boreal forest.

*Cerastium arvense* L.                                                         FIELD CHICKWEED

A low-growing perennial species 10–30 cm high, often with the stems prostrate at the base. Stems covered with hairs pointing downward; white flowers 15–20 mm across, blooming very early in the season. A very common spring flower; on prairie; throughout the Prairie Provinces.

*Cerastium nutans* Raf.                                             LONG-STALKED CHICKWEED

A bright green erect annual to 25 cm high, with leaves 6–30 mm long. Fruiting capsules much longer than sepals, usually decidely curved at maturity. Fairly plentiful; in moist woodlands; throughout Boreal forest, Rocky Mountains.

*Cerastium tomentosum* L.                                                    SNOW-IN-SUMMER

Perennial mat-forming plants, densely white lanate, up to 45 cm high, with leaves 10–30 mm long, 2–5 mm wide, the margins somewhat revolute. Inflorescence an elongated cyme; flowers about 30 mm across. Introduced as an ornamental; escaped into various locations.

*Cerastium vulgatum* L.                                              MOUSE-EARED CHICKWEED

A biennial or perennial plant 10–40 cm high, with glutinous hairy stems and oblong-spatulate leaves 10–25 mm long. Fairly common; in woodlands and fields; Boreal forest.

**Dianthus**        pink

1. Calyx pubescent. ........................................................................ *D. deltoides*
   Calyx glabrous. .......................................................................................... 2
2. Leaves 5–20 mm wide, obtuse; flowers in
   heads. ...................................................................................... *D. barbatus*
   Leaves 0.5–1 mm wide, acute; flowers soli-
   tary. ....................................................................................... *D. sylvestris*

*Dianthus barbatus* L.                                                        SWEET WILLIAM

Perennial plants to 60 cm high. Leaves lanceolate, the lower ones obtuse; flowers white or pink, several crowded in a head. Introduced as an ornamental plant; occasionally escaped and persistent in moist places.

*Dianthus deltoides* L.                                                       STEPPEN PINK

Perennial plants 20–45 cm high. Leaves linear-lanceolate, short pubescent; flowers solitary, terminating stems or peduncles, reddish purple. Probably introduced; found only once, in Meadow Lake Provincial Park.

*Dianthus sylvestris* Wulf. WOOD PINK

Perennial plants 10–30 cm high. Leaves filiform; flowers solitary on stems or peduncles, pink, with petals fringed. Introduced as ornamental; occasionally escaped.

### *Gypsophila*    baby's-breath

1. Annual plants 30–50 cm high. ...................................................................... *G. elegans*
   Perennial plants 50–150 cm high. ................................................................ 2
2. Inflorescence usually glandular pubescent; calyx 3–3.5 mm long. .................................. *G. acutifolia*
   Inflorescence usually glabrous; calyx 1.5–2 mm long. ................................................ *G. paniculata*

*Gypsophila acutifolia* Steven STICKY BABY'S-BREATH

Perennial plants with long-creeping rootstocks, up to 150 cm high. Inflorescence a rather dense panicle, with the pedicels 1–4 mm long; flowers white. Introduced ornamental; occasionally escaped and established as a roadside weed.

*Gypsophila elegans* Bieb. ANNUAL BABY'S-BREATH

An annual, much-branched plant, with narrow lanceolate leaves and many white flowers up to 12 mm across, with petals much longer than sepals. Escaped from gardens, sometimes found growing in waste places.

*Gypsophila paniculata* L. BABY'S-BREATH

A perennial, much-branched plant, with linear-lanceolate leaves and many white flowers about 6 mm across, with petals about the same length as the sepals. Escaped from gardens, locally abundant in waste areas, roadsides, and gravel pits.

### *Lychnis*    campion

1. Petals large, much protruding beyond the calyx; calyx enlarging in fruit; flowers of different sexes on separate plants. ........................................................ *L. alba*
   Petals small, not or little protruding beyond the calyx; calyx not or little enlarging; flowers perfect. ................................................................ 2
2. Flowers nodding, usually solitary, rarely 2 or 3. ........................................................ *L. apetala*
   Flowers erect, usually 3 or more, on stiffly ascending pedicels. ........................................ 3
3. Calyx not inflated, tightly enclosing the capsule; petals included or barely exserted. ........................................................ *L. drummondii*
   Calyx somewhat inflated; petals conspicuously exserted. ........................................ *L. affinis*

*Lychnis affinis* J. Vahl                                    MOUNTAIN COCKLE

A perennial, with stems 5–30 cm high, stiffly erect, more or less glandular viscid; flowers usually 3, on erect pedicels; calyx 8–10 mm long, glandular, somewhat inflated, with petals white, conspicuously exserted. Rare; rock crevices; Hudson Bay, Rocky Mountains.

*Lychnis alba* Mill.                                         WHITE COCKLE

A biennial or short-lived perennial 30–75 cm high, much-branched, and with many sticky-haired stems. Leaves opposite, ovate-oblong, 2–8 cm long; flowers white, about 2 cm in diam, with stamen-bearing flowers on some plants and pistillate flowers on others. Fruiting capsule becoming enlarged and swollen at maturity and bearing 10 teeth at the top. Resembles night-flowering catch-fly, which, however, has perfect flowers. Becoming increasingly common; in western parts of the Prairie Provinces.

*Lychnis apetala* L.                                        NODDING COCKLE

Small perennials 5–15 cm high; stems pubescent, often reddish viscid above. Flowers nodding; calyx 10–15 mm long; petals included or barely exserted, purple. Plants with the calyx conspicuously inflated, and petals not exserted are var. *arctica* (Fries) Cody; those with the calyx not inflated, and petals clearly exserted are ssp. *attenuata* (Farr) Mag. Alpine areas; Rocky Mountains, Hudson Bay.

*Lychnis drummondii* Wats.                                  DRUMMOND'S COCKLE

A tall, slender, erect perennial to 70 cm high with a sticky, hairy stem and opposite linear leaves up to 10 cm long. Flowers borne at head of stem on fairly long stalks, usually white or purplish and only slightly exceeding the sepals. Sepals about 12 mm long and joined to form a somewhat cylindric tube, usually pale yellow with green lines. Not plentiful but very widespread; on open prairie, especially on sandy soils; throughout Prairies and Parklands. Syn.: *Wahlbergella drummondii* (Wats.) Rydb.


**Paronychia**        whitlowwort

*Paronychia sessiliflora* Nutt.                             LOW WHITLOWWORT

A woody-based perennial, forming dense cushions, usually 20–75 mm high and 7–15 cm in diam. Leaves linear, spine pointed and very short, 5–6 mm long, bright green, and so closely overlapping that the stems or branches are usually hidden. Flowers yellow, 3–4 mm high, and dotted singly at intervals on the plant. Fairly common; on dry hillsides and eroded places; throughout Prairies. On casual inspection, this species may be mistaken for moss phlox, *Phlox hoodii*, although moss phlox is more open. Syn.: *P. depressa* Nutt.


**Sagina**        pearlwort

1. Plants annual. ...................................................................................... *S. decumbens*
   Plants perennial. ...................................................................................... 2

2. Stem leaves at the upper 2 nodes distinctly shorter than lower ones, and with axillary bulb-like leaf bundles. ..................................................... *S. nodosa*
   Stem leaves all about the same size. ................................................................. 3

3. Plants tufted; sepals with purple margins. ........................................... *S. caespitosa*
   Plants mat-forming; sepals with white margins. ................................................................. *S. saginoides*

*Sagina caespitosa* (J. Vahl) Lange            CUSHION PEARLWORT

Perennial plants growing in small cushions. Flowers singly or in pairs, 10–12 mm across. Known only from Northern Manitoba (Baralson Lake).

*Sagina decumbens* (Ell.) T. & G.            SPREADING PEARLWORT

Annual plants usually less than 10 cm high; stems ascending or decumbent. Leaves linear to awl-shaped; petals the same length as or shorter than the sepals, or lacking. Rare weed; introduced from the eastern USA.

*Sagina nodosa* (L.) Fenzl            PEARLWORT

Perennial plants 5–15 cm high; stems erect, ascending, or decumbent. Lower stem leaves 5–20 mm, the upper ones scale-like, with bulb-like sterile shoots replacing flowers. Rocky or sandy shores and beaches; Boreal forest.

*Sagina saginoides* (L.) Karsten            MOUNTAIN PEARLWORT

Perennial plants usually less than 10 cm high; rosette leaves, when present, to 20 mm long, the cauline leaves 5–10 mm. Flowers mostly solitary or occasionally in pairs; the petals about the same length as the sepals. Moist slopes and shores; Rocky Mountains.

**Saponaria**        soapwort

*Saponaria vaccaria* L. (Fig. 108)            COW COCKLE

A smooth, hairless annual plant 30–70 cm high. Leaves grayish green, smooth, entire, clasping, borne opposite each other in pairs on the stem, ovate-lanceolate, 2–8 cm long. Flowers in loose corymbose cymes at head of stem. Ovate calyx formed by the united sepals, about 12 mm long and 10 mm across, with the 5 pale pink petals forming a flat corolla about 12 mm across. Seeds borne in an ovoid capsule, round, dull black, about 2 mm in diam. Introduced from Europe, now common; in grainfields and on roadsides; throughout the southern part of the Prairie Provinces. Seeds are often in "Wild Flower Garden" packets available commercially. Syn.: *Vaccaria vulgaris* Host.

**Silene**        catchfly

Annual, biennial, or perennial plants with opposite, entire leaves and perfect flowers. Seeds contained in a capsule.

1. Annual or biennial plants. ................................................................. 2
   Perennial plants. ................................................................. 4

Fig. 108. Cow cockle, *Saponaria vaccaria* L.

*Dicots*

2. Plants not sticky. ............................................................................... *S. cserei*
   Plants sticky, at least around the nodes of
   the stem. ....................................................................................................... 3

3. Plants almost without hairs; flowers pink-
   ish white and about 5 mm across. ..................................................... *S. antirrhina*
   Plants sticky and hairy; flowers white and
   6–10 mm across. ........................................................................... *S. noctiflora*

4. Plants densely matted or cushion-
   forming. .................................................................. *S. acaulis* var. *exscapa*
   Plants with elongate, erect or decumbent
   stems. ........................................................................................................... 5

5. Calyx more or less glandular pubescent. ................................................. 6
   Calyx glabrous, or if pubescent, not glan-
   dular. ........................................................................................................... 7

6. Plants rather straggly, weak-stemmed;
   calyx 5–8 mm. ............................................................................. *S. menziesii*
   Plants with stems 20–50 cm high; calyx
   12–16 mm. ...................................................................................... *S. parryi*

7. Calyx to 8 mm long, with 10 simple veins. .................................... *S. sibirica*
   Calyx 10–15 mm long, with 20 main veins,
   strongly reticulate. ................................................................... *S. cucubalus*

*Silene acaulis* L. var. *exscapa* (All.) DC.                    MOSS CAMPION

Cushion plants or mat-forming 3–6 cm high; leaves 4–12 mm long, linear-lanceolate. Flowers solitary, often imperfect, purplish. Alpine meadows, scree fields; Rocky Mountains.

*Silene antirrhina* L.                                        SLEEPY CATCHFLY

An erect annual 40–50 cm high, with erect branches. Stem usually sticky near the nodes, sometimes slightly hairy. Leaves lanceolate, 2–5 cm long. Flowers pink, very small, about 5 mm across. Not common; may be found in sandy areas; Parklands, Boreal forest.

*Silene cserei* Baumgarten                                   SMOOTH CATCHFLY

A biennial plant 10–70 cm high somewhat resembling bladder campion but with a taproot. Leaves elliptic-lanceolate, stalkless, glaucous and thick, 5–10 cm long, and borne oppositely on the stem. Flowers numerous with pink-ish slightly inflated calyx and white petals cleft at the tip and borne in whorls around the stem. At maturity, calyx is somewhat inflated and ovoid, about 12 mm long. Becoming increasingly abundant; along railway grades; throughout Prairies and Parklands.

*Silene cucubalus* Wibel                                     BLADDER CAMPION

A tall erect perennial with hairless stems 15–60 cm high. Leaves opposite, lanceolate and smooth, 2–8 cm long. The flowers, 10–20 mm across, borne at the summit of the stem in loose open panicles; sepals united to form a blad-der-like calyx 10–12 mm long; petals white and 2-cleft. An introduced and very persistent weed. Not common, but generally plentiful where found. Syn.: *S. vulgaris* (Moench) Garcke.

*Silene menziesii* Hook.                                    MENZIES CATCHFLY

A straggly weak-stemmed perennial 15–40 cm high, with oval to lanceolate leaves 2–5 cm long. Flowers white with cleft petals, about 10 mm long.
Found sparingly; in Boreal forest, Rocky Mountains, and has been reported
from the Cypress Hills.

*Silene noctiflora* L.                                NIGHT-FLOWERING CATCHFLY

A stout, very sticky, hairy, erect annual weed 30–90 cm high. Basal leaves
short-stalked, 5–12 cm long, oblanceolate. Upper leaves stalkless, lanceolate,
2–8 cm long. Leaves hairy and slightly sticky. Sepals united to form an oval
tubular calyx up to 12 mm long, with light and dark green upright stripes.
White petals deeply cleft, opening only at night. An introduced weed; in
grainfields; throughout southern Prairie Provinces.

*Silene parryi* (Wats.) Hitchc. & Mag.                      MACOUN'S CAMPION

A perennial with erect stems 10–30 cm high. Leaves 5–7 cm long, linear-
lanceolate; stem and leaves finely puberulent and densely glandular above.
Flowers usually 3–7, with the pedicels of lateral flowers 2–3 cm long; calyx
8–10 mm long, prominently 10-nerved, glandular pubescent. Grassy slopes;
southern Rocky Mountains.

*Silene sibirica* (L.) Pers.                               SIBERIAN CAMPION

Perennial plants with erect subglabrous stems 40–60 cm high. Basal leaves
oblong-linear; stem leaves all with sterile shoots in the axils. Inflorescence
interrupted, the flowers in verticils, unisexual; calyx 4–5 mm long. Rare, an
introduced weed.

**Spergula**      spurry

*Spergula arvensis* L.                                      CORN SPURRY

An introduced annual, branching plant 15–40 cm high, with very narrow
leaves 20–30 mm long, borne in whorls around the stem. White flowers about 6
mm across, in branching cymes at end of stem. Introduced, not common;
sometimes found in fields and on roadsides; western Prairies.

**Spergularia**      sand spurry

*Spergularia marina* (L.) Griseb. var. *leiosperma* (Kindb.) Gurke
                                                   SALT-MARSH SAND SPURRY

A low, annual, clustered plant, much-branched, about 10–20 cm high,
with small ovate stipules at the bases of the opposite, linear, fleshy leaves.
Leaves circular in cross section, 6–20 mm long. Flowers pink and numerous;
petals shorter than sepals, which are about 3 mm long. This is a rather rare
plant; found in the margins of saline sloughs; throughout Prairies and Parklands. Syn.: *S. salina* J. & C. Presl.

**Stellaria**      stitchwort

Mostly low-growing or straggling plants with small white flowers, 5 petals
so deeply cleft that often they look like 10 petals.

1. Plants glandular pubescent throughout. ............................................. *S. americana*
   Plants glabrous. ............................................................................................................ 2

2. Plants annual; basal leaves ovate or rhombic-ovate, petioled. ................................................. *S. media*
   Plants perennial; all leaves sessile. .......................................................... 3

3. Bracts of the inflorescence, at least the upper ones, membranous. ................................................. 4
   Bracts of the inflorescence all green, not membranous. ...................................................................... 8

4. Upper part of stems and leaf margins papillate, rough. ........................................................ *S. longifolia*
   Upper part of stems and leaf margins smooth. ....................................................................... 5

5. Bracts and leaves or sepals ciliate. ................................................................ 6
   Bracts, leaves, and sepals not ciliate. ........................................................ 7

6. Bracts and leaves ciliate; leaves to 15 mm long. ................................................. *S. calycantha*
   Bracts and sepals ciliate; leaves to 10 mm long. ................................................. *S. ciliatisepala*

7. Inflorescence subumbellate; petals absent. ........................................................ *S. umbellata*
   Inflorescence cymose; petals present, larger than sepals. ....................................................... *S. longipes*

8. Leaves ovate-lanceolate, length-to-width ratio less than 4:1. ........................................................ 9
   Leaves linear- or elliptic-lanceolate, length-to-width ratio more than 4:1. ................................................. 10

9. Sepals 1.5–2.5 mm long, obtuse; leaves about 5 mm long. ........................................................ *S. obtusa*
   Sepals 2.5–3.5 mm long, acute; leaves 5–10 mm long. ............................................................ *S. crispa*

10. Petals shorter than the sepals, or lacking. ................................................. *S. calycantha*
    Petals the same length as or longer than the sepals. ....................................................... 11

11. Petals cleft more than halfway, almost to base. ...................................................................... 12
    Petals cleft less than halfway. ...................................................... 13

12. Leaves mostly 6–20 mm long, 2–6 mm wide; sepals 2–3 mm long. ....................................... *S. crassifolia*
    Leaves mostly 2–8 mm long, 1–4 mm wide; sepals 3.5–6 mm long. ....................................... *S. humifusa*

13. Plants usually less than 10 cm high; internodes short; leaves thick, rather rigid, green. ................................................. *S. laeta*
    Plants usually 10–20 cm high; internodes long; leaves not thick or rigid, glaucous. ................................................. *S. longipes*

*Stellaria americana* Porter

Plants with slender rootstocks, mat-forming; stems leafy, 10–20 cm long; leaves 1–3 cm long; cymes leafy; sepals 3–4 mm long; petals exceeding the sepals. Rocky alpine slopes; southern Rocky Mountains.

*Stellaria calycantha* (Ledeb.) Bong.                    NORTHERN STITCHWORT

A weak-stemmed trailing plant 15–50 cm long, with linear-lanceolate leaves 10–35 mm long. Not common; in shady woodlands; Rocky Mountains, Boreal forest, Cypress Hills. Syn.: *S. borealis* Bigel.

*Stellaria ciliatisepala* Trautv.

Plants with short creeping rootstocks; stems to 20 cm high; leaves to 10 mm long. Inflorescence few-flowered, or rarely a single flower. Open, rocky areas; Boreal forest.

*Stellaria crassifolia* Ehrh.                    FLESHY STITCHWORT

A small rather weak-stemmed plant 5–25 mm high, with short, fleshy oblong-lanceolate leaves 5–15 mm long. White flowers 4–6 mm across, not numerous. Uncommon; may be looked for in wet and very shaded places.

*Stellaria crispa* Cham. & Schl.

Weak-stemmed plants with creeping rootstocks; leaves ovate, sessile or subsessile; flowers axillary, solitary; sepals 3–4 mm long; petals minute or absent. Moist places in mountains; southern Rocky Mountains.

*Stellaria humifusa* Rottb.

Mat-forming plants with creeping rootstocks; often reproducing by vegetative buds in axils of leaves; flowers 8–16 mm across. Sandy beaches; Hudson Bay.

*Stellaria laeta* Richardson

Mat-forming plants with creeping rootstocks, and a few-flowered inflorescence. The sepals ciliate, puberulent; the stem often puberulent. Sandy beaches; Hudson Bay, southern Rocky Mountains.

*Stellaria longifolia* Muhl.                    LONG-LEAVED STITCHWORT

A weak-stemmed semierect plant 20–40 cm high, with bright green opposite linear leaves, tapered at both ends, 10–60 mm long. Numerous flowers 6–9 mm across, borne at the summit of the stem, with the petals deeply cleft and longer than the sepals. Fairly common; woodlands, moist and shady places; throughout the Prairie Provinces.

*Stellaria longipes* Goldie                    LONG-STALKED STITCHWORT

An erect plant 6–30 cm high, with short lanceolate leaves 10–25 mm long, broadest near the base and tapering upward. Flowers few, on long stalks at the head of the stem, the 2-cleft white petals slightly longer than the sepals. Petals not usually so deeply cleft as in the preceding species. Common; in moist places and woodlands; throughout the Prairie Provinces.

*Stellaria media* (L.) Cyrill.  COMMON CHICKWEED

A prostrate-growing trailing annual, with lines of fine white hairs on the stems. Leaves broadly ovate, 5–25 mm long, on short stalks. Flowers white, about 6 mm across, and deeply cleft. An introduced European weed, which has become very common; on lawns and in gardens; throughout the Prairie Provinces.

*Stellaria obtusa* Engelm.

Matted plants with many prostrate branching stems 5–15 cm long; flowers solitary in leaf axils; sepals 2–3 mm long; petals minute or absent. Very rare; wet areas; southern Rocky Mountains.

*Stellaria umbellata* Turcz.

Erect or ascending plants with creeping rootstocks, 10–30 cm high. Leaves 5–20 mm long, linear-lanceolate, slim; flowers in upper leaf axils and in terminal umbellate cymes; sepals 2–5 mm long; petals minute or absent. Rare; moist areas in mountains; southern Rocky Mountains.


## CERATOPHYLLACEAE—hornwort family

**Ceratophyllum**       hornwort

*Ceratophyllum demersum* L.  HORNWORT

A completely submersed aquatic plant with thread-like leaves, 2 or 3 times forked, 5–25 mm long, arranged in whorls of 6–9 leaves at intervals on the stem. Flowers without stalks, unisexual, and borne singly in the axils of the leaves. Fruit an achene about 5 mm long, with a persistent style about 6 mm long. May be mistaken for the water milfoil, *Myriophyllum*, from which it is distinguished by its 2- or 3-forked leaves, the leaves of *Myriophyllum* being pinnate and not forked. Fairly common; in ponds and still water; eastern Prairies and Parklands, rare along northern fringe of Parklands and in Boreal forest.


## NYMPHAEACEAE—water-lily family

Flowers yellow, with sepals petal-like, larger than the petals. ........................................................................................ *Nuphar*

Flowers white, with green sepals and large petals. ........................................................................................ *Nymphaea*

**Nuphar**       yellow pond-lily

Leaves 5–10 cm long; flowers 15–20 mm across. ........................................................................................ *N. microphyllum*

Leaves 10–25 cm long; flowers 40–50 mm across. ........................................................................................ *N. variegatum*

*Nuphar microphyllum* (Pers.) Fern.                    SMALL POND-LILY

Leaves submersed or floating, oval or elliptic in outline, deeply cordate at the base. Flowers yellow within; anthers 1.5–3 mm long, shorter than the filaments; the stigmatic disk red; fruit about 15 mm long. Ponds and lakes; eastern Parklands and Boreal forest.

*Nuphar variegatum* Engelm.                    YELLOW POND-LILY

Perennial aquatic plant with stout creeping rootstock. Large single cordate leaves that float on the surface borne on long slender stalks, arising from the rootstock. Leaves 10–25 cm long and 8–15 cm broad. Occasionally a few entirely submersed leaves, thin and membranous. Flowers reddish within; anthers 4.5–7 mm, shorter than the filaments; stigmatic disk green; fruit about 4 cm long. Ponds and lakes; Parklands, Boreal forest, Cypress Hills.

**Nymphaea**          water-lily

Leaves rotund in outline, 10–20 cm across;
  flowers 5–8 cm across, fragrant. ........................................................... *N. odorata*
Leaves elliptic in outline, 7–12 cm long, 4–8
  cm wide; flowers 2–5 cm across, odorless. ............................ *N. tetragona* ssp. *leibergii*

*Nymphaea odorata* Ait.                    FRAGRANT WATER-LILY

Leaves commonly purple or red below, deeply cordate, with the petiole inserted almost at the middle of the leaf. Flowers with 17–32 petals, opening in the morning. Lakes and ponds; southeastern Boreal forest.

*Nymphaea tetragona* Georgi ssp. *leibergii* (Morong) Porsild   SMALL WATER-LILY

Leaves with a V-shaped sinus. Flowers with 8–17 petals, opening in the afternoon. Ponds, lakes, and marshes; Boreal forest.


# RANUNCULACEAE—crowfoot family

A large and variable family, with leaves alternate, except in *Clematis*; sepals 3–15, often colored, and resembling petals. Petals sometimes absent, but, when present, the same number as sepals. Many stamens. Fruit in the form of achenes, follicles (dry pods), or berries.

1. Climbing plants with opposite leaves;
     without petals but with colored sepals;
     fruit with persistent feathery style. ........................................................... *Clematis*
   Plants not climbing; leaves either basal or
     alternate. ............................................................................................................ 2

2. Flowers irregular or spurred. ...................................................................................... 3
   Flowers regular, not spurred. ...................................................................................... 6

3. Flowers irregular, not spurred. ............................................................................ *Aconitum*
   Flowers spurred. ...................................................................................................... 4

4. Flowers with 1 spur. ................................................................ *Delphinium*
   Flowers with 5 spurs. ................................................................ 5

5. Plants tall; leaves compound; flowers
   15–25 mm long. ....................................................................... *Aquilegia*
   Plants small; leaves simple; flowers 3–5
   mm long. ................................................................................... *Myosurus*

6. Fruit berry-like. ........................................................................... *Actaea*
   Fruit not berry-like. ..................................................................... 7

7. Fruit dry pods (or follicles). ...................................................... 8
   Fruit achenes. ............................................................................... 10

8. Leaves simple, entire or incised. ............................................... 9
   Leaves compound with 3 leaflets. ............................................. *Coptis*

9. Leaves entire or toothed; flowers yellow. ................................ *Caltha*
   Leaves palmately incised; flowers white. .................................. *Trollius*

10. Leaves all basal, 3-lobed, with the lobes
    entire, rounded. ........................................................................ *Hepatica*
    Leaves not all basal; stem leaves present;
    leaf segments mostly toothed, acute. ...................................... 11

11. Petals usually present. .............................................................. *Ranunculus*
    Petals absent, but sepals colored and
    petal-like. ................................................................................. 12

12. Leaves all basal except involucre of leafy
    bracts some distance below
    inflorescence. ........................................................................... *Anemone*
    Leaves not basal; without involucral
    bracts below inflorescence. ..................................................... *Thalictrum*

## *Aconitum*    monkshood

*Aconitum delphinifolium* DC.                                   MONKSHOOD

Perennial plants 30–70 cm high, with short tubers. Stems finely pubescent; leaves glabrous, palmately lobed, with the segments lanceolate. Inflorescence a short raceme; flowers deep blue to purple. **Poisonous** to cattle. Mountain meadows; Rocky Mountains.

## *Actaea*    baneberry

*Actaea rubra* (Ait.) Willd.                                   RED BANEBERRY

Erect perennial herbs 30–100 cm high, with large compound leaves. Flowers small, white, in dense clusters at the ends of the stems, with the sepals falling off when the flower opens. Fruits large and berry-like, 6–10 mm long, and clustered in a raceme. **Poisonous** to humans, especially children. Fruit of the common typical species is bright red, but there is also a fairly common white-berried form, *Actaea rubra* forma *neglecta* (Gilman) Robins, white baneberry. Both baneberries common; in rich woodlands and in shady, wooded ravines; throughout the Prairie Provinces.

***Anemone***    anemone

Perennial plants from bulb-like taproots, with basal leaves long-stalked and palmately divided or dissected. Flowering stalks bearing a whorl of involucral leaves (large bracts resembling leaves) borne part way up the flowering stem. Flowers borne at the summits of long stems and consisting of 5 or 6 colored sepals but no petals. Fruits achenes borne in globular or cylindrical heads.

1. Styles long and feathery at maturity; sepals 15–40 mm long. ...................................................................... 2

   Styles short; sepals 5–25 mm long. ............................................................................................................... 3

2. Flowers purplish or bluish; bracts sessile; basal leaves expanding after flowering. .................... *A. patens* var. *wolfgangiana*

   Flowers white; bracts short-petioled; basal leaves fully expanded at flowering. ........................................................................ *A. occidentalis*

3. Carpels and achenes not woolly; glabrous or somewhat pubescent. ...................................... 4

   Carpels and achenes densely woolly. ......................................................................................... 7

4. Involucral leaves petioled; achenes fusiform, with the style short, hooked. .................................... *A. nemorosa* var. *bifolia*

   Involucral leaves sessile or subsessile; achenes flattened. ..................................................................... 5

5. Sepals yellow; styles very long in fruit, reflexed. ........................................................................ *A. richardsonii*

   Sepals white; styles straight or hooked. ...................................................................................... 6

6. Achenes stipitate, glabrous; styles hooked, short. ...................................................................... *A. narcissiflora*

   Achenes sessile, pubescent; styles straight, long. ...................................................................... *A. canadensis*

7. Involucral leaves long-petioled; heads of achenes cylindrical or narrowly ovoid. ...................................... 8

   Involucral leaves sessile or subsessile; heads of achenes globose or ovoid. ...................................... 9

8. Sepals 8–10 mm long, greenish white; fruiting heads long cylindrical. ...................................... *A. cylindrica*

   Sepals 10–20 mm long, white; fruiting heads oblong or oblong-ovoid. ...................................... *A. virginiana*

9. Leaves with 3 broad wedge-shaped segments. ...................................................................... *A. parviflora*

   Leaves 2–4 times ternate, with the segments linear or lanceolate. ...................................... 10

10. Leaves 3–4 times ternate; mostly 1-flowered, with mature styles 4–6 mm long. ...................................... *A. drummondii*

    Leaves 2–4 times ternate; mostly 1- to 4-flowered, with mature styles 1–2 mm long. ...................................... *A. multifida* and var. *richardsiana*

*Anemone canadensis* L. CANADA ANEMONE

A hairy-stemmed plant to 30 cm high, with several 5- to 7-parted basal leaves. In this species the flowering stems divide, and fresh stems appear, each bearing a whorl of involucral leaves and a flower at the end. Flowers white, 25–30 mm across. Fruiting head globular. One of the commonest anemones; found in large patches at the edges of woodlands, low moist places, and hollows; throughout the Prairie Provinces.

*Anemone cylindrica* A. Gray LONG-FRUITED ANEMONE

Plants with a long slender stem 15–50 cm high, branching at the involucral leaves into 2–6 flowering stems, each bearing a 5-sepaled greenish white flower almost 20 mm across. Fruiting head long, cylindric, often 20 mm high, and densely woolly. Common; on moist prairie; throughout Prairies and Parklands.

*Anemone drummondii* Wats. DRUMMOND'S ANEMONE

Plants with a stout woody caudex, silky hirsute stems 10–25 cm high, and long-petioled basal leaves. Flower usually solitary, appressed pubescent. Mountain meadows and alpine tundra; Rocky Mountains.

*Anemone multifida* Poir. CUT-LEAVED ANEMONE

Erect plants, usually with several silky-haired purplish stems 15–60 cm high. Flowering stalks 1–7, one usually having no involucral leaves. Leaves cleft several times into very narrow lobes. Flowers varying from reddish purple to white or yellowish green and the fruiting heads globular and very woolly. Common; in moist spots; throughout the Prairie Provinces. A var., *richardsiana* Fern., with longer sepals, usually whitish, but sometimes bright red. Fairly common; in moister spots; throughout the Prairie Provinces.

*Anemone narcissiflora* L. NARCISSUS ANEMONE

Plants with stout rootstocks; the stems pubescent, 20–40 cm high. Leaves deeply palmately divided, long-petioled. Inflorescence umbellate, 3- to 8-flowered. Mountain meadows; Rocky Mountains. Occurrence in Alberta not yet definitely established.

*Anemone nemorosa* L. var. *bifolia* (Farwell) Boiv. WOOD ANEMONE

Plants with long creeping rootstocks; sterile leaves 3 or 5 foliate; flowering stems 10–20 cm high, usually with a single flower, and a single verticil of 3-petioled stem leaves. Moist woods; Parklands and Boreal forest. Syn.: *Anemone quinquefolia* L. var. *interior* Fern.

*Anemone occidentalis* Wats. CHALICEFLOWER

Plants 10–60 cm high, at first densely villose, later often glabrate. Basal leaves few, 3-parted, with numerous linear-lanceolate segments; involucral leaves short-petioled. Flowering stalk elongating in fruit, the styles 2–4 cm long. Mountain meadows and slopes; Rocky Mountains. Syn.: *Pulsatilla occidentalis* (Wats.) Freyn.

*Anemone parviflora* Michx.                              SMALL WOOD ANEMONE

Plants with slender rootstocks; stems 10–30 cm high, sparingly pubescent. Basal leaves long-petioled; the segments crenately lobed into broad, obtuse divisions; peduncle solitary, with a single flower. Woods, meadows, and slopes; Boreal forest, Rocky Mountains.

*Anemone patens* L. var. *wolfgangiana* (Bess.) Koch (Fig. 109) CROCUS ANEMONE

An early spring flowering plant, from a thick woody taproot. Flower appears before leaves on an erect silky stem 8–20 cm high. Basal leaves much divided, on long stalks, with the involucral leaves also being cleft, but without stalks. Flowers mauve or pale blue, sometimes white, 4–6 cm across, usually the first flower to appear on the prairies in spring. Fruiting stems tall, often over 30 cm high, bearing many achenes, each having a long persistent tail-like feathery style about 3 cm long. Common on open prairie and hills; throughout southern and western parts of the Prairie Provinces, and particularly conspicuous on burned-over prairie and railway rights-of-way. Sheep may be **poisoned** by this plant and their digestive system may be impaired by its felty hairs. Very common on overgrazed pasture, where dense stands during the early spring indicate an overgrazed condition. Syn.: *Pulsatilla ludoviciana* (Nutt.) Heller; *P. patens* (L.) Miller ssp. *teklae* (Zamels) Zamels.

*Anemone richardsonii* Hook.                              YELLOW ANEMONE

Plants with slender yellow or brownish rootstocks; stems slender, delicate, 6–15 cm high. Basal leaves solitary, rotund or reniform in outline, 5-cleft into cuneate incised segments; involucral leaves 3, sessile or subsessile. Flowers sulfur yellow. Marshes, wet woods, and bogs; Boreal forest, Rocky Mountains.

*Anemone virginiana* L.                              TALL ANEMONE

A tall erect species 30–80 cm high, with hairy stems. Flowers greenish white, with sepals about 10 mm long. Head of fruit is ovoid or cylindric and woolly. A very variable species, found occasionally; in woods and shady places; in eastern part of the Prairie Provinces, but rare elsewhere. Some authorities consider the western plant to be *A. riparia* Fern.

*Aquilegia*        columbine

Erect, branching plants from rootstocks, with slightly hairy and glandular stems. Leaves compound and divided into 3 leaflets. Flowers with 5 sepals colored and petal-like, also 5 colored petals extending at the base forming a long tubular spur. Fruit a follicle or dry capsule containing numerous seeds.

1. Plants small, usually less than 20 cm high;
   leaves all basal; usually 1-flowered. ........................................................ *A. jonesii*
   Plants 20–80 cm high, with stem leaves;
   several-flowered. ........................................................................................ 2

2. Blade of petals longer than the spurs;
   flowers blue or purple. ............................................................. *A. brevistyla*
   Blade of petals shorter than the spurs. ................................................... 3

3. Flowers yellow; spurs of petals slightly
   hooked. ....................................................................................... *A. flavescens*

A.C. Budd

Fig. 109. Crocus anemone, *Anemone patens* L. var. *wolfgangiana* (Bess.) Koch.

Flowers scarlet; spurs of petals almost
straight. ..................................................................................................... 4

4. Sepals erect, shorter than the spur. ......................................................... *A. canadensis*
Sepals spreading, longer than the spur. ...................................................... *A. formosa*

### *Aquilegia brevistyla* Hook.   SMALL-FLOWERED COLUMBINE

A small plant 20–50 cm high; flowers blue or purple, nodding, with sepals about 15–20 mm long. Rare; in woodlands; Boreal forest, Rocky Mountains.

### *Aquilegia canadensis* L.   WILD COLUMBINE

A stout erect plant 30–80 cm high; flowers large, nodding, with scarlet sepals and sometimes yellow petals. Stamens and styles much longer than sepals and protruding conspicuously from the flower. Fairly common; in open woodlands; eastern Boreal forest, reported form Qu'Appelle Valley in Saskatchewan.

### *Aquilegia flavescens* S. Wats.   YELLOW COLUMBINE

A slender branching plant 30–80 cm high, with yellow or yellowish white flowers. In woodlands; Rocky Mountains.

### *Aquilegia formosa* Fisch.   CRIMSON COLUMBINE

A stout glabrous or sparingly pubescent plant 30–80 cm high, with sepals about 2 cm long; sepals and spurs crimson; the lamina of the petals yellow. Open woods; southern Rocky Mountains.

### *Aquilegia jonesii* Parry   BLUE COLUMBINE

Plants with a short caudex; the leaves crowded, finely pubescent, and often glaucous. Flowering stem with a single erect flower about 1.5 cm long; sepals blue; petals often whitish. Exposed slopes, rocky areas; southern Rocky Mountains.

### *Caltha*   marsh-marigold

1. Stems erect; bearing a single leaf or none;
flowers 1 or 2. ...................................................................................... *C. leptosepala*
Stems ascending, decumbent or floating;
bearing several leaves; flowers several. ................................................................ 2

2. Sepals white or pink; flowers not over 15
mm across; plants often floating. ........................................................... *C. natans*
Sepals yellow; flowers over 15 mm across;
plants usually rooted in mud. ................................................................. *C. palustris*

### *Caltha leptosepala* DC.   MOUNTAIN-MARIGOLD

Plants 10–40 cm high, from a short erect rootstock. Basal leaves long-petioled; blades oval to oblong-oval, cordate at base, with the margins toothed. Flowers 2.5–3.5 cm across, white or bluish. Alpine meadows; Rocky Mountains.

*Caltha natans* Pall. FLOATING MARSH-MARIGOLD

Plants of lakes and small ponds, with cordate to kidney-shaped leaves and stems rooting at the nodes. Flowers 10–15 mm across, with white or pinkish petal-like sepals; fruiting clusters about 6 mm across. Quite rare; has been found floating or rooted in the mud in woodland lakes and slow streams; Boreal forest.

*Caltha palustris* L. (Fig. 110) MARSH-MARIGOLD

Stout marsh plants with smooth hollow stems arising from coarse fleshy roots. Leaves round or kidney-shaped and heart-shaped at the base, the basal ones with long stalks and the upper ones stalkless. Flowers bright yellow, 20–30 mm across, with 5–9 sepals but no petals. Fruits many-seeded follicles. Common in wet places, more common in slightly moving water; Parklands and Boreal forest.

**Clematis** virgin's-bower

Climbing vines with more or less woody stems and opposite leaves. Plants climb by clasping the supporting plant with the stalks of their leaves and not by means of tendrils. Flowers have colored sepals but no obvious petals; fruit an achene with a long, persistent, feathery style.

1. Flowers solitary. ...................................................................................... 2
   Flowers in panicles or corymbs. .......................................................... 3

2. Flowers blue; leaves with 3 leaflets. ......................... *C. verticellaris* var. *columbiana*
   Flowers yellow; leaves with 5 leaflets. ....................................... *C. tangutica*

3. Flowers yellowish or greenish; leaves with
   3 leaflets. ................................................................................ *C. virginiana*
   Flowers white; leaves with 5–7 leaflets. ............................................ *C. ligusticifolia*

*Clematis ligusticifolia* Nutt. WESTERN VIRGIN'S-BOWER

A climbing plant, with opposite leaves divided into 5-stalked leaflets arranged pinnately on stalk, each leaflet pointed at tip, 2–8 cm long. Flowers borne in clusters, each flower about 10–15 mm across with white sepals and no petals. Fruit a cluster of achenes, each with a persistent feathery style. Common; in coulees and ravines, climbing over bushes and shrubs; Prairies.

*Clematis tangutica* (Max.) Korsh. CLEMATIS

Plant with greenish yellow flowers 3–5 cm across. Leaflets 2–5 cm long, coarsely toothed. Introduced ornamental, locally escaped and established; in coulees and shrubbery.

*Clematis verticellaris* DC. var. *columbiana* (Nutt.) A. Gray
PURPLE VIRGIN'S-BOWER

A climbing plant, with opposite leaves divided into 3 long-stalked leaflets, each 2–8 cm long, pointed at the tip. Flowers 5–10 cm across, having 4 (sometimes 5) pale blue petal-like sepals. Petals reduced, resembling sterile stamens. Achenes borne in clusters, each with a feathery style about 5 cm long. In shady woodlands; in Cypress Hills and Foothills region. Syn.: *C. columbiana* (Nutt.) T. & G.

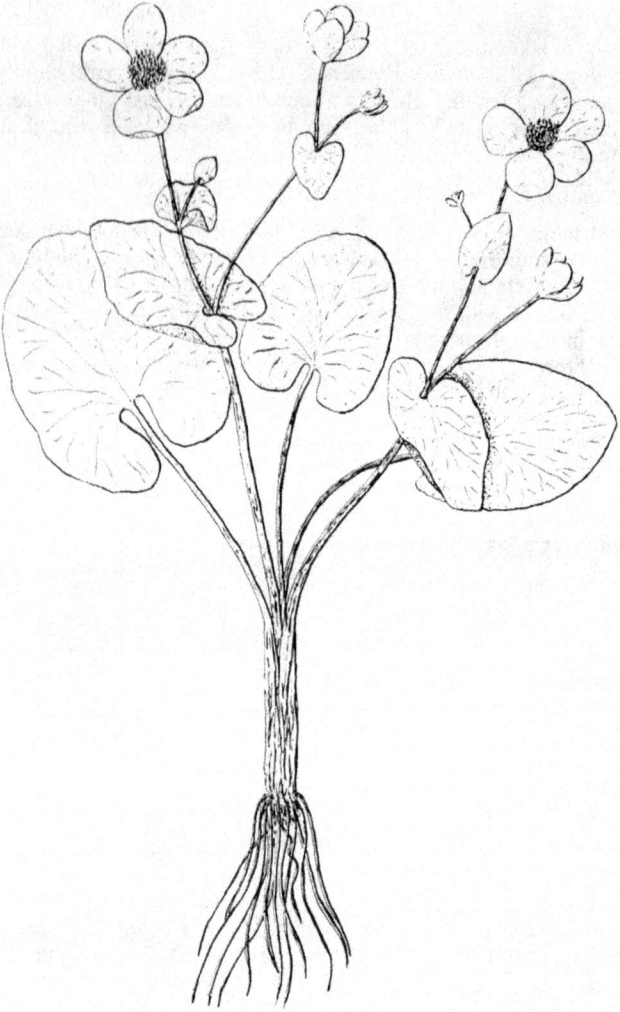

Fig. 110.   Marsh-marigold, *Caltha palustris* L.

*Clematis virginiana* L. VIRGIN'S-BOWER

Plants climbing, with stems 3–4 m long; leaflets usually 3, ovate, commonly coarsely toothed. Panicles from many of the leaf axils many-flowered. Woods; eastern Parklands and Boreal forest.

**Coptis**      goldthread

*Coptis trifolia* (L.) Salisb. GOLDTHREAD

Low, evergreen, perennial herbs 5–15 cm high. The leaves all basal, long-stalked, each bearing 3 leaflets. The single flowers about 6 mm across, on a long slender stem, and bearing 5–7 white sepals with yellowish bases, and smaller club-shaped petals, each with a tiny drop of nectar in its summit. Fruit a cluster of follicles, each about 6 mm long. Called goldthread because of its fine yellow rootstocks. In moist spots; Boreal forest.

**Delphinium**      larkspur

Perennial herbs with alternate much-divided or lobed leaves. Flowers in racemes, perfect, irregular, with 5 petal-like colored sepals, the upper one extended at the base into a spur. Petals 4, the upper pair extended into spurs projecting into the sepal spur, the lower pair small with short claws. Fruit a many-seeded follicle.

Plants tall and erect, 50–150 cm high; racemes
   of flowers spike-like; lower flower stalks
   shorter than flowers. ..................................................................... *D. glaucum*
Plants low, 15–50 cm high; racemes of flowers
   loose and spreading; lower flower stalks
   longer than flowers. ..................................................................... *D. bicolor*

*Delphinium bicolor* Nutt. (Fig. 111) LOW LARKSPUR

A perennial from a thick, fleshy, fibrous root, with hairy stems 15–50 cm. Leaves finely hairy, much cleft and dissected, on long stalks. Flowers dark blue, with a long blue spur at the rear. Flowers 15–35 mm across, borne on long stalks in a loose terminal raceme. Blooming in early spring, May, or June. Fruit a dry follicle 15–20 mm high, brownish, containing many seeds. Not common, but locally very abundant; in large patches in openings in woodlands, hillsides, and sheltered areas; on Cypress loam soils throughout the southwest, particularly in the Cypress Hills and Wood Mountains, also in the Rocky Mountains. Very **poisonous** to cattle and causing some heavy losses, but apparently harmless to sheep.

*Delphinium glaucum* Wats. (Fig. 112) TALL LARKSPUR

A tall erect plant 30–150 cm high, with smooth stem and deeply cut, dissected slightly hairy leaves. Dark blue flowers borne on very short stalks in a long close spike-like raceme. Fairly abundant; throughout the Foothills region. Very **poisonous** to cattle. Growing in association with open stands of aspen poplar and willows in the Rocky Mountains and western Boreal forest. Syn.: *D. brownii* Rydb.; *D. canmorense* Rydb.

Fig. 111.   Low larkspur, *Delphinium bicolor* Nutt.

*Dicots*

Fig. 112.   Tall larkspur, *Delphinium glaucum* Wats.

*Hepatica*       hepatica

*Hepatica nobilis* Miller                                                    LIVERLEAF

Plants with short thick brown rootstocks; scapes to 15 cm high. Leaves overwintering and persisting until flowering, reniform in outline, 5–15 cm broad, 3-lobed, the lobes obtuse-ovate; petioles 5–15 cm long, pubescent. Flowers 15–25 mm across, varying in color from white or pinkish to bluish purple; perianth with 6–9 segments, with 3 bract-like stem leaves appearing to form a calyx. Achenes hirsute. Very rare; dry to moist woods; eastern Parklands and Boreal forest.

*Myosurus*       mousetail

Small low-growing annual mud plants from fibrous roots. Leaves all basal and linear or thread-like. Flowering scapes bearing 1 flower with 5 sepals and 5 narrow greenish petals. Occasionally 6 or 7 sepals and petals, and in some instances the petals absent. Fruits numerous small achenes crowded on a tall narrow spike, from which the name mousetail was given to the genus.

*Myosurus minimus* L.                                                    LEAST MOUSETAIL

Small plants, 5–15 cm high. Sepals and petals 3–4 mm long; achenes 1–1.5 mm long, with a very short beak. Not common; in muddy slough margins and marshy areas; throughout Prairies.

*Ranunculus*       buttercup

A large genus, usually with alternate leaves. Flowers (Fig. 7) perfect and regular, with 5 sepals and usually 5 petals, either yellow or white. Fruits numerous achenes borne in a head or a short spike.

1. Plants aquatic; leaves submerged, finely
   dissected. ....................................................................................................... 2
   Plants not aquatic, mud or dryland
   plants; leaves not finely dissected. ........................................................... 3

2. Flowers white, axillary. ............................................................. *R. aquatilis*
   Flowers yellow, terminal or in cymes. ..................................... *R. gmelinii*

3. Basal leaves elliptic to linear, not deeply
   divided. ............................................................................................................. 4
   Basal leaves ovate to rhomboid, reniform,
   entire to compound. ..................................................................................... 6

4. Flowers white; leaves with a petiole
   almost as wide as the blade. ..................................................... *R. pallassii*
   Flowers yellow. ............................................................................................. 5

5. Basal leaves elliptic to lanceolate, entire;
   stem leaves lobed. ............................................................... *R. glaberrimus*
   Basal leaves linear-lanceolate to filiform. ....................... *R. flammula*

6. Achenes pubescent. ................................................................................... 7
   Achenes glabrous. ....................................................................................... 9

7. Basal leaves deeply 3-parted or lobed almost to base. ..................................................... *R. uncinatus*

   Basal leaves crenate or lobed to almost halfway. ................................................................ 8

8. Stems and leaves pilose-pubescent or villose. ................................................. *R. cardiophyllus*

   Stems and leaves subglabrous or puberulent. ...................................................... *R. inamoenus*

9. Basal leaves crenate or lobed to almost halfway. ............................................................... 10

   Basal leaves lobed almost to the base, or divided. ............................................................... 12

10. Plants stoloniferous, creeping; leaves cordate at the base. ........................................... *R. cymbalaria*

    Plants not stoloniferous, stems upright. ......................................................... 11

11. Leaves deltoid-ovate in outline, acute to rounded at base. ..................................... *R. rhomboideus*

    Leaves reniform in outline, more or less cordate at base. ..................................... *R. abortivus*

12. Stem leaves and leaves in the inflorescence sessile or subsessile. ........................ 13

    Stem and inflorescence leaves on well-developed petioles. ............................................. 15

13. Plants small, usually less than 10 cm high; petals as long as or shorter than sepals. ..................... *R. pygmaeus*

    Plants usually more than 10 cm high; petals as long as or longer than sepals. ............................. 14

14. Segments of stem leaves linear, 1–3 mm wide. ........................................ *R. pedatifidus* var. *leiocarpus*

    Segments of stem leaves lanceolate, 3–6 mm wide. ......................................................... *R. nivalis*

15. Plants small, rarely more than 5 cm high; leaves mostly 3- or 5-lobed. ................................. *R. hyperboreus*

    Plants larger, with the stems usually erect. ................................................................ 16

16. Leaves simple. ................................................................. 17

    Leaves compound. ............................................................... 22

17. Stems bearing a single leaf and a single flower. ............................................. *R. lapponicus*

    Stems bearing several leaves and flowers. ................................................. 18

18. Sepals reflexed at the middle or base. ................................................................ 19

    Sepals spreading or curved inward. ................................................................... 20

19. Petals and sepals about the same length, 3–5 mm. ........................................ *R. uncinatus*

    Petals about 3 times as long as the sepals. .................................... *R. occidentalis*

20. Plants villose-pubescent, especially the petioles. ........................................ *R. acris*

Plants glabrous to subglabrous or puberu-
lent. ..................................................................................................... 21

21. Beak of the achene very short, to 0.1 mm
    long. .................................................................................. *R. sceleratus*

    Beak of the achene developed, 0.5–1.0
    mm long. .............................................................................. *R. gmelinii*

22. Flowers 6–8 mm across, with the petals
    shorter than the sepals. ............................................. *R. pensylvanicus*

    Flowers larger, with the petals longer than
    the sepals. ........................................................................................... 23

23. Leaves divided into many narrow seg-
    ments. ................................................................................................. 24

    Leaves trifoliate. ................................................................................. 25

24. Plants subglabrous; petals about 5 mm
    long, 4–5 mm wide. ........................................................... *R. gelidus*

    Plants appressed pubescent; petals about
    10 mm long, 5 mm wide. ............................................... *R. fascicularis*

25. Sepals reflexed at base, 3.5–5 mm long;
    petals 4–7 mm long. ......................................................... *R. macounii*

    Sepals spreading or incurved; petals 7–15
    mm long. ............................................................................................. 26

26. Plants stoloniferous; beak of the achene
    about 1 mm long. ................................................................. *R. repens*

    Plants not stoloniferous; beak of the
    achene to 3 mm long. ................................................. *R. septentrionalis*

*Ranunculus abortivus* L.                      SMOOTH-LEAVED BUTTERCUP

A biennial plant 15–50 cm high, with a rather fleshy smooth stem. Basal
leaves long-stalked, usually cordate-based, and round with wavy margins;
stem leaves usually 3-cleft, the upper ones without stalks. Petals usually
shorter than the reflexed (turned downward) sepals; flowers yellow, 6–10 mm
across. Found in open woodlands, ravines, and moist places. Plants growing
on margins of streams or sloughs and wet spots have large and branched
inflorescence with numerous small flowers; when growing in less favorable
spots, more erect, with fewer flowers and less-branched inflorescence. The two
phases may be mistaken for different species. Fairly plentiful throughout Prairies and Parklands; rare in Boreal forest.

*Ranunculus acris* L.                                TALL BUTTERCUP

A tall erect perennial, branched above, with a hairy stem 30–80 cm high.
Basal leaves stalked and much cleft and divided; upper leaves divided into 3
lobes, with short stalks. Numerous yellow flowers 20–35 mm across, with petals more than twice as long as sepals. Achenes smooth, flattened, with a short
beak. Introduced from Europe. Not common; found occasionally in moist
places and along railway grades; throughout the Prairie Provinces. Very plentiful, however, in some irrigated areas, especially in southern Alberta.

*Ranunculus aquatilis* L.  LARGE-LEAVED WATERCROWFOOT

A fully submersed aquatic plant, with leaves finely dissected, about 10–25 mm long, growing from a sheathing stipule. Flowers either floating on the water or protruding above it, white, about 10–15 mm across. Achenes with a short beak not more than one-third the length of the body of the achene. In slowly moving water and brackish pools that are not saline; found throughout the Prairie Provinces. A very variable species, of which the following varieties occur:

1. Beak of the achene more than 0.5 mm
   long. ......................................................... var. *longirostris* (Godr.) Laws.
   Beak of the achene less than 0.5 mm
   long. ..................................................................................................... 2

2. Leaves stiff, with a grayish green petiole
   included in the sheath. ............................................ var. *subrigidus* (Drew) Breit.
   Leaves not stiff, with the green petiole to
   1 cm long. ............................................................................................. 3

3. Stem 1.0–1.5 mm thick; plants large. ......................... var. *capillaceus* (Thuill.) DC.
   Stem less than 1 mm thick; plants small
   and reduced. ....................................................................... var. *eradicatus* Laest.

*Ranunculus cardiophyllus* Hook. (Fig. 113)  HEART-LEAVED BUTTERCUP

An erect species 15–50 cm high, with few branches. Basal leaves stalked, round to ovate, sometimes lobed, and usually heart-shaped at the base; stem leaves with narrow, almost linear lobes, very short-stalked. Yellow flowers 10–20 mm across. Not common; in moist places, meadows, and stream banks; in western Parklands.

*Ranunculus cymbalaria* Pursh  SEASIDE BUTTERCUP

Low perennial plants with runners and predominantly basal leaves on thin stalks. Leaves small, rounded with cordate bases and wavy margins, 5–15 mm long. Flowers with a conical center, yellow, and quite small, usually 5–8 mm across; the petals slightly shorter than the sepals. Achenes numerous, with longitudinal grooves. Common; on margins of sloughs and lakes, on saline wet areas and stream banks; throughout the Prairie Provinces.

*Ranunculus fascicularis* Muhl.  EARLY BUTTERCUP

Plants with 2–5 stems 10–30 cm high, strigose. Leaves mostly basal, ovate in outline, with the segments lobed. Flowers 15–25 mm across, long-peduncled. Southern Parklands and Boreal forest.

*Ranunculus flammula* L. var. *ovalis* (Bigel.) Benson  CREEPING SPEARWORT

A low-growing trailing plant, with stems rooting at the nodes and linear or spatulate leaves 10–25 mm long. Bright yellow flowers about 8 mm across borne singly on short stems. Uncommon; on lakeshores and riverbanks; Boreal forest, Parklands. In var. *ovalis* the leaves lanceolate and about 5 mm wide. Plants with filiform leaves, usually 1–2 mm wide, are var. *filiformis* (Michx.) Hooker.

Fig. 113. Heart-leaved buttercup, *Ranunculus cardiophyllus* Hook.

*Ranunculus gelidus* Kar. & Kir.                    DWARF BUTTERCUP

Plants very small, with stems 5–10 cm high, glabrous or nearly so, bearing 1–3 leaves, these 3–5 foliate. Flowers 1–3, long-peduncled. Scree slopes; Rocky Mountains.

*Ranunculus glaberrimus* Hook.                    SHINY-LEAVED BUTTERCUP

A very early low-growing smooth-stemmed buttercup. Basal leaves stalked, some entire and not lobed, others 3-lobed. In the type the lower leaves usually rounded or kidney-shaped and wavy-margined, but the plants usually found are the var. *ellipticus* Greene (called *R. buddii* by Boivin), with entire long-elliptical basal leaves. Flowers 10–20 mm across, yellow, usually with a lavender or purplish tinge on the backs of the sepals. Occasionally large patches of this species with only one floral ring are found (called *R. buddii* f. *monochlamydeus* by Boivin). A very early spring flowering plant, local, and not usually abundant; plentiful in certain parts of the Prairies.

*Ranunculus gmelinii* DC.                    SMALL YELLOW WATERCROWFOOT

A semiaquatic perennial with underwater leaves divided into very narrow, flat segments, the floating leaves having wider lobes. Petals yellow, about 5 mm long; head of fruit globose, about 5 mm thick. Occasionally found in ponds and lakes; throughout the Prairie Provinces except the Prairies.

*Ranunculus hyperboreus* Rottb.                    BOREAL BUTTERCUP

Glabrous aquatic plants, with the stems floating, freely branching. Leaves with simple reniform blades, lobed or 3-parted. Petals 2–4 mm long, the same length as the sepals. Wet areas; Boreal forest. Large plants with cordate leaves and petals 3–4 mm long are var. *intertextus* (Greene) Boiv.; rare; Rocky Mountains.

*Ranunculus inamoenus* Greene                    GRACEFUL BUTTERCUP

A species 15–30 mm high, with stalked basal leaves varying from almost circular to fan-shaped and sometimes divided; stem leaves divided into 3 segments. Petals small, usually less than 6 mm long. Not common; has been found in the Cypress Hills and Rocky Mountains.

*Ranunculus lapponicus* L.                    LAPLAND BUTTERCUP

Plants with slender rootstocks; stems solitary, 10–20 cm high. Basal leaf 1 or rarely 2; cauline leaf solitary or absent. Flower solitary, with 3 sepals and 6–10 petals. Wet woods and muskeg; Boreal forest.

*Ranunculus macounii* Britt.                    MACOUN'S BUTTERCUP

A tall, branching hairy species 30–60 cm high, usually decumbent when partly grown. Leaves in broad divisions, with the leaf segments usually stalked. Yellow flowers 10–15 mm across; achenes having a sharp stout beak, about one-quarter the length of the body. Fairly common; in moist and wet places; throughout the Prairie Provinces.

*Ranunculus nivalis* L.                         SNOW BUTTERCUP

Plants usually unbranched, 10–20 cm high. Basal leaves orbicular-reniform, deeply lobed. Sepals villous; achenes with a beak about as long as the body. Tundra and alpine prairies. In var. *nivalis* the pubescence of the sepals brown to blackish; Hudson Bay; in var. *eschscholtzii* (Schlecht.) Benson the pubescence pale yellow; Rocky Mountains. Syn.: *R. verecundus* Rob.

*Ranunculus occidentalis* Nutt.                 WESTERN BUTTERCUP

Erect or somewhat decumbent plants, with stems 20–70 cm high, branching above. Basal leaves 3-parted, with the lobes cuneate, the petioles pubescent. Flowers large, 15–25 mm across. Meadows and prairies; Rocky Mountains. Several varieties have been described; most plants from Alberta are var. *brevistylis* Greene, with narrowly obovate petals, and the achene 2–2.5 mm long with a straight beak about half as long as the body.

*Ranunculus pallassii* Schlecht.                 PALLAS BUTTERCUP

Plants small, creeping, with stems 10–15 cm high. Leaves obovate-cuneate in outline, deeply 3-lobed. Flowers solitary, 20–25 mm across, white or somewhat violet-tinged; achenes 5–7 mm long. Wet shores; Hudson Bay, Boreal forest.

*Ranunculus pedatifidus* J. E. Smith var. *leiocarpus* (Trautv.) Fern.
                                   NORTHERN BUTTERCUP

An erect plant 10–30 cm high. Basal leaves deeply dissected into linear divisions. In the typical species, flowers 8–15 mm across with yellow petals and greenish yellow sepals. Fruit has a slender curved beak. The type is apparently a mountain and arctic plant, but the apetalous form has been found on open grassland in Prairies and Parklands. Formerly known as *R. affinis* R. Br. This form has only one floral ring of greenish yellow petal-like sepals.

*Ranunculus pensylvanicus* L. f.                 BRISTLY BUTTERCUP

An erect species, usually annual, very hairy, 30–60 cm high. Leaves divided into 3-stalked segments, again divided into 3 lobes. Yellow flowers 6–8 mm across, with petals shorter than reflexed sepals; achenes with a sharp beak about one-third the length of the body. Not common; occasionally found in wet places; Parklands and Boreal forest.

*Ranunculus pygmaeus* Wahl.                  PYGMEE BUTTERCUP

Small plants with a rootstock, and erect stems 1.5–5 cm high. Leaves about 1 cm long, glabrous. Flowers 5–10 mm across, with petals and sepals about the same length; achenes about 1 mm long. Rare; alpine tundra; Rocky Mountains.

*Ranunculus repens* L.                         CREEPING BUTTERCUP

Plants stoloniferous, with stems 30–45 cm high. Leaves 3-foliate. Flowers 20–30 mm across, yellow. Introduced species; occasionally found in wet areas.

*Ranunculus rhomboideus* Goldie               PRAIRIE BUTTERCUP

A short hairy-stemmed low-growing species 15–45 cm high. Rounded or oval wavy-margined basal leaves on long stalks; stem leaves deeply cleft or divided and without stalks. Flowers yellow, 10–20 mm across; petals rather narrow, appearing to be somewhat widely spaced. Flowering very early in the spring. Fruiting head globular; achenes with a short beak. Perhaps the most common species on the prairies; over all the open plains; the Prairie Provinces. Syn.: *R. ovalis* Raf.

*Ranunculus sceleratus* L. CELERY-LEAVED BUTTERCUP

A hollow-stemmed annual plant, with smooth stems 15–60 cm high. Leaves deeply 3- to 5-lobed and thick; basal leaves with long stalks but upper stem leaves stalkless. Yellow flowers small and numerous, 6–8 mm across, with petals about the same length as sepals. Common; along shores of lakes and sloughs and often the dominant plant of the hummocky bed of a drying slough; throughout the Prairie Provinces. Its acrid **poisonous** juice blisters the skin and produces intestinal inflammation if eaten. Sometimes called cursed crowfoot because of its toxic nature.

*Ranunculus septentrionalis* Poir. SWAMP BUTTERCUP

Stout-stemmed plants 30–60 cm high, erect or ascending, often rooting at the lower nodes of prostrate stems. Flowers 20–30 mm across; achenes 3–4 mm long, with a straight beak about half the length of the body. Wet areas; southeastern Parklands and Boreal forest.

*Ranunculus uncinatus* D. Don HAIRY BUTTERCUP

Plants with erect hirsute stems 30–60 cm high. Leaves uniform, cordate at the base, 3-parted. Flowers about 5 mm across, with the petals 2–3 mm long, the sepals reflexed; achenes 1.5–2 mm long, flattened, with a beak as long as the body, strongly recurved. Moist areas; Rocky Mountains, Peace River.

**Thalictrum** meadow-rue

Perennial plants from rootstocks, with alternate leaves divided into 3 leaflets, each 3-lobed. The segments and the leaflets are usually stalked. Flowers, either perfect or unisexual, with greenish white sepals and no petals, borne in rather open panicles. The yellow stamens of the male flowers make the inflorescence very attractive. Fruits achenes, usually ribbed, and borne in small clusters.

1. Flowers perfect; achenes flat and half
    obovoid with a straight back. .............................. *T. sparsiflorum* var. *richardsonii*
    Flowers unisexual. ................................................................................................ 2

2. Stem leaves barely stalked; leaves oblong,
    usually longer than wide, often very
    finely hairy beneath; stems purplish. ........................................... *T. dasycarpum*
    All leaves with stalks; leaflets almost
    round, pale, and smooth beneath. ....................................................................... 3

3. Leaves thin, not distinctly veined;
    achenes similarly pointed at both ends. ................. *T. occidentale* var. *palousense*
    Leaves thick and prominently veined;
    achenes obovoid and sharper at the
    upper end than at the lower. ................................................................. *T. venulosum*

*Dicots* RANUNCULACEAE – 381

*Thalictrum dasycarpum* Fisch. & Lall. (Fig. 114)     TALL MEADOW-RUE

A tall, erect purplish-stemmed plant 50–150 cm high. Leaves decompound, the long leaflets usually longer than wide, with 3 pointed lobes, dark green above, and pale and prominently veined beneath, often with very fine hairs beneath. Flowers borne in a long open panicle 30 cm long or longer. Flowering in midsummer. A fairly common species; in rich moist woodland; throughout the Prairie Provinces.

*Thalictrum occidentale* A. Gray var. *palousense* St. John  WESTERN MEADOW-RUE

An erect smooth-stemmed perennial 30–60 cm high, with thin pale green leaflets almost round but margins many-lobed. Flowers in an open panicle, appearing fairly early in the spring. In shady moist places, woodlands, and coulees; Cypress Hills, Rocky Mountains, Peace River.

*Thalictrum sparsiflorum* Turcz. var. *richardsonii* (Gray) Boiv.

FEW-FLOWERED MEADOW-RUE

A species with perfect flowers. Leaflets glandular below and scented. Rare; listed from several localities.

*Thalictrum venulosum* Trel.     VEINY MEADOW-RUE

Plants 15–200 cm high, with the leaflets rounded, strongly veined, bluish green, pale below. Panicle of flowers narrow and dense. Achenes ovoid. Fairly common; in woodlands; throughout most of the Prairie Provinces. Three varieties can be recognized:

1. Plants 15–40 cm high; filaments 3–4 mm
   long; fruit 3–4 mm long. ................................................................ var. *venulosum*
   Plants larger. ................................................................................................ 2
2. Plants 60–80 cm high; filaments 4–5 mm
   long; fruit 4–4.5 mm long. ............................................. var. *turneri* Boiv.
   Plants 1–2 m high, lax; filaments 5–6 mm
   long; fruit 5–6.5 mm long. ........................................... var. *lunellii* (Greene) Boiv.

**Trollius**     globeflower

*Trollius laxus* Salisb.     AMERICAN GLOBEFLOWER

Plants with a thick caudex, and erect stems 20–40 cm high, glabrous, 2- to 4-leaved. Flowers usually solitary, 2–4 cm across, whitish. Follicles about 1 cm long. Marshy areas and alpine meadows; Rocky Mountains.


# BERBERIDACEAE—barberry family

Plants shrubby, the leaves pinnate with spinu-
lose toothed leaflets. .................................................................................... *Berberis*
Plants herbaceous, with a single compound
leaf, the leaflets 5–8 cm long. ......................................................... *Caulophyllum*

Fig. 114.   Tall meadow-rue, *Thalictrum dasycarpum* Fisch. & Lall.

***Berberis*** barberry

*Berberis aquifolium* Pursh MAHONIA

    Shrubby plants 10–100 cm high, the leaves alternate with 5–7 leathery leaflets. Inflorescence a raceme with yellow flowers; fruit a blue berry with a few large seeds. In mountain woods; southern Rocky Mountains. Plants with trailing stems 10–30 cm high are forma *repens* (Lindl.) Boiv.

***Caulophyllum*** blue cohosh

*Caulophyllum thalictroides* (L.) Michx. BLUE COHOSH

    Plants erect, 30–80 cm high. The leaf inserted at about the middle of the stem, consisting of 3 long-petioled leaflets each of which is twice divided into 3 segments; the ultimate leaflets obovate-oblong. A smaller leaf subtends the inflorescence, consisting of a panicle of yellowish green or purplish flowers, about 1 cm across; seeds drupe-like, blue. Rare; moist woods; southeastern Parklands and Boreal forest.


## MENISPERMACEAE—moonseed family

***Menispermum*** moonseed

*Menispermum canadense* L. YELLOW PARILLA

    Plants woody climbers 2–3 m high; leaves 10–15 cm long, suborbicular, peltate near the margin. Inflorescence a panicle; flowers 3–5 mm across; fruit a blackish drupe. Woods; southeastern Parklands and Boreal forest.


## PAPAVERACEAE—poppy family

Petals usually 4; stems leafy; capsule opening
  by a circle of pores under the broad stigma. ...................................................... *Papaver*
Petals 8–12; leaf solitary, basal; capsule 2-
  valved. ................................................................................................. *Sanguinaria*

***Papaver*** poppy

1. Plants perennial; the leaves all basal. .......................................................................... 2
   Plants annual; cauline leaves present. ...................................................................... 3
2. Flowers sulfur yellow; leaves hirsute to
    subglabrous. ................................................................................ *P. pygmaeum*
   Flowers red or reddish orange; leaves
    glabrous and glaucous. ............................................................... *P. nudicaule*
3. Cauline leaves cordate-clasping, glabrous,
    not divided. ................................................................................ *P. somniferum*
   Cauline leaves not clasping, pubescent,
    pinnately divided. ........................................................................... *P. rhoeas*

*Papaver nudicaule* L.                                              ICELAND POPPY

Plants with a rosette of basal leaves, and scapose stems to 50 cm high bearing solitary flowers to 6 cm across. Introduced ornamental; in various locations escaped and established.

*Papaver pygmaeum* Rydb.                                            ALPINE POPPY

Plants with a rosette of basal leaves 2–3 cm long, subglabrous to pubescent. Scape 4–10 cm high, hirsute; flower 2–3 cm across; capsule about 1 cm long, densely bristly. Alpine areas; southern Rocky Mountains.

*Papaver rhoeas* L.                                                 CORN POPPY

Plants 30–60 cm high, hirsute; leaves pinnately divided, with the pinnae lobed or toothed. Flowers 5–7 cm across, scarlet. Introduced; occasionally escaped and established.

*Papaver somniferum* L.                                             COMMON POPPY

Plants 60–100 cm high, usually blue green, glabrous. Flowers 6–10 cm across, white to purplish or red, with the petals black at the base; capsule 5–8 cm long, to 5 cm across. Introduced as ornamental and for poppy seed; occasionally escaped and established.

**Sanguinaria**      bloodroot

*Sanguinaria canadensis* L.                                         RED PUCCOON

Plants with a stout rootstock, a single basal leaf, and scape with a solitary flower. Leaf orbicular in outline, 3- to 9-lobed, with the lobes undulate. Flowers white or pinkish, 2–5 cm across. Capsule 3–5 cm long. The name derives from the brilliant red sap, exuding when the plant is wounded. Rare; more or less open woods; southeastern Parklands and Boreal forest.


## FUMARIACEAE—fumitory family

1. Plants climbers; corolla persistent, the
   two outer petals saccate. ........................................................... *Adlumia*
   Plants not climbing; corolla deciduous,
   only one of the petals saccate. ........................................................... 2
2. Fruit an elongate capsule with more than
   one crested seed. ........................................................... *Corydalis*
   Fruit globose, with a single noncrested
   seed. ........................................................... *Fumaria*

**Adlumia**      fumitory

*Adlumia fungosa* (Ait.) Greene                                     CLIMBING FUMITORY

A biennial vine, stemless in the first year, growing to a climbing vine 2–3 m long. Leaves pinnate-ternate, with the rachis of the uppermost leaflets elongating and twining around supporting vegetation. Flowers pink, with the corolla becoming spongy and enclosing the capsule. Very rare; moist woods; southeastern Boreal forest.

### *Corydalis*    corydalis

Flowers yellow; plants usually somewhat prostrate. ................................................................ *C. aurea*

Flowers pink with yellow tips; plants fairly erect. ................................................................ *C. sempervirens*

### *Corydalis aurea* Willd.    GOLDEN CORYDALIS

A much-branched low, sometimes prostrate annual or biennial, with pale green much-dissected leaves. Plants sometimes spread about 45 cm. Racemes of narrow golden yellow flowers, each 10–15 mm long, with a spur at the base about half as long as the body of the corolla. Fruits narrow pods 15–25 mm long containing many small shiny seeds. Common; in edges of woodlands, moist spots, railway grades; throughout the Prairie Provinces. Reported to be somewhat **poisonous** to sheep.

### *Corydalis sempervirens* (L.) Pers. (Fig. 115)    PINK CORYDALIS

A slender-stemmed annual or biennial plant 30–75 cm high, with alternate twice-divided leaves. Flowers 10–20 mm long, in a raceme, pinkish with yellow tips. Fruits long, narrow cylindric capsules, turning brown at maturity, and containing many seeds. Often found; in rocky woodlands and disturbed areas; Boreal forest.

### *Fumaria*    fumitory

### *Fumaria officinalis* L.    EARTH-SMOKE

Annual plants 20–80 cm high, with the stems lax, diffusely branched. Inflorescence a many-flowered raceme 2–4 cm long; flowers 7–9 mm long, reddish purple; fruit about 2.5 mm in diam. Introduced; occasionally a weed.

## CAPPARIDACEAE—caper family

Pods borne on a stalk about 10 mm long above the calyx; stamens 6. ................................................................ *Cleome*

Pods on a very short stalk above the calyx, clammy to touch; stamens 8–32. ................................................................ *Polanisia*

### *Cleome*    spiderflower

### *Cleome serrulata* Pursh (Fig. 116)    SPIDERFLOWER, PINK CLEOME

An erect, branching annual plant 30–100 cm high. Stem smooth and hairless. Leaves trifoliolate with lanceolate leaflets 2–7 cm long. Flowers usually pale pink, sometimes white, stamens protruding above the flower. Pods 2–5 cm long, usually slightly curved, and containing one compartment with a single row of large seeds. Plant strong and unpleasant smelling. Very common; along roadsides and on light soil; Prairies. Syn.: *Peritoma serrulatum* (Pursh) DC.

Fig. 115.   Pink corydalis, *Corydalis sempervirens* (L.) Pers.

A.C. Budd

Fig. 116.  Spiderflower, *Cleome serrulata* Pursh.

*Dicots*

*Polanisia*        clammyweed

*Polanisia dodecandra* (L.) DC.                                   CLAMMYWEED

A sticky, glandular annual, branched, 15–50 cm high, with trifoliolate leaves, leaflets 10–25 mm long. Flowers yellowish white or pink, about 6 mm long, with pods 20–35 mm long, glandular. Not common; found on sandy soils; throughout Prairies and southeastern Parklands. Syn.: *P. graveolens* Raf. Large-flowered plants are distinguished as var. *trachysperma* (T. & G.) Iltis, large-flowered clammyweed, with flowers 8–12 mm long, pale yellow, the stamens protruding. Not common; on rocky banks, hillsides, and light soils; same locations as the species.

## CRUCIFERAE—mustard family

Annual, winter annual, biennial, or perennial plants with various-shaped alternate leaves. Flowers perfect and regular, with 4 sepals, 4 petals, and 6 stamens divided into 2 groups, 4 long and 2 short stamens. Fruit pods (siliques) (Fig. 117) divided into 2 compartments by a thin partition and usually containing many seeds (Fig. 117*A*). This family contains many troublesome weeds and also many edible plants.

1. Pods compressed at right angles to the
   central partition. ................................................................................. 2
   Pods not compressed at right angles to the
   central partition. ................................................................................. 9

2. Plants with basal leaves only; aquatic,
   usually submerged. ................................................................... *Subularia*
   Plants with leafy stems; mostly
   terrestrial. ........................................................................................... 3

3. Pods strongly inflated. ................................................................ *Physaria*
   Pods not strongly inflated. ................................................................... 4

4. One seed in each compartment of pod. ............................................... 5
   More than one seed in each compartment
   of pod. ................................................................................................ 6

5. Pods not winged or notched at top. ............................................. *Cardaria*
   Pods notched at apex, usually winged
   above. ............................................................................................ *Lepidium*

6. Pods strongly flattened. ...................................................................... 7
   Pods not strongly flattened. ................................................................. 8

7. Pods more or less winged, orbicular. ........................................... *Thlaspi*
   Pods not winged, triangular. ....................................................... *Capsella*

8. Stigma sessile; plants small, not fleshy. ................................. *Hymenolobus*
   Stigma not sessile; plants with fleshy
   leaves. ......................................................................................... *Cochlearia*

9. Pods usually not more than 3 times as
   long as wide. ..................................................................................... 10

*Dicots*                                    CRUCIFERAE (BRASSICACEAE) – 389

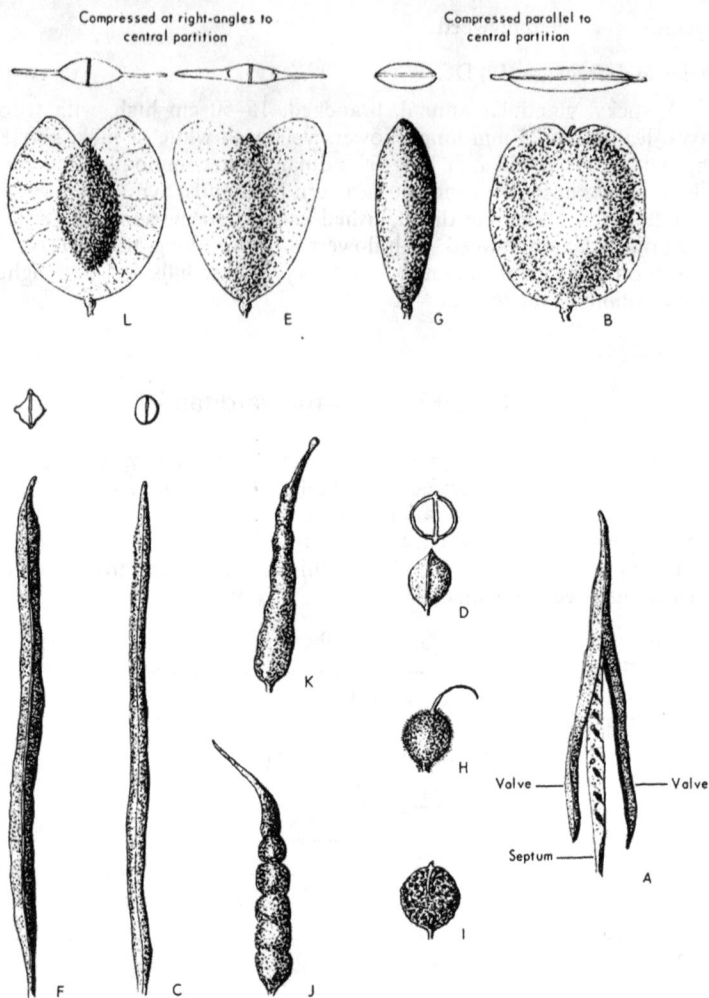

Compressed at right-angles to central partition

Compressed parallel to central partition

L   E   G   B

K   D   H   I   A

Valve —— —— Valve

Septum ——

F   C   J

A.C. Budd

Fig. 117.   Pods of the mustard family (Cruciferae): *A*, a dehiscent pod; *B*, small alyssum, *Alyssum alyssoides* L.; *C*, tower mustard, *Arabis glabra* (L.) Bernh.; *D*, false flax, *Camelina sativa* (L.) Crantz; *E*, shepherd's-purse, *Capsella bursa-pastoris* (L.) Medic.; *F*, hare's-ear mustard, *Conringia orientalis* (L.) Dum.; *G*, yellow whitlow-grass, *Draba nemorosa* L. var. *leiocarpa* Lindbl.; *H*, sand bladderpod, *Lesquerella ludoviciana* (Nutt.) Wats. var. *arenosa* (Rich.) Wats.; *I*, ball mustard, *Neslia paniculata* (L.) Desv.; *J*, wild radish, *Raphanus raphanistrum* L.; *K*, wild mustard, *Brassica kaber* (DC.) L. C. Wheeler; *L*, stinkweed, *Thlaspi arvense* L.

24. Pods flattened; seeds more or less in 2
    rows. ................................................................................ *Diplotaxis*
    Pods circular in cross section; seeds in a
    single row. ........................................................................ 25
25. Beak joint of capsule containing a seed. ................................ *Erucastrum*
    Beak joint of capsule seedless. ............................................. *Brassica*
26. Flowers yellow. ...................................................................... 27
    Flowers white to pinkish to purple, not
    yellow. ................................................................................ 34
27. Stem leaves entire to dentate, not lobed. .............................. 28
    Stem leaves deeply lobed, lyrate to
    tripinnate. ........................................................................... 31
28. Pubescence consisting of hairs, attached
    at the middle. ...................................................................... *Erysimum*
    Pubescence consisting of simple or sev-
    eral-forked hairs. ................................................................ 29
29. Stem leaves narrowly linear, long attenu-
    ate at base. .......................................................................... *Sisymbrium*
    Stem leaves broad, more or less auriculate
    at base. ................................................................................ 30
30. Stem leaves clasping, with the base cor-
    date; plants glabrous; flowers yellowish
    white. ................................................................................... *Conringia*
    Stem leaves with a sagittate base; plants
    pubescent below; flowers greenish to
    yellowish white. ................................................................... *Arabis*
31. Stem leaves bi- to tri-pinnately divided;
    pubescence of forked hairs. ................................................ *Descurainia*
    Stem leaves lyrate to pinnatifid; plants
    with simple hairs or glabrous. ............................................ 32
32. Pods 4-angled in cross section, with
    valves of pod nerveless; seeds globose. ............................. *Rorippa*
    Pods circular or nearly so in cross
    section. ............................................................................... 33
33. Leaf blades pinnatifid; seeds flat. ......................................... *Barbarea*
    Leaf blades entire or merely toothed;
    seeds plump. ....................................................................... *Erysimum*
34. Leaves deeply lobed to dissected. ........................................ 35
    Leaves entire to more or less deeply
    toothed. ............................................................................... 38
35. Plants densely grayish to silvery
    pubescent. ........................................................................... *Smelowskia*
    Plants glabrous to more or less
    pubescent. ........................................................................... 36
36. Pods flattened; leaves and stem usually
    with forked or starry hairs; leaves usu-
    ally simple and not deeply pinnately
    lobed. ................................................................................... *Arabis*

*Alyssum*    alyssum

*Alyssum alyssoides* L.                          SMALL ALYSSUM

A low annual 10–30 cm high, covered with starry hairs and bearing linear-oblong to spatulate leaves 6–12 mm long. Creamy white flowers about 5
mm across, in narrow racemes. Pods (Fig. 117*B*) yellowish, almost circular,
about 3 mm across, having a narrow flat margin around the edges. An introduced weed, not common; in sandy areas and disturbed places; Prairies,
southeastern Parklands, southern Rocky Mountains.

***Arabis***      rock cress

Biennial or perennial plants with straight erect stems. Basal leaves usually in a rosette, with stem leaves stalkless and clasping. Flowers usually small and in racemes; fruits long narrow pods containing many seeds.

1. Pods erect or ascending. ............................................................................ 2
   Pods spreading or reflexed. ....................................................................... 7

2. Pods almost circular in cross section. ...................................... *A. glabra*
   Pods flattened parallel to the partition. ................................................ 3

3. Plant tufted; basal leaves lyrately lobed;
   stem leaves not eared or clasping. .................... *A. lyrata* var. *kamchatica*
   Plant slightly tufted; basal leaves not
   lyrately lobed; stem leaves often clasp-
   ing or eared. ............................................................................................ 4

4. Stem leaves sessile, not auricled or
   sagittate. ................................................................................................ 5
   Stem leaves with the base auricled or
   sagittate. ................................................................................................ 6

5. Pods 0.5–1.5 mm wide, 1–3 cm long;
   flowers white. .................................................................... *A. nuttallii*
   Pods 2–3 mm wide, 2–5 cm long; flowers
   pinkish to purple. ................................................................. *A. lyallii*

6. Leaves coarsely hairy; stem leaves eared
   at base. ............................................................................ *A. hirsuta*
   Leaves not coarsely hairy, the hairs, if
   any, 2- or 3-forked. .................................................... *A. drummondii*

7. Pods   spreading   to   reflexed   and
   pendulous. ............................................................................................ 8
   Pods   spreading,   never   reflexed   or
   pendulous. .......................................................................................... 10

8. Pods   always   sharply   reflexed   and
   pendulous. ..................................................................... *A. retrofracta*
   Pods   in   part   spreading,   in   part
   descending. ............................................................................................ 9

9. Pods 2–5 cm long, 2–3.5 mm wide. ..................................... *A. lemmonii*
   Pods 5–10 cm long, 1–2 mm wide. ..................................... *A. divaricarpa*

10. Plants 10–20 cm high; pods 1–3 cm long,
    1.5–2.5 mm wide. ............................................................. *A. arenicola*
    Plants 20–80 cm high; pods 5–10 cm long,
    1–2 mm wide. .................................................................. *A. divaricarpa*


*Arabis arenicola* (Rich.) Gelert

Perennial plants with dentate basal leaves; stem leaves oblong, cuneate at the base, slightly fleshy. Arctic and subarctic gravelly or sandy areas. Glabrous plants are var. *arenicola,* those with lower stem leaves and basal leaves pubescent are var. *pubescens* (Wats.) Gelert. Boreal forest.

*Arabis divaricarpa* A. Nels.                    PURPLE ROCK CRESS

A biennial species, usually with a purplish stem 30–50 cm high. Basal leaves 20–40 mm long with starry white hairs; stem leaves stalkless and narrow-lanceolate. Flowers white or pinkish purple, about 6 mm across. Pods spreading or curved slightly downward, 35–60 mm long. Dry slopes, sandy areas; Prairies and Parklands. A form with the pods sickle-shaped and spreading at various angles, var. *dacotica* (Greene) Boiv., more common than the typical variety.

*Arabis drummondii* A. Gray                    DRUMMOND'S ROCK CRESS

A biennial species with a tall, smooth, erect stem 20–70 cm high. Leaves generally smooth, but basal ones sometimes having a few 2-forked hairs. Stem leaves have arrow-shaped clasping bases. Pink or white flowers borne in a terminal raceme. Pods usually fairly erect, 5–8 cm long. Fairly plentiful; on dry hillsides; throughout most of the Prairie Provinces.

*Arabis glabra* (L.) Bernh.                    TOWER MUSTARD

A tall biennial species with erect stems 20–60 cm high; stems quite smooth above but slightly hairy near the base. Basal leaves have short stalks and are 5–15 cm long and slightly hairy. Stem leaves stalkless with an arrow-shaped base. Flowers usually greenish white or yellowish white and quite small. Pods (Fig. 117C) erect and closely pressed against the stem, 5–10 cm long. Not particularly plentiful; in waste places and fields; throughout the Prairie Provinces and quite common in the Cypress Hills.

*Arabis hirsuta* (L.) Scop.                    HIRSUTE ROCK CRESS

An erect simple-stemmed species 30–60 cm high, either smooth or hairy-stemmed, with leaves coarsely hairy, the basal ones forming a rosette, the stem leaves stalkless and clasping. Flowers white or greenish white. The fairly erect pods are 2–5 cm long. Not plentiful; on dry, rocky places; throughout the Prairies and Parklands. A form with larger flowers, about 2 cm across, and only the lower part of the plant very coarsely pubescent, is var. *glabrata* T. & G. Southern Rocky Mountains.

*Arabis lemmonii* Wats.

Perennial plants with branching caudex, 5–30 cm high; pubescent, with branching hairs. Basal leaves spatulate; stem leaves oblong-lanceolate. The common form, var. *drepanoloba* (Greene) Rollins, usually 20–30 cm high, and with pods 3–5 cm long, 2.5–3.5 mm wide. Not common; alpine areas; southwestern Rocky Mountains.

*Arabis lyallii* Wats.

Perennial plants with branching caudex, few to several glabrous stems, 5–25 cm high. Basal leaves oblanceolate, glabrous or pubescent. Alpine areas; Rocky Mountains.

*Arabis lyrata* L. var. *kamchatica* Fisch.                    LYRE-LEAVED ROCK CRESS

A native, tufted perennial or biennial plant 10–30 cm high, with lyrately lobed basal leaves and linear to spatulate stem leaves. Flowers usually white but occasionally pinkish. Found sparingly; on sandy soils in lightly forested areas; Boreal forest.

*Arabis nuttallii* Robins.

Perennial plants with several to many stems from a branching caudex, 5–30 cm high, glabrous above, usually hirsute below. Alpine areas; southern Rocky Mountains.

*Arabis retrofracta* Graham                                    REFLEXED ROCK CRESS

A biennial or perennial 10–60 cm high, with one or several stems from the base, usually with appressed hairs. Leaves usually covered with fine hairs, the lower leaves forming a rosette and the stem leaves often clasping the stem. Flowers purplish pink to white and borne in a terminal raceme. Pods usually straight, 2–6 cm long. Common; dry slopes, open areas; Prairies and Parklands. The var. *collinsii* (Fern.) Boiv. is similar, but has long spreading hairs at the base of the stem instead of appressed hairs as in the typical variety; similar locations. The var. *multicaulis* Boiv. is many-stemmed and has basal leaves about twice as long as the stem leaves; pods somewhat longer and wider. Western Parklands, Rocky Mountains.

*Armoracia*          horse-radish

*Armoracia rusticana* P. Gaertner, B. Meyer & Scherb.          HORSE-RADISH

Perennial plants with stems 20–100 cm high, glabrous, much-branched. Basal leaves 10–40 cm long, crenate. Flowers white, 10–15 mm across. Introduced; cultivated for its roots for making table relish; occasionally escaped and established along roadsides and in moist areas.

*Barbarea*          winter cress

*Barbarea orthoceras* Ledeb.                              AMERICAN WINTER CRESS

A smooth-stemmed biennial 20–50 cm high. Leaves alternate and usually lyrate with one or two pairs of narrow lobes and a large terminal lobe on each leaf. Streams and wet places; in the Cypress Hills, Rocky Mountains, Boreal forest. Plants with large flowers 10–15 mm across and the beak of the pod 2–3 mm long are *B. vulgaris* R. Br., an introduced species.

*Berteroa*          hoary-alyssum

*Berteroa incana* (L.) DC.                                    HOARY-ALYSSUM

A starry-haired plant 30–60 cm high, with numerous lanceolate leaves 10–25 mm long. Flowers about 3 mm across with white deeply notched petals. Pods starry-hairy, broadly oval, about 6 mm long. An introduced weed reported at various localities across the Prairies and Parklands.

*Brassica*          mustard

1. Pods with a large flat or angled beak,
   often containing a seed. .................................................................................. 2

Pods with a slender round or conical beak
without seeds. ...................................................................................................... 3

2. Pods densely pubescent; the beak usually
longer than the body. ................................................................................ *B. hirta*

Pods glabrous; the beak half as long as
the body. ................................................................................................ *B. kaber*

3. Upper stem leaves cordate and clasping. ........................................... *B. campestris*

Upper stem leaves cuneate at the base,
not clasping. ........................................................................................ *B. juncea*

### *Brassica campestris* L. BIRD-RAPE

Annual or biennial plants 40–80 cm high. Inflorescence not elongating, the opened flowers overtopping the buds; pods 3–7 cm long, with seeds reddish brown to blackish. Introduced; occasionally grown for seed; weedy in several areas. Plants in which the inflorescence elongates have been distinguished as *B. napus* L.

### *Brassica hirta* Moench WHITE MUSTARD

Annual plants 30–70 cm high, more or less pubescent. Flowers about 2 cm across; pods about 3 cm long, spreading, hirsute, with seeds yellowish to pale brown. Occasionally grown for seed; rare as a weed.

### *Brassica juncea* (L.) Cosson INDIAN MUSTARD

An annual erect, weedy plant 30–100 cm high, with almost hairless smooth stems and leaves. Lower leaves lyrate-pinnatifid with a large end lobe, stalked, 10–15 cm long; upper leaves slightly stalked and often not lobed, much smaller. Flowers 10–15 mm across, yellow; pods 2–5 cm long, with a conic beak. Introduced, common; in grainfields and waste places; throughout the Prairies and Parklands.

### *Brassica kaber* (DC.) L. C. Wheeler WILD MUSTARD

A more or less hirsute annual 20–80 cm high; the lower leaves usually lyrate-pinnatifid, with the terminal lobe large, rounded. Flowers 10–15 mm across; pods (Fig. 117*K*) 2–5 cm long, with the beak half as long as the body and seeds black or brown. Introduced weed; in waste places and disturbed areas.

### *Braya*

*Braya humilis* (C. A. Mey.) Robins.

Perennial plants with several stems arising from the caudex, 5–30 cm high, more or less pubescent. Basal leaves linear oblanceolate, 1–3 cm long, thickish, entire or somewhat toothed; stem leaves smaller, distant. Flowers white or purplish-tinged, 6–10 mm across; pods 1–3 cm long, cylindric, pubescent, swollen at intervals. Gravelly or sandy areas; Boreal forest, Rocky Mountains. A very variable species, in which several varieties have been described that overlap in characters.

### *Camelina*   false flax

Introduced annual or winter-annual weeds, with small yellow flowers and long-stalked obovoid (or pear-shaped) pods containing numerous seeds, which have a very short dormant period.

1. Stems and leaves hairy; lower leaves not
     stalked. ........................................................................................ *C. microcarpa*
   Stems and leaves smooth or slightly hairy;
     lower leaves with stalks. ................................................................................. 2
2. Lower leaves entire. ......................................................................... *C. sativa*
   Lower leaves dentate or lobed. ................................................... *C. dentata*

### *Camelina dentata* Pers.                    FLAT-SEEDED FALSE FLAX

Stem erect and smooth, 30–50 cm high. Lower leaves lobed, with winged stalks, 35–70 mm long; upper leaves smaller and entire. Flowers yellow; pod obovoid, 5–10 mm long. An introduced weed; sometimes found in grainfields and waste places.

### *Camelina microcarpa* Andrz.                SMALL-SEEDED FALSE FLAX

A densely hairy plant 30–50 cm high. Leaves stalkless, with simple starry hairs. Flowers yellow; pods ovoid, to 5 mm long. An introduced weed, fairly common; in grainfields and waste places.

### *Camelina sativa* (L.) Crantz                FALSE FLAX

An introduced annual or winter annual 30–100 cm high, with a few hairs on the lower leaves and the lower part of the stem. Upper leaves stalkless and clasping; lower ones tapering to a winged stalk. Flowers yellow, about 3 mm across; pods (Fig. 117*D*) 5–10 mm long, ovoid, containing about 10 seeds. Introduced from Europe, found occasionally; a weed in grainfields and waste places.

### *Capsella*   shepherd's-purse

### *Capsella bursa-pastoris* (L.) Medic. (Fig. 118)    SHEPHERD'S-PURSE

An introduced annual or winter-annual plant, with branched stems 15–50 cm high. Basal leaves forming a rosette and often deeply cut and lobed; stem leaves usually clasping with ears at the base. Flowers in terminal racemes, small and white. The seedpods (Fig. 117*E*) are characteristic of this species; they resemble an inverted triangle, the flat base uppermost and notched, each pod containing about 20 seeds. An introduced weed, common; in gardens and waste places; throughout the Prairie Provinces.

### *Cardamine*   bitter cress

1. Stem leaves entire or dentate, not lobed
     or divided. ......................................................................................................... 2
   Stem leaves deeply divided to pinnate. .......................................................... 3
2. Stem leaves entire. ...................................................................... *C. bellidifolia*
   Stem leaves coarsely dentate. ........................................................... *C. bulbosa*

Fig. 118.  Shepherd's-purse, *Capsella bursa-pastoris* (L.) Medic.

3. Flowers 15–25 mm across. ................................................................. *C. pratensis*
   Flowers 3–10 mm across. ................................................................................ 4

4. Plants perennial, with rootstocks; inflorescence corymbose, appearing umbellate; flowers 5–10 mm across. .............................................. *C. umbellata*
   Plants annual or perennial, without rootstocks; inflorescence elongate; flowers 3–5 mm across. ................................................................................ 5

5. Stems mostly pubescent at the base; terminal leaflets obovate. ....................................................................... *C. pensylvanica*
   Stems glabrous throughout; terminal leaflets linear-oblanceolate. ............................................................. *C. parviflora*

### *Cardamine bellidifolia* L.                                    ALPINE BITTER CRESS

Perennial plants with branched caudex; stems 3–10 cm high, glabrous. Flowers 1–5, white. Rare; alpine areas; southern Rocky Mountains.

### *Cardamine bulbosa* (Schreb.) BSP.                                    SPRING CRESS

Perennial plants with a fleshy bulb; stem solitary, 15–30 cm high; leaves mostly coarsely toothed. Rare; wet areas; southeastern Boreal forest.

### *Cardamine parviflora* L.                                    SMALL BITTER CRESS

Annual or biennial plants; stems usually solitary, 10–30 cm high, simple to branched, glabrous. Flowers about 4 mm across. Rare; dry, usually rocky areas; southeastern Boreal forest, Rocky Mountains.

### *Cardamine pensylvanica* Muhl.                                    BITTER CRESS

A biennial or short-lived perennial plant 10–50 cm high. Leaves very deeply pinnately lobed with 2–8 pairs of lateral segments and a larger terminal segment. Stem usually bearing a few scattered hairs. Flowers small and white. Fruits linear pods 10–30 mm long. Uncommon; in wet or very moist places; Boreal forest, Rocky Mountains, Cypress Hills. Syn.: *C. scutata* Thunb.

### *Cardamine pratensis* L.                                    MEADOW BITTER CRESS

A hairless perennial 20–50 cm high, with deeply pinnately lobed leaves and white or pink flowers 15–25 mm across. Rare; a plant of bogs and swampy places; Boreal forest, Peace River. Mostly represented by var. *angustifolia* Hook., with stem leaves narrow, not decurrent.

### *Cardamine umbellata* Greene                                    MOUNTAIN CRESS

Perennial with rootstocks; stems erect, 20–50 cm high, glabrous or nearly so. Flowers 5–10 mm across; pods crowded, 2–2.5 cm long. Rare; wet areas; southern Rocky Mountains.

### **Cardaria**          hoary cress

Introduced perennial weeds spreading by seeds and by running roots penetrating deeply into the soil. Stem leaves alternate and clasping; basal ones with a short stalk. Flowers white in a close raceme. Seedpods varying in shape, a character useful in distinguishing the species.

1. Pods with downy hairs, globular and
    inflated. ........................................................... *C. pubescens* var. *elongata*
    Pods smooth, flattened, not hairy. ........................................................................... 2
2. Pods heart-shaped at base, not inflated. ...................................................... *C. draba*
    Pods somewhat inflated, not heart-shaped
    at base. .................................................................................... *C. draba* var. *repens*

*Cardaria draba* (L.) Desv.                    HEART-PODDED HOARY CRESS

Perennial from deep running roots, with dense heads of white flowers and heart-shaped smooth seedpods. Introduced from Europe, found occasionally; in gardens and shelterbelts; throughout the Prairie Provinces.

The var. *repens* (Schrenk) O. E. Schulz, lens-podded hoary cress, is similar to the species, but with seedpods slightly inflated and almost round instead of heart-shaped. Introduced, probably from Afghanistan; found in the same locations as the species.

*Cardaria pubescens* (Meyer) Rollins var. *elongata* Rollins
                                               GLOBE-PODDED HOARY CRESS

Plants similar to *C. draba*, but with leaves more hoary in appearance and seedpods globular, inflated with fine, downy hairs. Apparently introduced from Asia, becoming quite common; in gardens and shelterbelts and some-times in fields and roadsides; southern part of the Prairie Provinces.

**Cochlearia**          scurvy-grass

*Cochlearia officinalis* L.                              SCURVY-GRASS

Annual plants of very variable size, 1–40 cm high; basal leaves reniform, usually fleshy, glabrous. Flowers white, with petals 3–7 mm long; silicle ovoid to globose, 4–7 mm long. Shores and beaches. Rare; Hudson Bay.

**Conringia**          hare's-ear mustard

*Conringia orientalis* (L.) Dum.                    HARE'S-EAR MUSTARD

An introduced annual or winter-annual weed with taproots and a per-fectly smooth stem 15–60 cm high. Seedling leaves round, on stalks, but subse-quent leaves clasping and elliptical, entire, eared at the base, quite smooth, and the whole plant covered with a bluish bloom. Flowers creamy white, about 6 mm across. Pods (Fig. 117*F*) narrow, 4-angled, erect, 7–10 cm long. Common; roadside and field weed.

**Descurainia**          tansy mustard

Annuals or biennials from taproots, usually with much-dissected leaves, yellow flowers, and long narrow seedpods.

1. Both upper and lower leaves 2 or 3 times divided. ........................................................................ 3

   Upper leaves simply pinnate. ........................................................................ 2

2. Seeds in 1 row in each compartment of pod; pods erect and closely pressed to the stem; plant grayish. ........................................................................ *D. richardsonii*

   Seeds in 2 rows in each compartment of pod; pods divergent from stem. ........................................ *D. pinnata* var. *brachycarpa*

3. Pubescence of simple hairs; plants glandular. ........................................................................ *D. sophioides*

   Pubescence of mostly stellate hairs; plants not glandular. ........................................................................ *D. sophia*

*Descurainia pinnata* (Walt.) Britt. var. *brachycarpa* (Richards.) Fern.

SHORT-FRUITED TANSY MUSTARD

Annual plants 20–60 cm high, erect, with a hairy and glandular stem. Leaves dark green; the upper ones pinnate and the lower ones often 2 or 3 times divided, glandular. Flowers yellow. Pods somewhat club-shaped, 5–12 mm long, on stalks 8–15 mm long. Not very common; has been found in southern part of the Prairie Provinces. Syn.: *Sophia brachycarpa* (Richards.) Rydb.

*Descurainia richardsonii* (Sweet) O. E. Schulz          GRAY TANSY MUSTARD

Tall, erect biennial plants 30–90 cm high, with pinnate leaves, sometimes doubly pinnate. Whole plant covered with very fine hairs, giving it a grayish appearance. Flowers small, pale yellow. Pods small, 5–10 mm long, tightly compressed against stem, and crowded. Common; in fields and waste places; throughout the Prairie Provinces. Syn.: *Sophia richardsonii* (Sweet) Rydb.

*Descurainia sophia* (L.) Webb.          FLIXWEED

An annual or biennial, branched, with all leaves 2 or 3 times dissected, and with some star-like hairs. Flowers yellow. Pods linear, 10–30 mm long, on stalks 6–12 mm long, and borne at an angle from the stem. An introduced weed, which has become very common; in fields, waste places, and even on prairie land; one of the commonest weeds in Western Canada. Syn.: *Sophia multifida* Gilib.

*Descurainia sophioides* (Fisch.) O. E. Schulz          NORTHERN FLIXWEED

Much like *D. sophia*, but brighter green and glandular. Inflorescence tardily elongating. Pods sickle-shaped, irregularly spreading to erect. Dry gravel beds; Hudson Bay. Syn.: *Sisymbrium sophioides* Fisch.

**Diplotaxis**          sand-rocket

*Diplotaxis muralis* (L.) DC.          SAND-ROCKET

Annual plants, branched from base, 30–60 cm high. Stems leafy only near the base, with oblanceolate lobed leaves 5–10 cm long, usually with a slender stalk. Flowers yellow. Pods 20–25 mm long with a short beak, on stalks 10–15 mm long. Introduced from Europe, but occasionally found; in waste places; throughout the Prairies and Parklands.

***Draba***      whitlow-grass

Low tufted plants, usually with a rosette of basal leaves, scapose or with a few stem leaves. Flowers small, white or yellow, usually in racemes. Pods (Fig. 117*G*) either oval or linear and flat, with seeds in 2 rows in each cell.

14. Leaves ciliate, at least toward the base. ................................................................ 15
    Leaves densely stellate throughout, not
    ciliate. ...................................................................................................... *D. nivalis*

15. Plants densely stellate, pubescent
    throughout. ............................................................................................. *D. cinerea*
    Plants not densely pubescent. ................................................................................. 16

16. Basal leaves rather thick, oblong-spatu-
    late, subglabrous. ............................................................................. *D. crassifolia*
    Basal leaves oblong-obovate, usually
    sparsely pubescent, not thick. ...................................................... *D. fladnizensis*

17. Basal leaves glabrous, but ciliate toward
    the tip. .......................................................................................... *D. crassifolia*
    Basal leaves more or less pubescent, at
    least below. ................................................................................................. 18

18. Stems glabrous; leaves very narrow,
    mostly less than 1 mm,
    stellate-pubescent. ....................................................................... *D. oligosperma*
    Stems pubescent; leaves mostly 1 mm
    wide or wider. ............................................................................................ 19

19. Leaf margins conspicuously long-ciliate,
    pubescent with long tangled hairs. ............................................... *D. stenopetala*
    Leaf margins not long-ciliate. ................................................................................ 20

20. Pubescence of simple or forked hairs. ........................................................ *D. alpina*
    Pubescence mostly of stellate hairs. ......................................................... *D. incerta*

### *Draba alpina* L.                                         ALPINE WHITLOW-GRASS

Plants with a dense rosette, the leaves elliptic-lanceolate, densely pubescent, entire. Scape to 20 cm high, slightly elongating in fruit. Flowers 6–10 mm
across, bright yellow; pods glabrous. Open places in rocky arctic tundra;
Boreal forest, Manitoba.

### *Draba aurea* Vahl                                        GOLDEN WHITLOW-GRASS

Plants with basal rosettes and erect or decumbent leafy stems. Stem leaves
entire or somewhat dentate. Flowers 8–12 mm across, yellow; pods 8–16 mm
long, puberulent, often twisted. Open forest and alpine meadows; Rocky
Mountains. Plants with glabrous pods are var. *leiocarpa* (Payson & St. John) C.
L. Hitchc.

### *Draba cinerea* Adams

Plants densely stellate-pubescent in all parts; stems 5–25 cm high, with
1–4 stem leaves. Flowers 6–10 mm across; pods 4–8 mm long, inflated. Sandy
or gravelly areas; southern Rocky Mountains.

### *Draba crassifolia* R. C. Graham                  THICK-LEAVED WHITLOW-GRASS

Dwarf perennial plants with dense rosettes; leaves rather thick, sparsely
pubescent. Stems usually leafless, 5–15 cm high, glabrous. Flowers 4–6 mm
across, with the petals little larger than the sepals; pods 4–7 mm long, glabrous. Alpine meadows and scree fields; southern Rocky Mountains.

*Draba fladnizensis* Wulf.

Small, often dwarf perennials, with stems rarely more than 10 cm high. Stem leaves 0–2; basal leaves oblong-obovate, glabrous except the ciliate margins. Inflorescence 2- to 10-flowered; flowers 5–8 mm across; pods 5–7 mm long. Arctic and alpine, usually stony areas; Boreal forest, southern Rocky Mountains.

*Draba hirta* L.                                                          HAIRY WHITLOW-GRASS

Plants with stems to 25 cm high, and 1–4 stem leaves, usually unbranched. Basal leaves to 2 cm long, narrowly lanceolate, subentire, ciliate, and more or less densely stellate-pubescent. Flowers 6–10 mm across, creamy white or light yellow; pods 6–12 mm long, glabrous, often somewhat twisted. Gravelly shores and riverbanks; Boreal forest, Rocky Mountains.

*Draba incana* L.

Plants biennial or short-lived perennial; stems to 35 cm high, erect, simple or branched, with 6–25 stem leaves. Basal leaves 1–2.5 cm long, pubescent with simple or branched hairs. Stem, stem leaves, and inflorescence stellate-pubescent. Flowers 6–10 mm across, white; pods 6–10 mm long, often somewhat twisted. Dry gravel beds; Boreal forest, Manitoba.

*Draba incerta* Payson

Loosely tufted perennial; scapes 2–10 cm high, often with a reduced leaf below the first flower, pubescent with branched hairs. Leaves linear-oblanceolate, stellate-pubescent below. Flowers 8–10 mm across, yellow; pods 5–10 mm long, pubescent. Alpine meadows and slopes; Rocky Mountains.

*Draba lanceolata* Royle

Plants perennial, with leafy simple or branched stems 5–25 cm high, pubescent; basal leaves 1–3 cm long, stem leaves 5–25 mm long, dentate, pubescent with stellate or bushy hairs. Flowers 6–10 mm across, white; pods 4–12 mm long, often twisted, pubescent. Meadows and slopes; Rocky Mountains.

*Draba mccallae* Rydb.

Plants to 25 cm high, with 3–5 or sometimes more stem leaves. Stem long pilose below; leaves stellate pubescent. Flowers 6–10 mm across; pods 7–10 mm long, densely puberulent. Meadows, slopes, and gravel banks; southern Rocky Mountains.

*Draba nemorosa* L.                                                     YELLOW WHITLOW-GRASS

A low plant, 10–30 cm high, branching from the rosette of basal leaves. Flowers pale yellow, occasionally fading to white; pods 6–8 mm long, on oval stalks 6–20 mm long. Fairly common; throughout southern part of the Prairie Provinces. The var. *leiocarpa* Lindbl. differing from the type by having smooth pods (Fig. 117G) and being probably more plentiful, especially in the northern and eastern parts. Both the species and the variety flowering quite early in the spring. Syn.: *D. lutea* Gilib.

*Draba nivalis* Liljeb.

Perennial plants with rosettes on the caudex branches; leaves densely stellate-pubescent, with branched hairs mixed in. Stems 2–10 cm high, leafless or with a single reduced leaf. Flowers 5–10 mm across, white. In var. *nivalis,* the pods less than 1 cm long, flat to slightly twisted; rock outcrops; Boreal forest and southern Rocky Mountains; in var. *elongata* Wats., the pods 10–15 mm long, and conspicuously twisted; alpine areas; southern Rocky Mountains.

*Draba oligosperma* Hook.                              FEW-SEEDED WHITLOW-GRASS

Small perennials with rosettes on the caudex branches. Scapes 1–10 cm high; flowers 6–10 mm across, yellow; pods 3–5 mm long, ovate, pubescent to glabrous. Rocky alpine areas; southern Rocky Mountains.

*Draba praealta* Greene

Biennial or perennial plants with stems 10–30 cm high, and compact rosettes. Basal leaves 1–3 cm long, entire or somewhat dentate, densely pubescent with branched hairs. Stem leaves 2–6, densely pubescent. Flowers 4–8 mm across, white; pods 8–15 mm long, soft pubescent. Slopes and rocky areas; Rocky Mountains.

*Draba reptans* (Lam.) Fern. var. *micrantha* (Nutt.) Fern.

CREEPING WHITLOW-GRASS

A small plant 2–15 cm high, with a leafy stem. Leaves usually 8–20 mm long. White flowers, 4–6 mm across, but often without petals; pods linear, 8–12 mm long, clustered near the top of the stem, usually covered with fine, stiff hairs. Flowering in early June. Not common; has been found in disturbed sandy areas; Prairies.

*Draba stenoloba* Ledeb.

Biennial or perennial plants 5–30 cm high, with stems arising from a simple caudex, glabrous above, somewhat pubescent below, with 1–4 pubescent leaves. Basal leaves 1–4 cm long, denticulate, stiffly pubescent. Flowers 6–10 mm across, yellow; pods 8–20 mm long. Alpine slopes; southern Rocky Mountains. Plants with the pubescence mostly of simple hairs and the stem coarsely pilose are var. *nana* (Schulz) C. L. Hitchc.; southern Rocky Mountains.

*Draba stenopetala* Trautv.

Perennial mat-forming plants with leafless stems 1–6 cm high; leaves crowded, linear to linear-oblanceolate, the margins ciliate with simple or branched hairs, pubescent below with simple or branched hairs. Flowers 5–9 mm across, light yellow; pods 3–5 mm long, ovate, pubescent to subglabrous. Alpine slopes; southern Rocky Mountains. Syn.: *D. paysonii* Macbr. var. *treleasii* (Schulz) C. L. Hitchc.

**Eruca**      rocket

*Eruca sativa* L.                                        GARDEN-ROCKET

Annual plants, with stems 20–100 cm high; leaves lyrate-pinnatifid, with the terminal lobe large, and 2–5 pairs of lateral lobes. Flowers 30–40 mm across, whitish or yellowish, purplish-veined; pods 12–25 mm long. Introduced; occasionally escaped from cultivation.

*Erucastrum*    dog mustard

*Erucastrum gallicum* (Willd.) O. E. Schulz                DOG MUSTARD
An annual plant 20–50 cm high, erect, with stem hairs pointing downward. Leaves deeply cut, often to the midrib, making them appear pinnate, and often deeply lobed. Leaves varying to 25 cm long. Flowers pale yellow; pods 25–40 mm long, linear, circular in cross section, on a short stalk tipped with a slender style, with one row of seeds in each compartment. An introduced weed; in grainfields; throughout Prairies and Parklands.

*Erysimum*    treacle mustard

Annual or biennial plants with leafy stems and 2-branched hairs, yellow flowers and linear 4-angled seedpods with seeds in one row in each compartment.

1. Flowers large, 2–4 cm across, purple. ....................................................... *E. pallassii*
   Flowers smaller, yellow. ................................................................................ 2
2. Petals longer than 10 mm. ........................................................ *E. asperum*
   Petals shorter than 10 mm. .......................................................................... 2
3. Stalks of seedpods almost half as long as
   pods. ....................................................................................... *E. cheiranthoides*
   Stalks of seedpods not one-quarter as
   long as pods. ................................................................................................. 4
4. Pubescence mostly or entirely of 2-
   branched hairs. ................................................................... *E. inconspicuum*
   Pubescence mostly or entirely of stellate
   hairs. ...................................................................................... *E. hieracifolium*

*Erysimum asperum* (Nutt.) DC. (Fig. 119)              WESTERN WALLFLOWER
A rough, usually erect plant 20–60 cm high, often much branched and coarse, with appressed small white hairs. Flowers pale yellow, 10–15 mm across; pods 3–10 cm long, narrow, rough, and 4-angled, with a short, thick style at the end, and a very short, stout stalk. Spreading out considerably from the stem in all directions. Common; on light and sandy soil; in Prairies, rare in Parklands. Syn.: *Cheirinia aspera* (Nutt.) Rydb.

*Erysimum cheiranthoides* L.                            WORMSEED MUSTARD
An erect plant 20–60 cm high, with lanceolate or oblong-lanceolate dark green leaves 2–10 cm long. Flowers yellow and small, about 5 mm across, in dense terminal clusters 2–3 cm across. Seedpods linear, 10–25 mm long, on slender stalks 6–12 mm long. Not common; a weedy plant found in moist places and fields; throughout the Prairie Provinces. Syn.: *Cheirinia cheiranthoides* (L.) Link.

Fig. 119. Western wallflower, *Erysimum asperum* (Nutt.) DC.

*Erysimum hieracifolium* L. GRAY ROCKET

Biennial or perennial plants 30–80 cm high; leaves linear or oblong, wavy-dentate, greenish gray. Flowers 15–20 mm across; pods 30–55 mm long, 4-angled. Rare, introduced weed; southeastern Parklands.

*Erysimum inconspicuum* (S. Wats.) MacM. SMALL-FLOWERED PRAIRIE-ROCKET

An erect grayish green perennial 20–50 cm high, and not much branched. Leaves narrow, 2–8 cm long, the lower ones having stalks but the upper ones stalkless. Yellow flowers about 6 mm across, on the top of the stem. Linear seedpods 2–6 cm long on a very short stalk and borne erect on the stem. Common; on dry and sandy prairie; throughout Prairies and Parklands. Syn.: *E. parviflorum* Nutt.

*Erysimum pallassii* (Pursh) Fern. PURPLE ROCKET

Biennial plants 10–20 cm high, with numerous linear leaves. Pods 6–10 cm long, ascending. Shale and gravel slides; southern Rocky Mountains.

## Eutrema

*Eutrema edwardsii* R. Br.

Rhizomatous perennial to 40 cm high, with rather fleshy ovate to orbicular leaves, truncate to cordate at the base. Flowers 5–8 mm across; pods 6–20 mm long, linear to oblong, erect or appressed. Arctic and alpine tundra; Boreal forest, southern Rocky Mountains.

## Halimolobos

*Halimolobos virgata* (Nutt.) O. E. Schulz

Perennial plants 15–40 cm high, much resembling *Arabis* ssp., densely pubescent with branched and stellate hairs. Flowers 4–8 mm across, white; pods 1–3 cm long, round, slightly constricted between the seeds, strongly ascending. Dry grasslands; western Prairies.

## Hesperis    rocket

*Hesperis matronalis* L. DAME'S-ROCKET

A tall perennial species with erect stems 30–100 cm high, simple or branched above. Leaves simple oblong- to ovate-lanceolate, dentate, pubescent on both sides. Flowers 3–5 cm across, purple, fragrant; pods 5–10 cm long, linear, ascending or spreading. Introduced ornamental, occasionally escaped and established; in tree groves or shrubbery.

## Hymenolobus

*Hymenolobus procumbens* (L.) Nutt.

Small annual or biennial plants 3–30 cm high, with stems procumbent or erect. Lower leaves entire to deeply lyrate-pinnatifid; stem leaves entire, pubescent throughout with sparse simple hairs. Inflorescence many-flowered, the flowers 3–6 mm across; petals white, spatulate, the same length as or somewhat exceeding the sepals; pods 1.5–5 mm long, elliptic to obovate, with the valves translucent. Very rare; wet, often saline areas; Hudson Bay, Prairies. Syn.: *Hutchinsia procumbens* (L.) Desv.

*Lepidium*        pepper-grass

Annual or perennial herbs, glabrous or pubescent with simple hairs. Inflorescence in dense, bractless racemes; flowers small, usually with whitish petals, or apetalous. Fruit a silicle, strongly flattened at right angles to the septum, orbicular to ovate or obovate; seeds one in each locule. Weedy plants of roadsides, waste places, and disturbed areas.

1. Upper leaves with a sagittate or cordate-
    clasping base. ...................................................................................................... 2
   Upper leaves tapering at the base. ....................................................................... 3

2. Auricles of the leaf base acute, much
    shorter than the leaf blade; lower
    leaves similar to the upper leaves. ..................................... *L. campestre*
   Auricles of the leaf base rounded, about
    the same length as the leaf blade; lower
    leaves pinnate. ................................................................... *L. perfoliatum*

3. Silicle not notched at the apex; perennial
    plant. ................................................................................ *L. latifolium*
   Silicle notched at the apex; annual plant. ......................................................... 4

4. Silicle 5–7 mm long, distinctly winged;
    stamens 6; petals 2 mm long. ............................................... *L. sativum*
   Silicle 2–4 mm long, not winged; stamens
    2 or 4; petals minute or none. .......................................................................... 5

5. Plants with short axillary racemes as well
    as terminal ones; petals linear, shorter
    than the sepals. ............................................................. *L. ramosissimum*
   Plants with elongated terminal racemes;
    petals none or rudimentary. .............................................................................. 6

6. Silicles oval or elliptic-ovate, gradually
    narrowed toward the apical teeth;
    basal leaves bipinnatifid. ....................................................... *L. ruderale*
   Silicles round-cordate or round-obovate,
    rounded at the apex; basal leaves at
    most pinnatifid. .............................................................. *L. densiflorum*

*Lepidium campestre* (L.) R. Br.                    PEPPERWORT

An annual or biennial species, with simple or branched stems 20–50 cm high, grayish green, densely pubescent. Basal leaves entire or lyrate; lower stem leaves petiolate, the upper ones clasping the stem with sagittately lobed leaves. Flowers about 3 mm across, white; silicle about 5 mm long. Rare; introduced weed, roadsides, railway grades; western Prairies.

*Lepidium densiflorum* Schrad. COMMON PEPPER-GRASS

An annual or winter annual with erect stem, much branched above, 15–60 cm high. Stem leaves lanceolate, with a few coarse teeth; basal leaves often deeply incised and divided. Flowers minute, with petals missing or very rudimentary. Seedpods borne on short stalks and very numerous on the stem, the spike of pods often 10–15 cm long; each pod about 2–3 mm wide, heart-shaped with a notch at the top, and containing a single seed in each compartment. Common; in fields, waste places, and roadsides; throughout the Prairie Provinces. Syn.: *L. apetalum* A. Gray. Plants with flattened pedicels and some axillary racemes are var. *bourgeauanum* (Thell.) C. L. Hitchc.

*Lepidium latifolium* L. BROAD-LEAVED PEPPER-GRASS

Tall plants 50–100 cm high, with a thick, branching rootstock, glabrous or nearly so. Basal leaves petiolate, to 30 cm long; stem leaves subsessile, much reduced. Racemes numerous in a large panicle; flowers about 3 cm across, with the sepals white-margined. Very rare; introduced weed; southwestern Prairies.

*Lepidium perfoliatum* L. PERFOLIATE PEPPER-GRASS

An annual weed with a stem much branched and bushy above, 20–50 cm high. Basal leaves pale green, very finely dissected and divided; subsequent leaves more entire, the upper ones entire with a pointed end and a cordate to entirely clasping base. Flower terminal in racemes and very small with yellow petals. Numerous seedpods about 3 mm long, oval, and containing a seed in each compartment. When growing, the plants have a reddish tinge to the leaves, which, with the yellow petals, makes them easy to notice. An introduced weed, common in dry areas of western USA but rare in Canada; western Prairies.

*Lepidium ramosissimum* A. Nels. BRANCHED PEPPER-GRASS

An annual or biennial, very similar to *L. densiflorum,* but branched diffusely from the base, and with white petals shorter than the sepals. Seedpods rather pointed at the notched apex, whereas in *L. densiflorum* the pods are more rounded. Probably fairly common throughout southern part of area, but distinction between the two species is obscure.

*Lepidium ruderale* L. ROADSIDE PEPPER-GRASS

Annual or biennial with a single erect or ascending stem, 10–30 cm high, sparsely pubescent. Basal leaves 5–7 cm long, long-petioled, pinnatifid to bipinnatifid. Flowers about 1.5 mm across; silicle 2–2.5 mm long, deeply notched. Introduced weed, rare; southeastern Parklands.

*Lepidium sativum* L. GARDEN CRESS

Annual with a single erect stem, glabrous. Basal leaves long-petioled, lyrate, with the lobes toothed; upper stem leaves linear, entire. Flowers 3–5 mm across; silicle 5–6 mm long, deeply notched. Introduced, occasionally escaped from cultivation.

***Lesquerella***     bladderpod

Low tufted annual or perennial plants with clustered basal leaves and a few stem leaves. Plants with starry hairs. Flowers perfect, in racemes, yellow or purple. Seedpods (Fig. 117*H*) inflated, globose or oval, with 2 to many seeds in each compartment.

1. Pods ovate; plants densely tufted; flowers
    bright yellow. ........................................................................ *L. alpina* var. *spathulata*
    Pods globose; plants not densely tufted;
    flowers reddish yellow. ........................................................................ 2

2. Pedicels ascending in fruit; pods glabrous
    or nearly so. ........................................................................ *L. arctica*
    Pedicels recurved in fruit; pods densely
    pubescent. ........................................................................ *L. ludoviciana* var. *arenosa*

*Lesquerella alpina* (Nutt.) S. Wats. var. *spathulata* (Rydb.) Payson
                                                            SPATULATE BLADDERPOD

A deep-rooted tufted perennial 5–15 cm high, with several stems. Numerous basal leaves 1–3 cm long; very few linear stem leaves. Flowers yellow, about 6 mm across; pods ovoid, very slightly compressed. Not common; dry hillsides and badlands; Prairies, southern Rocky Mountains.

*Lesquerella arctica* (Wormsk.) Wats.                    NORTHERN BLADDERPOD

Small plants with spatulate or oblanceolate basal leaves. Flowers long-pedicellate, yellow; pods on straight pedicels, 1–2 cm long. Gravelly areas; Hudson Bay. Plants with the pods scurfy and few white stellate hairs are var. *purshii* Wats.; southern Rocky Mountains.

*Lesquerella ludoviciana* (Nutt.) Wats. var. *arenosa* (Rich.) Wats. (Fig. 120)
                                                            SAND BLADDERPOD

A very slender stemmed spreading and decumbent plant 5–30 cm across, with mostly basal linear-oblanceolate leaves. Flowers dull yellow or reddish-tinged, about 6 mm across. Pods globose (Fig. 117*H*), 3–5 mm across, on curved stalks. Common; on dry hillsides and prairie; throughout the southern part of the Prairie Provinces, not necessarily on sandy soil, specimens having been collected on Regina Plains.

***Neslia***     ball mustard

*Neslia paniculata* (L.) Desv.                    BALL MUSTARD

A tall erect annual or winter annual 30–60 cm high, on much-branched stems. The whole plant is pale yellowish green, with starry hairs. Lower leaves somewhat stalked and lanceolate; stem leaves arrow-shaped and clasping the stems at the base. Flowers small, orange yellow, about 3 mm across, and clustered on the ends of the stems. Fruit pods (Fig. 117*I*) in long racemes, small, round, about 2 mm long. Pods usually contain only 1 seed and remain on the seed when ripe. Introduced from Europe, a weed of crop lands and waste places; throughout the Prairie Provinces.

Fig. 120.   Sand bladderpod, *Lesquerella ludoviciana* (Nutt.) Wats. var. *arenosa* (Rich.) Wats.

*Physaria*        bladderpod

*Physaria didymocarpa* (Hook.) Gray        TWIN BLADDERPOD

A tufted perennial with dense rosettes and several stems 5–15 cm long. Rosette leaves broadly obovate or spatulate, 1–4 cm wide, densely silvery stellate; stem leaves few and small. Flowers 1–2 cm across, bright yellow; pods 2-lobed, strongly inflated, often 2 cm or more across; seeds 4 in each locule. Slopes and rocky areas; Rocky Mountains.

*Raphanus*        radish

*Raphanus raphanistrum* L.        WILD RADISH

An erect annual or winter-annual weed 30–70 cm high. Basal and lower leaves 10–20 cm long, deeply lobed, with a large terminal lobe; upper leaves smaller. Flowers yellow with purplish veins, 10–20 mm across. Seedpods (Fig. 117*J*) 3–4 cm long, constricted between each seed, and breaking rapidly into 1-seeded sections when handled. Introduced from Europe, becoming increasingly common; a weed of abandoned fields and waste places; in eastern Prairies and Parklands.

*Rapistrum*

*Rapistrum perenne* (L.) All.        PERENNIAL MUSTARD

A biennial or perennial, with stems 30–80 cm high, densely hispid below, glabrous above. Lower leaves pinnate, coarsely serrate. Flowers 10–15 mm across, bright yellow; pods 7–10 mm long, the lower part cylindrical, the upper part ovoid, strongly ribbed. Introduced, occasional weed; southeastern Parklands.

*Rorippa*        yellow cress

1. Creeping-rooted perennials. ............................................................................ 2
   Fibrous-rooted plants without creeping
   roots; winter annual or biennial. ......................................................... 4

2. Pods not more than 3 times as long as
   wide; leaves entire or merely toothed. ...................................... *R. austriaca*
   Pods 4 or more times as long as wide;
   leaves dissected or lobed. ........................................................................ 3

3. Leaves with sharply toothed or incised
   divisions. ........................................................................................ *R. sylvestris*
   Leaves with blunt lobes. .......................................................................... *R. sinuata*

4. Stems erect, branched above; petals 1.5–2
   mm long; sepals 2 mm. ........................................................................ *R. islandica*
   Stems branched from the base; petals
   1.0–1.4 mm long; sepals 1.5 mm. ...................................................... *R. tenerrima*

*Rorippa austriaca* (Crantz) Besser        AUSTRIAN CRESS

A tall-growing introduced perennial 40–90 cm high, with running roots and smooth stem. Leaves alternate, almost entire or merely toothed, not dissected; the lower leaves with stalks and the upper ones without. Flowers yellow and small, in racemes. Pods small, almost globular, 2–3 mm long, on stalks 5–15 mm long. An introduced weed found occasionally; in grainfields in Prairies and Parklands, but uncommon at present. However, it should be watched to prevent its spread.

*Rorippa islandica* (Oeder) Borbas               MARSH YELLOW CRESS

Native annual or biennial branching plants 20–60 cm high, with deeply dissected leaves. The lower leaves stalked, 6–15 cm long, and the upper ones less dissected or lobed and without stalks. Yellow flowers, small, usually less than 6 mm across. Pods either globose or linear, with two rows of seeds in each compartment. Authorities have split this species into several varieties, two of which are common in the Prairie Provinces. The var. *fernaldiana* Butt. & Abbe with almost or entirely hairless stems and linear to linear-oblong pods 6–9 mm long. It was formerly called *R. palustris* (L.) Besser. The var. *hispida* (Desv.) Butt. & Abbe has the stem covered with bristly hairs and the pods globose or oval, not more than 6 mm long. This was formerly known as *R. hispida* (Desv.) Britt. Both varieties common on lakeshores, sloughs, and wet places; throughout the Prairie Provinces.

*Rorippa sinuata* (Nutt.) Hitchc.               SPREADING YELLOW CRESS

A native perennial with smooth stems 10–40 cm high, and cleft or divided leaves with rather obtuse lobes. Leaves 3–8 cm long. Yellow flowers about 4 mm across. Pods 8–12 mm long, linear-oblong. Not particularly common; but has been found in several locations throughout the Prairie Provinces as far north as northern Alberta.

*Rorippa sylvestris* (L.) Besser               CREEPING YELLOW CRESS

An introduced perennial with creeping roots, smooth stem, and pinnately-divided leaves with sharp-toothed divisions. Yellow flowers 6–8 mm across. Pods linear, 8–12 mm long, on slender stalks, with 1 row of seeds in each compartment. Found in a few isolated places in the Prairie Provinces. Another potential pest that should be watched to prevent its spread.

*Rorippa tenerrima* Greene               SLENDER CRESS

An annual species 10–30 cm high; stems numerous from the root crown, slender, prostrate or erect. Leaves deeply pinnatifid, with the lobes rounded, sinuately toothed. Wet places; southern Rocky Mountains.

***Sisymbrium***       sisymbrium

1. Pods 1–2 cm long, conical, closely appressed to the stem. ............................................................ *S. officinale*
   Pods 2–10 cm long, linear, spreading or ascending. ................................................................................ 2

2. Leaves all or mostly entire, very narrow. ............................................. *S. linifolium*
   Leaves all pinnatifid to pinnate. ................................................................................ 3

3. Stem with spreading hairs; upper leaves with narrow, linear divisions; pods 5–10 cm long. ................................................................. *S. altissimum*

Stem with reflexed hairs; upper leaves
with lanceolate divisions; pods about 4
cm long. .................................................................................................. *S. loeselii*

### Sisymbrium altissimum L. TUMBLING MUSTARD

An introduced annual or winter-annual weed 30–100 cm high. Basal leaves a rosette of pale green soft hairy divided leaves; the stem leaves of various shapes from deeply lobed or pinnate to entire. Flowers pale yellow, about 8 mm across. Very thin pods 5–10 cm long, each containing over 100 small light brown seeds. At maturity the plant dries and breaks off, acting as a tumbling weed and rolling with the wind, leaving seeds in its wake. A very common weed; on cultivated land; throughout Prairie Provinces, but more prevalent on open plains where it can travel with the wind.

### Sisymbrium linifolium Nutt. NARROW-LEAVED MUSTARD

Perennial plants, with much-branched glabrous stems and narrow mostly entire leaves. Pods spreading, 3–6 cm long. Rare weed, introduced from the west; western Parklands.

### Sisymbrium loeselii L. TALL HEDGE MUSTARD

A tall branched annual weed 40–100 cm high. Lower parts with downward-pointed hairs; upper parts usually smooth. Leaves deeply lobed. Flowers bright yellow; seedpods linear, 2–4 cm long. Introduced, fortunately not common; found in disturbed areas; Prairies and Parklands.

### Sisymbrium officinale (L.) Scop. HEDGE MUSTARD

Annual or biennial plants 5–80 cm high, erect, branched above. Lower leaves petiolate, pinnatifid, with the terminal lobe rounded. Flowers yellow, 4–6 mm across; pods 10–20 mm long, pubescent, or glabrous as in var. *leiocarpum* DC. Introduced, not common; a weed of waste areas.

### **Smelowskia** rockcress

### Smelowskia calycina (Stephan) C. A. Mey.
var. *americana* (Regel & Herder) Drury & Rollins SILVER ROCKCRESS

Densely tufted perennial, with branched caudex. Stems 5–20 cm high, more or less pubescent; leaves pinnately divided, silvery gray pubescent, with the hairs mostly branched; basal leaf bases ciliate. Flowers white or cream-colored, about 1 cm across; pods 5–10 mm long, linear or narrowly oblong. Rock slides, stony areas; southern Rocky Mountains.

### **Subularia** awl-wort

### Subularia aquatica L. WATER AWL-WORT

A glabrous annual, with the leaves all basal, 2–7 cm long, terete, awl-shaped, entire. Flowers 2–12, white; silicle 2–5 mm long. In shallow waters; eastern Boreal forest.

## Thellungiella

*Thellungiella salsuginea* (Pall.) O. E. Schulz        MOUSE-EAR CRESS

An erect divaricately branched annual 10–35 cm high. Stem leaves oblong-ovate, with rounded auricles. Flowers 6–8 mm across, white; pods linear, 12–16 mm long, ascending or erect. Saline areas; Prairies and Parklands.

## Thlaspi        pennycress

*Thlaspi arvense* L.        STINKWEED

An introduced annual or winter-annual weed 2–40 cm high, with hairless stems and smooth leaves. Basal leaves stalked and oblanceolate, soon withering and falling off; stem leaves oblong to lanceolate, eared at the base and clasping the stems. Flowers small and white, about 3 mm across, in a cluster at the head of the stem. Pods (Fig. 117*L*) flat, oval, and broadly winged, 6–12 mm across, deeply notched at the top, on stalks 1–2 cm long. They form large racemose clusters, at first bright green but turning yellow or orange when mature. Seeds purplish chocolate brown, about 1 mm across, and bearing concentric grooves like a fingerprint on each side. A very common unpleasant-smelling weed, which when eaten by cattle taints their milk, butter, and meat. Often flowering and producing seed when only 3–6 cm high in early spring. Because of its early spring growth, it depletes soil of moisture. In fields and waste places; throughout the Prairie Provinces.

# RESEDACEAE—mignonette family

## Reseda        mignonette

Introduced perennials with lobed or pinnatifid alternate leaves bearing small glandular stipules. Flowers in narrow racemes, with the stamens on one side of the flower. Seeds contained in numerous globose capsules.

Petals greenish yellow, 3 or 4 of them divided. ...................................................... *R. lutea*
Petals white, all cleft or divided. ........................................................................ *R. alba*

*Reseda alba* L.        WHITE CUT-LEAVED MIGNONETTE

An erect hairless plant 30–50 cm high, with deeply cut or pinnate leaves, and spikes of white flowers. Rarely found, but has been introduced in seeds in some places.

*Reseda lutea* L.        YELLOW CUT-LEAVED MIGNONETTE

An erect plant, sometimes slightly hairy, 20–60 cm high, with deeply cut leaves, which have spatulate lobes. Flowers yellowish green in a long spike-like raceme. Occasionally found as a weed in imported lawn grass seed.

# DROSERACEAE—sundew family

Perennial bog herbs with basal leaves bearing glandular sticky hairs that close over insects and entrap them. Flowers in a 1-sided raceme on an elongated stem, and numerous seeds enclosed in a capsule.

***Drosera***      sundew

1. Leaf blades almost round, broader than
   long. ............................................................................ *D. rotundifolia*
   Leaf blades longer than broad. ................................................................ 2
2. Leaf blades spatulate or oblanceolate. ...................................... *D. anglica*
   Leaf blades linear. .......................................................................... *D. linearis*

*Drosera anglica* Huds.                              OBLONG-LEAVED SUNDEW

Leaves sticky-haired, spatulate, 15–25 mm long. Flowers white, on a stem 4–10 cm high. Found in cold bogs; Boreal forest.

*Drosera linearis* Goldie                           SLENDER-LEAVED SUNDEW

Leaves linear, 1–6 cm long. Flowers white, solitary or a few on a stem 4–10 cm high. Rare; in bogs; Boreal forest.

*Drosera rotundifolia* L.                           ROUND-LEAVED SUNDEW

Leaves roundish, 6–10 mm long, broader than long, with the upper surface covered with fine glandular insect-catching hairs. Flowers white, about 4 mm wide, on a stem 10–20 cm high. The commonest of the sundews; in moist swamps and bogs; Boreal forest.

# SARRACENIACEAE—pitcherplant family

***Sarracenia***      pitcherplant

*Sarracenia purpurea* L.                                        PITCHERPLANT

A peculiar and conspicuous bog plant, with pitcher-shaped leaves 10–30 cm long, erect, greenish yellow with purple veins, and hood-like top containing downward-pointing bristly hairs that trap insects. The hollow lower part of the leaf usually containing water in which insects drown and are absorbed by the plant. Flower large, about 5 cm across, nodding, yellow and purple, and borne on a long stalk 20–50 cm high. Not common; occasionally in cold bogs; in Boreal forest.

# CRASSULACEAE—orpine family

***Sedum***      stonecrop

Succulent or fleshy-leaved perennials with perfect flowers borne in cymes. They have 4 or 5 sepals and 4 or 5 petals and their numerous seeds are contained in follicle-like capsules. Plants of rocky places, usually associated with mountainous locations.

1. Leaves very thick, less than 3 mm wide. ................................................................ 2
   Leaves flat, mostly more than 5 mm wide. ................................................................ 4

2. Stem leaves less than 5 mm long, persistent in drying. ................................................................ *S. acre*
   Stem leaves about 10 mm long, deciduous in drying. ................................................................ 3

3. Leaves narrowed at the base, not bulbiferous. ................................................................ *S. lanceolatum*
   Leaves broadened at the base, bulbiferous in upper part of the stem. ................................................................ *S. stenopetalum*

4. Flower parts mostly in 4's; flowers dark purplish. ................................................................ *S. rosea*
   Flower parts mostly in 5's; flowers yellow or pinkish. ................................................................ 5

5. Flowers pinkish red. ................................................................ *S. telephium*
   Flowers yellow. ................................................................ 6

6. Leaves lanceolate, with the margins serrate. ................................................................ *S. aizoon*
   Leaves spatulate, with the margins dentate toward the apex. ................................................................ *S. hybridum*

*Sedum acre* L.                                                                    MOUNTAIN MOSS

   Glabrous, tufted, mat-forming plants with numerous short, sterile shoots, and flowering stems 5–12 cm high. Leaves 3–6 mm long, elliptic in cross section. Flowers about 15 mm across, yellow, in small cymes. Introduced, rarely escaped from cultivation; dry rocky places.

*Sedum aizoon* L.

   Plants with erect stems 30–40 cm high; leaves 5–8 cm long, lanceolate. Flowers 15–20 mm across, yellow, in dense compound cymes. Introduced ornamental, occasionally escaped from cultivation.

*Sedum hybridum* L.

   Plants with creeping woody stems, short sterile shoots, and flowering stems 15–20 cm high. Leaves 2–3 cm long, oblong-cuneate. Flowers 10–20 mm across, golden yellow, numerous in terminal corymbs. Introduced ornamental, and rarely escaped; dry rocky places.

*Sedum lanceolatum* Torr.                                          LANCE-LEAVED STONECROP

   Plants mat-forming, with numerous sterile shoots, and flowering stems 5–15 cm high. Leaves 6–15 mm long, linear, round in cross section. Flowers about 1 cm across, yellow, borne in a terminal cyme. Not common; dry stony slopes and hillcrests; western Prairies, Cypress Hills.

*Sedum rosea* (L.) Scop. var. *integrifolium* (Raf.) Berger.          ROSE-ROOT

Plants with a thick, fleshy rootstock, fragrant when cut. Flowering stems 5–35 cm high, erect; leaves linear-oblong to orbicular-ovate, entire. Flowers 6–10 mm across, dark purple to dull reddish yellow, borne in a terminal cyme. Alpine meadows and scree fields; Rocky Mountains.

*Sedum stenopetalum* Pursh          NARROW-PETALED STONECROP

A low fleshy-leaved tufted perennial, branched, 5–15 cm high. Leaves round in cross section (terete), 6–15 mm long, usually crowded around the base, and overlapping. Yellow flowers borne at the head of a stem 5–15 cm long, fruiting follicles erect and close to the stem. On gravelly slopes and rocky places; southern Rocky Mountains.

*Sedum telephium* L.          STONECROP

Plants with tuberous roots, usually simple stems 15–60 cm high, with leaves 2–7 cm long, suborbicular to oblong. Flowers 6–10 mm across, pinkish to reddish, borne in large terminal corymbs. Introduced, rarely escaped from cultivation.

## SAXIFRAGACEAE—saxifrage family

Shrubs and herbs with opposite or alternate leaves with no stipules. Usually 5 sepals and 5 petals, and flowers perfect and regular. Stamens 5 or 10. The fruit usually, but not always, a capsule.

1. Shrubs. ..................................................................................................... 2
   Herbs. ....................................................................................................... 3
2. Fruit a berry; stamens 5. ............................................................... *Ribes*
   Fruit a capsule; stamens 20–40. .......................................... *Philadelphus*
3. Petals missing; low plants. ................................................................ 4
   Petals present. ......................................................................................... 5
4. Leaves reniform, crenate; leaves and
   flowers clustered at tip of stem. ..................................... *Chrysosplenium*
   Leaves lanceolate, serrate; flowers in ter-
   minal cymes. ...................................................................... *Penthorum*
5. Bundles of sterile stamens (staminodia)
   alternating with stamens; flowers singly
   on long stems. ......................................................................... *Parnassia*
   Sterile stamens not present. ............................................................ 6
6. Petals divided. ........................................................................................ 7
   Petals entire. ........................................................................................... 8
7. Petals 3- to 5-lobed; leaves deeply lobed
   or divided. ........................................................................ *Lithophragma*
   Petals finely lobed, filament-like; leaves
   shallowly to more or less deeply lobed. ................................... *Mitella*
8. Inflorescence a simple raceme. ................................... *Conimitella*
   Inflorescence not a raceme. ............................................................ 9

9. Stamens 5. ........................................................................................... 10
   Stamens 10. ........................................................................................ 11

10. Inflorescence a spike-like panicle; root-
    stock elongate. ................................................................. *Heuchera*
    Inflorescence a loose panicle; rootstock
    corm-like. ....................................................................... *Suksdorfia*

11. Petals linear, filiform. .......................................................... *Tiarella*
    Petals spatulate or wider. ................................................................ 12

12. Leaves leathery, 3–8 cm long, glossy
    green above. .................................................................. *Leptarrhena*
    Leaves not leathery, smaller, not glossy
    green. ................................................................................................ 13

13. Flowers pink to purple; leaves reniform;
    styles slender, elongate, partly fused. ........................... *Telesonix*
    Flowers white or yellowish; leaves vari-
    ous; styles short, not fused. ...................................... *Saxifraga*

### *Chrysosplenium*     golden saxifrage

*Chrysosplenium tetrandrum* (Lund) Fries.     GOLDEN SAXIFRAGE

Stems erect, 5–15 cm high, with alternate round leaves. Flowers greenish, mostly at the end of the stems, with no petals, but the sepals usually orange yellow inside. Rare; found in wet coniferous forests; Boreal forest. Plants with outer sepals larger and wider, and 4–8 stamens are often distinguished as *C. iowense* Rydb.

### *Conimitella*

*Conimitella williamsii* (Eat.) Rydb.

Plants with basal leaves only, these reniform, glandular puberulent. Scapes 15–30 cm high, bearing 5–10 flowers, with white petals 4–5 mm long. Montane forests; southern Rocky Mountains.

### *Heuchera*     alumroot

Perennial plants from scaly rootstocks, with broad, rounded cordate or reniform, stalked, all-basal leaves; the flowers in a narrow panicle on a long stem. Fruits are capsules opening at top, with spreading beaks.

1. Flowers small; the calyx 2–4 mm long. ................................................ 2
   Flowers large; the calyx 5–12 mm long. ............................................. 3

2. Panicle open; teeth of the leaves acute. ........................... *H. glabra*
   Panicle very narrow; teeth of the leaves
   rounded. ......................................... *H. parvifolia* var. *dissecta*

3. Stamens exserted; the calyx strongly
   oblique. ..................................................... *H. richardsonii*
   Stamens included; the calyx not strongly
   oblique. ........................................................ *H. cylindrica*

*Heuchera cylindrica* Dougl.                           STICKY ALUMROOT

Plants 20–50 cm high; leaves 2–5 cm long, to 4 cm wide, deeply cordate at the base, glabrous, 5- to 7-lobed. Inflorescence a narrow panicle; petals included and inconspicuous or absent; base of the calyx glandular pubescent. Plants with the leaves crenate, with broad, somewhat mucronate teeth are var. *glabella* (T. & C.) Wheelock. Not common; rocky slopes, ledges, and ridges; southern Rocky Mountains.

*Heuchera glabra* Willd.                               ALPINE ALUMROOT

Leaves ovate to rounded in outline, deeply 5- to 7-lobed, cordate at the base, sparingly pubescent when young, soon glabrous, 3–10 cm long, almost as wide as long. Stems 10–60 cm high, with the panicle 5–20 cm long, lax, glandular puberulent, open. Rare; riverbanks, rock slides; southern Rocky Mountains.

*Heuchera parvifolia* Nutt. var. *dissecta* M. E. Jones   SMALL-LEAVED ALUMROOT

Leaves 15–30 mm broad, rounded cordate, but deeply cleft one-third to half their length in 7–9 wedge-shaped lobes. The space between the lower lobes of the leaves is very narrow and the lobes frequently overlap. Flowering stalk 20–30 cm high; flowers greenish. Found only in southern Rocky Mountains and in Cypress Hills.

*Heuchera richardsonii* R. Br. (Fig. 121)             ALUMROOT

Leaves coarse, basal, rounded cordate with broadly ovate teeth, 2–6 cm across, dark green. Flowering stalks 30–50 cm high with a few glandular hairs, especially near the top. Petals purplish, a trifle longer than sepals. Indians and early settlers chewed the scaly rootstocks of this plant as a cure for diarrhea. Common; on the prairie, especially in lower or moister places; throughout Prairies, Parklands, and southern fringe of Boreal forest.


**Leptarrhena**     leather-leaved saxifrage

*Leptarrhena pyrifolia* (D. Don) R. Br.               LEATHER-LEAVED SAXIFRAGE

Plants with creeping or ascending rootstocks. Stems 10–40 cm high, glandular pubescent above, with 1 or 2 clasping leaves. Leaves alternate and rather crowded on the caudex, 3–8 cm long, glabrous, deep green above, pale green or brownish below; stem leaves few and distant. Flowers arranged in dense cymules in a short panicle, whitish or pinkish, 4–5 mm across. Not common; wet areas; southern Rocky Mountains.


**Lithophragma**     starflower

Upper leaf axils usually bearing bulblets. ........................................................ *L. bulbifera*
Upper leaf axils not bearing bulblets. .............................................................. *L. parviflora*

*Lithophragma bulbifera* Rydb.                        ROCKSTAR

A small plant from bulblet-bearing fibrous rootstocks. Stems erect, 10–20 cm high, glandular hairy. Leaves small, thrice-divided, stalked, 5–10 mm

A.C. Budd

Fig. 121.   Alumroot, *Heuchera richardsonii* R. Br.

across; stem leaves usually having several very tiny bulblets in the axils. Flowers white or rose, with petals 3- to 5-cleft, 3–6 mm long. Very uncommon; on dry hills; southern Rocky Mountains and Cypress Hills.

*Lithophragma parviflora* (Hook.) Nutt.    SMALL-FLOWERED ROCKSTAR

Plants with more or less densely glandular hairy stems 10–30 cm high. Basal leaves white pubescent, 3- to 5-lobed to the base, with the divisions ternately divided into entire or toothed segments. Raceme 3- to 9-flowered; the calyx densely glandular hairy; petals 4–10 mm long, 3- to 5-cleft. Eroded, gravelly slopes; southern Rocky Mountains.

## *Mitella*    miterwort

1. Petals 3-cleft at the tip, white or purplish. .................................................... *M. trifida*
   Petals pinnatifid or fringed, greenish or
   yellow. ........................................................................................................................ 2
2. Stamens 10; rootstocks creeping, very
   slender. ................................................................................................................ *M. nuda*
   Stamens 5; rootstocks short, thick. ...................................................................... 3
3. Stamens opposite sepals; leaves reniform,
   obscurely lobed. ............................................................................................... *M. breweri*
   Stamens opposite petals; leaves broadly
   cordate, shallowly lobed. ............................................................... *M. pentandra*

*Mitella breweri* Gray    BREWER'S MITERWORT

Plants with brown scaly rootstocks, and scapes 10–25 cm high. Leaves 4–8 cm wide, orbicular to reniform, with a deep basal sinus; petioles long pubescent. Petals usually with 3 pairs of filiform divisions. Moist woods; southern Rocky Mountains.

*Mitella nuda* L.    BISHOP'S-CAP

A low-growing perennial with scaly rootstocks and long-stalked, mostly basal, rounded cordate leaves 2–5 cm across. Flowers borne in a raceme on a stalk 5–20 cm high, 5–6 mm across, with 4 greenish sepals and 5 greenish white petals, very finely divided and branched, pinnatifid. Fairly common; in cold, wet woodlands; throughout Rocky Mountains, Boreal forest, and also in the Cypress Hills.

*Mitella pentandra* Hook.

Much like *M. breweri*, but with the petioles glabrous to subglabrous and the calyx often purplish inside. Mountain forests; Rocky Mountains.

*Mitella trifida* Graham

Perennials with scaly rootstocks, 1 to several scapes, 10–40 cm high. Leaves round-reniform to cordate, 2–4 cm wide, 7- to 9-lobed, with the lobes shallow, rounded, obscurely glandular-toothed, pubescent with white hairs. Inflorescence a one-sided raceme; the flowers subtended by small, fringed bracts; petals white, wedge-shaped, 3-cleft. Mountain woods; southern Rocky Mountains.

***Parnassia***      grass-of-Parnassus

Perennials with rootstocks and stalked, entire, basal leaves. Flowers single on a long stalk, with sometimes a leaf on the stalk, and 5 sepals and 5 white petals with green veins. There are 5 fertile stamens, alternating with 5 clusters of gland-tipped infertile stamens termed staminodia. Fruit a 1-celled capsule opening at the top.

1. Petals fringed on margins, at least below;
   leaves reniform to reniform-cordate. ................................................. *P. fimbriata*

   Petals not fringed; leaves ovate to
   cordate. ......................................................................................................... 2

2. Petals shorter than, to as long as, the sep-
   als; stem leafless or with a leaf at base. ............................................ *P. kotzebuei*

   Petals much larger than sepals; stem
   leafless or with a leaf near the middle. ................................................................ 3

3. Petals 3 or more times longer than sepals;
   stem leafless. ................................................................................. *P. glauca*

   Petals 1.5 times longer than sepals; stem
   bearing 1 leaf near the middle. .......................................... *P. palustris* var. *tenuis*

*Parnassia fimbriata* Konig          FRINGED GRASS-OF-PARNASSUS

Plants with leaves 2–4 cm wide, reniform, with a broad sinus. Stems 10–30 cm high, with a single ovate leaf near the middle. Flowers about 20 mm across; petals 5-nerved. Moist, springy areas and banks; Rocky Mountains.

*Parnassia glauca* Raf.          GLAUCOUS GRASS-OF-PARNASSUS

Low plants with broadly ovate basal leaves 2–5 cm long. Flowers white, strongly veined, 15–30 mm across, and borne singly on stems 20–40 cm high; 3–5 staminodia in each cluster. In cold bogs; Boreal forest. Syn.: *P. americana* Muhl.

*Parnassia kotzebuei* Cham. & Schl.          SMALL GRASS-OF-PARNASSUS

Plants about 10 cm high, with leaves ovate or oval and the base rounded or subcordate. Stem leafless or sometimes with an oval leaf near the base. Flowers 8–12 mm across; petals 3-nerved. Wet areas, bogs; Boreal forest and Rocky Mountains.

*Parnassia palustris* L. var. *tenuis* Wahl. (Fig. 122)

NORTHERN GRASS-OF-PARNASSUS

Cordate leaves 10–25 mm wide. The flowering stalk 10–30 cm high with a single clasping stem leaf and at the top the white flower 15–25 mm across. Petals much longer than sepals, distinguishing this variety from var. *montanensis.* The most common species of the genus; found in wet shady places; throughout Boreal forest and less plentiful in favorable sites in Parklands. In var. *montanensis* (Fern. & Rydb.) C. L. Hitchc., the petals slightly longer than the sepals, the plants generally smaller; wet areas; Rocky Mountains. Still smaller is var. *parviflora* (DC.) Boiv., usually less than 20 cm high, and with most of the flowers less than 15 mm across; wet areas; Boreal forest and Rocky Mountains.

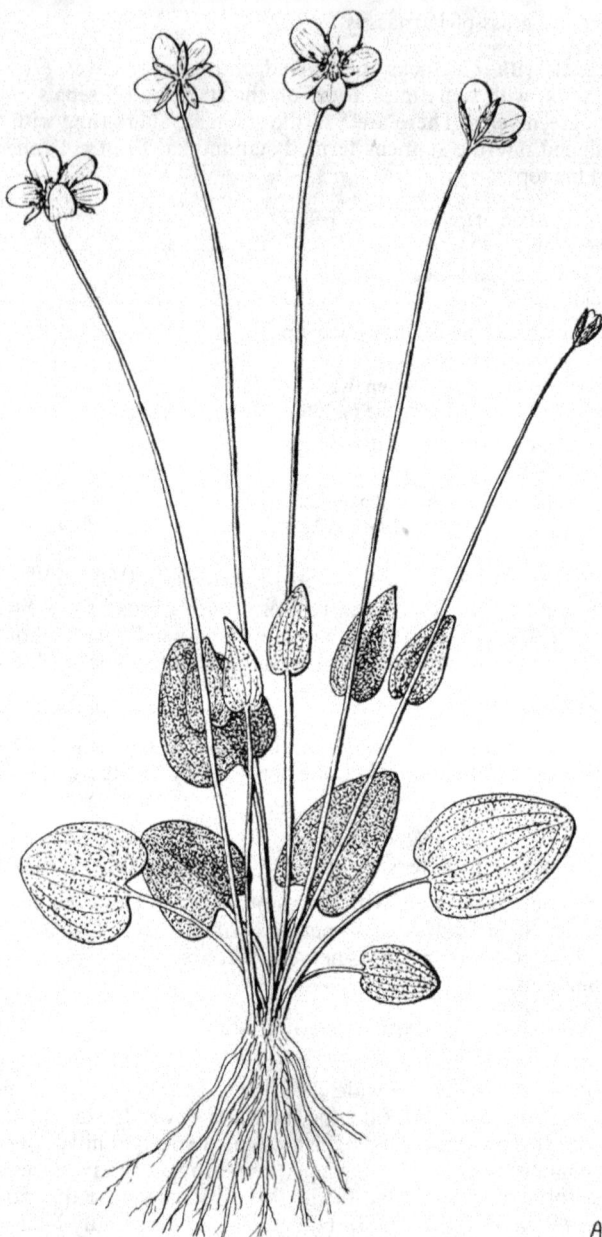

Fig. 122.   Northern grass-of-Parnassus, *Parnassia palustris* L. var. *tenuis* Wahl.

*Dicots*

***Penthorum***     stonecrop

*Penthorum sedoides* L.                       DITCH STONECROP

Perennial plants with stems 20–70 cm high arising from a creeping base. Leaves 5–10 cm long, lanceolate, serrate. Inflorescence terminal, glandular; flowers 4–6 mm across; petals none, in cymes 2–8 cm long. Rare; shores and ditches; southeastern Boreal forest.

***Philadelphus***     mock orange

*Philadelphus lewisii* Pursh                    MOCK ORANGE

Shrubs with freely branching stems 1–3 m high; young twigs reddish, glabrous; leaves 2–6 cm long, ovate-lanceolate, opposite. Flowers showy, fragrant, in racemose cymes; 4 or 5 white petals 1–2 cm long; sepals usually 4, rarely 5, pubescent at the inside of the tips; stamens 20–40. Rare; mountain slopes; southern Rocky Mountains.

***Ribes***     currant

Shrubs, with palmately divided and veined leaves, and regular perfect flowers. Fruit a globose or ovoid pulpy berry. Some plants armed with prickles, especially at the nodes of the stem. Fruit edible, but sometimes not very tasty.

1. Inflorescence a reduced raceme, 1- to 3-flowered; most of the stems spiny. ........... *R. oxyacanthoides* var. *oxyacanthoides*
   Inflorescence an elongated raceme, flowers numerous. ................................................................................ 2

2. Stems densely spiny along the internodes. ................................................ *R. lacustre*
   Stems not spiny, or spines only at the nodes. ................................................................................ 3

3. Flowers golden yellow to somewhat reddish; leaves rolled in buds. ................................. *R. aureum*
   Flowers not golden yellow; leaves folded in buds. ................................................................................ 4

4. Ovary and fruit more or less densely glandular pubescent. ................................................................... 5
   Ovary and fruit glabrous, or with a few sessile glands. ................................................................................ 7

5. Leaves and young stems glandular puberulent. ....................................................... *R. viscosissimum*
   Leaves and young stems not glandular. ................................................................................ 6

6. Fruit red, glandular only. ................................................................... *R. glandulosum*
   Fruit purplish black, pubescent and glandular. ............................................................... *R. laxiflorum*

7. Leaves abundantly glandular dotted on the underside. ................................................................................ 8
   Leaves without glands. ................................................................................ 9

8. Pedicels much longer than the bracts. .............................................. *R. hudsonianum*
   Pedicels much shorter than the bracts. .............................................. *R. americanum*

9. Lobes of leaves closely and uniformly ser-
     rate from base to tip. ...................................................... *R. rubrum* var. *propinquum*
   Lobes of leaves coarsely dentate above
     the middle. ........................................................................................ *R. diacanthum*

### Ribes americanum Mill.                                    WILD BLACK CURRANT

A low shrub 1–2 m high, with unarmed stems and 3- to 5-lobed leaves 2–7 cm across and somewhat hairy and resinous-dotted on the underside. Flowers greenish white, with a tubular calyx 3–5 mm long, and borne in drooping racemes. Fruit black, 6–10 mm across, and smooth. Fairly common; in moist woodlands; throughout the Prairie Provinces. Syn.: *R. floridum* L'Her.

### Ribes aureum Pursh                                        GOLDEN CURRANT

An erect shrub 1–2 m high, with 3-lobed leaves 2–5 cm across. Flowers bright yellow tipped with red, with a long cylindric calyx tube 6–10 mm long, clove-scented, in small racemes. Fruit, about 6 mm in diam, varying from pale yellow through shades of red to black. Often grown as an ornamental garden shrub. Only found in the Prairies; plentiful in some localities on the southern slope of Cypress Hills. Syn.: *Chrysobotrya aurea* (Pursh) Rydb.

### Ribes diacanthum Pall.                                    RED CURRANT

Shrubs 1–2 m high; leaves 2–6 cm long, rather deeply 3-lobed. Flowers in racemes, functionally dioecious, with the organs of one sex rudimentary; each flower subtended by a conspicuous bract; corolla greenish; fruit scarlet. Introduced and cultivated, occasionally escaped.

### Ribes glandulosum Grauer                                  SKUNKBERRY

A low spreading or reclining shrub about 1 m long, with no thorns or prickles. Leaf blades 5- to 7-lobed, cordate at base, hairy on the veins below, 2–7 cm across, and having a somewhat skunk-like odor. Flowers in a short raceme. Fruit a red, glandular, bristly berry about 6 mm across. Common; in damp woodlands; throughout Parklands and Boreal forest.

### Ribes hudsonianum Richards.                               NORTHERN BLACK CURRANT

An erect shrub 1–1.5 m high, with unarmed stems and 3-lobed (occasionally 5-lobed) leaves, wider than long, 2–10 cm wide, more or less hairy, and dotted with resinous glands on the undersides. Flowers white, in racemes. Fruit black, 5–10 mm in diam. Common in shady woodland; in the northern parts of the Prairie Provinces and in the Cypress Hills.

### Ribes lacustre (Pers.) Poir.                              SWAMP GOOSEBERRY

Shrubs 1–2 m high, with clusters of slender spines; stems and branches densely bristly. Leaves 2–5 cm long, deeply 5- to 7-lobed. Flowers in racemes, green or purplish. Fruit a densely hairy, glandular reddish berry about 6 mm in diam. Found occasionally; in swamps; in Boreal forest and in the Riding Mountain. Syn.: *Limnobotrya lacustris* (Pers.) Rydb.

*Ribes laxiflorum* Pursh                                      MOUNTAIN CURRANT

Shrubs with ascending or spreading stems; leaves more or less orbicular in outline, 5- to 7-lobed, cordate at base. Flowers in ascending racemes, with the pedicels glandular, the corolla red, the petals 3–4 mm long, pubescent on the outside; berry purplish black to blue black. Swamps and woods; southern Rocky Mountains.

*Ribes oxyacanthoides* L. var. *oxyacanthoides* (Fig. 123)  NORTHERN GOOSEBERRY

A low, bristly shrub or bush to 1 m high, with lobed leaves 1–4 cm broad and greenish purple or white sepals and petals; bracts glandular ciliate. Calyx tube not longer than lobes. Fruit a globose berry 10–15 mm across, turning reddish purple when ripe. Common in woodlands and shrubbery, especially in Boreal forest. Syn.: *R. setosum* Lindl. The var. *saxosum* (Hook.) Cov. having the bracts long-ciliate but not glandular, generally less spiny, particularly at the nodes and on older stems. Woodlands; in northern Parklands and Boreal forest. Syn.: *R. hirtellum* Michx.

*Ribes rubrum* L. var. *propinquum* Trautv. & Mey.      SWAMP RED CURRANT

A shrub about 1 m high, unarmed; leaves paler beneath, usually 3-lobed, sometimes 5-lobed, 3–8 cm across. Flowers usually purplish; fruit a smooth red berry, very similar to the garden red currant, about 6 mm across. Found occasionally; in rich poplar woods; especially in Parklands, Boreal forest, and Riding Mountain. Syn.: *R. triste* Pall.

*Ribes viscosissimum* Pursh                                   STICKY CURRANT

A rather stout shrub 50–100 cm high, with spreading branches. Leaves 4–7 cm across, orbicular or reniform in outline, 3- or sometimes 5-lobed, cordate, sticky. Racemes 3- to 7-flowered, glandular puberulent; flowers greenish white tinged with pink; fruit black, glandular hairy, about 1 cm long. Mountain forests and slopes; southern Rocky Mountains.

## *Saxifraga*      saxifrage

1. Leaves opposite, densely crowded, 4-
     ranked on matted stems; petals 6–8
     mm long, purple. ................................................................ *S. oppositifolia*
   Leaves alternate or basal. ........................................................................ 2

2. Stems leafless, except for reduced leaves
     subtending branches of inflorescence. ............................................... 3
   Stems with several well-developed leaves. ............................................ 9

3. Basal leaves 10–20 cm long; scapes 30–80
     cm high. ............................................................................... *S. pensylvanica*
   Basal leaves much smaller; scapes lower. .............................................. 4

4. Leaves subcordate to deeply cordate at
     base. .................................................................................................... 5
   Leaves cuneate to acute at the base. .................................................... 6

5. Flowers in part replaced by clusters of
     bulblets. .............................................................................. *S. mertensiana*

Fig. 123.  Northern gooseberry, *Ribes oxyacanthoides* L. var. *oxyacanthoides*.

Flowers normal, bulblets absent. ....................................... *S. punctata* var. *porsildiana*

6. Sepals sharply reflexed, pendent. .......................................................................... 7
   Sepals spreading or ascending. ............................................................................... 8

7. Plants glabrous or somewhat puberulent
   above. ........................................................................................................ *S. lyallii*
   Plants densely glandular pubescent
   throughout. ............................................................................................ *S. ferruginea*

8. Petals 2–4 mm long, obovate to oblong. .......................................... *S. occidentalis*
   Petals 4–5 mm long, oblanceolate to
   lanceolate. ............................................................................................ *S. virginiensis*

9. Leaves 3-parted to palmately lobed. ...................................................................... 10
   Leaves shallowly 3-lobed to entire. ....................................................................... 12

10. Plants with bulblets in the upper axils. ............................................... *S. cernua*
    Plants not bulbiferous. ....................................................................................... 11

11. Lobes of leaves linear to
    linear-lanceolate. .......................................................................... *S. cespitosa*
    Lobes of leaves ovate to rounded. ...................................................... *S. rivularis*

12. Flowers white, the petals often
    purplish-dotted. ................................................................................................ 13
    Flowers yellow, the petals not dotted. ................................................................. 15

13. Leaves soft, not spine-tipped. ....................................... *S. adscendens* ssp. *oregonensis*
    Leaves stiff, thick, spine-tipped. ........................................................................ 14

14. Leaves entire, spinulose on the margins. .............. *S. bronchialis* var. *austromontana*
    Leaves 3-toothed at the apex, the teeth
    spine-tipped. ...................................................................................... *S. tricuspidata*

15. Plants long-stoloniferous, the stolons
    whip-like. ................................................................................................ *S. flagellaris*
    Plants not stoloniferous. ................................................................................... 16

16. Leaves all sessile. ..................................................................................... *S. aizoides*
    Basal leaves long-petioled, stem leaves
    subsessile. ............................................................................................... *S. hirculus*

*Saxifraga adscendens* L. ssp. *oregonensis* (Raf.) Bacig.

WEDGE-LEAVED SAXIFRAGE

A small perennial, with stems 2–8 cm high; several stem leaves. Basal leaves in compact rosettes 5–15 mm long, oblong-spatulate or cuneate, entire or 3-toothed; stem leaves cuneate, 3-toothed. Petals 3–5 mm long, white. Moist slopes and ledges; southern Rocky Mountains.

*Saxifraga aizoides* L.                     YELLOW MOUNTAIN SAXIFRAGE

Perennial plants with stems from a multiple caudex; flowering stems 5–20 cm high, with several stem leaves. Lower leaves 1–2 cm long, linear-oblong, thick, often ciliate. Cymes usually several-flowered; flowers erect, 5–8 mm across, yellow, predominantly orange-dotted. Dry, gravelly slopes; southern Rocky Mountains.

*Saxifraga bronchialis* L. var. *austromontana* (Wieg.) G. N. Jones

Plants forming dense mats or cushions; the flowering stem 5–15 cm high. Leaves 8–15 mm long, loosely imbricate, linear to narrowly lanceolate, with the margin spinulose. Inflorescence a corymbose cyme; flowers 6–12 mm across, pale yellow or whitish, often red-spotted. Usually on rock outcrops; Rocky Mountains.

*Saxifraga cernua* L.

Plants with usually simple stems 10–20 cm high, more or less glandular puberulent. Basal leaves reniform, 15–25 mm wide, 5- to 7-lobed; the cauline leaves similar but progressively smaller, all with reddish bulblets in the axils. Usually a solitary terminal flower 15–20 mm across, white. Alpine rockslides; southern Rocky Mountains.

*Saxifraga cespitosa* L.

Plants with short leafy shoots, more or less erect, forming lax to rather dense cushions. Stems 4–10 cm high. Leaves 5–15 mm long, mostly 3-lobed, cuneate, glandular pubescent. Flowers 8–10 mm across, dull white or slightly greenish. Arctic and alpine gravel slopes; northeastern Boreal forest, southern Rocky Mountains.

*Saxifraga ferruginea* Graham

Plants with stems 10–30 cm high, more or less glandular pubescent. Basal leaves crowded, 2–6 cm long, spatulate to oblanceolate, densely pubescent. Flowering stems with branches ascending, flowers white, the 3 upper petals with 2 yellow spots at the base. In f. *vreelandii* (Small) St. John & Thayer, some flowers replaced by bulblets. Moist ledges and banks; southern Rocky Mountains.

*Saxifraga flagellaris* Stemb. & Willd.

Plants with erect stems 5–20 cm high, with whip-like stolons to 15 cm long from the base, each terminating in a rosette. Leaves mainly in a basal rosette, more or less glandular pubescent. Flowers few, in a lax cyme 15–20 mm across, bright yellow. High alpine soils; southern Rocky Mountains.

*Saxifraga hirculus* L.

Plants cespitose or with short stolons; stems 15–25 cm high; leafy below. Leaves 10–25 mm long, lanceolate, pubescent with reddish brown hairs. Flowers solitary or 2–4 in a loose corymb 25–30 mm across, bright yellow, sometimes red-spotted. Wet tundra; northeastern Boreal forest.

*Saxifraga lyallii* Engler

Plants with a prominent caudex; leaves fan-shaped to cuneate-obovate, 1–4 cm long, usually coarsely toothed above the middle. Scapes 5–10 cm high, glabrous or sparsely glandular pubescent above. Inflorescence open; flowers 5–8 mm across, white, with deep red sepals. In var. *hultenii* Calder & Savile, plants 10–30 cm high, with larger basal leaves and paniculate inflorescence. In

*Dicots*

var. *laxa* Engler, plants to 40 cm high, with the basal leaves orbicular, cuneate to subtruncate at base, and paniculate inflorescence. Along alpine creeks and rivulets; southern Rocky Mountains.

*Saxifraga mertensiana* Bong.

Plants with stout, erect, often bulb-forming rootstocks. Leaves basal, glandular pubescent, with the blades orbicular or ovate-reniform, cordate at the base, shallowly lobed; the lobes 3-toothed. Flowering stems 10–30 cm high, minutely glandular pubescent, with 1 or 2 leaves at the base, bracteose above. Inflorescence few-flowered; often all but the terminal flower replaced by bulblets; flowers white, with the filaments conspicuously club-shaped. Moist cliffs; southern Rocky Mountains.

*Saxifraga occidentalis* Wats.    RHOMBOID-LEAVED SAXIFRAGE

A low plant with fleshy basal leaves, usually ovate, 1–4 cm long, wavy-margined, and smooth. Flowers borne in a branched inflorescence at the head of a glandular haired stalk 8–20 cm high. Flowers about 5 mm across, with white petals. Fruits are follicle-like capsules. Moist meadows and slopes; southern Rocky Mountains, Cypress Hills. Syn.: *Micranthes rhomboidea* (Greene) Small; *S. rhomboidea* Greene.

*Saxifraga oppositifolia* L.    RED-FLOWERED SAXIFRAGE

Mat-forming plants, with stems procumbent or ascending. Leaves 2–6 mm long, suborbicular to obovate-lanceolate, dull bluish green, thick, keeled below. Flowering stems very short, leafy, glandular pubescent; flowers solitary, pale pink to deep purple. Rocky and gravelly areas, arctic–alpine; northeastern Boreal forest, Rocky Mountains.

*Saxifraga pensylvanica* L. (Fig. 124)    MARSH SAXIFRAGE

Stout plants with erect stems 30–80 cm high. Basal leaves 10–20 cm long, oblong to lanceolate or oblong-lanceolate, entire to sparsely glandular serrulate, sparingly pubescent. Inflorescence at first compact, becoming elongate, lax; flowers 5–8 mm across, greenish white. Rare; bogs and wet meadows; southeastern Boreal forest.

*Saxifraga punctata* L. var. *porsildiana* (Calder & Savile) Boiv.

SPOTTED SAXIFRAGE

Plants with scapes 8–15 cm high. Basal leaves long-petioled, with the blades nearly orbicular, deeply cordate at base, 2–5 cm across, crenate-dentate, glabrous or nearly so. Scapes solitary or few tufted; inflorescence closely corymbose, glandular puberulent; flowers 6–10 mm across, white; sepals reflexed, ciliate. Moist areas; Rocky Mountains.

*Saxifraga rivularis* L.    ALPINE BROOK SAXIFRAGE

Plants with short creeping rootstocks; erect stems 6–15 cm high, glabrous to glandular puberulent. Basal leaves 10–15 mm wide, 3- to 5-lobed; cauline leaves soft pubescent. Flowers 6–10 mm across, white or pale pink. Moist, springy areas; Rocky Mountains, northeastern Boreal forest.

Fig. 124.  Marsh saxifrage, *Saxifraga pensylvanica* L.

*Saxifraga tricuspidata* Rottb.                              THREE-TOOTHED SAXIFRAGE

Plants with branched caudex, the branches mat-forming; leaves closely crowded, 1–2 cm long, narrowly oblong, ciliate, the apex with 3 sharp erect teeth. Stems 10–20 cm high, with several stem leaves 3–10 mm long; flowers in a cyme 12–15 mm across, white with light purplish dots. Rock outcrops and rocky slopes; Boreal forest, Rocky Mountains.

*Saxifraga virginiensis* Michx.                              EARLY SAXIFRAGE

Plants with basal leaves 2–5 cm long, oblong to ovate, entire or serrate, the base narrowed to a petiole, glandular pubescent. Scapes 10–20 cm high, glandular pubescent; inflorescence branched, at first compact, later elongating and open; flowers 8–12 mm across; sepals spreading to ascending. Moist or dry open woods; southeastern Boreal forest.

**Suksdorfia**          suksdorfia

*Suksdorfia violacea* Gray                                   BLUE SUKSDORFIA

A slender perennial, with more or less glandular pubescent stems 10–20 mm high. Basal leaves kidney-shaped, 1–3 cm wide, with 5–7 rounded lobes, on petioles 2–8 cm long, glandular puberulent; upper leaves sessile. Inflorescence a few-flowered panicle; flowers long-pediceled, 10–15 mm across, violet. Wet rocks along streams; southern Rocky Mountains.

**Telesonix**

*Telesonix jamesii* (Torr.) Raf. var. *heucheriformis* (Rydb.) Bacig.

Low perennial plants with a thick scaly rootstock; stems 5–15 cm high, glandular pubescent. Basal leaves reniform to orbicular, 2–4 cm wide, deeply crenate-dentate, with sessile glands; upper cauline leaves fan-shaped to cuneate, short-petioled or sessile. Inflorescence a short, leafy panicle; flowers 3–5 mm long, deep violet. Rock crevices; southern Rocky Mountains.

**Tiarella**      false miterwort

Leaves compound, usually trifoliate. ................................................................ *T. trifoliata*
Leaves simple, shallowly 3- to 5-lobed. ........................................................ *T. unifoliata*

*Tiarella trifoliata* L.                                     LACEFLOWER

Perennial plants with slender, sparsely hirsute stems 20–60 cm high, 2- to 4-leafed, glandular hirsute above. Leaves with 3 leaflets, on petioles 5–15 cm long, glabrous to villous-hirsute; middle leaflet more or less 3-lobed, 2–8 cm long. Inflorescence an elongated, narrow panicle; sepals about 1.5 mm long, white; petals white, linear, about 5 mm long. Moist coniferous forests; Rocky Mountains.

*Tiarella unifoliata* Hook.                                  SUGARSCOOP

Slender perennial, with stems 10–50 cm high, glabrous or hirsute with white hairs, glandular puberulent above. Basal leaves cordate in outline, 4–10 cm wide, 3- to 5-lobed, with the lobes broadly ovate, acute or obtuse, crenate-dentate, and the petioles 5–15 cm long. Stem leaves 1–4, similar to but smaller than the basal leaves. Inflorescence a narrow panicle; sepals white to pinkish, 1.5–2.5 mm long; petals white, linear, about 6 mm long. Moist woods; Rocky Mountains and northwestern Boreal forest.

## ROSACEAE—rose family

An extremely variable family, including herbs, shrubs, and small trees, all having alternate leaves with stipules. Flowers in racemes or cymes, perfect and regular, with 5 petals and 5 sepals, often with 5 bracts below, alternating with the sepals. Many stamens and from one to many pistils. Fruits varying: dry achenes, follicles, fleshy receptacles (strawberry), fleshy drupes or drupelets (raspberry), and berry-like pomes (saskatoon).

1. Shrubs or trees. ......................................................................................... 2
   Herbs or half-shrubs. ............................................................................. 14
2. Leaves simple. ........................................................................................ 3
   Leaves compound or lobed. .................................................................... 9
3. Flowers solitary on a long peduncle;
   creeping dwarf shrubs. ............................................................... *Dryas*
   Flowers usually more numerous; plants
   taller. ..................................................................................................... 4
4. Spiny shrubs or small trees. .................................................................. 5
   Shrubs or trees not spiny. ...................................................................... 6
5. Spines never leafy. .......................................................... *Crataegus*
   Spines leafy in the first year, leafless
   thereafter. ........................................................................... *Prunus*
6. Leaves entire; ovary inferior. ................................... *Cotoneaster*
   Leaves serrate or toothed. ...................................................................... 7
7. Fruit a leathery follicle. ............................................... *Spiraea*
   Fruit fleshy. ............................................................................................ 8
8. Fruit a red or black drupe with a single
   stone. ....................................................................................... *Prunus*
   Fruit a berry-like pome with several
   seeds. ................................................................................ *Amelanchier*
9. Leaves lobed, not distinctly divided. .................................................. 10
   Leaves compound, distinctly divided into
   leaflets. ................................................................................................ 11
10. Shrubs tall; flowers in dense racemose
    clusters. ............................................................................ *Physocarpus*
    Shrubs usually low; flowers in loose
    racemes or solitary. ........................................................... *Rubus*
11. Flowers yellow; fruit dry; leaves with 3–7
    small leaflets. ..................................................................... *Potentilla*

## *Agrimonia*    agrimony

*Agrimonia striata* Michx.                    AGRIMONY

An erect-stemmed perennial herb 30–80 cm high, sometimes branched above, with soft, fine brownish hairs. Leaves pinnate, with 7–9 leaflets, deeply indented, 2–7 cm long, smooth above, and softly hairy below. Small yellow flowers borne in a spike-like raceme. The fruiting calyx reflexed (pointing downward), with short hooked bristles at the top, and containing 2 seeds. Found occasionally; in small poplar woods and along roadsides; throughout the Prairie Provinces, but very rare in Prairies.

## *Amelanchier*    juneberry

Small trees or bushes without spines or prickles, with simple alternate leaves, and with white flowers in racemes. Fruit a berry-like, sweet-flavored, several-seeded pome.

1. Pedicels short, mostly 6–8 mm. ................................................................ *A. alnifolia*
   Pedicels long, the lower ones 15–20 mm
   long. .............................................................................................................. 2
2. Leaves glabrous or nearly so, rounded to
   cordate. ............................................................................................... *A. florida*
   Leaves white pubescent below, obtuse to
   rounded. ......................................................................................... *A. sanguinea*

*Amelanchier alnifolia* Nutt. (Fig. 125)                                    SASKATOON

A tall shrub or small tree 1–4 m high, with smooth chocolate brown stems and branches. Leaves simple, rounded at the ends, 1–5 cm long, with a few serrate teeth at the apex. Flowers white, about 8–12 mm across, in racemes quite early in the season. Fruits purple and berry-like, very sweet when ripe, globular, 6–10 mm across, and used for preserves. Used by the Indians and early settlers as a constituent of pemmican. Very common; in coulees, bluffs, and open woodlands; throughout the Prairie Provinces. The most commonly found *Amelanchier* in the area.

*Amelanchier florida* Lindl.                                    SERVICEBERRY

A shrub or small tree 2–5 m high, with reddish brown branchlets. Leaves broadly oblong to suborbicular, 3–4 cm long, subcordate to rounded at base, coarsely toothed above the middle, more or less tomentose below. Flowers 20–30 mm across; fruit dark purple. Slopes of valleys and along rivers; Rocky Mountains, Peace River, and Cypress Hills.

*Amelanchier sanguinea* (Pursh) DC.                                    EASTERN SERVICEBERRY

An erect or straggling shrub or small tree 1–3 m high, usually in clumps of several stems. Leaves densely tomentose at flowering time, only half-grown; later becoming glabrous, oblong to subrotund, to 7 cm long, finely to coarsely toothed above the middle. Racemes loose; flowers 20–30 mm across; fruit purple. Forest margins and openings; southeastern Boreal forest.

**Chamaerhodos**      chamaerhodos

*Chamaerhodos erecta* (L.) Bunge var. *parviflora* (Nutt.) C. L. Hitchc.

CHAMAERHODOS

Low-growing perennial or biennial glandular herbs 10–30 cm high, from a single stem, but sometimes branching above; the usual form, however, being a very narrow pyramid. Basal leaves, from a stout root, usually forming a rosette, and later becoming a small clump; leaves small, with 3 leaflets, each leaflet again divided into 3 lobes, and sometimes these lobes again 3 times cleft. Stem leaves similar to the basal ones, but less divided. Inflorescence much-branched, of many small white flowers about 3 mm across. Very common; on dry hillsides and light soil; throughout Prairies and Parklands. Syn.: *C. nuttallii* Pickering.

**Cotoneaster**      cotoneaster

Young shoots and leaves yellowish tomentose
   on underside, later pubescent. .......................................................... *C. acutifolia*

A.C. Budd

Fig. 125. Saskatoon, *Amelanchier alnifolia* Nutt.

Young shoots and leaves white tomentose on
underside, remaining so. .................................................................. *C. melanocarpa*

### *Cotoneaster acutifolia* Turcz.           COTONEASTER

Shrubs with branches and leaves 2-ranked, forming flattish sprays. Leaves 3–4 cm long, ovate, dark green above, pale green below; tomentum of young leaves bright to brownish yellow. Inflorescence a small corymb; flowers yellow; fruit black with 2 nutlets. Introduced and cultivated as an ornamental or hedge; occasionally spreading into native woodlands or shrubbery.

### *Cotoneaster melanocarpa* Lodd.           COTONEASTER

Similar to the previous species, but smaller, with the tomentum denser, pure white, and persistent; the fruit with 4 nutlets. A cultivated introduction, occasionally escaped.

### *Crataegus*       hawthorn

Shrubs or small trees with sharp thorns or spines on the stem, and alternate, sometimes lobed, leaves. Flowers white, in clusters. Fruit a round berry-like pome. A difficult genus to separate into species because of apparent hybridizing between species, thereby obscuring the characters.

1. Sepals short, triangular, acute but not
    attenuate; fruit black or very dark
    purple. ................................................................................ *C. douglasii*
   Sepals attenuate, long-pointed; fruit pur-
    ple or red. .......................................................................................... 2
2. Teeth of leaves not gland-tipped; nutlets
    deeply pitted on faces. ...................................................... *C. succulenta*
   Teeth of leaves gland-tipped; nutlets not
    pitted. ...................................................................................... *C. rotundifolia*

### *Crataegus douglasii* Lindl.           DOUGLAS HAWTHORN

A tall shrub or small tree 3–6 m high with short spines 6–25 mm long. Leaves ovate to obovate, 15–50 mm long; fruit black or very dark purple. Found occasionally; in western part of the Prairie Provinces and in Cypress Hills.

### *Crataegus rotundifolia* Moench (Fig. 126)       ROUND-LEAVED HAWTHORN

A round-topped shrub 1–3 m high with stout thorns, usually about 3 cm long, on the stems and branches. Leaves usually almost round, 2–5 cm across, sometimes lobed, but usually double-toothed. Flowers white, 10–15 mm across, borne in clusters at the ends of the branches, and followed by red berry-like fruits about 1 cm across containing several bony carpels. Fairly common; in coulees, stream banks, and open woods; throughout the Prairie Provinces. Syn.: *C. chrysocarpa* Ashe; *C. columbiana* Howell.

### *Crataegus succulenta* Link           LONG-SPINED HAWTHORN

A shrub very similar to *C. rotundifolia* with no dark glandular tips on the leaf teeth and with deeply pitted nutlets. Plentiful in eastern parts but appears to be replaced by *C. rotundifolia* in the west.

Fig. 126. Round-leaved hawthorn, *Crataegus rotundifolia* Moench.

### Dryas    mountain-avens

1. Leaves entire or nearly so. ................................................................. *D. integrifolia*
   Leaves crenate to the tip. ...................................................................... 2
2. Flowers yellow; leaves tapering to base. ........................................... *D. drummondii*
   Flowers white; leaf base blunt. ................................... *D. octopetala* var. *hookeriana*

*Dryas drummondii* Richardson                    YELLOW MOUNTAIN-AVENS

Depressed, mat-forming shrubs; the young branches white tomentose. Leaves elliptic to obovate, 1–3 cm long, densely white tomentose below, dark green and rugose above. Scape 5–20 cm long, tomentose; calyx black glandular pubescent; petals yellow, about 1 cm long; styles plumose in fruit, 3–5 cm long. Gravel banks, scree fields, and slopes; Rocky Mountains, northern Boreal forest.

*Dryas integrifolia* Vahl                         WHITE MOUNTAIN-AVENS

Depressed, densely mat-forming shrub, very leafy. Leaves 8–15 mm long, oblong-lanceolate, entire, often cordate at base. Scape 3–10 cm high; calyx tomentose; petals about 1 cm long, white; styles 15–25 mm long in fruit. Gravel banks along rivers and rocky slopes; northeastern Boreal forest, Rocky Mountains.

*Dryas octopetala* L. var. *hookeriana* (Juz.) Breit.          MOUNTAIN-AVENS

Similar to *D. drummondii,* but forming small mats. Leaves strongly rugose above, truncate to subcordate at base, and black glandular as well as tomentose; calyx white tomentose and black pubescent; petals white. Alpine gravels and rock outcrops; Rocky Mountains.

### Fragaria    strawberry

Perennial low-growing herbs with 3-foliolate leaves, and running stems rooting at tips and producing new plants. Flowers white, with 5 petals, 5 sepals, and 5 sepal-like bracts. Fruit is the enlarged fleshy receptacle of the flower, bearing very juicy and sweet achenes on the surface.

Hairs on the stems ascending and closely
  pressed to stems. ............................................................................... *F. virginiana*
Hairs on the stems spreading in various
  directions. ................................................................... *F. vesca* var. *americana*

*Fragaria vesca* L. var. *americana* Porter          AMERICAN WILD STRAWBERRY

Similar to the following species, but with leaves paler, flowers smaller, and leaflets stalkless (leaflets of *F. virginiana* having short stalks). Fruit ovoid or conical, with seeds borne on the surface. Rocky woodlands; in Boreal forest. Syn.: *F. americana* (Porter) Britt.

*Fragaria virginiana* Dcne.                       SMOOTH WILD STRAWBERRY

A low-growing herb with coarsely toothed, broadly ovate leaflets, sometimes silky beneath. Flowers white, 15–20 mm across, appearing fairly early in the season. Fruit almost round, 10–15 mm in diam, with seeds sunk in shallow pits. The commonest strawberry; in low spots on prairie, open woodlands, and moist areas; throughout the Prairie Provinces. Syn.: *F. virginiana* Dcne. var. *glauca* Wats.; *F. glauca* (S. Wats.) Rydb.

**Geum**         avens

Perennials from stout rootstocks, with lyrate or pinnate leaves, perfect regular flowers, and many achenes. The long styles, remaining on the achenes, either jointed or feathery.

1. Styles feathery and not jointed; leaves
   pinnate. ................................................................................................ *G. triflorum*
   Styles not feathery, but jointed; leaves
   lyrately pinnate. ................................................................................................ 2

2. Sepals erect or spreading; flowers purple
   or flesh-colored; upper part of style at
   least half as long as lower part. ................................................................ *G. rivale*
   Sepals reflexed; flowers yellow; upper
   portion of style less than one-third the
   length of lower part. ................................................................................................ 3

3. Upper portion of style hairy. ................................................................ *G. aleppicum*
   Upper portion of style not hairy, or with
   only a few stiff hairs. ................................................................ *G. perincisum*

*Geum aleppicum* Jacq.                              YELLOW AVENS

A hairy erect plant 40–120 cm high, with broad leafy stipules and basal leaves lyrately pinnate (terminal lobe longer than others) with 5- to 7-toothed or divided leaflets and often a few very small interspersed leaflets. Stem leaves with 3–5 leaflets, and either with or without very short stalks. Flowers bright yellow, 10–25 mm across, and followed by a fruiting head about 15 mm in diam, on which the characteristic hooked or bent styles are located. Common; in wet or moist locations on prairies and in meadows and open woods; throughout the Prairie Provinces. Syn.: *G. strictum* Ait.

*Geum perincisum* Rydb.                              LARGE-LEAVED AVENS

Very similar to *G. aleppicum,* but the terminal segment of the basal leaves large, often 3-lobed, and having 5–15 leaflets. Flowers yellow and the fruiting head similar to that of the preceding species except that the upper portion of the style on the fruit is entirely or almost hairless. Not so common as yellow avens; found in moist meadows and open woods; throughout the Prairie Provinces.

*Geum rivale* L.                              PURPLE AVENS

An erect, simple, and little-branched herb, more or less hairy, 30–60 cm high, with lyrate-pinnate basal leaves and thrice-divided stem leaves. Sepals not bent abruptly downward as those of the two preceding species. Flowers 15–20 mm across, flesh-colored or purple, often with a yellowish tinge. Fruiting heads very similar to those of the preceding species, but the upper portion of the style at least half as long as the lower part. Found occasionally; in moist places and wet, swampy ground; Boreal forest and the Cypress Hills.

*Geum triflorum* Pursh (Fig. 127)    THREE-FLOWERED AVENS, TORCHFLOWER

Erect perennial herbs 15–40 cm high from a thick, coarse, almost black rootstock, with coarse black roots. The many basal leaves pinnate, with many wedge-shaped lobed leaflets, often with smaller leaflets mixed with the larger ones. Flowering stem usually with a tuft of small leaves halfway up, and some thin leafy bracts at summit. Flowers usually 3, 12–20 mm across; sepals purplish pink; petals pink, yellowish, or flesh-colored. Fruiting head bearing long, persistent, feathery styles 2–5 cm long. A very common spring flower on the prairies, the bright green leaves being among the first new foliage to show in spring and the flowers among the very early spring blooms. Plentiful; everywhere on open prairie; Prairies, Parkland, and Rocky Mountains. Syn.: *Sieversia triflora* (Pursh) R. Br.

**Luetkea**    luetkea

*Luetkea pectinata* (Pursh) Ktze.    LUETKEA

A cespitose plant, semishrubby, with prostrate, stoloniferous branches. Leaves dissected into linear divisions, 10–15 mm long, acute, grooved above. Raceme narrow, 1–5 cm long; petals 3–3.5 mm long, white; sepals 2 mm long, ovate, acute. Moist slopes and mountain meadows; southwestern Rocky Mountains.

**Physocarpus**    ninebark

*Physocarpus malvaceus* (Greene) Ktze.    MALLOW-LEAVED NINEBARK

A shrub 1–2 m high, with brown branches, these glabrous or stellate pubescent. Leaves 2–6 cm long, round-ovate, 3- to 5-lobed, more or less doubly crenate, mostly cordate at base, and stellate pubescent on both sides. Inflorescence of half-round corymbs, 3–5 cm wide; flowers white, about 1 cm across. Rare; thickets and forest openings; southern Rocky Mountains.

**Potentilla**    cinquefoil

Generally perennial, but some species annual or biennial, with alternate leaves, either palmately or pinnately compound. Flowers perfect, with 5 sepals, 5 petals, and many stamens. Fruit a head with many achenes.

1. Flowers reddish purple; leaves with 5–7
   leaflets. ............................................................................................ *P. palustris*
   Flowers white or yellow. ........................................................................................ 2
2. Plant shrubby or with woody bases;
   achenes hairy. ........................................................................................................ 3
   Plant herb-like, not woody based;
   achenes generally smooth. ................................................................................... 4
3. Shrub 15–150 cm high; flowers yellow;
   pinnate leaves. ................................................... *P. fruticosa* ssp. *floribunda*
   Low plant, woody at base; flowers white;
   palmate leaves. .................................................................................. *P. tridentata*

A.C. Budd

Fig. 127.   Three-flowered avens, *Geum triflorum* Pursh.

4. Low creeping plant with runners; leaves basal and pinnate; yellow flowers singly on long stalks from base of plant. ............................................................................ *P. anserina*

   Plant without runners; flowers borne in terminal cymes and not singly. ............................................................................ 5

5. Annual or biennial plant without perennial rootstocks; inflorescence many-flowered and leafy. ............................................................................ 6

   Perennial plant with stout rootstocks; often showing old bases of leaves. ............................................................................ 8

6. Lower leaves pinnate, with 5–11 leaflets. ............................................................................ 7

   Lower leaves digitate, with 3 leaflets. ............................................................................ *P. norvegica*

7. Lower leaves with 7–11 leaflets; achenes ribbed with a corky enlargement on one side. ............................................................................ *P. paradoxa*

   Lower leaves with 5–7 leaflets; achenes smooth. ............................................................................ *P. rivalis*

8. Plants glandular pubescent; petals white or yellowish; stamens 25–30. ............................................................................ 9

   Plants not usually glandular; petals yellow; stamens 5–20. ............................................................................ 10

9. Cymes short, crowded; petals white, slightly longer than the sepals. ............................................................................ *P. arguta*

   Cymes open; petals white to yellowish, as long as the sepals. ............................................................................ *P. glandulosa* var. *intermedia*

10. Leaves all or mostly pinnate, with the upper stem leaves sometimes trifoliate. ............................................................................ 11

    Leaves all trifoliate or digitate with 5–7 leaflets. ............................................................................ 21

11. Leaflets serrate to shallowly lobed. ............................................................................ 12

    Leaflets dissected or incised more than halfway to midrib. ............................................................................ 15

12. Leaflets green on both sides. ............................................................................ 13

    Leaflets gray or white tomentose below. ............................................................................ 14

13. Plants with 0–2 stem leaves. ............................................................................ *P. drummondii*

    Plants with 4–7 stem leaves. ............................................................................ *P. paradoxa*

14. Leaves white tomentose below. ............................................................................ *P. hippiana*

    Leaves grayish pilose to hirsute below. ............................................................................ *P. pensylvanica*

15. Leaflets green on both sides. ............................................................................ *P. plattensis*

    Leaflets white or a different shade of green below. ............................................................................ 16

16. Leaflets pale green or grayish pilose to glandular below. ............................................................................ *P. pensylvanica*

    Leaflets white tomentose below. ............................................................................ 17

17. Leaflets pectinately toothed or lobed; leaf margins revolute. ............................................................................ 18

31. Leaflets entire below the middle, dentate
    above. ........................................................ *P. diversifolia* var. *glaucophylla*
    Leaflets dentate along the margins. ........................................................ *P. gracilis*

*Potentilla anserina* L.                                                    SILVERWEED

A low tufted perennial plant, spreading by runners. Leaves pinnate, 10–45
cm long, with 7–25 leaflets, often with some smaller interspersed leaflets. In the
typical form, the leaflets are green above and silky white woolly beneath, but
in forma *sericea* (Hayne) Hayek the leaves are silvery white on both sides.
Both the type and the form have been found in the same clump of plants.
Flowers bright yellow, 20–25 mm across, borne singly on a long stalk, blooming early in season until fall. Very common; in low, wet places and slough
margins; throughout the Prairie Provinces.

*Potentilla argentea* L.                                          SILVERY CINQUEFOIL

Perennial with white woolly freely branching stems 10–50 cm high,
ascending or erect. Lower leaves mostly 5-digitate, with the leaflets white
tomentose on underside, toothed to pinnatifid above the middle and the margins revolute. Cyme leafy, many-flowered; petals yellow, about the same
length as the sepals. Roadsides and waste areas; eastern Prairies, Parklands,
and Boreal forest.

*Potentilla arguta* Pursh                                          WHITE CINQUEFOIL

Perennial glandular erect herbs 30–90 cm high, with pinnate leaves; the
lower leaflets smaller than the upper ones. Leaflets 7–11 on the stalked basal
leaves, fewer on the stem leaves. Flowers in a rather dense inflorescence, 12–20
mm in diam; petals white. Common; in moist places, slough margins on prairie; throughout Prairies and Parklands. Syn.: *Drymocallis agrimonioides*
(Pursh) Rydb.

*Potentilla bipinnatifida* Dougl.                                 PLAINS CINQUEFOIL

A many-stemmed perennial 20–50 cm high, with pinnate leaves of 3–7
leaflets. Leaflets silky and green above, snowy white woolly beneath, deeply
cut into narrow lobes. Fairly well distributed; on dry prairies; across Prairies
and Parklands. Syn.: *P. pensylvanica* L. var. *bipinnatifida* (Dougl.) T. & G.

*Potentilla concinna* Richardson                                   EARLY CINQUEFOIL

A low-growing perennial from a very coarse, woody rootstock, with 5
(rarely 7) oblong or obovate leaflets, either palmately or pinnately arranged.
Leaflets 12–30 mm long and toothed, greenish silky above and whitish below.
Flowers yellow, 6–12 mm across, opening early in the season. Common; on
dry hillsides and prairie; throughout Prairies, Parklands, and southern Rocky
Mountains.

*Potentilla diversifolia* Lehm. var. *glaucophylla* Lehm.
                                               SMOOTH-LEAVED CINQUEFOIL

A perennial plant 15–50 cm high, with palmately lobed leaves; the leaflets
green and almost hairless on both sides. Leaflets usually coarsely toothed
along the upper two-thirds of their margins, with forward-pointing teeth. A

mountain species found in the southern Rocky Mountains and occasionally in the Cypress Hills.

*Potentilla drummondii* Lehm.                                    DRUMMOND'S CINQUEFOIL

Plants with a short caudex; the stems 30–60 cm high, more or less hirsute. Basal leaves 5–10 cm long, more or less hirsute; leaflets 2–5 cm long, deeply incised. Flowers in a cyme, long-pedicelled; petals 6–10 mm long. Mountain meadows, riverbanks; southern Rocky Mountains.

*Potentilla flabellifolia* Hook.                                  MOUNTAIN CINQUEFOIL

Plants with branching, scaly rootstocks; the stems slender, 15–30 cm high, minutely puberulent. Basal leaves 3-foliate, very thin; leaflets fan-shaped, deeply incised to serrate, short-pubescent. Cymes few-flowered; petals 8–10 mm long. The var. *emarginata* (Pursh) Boiv. includes more densely pubescent plants, about 10 cm high, and the marginal teeth with tufts of hairs. Moist mountain meadows; southern Rocky Mountains.

*Potentilla fruticosa* L. ssp. *floribunda* (Nutt.) Elk. (Fig. 128)
                                                                 SHRUBBY CINQUEFOIL

A much-branched shrub from branching rootstocks, 15–150 cm high. Leaves pinnate, leathery, with 5–7 leaflets, 12–25 mm long, linear-oblong and pointed at both ends. Flowers yellow, 15–25 mm across. Achenes densely hairy. Much used as a garden ornamental. Very abundant on low moist ground; especially on Cypress Hills – Wood Mountain bench and in the Rocky Mountains, but also found across the Prairie Provinces. Syn.: *Dasiphora fruticosa* (L.) Rydb.; *P. fruticosa* Auct.; *P. floribunda* Pursh.

*Potentilla glandulosa* Lindl. var. *intermedia* (Rydb.) C. L. Hitchc.
                                                                 STICKY CINQUEFOIL

Plants with slender, viscid, and glandular hirsute stems 15–30 cm high. Basal leaves with 5–9 leaflets, glabrous above, or nearly so, sparingly glandular hirsute below; leaflets 1–3 cm long, obovate, serrate; stem leaves reduced. Cymes many-flowered; sepals 6–7 mm long, glandular hirsute; petals about the same length as the sepals. Alpine slopes and mountain meadows; southern Rocky Mountains.

*Potentilla gracilis* Dougl.                                     GRACEFUL CINQUEFOIL

A somewhat tufted species with root crowns bearing brownish remains of old stipules, and several stems 30–60 cm high. Basal leaves long-stalked and digitate, of 5–7 narrowly oblanceolate leaflets. Flowers yellow, 15–20 mm across. A very variable plant, some of its varieties have been considered separate species. Two varieties are distinguished here.

Leaflets oblanceolate, serrate to pinnatifid,
  green silky to white tomentose below. ........................................ var. *gracilis*
Leaflets fan-shaped, deeply incised, densely
  tomentose below. ........................................ var. *flabelliformis* (Lehm.) Nutt.

A.C. Budd

Fig. 128.   Shrubby cinquefoil, *Potentilla fruticosa* L. ssp. *floribunda* (Nutt.)
Elk.

Both varieties occur throughout the Prairie Provinces, north to the southern parts of the Boreal forest; var. *flabelliformis* is usually found in swampy or somewhat alkali meadows; var. *gracilis* in moist grasslands, slopes, and thickets.

*Potentilla hippiana* Lehm.                                                   WOOLLY CINQUEFOIL

A perennial from a stout rootstock, with several woolly stems 10–25 cm high, and mostly basal pinnate leaves with 7–11 leaflets 1–5 cm long, toothed, densely silky above and white woolly below. Stem leaves smaller, with fewer leaflets. Flowers fairly numerous, in a terminal cyme 6–10 mm across. Found fairly often on prairie land and in valleys in southwestern parts of the Prairie Provinces. The var. *argyrea* (Rydb.) Boivin, silvery cinquefoil, has leaflets white woolly on both upper and lower sides; found occasionally on dry hillsides; in the Prairies and in the Cypress Hills.

*Potentilla multifida* L.                                                   BRANCHED CINQUEFOIL

Plants with strigose stems 20–40 cm high. Basal leaves with 7 leaflets; stem leaves with 5–7 leaflets; the leaflets deeply incised, green below, with revolute margins. Cymes few- to many-flowered; the petals 6–10 mm long. Open areas and gravel banks; Boreal forest.

*Potentilla nivea* L.                                                   SNOW CINQUEFOIL

A cespitose perennial with a stout caudex; the flowering stems 10–20 cm high. Leaves mostly basal, 3-foliolate or rarely 5-digitate; leaflets 1–3 cm, oblong or obovate, densely white tomentose beneath, deeply 3- to 5-dentate on each side. Flowers 10–15 mm across. Alpine, subarctic, and arctic meadows; Boreal forests, Rocky Mountains. Besides the typical variety described, two other varieties occur: var. *villosa* (Pall.) Regel & Tiling, more coarsely and densely villous throughout, including the upper surface of the thick, rugose leaflets; and var. *pulchella* (R. Br.) Durand, with the leaves mostly short-pinnate with 5 coarse leaflets, and rather large, brown stipules.

*Potentilla norvegica* L.                                                   ROUGH CINQUEFOIL

A very coarse hairy annual or biennial plant, with erect branched stems 15–60 cm high. Leaves digitately 3-foliolate, with leaflets 2–10 cm long, obovate or elliptic, and much-toothed. Flowers numerous, 6–12 mm across, yellow, in a fairly dense cyme; the sepals a little longer than the petals. A common plant; moist meadows and waste places, a bad weed in gardens; throughout the Prairie Provinces. Syn.: *P. monspeliensis* L.

*Potentilla palustris* (L.) Scop.                                                   MARSH CINQUEFOIL

A somewhat decumbent herb 20–50 cm high, from long, creeping, weedy rootstocks. Leaves pinnate, often purple; the lower ones long-stalked, with 5–7 oblong or oval leaflets 2–7 cm long, tapering to the base and toothed. Upper leaves short-stalked, with 3–5 leaflets. Flowers conspicuous, 15–35 mm across, purple or maroon; the sepals much longer than the petals and also purplish. Found in shallow water and bogs; in Boreal forest and rarely in Parklands. Syn.: *Comarum palustre* L.

*Potentilla paradoxa* Nutt.                                    BUSHY CINQUEFOIL

An annual, biennial, or short-lived perennial species, with spreading or somewhat ascending stems 20–50 cm high. The pinnate leaves of 5–11 almost smooth leaflets. Flowers about 6 mm across, in a leafy cluster. Achenes or fruits ribbed lengthwise and having a corky enlargement on one side. Found occasionally in low moist places; throughout the Prairie Provinces.

*Potentilla pensylvanica* L. (Fig. 129)                      PRAIRIE CINQUEFOIL

A low-growing tufted species, 10–40 cm high, with the crown usually bearing the brown remains of previous leaf stipules. Two varieties of this species occur throughout the Prairies and Parklands.

Hairs on stem very short and appressed.
  Leaflets grayish and hairy on underside. ............................................. var. *pensylvanica*
Hairs on stem obvious and spreading. Leaflets
  lobed to about halfway to midrib. ................................. var. *atrovirens* (Rydb.) T. Wolf

The var. *atrovirens* common in dry grasslands in the Prairies; var. *pensylvanica* common in moister grasslands of the Prairies, Parklands, and Cypress Hills.

*Potentilla plattensis* Nutt.                                  LOW CINQUEFOIL

A low decumbent or spreading perennial 10–20 cm high, with many basal pinnate leaves of 9–17 leaflets. Leaflets light green, deeply divided into oblong to nearly linear divisions. Flowers few and in rather open terminal cymes; petals yellow. Found occasionally in valleys; Prairies.

*Potentilla quinquefolia* Rydb.                          FIVE-FINGERED CINQUEFOIL

A low tufted perennial 10–20 cm high, with small leaves, palmately divided into 3–5 leaflets. Leaflets green above and white woolly beneath, deeply divided into narrow lobes. Flowers about 1 cm across, in a fairly dense cluster. Quite rare; but has been found in Parklands and Boreal forest.

*Potentilla recta* L.                                     ROUGH-FRUITED CINQUEFOIL

A leafy-stemmed erect plant 15–50 cm high, with loosely hairy stems and leaf stalks. The deeply toothed leaves are digitate, usually of 5–7 leaflets, more or less hairy, but paler on the underside. Flowers on erect stalks 15–25 mm across, pale yellow or sulfur. This introduced plant has been found at various places throughout the Prairie Provinces.

*Potentilla rivalis* Nutt.                                     BROOK CINQUEFOIL

An erect annual or biennial species, branching above, and with finely hairy, sometimes sticky, stems 20–40 cm high. The lower leaves have 5 obovate coarsely toothed leaflets 2–5 cm long; upper stem leaves have 3 leaflets. Yellow flowers less than 5 mm across, with petals shorter than sepals. Many authorities consider this species and *P. millegrana* as forms of the same species, differing mainly in the number of leaflets on the basal leaves. Found occasionally; in river valleys; Prairies.

Fig. 129.   Prairie cinquefoil, *Potentilla pensylvanica* L.

*Potentilla saximontana* Rydb.                                    ALPINE CINQUEFOIL

A low perennial with stout caudex; the stems 10–20 cm high, glabrous to pubescent. Leaves mostly basal, pinnate, usually with 5 leaflets 1–5 cm long, deeply cleft, the divisions narrow, white tomentose or strigose below, silky hairy above. Flowers few to several, 10–15 mm across. Alpine slopes; southern Rocky Mountains.

*Potentilla sibbaldii* Hall. f.                                         SIBBALDIA

A low densely tufted herb with short decumbent or creeping woody stems. Leaves 3-foliolate, with slender petioles; leaflets 3- to 5-toothed at the apex, obovate, cuneate at the base, sparsely pubescent on both surfaces. Flowers on axillary peduncles; sepals 2–3 mm long; petals minute, yellow. Alpine slopes and scree fields; Rocky Mountains.

*Potentilla tridentata* Ait.                            THREE-TOOTHED CINQUEFOIL

A tufted, low-growing, woody-based subshrub, much-branched, 10–20 cm high. Stems have silky, appressed hairs. Leaves mostly stalked with 3 narrow wedge-shaped leaflets 12–25 mm long, dark green and shiny on the upper side and paler beneath, having 3 teeth at their apex. Flowers 1–6 in a cyme, white, about 6–8 mm across. Fairly common; in dry sandy places, in pine forests; Boreal forest. Syn.: *Sibbaldiopsis tridentata* (Ait.) Rydb.


**Prunus**      plum

Shrubs or trees with bark almost black; alternate leaves; perfect regular flowers with 5 petals and 5 sepals; fruit a fleshy drupe with a smooth bony stone.


1. Flowers in long racemes at the ends of the
     branches. ........................................................................................ 2
   Flowers in small umbels or corymbs. ........................................................ 3

2. Fruit red or purplish, very astringent;
     leaves thin. ............................................................. *P. virginiana*
   Fruit black and sweetish; leaves thick. ................ *P. virginiana* var. *melanocarpa*

3. Stone more or less flattened, with a
     groove on the end. ............................................................................ 4
   Stone not flattened or grooved. ............................................................. 5

4. Teeth of leaves not very deep and tipped
     with a small gland; lobes of sepals
     toothed and glandular; fruit oval. .............................................. *P. nigra*
   Teeth of leaves deeper and not gland-tip-
     ped; lobes of sepals not toothed or glan-
     dular; fruit round. ...................................................... *P. americana*

5. Tall shrub or small tree; flowers 12–25
     mm across, in corymbs. ..................................................... *P. pensylvanica*
   Low shrub; flowers 6–12 mm across, in
     umbels. ......................................................................... *P. pumila*

*Prunus americana* Marsh.                                    AMERICAN PLUM

A tree 3–8 m high, with more or less thorny branches. Leaves narrowly obovate, 3–10 cm long, with double teeth and a pointed apex. Flowers white, 15–25 mm across, appearing before the leaves. Fruit a red or yellow plum, almost round, 15–25 mm long. Found occasionally; in moist woods and along riverbanks; southeastern Parklands and Boreal forest.

*Prunus nigra* Ait.                                          CANADA PLUM

A tree or shrub 2–8 m high, with oval or obovate leaves 7–12 cm long, pointed at the apex, dark green with gland-tipped teeth. Flowers white, turning pink, about 15–30 mm across, opening before the leaves. Fruit a yellow to orange plum, oval, 25–35 mm long. Found occasionally in woodlands and bluffs; southeastern Parklands and Boreal forest.

*Prunus pensylvanica* L. f.                                  PIN CHERRY

A small tree 4–8 m high, with lanceolate finely toothed leaves 3–10 cm long. Flowers small, white, on long stalks in corymbose clusters 6–10 mm across, opening about the same time as the leaves. Fruit a small, sour, bright red cherry, 5–8 mm in diam. Fairly common; in bluffs, ravines, and hillsides; throughout the Prairie Provinces.

*Prunus pumila* L.                                           LOW SAND CHERRY

A low, spreading bush with oblanceolate to spatulate leaves, dark green above and pale beneath. Flowers in clusters of 2–4, white, about 1 cm across. Fruit a dark brown to purplish cherry, 6–12 mm across. Found on dry prairie and in sandhills; southeastern Parklands and Boreal forest.

*Prunus virginiana* L.                                       RED-FRUITED CHOKE CHERRY

A shrub 1–3 m high, with grayish stems and thin ovate leaves 5–10 cm long. Flowers white, about 12 mm across, in loose racemes. Fruit a red, very astringent cherry 8–10 mm in diam. Quite common; on riverbanks and open woodlands; in the southeastern Parklands. It is replaced further west by the var. *melanocarpa*.

*Prunus virginiana* L. var. *melanocarpa* (A. Nels.) Sarg. (Fig. 130)
                                                             BLACK-FRUITED CHOKE CHERRY

A small tree or large shrub 2–6 m high, with smooth reddish brown stems. Leaves obovate or oval, rather thick, 2–8 cm long, smooth on both sides. Flowers numerous, white or pale cream, about 1 cm across, in dense racemes. Fruit a black cherry, slightly astringent, round, 6–8 mm in diam. Very common; in bluffs, ravines, sandhills, and open woodlands; throughout the Prairie Provinces. Leaves of choke cherry injured by frost or extreme drought are reported to be **poisonous** to sheep and cattle.

**Rosa**      rose

From low shrubs or herbaceous plants to large bushes, usually with prickly stems. Leaves pinnate, with odd number of small leaflets. Flowers large and showy, generally pinkish and fragrant. Fruit a hip, or berry-like enlarged

*Dicots*                                          ROSACEAE — 455

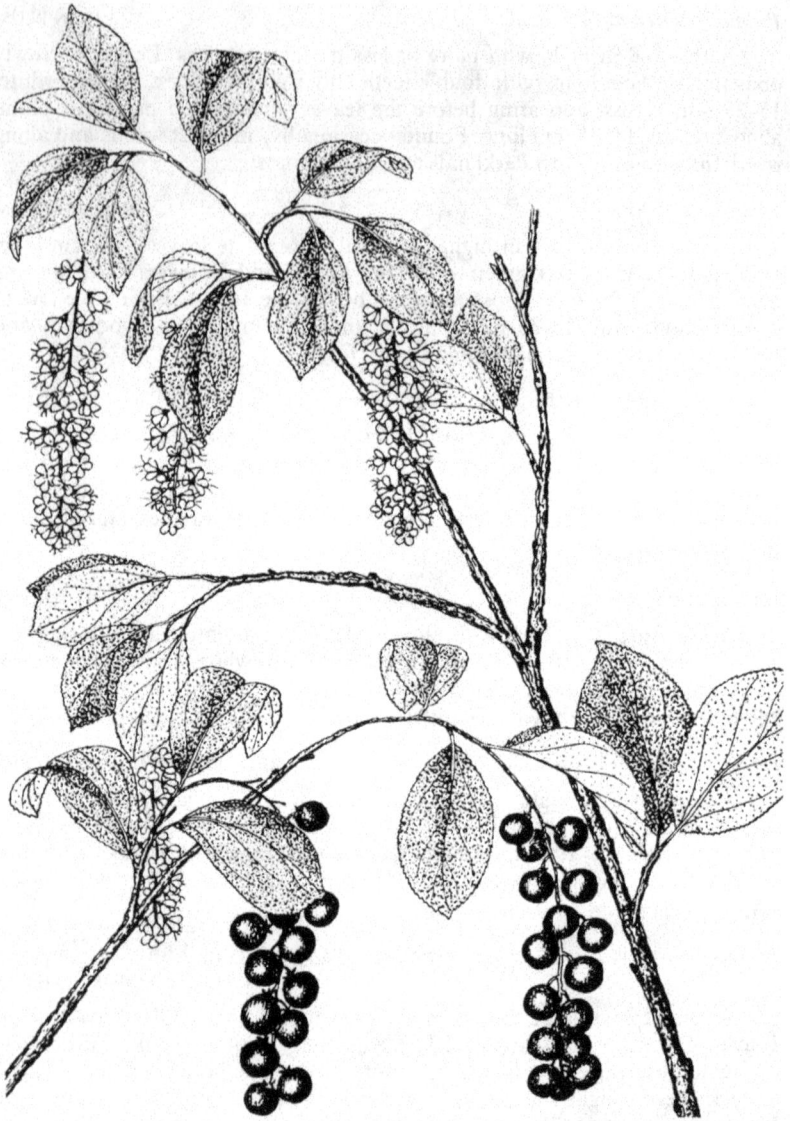

Fig. 130.   Black-fruited choke cherry, *Prunus virginiana* L. var. *melanocarpa* (A. Nels.) Sarg.

calyx tube containing numerous achenes. A difficult genus to identify, local variations in form being so common that authorities do not agree on nomenclature, and some specialists have separated them into many species. Probably only about 4 species, with many slight variations, occur in the Prairie Provinces.

1. Plant with few, if any, bristles. ........................................................ *R. blanda*
   Plant definitely bristly. ................................................................................ 2

2. Thorns present below the stipules; bristles broad and usually flattened at the base; fruit globose without constricted neck. ...................................................... *R. woodsii*
   Thorns not present; bristles usually not broad or flattened at base. ......................................................... 3

3. Plant shrubby; leaflets usually 5–9 (9 on young shoots); flowers usually borne singly; fruit oval with distinct neck. ...................................................... *R. acicularis*
   Plant partly shrubby, often dying back close to the ground in winter; leaflets 9–11. ................................................................................ *R. arkansana*

*Rosa acicularis* Lindl.                                          PRICKLY ROSE

A low bushy plant 30–120 cm high, with stems densely covered with straight weak bristles. Leaves usually of 5–7 hairy leaflets 12–50 mm long, elliptic or oval. Stipules broad, usually both hairy and glandular. Flowers usually borne singly, 5–7 cm across. Fruit usually ovoid with a constricted neck, about 1.5 cm long. Common; in bluffs, around woods, in fields, and along roadsides; throughout the Prairie Provinces.

*Rosa arkansana* Porter                                     LOW PRAIRIE ROSE

Low shrubs 20–30 cm high with densely prickly stems, which usually die off annually close to the root. Leaflets 9–11, usually without hairs, smooth and shiny, 2–5 cm long. Flowers in corymbs of 2 or 3 flowers. Fruit almost globular, about 12 mm across. Common; on hills and sandy open prairie; throughout the Prairies.

*Rosa blanda* Ait.                                              SMOOTH ROSE

A low bush 60–120 cm high, with an unarmed stem or with a very few straight prickles. Stipules rather broad; leaves of 5–7 leaflets 25–40 mm long. Flowers pink, usually borne singly, may be up to 7 cm across. Not common; but has been found in eastern Parklands.

*Rosa woodsii* Lindl.                                           WOOD'S ROSE

A bush 50–200 cm high, with the stems armed with straight or slightly curved prickles, often broad and flattened at the base. Spines below the stipules usually well-defined. Leaves of 5–9 oval or obovate leaflets 15–35 mm long. Flowers 2–5 cm across. Fruit globular without a constricted neck, about 1 cm across. In bluffs, on ravines, and in sandhills; throughout the Prairies and Parklands.

***Rubus***    raspberry

Perennial or biennial, shrubby or herbaceous, from rootstocks, often with prickly stems. Leaves various, from simply lobed to 3–5 times pinnately or palmately compound. Flowers regular, unisexual or perfect, usually with 5 sepals and 5 petals, either pink or white. Fruit a berry composed of many fleshy drupelets.

1. Stems more or less woody; leaves 3–5 foliolate, usually pinnate or lobed. ........................................................................ 2

   Stems herbaceous, dying down annually, not prickly; leaves with 3 leaflets or merely lobed. .................................................................................. 3

2. Stems prickly as well as bristly; leaves compound. .................................................. *R. idaeus* var. *aculeatissimus*

   Stems not prickly; leaves lobed, 10–20 cm wide. .............................................................................. *R. parviflorus*

3. Leaves merely lobed; petals spreading; flowers unisexual. ........................................................ *R. chamaemorus*

   Leaves compound; petals erect. .................................................................................. 4

4. Leaves 5-digitate. ................................................................................ *R. pedatus*

   Leaves 3-foliolate. .................................................................................. 5

5. Central leaflet with rounded tip. ........................................................ *R. acaulis*

   Central leaflet with pointed tip. ........................................................ *R. pubescens*


*Rubus acaulis* Michx.                                    STEMLESS RASPBERRY

A low, herbaceous, unarmed perennial 5–20 cm high, with pinnately divided leaves of 3 broadly ovate leaflets 15–35 mm long. Flowers usually borne singly, pink; fruit red. Only found in bogs and meadows; in Boreal forest.


*Rubus chamaemorus* L.                                    CLOUDBERRY

A herbaceous perennial 5–20 cm high, with 2 or 3 stalked round leaves 2–7 cm across, each with 5–9 rounded lobes. Flowers borne singly, terminal, white, 10–25 mm across; fruit reddish, turning golden yellow when ripe. Found in bogs and swamps on Boreal forest.


*Rubus idaeus* L. var. *aculeatissimus* Regel & Tiling (Fig. 131)

WILD RED RASPBERRY

A large bush 1–2 m high, with brownish more or less bristly stems and pinnate leaves with 5 leaflets (the floral branches with 3 leaflets). Leaflets ovate, the terminal one often being 3-lobed, 5–10 cm long, dark green above and white woolly beneath. New shoots fairly bristly, but rarely glandular. Flowers white, 8–12 mm across; fruit round, light red, about 1 cm across. Probably the most common raspberry; in shady, wooded places, on burned-over woodlands, bluffs, riverbanks; throughout the Prairie Provinces. Syn.: *R. melanolasius* Focke; *R. strigosus* Michx.

A.C. Budd

Fig. 131. Wild red raspberry, *Rubus idaeus* L. var. *aculeatissimus* Regel & Tiling.

*Rubus parviflorus* Nutt.                                               THIMBLEBERRY

An erect shrub 50–200 cm high, unarmed, with the young growth more or less glandular hairy. Leaves 10–20 cm wide, simple, round to reniform in outline, few-lobed, with the lobes triangular, glabrous or pubescent below. Flowers 2–4 cm across, white; fruit red, 15–20 mm across. Forest margins and openings; common in southern Rocky Mountains, rarer in northern Rocky Mountains and Cypress Hills.

*Rubus pedatus* J. E. Smith                                            DWARF BRAMBLE

A low or creeping unarmed herbaceous species, with glabrous stems rooting at the nodes, to 1 m long. Branches short; leaves 1–4, digitately 5-foliolate, or rarely 3-foliolate; leaflets 1–5 cm long, obovate, irregularly incised, glabrous or very sparsely pubescent. Flowers solitary on slender peduncles 4–10 cm long; sepals green, 8–10 mm long; petals white, the same length as the sepals; fruit reddish, 8–10 mm long. Moist woods; Rocky Mountains, western Parklands, and Boreal forest.

*Rubus pubescens* Raf.                                                 DEWBERRY

A trailing or climbing herbaceous plant, with pinnate leaves usually of 3 leaflets (rarely 5), ovate or rhombic, 2–10 cm long, sharply toothed, and green on both sides. Flowers borne in groups of 2 or 3 with white or pink petals, 8–12 mm across, with reflexed sepals. Fruit reddish purple, about 1 cm across. Common; in rich woods and bluffs; Parklands, Boreal forest, and Cypress Hills.


**Sorbaria**       false spiraea

*Sorbaria sorbifolia* (L.) A. Braun                                    ASH-SPIRAEA

Shrubs to 2 m high; young branches tomentose with stellate hairs. Leaves pinnate; leaflets 3–7 cm long, acuminate, sharply double serrate. Inflorescence a panicle to 30 cm long, flowers white, 5–6 mm across. Introduced ornamental, occasionally escaped and persisting.


**Sorbus**       rowan

1. Trees 6–10 m high. .................................................................... *S. aucuparia*
   Shrubs or small trees less than 5 m high. .................................................. 2

2. Leaflets usually 11–13, serrate to close to
   the    base;    inflorescence    broad,
   flat-topped. ................................................................. *S. americana*
   Leaflets usually 7–11, entire toward the
   base; inflorescence small, rounded. ...................................... *S. sitchensis*

*Sorbus americana* Marsh.                                    WESTERN MOUNTAIN ASH

A small tree 1–4 m high, with pinnate leaves of 11–13 elliptic-lanceolate leaflets 3–6 cm long, bright green on both sides. Flowers white, in dense terminal clusters, 6–10 mm across, with the flower clusters up to 10 cm broad; fruit globose, red, about 1 cm across. Not common; but found in Parklands, Boreal forests, Riding Mountain, and Cypress Hills. Syn.: *Sorbus scopulina* Greene.

*Sorbus aucuparia* L.                                              ROWAN TREE

A tree to 10 m high, round-topped, with young branches more or less pubescent; winter buds white villose. Leaflets 9–15, oblong, 3–5 cm long, serrate, usually long-pubescent below. Inflorescence to 20 cm broad, flat-topped; fruit bright orange red. Introduced from Europe, extensively planted, and occasionally escaped.

*Sorbus sitchensis* Roemer                                    MOUNTAIN ASH

A tall shrub, usually 1–3 m high, with stems dull brown, sparsely branched; twigs pubescent with red brown hairs. Leaves commonly with 7–11 leaflets, 2–4 cm long, serrate above the middle. Inflorescence small, round-topped; fruit orange, red, or purplish. Woods and openings in mountains; southern Rocky Mountains.

**Spiraea**        meadowsweet

Shrubs with simple leaves and no stipules. Flowers perfect, with 5 sepals and 5 petals; fruit a follicle opening along one side.

1. Inflorescence flat-topped. ................................................................................ 2
   Inflorescence elongate. .................................................................... *S. alba*
2. Petals pink or roseate; inflorescence
   small, dense. ................................................................... *S. densiflora*
   Petals white; inflorescence broad, open. ...................... *S. lucida*

*Spiraea alba* Du Roi              NARROW-LEAVED MEADOWSWEET

An erect shrub 50–100 cm high, with brown twigs and narrowly oblanceolate leaves, pointed at both ends, 3–6 cm long, sharply toothed, and sometimes with fine hairs beneath on the veins. Inflorescence finely hairy in dense terminal panicles; flowers small and white. Fairly common; Parklands and Boreal forest.

*Spiraea densiflora* Nutt.                            PINK MEADOWSWEET

A slender shrub 50–150 cm high, with branches ascending, reddish brown. Leaves oval to elliptic, 15–30 mm long, rounded at base and tip, crenate or serrate above the middle, glabrous or nearly so. Inflorescence 2–5 cm broad, dense; flowers showy, very sweet-scented, with petals about 1.5 mm long. Moist thickets and meadows; southern Rocky Mountains.

*Spiraea lucida* Dougl.                    SHINING-LEAVED MEADOWSWEET

A low shrub 30–100 cm high; stems and branches erect, usually dying down to close to the base annually. Leaves 2–5 cm long, obovate to oval, on short stalks, shiny green above, paler below. Flowers white, small, in a flat-topped panicle. Common; in the Cypress Hills and Rocky Mountains.

# LEGUMINOSAE—pea family

Shrubs or herbs with alternate compound leaves with stipules. Flowers (*see* Fig. 7) perfect and irregular, having 5 more or less united sepals and 5 petals. The upper petal, larger than the others, is called the standard; the two side petals are called wings, the two lower ones, united, form the keel. The pistil and stamens are contained in the keel. The stamens, usually 10, are arranged in various groupings, their arrangement being one of the characters used in separating the genera of this large family. They can all be distinct and separate, or all joined in one group, or diadelphous (separated into two groups of 9 and 1). The fruit is either a 1- or 2-compartment pod (legume) or a pod constricted between the seeds (loment).

1. Shrubs. ................................................................................................................. 2
   Herbs. .................................................................................................................. 3
2. Flowers yellow; leaves pinnate, leaflets even-numbered, the terminal leaflet lacking. .......................................................... *Caragana*
   Flowers purple; leaves pinnate, leaflets odd-numbered, the terminal leaflet present. ............................................................. *Amorpha*
3. The terminal leaflet replaced by a tendril. .................................................. 4
   The terminal leaflet normal. ......................................................................... 5
4. Styles filiform, bearded at the tip; wings of flower coherent with the keel. ............................................. *Vicia*
   Styles flattened, bearded along the inside; wings free from the keel. ................................................. *Lathyrus*
5. Leaves palmately divided. ............................................................................ 6
   Leaves pinnately divided. ............................................................................ 16
6. Plants very low cushion plants. ................................................................... 7
   Plants not cushion plants. ............................................................................ 8
7. Flowers white, 2 cm long. ........................................... *Astragalus gilviflorus*
   Flowers purple, 6–8 mm long. .................................... *Astragalus spatulatus*
8. Leaves glandular-dotted. .......................................................... *Psoralea*
   Leaves not glandular-dotted. ........................................................................ 9
9. Leaflets clearly toothed. ............................................................................. 10
   Leaflets entire. ............................................................................................. 12
10. Pods strongly curved or coiled. ............................................... *Medicago*
    Pods straight or almost straight. ............................................................... 11
11. Flowers in long spike-like racemes; the deciduous petals free from the stamen tube. ...................................................................... *Melilotus*
    Flowers in short loose or dense heads; the petals more or less adhering to the stamen tube. ................................................................. *Trifolium*
12. Stems twining on vegetation; upper flowers with petals, the lower ones apetalous. ................................................................ *Amphicarpa*

Stems not twining; flowers all alike. .......................................................... 13

13. Pods usually with 3–5 one-seeded segments,
    which separate at maturity. ...................................................... *Desmodium*
    Pods not segmented. ........................................................................ 14

14. Flowers about 4 mm long, usually solitary
    in the leaf axils. ............................................................................. *Lotus*
    Flowers 10–20 mm long, in spikes or racemes. ............................................ 15

15. Plants with rootstocks; flowers yellow;
    pods 4–7 cm long, sickle-shaped. .............................................. *Thermopsis*
    Plants tufted; flowers purple or yellow;
    pods 2–4 cm long, straight. ...................................................... *Lupinus*

16. Pods segmented or jointed, breaking up at maturity. ..................................... 17
    Pods not segmented or jointed. ........................................................... 18

17. Inflorescence a loose few-flowered head-like umbel. ............................. *Coronilla*
    Inflorescence an elongated raceme. ............................................... *Hedysarum*

18. Leaves glandular-dotted. ................................................................. 19
    Leaves not glandular. ..................................................................... 20

19. Pods with hooked prickles; flowers 10–15 mm long. ......................... *Glycyrrhiza*
    Pods not prickly; flowers small
    to 5 mm; stamens 5. ........................................................... *Petalostemon*
    Pods not prickly; flowers 10–12 mm
    long, purple with yellowish base; stamens 10. ....................... *Oxytropis viscida*

20. Keel tipped with a point. ............................................................. *Oxytropis*
    Keel not tipped with a point. ....................................................... *Astragalus*

### *Amorpha*      false indigo

Shrubs with pinnate leaves, bearing odd number of glandular-dotted leaflets; midrib projecting from end of leaflets. Flowers having the standard petal only, other petals missing. Flowers borne in long narrow spikes. Fruit a short, oblong, curved, 1- or 2-seeded pod.

1. Plants densely pubescent. ...................................................... *A. canescens*
   Plants sparsely pubescent to glabrous. ............................................... 2

2. Low shrubs; leaflets to 1 cm long. ................................................. *A. nana*
   Tall shrubs; leaflets to 3 cm long. .............................................. *A. fruticosa*

### *Amorpha canescens* Pursh                               LEADPLANT

A bushy shrub, densely white hairy, 30–100 cm high. Leaves very dense, 5–10 cm long, with 21–51 oval leaflets, each 8–12 mm long. Flowers in dense, clustered, spike-like racemes 5–15 cm long, small, with a bluish purple petal. Pods single-seeded, about 3–5 mm long. Found occasionally on dry prairies; southeastern Parklands.

### *Amorpha fruticosa* L.                                FALSE INDIGO

A fairly tall shrub 2–5 m high, with stalked leaves 15–40 cm long and 11–25 oval leaflets 10–30 mm long. The spike-like racemes of flowers either clustered or borne singly, 7–15 cm long, with a violet purple petal. Found occasionally on riverbanks; southeastern Parklands.

*Amorpha nana* Nutt.                                                    DWARF FALSE INDIGO

A low shrub, usually less than 30 cm high, almost hairless, with numerous leaves 2–8 cm long, each having 13–31 stiff elliptic leaflets 6–12 mm long. Spike-like racemes of flowers borne singly, 2–7 cm long, fragrant with a purplish petal. Not common; moist prairie and waste places; southeastern Parklands.

**Amphicarpa**      hog-peanut

*Amphicarpa bracteata* (L.) Fern.                                       HOG-PEANUT

Twining plants with stems to 1 m long, pubescent with retrorse appressed hairs. Leaflets thin, somewhat pubescent, with the terminal one 2–6 cm long. Racemes with 2–10 flowers, white to pale purplish; pods sparsely pubescent, mainly at maturity, those of the upper flowers 3-seeded, those of the basal ones 1-seeded and often underground. Not common; woods along rivers and in ravines; southeastern Parklands and Boreal forest.

**Astragalus**      milk-vetch

Perennial herbs with pinnate leaves of 3 to many leaflets, and perfect irregular flowers having a blunt keel. Stamens 10, diadelphous (9 being united in a bundle and 1 separate). Flowers in spikes or racemes. Fruit a pod, sometimes only 1-celled and often appearing 2-celled because of the ingrowth of one or both of the joints between the halves of the pods. *Astragalus* is of considerable importance in the Prairie Provinces, most of the species occurring in the native grasslands and woodlands, where they are accessible to grazing livestock. Several species are **poisonous** or potentially poisonous to livestock, and although poisoning by these species seems to be rare, in periods of drought or under other unfavorable conditions they can be dangerous. In the Rocky Mountains, *A. miser*, and possibly some other species, produce an alkaloid that causes partial paralysis, which has led to the name "locoweed," because of the erratic movements of affected animals. Other species, such as *A. pectinatus* and *A. bisulcatus*, can accumulate the element selenium, which can cause sickness or death when eaten.

Some authors have divided the genus *Astragalus* into several small genera, or sections, by using the characters of the legumes. The legumes (or pods) vary from short to long, compressed to inflated, papery to woody, 1-celled to 2-celled, and few to many seeds. The inflorescence is a raceme with few to many flowers more or less long peduncled, and axillary in the leaf axils.

3. Flowers white, borne in the crown of the plant; leaves with 3 leaflets, densely villous. ........................................................................................ *A. gilviflorus*

   Flowers purplish or yellowish, borne on more or less well developed peduncles. ................................................................... 4

4. Leaves usually with a single leaflet, occasionally with 3 or 5 leaflets; flowers purple. .................................................... *A. spatulatus*

   Leaves with at least 7 leaflets. ....................................... 5

5. Flowers yellowish, with or without purplish tips. ................................................................ 6

   Flowers purple or blue, occasionally whitish. ............................................................... 8

6. Plants cushion-like; leaves silky villous, tomentose; flowers about 2 cm long. ........................... *A. purshii*

   Plants not cushion-like; stems spreading. ......................... 7

7. Plants large; flowers about 2 cm long; leaves with 17–25 leaflets; pods glabrous. ................................................... *A. crassicarpus*

   Plants small; flowers about 1 cm long; leaves with 7–15 leaflets; pods densely pubescent. .......................................................... *A. lotiflorus*

8. Plants with taproots and caudex. .................................. 9

   Plants with rootstalks. ................................................. 11

9. Flowers about 2 cm long; leaves silvery pubescent. .............................................. *A. missouriensis*

   Flowers about 1 cm long; leaves not silvery pubescent. .............................................. 10

10. Stems long, few-leafed; leaflets 15–21; inflorescence elongating at maturity; pods cylindric. ........................................... *A. flexuosus*

    Stems short, densely leafed; leaflets 7–11; inflorescence not elongating; pods flattened. ....................................... *A. vexilliflexus*

11. Inflorescence a dense, head-like raceme; flowers and pods erect; pods densely pubescent. ............................................ *A. danicus*

    Inflorescence a short, loosely flowered raceme; flowers spreading; pods reflexed, black hairy. ................................ *A. alpinus*

12. Flowers white, yellow, or greenish. ............................ 13

    Flowers at least partly reddish, purplish, or bluish. ........................................................... 21

13. Leaflets very narrowly linear. ........................... *A. pectinatus*

    Leaflets not narrowly linear. ....................................... 14

14. Plants densely hairy. ..................................... *A. drummondii*

    Plants not densely hairy. ............................................ 15

Pods drooping, 10–15 mm long; leaflets
13–23. ................................................................................................ *A. bourgovii*

27. Flowers in a dense raceme, somewhat
    reflexed; pods 2-grooved on the upper
    side. ............................................................................................ *A. bisulcatus*
    Flowers in loose racemes; pods not
    2-grooved. ................................................................................................ 28

28. Pods stipitate, 1.5–2 cm long, spreading or
    arched downward. ........................................................................ *A. robbinsii*
    Pods sessile, 1 cm long, drooping. ...................................... *A. eucosmus*


*Astragalus aboriginum* Richardson                INDIAN MILK-VETCH

Plants with a stout taproot, few- to many-stemmed from a branched caudex. Stems 20–40 cm long, ascending or erect, densely and finely pubescent
with spreading hairs. Leaves with 7–15 leaflets 1–3 cm long, elliptic to linear-
lanceolate, pubescent on both sides. Inflorescence 2–3 cm long in flower, elongating to 10–15 cm in fruit; flowers 6–10 mm long, whitish, bluish purple tipped; calyx 3–4 mm long, black pubescent. Legumes 1.5–2.5 cm long, straight
to somewhat falcate, flattened, glabrous to white pubescent, drooping, borne
on a stipe 5–8 mm long. An almost glabrous form has been named var. *major*
Gray. Gravel banks along rivers and on slopes; throughout the Prairie Provinces. Syn.: *A. aboriginorum* Richardson; *Atelophragma aboriginum* (Richardson) Rydb.


*Astragalus alpinus* L.                ALPINE MILK-VETCH

Plants with extensively creeping roots; stems 10–40 cm high, solitary or in
small tufts, slender, decumbent or ascending, glabrous or nearly so. Leaves
with 11–25 leaflets 1–2 cm long, oblong to oval, with the apex retuse or obtuse,
glabrous above or somewhat pubescent on both sides. Inflorescence 3–4 cm
long, at first dense, soon elongating to 5–8 cm in fruit; flowers 7–13 mm long,
light purplish blue to almost white; calyx 3–4 mm long, black pubescent.
Legume 8–12 mm long, straight or falcate, compressed, black pubescent,
reflexed, stipitate. The var. *brunetianus* Fern. is distinguishable by having a
short glabrescent calyx, 2–2.5 mm long, and a glabrescent pod. Forest margins, open and disturbed areas; Boreal forest and Rocky Mountains. Syn.:
*Atelophragma alpinum* (L.) Rydb.


*Astragalus bisulcatus* (Hook.) Gray (Fig. 132*A*)        TWO-GROOVED MILK-VETCH

Plants densely tufted, often many-stemmed, erect; stems stout, 30–80 cm
high, usually reddish to purplish, finely pubescent to glabrous. Leaves with
13–29 leaflets 1–3.5 cm long, oblong-elliptic, somewhat pubescent to glabrous.
Inflorescence 10–18 cm long, dense; flowers 10–15 mm long, reddish purple,
usually somewhat pendent; calyx about 4 mm long, inflated at base. Legume
18–22 mm long, pendent, linear oblong, the upper side deeply 2-grooved. Plant
with an unpleasant odor. Prairies and Parklands. Syn.: *Diholcos bisulcatus*
(Hook.) Rydb.

Fig. 132.   *A*, Two-grooved milk-vetch, *Astragalus bisulcatus* (Hook.) Gray; *B*, narrow-leaved milk-vetch, *Astragalus pectinatus* Dougl.

*Astragalus bourgovii* Gray

Plants densely tufted from a stout caudex. Stems erect or ascending, 10–30 cm high. Leaves with 13–25 leaflets 5–12 mm long, lanceolate to oblong-elliptic, pubescent on both sides. Inflorescence lax, loosely flowered; flowers 8–11 mm long, bluish purple or violet; calyx 4–5 mm long, black pubescent. Legume 1–1.5 cm long, oblong elliptic, flattened, black pubescent. Montane and alpine meadows; Rocky Mountains.

*Astragalus canadensis* L.                                    CANADIAN MILK-VETCH

Plants with creeping roots; stems single or a few together. Stems stout, 40–100 cm high, glabrous or thinly pubescent. Leaves with 13–27 leaflets 2–4.5 cm long, elliptic to oblong, glabrous to thinly pubescent. Inflorescence dense, 8–10 cm long; flowers greenish white to white, 1.0–1.5 cm long, ascending to spreading; calyx 4–5 mm long, strigose. Legume 1.2–1.6 cm long, woody, glabrous or nearly so, terete, sessile. Moist areas, woodland; throughout the Prairie Provinces.

*Astragalus cicer* L.                                            CICER MILK-VETCH

Plants with creeping roots; stems single, 40–60 cm high. Leaves with 23–33 leaflets 5–20 mm long, pubescent on both sides. Inflorescence dense, 4–6 cm long, black pubescent; flowers yellow, 10–15 mm long; calyx 6–8 mm long, black pubescent. Legume 10–15 mm long, ovoid to globose, inflated, thin-walled, black pubescent. An introduced species, rare as a weed; Manitoba, Alberta.

*Astragalus crassicarpus* Nutt.                                    GROUND-PLUM

Plants with a stout caudex; many-stemmed; stems procumbent, straggling, 20–35 cm long, finely pubescent. Leaves with 13–27 leaflets 8–20 mm long, oblong to linear, strigose below, glabrous above. Inflorescence 4–5 cm long, few-flowered; flowers 1.5–2.0 cm long, whitish with purplish tip or bluish purple; calyx 5–8 mm long, pubescent with black and white hairs. Legume subglobose, 1.5–2.5 cm in diam, deep reddish purple when ripe. Grasslands; Prairies and Parklands. Syn.: *A. caryocarpus* Ker; *Geoprumnon crassicarpum* (Nutt.) Rydb.

*Astragalus danicus* Retz.                                    PURPLE MILK-VETCH

Plants with slender creeping roots; stems single or tufted, slender, 10–30 cm high, sparsely pubescent to glabrous. Leaves with 11–21 leaflets 1–2 cm long, lanceolate to linear-oblong, sparsely pubescent on both sides. Inflorescence dense, 2–4 cm long; flowers 14–17 mm long, purplish, erect; calyx black pubescent, 5–6 mm long. Legume about 1 cm long, densely pubescent. A Eurasian steppe species. North American plants are considered to be var. *dasyglottis* (Fisch.) Boiv. A white-flowered form is f. *virgultulus* (Sheld.) Boiv. Grasslands; Prairies and Parklands. Syn.: *A. goniatus* Nutt.; *A. hypoglottis* Richardson.

*Astragalus drummondii* Dougl.                                 DRUMMOND'S MILK-VETCH

Plants with well-developed caudex; many-stemmed; stems 30–60 cm high, more or less densely pubescent with long, spreading hairs. Leaves with 23–33 leaflets 10–15 mm long, linear-oblong to elliptic, glabrous or nearly so above, densely soft pubescent below. Inflorescence 5–15 cm long, dense at first, soon elongating; flowers 1.5–2.0 cm long, white, pendent; calyx 5–6 mm long, pubescent with mixed white and black hairs. Legumes pendulous, 2–3 cm long, linear, glabrous, with the stipe 6–8 mm long. Grassland; western Prairies and Parklands. Syn.: *Tium drummondii* (Dougl.) Rydb.

### *Astragalus eucosmus* Robinson

Plants with creeping roots; stems solitary or in small tufts, 30–60 cm high, glabrous or nearly so. Leaves with 11–17 leaflets 8–25 mm long, elliptic to oblong, glabrous or nearly so above, sparsely pubescent on underside. Inflorescence at first dense, soon elongating to 10–15 cm long; flowers 7–10 mm long, purple to whitish purple or almost white; calyx 3–4 mm long, black pubescent. Legume 8–10 mm long, obliquely elliptic, reflexed, sessile, densely pubescent. The var. *eucosmus* has black hairy pods; plants with white hairy pods are f. *leucocarpus* Lepage. Open woods, riverbanks; Boreal forest and Rocky Mountains.

### *Astragalus flexuosus* Dougl.                    SLENDER MILK-VETCH

Plants with stout creeping roots; stems solitary or few together, 30–50 cm long, straggling, thinly pubescent. Leaves with 13–23 leaflets 5–15 mm long, linear to oblong, glabrous above, somewhat pubescent below. Inflorescence 5–10 cm long in flower, elongating to 15 cm in fruit; flowers white, tipped purplish to reddish purple; calyx 3–3.5 mm long, pubescent. Legume 1–2 cm long, almost terete, linear, spreading or reflexed, with the stipe shorter than the calyx tube. Prairies and Parklands throughout the Prairie Provinces. Syn.: *Pisophaca flexuosa* (Dougl.) Rydb.

### *Astragalus frigidus* (L.) Gray                    AMERICAN MILK-VETCH

Plants with a stout caudex; stems usually several, erect, seldom branching, glabrous or nearly so. Leaves with 9–15 leaflets 2–5 cm long, oblong-elliptic to elliptic, glabrous above, somewhat pubescent below. Inflorescence 5–7 cm long in flower, elongating to 10 cm in fruit; flowers 13–15 mm long, yellowish white, at first ascending, later reflexed; calyx oblique, about 5 mm long, with very short teeth. Legume 1.5–2.0 cm long, ellipsoid, inflated, pendent, with the stipe about 5 mm long. In the var. *frigidus* of Eurasia, the legume is at first black pubescent, and later glabrescent. In Canada, the legume is glabrous, or very sparsely pubescent; the plants are distinguished as var. *americanus* (Hook.) Wats. (Fig. 133). Moist woods, riverbanks, and openings; Boreal forest, Rocky Mountains. Syn.: *A. americanus* (Hook.) M. E. Jones; *Phaca americana* (Hook.) Rydb.

### *Astragalus gilviflorus* Sheld.                    CUSHION MILK-VETCH

Plants with a caudex and deep taproot, stemless, forming a dense cushion. Leaves with 3 leaflets 1–2.5 cm long, densely silvery silky villous on both sides. Inflorescence 1 or 2 flowers, subsessile at base of leaves; flowers white, with purplish-spotted keel, 1.5–3.0 cm long; calyx 15 mm long, densely pubescent. Legume about 15 mm long, oblong-ovate, terete, silky pubescent. Dry eroded hills, slopes, and disturbed areas; western Prairies. Syn.: *A. triphyllus* Pursh.; *Orophaca caespitosa* (Nutt.) Britt.

Fig. 133.   American milk-vetch, *Astragalus frigidus* (L.) Gray var. *americanus* (Hook.) Wats.

*Astragalus iochrous* Barneby (Fig. 134)                    AUSTRIAN FIELD-PEA

Plants with long creeping roots; stems solitary or few together, 40–90 cm high. Leaves with 11–17 leaflets 5–25 mm long, oblong to elliptic, glaucous green. Inflorescence 5–10 cm long, loosely flowered; flowers 10–15 mm long, brick red; calyx 3–5 mm long. Legume 15–25 mm long, ovoid, glabrous, with the stipe about twice as long as the calyx. An introduced species, rare; known from Maple Creek, Sask. Syn.: *Swainsona salsula* (Pall.) Taub.

*Astragalus kentrophyta* Gray                    PRICKLY MILK-VETCH

Plants with a deep taproot; stems pubescent, straggling. Leaves 2–3 cm long; leaflets 1–2 cm long, tipped with a sharp spine. Inflorescence 1–2 cm long, almost hidden among the leaves; flowers 4–5 mm long, white with purplish tinge; calyx 2–3 mm long, pubescent. Legume about 5 mm long, ovoid, somewhat compressed. Rare; sand dunes and other sandy areas; southwestern Prairies. Syn.: *Kentrophyta montana* Nutt.

*Astragalus lotiflorus* Hook.                    LOW MILK-VETCH

Plants with a deep taproot; stems erect, pubescent, 10–15 cm high. Leaves with 9–13 leaflets 8–15 mm long, elliptic to oblong, thinly pubescent above, densely pubescent below. Inflorescence 2–3 cm long; flowers 8–10 mm long, yellow white, often purplish-tipped; calyx 3–3.5 mm long, densely pubescent. Legumes 1.5–2.0 cm long, long-pointed, densely pubescent, usually sessile amongst the leaf bases, sometimes on stems about 5 cm long. Not common; grassland and openings on light soils; Prairies and Parklands. Syn.: *Batidophaca lotiflora* (Hook.) Rydb.

*Astragalus miser* Dougl.                    TIMBER MILK-VETCH

Plants with several stems arising from caudex. Stems slender, 5–20 cm high, ascending, somewhat pubescent or glabrous. Leaves with 9–15 leaflets 3–10 mm long, linear to linear-oblong, pubescent on both sides. Inflorescence 6–10 cm long; flowers 8–10 mm long, white to yellowish, purple-tipped; calyx 3–4 mm long, pubescent with mixed white and black hairs. Legume 15–20 mm long, linear, subsessile, drooping, somewhat pubescent. Rare; in Rocky Mountains. The var. *serotinus* (Gray) Barneby, with the calyx 2–2.5 mm long, flowers 6–8 mm long, purple. More common; open woods, slopes; Rocky Mountains. Syn.: *Homalobus serotinus* (A. Gray) Rydb.; *A. serotinus* A. Gray.

*Astragalus missouriensis* Nutt.                    MISSOURI MILK-VETCH

Plants with a deep stout taproot and caudex. Stems tufted, to 20 cm long, usually ascending, more or less densely gray or silvery pubescent. Leaves with 9–21 leaflets 5–15 mm long, elliptic to obovate, densely silvery pubescent. Inflorescence 1.5–5 cm long, dense at first, and elongating in fruit; flowers 15–20 mm long, deep reddish purple to bluish purple; calyx 7–8 mm long, silvery pubescent. Legume 2.5–3 cm long, oblong, pubescent, erect. Slopes and eroded hillsides; Prairies. Syn.: *Xylophacos missouriensis* (Nutt.) Rydb.

*Astragalus pectinatus* Dougl. (Fig. 132*B*)                    NARROW-LEAVED MILK-VETCH

Plants with a deep taproot, and few-stemmed caudex. Stems 20–50 cm long, decumbent to ascending, glabrous, often reddish at base. Leaves with

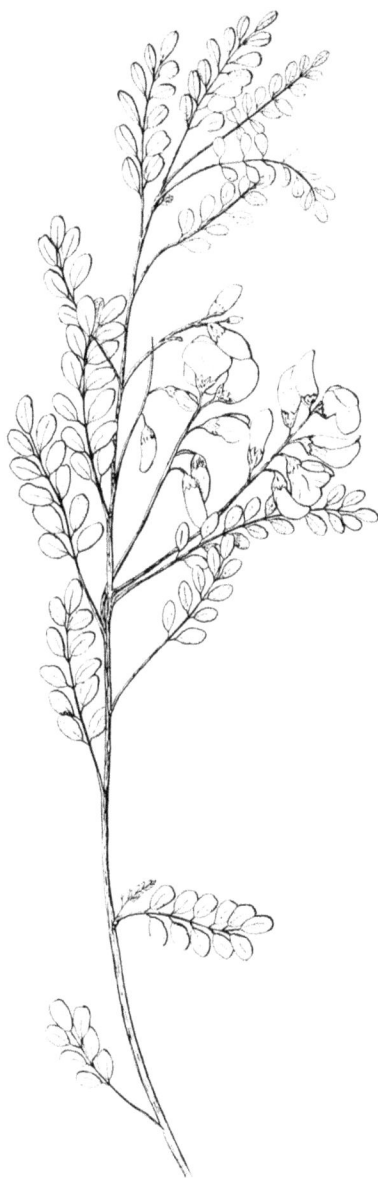

Fig. 134. Austrian field-pea, *Astragalus iochrous* Barneby.

9-17 leaflets 2-6 cm long, narrowly linear, 1-2 mm wide, sparingly pubescent. Inflorescence 5-8 cm long; flowers 1.5-2.5 cm long, white to yellowish; calyx 7-8 mm long. Legume 10-20 mm long, oblong, terete, glabrous, fleshy at first, becoming woody when dry. Grasslands and slopes; Prairies. Syn.: *Cnemidophacos pectinatus* (Dougl.) Rydb.

*Astragalus purshii* Dougl.                                    PURSH'S MILK-VETCH

Cushion-like plants with a deep taproot and caudex. Stems 5-8 cm long, densely pubescent. Leaves with 9-15 leaflets 8-12 mm long, oblanceolate to ovate, densely long pubescent. Inflorescence 2-3 cm long; flowers 2-2.5 cm long, yellowish white, purple-tipped; calyx to 10 mm long, densely pubescent. Legume 15-20 mm long, ovoid, densely long pubescent. Rare; dry prairie slopes and eroded hills; western Prairies. Syn.: *Xylophacos purshii* (Dougl.) Rydb.

*Astragalus racemosus* Pursh                              RACEMOSE MILK-VETCH

Plant several-stemmed from a stout taproot and caudex. Stems 40-80 cm high, sparingly pubescent. Leaves with 15-27 leaflets 1-2 cm long, glabrous above, somewhat pubescent below, linear to linear-oblong. Inflorescence 5-7 cm in flower, elongating to 10 cm in fruit; flowers 15-20 mm long, yellowish white; calyx 2.5-3 mm long. Legume 1.5-2.5 cm long, glabrous, triangular in cross section, drooping, with the stipe 5-7 mm long. Rare; dry prairie slopes and hillsides; Saskatchewan. Syn.: *Tium racemosum* (Pursh) Rydb.

*Astragalus robbinsii* (Oakes) Gray

Plants with several stems arising from caudex. Stems 30-50 cm high, erect, pubescent. Leaves with 9-17 leaflets 1.5-2.5 cm long, elliptic to oval or oblong, glabrous above, pubescent below. Inflorescence 5-10 cm long in flower, elongating to 15 cm in fruit; flowers 8-10 mm long, bluish purple to almost white; calyx 3-4 mm long, black hairy. Legume 15-20 mm long, flattened, black pubescent, drooping, stipitate. Riverbanks, lakeshores; Rocky Mountains.

*Astragalus spatulatus* Sheld.                              TUFTED MILK-VETCH

Cushion-like plant with a deep taproot and caudex. Stems 5-7 cm long, silvery gray pubescent. Leaves usually with a single leaflet, occasionally 3 leaflets 1-4 cm long, oblong to linear oblong, densely silvery gray pubescent. Inflorescence 2-3 cm long in flower, the peduncle elongating to 7-8 cm in fruit; flowers 6-8 mm long, bluish purple; calyx 2-2.5 mm long, pubescent. Legume 6-12 mm long, flattened, elliptic, glabrous, erect. Locally moderately common; eroded hillsides and slopes; western Prairies. Syn.: *A. caespitosus* (Nutt.) Gray; *Homalobus caespitosus* Nutt.

*Astragalus striatus* Nutt.                        ASCENDING PURPLE MILK-VETCH

Plants with several to many stems arising from a stout taproot. Stems decumbent or ascending, 15-40 cm high, pubescent. Leaves with 9-19 leaflets 1-2 cm long, elliptic to oblong, pubescent, with the hairs attached in the middle. Inflorescence 4-5 cm long, dense; flowers 12-15 mm long, purplish, rarely white; calyx 5 mm long, pubescent with black and white hairs. Legume 7-15 mm long, ovoid, pubescent. Open grasslands and slopes; Prairies and Parklands. Syn.: *A. adsurgens* Hook.

*Astragalus tenellus* Pursh                    LOOSE-FLOWERED MILK-VETCH
   Plants with few to several stems from a stout caudex. Stems 20–50 cm high, somewhat pubescent, erect. Leaves with 11–21 leaflets 8–12 mm long, linear to linear-oblong, glabrous above, somewhat pubescent below. Inflorescence 4–5 cm long in flower, elongating to 7 cm in fruit; flowers 8–10 mm long, whitish; calyx 2–2.5 mm long, pubescent. Legume 8–12 mm long, flattened, glabrous, drooping, stipitate. Coulees, forest margins, lakeshores; Prairies and Parklands. Syn.: *Homalobus tenellus* (Pursh) Britt.

*Astragalus vexilliflexus* Sheld.                    FEW-FLOWERED MILK-VETCH
   Plants matted, many-stemmed from a deep taproot and caudex. Stems 10–30 cm long, straggling, pubescent. Leaves with 7–11 leaflets 5–18 mm long, oblong to oblong-lanceolate, pubescent. Inflorescence 1.5–3.0 cm long; flowers 6–8 mm long, purplish, occasionally whitish; calyx 2 mm long, pubescent. Legume 7–10 mm long, somewhat pubescent, flattened, sessile. Rare, locally moderately common; eroded slopes and hills; western Prairies, Rocky Mountains. Syn.: *Homalobus vexilliflexus* (Sheld.) Rydb.

*Astragalus yukonis* M. E. Jones
   Plants with several to many stems arising from a stout taproot and caudex. Stems 5–30 cm high, slender, decumbent to ascending, glabrous. Leaves with 7–15 leaflets 4–12 mm long, oblong, apex often retuse, glabrous above, pubescent below. Inflorescence 2–4 cm long on a long peduncle; flowers 7–10 mm long, bluish- or purple-tipped; calyx 2–2.5 mm long, black pubescent. Legume 5–10 mm long, ellipsoid, black pubescent, erect. Grassy openings; Boreal forest.

**Caragana**      Siberian peatree

*Caragana arborescens* Lam.                    COMMON CARAGANA
   A bush 3–4 m high; leaves pinnate with 8–12 pale green leaflets 10–25 mm long; the leaf stem ending in a short spine. Flowers bright yellow, 15–25 mm long, followed by dark brown linear pods 3–5 cm long. Introduced as a hedge and ornamental plant from Siberia and now established in many places.

**Coronilla**      crown-vetch

*Coronilla varia* L.                    FIELD CROWN-VETCH
   Perennial plants with stems 20–100 cm high. Leaves with 6–10 pairs of oblong or elliptic leaflets 6–20 mm long. Heads mostly 10- to 20-flowered; corolla 10–15 mm long, white to purplish; pods 20–60 mm long, with 3–7 joints, 4-angled. Introduced and occasionally spreading from cultivation; southeastern Parklands.

***Desmodium***      tick-trefoil

*Desmodium canadense* (L.) DC.      BEGGAR'S-LICE

Plants perennial, with erect pubescent stems, mostly about 1 m, occasionally to 2 m high. Leaves trifoliate, with the leaflets oblong to oblong-lanceolate, the terminal one petioled, 5–9 cm long; racemes densely flowered; flowers 10–15 mm long, purplish; legume 2–3 cm long, with 3–5 joints. Rare; moist open areas; southeastern Parklands and Boreal forest.

***Glycyrrhiza***      wild licorice

*Glycyrrhiza lepidota* (Nutt.) Pursh (Fig. 135)      WILD LICORICE

A coarse erect branching herb 30–100 cm high, from a thick sweet-tasting rootstock having a slight licorice flavor. Leaves of 11–19 lanceolate or oblong leaflets, pale green, glandular-dotted, 20–35 mm long, and pointed at both ends. Flowers yellowish white, 10–15 mm long, and borne in rather dense racemes 2–6 cm long, arising from the axils of the leaves. Fruit in clusters of oblong reddish brown pods 10–15 cm long, densely covered with long hooked prickles, and containing several large seeds. Rootstocks were chewed by Indians. Palatable as hay, but seldom grazed. Very common in southern portion, but becoming scarcer toward the north; found in low spots on the prairie, slough margins, riverbanks, and coulees; throughout the Prairies and Parklands.

***Hedysarum***      sweet-broom

Perennial herbs with pinnate leaves; the flowers usually reflexed in long, spike-like racemes, and the pods flat and jointed or constricted between the seeds. The plants make good forage.

1. Calyx teeth very unequal, the upper ones nearly triangular, shorter than the calyx tube; veins of leaflets conspicuous. ............................................................................................................ 2

   Calyx teeth almost equal, linear, about as long as the calyx tube; veins of leaves not conspicuous. ........................................................................ *H. boreale*

2. Flowers violet or pale pink. ............................................ *H. alpinum* var. *americanum*

   Flowers sulfur yellow. ........................................................................ *H. sulphurescens*

*Hedysarum alpinum* L. var. *americanum* Michx. (Fig. 136)

AMERICAN HEDYSARUM

An erect plant 15–80 cm high, usually with few branches. Leaves of 11–21 oblong leaflets 10–30 mm long. Flowers on a long raceme, pinkish or violet, 10–15 mm long, usually pointing downward. Pods usually have 3–5 internodes or enlargements, hairless except perhaps on the margins. Common; in semi-open prairie and open woods; especially in the Boreal forest and the Cypress Hills. The most common species of the genus in the Prairie Provinces, and readily eaten by livestock. A variety with hairy pods has been found in the southwestern part of the Prairie Provinces. Syn.: *H. americanum* (Michx.) Britt.

Fig. 135.   Wild licorice, *Glycyrrhiza lepidota* (Nutt.) Pursh.

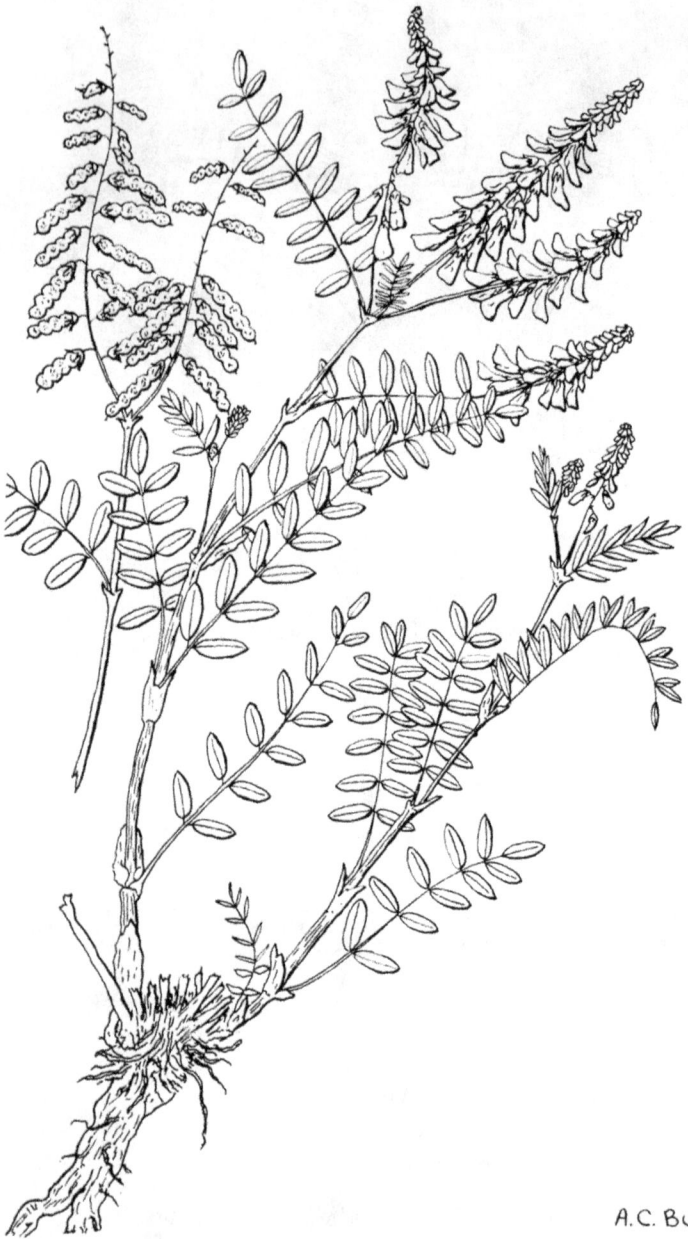

Fig. 136.   American hedysarum, *Hedysarum alpinum* L. var. *americanum* Michx.

*Hedysarum boreale* Nutt.                               NORTHERN HEDYSARUM

Somewhat similar to *H. alpinum* but the lobes of the calyx longer and very narrow and the veins on the leaflets not distinct. Leaflets almost smooth and flowers reddish purple and somewhat reflexed.

Three forms of this species occur:

1. Inflorescence short and dense; flowers to
     20 mm long; stipules whitish. ........................................................ var. *mackenzii*
   Inflorescence elongate and loose; flowers
     about 15 mm long; stipules brownish. ........................................................ 2
2. Leaflets glabrous or sparsely pubescent;
     pods rather smooth. ........................................................ var. *boreale*
   Leaflets densely silky hairy; pods rugose. ........................................ var. *cinerascens*

Both var. *boreale* and var. *cinerascens* (Rydb.) Rollins occur on slopes of ravines and coulees in Prairies; var. *cinerascens* in drier areas, var. *boreale* in moister situations and also occasionally in Parklands. The var. *mackenzii* (Richardson) C. L. Hitchc., Mackenzie's hedysarum, is fairly common in meadows and on slopes in Parklands and Boreal forest.

*Hedysarum sulphurescens* Rydb.                         YELLOW HEDYSARUM

An erect species 30–60 cm high; leaves with 11–15 oblong to oval leaflets 10–85 mm long. Flowers 12–15 mm long, sulfur yellow, in a long loosely-flowered raceme. A species readily distinguishable by the flower color. Open woods, slopes, and meadows; southern Rocky Mountains, rare farther north.

*Lathyrus*          vetchling

Perennial twining vines, dying to the ground each year and bearing pinnate leaves with a tendril taking the place of the terminal leaflet. The flowers perfect, the stamens diadelphous. Style hairy along its inner side, and somewhat flattened, distinguishing this genus from *Vicia* (the true vetch), with merely a tuft of hairs at the end of an unflattened style. Pods somewhat flattened containing several seeds. Very palatable, a valuable native fodder in many localities. Grazed out readily, but making good growth again when protected from livestock. The sweet pea, *Lathyrus odoratus* L., occurs occasionally as an escape, but is not persistent.

1. Leaves with only 2 leaflets. ........................................................ 2
   Leaves with more than 2 leaflets. ........................................................ 3
2. Stem clearly winged. ........................................................ *L. sativus*
   Stem wingless. ........................................................ *L. tuberosus*
3. Flowers creamy white or yellowish. ........................................ *L. ochroleucus*
   Flowers purplish. ........................................................ 4
4. Racemes with 15–20 or more flowers. ........................................ *L. venosus*
   Racemes with fewer than 15 flowers. ........................................ 5
5. Leaflets linear to linear-lanceolate; stem
     narrowly winged, at least above. ........................................ *L. palustris*
   Leaflets oblong to ovate; stem wingless. ........................................ *L. japonicus*

*Lathyrus japonicus* Willd. BEACH-PEA

A somewhat glaucous glabrous or pubescent perennial, usually straggling with stems to 1 m long. Leaves with 4–10 leaflets, often without a tendril. Racemes 2- to 12-flowered; corolla 15–25 mm long, purplish; legume 3–5 cm long, with 4–10 seeds. Not common; lakeshores and beaches; Hudson Bay, Lake Winnipeg, Lake Manitoba. The typical form has 2- to 7-flowered racemes, and a pubescent calyx; plants with 5- to 12-flowered racemes and glabrous calyx distinguishable as subsp. *maritimus* (L.) R. W. Ball.

*Lathyrus ochroleucus* Hook. (Fig. 137) CREAM-COLORED VETCHLING

A slender smooth climber up to 1 m long, with a somewhat angled stem. Stipules large and almost cordate. Leaves of 6–10 broad oval leaflets 2–5 cm long. Flowers cream-colored, about 15 mm long, in racemes of 5–10 flowers. Pods about 4 cm long. Very common; in bluffs, open woodlands, and among bushes; throughout the Prairie Provinces.

*Lathyrus palustris* L. MARSH VETCHLING

A smooth climber 30–100 cm long, with a somewhat winged stem and leaves of 4–8 linear or linear-oblong leaflets 10–65 mm long. Stipules small and almost linear. Flowers purple, 10–15 mm long, 2–8 in each raceme. Pods 3–5 cm long. Fairly common; in moist places and damp woodlands; Boreal forest.

*Lathyrus sativus* L. CHICKEN VETCH

Annual plants with winged stems to 1 m high; leaflets 2–15 cm long, linear-lanceolate. Flowers solitary, white, pink, or bluish; legume 2–4 cm long, 1–2 cm wide. Introduced, occasionally sown for fodder, and rarely escaped.

*Lathyrus tuberosus* L. TUBEROUS VETCHLING

Perennial with a tuberous root; stems 30–100 cm high, glabrous or subglabrous. Leaflets 15–45 mm long, elliptic to oblong. Racemes 2- to 7-flowered; corolla 12–20 mm long, bright reddish purple; legume 2–4 cm long. Introduced, occasionally sown for fodder, and spreading from cultivation.

*Lathyrus venosus* Muhl. WILD PEAVINE

A climbing plant 50–100 cm long, sometimes somewhat finely hairy. Stems strongly 4-angled. Leaves having 8–12 oblong-ovate blunt-tipped leaflets 2–5 cm long. Flowers purple, 10–18 mm long, 15–20 in each raceme. Pods 3–5 cm long and veiny. This wild peavine provided a valuable source of forage and hay in the earlier days of settlement of northern bushlands. Common; around bushes and woodlands; eastern Parklands and Boreal forest. The pubescent form is var. *intonsus* Butt. & St. John.

**Lotus**    trefoil

Annual or perennial herbs; leaves pinnate, with odd-numbered leaflets. Flowers solitary or in heads; legume cylindric. Mostly Eurasian; several species useful as forage crops, two of these introduced in Canada.

Fig. 137. Cream-colored vetchling, *Lathyrus ochroleucus* Hook.

1. Plants annual; flowers solitary in upper leaf axils. ................................................................................ *L. americanus*

   Plants perennial; flowers in umbellate heads. ...................................................................................... 2

2. Stems solid; calyx lobes incurved in bud, 1.5–2.0 mm long; leaflets 5–15 mm long. ................................................................................ *L. corniculatus*

   Stems hollow; calyx lobes recurved in bud, 2–4 mm long; leaflets 10–25 mm long. ................................................................................ *L. pedunculatus*

### *Lotus americanus* (Nutt.) Bisch.      SPANISH CLOVER

Plants 10–40 cm high, with stems erect, branched or simple, pubescent. Leaves subsessile, 3-foliate, 1–2 cm long, with the terminal one stalked. Flowers pinkish, 5–7 mm long; pods 2–3 cm long, deflexed. Not common; dry to moist grasslands; southeastern Parklands. Syn.: *Trigonella americana* Nutt.; *Lotus purshianus* (Benth.) Clem. & Clem.

### *Lotus corniculatus* L.      BIRD'S-FOOT TREFOIL

Stems prostrate, erect or ascending, to 60 cm long, solid or nearly so. Inflorescence 1- to 5-flowered; flowers yellow, often tinged with red; pods 2–4 cm long. Introduced as forage crop and locally established.

### *Lotus pedunculatus* Cav.      TREFOIL

Resembling *L. corniculatus*, but the stem with a large cavity, the inflorescence 5- to 10-flowered, and the leaflets usually larger. Legume 2–4 cm long. Probably introduced with *L. corniculatus*; rare as an escape.

### *Lupinus*      lupine

Annual or perennial showy herbs with alternate palmate leaves, each bearing 5–12 leaflets. Flowers perfect and borne in terminal racemes. Ten stamens united into one bundle with the anthers alternately elongated and short. Pods flattened, consisting of 2 cells, and containing 1–6 seeds. Some species of lupines **poisonous** to stock, particularly sheep; the seedpods apparently being the most dangerous part of the plants.

1. Plants annual; stem and leaves densely pilose to hirsute; flowers pale blue to white. ................................................................................ *L. pusillus*

   Plants perennial. ................................................................................ 2

2. Standard (or banner) pubescent on the back over most of its surface. ................................................................................ *L. sericeus*

   Standard glabrous or with a few cilia. ................................................................................ 3

3. Leaflets sericeous above; plants silky-sericeous, up to 20 cm high; the leaves mostly basal. ................................................................................ *L. minimus*

   Leaflets glabrous above, or occasionally sparsely puberulent or strigose; plants more than 20 cm high. ................................................................................ 4

4. Keel densely ciliate along most of its
   upper edge; petioles usually less than
   twice the length of the leaflets. .......................................................... *L. nootkatensis*

   Keel ciliate only toward the tip, or glab-
   rous; petioles of various lengths. ........................................................................... 5

5. Leaves mostly basal, long-petioled;
   leaflets 10–17, glabrous above, sparsely
   strigose below, the largest to 12 cm
   long, to 25 mm wide. .......................................................... *L. polyphyllus*

   Leaves mostly cauline, short-petioled;
   leaflets 7–9, glabrous above, sericeous
   below, the largest to 4.5 cm long, to 5
   mm wide. ................................................................................... *L. argenteus*

*Lupinus argenteus* Pursh (Fig. 138)                    SILVERY LUPINE

A rather shrubby much-branched herb 30–60 cm high, with stems covered
with appressed silky hairs. Leaves of 6–9 narrowly oblanceolate leaflets 2–5 cm
long, sometimes silvery hairy or smooth above. Flowers varying from light violet or purplish to almost white, in long terminal racemes. Pods densely silky
hairy, 15–25 mm long, containing up to 5 seeds. Plentiful; on submontane
prairie; southern Rocky Mountains, Cypress Hills, and Wood Mountain.

*Lupinus minimus* Dougl.                                ALPINE LUPINE

Stems erect to decumbent, unbranched, 15–20 cm high, silky pubescent,
with 1 or 2 stem leaves. Basal leaves long-petioled, with 5–9 leaflets 2–3 cm
long. Very rare; alpine slopes; southern Rocky Mountains.

*Lupinus nootkatensis* Donn                             NOOTKA LUPINE

Plants 20–60 cm high, with stems erect, densely villose pubescent. Leaves
with 7–15 leaflets; the petioles 1.5–2 times as long as the leaflets. Racemes to
20 cm long; flowers 12–16 mm long. Pods loosely pubescent, 3–4 cm long. Not
common; moist alpine meadows; southern Rocky Mountains.

*Lupinus polyphyllus* Lindl.                            LARGE-LEAFED LUPINE

Plants 50–120 cm high, stout-stemmed, appressed to spreading pubescent.
Leaves long-petioled, with up to 17 leaflets. Inflorescence 15–30 cm long,
rather loosely flowered; petals blue to reddish or yellowish. Pods pubescent,
3–5 cm long. Not common; grassland, open woods, and meadows; southern
Rocky Mountains.

*Lupinus pusillus* Pursh                                SMALL LUPINE

A low-growing annual plant 10–25 cm high, with decumbent branches
and very hairy. Leaves usually of 5 oblong leaflets, smooth above but with
long hairs beneath, usually rounded at the ends, 20–35 mm long. Flowers
tinged with purple or rose, sometimes almost white, in short dense racemes on
very short stalks. Pods about 2 cm long with 1 or 2 seeds and somewhat constricted between the seeds. Found locally in sandhills and among sand dunes,
but not common.

Fig. 138.   Silvery lupine, *Lupinus argenteus* Pursh.

*Lupinus sericeus* Pursh                                   FLEXILE LUPINE

A branching plant 40–80 cm high, with grayish appressed silky hairs. Leaves of 6–10 narrowly oblanceolate leaflets 2–5 cm long, densely appressed silky hairy on both sides. Dark blue flowers in dense terminal spikes. Densely white hairy pods 20–35 mm long, containing 2–5 seeds. A very showy species. In grasslands and open woods; southern Rocky Mountains. Syn.: *L. flexuosus* Lindl.

## *Medicago*        medick

Annual or perennial plants, not native to the prairie, with trifoliolate leaves, toothed only beyond the center. Flowers perfect, borne in racemes or spikes; stamens diadelphous. Pods curved or spirally twisted.

1. Plants annual or biennial; flowers yellow;
   leaflets about 10 mm long, obovate. ........................................................................ 2

   Plants perennial; flowers yellow or some
   shade of purple; leaflets about 15 mm
   long, linear. ........................................................................ 3

2. Flowers in dense head-like racemes, 10-
   to 30-flowered; legume kidney-shaped,
   to 3 mm across, black. ........................................................ *M. lupulina*

   Flowers in a raceme with 1–5 flowers;
   legume spirally coiled, to 8 mm across,
   spiny. ........................................................ *M. polymorpha*

3. Flowers some shade of purple, sometimes
   whitish; legume coiled; corolla 7–11
   mm. ........................................................ *M. sativa* ssp. *sativa*

   Flowers yellow; legume falcate to almost
   straight; corolla 5–8 mm. ........................................................ *M. sativa* ssp. *falcata*

*Medicago lupulina* L.                                   BLACK MEDICK

A prostrate branched annual weed 10–60 cm across, with trifoliolate leaves. Leaflets obovate, toothed above the middle, 3–10 mm long. Flowers yellow, about 3 mm long, in dense head-like racemes, less than 1 cm long. Pods small and black, containing a single seed. Introduced; common in disturbed areas in Parklands and Boreal forest; not common in Prairie.

*Medicago polymorpha* L.                                   BUR-CLOVER

Annual plants 15–40 cm high, with glabrous or sparingly pubescent stems. Racemes of 1–8 yellow flowers 3–5 cm long. Pods 4–8 mm in diam, spirally coiled, shiny. Introduced, rarely weedy.

*Medicago sativa* L. (Fig. 139)                                   ALFALFA

A fairly erect perennial 30–80 cm high, much-branched with trifoliolate leaves. Leaflets 10–35 mm long, obovate, and sharply toothed toward the apex. Flowers 5–11 mm long, in a dense oblong raceme 10–45 mm long. Introduced as a fodder crop from Europe and now very common; along roadsides and waste places; throughout most of the Prairie Provinces.

A.C. Budd

Fig. 139.   Alfalfa, *Medicago sativa* L. ssp. *sativa*.

This species consists of several entities that hybridize quite freely and are often considered as separate species. Two of these are commonly used as forage, along with many cultivated varieties (cultivars) resulting from the breeding program.

Corolla blue to purple, 7–11 mm; pods coiled
in 1.5–3.5 turns. ............................................................................... ssp. *sativa*
Corolla yellow, 5–8 mm; pods straight to
sickle-shaped. ................................................................ ssp. *falcata* (L.) Arcangeli

Cultivars of the two subspecies have flower colors and pod shapes between those of the parents.

## *Melilotus*    sweet-clover

Annual or biennial introduced legumes with trifoliolate leaves; the leaflets toothed almost to the base. Flowers perfect, small, and in elongate, spike-like racemes. Pods short, thick, and straight with 1 or few seeds. Excellent for honey production and also as pasture for stock. Very common roadside weeds.

1. Flowers white. ....................................................................................................... 2
   Flowers yellow. ..................................................................................................... 3
2. Corolla 3–3.5 mm long, with pedicels 2–4
   mm long; racemes 5–10 cm long. ........................................................ *M. wolgica*
   Corolla 4–5 mm long, with pedicels 1–1.5
   mm long; racemes to 20 cm long. ........................................................... *M. alba*
3. Corolla 2–3 mm long; legume 1.5–3 mm
   long. ............................................................................................................ *M. indica*
   Corolla 4–7 mm long; legume 3–5 mm
   long. ..................................................................................................... *M. officinalis*

*Melilotus alba* Medic. (Fig. 140)                    WHITE SWEET-CLOVER
An erect plant 50–250 cm high, with palmately or pinnately trifoliolate leaves; the leaflets 10–25 mm long, toothed almost to the base. Flowers white, in long narrow spike-like racemes. Introduced as a forage plant from Europe and Asia.

*Melilotus indica* (L.) All.                          YELLOW SWEET-CLOVER
Erect plants to about 50 cm high. Racemes dense; flowers pale yellow. Occasionally cultivated, rare as a weed; southeastern Parklands.

*Melilotus officinalis* (L.) Pall.                    YELLOW SWEET-CLOVER
Similar to *M. alba*, but with yellow flowers. Introduced as a forage plant; common along roadsides and in disturbed areas.

*Melilotus wolgica* Poir.                             WHITE SWEET-CLOVER
Also similar to *M. alba*, but with longer pedicels, 2–4 mm, and smaller flowers. Introduced as a forage plant, and rarely escaped.

Fig. 140.   White sweet-clover, *Melilotus alba* Medic.

*Oxytropis*    locoweed

Perennial herbs, usually with no apparent stem (except in *O. deflexa*), pinnate leaves with odd-numbered leaflets, and perfect flowers borne in a spike or a raceme. Stamens diadelphous; keel extended to a protruding point, distinguishing this genus from *Astragalus*. Several species of this genus **poisonous** to livestock, causing well-known loco disease, affecting the nervous system.

1. Plant caulescent; stipules adnate to the base of the petiole only; pods pendulous. ............................................................ *O. deflexa*
   Plants acaulescent; stipules adnate for half their length; pods erect. ............................................................ 2

2. Leaflets whorled on the rachis or appearing so; few leaflets in pairs. ............................................................ 3
   Leaflets all paired or scattered on the rachis. ............................................................ 5

3. Inflorescence capitate, few-flowered. ............................................................ *O. bellii*
   Inflorescence elongate, many-flowered. ............................................................ 4

4. Corolla purple; plants silky pubescent. ............................................................ *O. splendens*
   Corolla yellow; plants appressed-pubescent. ............................................................ *O. campestris*

5. Corolla purple to bluish. ............................................................ 6
   Corolla yellow to whitish. ............................................................ 11

6. Pubescence in part composed of malpighian hairs. ............................................................ *O. lambertii*
   Pubescence entirely composed of basi-fixed hairs. ............................................................ 7

7. Plants more or less glandular-viscid. ............................................................ *O. viscida*
   Plants not glandular-viscid. ............................................................ 8

8. Pods ovoid, stipitate, strongly inflated. ............................................................ *O. podocarpa*
   Pods not ovoid, not strongly inflated. ............................................................ 9

9. Corolla small, the keel to 13 mm long. ............................................................ *O. campestris*
   Corolla large, the keel 15–25 mm long. ............................................................ 10

10. Plants very densely silky and spreading pubescent; bracts of the inflorescence linear-lanceolate, membranous with involute blades. ............................................................ *O. lagopus*
    Plants hispid-hirsute to subappressed-pilose; bracts of the inflorescence rhombic-lanceolate, herbaceous with flat blades. ............................................................ *O. besseyi*

11. Keel of the corolla to 13 mm long; leaves with 17–33 leaflets; pods not rigid at maturity. ............................................................ *O. campestris*
    Keel of the corolla to 20 mm long; leaves with 11–19 leaflets; pods rigid at maturity. ............................................................ *O. sericea*

*Oxytropis bellii* (Britt.) Palibine

Plants 10–15 cm high. Leaves with 17–35 leaflets 5–10 mm long, linear-lanceolate, acute, silky pubescent, verticillate, mostly 3 or 4 leaflets per verticil. Inflorescence with 4–7 flowers in a dense raceme; flowers 15–20 mm long, large, dark blue purple; calyx 4–5 mm long, villous. Legume 8–10 mm long, black pubescent, ovoid, erect. Very rare; Hudson Bay.

*Oxytropis besseyi* (Rydb.) Blank. (Fig. 141)          BESSEY'S LOCOWEED

A tufted silvery hairy plant 10–20 cm high. Leaflets 5–20 mm long. Inflorescence a short dense spike 2–5 cm high; flowers reddish purple. The plant is easily mistaken for Missouri milk-vetch. Pods about 2 cm long, covered with long silky hairs. This is a southern species, which has been found, but very rarely; on dry hillsides; Wood Mountain.

*Oxytropis campestris* (L.) DC.          LATE YELLOW LOCOWEED

Plants with a deep taproot and branching caudex, 15–40 cm high. Leaves with 7–33 leaflets 1–2 cm long, oblong-lanceolate, silky pubescent. Inflorescence 5–10 cm long; flowers 12–15 mm long, variable in color; calyx 5–7 mm long, pubescent with both white and black hairs. Legume 16–20 mm long, oblong-ovate, pubescent with both white and black hairs, semimembranous.

A circumboreal species, very variable. It has been divided into a large number of "small" species by various authors. North American plants may be considered as varieties of ssp. *gracilis* (Nelson) Boiv., in which the legumes lack a fully developed septum, whereas in the Eurasian ssp. *campestris* the septum is present. Some authorities consider all these forms as varieties of *O. campestris*.

1. Flowers bluish or purplish. .................................................................................... 2
   Flowers yellow or creamy white. ........................................................................... 3

2. Leaves of two sizes, the long ones about
      twice as long as the short ones; leaflets
      often in part verticillate. ................................................. var. *dispar* (Nels.) Barneby
   Leaves all about the same length; leaflets
      all paired. ............................................................................ var. *johannensis* Fern.

3. Stipules villous, often with club-shaped
      processes on margins. ............................................... var. *varians* (Rydb.) Barneby
   Stipules glabrous or glabrate, without
      club-shaped processes. .................................................................................... 4

4. Leaves with 17–33 leaflets; scapes 15–30
      cm; inflorescence many-flowered. ............................................... var. *gracilis*
   Leaves with 7–17 leaflets; scapes 1.5–15
      cm; inflorescence few-flowered. ......................... var. *cusickii* (Greenm.) Barneby

Of these forms, var. *gracilis* is common in moist grassland, open woods, and openings throughout the Prairie Provinces; var. *cusickii* occurs in alpine and subalpine meadows and rockslides in the Rocky Mountains; var. *dispar*, rare, on slopes of ravines and margins of groves, in southeastern Parklands; var. *johannensis*, gravel bars, Hudson Bay; var. *varians*, rocky hillsides and meadows, Hudson Bay.

Fig. 141.  Bessey's locoweed, *Oxytropis besseyi* (Rydb.) Blank.

*Oxytropis deflexa* (Pall.) DC.  REFLEXED LOCOWEED

Plants with a deep taproot and caudex. Stems well-developed, 10–40 cm long, decumbent to ascending, or acaulescent with the scape 10–20 cm long, loosely pubescent. Leaves appearing flattened, with 17–41 leaflets 5–15 mm long, lanceolate, pubescent on both sides. Inflorescence loosely flowered, elongating in fruit; flowers 6–10 mm long, bluish or creamy white with purple tip; calyx 3–4 mm long, black pubescent, reflexed.

This species has been divided into the following varieties:

1. Lateral sinuses of the calyx similar to the ventral pair, broad and obtuse; petals ample, purple, with the banner broadly obcordate, about twice as long as broad. .................................................................................... 2

   Lateral sinuses similar to the dorsal pair, narrow and acute, rarely a little broader and rounded; petals narrower, mostly blue- or purplish-tinged, with the banner oblanceolate, about 3 times as long as broad. ........................ var. *sericea* T. & G.

2. Plants copiously and loosely villous-pilose; stems commonly developed, with at least 1 apparent internode at maturity; racemes 5–10 to 20-flowered, usually elongating in fruit. ........................... var. *deflexa*

   Plants sparingly pilose, green, with the hairs appressed; stems usually none, occasionally with 1 or 2 internodes; racemes 2–7 to 10-flowered, almost always compact in fruit. ........................ var. *foliolosa* (Hook.) Barneby

Of these forms, var. *sericea* is found in grasslands and openings throughout Parklands and Boreal forest; var. *foliolosa* in openings and on slopes in Rocky Mountains; var. *deflexa* does not occur in the Prairie Provinces.

*Oxytropis lagopus* Nutt.

Plants with a stout taproot, densely pubescent. Leaves with 9–15 leaflets 5–15 mm long, elliptic to lanceolate. Scape 15–20 cm high; inflorescence 3–5 cm long; flowers 15–20 mm long, dark bluish purple; calyx 5–6 mm long. Legume 12–15 mm long, oblong ovoid, densely long villous. Very rare; dry hills; southwestern Prairies.

*Oxytropis lambertii* Pursh  PURPLE LOCOWEED

Plants with a stout taproot, 15–25 cm high. Leaves with 11–17 leaflets 15–30 mm long, lanceolate, somewhat pubescent on both sides. Scape to 25 cm high; inflorescence 7–12 cm long; flowers 15–20 mm long, dark bluish purple; calyx about 7 mm long. Legume 15–20 mm long, lanceolate, pubescent. Some of the pubescence of this species is malpighian, which makes it distinguishable from the other purple-flowered species. Grasslands; southeastern Parkland.

*Oxytropis podocarpa* Gray

Plants with a deep taproot and branched caudex, somewhat cushion-like. Leaves with 11–25 leaflets 3–10 mm long, linear to oblong, silky pubescent. Scape 3–6 cm high; inflorescence with 1–3 dark bluish purple flowers; calyx 6–7 mm long, villose pubescent. Legume 15–20 mm long, ovoid-ellipsoid, inflated, black pubescent, stipitate. Arctic–alpine; Rocky Mountains, Alberta.

*Oxytropis sericea* Nutt.                                        EARLY YELLOW LOCOWEED

Plants with a stout taproot and branched caudex. Leaves with 7–15 leaflets 10–30 mm long, elliptic or oblong, silky pubescent on both sides. Scape 10–20 cm high; inflorescence 5–7 cm long; flowers 18–20 mm long, yellowish; calyx 6–7 mm long, silky pubescent, black hairy on the calyx lobes. Legume 20 mm long, oblong, pubescent with both black and white hairs, rigid. The var. *spicata* (Hook.) Barneby (Fig. 142) is common in grasslands in Prairies, less common in Parklands; var. *sericea* with flowers light purple to whitish is more southern.

*Oxytropis splendens* Dougl.                                         SHOWY LOCOWEED

Plants with a deep taproot and branched caudex. Leaves with up to 60 leaflets arranged in verticils of 3–6; leaflets 10–25 mm long, linear-lanceolate, long silky pubescent. Scape 20–30 cm high; inflorescence 4–10 cm long; flowers 12–15 mm long, dark blue; calyx 6–7 mm long, densely long silky pubescent. Legume 8–12 mm long, ovoid, densely long pubescent. Grasslands and open woods; Boreal forest, Parklands, Rocky Mountains. The var. *richardsonii* Hook., wooly locoweed, differs from the species in the type of hairiness: the variety has hairs appressed (lying flat) and silky, whereas the typical plant has long, soft, spreading hairs. Flowers dark purple; leaflets lanceolate. Common in southern Rocky Mountains, decreasing in abundance toward the east.

*Oxytropis viscida* Nutt.                                          VISCID LOCOWEED

Plants with a deep taproot and branched caudex. Leaves with 21–35 leaflets 5–12 mm long, lanceolate, somewhat pubescent, and glandular. Scape 5–20 cm high; inflorescence 3–8 cm long; flowers 10–12 mm long, purple with yellowish base; calyx 4–6 mm long, densely glandular pubescent. Legume 10–15 mm long, finely black pubescent, ovate. Southwestern Alberta, Prairies, and Parklands.

**Petalostemon**        prairie-clover

Rather low-growing, often prostrate perennial plants, with glandular-dotted pinnate leaves having odd-numbered leaflets. Flowers perfect, borne in dense spikes at head of flowering stalks. Five stamens united into one bundle. Fruits short; pods containing 1 or 2 seeds. Many authorities call this genus *Petalostemum*; both spellings are acceptable.

1. Leaves with 7–17 leaflets; plants densely
     hairy throughout. ................................................................. *P. villosum*
   Leaves with 3–5 (sometimes 7) leaflets;
     plants glabrous or sparsely pubescent. .................................... 2

A.C. Budd

Fig. 142.   Early yellow locoweed, *Oxytropis sericea* Nutt. var. *spicata* (Hook.) Barneby.

2. Flowers purple; leaflets narrowly linear,
about 15–25 mm long and 1–1.5 mm
wide. ........................................................................................ *P. purpureum*

Flowers white; leaflets linear-oblong,
about 15–25 mm long and 2–3 mm
wide. ........................................................................................ *P. candidum*

*Petalostemon candidum* (Willd.) Michx.         WHITE PRAIRIE-CLOVER

Stems 20–50 cm high, usually erect. Leaves of 7–9 linear-oblong leaflets
1–2 cm long. Flowers in a compact spike 2–8 cm long, white, and a little under
6 mm long. Common; on dry prairie and hillsides; throughout Prairies and
Parklands.

*Petalostemon purpureum* (Vent.) Rydb. (Fig. 143)         PURPLE PRAIRIE-CLOVER

A several-stemmed plant, erect or decumbent, 20–50 cm high, but usually
prostrate. Leaves of 3–7 linear leaflets 5–20 mm long, sparingly hairy or glabrous. Flower spikes dense and cylindric, 1–5 cm long, with red or purple
flowers. Common; on hillsides, dry banks, and prairie; throughout Prairies. A
densely hairy form, var. *pubescens* (Gray) Fassett, is sometimes found in the
southwest.

*Petalostemon villosum* Nutt.         HAIRY PRAIRIE-CLOVER

A densely hairy plant 20–50 cm high, branching from the base; leaves
bearing 7–17 closely packed leaflets 5–15 mm long, covered with silky hairs.
Flower spikes either single or clustered, 2–10 cm long, flowers reddish purple
or pink. Found occasionally; in sandhills; southeastern Parklands.

*Psoralea*         breadroot

Perennial herbs with glandular-dotted foliage and leaves. Leaves palmately compound, of 3–7 leaflets. Flowers in spikes or racemes, perfect, with
stamens in 1 or 2 bundles. Pods ovoid, short, 1-seeded, not splitting open at
maturity, but opening very irregularly. Roots, especially those of *P. esculenta*,
were used by Indians for food.

1. Plants densely long-hairy; flowers in a
short, dense spike. ........................................................................................ *P. esculenta*

Plants not densely long-hairy; flowers in
racemes or loose spikes. ........................................................................................ 2

2. Plants silvery hairy; leaves with 3–5
leaflets; inflorescence loose. ........................................................................................ *P. argophylla*

Plants not silvery hairy; leaves and stem
glandular-dotted; leaves with 3 leaflets;
inflorescence dense. ........................................................................................ *P. lanceolata*

*Psoralea argophylla* Pursh         SILVERLEAF PSORALEA

An erect much-branched plant 30–60 cm high with silvery whitish hairiness throughout. Leaves of 3–5 obovate silvery-haired leaflets 10–35 mm long.
Flowers borne on interrupted spikes in clusters of 2 or 4, about 6 mm long,
blue fading during drying. Common; on dry to moist grassland; throughout
Prairies and Parklands. Syn.: *Psoralidium argophyllum* (Pursh) Rydb.

Fig. 143.  Purple prairie-clover, *Petalostemon purpureus* (Vent.) Rydb.

*Psoralea esculenta* Pursh (Fig. 144)                          INDIAN BREADROOT

A low, stout short-stemmed plant 10–50 cm high, densely covered with loose white hairs, growing from a large tuberous starchy root or cluster of roots. Leaves of 5 leaflets 2–5 cm long. Flowers in a dense oblong spike 3–8 cm long, blue, a little longer than the greenish sepals. Fairly common on prairie and in sheltered places, or sandy banks throughout Prairies. Roots edible, raw or cooked. Syn.: *Pediomelum esculentum* (Pursh) Rydb.

*Psoralea lanceolata* Pursh (Fig. 145)                    LANCE-LEAVED PSORALEA

A low-growing semiprostrate or erect plant 20–50 cm high, with glandular-dotted stems; the whole plant pale yellowish green. Very long stringy roots with ramifications extending for many meters. Leaves of 3 linear-lanceolate leaflets 10–35 mm long. Flowers about 6 mm long, pale bluish white, in short dense spikes. Fruit a globular dotted lemon-shaped pod 5–8 mm in diam containing a single seed. A species of sandhills and sandy land, often the dominant plant in some areas, especially on partly stabilized dunes. The rough roots often exposed for long distances, bridging the gap where sand has been blown out between dunes. Very common in suitable sandy sites throughout the southwestern portion of the area, but not found on heavier soils. Not palatable to livestock. Syn.: *Psoralidium lanceolatum* (Pursh) Rydb.

*Thermopsis*          golden-bean

*Thermopsis rhombifolia* (Nutt.) Richardson (Fig. 146)          GOLDEN-BEAN

An erect, branched perennial 15–50 cm high, usually in large patches, from running rootstocks. Leaves of 3 obovate leaflets 2–4 cm long, with appressed silky gray hairs, leaf stalks large and leaf-like, stipules at junction of stem. Very bright golden yellow flowers, 1–2 cm long, in rather dense racemes; stamens separate. Pods 3–7 cm long, curved, grayish hairy, and containing 10–13 seeds. An early blooming plant and one of the most striking and colorful early spring flowers. The milk from cows that have eaten the flowers of this species is said to have a peculiar odor and flavor. The fruit of golden-bean has caused **severe sickness** in children. Very common; in great masses along roadsides, on edges of buffalo wallows, and on hillsides; throughout Prairies, also in sandy areas in Parklands.

*Trifolium*          clover

Perennial or biennial herbs, with leaves of 3 leaflets and flowers in short very dense head-like racemes. Flowers perfect; stamens diadelphous. European plants much used for forage and lawns, but escaped from cultivation.

1. Flowers yellow or white. ............................................................................ 2
   Flowers purple or pinkish. ...................................................................... 4
2. Flowers white; perennial, with extensively
   creeping stems, rooting at the nodes. ....................................... *T. repens*
   Flowers yellow; annual, with erect or
   ascending, not creeping, stems. ..................................................... 3

Fig. 144.   Indian breadroot, *Psoralea esculenta* Pursh.

Fig. 145.   Lance-leaved psoralea, *Psoralea lanceolata* Pursh.

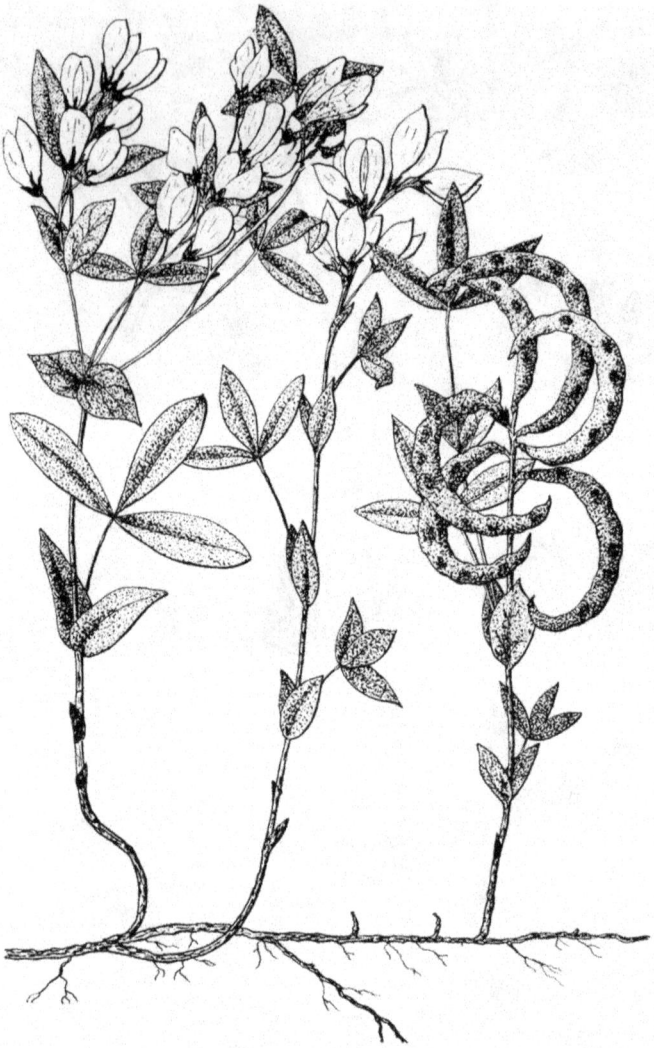

A.C. Budd

Fig. 146.   Golden-bean, *Thermopsis rhombifolia* (Nutt.) Richardson.

3. Leaflets to 15 mm long, with the terminal
   one nearly sessile; stipules not dilated. ...................................................... *T. aureum*

   Leaflets to 10 mm long, with the terminal
   one petiolulate; stipules dilated below. ............................................ *T. campestre*

4. Plants pubescent; flowers sessile in the
   heads. ............................................................................................................ *T. pratense*

   Plants glabrous or nearly so; flowers pedi-
   celed in the heads. ............................................................................. *T. hybridum*


*Trifolium aureum* Poll.                                    YELLOW CLOVER

Plants biennial or annual; stems 15–30 cm high, erect, branched. Leaflets to 15 mm long, oblong-lanceolate; petiole of terminal leaflet as long as that of the lateral ones. Flowers 5–8 mm long, golden yellow; pods 1-seeded. Introduced, cultivated, and occasionally escaped.


*Trifolium campestre* Schreb.                            YELLOW FIELD CLOVER

Similar to *T. aureum*, but the petiole of the terminal leaflet distinctly longer than that of the lateral ones, and the stipules larger. Flowers 4–5 mm long, lemon yellow. Introduced for cultivation, rarely escaped.


*Trifolium hybridum* L.                                    ALSIKE CLOVER

An erect species 30–60 cm high, with long-stalked leaves of 3 obovate leaflets 10–25 mm long, smooth. Flowers pink, in globose head-like racemes. Common; in waste places and roadsides, where it has escaped from cultivation; Parklands and Boreal forest.


*Trifolium pratense* L.                                    RED CLOVER

An erect somewhat hairy biennial or perennial species. Leaflets large, 1–5 cm long, often with a reddish inverted V on the upper surface of the leaflets. Flowers red, in globose heads. Rarely found in Prairies, but fairly common in waste places in Parklands and Boreal forest.


*Trifolium repens* L.                                    WHITE CLOVER

A creeping perennial with smooth hairless leaflets 5–20 mm long, often having a whitish or pale inverted V on the upper surface of the leaflets. Flowers white or somewhat pinkish-tinged, in round head-like racemes. Often used for lawns and occasionally found along roadsides, in meadows, and throughout forest areas, where it has escaped from cultivation.


*Vicia*     vetch

Annual or perennial herbaceous vines, with pinnate leaves and even-numbered leaflets, the terminal leaflet being replaced by tendrils. Flowers perfect, in spikes or racemes; stamens diadelphous. Vetches distinguishable from vetchlings (*Lathyrus*) by having the style or female organ not flattened and merely a tuft of hairs at its summit instead of down one side. Good forage for livestock.

1. Flowers 1 or 2 (seldom 4) in the upper
   leaf axils. ........................................................................ *V. sativa* ssp. *nigra*
   Flowers many in a more or less one-sided
   raceme. ........................................................................................................ 2

2. Plants annual; villous with more or less
   long-spreading pubescence. ........................................................ *V. villosus*
   Plants perennial; glabrous or sparsely
   appressed pubescent. ................................................................................. 3

3. Inflorescence with 10–40 flowers in a
   dense raceme. ...................................................................................... *V. cracca*
   Inflorescence with 3–9 flowers in a loose
   raceme. .......................................................................................... *V. americana*

*Vicia americana* Muhl. (Fig. 147)                          AMERICAN VETCH

A smooth trailing or climbing plant 40–80 cm long. Leaves of 8–14 ovate or elliptic leaflets 15–35 mm long, very strongly veined. Flowers 15–20 mm long, bluish purple, in loose 3- to 9-flowered racemes. Pods smooth, 2–4 cm long. Very common; around bluffs and shady parts of prairie; throughout the Prairie Provinces.

The var. *truncata* (Nutt.) Brewer, Oregon vetch, differs from the species by being somewhat hairy below, especially when young, and having the leaflets abruptly flattened at the apex and sometimes toothed. Sometimes found in southeastern Parkland. Syn.: *V. oregana* Nutt.

The var. *minor* Hook., narrow-leaved vetch, a prostrate trailing plant with 8–12 narrowly linear strongly-veined leaflets 10–35 mm long and very narrow. Racemes 2- to 6-flowered; flowers almost 2 cm long, bluish purple. Pods a little over 25 mm long. Common; on open prairie and dry soil; throughout the Prairies, often persisting after cultivation. Syn.: *V. americana* Muhl. var. *angustifolia* Nees.

*Vicia cracca* L.                                            TUFTED VETCH

A tufted weak-stemmed vetch 50–125 cm long, with leaves formed of 8–24 linear-oblong leaflets. Flowers in a dense 1-sided raceme. Rarely found, but has escaped from cultivation; in some localities; Parklands and Boreal forest.

*Vicia sativa* L. ssp. *nigra* (L.) Erhr.                     VETCH

A pubescent annual 50–80 cm high. Leaflets 3–8 pairs, linear, 6–25 mm long. Flowers 1 or 2 together (or rarely 4), 10–30 mm long, purple; pods 25–70 mm long. Introduced, occasionally escaped.

*Vicia villosus* Roth                                        HAIRY VETCH

An annual with stems 30–150 cm high, villose. Leaves with 4–12 pairs of linear to elliptic leaflets. Flowers 10–20 mm long, purple to violet, occasionally the wing white or yellowish; the calyx strongly swollen at the base; pods 2–4 cm long. Introduced, occasionally escaped from cultivation.

A.C. Budd

Fig. 147.   American vetch, *Vicia americana* Muhl.

# GERANIACEAE—geranium family

Herbs with opposite stipulate leaves. Flowers perfect and regular, with 5 petals and 5 sepals. Stamens either 5 or 10. Style extending into a long beak-like column with short appendages at tip, and splitting from bottom upward into 5 sections, each with a single-seeded capsule bearing a long tail. Plants of this family not palatable to livestock.

Leaves pinnately lobed or dissected; tails of
    capsules twisted at maturity; capsule
    spindle-shaped. ............................................................................................... *Erodium*
Leaves palmately lobed or divided; tails of
    capsules merely curved; capsule round. ......................................... *Geranium*

## *Erodium*    stork's-bill

*Erodium cicutarium* (L.) L'Hér.        STORK'S-BILL

A low prostrate annual with pinnately divided leaves. Sepals 5, somewhat awn-tipped. Flowers about 1 cm across, pink or purplish, and borne in clusters of 2–12 on long flower stalks. Stamens 5 fertile and 5 sterile. Style column very long, 2–4 cm, and splitting into 5 segments, each with a long spirally twisted tail. Introduced from Europe, and occasionally found; around towns; in southern and eastern Prairie Provinces.

## *Geranium*    geranium

1. Flowers not over 12 mm across; petals
   scarcely longer than sepals; annuals or
   biennials. .......................................................................................... 2
   Flowers over 12 mm across; petals much
   longer than sepals; perennials. .................................................. 4
2. Sepals without bristle tips; seeds smooth. ........................ *G. pusillum*
   Sepals bristle-tipped; seeds rough. ........................................ 3
3. Stalks of individual flowers more than
   twice as long as calyx; inflorescence
   loose. ............................................................................................. *G. bicknellii*
   Stalks of individual flowers not more than
   twice as long as calyx; inflorescence
   compact. ..................................................................................... *G. carolinianum*
4. Petals white. ........................................................................... *G. richardsonii*
   Petals rose or purple. ......................................................... *G. viscosissimum*

*Geranium bicknellii* Britt.        BICKNELL'S GERANIUM

A fairly erect annual or biennial plant 15–50 cm high, with a loosely hairy stem. Leaves very deeply dissected into narrow oblong segments 2–6 cm broad. Stalks of leaves and flowers hairy and glandular; petals rose-colored, about the same length as sepals. Mature style column about 25 mm long, with narrow beak about 6 mm long. Inflorescence loose. Fairly common; in Park-lands and Boreal forest and on pathways and roads in Cypress Hills.

*Geranium carolinianum* L. CAROLINA WILD GERANIUM

An erect annual species 15–40 cm high, with stems loosely hairy and somewhat glandular. Leaves deeply cut into wedge-shaped lobed segments 2–6 cm across. Flowers pale pink or whitish, about the same length as sepals, and borne in rather compact clusters. Style column about 25 mm long with a very short narrow beak. Not so plentiful as *G. bicknellii*; found in meadows and waste places; throughout the Prairie Provinces.

*Geranium pusillum* L. SMALL-FLOWERED CRANE'S-BILL

A weak-stemmed annual, spreading and hairy, 10–50 cm long. Leaves 1–5 cm long, 5- to 7-lobed, the lobes either entire or 3-toothed. Sepals not bristle-tipped. Flowers pale purple, 6–10 mm across, with often only 5 fertile stamens. Introduced from Europe and occasionally found; in waste places near towns; in eastern Parklands and Boreal forest.

*Geranium richardsonii* Fisch. & Trautv. WILD WHITE GERANIUM

An erect perennial species 30–80 cm high, often with spreading hairs. Leaves 3–10 cm across, deeply 3- to 5-lobed, with cut and toothed segments. Flowers 2–4 cm across, white with pink veins; style column 20–35 mm long. Plentiful; in meadows and open forest; Rocky Mountains and in Cypress Hills.

*Geranium viscosissimum* Fisch. & Mey. (Fig. 148) STICKY PURPLE GERANIUM

An erect branching perennial species 30–60 cm high, with the stem and leaf stalks sticky glandular. Leaves 4–10 cm across, hairy, 3–5 times cleft into sharply toothed segments. Sepals hairy and awn-tipped; flowers pink purple, 3–4 cm across, and very showy. Common; in open woodlands; in southern Rocky Mountains and also on south slope of Cypress Hills, but rare elsewhere in the area. *G. strigosum* Rydb., a very similar plant, having stems and leaf stalks with reflexed hairs and not glandular, occasionally found in the north-western Boreal forest and in the Peace River District.

## OXALIDACEAE—wood-sorrel family

Low herbs with rootstocks, and palmately divided leaves with 3-leaflets; the leaflets broadly inverted heart-shaped, indented at the apex. Flowers perfect and regular, with 5 petals, 5 sepals, and 10 stamens. Fruit a capsule.

**Oxalis** wood-sorrel

1. Stem creeping, often rooting at nodes. ................................................. *O. corniculata*
   Stem erect or decumbent in age, not
   rooting. ................................................................................................. 2
2. Stem with appressed hairs; capsule hairy. ................................................. *O. stricta*
   Stem with loose spreading hairs or nearly
   smooth; capsule smooth. ....................................................... *O. europaea*

*Dicots*                                                      OXALIDACEAE – 505

Fig. 148. Sticky purple geranium, *Geranium viscosissimum* Fisch. & Mey.

*Oxalis corniculata* L.

A decumbent creeping plant, branched at base, with sparse loose hairiness. Leaves trifoliolate, with leaflets 3–10 mm wide. Petals pale yellow, 5–10 mm long. Escaped from cultivation and occasionally found around buildings. Syn.: *Xanthoxalis corniculata* (L.) Small.

*Oxalis europaea* Jordan                    BUSH'S YELLOW WOOD-SORREL

A slender-stemmed erect plant 10–20 cm high, loosely hairy. Leaves trifoliolate, with leaflets 5–20 mm long, bright green. Flowers bright yellow, 10–15 mm across. Capsule smooth, 10–15 mm long. Found in dry soil; has been reported, though seldom, from Boreal forest.

*Oxalis stricta* L.                    YELLOW WOOD-SORREL

A low pale green plant, somewhat decumbent, usually branched from the base. Leaves trifoliolate, with leaflets 1–2 cm wide. Flowers pale yellow, in umbel-like cymes; petals almost 12 mm long. Capsule hairy, 15–25 mm long, rather abruptly pointed at the tip, and borne erect on reflexed stems. Fairly common locally; along roadsides, gardens, and waste places; in Prairies. Syn.: *Xanthoxalis stricta* (L.) Small.


# LINACEAE—flax family

Annual or perennial plants having simple leaves without stalks and perfect regular flowers with 5 sepals, 5 petals, and 5 stamens. Fruit a round capsule divided into 4 or 5 cells, each containing 2 flat seeds. Petals fall very readily from the plant, usually not lasting more than a day.

***Linum***        flax

1. Flowers blue; sepals without glands. ........................................................ 2

   Flowers yellow; inner sepals with marginal glands. ................................................................. 3

2. Sepals more than 5 mm long when mature, more than half as long as capsule. ............................................ *L. lewisii*

   Sepals less than 5 mm long when mature, less than half as long as capsule. ......................................... *L. pratense*

3. Sepals persistent; capsule not thickened at base. ...................................................... *L. sulcatum*

   Sepals falling off; capsule with firm thickenings at base. ...................................................... *L. rigidum*

*Linum lewisii* Pursh                    LEWIS WILD FLAX

An erect perennial plant, sometimes branched from base, growing 20–60 cm high from a woody root, hairless throughout. Leaves linear, 10–25 mm long, somewhat crowded on stem. Flowers blue, 20–35 mm across; the petals soon falling off. Capsules round, 5 mm in diam. Common; on dry prairie; throughout Prairies and Parklands.

*Linum pratense* (Norton) Small                              MEADOW WILD FLAX

Very similar to *L. lewisii*, but the sepals are less than 5 mm long and less than half the length of the capsule, and the flowers are smaller, 15–25 mm across. Occasionally found but not common on dry prairie; Prairies and Parklands.


*Linum rigidum* Pursh (Fig. 149)              LARGE-FLOWERED YELLOW FLAX

An erect pale green species 15–40 cm high, with the stem simple below and somewhat branched above. Leaves few, linear, 10–25 mm long, and very easily knocked off. Flowers yellow, 20–25 mm across. Common locally; in sand hills and sandy or very light soils; throughout Prairies and Parklands. Not generally common, but very plentiful where found. Syn.: *Cathartolinum rigidum* (Pursh) Small; *L. compactum* Nelson.


*Linum sulcatum* Riddell                              GROOVED YELLOW FLAX

An annual 5–50 cm high, with the stem angled and somewhat winged, simple below but much-branched above. Leaves linear, 10–25 mm long, soon falling off. Flowers yellow, 8–12 mm across. Not common, but found occasionally; in dry sandy soils; eastern Parklands. Syn.: *Cathartolinum sulcatum* (Riddell) Small.


# BALSAMINACEAE—touch-me-not family

Somewhat succulent herbs having simple leaves without stipules. Flowers perfect, but irregular; 3 sepals, 2 of them small and green, the other large, petal-like, and extended back into a bag-like pouch terminating in a nectar-filled spur. Only 3 apparent petals, 2 pairs of petals being united; 5 stamens. Fruit a 5-celled capsule, springing open at maturity and forcibly expelling the seeds. Plants found in moist places.


***Impatiens***        touch-me-not

Flowers orange, spotted, and sharply contracted to spur. ................................................................................................ *I. biflora*
Flowers light yellow, not spotted, gradually tapering to spur. ........................................................................................ *I. noli-tangere*


*Impatiens biflora* Walt.                              SPOTTED TOUCH-ME-NOT

An annual, branched, rather succulent plant 30–150 cm high, with stems often tinged with red, and leaves ovate, 2–10 cm long, green or purplish. Petals and the large pouch-like sepals orange, copiously dotted with reddish brown or purplish spots; flower 20–25 mm long. Found on banks of rivers and lakes; throughout Parklands and Boreal forest and also in Cypress Hills. Syn.: *I. capensis* Meerb.

Fig. 149.　Large-flowered yellow flax, *Linum rigidum* Pursh.

*Impatiens noli-tangere* L. WESTERN JEWELWEED

An annual very similar to *I. biflora*, but with light green or straw-colored stems and pale green leaves. Flowers pale yellow, unspotted, tapering to a spur about 10 mm long. Rare in the Prairie Provinces, but has been found in wet places in woodlands; Boreal forest. Syn.: *I. occidentalis* Rydb.


# POLYGALACEAE—milkwort family

Herbs with simple leaves and no stipules. Irregular flowers with 5 sepals, 2 of which are large, colored, and petal-like; either 3 or 5 more or less united petals; keel of petals with a fringed crest.

***Polygala***    milkwort

1. Leaves whorled; plants annual. ............................................................... *P. verticillata*
   Leaves alternate; plants perennial. ...................................................................... 2
2. Flowers, showy, rose purple. ....................................................... *P. paucifolia*
   Flowers white. ...................................................................................... 3
3. Leaves linear or linear-oblanceolate;
      inflorescence conic. ........................................................................... *P. alba*
   Leaves lanceolate to ovate; inflorescence
      oblong, cylindric. ................................................................... *P. senega*


*Polygala alba* Nutt. WHITE MILKWORT

An erect plant with several stems growing from the base, 15–40 cm high, with alternate linear leaves 5–25 mm long. Flowers borne in spike-like racemes 2–5 cm long, white or somewhat green-tinged. Found occasionally; in dry ground; along extreme southern border of the Prairie Provinces, but its natural habitat is farther south.

*Polygala paucifolia* Willd. FRINGED MILKWORT

An erect plant 5–20 cm high, usually branched, with small leaves near the lower part of the stem and larger ones above; upper leaves oval, 1–3 cm long. Flowers few, 3 or 4, rose purple or pink, 1–2 cm long; keel of corolla with a fringed crest. Found occasionally in Boreal forest.

*Polygala senega* L. SENECA SNAKEROOT

An erect plant with several stems growing from a thick rootstock to 10–50 cm high. Leaves numerous on the stems, lanceolate, 2–5 cm long. Flowers greenish white, borne in terminal spike-like racemes 2–6 cm long. Roots used for medicinal purposes. Fairly common; around edges of bluffs and in semi-wooded prairie; throughout Parklands.

*Polygala verticillata* L. WHORLED MILKWORT

Erect plants with stems 10–40 cm high, divergently branched. Leaves 1–2 cm long, linear to linear-oblong, with the lower ones mostly in whorls of 2–5.

Inflorescence with the lower branches opposite or whorled; racemes 5–15 mm long; flowers 2–3 mm long, whitish to pinkish. Moist grasslands; southeastern Parklands.

## EUPHORBIACEAE—spurge family

Erect or prostrate herbs with an acrid milky juice. Annual or perennial with simple, entire opposite or alternate leaves. Flowers unisexual (but both sexes on the same plant), with sepals reduced to a minute scale. Involucre resembling a calyx, with numerous male flowers consisting of a single stamen and a minute bract and one female flower consisting of a 3-lobed ovary, which, when fertilized, extends upward on a thin stalk and bears the capsule containing the seeds. In some cases stalks arising from the involucre bear still more involucres and flowers, thus forming a large branched inflorescence.

***Euphorbia***    spurge

1. Prostrate mat-like annual plants; inflorescence in leaf axils. ................................................... 2
   Erect plants; inflorescence in terminal umbels. ................................................................... 3

2. Leaves entire; seeds not compressed, 3-angled. ...................................................... *E. geyeri*
   Leaves denticulate; seeds compressed, 4-angled. ................................................... *E. serpyllifolia*

3. Upper leaves with a conspicuous white margin. ....................................................... *E. marginata*
   Upper leaves not with a white margin. ...................................................................... 4

4. Stem leaves finely serrulate; inflorescence very leafy. ............................................. *E. helioscopia*
   Stem leaves entire; inflorescence not very leafy. ....................................................... 5

5. Leaves obovate, rounded or blunt at end. ............................................................. *E. peplus*
   Leaves linear or lanceolate, pointed at end. ............................................................. 6

6. Leaves less than 3 mm wide, crowded on stem. ..................................................... *E. cyparissias*
   Leaves more than 3 mm wide, not very crowded. ..................................................... 7

7. Leaves usually not over 1 cm wide. ...................................................................... *E. esula*
   Leaves usually 1–3 cm wide. ................................................................................ *E. lucida*

*Euphorbia cyparissias* L.                                            CYPRESS SPURGE

A tufted introduced perennial plant growing from a rootstock to 15–60 cm high, with numerous linear pale green leaves 1–2 cm long and less than 3 mm wide densely crowded toward the upper part of the stems. Flowers in a terminal umbel-like inflorescence and also in the upper leaf axils. Escaped from cultivation and becoming weedy in several localities.

*Euphorbia esula* L. (Fig. 150)    LEAFY SPURGE

An erect bluish green perennial growing from running roots to 15–75 cm high. A few scattered linear or oblong stem leaves 10–35 mm long. A whorl of narrow leaves is located below the inflorescence. Inflorescence somewhat umbel-like; flowers borne on a pair of pale yellowish green leaf-like bracts, with many inconspicuous male flowers and one female flower for each pair of bracts. Female flowers extend upward on a short stalk, forming 3-seeded capsules. The capsules bursting and expelling the seeds, sometimes for a long distance. It is extremely difficult to eradicate this pernicious weed, which spreads by creeping underground rootstocks and also by seed. Becoming very plentiful in many localities in the Prairie Provinces. An introduced plant, native of Europe and Asia.

*Euphorbia geyeri* Engelm.    PROSTRATE SPURGE

An annual plant, prostrate in a mat on the ground, 5–25 cm across, with small oblong leaves 5–10 mm long, opposite and pale green. Inflorescence minute, in axils of leaves; seeds smooth, reddish. Rare; sandy soils; southeastern Parklands.

*Euphorbia helioscopia* L.    SUN SPURGE

An erect introduced annual, often branching from the base, 15–60 cm high. Leaves obovate or spatulate, 2–5 cm long, bluntly rounded at tip and narrowed at base. Flowers borne in leafy inflorescence at head of stem. Very scarce; reported as a weed from several locations in the Prairie Provinces.

*Euphorbia lucida* Waldst. & Kit.    SHINING SPURGE

An introduced perennial 20–70 cm high, from a thick rootstock. Leaves 15–30 mm wide and 5–10 cm long. Very similar to *E. esula*, but differing in the wider leaves and thicker rootstocks. At present very scarce; but has been reported from east central Alberta.

*Euphorbia marginata* Pursh    SNOW-ON-THE-MOUNTAIN

Erect annual plants 30–80 cm high, softly pubescent. Stem leaves 4–10 cm long, broadly ovate to elliptic. Inflorescence subtended by whorled white-margined leaves. Introduced as an ornamental plant; occasionally escaped and reseeding itself.

*Euphorbia peplus* L.    PETTY SPURGE

Annuals with erect stems 10–30 cm high; stem leaves 1–2 cm long, obovate to suborbicular, rounded at the tip. Inflorescence with 3–5 rays; capsule winged; seeds pitted. Introduced, and locally weedy.

*Euphorbia serpyllifolia* Pers.    THYME-LEAVED SPURGE

A native annual plant, usually prostrate and forming mats 5–50 cm across, usually smooth with somewhat reddish stems. Leaves opposite, small, 5–15 mm long, dark green, usually with a conspicuous red line down the center. Flowers inconspicuous in the leaf axils; seeds pitted and wrinkled. The var. *hirtella* (Engelm.) L. C. Wheeler is more or less hairy, but the typical variety is quite smooth. Common; on dry soil and in yards and waste places; throughout the Prairie Provinces. Includes *E. glyptosperma* Engelm.

Fig. 150.   Leafy spurge, *Euphorbia esula* L.

# CALLITRICHACEAE—water-starwort family

A small slender-stemmed aquatic perennial with opposite entire leaves. Leaves linear but some floating leaves obovate or spatulate. Flowers unisexual (but both sexes on the same plant), growing either singly or in groups of 2 or 3 in the axils of leaves. Male flower having 1 stamen and female flower a 4-celled ovary with 2 stigmas or style branches. Fruit small, nut-like, pendulous, and divided into 4 single-seeded sections.

*Callitriche*     water-starwort

All leaves submersed, linear, and 1-nerved. ........................................ *C. hermaphroditica*
Upper floating leaves obovate or spatulate, 3-
   nerved, but submersed leaves linear and
   1-nerved. ................................................................................................ *C. palustris*

*Callitriche hermaphroditica* L.                NORTHERN WATER-STARWORT

A completely submersed aquatic perennial, with stems 10–40 cm long. Leaves opposite, linear, 5–15 mm long. Usually rather crowded on stem and sometimes slightly indented at apex. Found occasionally; in flowing water; throughout the Prairie Provinces, but more frequently in the northern portion. Syn.: *C. autumnalis* L.

*Callitriche palustris* L.                VERNAL WATER-STARWORT

An aquatic perennial, usually with floating stems 2–30 cm long. Submersed leaves stalkless, up to 2 cm long, with a single nerve. Floating leaves obovate. Some plants with all leaves submersed and linear. Found occasionally; in ponds and ditches; throughout the Prairie Provinces.

# EMPETRACEAE—crowberry family

*Empetrum*     crowberry

*Empetrum nigrum* L.                BLACK CROWBERRY

A bushy-branched prostrate species, with stems to 40 cm long. Leaves numerous, crowded, 4–8 mm long, widely spreading or reflexed. Flowers small, purple; berries black. Bogs, tundra, and rocky areas; Boreal forest, Rocky Mountains.

# ANACARDIACEAE—sumach family

Small trees, shrubs, or very low herbs, with a somewhat acrid sap. Leaves consisting of from 3 to many leaflets. Both perfect and imperfect flowers, with 5 sepals, 5 petals, and 5 stamens. Fruit a drupe.

*Rhus*     sumach

1. Leaflets 13–31 ................................................................................ *R. glabra*
   Leaflets 3. ................................................................................................ 2

2. Low single-stemmed shrubs; leafing
   before flowering; leaflets 3–10 cm long. .......................... *R. radicans* var. *rydbergii*
   Large shrubs; flowering before leaves
   appear; leaflets not over 3 cm long. .............................. *R. aromatica* var. *trilobata*

*Rhus aromatica* Ait. var. *trilobata* (Nutt.) Gray                    SKUNKBUSH

A shrub 1–2 m high, much-branched, and usually flowering before the leaves appear. Leaves 3-foliate, with dark green leaflets, somewhat paler beneath, often 3-cleft, 1–3 cm long. Flowers yellowish green, minute, in clusters. Fruit a red globular drupe about 6 mm in diam. Found occasionally; in coulees, thickets, and open-wooded places; throughout Prairies. The unpleasant-smelling bush is not known to be poisonous.

*Rhus glabra* L.                                                      SMOOTH SUMACH

A shrub 2–4 m high, with pinnate leaves having 13–31 lanceolate leaflets, dark green above, paler beneath. Flowers in terminal clusters, bright green. Bright red fruits borne in dense clusters. Occasionally found on hillsides and dry soils; southeastern Parkland and Boreal forest.

*Rhus radicans* L. var. *rydbergii* (Small) Rehder (Fig. 151)          POISON-IVY

A single-stemmed erect shrub from a creeping rootstock, 10–30 cm high, with a rather woody stem. Leaves with 3 large bright green strongly veined leaflets 3–10 cm long. Flowers whitish yellow, in dense panicles from axils of leaves. Fruits somewhat globose berries, dull whitish color, about 6 mm in diam. Cattle appear to eat it with impunity, but pollen, sap, and even exhalations from the plant affect susceptible persons and cause severe skin eruptions and other troubles. Found in ravines, shady woodlands; Prairies and Parklands, but rare in Boreal forest. Syn.: *Toxicodendron rydbergii* (Small) Greene.


## ACERACEAE—maple family

Trees with a sweet sap and opposite, lobed, or pinnately compound leaves. Flowers either perfect or unisexual, sometimes appearing before the leaves. Fruit (a samara) consisting of 2 carpels joined at the base.

*Acer*          maple

1. Leaves pinnate with 3–5 leaflets; flowers
   unisexual, with male and female
   flowers on separate trees. ..................................... *A. negundo* var. *interius*
   Leaves lobed, not pinnate. .................................................................... 2
2. Leaves with lobes rhomboid, the base
   cordate. ........................................................................ *A. saccharinum*
   Leaves with lobes deltoid, the base not
   cordate. ........................................................................................... 3
3. Inflorescence a many-flowered racemose
   panicle. ........................................................................... *A. spicatum*
   Inflorescence a few-flowered corymb. ................................ *A. glabrum* var. *douglasii*

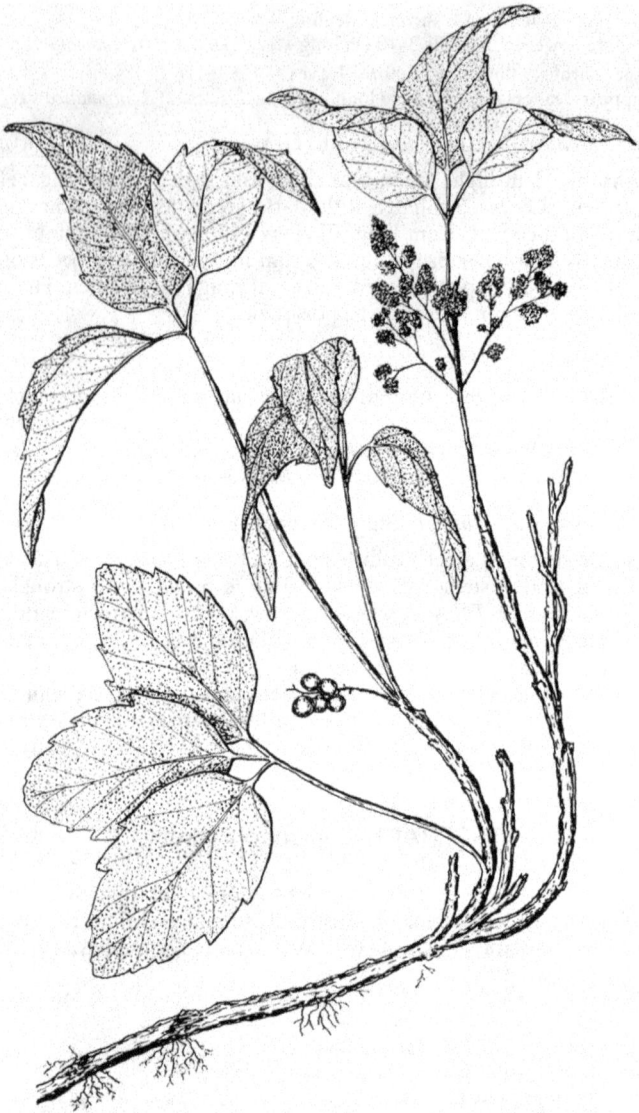

A.C. Budd

Fig. 151.   Poison-ivy, *Rhus radicans* L. var. *rydbergii* (Small) Rehder.

*Acer glabrum* Torr. var. *douglasii* (Hook.) Dippel        MOUNTAIN MAPLE

A small tree or tall shrub, with leaves palmately 3- to 5-lobed, coarsely and irregularly serrate. Panicle with fascicles of 2–4 flowers; samaras 15–25 mm long. Found occasionally; in deciduous and mixed forests; southern Rocky Mountains.

*Acer negundo* L. var. *interius* (Britt.) Sarg.        MANITOBA MAPLE, BOX ELDER

A tree, occasionally 6–7 m high, with rough grayish bark. Leaves of 3–5 lanceolate or ovate toothed leaflets, pale green, 5–12 cm long, each with a short stalk. Flowers appearing slightly before the leaves; unisexual, with male and female flowers on separate trees. Female flowers in small racemes, very small, and greenish. Male flowers consisting of 4 or 5 very small sepals and 4 or 5 stamens, reddish, borne in drooping clusters. Fruit a samara consisting of 2 oval carpels about 10 mm long, joined at the base and each terminating in a broad membranous wing about 25 mm long and 10 mm wide. Found along streams, in ravines, and wooded valleys; on Prairies and Parklands. Syn.: *Negundo interius* (Britt.) Rydb.

*Acer saccharinum* L.        SILVER MAPLE

Trees 10–20 m high, with leaves deeply 5-lobed, the base distinctly cordate. Flowers in clusters from lateral buds, either staminate or pistillate. Samaras 3.5–5 cm long. Planted as shade trees, and occasionally escaped.

*Acer spicatum* Lam.        WHITE MAPLE

Small trees 3–10 m high; the leaves clearly 3-lobed with 3 obscure basal lobes. Flowers in 3- to 7-flowered corymbs; samaras 3–4 cm long. Mixed woods and copses; eastern Parklands and Boreal forest.

## CELASTRACEAE—staff-tree family

Leaves alternate. ................................................................................................ *Celastrus*
Leaves opposite. .................................................................................................. *Pachystima*

**Celastrus**        staff-tree

*Celastrus scandens* L.        BITTERSWEET

Plants climbing to several meters high, with the stems becoming woody and eventually strangling the supporting trees. Leaves elliptic to ovate, acute, serrulate, 5–10 cm long. Panicles terminal, 3–8 cm long; fruits orange, clustered, subglobose, to 1 cm long. Woods, especially in sandy areas, river valleys, and ravines; southeastern Parklands and Boreal forest.

**Pachystima**        boxwood

*Pachystima myrsinites* (Pursh) Raf.        MOUNTAIN BOXWOOD

Small much-branched leafy shrubs 30–100 cm high; sometimes spreading to prostrate. Leaves ovate to oblanceolate, 15–30 mm long, serrulate, evergreen. Flowers axillary, solitary or in clusters of 2 or 3; petals reddish brown; capsule 4–5 mm long. Coniferous or mixed woods; Rocky Mountains.

# RHAMNACEAE—buckthorn family

Flowers numerous, in terminal umbels. .................................................. *Ceanothus*

Flowers in few-flowered axillary umbels or
  solitary. ........................................................................................ *Rhamnus*

## *Ceanothus*

Leaves lanceolate; panicles hemispheric or
  short-ovate. ...................................................................... *C. ovatus*

Leaves ovate or elliptic, glandular serrate;
  panicles compound with numerous umbels. ................................. *C. velutinus*

### *Ceanothus ovatus* Desf.                                    NEW JERSEY TEA

A bushy shrub about 1 m high; leaves 2–6 cm long, obtuse to subacute. Panicles on peduncles 2–5 cm long, with the umbels crowded. Rare; semiopen woods on light soils; southeastern Boreal forest.

### *Ceanothus velutinus* Dougl.                                STICKY LAUREL

Shrubs to 2 m high, with stout branches. Leaves evergreen, 2–8 cm long, obtuse or rounded at the tip, with the base subcordate, closely glandular denticulate. Panicles usually compound, with the umbels not crowded. Rare; open woods and slopes; southern Rocky Mountains.

## *Rhamnus*          buckthorn

Trees or shrubs, with leaves simple, usually alternate; flowers small, greenish or white; fruit a capsule or berry-like drupe.

1. Some branches ending in a short thorn. ................................. *R. catharticus*
   Branches not thorny. ........................................................................ 2

2. Flowers unisexual, apetalous; leaves
     serrate. ........................................................................ *R. alnifolius*
   Flowers perfect, with petals present;
     leaves entire. ................................................................ *R. frangula*

### *Rhamnus alnifolius* L'Hér.                         ALDER-LEAVED BUCKTHORN

A small shrub 1–2 m high, with grayish, very finely hairy branches. Leaves ovate to elliptical, strongly veined, 2–6 cm long. Flowers small and greenish, either single or in umbels of 2 or 3 on short stalks in axils of leaves. Fruit berry-like, black, 6–8 mm in diam, and **poisonous**. Fairly common; in moist woodlands and swamps; in northern and eastern fringes of the Prairie Provinces.

### *Rhamnus catharticus* L.                                    BUCKTHORN

Shrubs or small trees to 3 m high; lateral branches usually ending in a thorn. Leaves 3–6 cm long, broadly elliptic. Flowers having 4 parts; drupe black, 5–6 mm in diam. Introduced; occasionally establishing in shrubbery.

*Rhamus frangula* L. BLACK ALDER

Tall shrubs, with leaves 5–8 cm long, usually obovate or oblong. Flowers in axillary umbels; drupe at first red, becoming black. Introduced; occasionally establishing in shrubbery.


# VITACEAE—grape family

Woody vines climbing by means of tendrils. Leaves alternate, either lobed or palmately divided. Flowers in panicles, usually unisexual, sometimes both perfect and unisexual. Fruit a several-seeded berry.

Leaves simple, lobed. ................................................................................. *Vitis*
Leaves palmately compound, with 5–7 leaflets. ......................................... *Parthenocissus*


**Parthenocissus**      Virginia creeper

Leaves dull above, paler beneath; tendrils adhesive with disks; berry about 6 mm in diam. ................................................................................. *P. quinquefolia*

Leaves glossy above, slightly paler beneath; tendrils not adhesive, without disks; berry 8–10 mm in diam. ................................................................................. *P. inserta*

*Parthenocissus inserta* (Kerner) Fritsch      LARGE-TOOTHED VIRGINIA CREEPER

A straggling vine with smooth bark and long tendrils. Leaves of 5 or 6 lanceolate leaflets 4–10 cm long, large-toothed, glossy on upper side, and turning red in the fall. Tendrils branched with few, if any, adhesive disks at ends. Berries bluish black. Not common; but found in moist woods and shady banks; in eastern Parklands and Boreal forest as far north as Riding Mountain. Syn.: *Psedera vitacea* (Knerr) Greene.

*Parthenocissus quinquefolia* (L.) Planch.      VIRGINIA CREEPER

A tall climbing vine, usually with rather warty branches. Tendrils bearing an adhesive disk at end. Leaves usually of 5 ovate dull green leaflets 5–12 cm long, and coarsely toothed. Fruit blue with a slight bloom. Found occasionally; in moist woodlands; in southeastern Boreal forest. Syn.: *Psedera quinquefolia* (L.) Greene.


**Vitis**      grape

*Vitis riparia* Michx.      RIVERBANK GRAPE

A climbing or trailing vine, with greenish, somewhat angled branches. Leaves not lobed or sometimes 2-lobed near the apex and cordate at the base. The inflorescence with a fairly compact panicle; the berries blue, 6–12 mm in diam, and covered with a heavy bloom. In moist woods; in the eastern Parklands and Boreal forest, as far north and west as Riding Mountain.

*Dicots*                                              VITACEAE — 519

# TILIACEAE—linden family

*Tilia*      basswood

*Tilia americana* L.                            BASSWOOD

Trees to 40 m high. Leaves broadly ovate to suborbicular, sharply serrate, cordate to truncate at the base, glabrate to stellate pubescent above, and with conspicuous tufts of hairs in the vein axils below. Flowers in axillary cyme-like clusters; the petiole with a large, adnate foliaceous bract. Woods in ravines and along rivers; southeastern Parklands and Boreal forest.

# MALVACEAE—mallow family

Herbs with alternate, lobed or dissected, nearly round leaves. Flowers perfect, regular, and either single in clusters or in racemes in axils of leaves. Five partly united sepals and 5 petals united at base. Numerous stamens; fruit a many-segmented capsule. Velvetleaf, *Abutilon theophrasti* Medic., hollyhock, *Althea rosea* Cav., and gay mallows, *Lavatera thuringiaca* L., are used as ornamentals, and occasionally escape but are not long persistent.

1. Leaves palmately divided into 5 lobes;
   these entire or divided. ........................................................... *Malvastrum*
   Leaves more or less lobed, but not
   divided. ........................................................................................... 2
2. Flowers in axillary racemes. ............................................................. *Iliamna*
   Flowers in axillary clusters or solitary. ............................................. *Malva*

*Iliamna*      wild hollyhock

*Iliamna rivularis* (Dougl.) Greene      MOUNTAIN HOLLYHOCK

A stout perennial 50–200 cm high, sparsely stellate-pubescent. Leaves 5–15 cm across, cordate or reniform in outline, irregularly 4- to 7-lobed, the lobes toothed. Flowers about 5 cm across, pink to purplish or sometimes white; staminal column stellate-pubescent at the base; fruit segments 2- or 3-seeded. Mountain slopes and meadows; southern Rocky Mountains.

*Malva*      mallow

1. Petals 15–35 mm long; erect perennial
   plants. ............................................................................................... 2
   Petals 4–15 mm long; erect or prostrate
   annuals. ............................................................................................ 4
2. Flowers mostly in a terminal corymb. ........................................ *M. moschata*
   Flowers in the leaf axils. ................................................................. 3
3. Lower flowers solitary in the leaf axils. ........................................ *M. alcea*
   Lower flowers 2 or more in the leaf axils. .................................... *M. sylvestris*

4. Petals about twice as long as sepals. ........................................................................ 5
   Petals scarcely longer than sepals. ........................................................................... 6
5. Plants erect; carpels net-veined. ................................................................ *M. crispa*
   Plants decumbent, prostrate; carpels
      smooth. ................................................................................................. *M. neglecta*
6. Lower part of petals hairless; 8–12 car-
      pels, with net-like venation on backs;
      calyx reflexed at maturity. ....................................................... *M. parviflora*
   Lower parts of petals hairy-margined;
      12–15 carpels with smooth backs; calyx
      incurved at maturity. .................................................................... *M. pusilla*

*Malva alcea* L.                                                          PINK MALLOW

   Plants 50–150 cm high, with leaves 3- to 5-lobed, 5–10 cm long. Flowers
5–8 cm across, pale pink or mauve. Introduced as an ornamental; occasionally
escaped and established.

*Malva crispa* L.                                                         CRISP MALLOW

   An erect annual 50–150 cm high, with almost circular or kidney-shaped
leaves 5–20 cm across, wavy and lobed at the edges. Flowers purplish or white,
10-15 mm across, without stalks, and crowded in leaf axils. Introduced from
Europe; occasionally escaped.

*Malva moschata* L.                                                       MUSK MALLOW

   Plants 50–150 cm high, with leaves palmately 5- to 7-lobed, the lobes pinnatifid. Flowers in a terminal corymb, the lower ones solitary in the leaf axils,
4–5 cm across, white to pink. Introduced as an ornamental; occasionally
escaped and established.

*Malva neglecta* Wallr.                                                  COMMON MALLOW

   A prostrate annual weed with lobed, wavy-margined leaves, roughly kidney-shaped, 2–7 cm across. Flowers blue, lilac, or whitish, 10–15 mm across,
petals about twice the length of sepals. Fruit a circular series of 12–15 smooth
carpels. Introduced from Europe; becoming an increasingly abundant weed of
gardens and roadsides.

*Malva parviflora* L.                                              SMALL-FLOWERED MALLOW

   A prostrate mat-forming annual weed with many branches. Leaves 2–6
cm across, roughly kidney-shaped in outline, with about 7 wavy-margined
lobes. Flowers pink or lilac, about 10 mm across, with sepals almost as long as
petals. Fruit consisting of a series of 8–11 carpels or small capsules arranged in
a circle. Introduced from Europe, but becoming a common weed; of roadsides
and waste places; throughout most of the Prairie Provinces.

*Malva pusilla* Sm.                                                ROUND-LEAVED MALLOW

   A much-branched prostrate annual or biennial plant similar to *M.
parviflora*, with pale lilac flowers and wavy-margined, somewhat kidney-
shaped leaves. An introduced weed, becoming common; on wasteland and
roadsides.

*Malva sylvestris* L. (Fig. 152)     PURPLE MALLOW

Plants 50–150 cm high; leaves reniform to orbicular in outline, 5–15 cm across, usually with 5 acute lobes. Flowers 25–60 mm across, dark pink to purple. Introduced as an ornamental; occasionally escaped and established.

**Malvastrum**     false mallow

*Malvastrum coccineum* (Pursh) Gray (Fig. 153)     SCARLET MALLOW

A native perennial, with a woody base and running rootstocks. Grows 5–20 cm high, often forming large patches along roadsides and in disturbed prairie. Leaves roughly round in outline and divided to the base into wedge-shaped, lobed, and cleft leaflets. Leaves covered with fine white star-shaped hairs, giving the plant a grayish appearance. Flowers borne in a dense, short, raceme-like inflorescence, brick red (a shade rarely found in prairie flowers), 10–25 mm across. Common on dry prairie in lighter soils throughout Prairies and Parkland. Though common on virgin prairie, it is not generally noticeable because it does not flower profusely or form large clumps, but on disturbed soil or roadside cuts it often takes full possession of large areas and flowers freely.

# HYPERICACEAE—St. John's-wort family

**Hypericum**     St. John's-wort

1. Leaves narrowly lanceolate. ........................................................................ 2
   Leaves ovate or elliptic. ............................................................................ 3

2. Leaves 1- to 3-nerved, linear to oblanceolate; the margins of opposite leaves not meeting. ................................................... *H. canadense*
   Leaves 5- to 7-nerved, lanceolate; the margins of opposite leaves meeting around the stem. ......................................................... *H. majus*

3. Petals yellow, at least twice as long as the sepals; leaves, sepals, and petals black glandular. ........................................ *H. formosum* var. *nortoniae*
   Petals pinkish to greenish; plants with translucent glands. ............................................. *H. virginicum* var. *fraseri*

*Hypericum canadense* L.     CANADA ST. JOHN'S-WORT

Slender-stemmed annuals 10–40 cm high. Leaves 1–3 cm long, usually less than 3 mm wide, 1- or 3-nerved. Inflorescence few-flowered, with yellow petals the same length as or shorter than the sepals. Rare; lakeshores, wet meadows; eastern Boreal forest.

Fig. 152.  Purple mallow, *Malva sylvestris* L.

Fig. 153.   Scarlet mallow, *Malvastrum coccineum* (Pursh) Gray.

*Dicots*

*Hypericum formosum* HBK. var. *nortoniae* (M. E. Jones) C. L. Hitchc.

WESTERN ST. JOHN'S-WORT

Perennials with rootstocks; stems 15–60 cm high, few or solitary, simple or branched at the summit. Leaves ovate to oval-lanceolate, sessile, the base more or less clasping. Flowers 15–20 mm across, yellow, in more or less paniculate cymes. Wet mountain meadows and shores; southern Rocky Mountains.

*Hypericum majus* (Gray) Britt.

LARGE CANADA ST. JOHN'S-WORT

Similar to *H. canadense*, but with leaves to 4 cm long, 3–9 mm wide, 5- to 7-nerved, and somewhat clasping the stem; the lower leaf margins meeting or overlapping. Not common; wet meadows and lakeshores; Boreal forest.

*Hypericum virginicum* L. var. *fraseri* (Spach) Fern.

MARSH ST. JOHN'S-WORT

Perennial plants with erect stems 30–60 cm high. Leaves ovate-oblong to elliptic, 3–6 cm long, cordate to subcordate at the base. Flowers 15–20 mm across, with petals the same length as or somewhat longer than the sepals. Lakeshores and marshy or boggy areas; eastern Boreal forest.


## ELATINACEAE—waterwort family

*Elatine*        waterwort

*Elatine triandra* Schkuhr var. *americana* (Pursh) Fassett        MUD-PURSLANE

A tufted annual plant growing in either mud or water, rooting along the stems, 10–35 mm long. Leaves obovate, 2–6 mm long. Minute flowers occurring singly in the leaf axils. Neither leaves nor flowers stalked, but attached directly to the stem. Found occasionally along margins of ponds or slow-moving streams; in Boreal forest.


## CISTACEAE—rock-rose family

1. Petals 3, dark red, minute. ................................................................. *Lechea*
   Petals 5, yellow, conspicuous. ................................................................. 2
2. Leaves lanceolate or oblong, the main
   stem leaves 20–30 mm long. ................................................. *Helianthemum*
   Leaves narrowly linear or scale-like, 1–4
   mm long. ................................................................. *Hudsonia*

*Helianthemum*        rock-rose

*Helianthemum bicknellii* Fern.        FROSTWEED

Perennials with stems solitary or few together from rootstocks, erect or nearly so. Early flowers 5–12 in a loose terminal raceme, 15–25 mm across; later flowers apetalous, crowded on short axillary branches. Not common; open sandy or stony soils; southeastern Boreal forest.

*Hudsonia*     false heather

*Hudsonia tomentosa* Nutt.       SAND-HEATHER

A low, densely tufted, shrubby plant 10–20 cm high. Leaves 1–4 mm long, oval, densely imbricated (overlapping like shingles), and almost scale-like, pale grayish hoary. Flowers small, yellow, about 1 cm across, and borne near ends of branches. Found occasionally; on sandy shores and pine lands; Boreal forest.

*Lechea*     pinweed

Underside of leaves glabrous except on the midrib and margins; seeds 4–6, shaped like a section of orange. ...................................................................................... *L. intermedia*

Underside of leaves pubescent over entire surface; seeds 3 or 4, 2-sided or obscurely 3-sided. ............................................................................................................. *L. stricta*

*Lechea intermedia* Leggett       PINWEED

Perennial plants with pubescent stems 20–60 cm high. Leaves of the basal shoots lanceolate, 3–7 mm long; stem leaves linear-oblong, sparsely pubescent. Panicle slender, cylindric about half the height of the plants, branches to about 5 cm long. Flowers about 3 mm long, deep red. Rare; sandy forest openings; southeastern Boreal forest.

*Lechea stricta* Leggett       HAIRY PINWEED

Resembling *L. intermedia*, but coarser, with larger leaves, and more pubescent. Rare; sandy forest openings; southeastern Boreal forest.


# VIOLACEAE—violet family

*Viola*     violet

Perennial (rarely annual) herbs with either basal or alternate simple leaves with stipules. Flowers of two kinds: early spring flowers, showy and cross-fertilized; summer flowers, remaining closed and self-fertilized (cleistogamous). The spring flowers perfect, but irregular in shape, having 5 sepals and 5 petals, the lowest petal prolonged into a spur. Fruit an ovoid or cylindric capsule containing 20–60 obovate seeds.

1. Plants with well-developed leafy stems. ........................................................ 2
   Plants stemless. ............................................................................................. 9

2. Stipules large, about the same length as the leaf blades; plants annual. ............................................................ 3
   Stipules much smaller than the leaf blades; plants perennial. ......................................................................... 4

3. Corolla the same length as or shorter than the calyx, 10–15 mm across. ............................................... *V. arvensis*

*Viola adunca* J. E. Smith (Fig. 154*A*)          EARLY BLUE VIOLET

A plant 5–30 cm high, from a woody rootstock, usually showing the remains of the previous season's growth of foliage. Leaves ovate with somewhat cordate bases, 10–20 mm wide. Flowers violet or purple, 10–20 mm long,

A.C. Budd

Fig. 154.  Violets: *A*, early blue violet, *Viola adunca* J. E. Smith; *B*, northern bog violet, *Viola cucullata* Ait.; *C*, Nuttall's yellow violet, *Viola nuttallii* Pursh; *D*, Western Canada violet, *Viola rugulosa* Greene.

with somewhat bearded side petals and a spur almost as long as the petals. Flowering early in May. Fairly common; on prairie that is not too dry, and in shady places; throughout Prairies, Parklands, and southern fringe of Boreal forest.

*Viola arvensis* Murr.                                           WILD PANSY

An irregularly branched annual with leafy stems 10–30 cm high. Leaf blades varying from ovate or round lower leaves to oblong or oblanceolate upper ones. Stipules large, toothed, and leaf-like; leaves and center sections of stipules round-toothed at apex. Flowers pale yellow, with upper petals sometimes violet-tipped. Introduced from Europe, has been found as a field weed.

*Viola blanda* Willd.                                    SWEET WHITE VIOLET

Plants with slender creeping rhizomes. Leaf blades cordate-ovate, dark green. Flowers fragrant; petals white, with the 3 lower ones veined brownish purple near the base. Rich woods; eastern Boreal forest, southern Rocky Mountains.

*Viola cucullata* Ait. (Fig. 154*B*)              NORTHERN BOG VIOLET

A stemless species with all leaves basal. Earliest leaves round, later ones broadly ovate to reniform, heart-shaped at base, wavy-margined, and 3–6 cm across. Flowers fairly large and violet-colored; the side petals bearded and the spurred petal hairy. Flowering from mid-May until late July. Probably the most common of the blue violets; found in moist places on the prairies, woodlands, slough margins, and bogs; throughout the Prairie Provinces.

*Viola glabella* Nutt.                                   YELLOW WOOD VIOLET

A species usually with both basal and stem leaves. Plants smooth, with the only hairiness at the upper part of the stem and on the underside of the leaf veins. Flowers yellow, springing from the leaf bases. Found in woodlands; in southern Rocky Mountains.

*Viola nuttallii* Pursh (Fig. 154*C*)         NUTTALL'S YELLOW VIOLET

A somewhat hairy species with many basal leaves and a few stem leaves. Leaves mostly lanceolate, tapering to the stalk, and somewhat wavy-margined. Flowers yellow; petals occasionally somewhat bearded. Very early flowering. Common; on open prairie and hillsides; throughout Prairies and Parklands.

*Viola orbiculata* Geyer                      ROUND-LEAVED WOOD VIOLET

Plants small, with 1–3 small stem leaves; the basal leaves 2–5 cm across, orbicular in outline, deeply cordate. Petals pale yellow with the centers purplish-veined and the lateral petals bearded. Moist woods; in Rocky Mountains.

*Viola palustris* L.                                            MARSH VIOLET

A stemless species growing from creeping rootstocks. Stalks and leaves smooth and hairless; the leaf blades broadly ovate or heart-shaped, 2–6 cm wide. Flowers pale lilac, occasionally nearly white, with darker veins; the side petals may be somewhat bearded and the spur short and thick. Quite rare; in cold bogs and along stream banks; Boreal forest and Rocky Mountains.

*Viola pedatifida* G. Don                                              CROWFOOT VIOLET

A plant with all leaves basal and leaf blades cleft almost to the base into 3 divisions, each further cleft into 2–4 lobes. Leaves varying to 10 cm wide, and slightly hairy on margins. Flowers violet, very showy. Never abundant; found on prairies and exposed banks; Parklands, Wood Mountain, Cypress Hills.

*Viola pubescens* Ait.                                              DOWNY YELLOW VIOLET

A stout-stemmed species 15–30 cm high, covered with soft, downy hairs. Rarely some basal leaves and a few stem leaves near the top of the stem. Leaf blades broadly ovate with heart-shaped bases; stipules fairly large and ovate-lanceolate. Flowers bright yellow, growing on stalks from axils of stem leaves, with bearded side petals and a short spur. Found often; in moist woodlands; eastern Parklands and southeastern Boreal forest.

*Viola renifolia* Gray                                              KIDNEY-SHAPED VIOLET

A low-growing plant with running rootstocks and kidney-shaped, wavy-margined leaves with heart-shaped bases. Flowers white, without bearded petals. Cleistogamous flowers purple, on horizontal stalks. Flowering very early in spring. In cold woodlands, forests, and swamps; eastern Parklands, Boreal forest, and Cypress Hills.

*Viola rugulosa* Greene (Fig. 154*D*)                          WESTERN CANADA VIOLET

A tall woodland species with stems 20–60 cm high growing from numerous stolons. Cordate or heart-shaped leaves pointed at apex, borne on long stalks, and often up to 10 cm across, although becoming small toward the top of stems; often densely hairy beneath. Flowers white with pinkish or purplish veins, although the complete flower is sometimes pale pink. The common white violet of woodlands throughout Western Canada. Transplants readily into gardens and then forms a low, dense mass of plants flowering copiously quite early in the season. Spreads rapidly by its slender, numerous, white roots and may become too aggressive. Very common; in shady woodlands; throughout the Prairie Provinces.

*Viola selkirkii* Pursh                                              LONG-SPURRED VIOLET

A delicate species 6–15 cm high, tufted from elongate slender rhizomes. Leaf blades 15–30 mm long, elongating after flowering. Flowers about 15 mm across, pale violet, with the blunt spur 5–8 mm long. Moist or wet woods; Parklands, Boreal forest, and Rocky Mountains.

*Viola tricolor* L.                                                            PANSY

Annual, biennial, or perennial plants 10–40 cm high, glabrous or somewhat pubescent. Flowers usually 10–25 mm across, sometimes to 35 mm, violet, yellowish, or 3-colored, with a spur 3–6.5 mm long. An introduced species, very variable; occasionally escaped from cultivation, and established.

# LOASACEAE—loasa family

***Mentzelia***        sand-lily

*Mentzelia decapetala* (Pursh) Urban & Gilg. (Fig. 155)        EVENINGSTAR

Stout, erect, biennial plants 15–60 cm high, with rough, pale gray stems. Leaves alternate, oblanceolate to lanceolate, 5–15 cm long; the lower ones with short stalks, but the upper ones stalkless. Blades sharply and coarsely toothed and covered with tiny, white, stiff bristles, making them very rough to touch. Flowers borne singly or in clusters of 2 or 3 at ends of stems or branches; sharp-pointed sepals about 25 mm long; and petals 35–50 mm long, creamy white, narrow, and pointed at apex. Only 5 petals appearing to be 10, the inner ones being petal-like sterile stamens. Flowers opening only in evening. Fruit an oblong capsule about 35 mm long and up to 12 mm thick, opening at the top, and containing many seeds. Occasionally leafy bracts are attached to base or side of covering of capsule. Plentiful locally, but not common; on eroded hillsides and clay banks; throughout Prairies and on badlands. Syn.: *Nuttallia decapetala* (Pursh) Greene.

# CACTACEAE—cactus family

Almost leafless perennial plants with stems fleshy, thickened, succulent, and covered with spines. Flowers perfect and regular with many sepals and petals. Fruit a fleshy berry relished by antelope and sheep.

Globose or cushion-like plants; flowers red or
   purple. ................................................................................................. *Mamillaria*
Branching or jointed plants; stems flattened;
   flowers large, yellow or orange. ........................................................ *Opuntia*

***Mamillaria***        ball cactus

*Mamillaria vivipara* (Nutt.) Haw.        PURPLE CACTUS

A cushion-like cactus 3–20 cm high and 3–30 cm across, covered with somewhat cone-shaped tubercles each bearing a cluster of 3–8 reddish brown spines 12–20 mm long. Flowers borne between tubercles, 3–5 cm across, with numerous purple or dark red narrow petals and a yellow center of many stamens. Fruit a pale green fleshy berry 1–2 cm long, turning brown with age, and very sweet and edible when ripe. Very common; on open prairie and hillsides; throughout Prairies and southern fringe of Parklands. Syn.: *Neomamillaria vivipara* (Nutt.) Britt. & Rose.

***Opuntia***        prickly-pear

Cacti with flattened stems jointed and divided into somewhat plate-like sections called internodes. Plant often bearing small, scale-like, reddish leaves, soon falling off. Spines long, with a tuft of barbed bristles at the base. Flowers very showy, with large waxy petals. Fruit quite sweet and edible.

A.C. Budd

Fig. 155.   Eveningstar, *Mentzelia decapetala* (Pursh) Urban & Gilg.

Internodes very fleshy, often circular in cross
section, 3–5 cm long, the terminal one easily
breaking loose. ..................................................................................... *O. fragilis*
Internodes never circular in cross section, 5–15
cm long, always broader than thick. ......................................... *O. polyacantha*

*Opuntia fragilis* (Nutt.) Haw.                                   BRITTLE PRICKLY-PEAR

A low-growing decumbent cactus, often forming very large mats, red or
reddish green. Spines in divaricate groups, 10–25 mm long. Flowers pale yellow, about 5 cm across. Fruit a fleshy berry 15–25 mm long. Common on dry
prairie throughout Prairies; rare in western Parklands and Peace River.

*Opuntia polyacantha* Haw.                                              PRICKLY-PEAR

A prostrate bright green plant growing in large clumps. The internodes
large and much flattened, 5–15 cm long with reddish brown spines 1–5 cm
long. Flowers showy, yellow to pinkish orange, 5–8 cm across. Fruit a prickly
berry 25–35 mm long containing numerous seeds. Very common; on dry
prairies and light soils, increasing greatly through overgrazing and erosion;
Prairies.

# ELAEAGNACEAE—oleaster family

Shrubs or small trees with silvery scurfy leaves. Flowers either perfect or
unisexual, sometimes both kinds of flowers on the same plant and sometimes
only one sex on a plant. Flowers with 4 sepals but no petals, and 4 or 8 stamens. Fruit drupe-like. All species in this family have nodules on the root system in which nitrogen compounds are formed.

Leaves alternate; flowers perfect, with 4
stamens. ....................................................................................... *Elaeagnus*
Leaves opposite; flowers unisexual, the sexes
being on different plants, with 8 stamens. ....................................... *Shepherdia*

**Elaeagnus**      oleaster

Leaves lanceolate; shrubs spiny; twigs white
stellate-pubescent. ................................................................. *E. angustifolia*
Leaves elliptic; shrubs not spiny; twigs brown
stellate-pubescent. ................................................................... *E. commutata*

*Elaeagnus angustifolia* L.                                            RUSSIAN OLIVE

Shrubs or small trees 4–7 m high. Leaves lanceolate, 3–10 cm long. Flowers inconspicuous, in small clusters; sepals spreading, yellow within, about 4
mm across; fruit about 1 cm long, yellowish with silvery scales. An introduced
species, planted as an ornamental and occasionally escaped.

*Elaeagnus commutata* Bernh.                              SILVERBERRY, WOLF-WILLOW

Shrubs or small trees 2–5 m high, with brown scurfy twigs. Alternate
leaves silvery, scurfy on both sides, oblong or elliptic, 2–6 cm long. Flowers in

clusters of 2 or 3 in axils of leaves, yellowish, very fragrant. Fruit oval, drupe-like, silvery, about 1 cm long, containing a large stony seed. Very common on lighter soils (where moisture is plentiful); throughout the Prairie Provinces. Spreads rapidly in overgrazed pastures throughout Parklands. In sparse stands, the plant is kept grazed and does not spread.

***Shepherdia***    buffaloberry

Leaves oblong, silvery on both sides; shrub or
  small tree bearing long thorns. .......................................................................... *S. argentea*
Leaves oval, green above, silvery below; low
  undershrub without thorns. ......................................................................... *S. canadensis*

*Shepherdia argentea* Nutt. (Fig. 156)                BUFFALOBERRY
    A thorny shrub 1–5 m high, with whitish branches. Leaves oblong, 2–5 cm long, and densely silvery scurfy on both sides. Flowers unisexual, all flowers on a plant being the same sex, brownish, in small clusters at nodes formed in the preceding season. Fruit rounded, 3–5 mm across, orange and very sour, but after a hard frost being a good jelly fruit. Common around sloughs, in cou-lees, and on light soils; Prairies.

*Shepherdia canadensis* (L.) Nutt.                CANADA BUFFALOBERRY
    An unarmed undershrub 0.5–3 m high, with brown scurfy branches. Oval or ovate leaves 2–4 cm long, with green upper surface, but with silvery star-shaped hairs on underside. Flowers yellowish, borne at leaf nodes. Female plant producing round or oval fruit 3–5 mm long, reddish or yellowish, and insipid tasting. Fairly common in wooded places and riverbanks in Parklands and Boreal forest, rare in Cypress Hills, along South Saskatchewan River breaks, and adjacent areas.


# LYTHRACEAE—loosestrife family

***Lythrum***    loosestrife

*Lythrum salicaria* L.                PURPLE LOOSESTRIFE
    An erect perennial herb with stems 50–100 cm high; foliage glabrous to sparsely pubescent; leaves opposite or whorled, 3–10 cm long. Inflorescence 10–40 cm long, with foliaceous bracts; flowers 15–20 mm across, purple. Intro-duced, marsh species; Boreal forest.


# ONAGRACEAE—evening-primrose family

    Annual or perennial herbs with either opposite or alternate leaves. Flow-ers perfect and generally regular, with usually 4 sepals and 4 petals, and with as many, or twice as many, stamens as sepals. Style slender, either knobbed or

Fig. 156.  Buffaloberry, *Shepherdia argentea* Nutt.

4-lobed at the summit. Fruit either a capsule or a small nut. In some genera
seeds silky-tufted.

1.  Parts of flowers in 2's; fruit covered with
    hooked bristles; leaves opposite; plants
    with rootstocks. ........................................................................ *Circaea*

    Parts of flowers in 4's; fruit not covered
    with hooks. ........................................................................................ 2

2.  Fruit not splitting open when mature, 1-
    to 4-seeded; short and ribbed. ........................................ *Gaura*

    Fruit splitting open either at sides or sum-
    mit when mature. .......................................................................... 3

3.  Seed with a tuft of silky hairs. ................................... *Epilobium*

    Seed without a tuft of silky hairs. .................................... 4

4.  Capsules opening with 2 valves. ........................ *Gayophytum*

    Capsules opening with 4 valves. ..................................... 5

5.  Petals purple; anthers attached near their
    base. .............................................................................. *Boisduvalia*

    Petals white or yellow; anthers attached
    near their center. ............................................... *Oenothera*

### *Boisduvalia*     boisduvalia

*Boisduvalia glabella* (Nutt.) Walp.         SMOOTH BOISDUVALIA

Native annual plants 10–30 cm high, with leaves alternate, stalkless, lan-
ceolate-ovate, 10–20 mm long. Flowers small and in leafy terminal spikes, purple or violet, with petals almost 3 mm long. Rare; found in moist, mostly alkaline areas; Prairies.

### *Circaea*     enchanter's-nightshade

Leaves ovate, mostly less than twice as long as
  wide; fruit oblanceolate, not furrowed. ............................................... *C. alpina*

Leaves oblong-ovate, usually more than twice
  as long as wide; fruit obovoid, deeply
  furrowed. ........................................................ *C. quadrisulcata* var. *canadensis*

*Circaea alpina* L.         SMALL ENCHANTER'S-NIGHTSHADE

A slender plant 5–20 cm high, with leaves opposite, stalked, heart-shaped,
coarsely toothed, and 2–5 cm long. Flowers borne in a terminal raceme, very
small and white, on slender stalks; 2 sepals and 2 notched white petals, with 2
stamens. Fruit an ovoid or club-shaped capsule 2–3 mm long covered with fine
hooked hairs. In moist woodlands; Boreal forest, eastern Parklands.

*Circaea quadrisulcata* (Max.) Franch. & Sav. var. *canadensis* (L.) Hara
                    LARGE ENCHANTER'S-NIGHTSHADE

Plants with erect stems to 60 cm high or more, glabrous below, minutely
pubescent above. Leaves 6–12 cm long, rounded or subcordate at the base.

Racemes to 20 cm long; fruit 3.5–5.0 mm long. Rare; moist woods; southeastern Boreal forest.

**Epilobium**     willowherb

Flowers usually in spikes or racemes, with 4 sepals and 4 petals. Leaves opposite or alternate, often with lobed blades. Fruit a long, linear, nearly 4-angled capsule containing many seeds; each seed bearing a tuft of silky hairs at the upper end; capsule splitting lengthwise at maturity.

1. Flowers large, more than 12 mm across. ................................................................... 2
   Flowers small, less than 12 mm across. .................................................................. 3

2. Racemes long and narrow, not leafy;
   leaves with a vein running parallel to
   leaf margin; bracts small. ................................................................ *E. angustifolium*
   Racemes very leafy; leaves without a vein
   parallel to leaf margin; bracts leaf-like. ............................................. *E. latifolium*

3. Annual with shreddy, straw-colored bark;
   stigma 4-cleft. ................................................................................. *E. paniculatum*
   Perennial; stigmas entire or merely cleft. ................................................................. 4

4. Leaves linear or narrowly lanceolate, not
   toothed, often with rolled margins;
   stems not angled. ................................................................................. *E. palustre*
   Leaves lanceolate and toothed, not rolled
   at margins. ................................................................................................. 5

5. Plants low, often decumbent; leaves ovate
   to elliptic. ......................................................................................... *E. alpinum*
   Plants mostly erect; leaves lanceolate to
   oblong. ............................................................................................. *E. ciliatum*

*Epilobium alpinum* L.                                      ALPINE WILLOWHERB

Stems usually flexuous, 5–15 cm high. Basal leaves small, 3–10 mm long; stem leaves 10–25 mm long. Flowers solitary or in pairs, 8–12 mm across, pink or rose. Alpine meadows, moist areas; Rocky Mountains.

*Epilobium angustifolium* L. (Fig. 157)                              FIREWEED

An erect, fairly stout perennial plant 60–150 cm high, with alternate, very short-stalked, entire lanceolate leaves 5–15 cm long. Venation of leaves of this species interesting, because of lateral veins running parallel to leaf margins. Leaves slightly paler below than on upper side. Flowers pink to purple, 15–35 mm across, in a long terminal raceme, with a small bract below each flower stalk. Fruit a long, linear, somewhat 4-angled capsule 5–8 cm long, splitting lengthwise to release the numerous tufted seeds. One of the most ubiquitous of plants, being found over most of the northern hemisphere. A good source of honey, and fairly palatable to livestock, especially the upper and more tender tips. Very common; in woodlands, edges of forest, burned-over forests; throughout Boreal forest and Parklands; occasionally along roadsides and in moist places in Prairies. Syn.: *Chamaenerion spicatum* (Lam.) S. F. Gray; *C. angustifolium* (L.) Scop.

A.C. Budd

Fig. 157.   Fireweed, *Epilobium angustifolium* L.

*Epilobium ciliatum* Raf.                    NORTHERN WILLOWHERB

An erect perennial plant 30–100 cm high, sometimes rather sticky. Leaves mostly opposite, lanceolate or ovate-lanceolate, 2–7 cm long. Flowers pink, usually nodding in their early stage, about 6 mm across, with a long ovary tube beneath the flower. Fruit many-seeded, 3–5 cm long; seeds with a white tuft of hairs. Common; in sloughs and wet places; throughout the Prairie Provinces. Syn.: *E. adenocaulon* Hausskn.; *E. glandulosum* Lehm. var. *adenocaulon* (Hausskn.) Fern.

*Epilobium latifolium* L.                    BROAD-LEAVED FIREWEED

An erect or somewhat decumbent perennial plant, usually rather branched, 15–50 cm high. Leaves ovate to ovate-lanceolate, entire, 2–5 cm long, some opposite, some alternate. Flowers large and showy, purple, 2–5 cm across, in short leafy racemes; leaves intermixed with flowers. Found along banks of streams; in southern Rocky Mountains and northeastern Boreal forest. Syn.: *Chamaenerion latifolium* (L.) Sweet.

*Epilobium palustre* L.                    MARSH WILLOWHERB

An erect, branched perennial plant 30–60 cm high. Leaves very narrow and linear, 2–5 cm long, glabrous or covered with curled hairs, too fine to be seen without a small lens. Flowers few, pink or whitish, about 6 mm across. Fruit a capsule, 3–5 cm long; seeds bearing a dingy white tuft of hairs. Found in wet places and swamps; throughout the Prairie Provinces, but nowhere very common. Syn.: *E. lineare* Muhl.; *E. leptophyllum* Raf.

*Epilobium paniculatum* Nutt.                    ANNUAL WILLOWHERB

An annual with a pale, straw-colored, somewhat shreddy stem, erect, branched, 25–50 cm high. Leaves linear or linear-lanceolate, 20–35 mm long. Flowers pink, about 1 cm across, at the end of a tube-like ovary about 1 cm long. Capsule 10–25 mm long, splitting open to release the seeds, each bearing a tuft of white silky hairs. Common; on light and sandy soil; throughout southwestern part of the Prairie Provinces. Syn.: *E. adenocladon* (Hausskn.) Rydb.

**Gaura**      butterflyweed

Perennial, branching herbs with alternate, narrow leaves. Flowers have a long calyx tube, and 4 reflexed sepal lobes, 4 petals, and 8 stamens. Fruit a nut-like capsule, containing 1–4 seeds, without hairy tufts.

*Gaura coccinea* Pursh                    SCARLET GAURA

A perennial, decumbent to erect, 10–30 cm high, almost entirely covered with fine grayish hairs. Leaves numerous, oblong or lanceolate, alternate, without stalks, sometimes wavy-margined or shallowly toothed, 10–30 mm long. Flowers in terminal racemes, white when they emerge, but becoming scarlet in a few hours, about 1 cm across. Capsule about 6 mm long containing 1–4 seeds. Common; on dry prairies and hillsides; throughout Prairies and Parklands. The var. *glabra* (Lehm.) Torr. & Gray very similar to the species, but hairless, except for basal parts of flowers.

*Dicots*

***Gayophytum***     willowherb

*Gayophytum humile* Juss.                                    LOW WILLOWHERB

Plants 5–15 cm high, glabrous, branched from the base; leaves linear to linear-lanceolate, 1–3 cm long, entire. Flowers axillary, white, 2–3 mm across; capsule 10–15 mm long, flattened. Rare; disturbed areas; southern Rocky Mountains.

***Oenothera***     evening-primrose

A variable genus, some species being tall and erect, others low and stemless. Flowers having the tube of the calyx extended beyond the ovary, with 4 lobes turning downward, 4 petals, and 8 stamens. Flowers very variable in color and size; fruit a capsule opening at the top with 4 lobes and containing many seeds not bearing a hairy tuft.

1. Plants distinctly stemmed. ........................................................................ 2
   Plants stemless or nearly so. ..................................................................... 7
2. Much-branched perennials with brown
     woody stems; flowers yellow; stigma
     not deeply 4-cleft. ................................................................ *O. serrulata*
   Plants without brown woody stems;
     stigma deeply 4-cleft. ............................................................................ 3
3. Flowers white or pinkish; stems white
     with   shreddy   bark,   usually
     much-branched. ..................................................................... *O. nuttallii*
   Flowers  yellow;  plant  usually  not
     much-branched. ...................................................................................... 4
4. Petals 1–3 mm long. .................................................................. *O. andina*
   Petals more than 5 mm long. .................................................................... 5
5. Capsules  rounded  on  the  angles,  not
     winged. ................................................................................ *O. biennis*
   Capsules 4-angled, winged. ..................................................................... 6
6. Petals  5–9  mm  long,  inflorescence
     nodding. ............................................................................. *O. perennis*
   Petals 10–25 mm long, inflorescence erect. ........................... *O. fruticosa* var. *linearis*
7. Stigma not 4-cleft. ............................................................... *O. breviflora*
   Stigma divided into 4 linear lobes. ......................................................... 8
8. Flowers yellow, turning pink with age;
     petals less than 25 mm long; capsule
     narrowly winged at angles. ........................................................ *O. flava*
   Flowers white, turning pink with age; pet-
     als usually more than 25 mm long; cap-
     sule with double crests at angles. ....................................... *O. caespitosa*

*Oenothera andina* Nutt.                          UPLAND EVENING-PRIMROSE

Low, slender-stemmed annuals 2–15 cm high, with branches spreading from the base, finely pubescent. Leaves alternate, linear to narrowly lanceolate. Flowers axillary, yellow; capsule 5–6 mm long. Rare; dry sandy soils; western Prairies.

*Oenothera biennis* L.                                                 YELLOW EVENING-PRIMROSE

An erect biennial from a taproot, 30–200 cm high. Leaves lanceolate or ovate-lanceolate, 2–15 cm long, stalkless except for short stalks on lower leaves. Flowers yellow, 2–5 cm across, opening in evening, borne in a leafy terminal spike. Capsules 2–3 cm long, finely hairy, and opening at the top when mature. The type species is fairly common in the southeast, where it is an introduced weed. However, the common forms found over most of the Prairie Provinces are var. *canescens* Torr. & Gray and var. *hirsutissima* Gray.

The var. *canescens* Torr. & Gray, western yellow evening-primrose, very similar to the typical form, but usually distinguishable by the length of the free tips or lobed portions of the sepals: 3 mm or more in the typical forms, but only slightly over 1.5 mm in the variety. Very common; on lighter soils; throughout southwestern portion of the Prairie Provinces, and occasionally found in Boreal forest.

*Oenothera breviflora* (Nutt.) Torr. & Gray                               TARAXIA

A stemless perennial growing from a taproot. Leaves finely hairy, deeply incised, 5–12 cm long. Flowers yellow, turning reddish when dried, each 10–15 mm across with 4 sepals, 4 petals, and 8 stamens. Stigma or female portion of flower knobbed at end and not divided into 4 linear lobes. Fruit a 4-winged capsule about 15 mm long. Very rare; has been found on heavy soil on a slough margin; in western Prairies. Syn.: *Taraxia breviflora* Nutt.

*Oenothera caespitosa* Nutt. (Fig. 158)                        GUMBO EVENING-PRIMROSE

A low, stemless perennial from a thick woody root. Leaves oblanceolate to lanceolate, 7–20 cm long, growing from short, winged stalks, sometimes toothed and sometimes wavy-margined. Flowers 3–8 cm across, borne on stalks from the root crown, sweet-scented, white, opening in early morning, but soon fading to a pale pink. Capsules about 3 cm long, without stalks, in cluster on root crown. Found on dry hillsides of gumbo or clay soil and sometimes on gumbo flats; throughout Prairies. Syn.: *Pachylophus caespitosus* (Nutt.) Raim.

The var. *montana* (Nutt.) Durand is similar to the typical form but with smaller flowers. Capsules only about 2 cm long and wavy-margined leaves with fine white hairs around their margins. Flowers very early in the season. One of the sweetest-smelling flowers on the prairie, and a good garden plant. Found more often than the typical form; on clay hillsides; in western Prairies. Syn.: *Pachylophus montanus* (Nutt.) A. Nels.

*Oenothera flava* (A. Nels.) Garrett                                YELLOW LAVAUXIA

A low-growing, stemless perennial from a fleshy taproot. Leaves long and narrow, oblong-lanceolate, very deeply incised, medium green, 10–25 cm long, midrib very prominent on the underside. Flowers yellow, turning pink; petals 10–20 mm long. Capsules winged at the 4 angles, 2–3 cm long. Not common, but found occasionally; in valleys, slough margins, and drainage channels; in western Prairies. Syn.: *Lavauxia flava* A. Nels.

Fig. 158.   Gumbo evening-primrose, *Oenothera caespitosa* Nutt.

*Oenothera fruticosa* L. var. *linearis* (Michx.) Wats.　　　　LARGE SUNDROPS

Stems spreading or erect, pubescent, 30–50 cm high. Stem leaves to 6 cm long, linear to lanceolate. Flowers 2–5 cm across; sepals pubescent; capsules 6–10 mm long. Rare; introduced from the east; southeastern Parklands.

*Oenothera nuttallii* Sweet　　　　WHITE EVENING-PRIMROSE

An erect, often branched perennial 30–100 cm high from a white, fleshy rootstock. Stems white, somewhat shiny, and with a shreddy bark. Leaves entire, linear, 2–10 cm long, pale green, and with wavy margins. Unpleasantly scented flowers borne from axils of upper leaves, white but turning pinkish as they fade, about 4 cm across. Usually they open in the morning. Capsules narrow, somewhat curved, 4-angled, and about 25 mm long. Very common; on roadsides in light soil areas, and on sandy land, persistent in cultivated fields; Prairies and Parklands. Syn.: *Anogra nuttallii* (Sweet) A. Nels.

*Oenothera perennis* L.　　　　SUNDROPS

Stems 20–60 cm high, erect, usually simple. Leaves 3–6 cm long, oblanceolate to elliptic. Inflorescence nodding, becoming erect during flowering; flowers 1–2 cm across; capsules 5–10 mm long. Rare; gravelly soils; southeastern Parklands and Boreal forest.

*Oenothera serrulata* Nutt.　　　　SHRUBBY EVENING-PRIMROSE

An erect or decumbent, woody-crowned, brown, shrubby perennial 10–40 cm high. Leaves alternate, spatulate to linear-oblong or linear, 1–5 cm long, often entire but sometimes with small teeth, pale green. Flowers bright yellow, 10–25 mm across, with an almost disk-like stigma. Capsules 15–20 mm long. Not common; found on dry prairie and hillsides; Prairies and Parklands. Syn.: *Meriolix serrulata* (Nutt.) Walp.

## HALORAGACEAE—water-milfoil family

Perennial aquatic or marsh plants with creeping roots. Leaves alternate, or in whorls, with the inconspicuous flowers borne in the axils of the leaves. Petals 4 or absent; fruit a nutlet with 2 or 4 single-seeded sections.

Leaves entire and in whorls; flowers perfect,
　with 1 stamen and a pistil. ............................................................. *Hippuris*
Submersed leaves dissected; flowers perfect or
　unisexual, with 2–8 stamens. ..................................................... *Myriophyllum*

**Hippuris**　　　mare's-tail

*Hippuris vulgaris* L.　　　　MARE'S-TAIL

An aquatic plant with creeping rootstocks, 20–50 cm high, with simple unbranched stems. Leaves in whorls of 6–12 around the stem, linear, 1–3 cm long; those under water are soft and flaccid, the ones above water firm. Flowers lack sepals and petals, but are usually perfect, and borne in the axils of the leaves. Very common; in streams and sloughs, and in water and mud along the banks; throughout the Prairie Provinces.

*Myriophyllum*        water-milfoil

1. Most or all of the bracts and flowers
   alternate. ............................................................ *M. alterniflorum*

   Most or all of the bracts and flowers
   whorled. ........................................................................................ 2

2. Leaves all in whorls; stamens 8; floral
   leaves shorter than flowers, and usually
   entire. ......................................................................... *M. spicatum*

   Some leaves alternate; stamens 4; floral
   leaves longer than flowers, and
   toothed. ...................................................................... *M. pinnatum*

*Myriophyllum alterniflorum* DC.                    WATER-MILFOIL

Plants with very slender stems 15–20 cm long; leaves whorled, usually 5–10 mm long. Spikes emersed, 2–5 cm long; male flowers solitary or in opposite pairs, female flowers whorled at the base of the spike. Shallow water; Boreal forests.

*Myriophyllum pinnatum* (Walt.) BSP.            PINNATE WATER-MILFOIL

A smaller aquatic perennial 10–20 cm high, with whorled submersed leaves in 3–5 pairs of thread-like divisions. Floral leaves longer than flowers and toothed. Male flowers purplish, with 4 stamens, on spikes 10–20 cm long. A southern species, very uncommon; found in extreme southern part of the Prairie Provinces. Syn.: *M. scabratum* Michx.

*Myriophyllum spicatum* L.                    SPIKED WATER-MILFOIL

An aquatic perennial with branched stems 30–150 cm long. Leaves in whorls, 10–30 cm long, divided into many thread-like divisions. Flowers purplish, small, with 4 petals and 8 stamens, borne in whorls on a spike 2–8 cm long protruding above the water. Common; in streams, sloughs, and lakes; throughout the Prairie Provinces. Syn.: *M. spicatum* L. var. *exalbescens* (Fern.) Jepson; *M. exalbescens* Fern.


# ARALIACEAE—ginseng family

Leaves compound. .......................................................................... *Aralia*

Leaves simple. .......................................................................... *Oplopanax*

*Aralia*        wild sarsaparilla

1. Plants scapose; leaves and peduncle
   arising from the rootstock, glabrous. ................................. *A. nudicaulis*

   Plants with leafy stems. ....................................................................... 2

2. Umbels 2–10 in a loose cluster; stems
   bristly below. ................................................................ *A. hispida*

   Umbels numerous in a compound pani-
   cle; stems not bristly. ........................................................ *A. racemosa*

*Aralia hispida* Vent. (Fig. 159)                    BRISTLY SARSAPARILLA

Plants with stout rootstocks; stems up to 1 m high, bristly at the base with slender spines. Leaves few, with the petioles usually shorter than the blades; leaflets 2–6 cm long, oblong to ovate. Inflorescence terminal, with 2–10 umbels; berries globose. Dry woods and rock outcrops; Boreal forests.

*Aralia nudicaulis* L. (Fig. 160)                    WILD SARSAPARILLA

A perennial from a creeping rootstock, usually with a single leaf on a stalk 15–30 cm high. This stalk divides into 3 parts, each of which is again divided into 3–5 leaflets. Leaflets oval, pointed at apex, 5–15 cm long, finely toothed on margins, dark green above, and very pale green below. Flowers greenish, usually in 3 umbels 2–5 cm across, on a flowering stalk 10–40 cm high, growing from the rootstock. Fruit globular, purplish black, about 6 mm long. Common; in shady, rich woodlands and deep wooded ravines; throughout the Prairie Provinces. Occasionally forming the major part of herbaceous vegetation in forest.

*Aralia racemosa* L.                    SPIKENARD

Stout herbs to 2 m high. Leaves few, to 80 cm long, the three primary segments pinnately compound; leaflets up to 15 cm long, serrate. Inflorescence a large panicle with numerous umbels; fruit dark purple. Rare; rich woods; southeastern Boreal forest.

**Oplopanax**        devil's-club

*Oplopanax horridum* (J. E. Smith) Miq.                    DEVIL'S-CLUB

A densely prickly shrub 2–4 m high, with an unpleasant odor. Leaves rounded in outline, 15–50 cm broad, cordate at base, palmately 5- to 7-lobed, petioles and veins spiny. Inflorescence terminal, 10–30 cm long, pubescent and prickly; berry scarlet, 4–5 mm long. Moist woods; Rocky Mountains, western Boreal forest.

## UMBELLIFERAE—parsley family

Hollow-stemmed herbs with alternate usually divided leaves. Bases of leaf stalks usually enlarged and sheathing stem. Flowers small, in umbels, simple or compound. Sepals minute, 5 petals, 5 stamens, and 2 styles. Fruit a mericarp consisting of 2 single-seeded carpels united into one capsule, usually ribbed and sometimes winged. A large and widespread family, often difficult to identify positively without the fruit.

(Figs. 159 and 160 overleaf)

Fig. 159.   Bristly sarsaparilla, *Aralia hispida* Vent.

A.C. Budd

Fig. 160.   Wild sarsaparilla, *Aralia nudicaulis* L.

1. Foliage spiny; inflorescence bluish; invo-
   lucre spiny. .................................................................... *Eryngium*

   Foliage not spiny; inflorescence not
   bluish. ...................................................................................... 2

2. Fruit with hooked spines; foliage digi-
   tately compound. ............................................................ *Sanicula*

   Fruit not spiny; foliage not digitately
   compound. .............................................................................. 3

3. Leaves simple or divided into few well-de-
   veloped leaflets. .................................................................. 4

   Leaves much divided, with leaflets or
   segments numerous. ......................................................... 16

4. Leaves simple or with 3 leaflets. ...................................... 5

   Leaves with more than 3 leaflets. ..................................... 8

5. Leaves simple, entire. ................................................ *Bupleurum*

   Leaves serrate to trifoliate. ............................................... 6

6. Leaflets very large, 10 cm or more wide. ............... *Heracleum*

   Leaflets much smaller. ........................................................ 7

7. Flowers white; rays of umbel very
   uneven. ........................................................................ *Cryptotaenia*

   Flowers yellow; rays rather even. ............................... *Zizia*

8. Leaves pinnately divided. .................................................. 9

   Leaves ternately divided. ................................................ 10

9. Leaflets linear- to ovate-lanceolate;
   flowers white. ......................................................................... *Sium*

   Leaflets ovate to oblong; flowers yellow. ............... *Pastinaca*

10. Leaflets entire or with a few lobes. ............................... 11

    Leaflets finely to deeply serrate. .................................... 12

11. Plants tall, 1–2 m high; leaflets cuneate. ............. *Levisticum*

    Plants low, scapose; leaflets linear or
    lanceolate. ................................................................... *Lomatium*

12. Fruit strongly flattened, broadly winged. ............. *Angelica*

    Fruit flattened but wingless. ......................................... 13

13. Leaves divided into leaflets of about equal
    size, usually 3 segments each with 3
    leaflets. ................................................................... *Aegopodium*

    Leaves divided into leaflets of unequal
    size, often with 5 or more segments. ........................... 14

14. Fruit linear, more than 1 cm long, usually
    pubescent. .................................................................. *Osmorhiza*

    Fruit not linear, less than 1 cm long,
    glabrous. ................................................................................ 15

15. Flowers yellow; the central flower in the
    umbellets sessile. .............................................................. *Zizia*

    Flowers white; the central flower
    pedicellate. ....................................................................... *Cicuta*

16. Flowers in part replaced by axillary
    bulblets. ................................................................................ *Cicuta*

    Flowers not replaced by bulblets. ........................................ 17

17. Involucral bracts large, pinnately
    dissected. ............................................................................. *Daucus*

    Involucral bracts small, slightly or not
    dissected. .................................................................................... 18

18. Umbel simple, few-flowered. ............................................. *Scandix*

    Umbel compound, many-flowered. ........................................ 19

19. Plants scapose. ............................................................................. 20

    Plants with stem leaves. ........................................................... 22

20. Flowers white; fruit conspicuously
    winged on marginal and dorsal nerves. .................... *Cymopterus*

    Flowers yellow; fruit winged on the mar-
    ginal nerves only or wingless. ................................................ 21

21. Fruit wingless. ................................................................ *Musineon*

    Fruit winged. .................................................................... *Lomatium*

22. Lower stem leaves opposite; most leaves
    basal. ............................................................................... *Musineon*

    Lower stem leaves alternate; stem leaves
    sometimes opposite in the inflor-
    escence. ...................................................................................... 23

23. Leaf segments few, usually 5, long and
    linear. ............................................................................ *Perideridia*

    Leaf segments more than 5. .................................................... 24

24. Flowers yellow. ............................................................................ 25

    Flowers white or pinkish. ........................................................ 26

25. Plants strongly scented; introduced
    annuals. ............................................................................ *Anethum*

    Plants not scented; native perennials. ........................... *Lomatium*

26. Stout plants with purplish-spotted stems,
    to 3 m high. ......................................................................... *Conium*

    Plants with stems not purplish-spotted, to
    1 m high. ............................................................................... *Carum*


***Aegopodium***  goutweed

*Aegopodium podagraria* L.  GOUTWEED

A perennial herb with long creeping rootstocks; stems erect, 40–90 cm
high. Lower leaves long-petioled, mostly with 9 leaflets, very similar in size and
shape; leaflets 3–8 cm long, oblong to ovate, sharply serrate. Umbels 6–15 cm
wide, dense, with 15–25 primary rays; fruit 3–4 mm long. Introduced; formerly
much cultivated for medicinal purposes and as a vegetable; occasionally
escaped.

*Anethum*    dill

*Anethum graveolens* L.    DILLSEED

Glabrous and more or less glaucous herbs, to 1.5 m high, often much branched above. Leaves ovate in outline, pinnately dissected into numerous filiform segments, 5–20 mm long. Umbels to 15 cm across, usually with 30–40 primary rays; fruit 3–5 mm long. Introduced; occasionally escaped from cultivation.

*Angelica*    angelica

1. Flowers yellow; the involucral bracts
   large. ............................................................................. *A. dawsonii*
   Flowers white or pinkish; the involucral
   bracts small or none. ...................................................................... 2
2. Ovaries glabrous; pedicels conspicuously
   webbed. ......................................................................................... *A. arguta*
   Ovaries hispid; pedicels glabrous. ............................................. *A. genuflexa*

*Angelica arguta* Nutt.    LYALL'S ANGELICA

A tall, stout, smooth-stemmed plant 60–150 cm high, with a woody root. Leaves several times pinnately compound; leaflets ovate to lanceolate, 2–10 cm long, with toothed margins. Flowers white, in large compound umbels. Fruit oblong and smooth, 4–7 mm long, often reddish or purplish, making the fruiting plant very conspicuous. Whole plant with a pleasing odor. Found only in wet meadows and along mountain streams; southern Rocky Mountains. Syn.: *A. lyallii* S. Wats.

*Angelica dawsonii* Wats.    MOUNTAIN PARSNIP

A stout plant, 40–80 cm high, with a thick, woody, strongly scented root, glabrous. Leaves with 9–15 leaflets, these 3–6 cm long, sharply serrate. Umbels solitary on long peduncles; the involucre of bracts 1–3 cm long, toothed or divided, sometimes rather inconspicuous. Moist woods; southern Rocky Mountains.

*Angelica genuflexa* Nutt.    KNEELING ANGELICA

Stout plants 40–100 cm high, with the foliage glabrous to somewhat scabrous. Leaves ternate-pinnate or biternate, with the main divisions commonly reflexed, the rachis geniculate; leaflets 4–10 cm long, serrate to incised. Inflorescence hispid to pilose; umbels with 20–40 uneven rays; bractlets linear; fruit 3–4 mm long. Moist forests; western Parkland, Boreal forest, and Peace River.

*Bupleurum*    thorough-wax

*Bupleurum americanum* Coult. & Rose    THOROUGH-WAX

Perennial plants with a woody caudex; stems 10–30 cm high. Basal leaves entire, linear-lanceolate; stem leaves clasping, oblong to linear. Flowers yellow, in small umbels; bractlets conspicuous. Rare; gravelly and rocky areas; southern Rocky Mountains.

*Carum*     caraway

*Carum carvi* L.

Biennial plants, with glabrous stems to 1 m high. Leaves ovate, pinnate; the leaflets once or twice pinnately divided into narrow segments, 5–15 mm long. Umbels with 7–15 rays, bractless or with a few linear bracts; fruit 3–4 mm long. Introduced; occasionally escaped from cultivation.


*Cicuta*     water-hemlock

Marsh plants with smooth stems and pinnately compound leaves, growing from stout, often tuberous, rootstocks. Flowers white, in compound umbels; fruit oblong and slightly flattened. All species **very poisonous.** Genus contains some of Canada's most **poisonous** plants, and as yet no antidote is known. Although all parts of the plant contain toxic substances, the heaviest concentration occurs in the root. Humans as well as livestock are affected.

1. Leaflets very narrow; bulblets in axils of
   upper leaves. ................................................................................ *C. bulbifera*
   Leaflets lanceolate; no bulblets in axils of
   leaves. ................................................................................................ 2

2. Leaflets very narrowly lanceolate, length-
   to-width ratio 10:1 or more. ......................................... *C. mackenzieana*
   Leaflets lanceolate, length-to-width ratio
   about 5:1. .................................................... *C. maculata* var. *angustifolia*


*Cicuta bulbifera* L.                                    BULB-BEARING WATER-HEMLOCK

An erect, rather slender plant 30–90 cm high, with a few fleshy tuberous roots. Leaves two or three times pinnate, with leaf segments very narrowly linear, 2–5 cm long, sparsely toothed. Axils of upper leaves bearing bulblets. Fruits few, orbicular, less than 3 mm long. Common; in swamps and wet meadows; in Boreal forest and Parklands; **very poisonous** to all forms of livestock and to humans.

*Cicuta mackenzieana* Raup.                                    WATER-HEMLOCK

Stout marsh plants, with stems to 1.5 m high, often more than 1 cm thick; rootstock short, rather thin; tuberous roots hardly developed. Leaflets 5–8 cm long, usually less than 5 mm wide. Umbels several, 3–8 cm across; fruit 2–2.5 mm long. Lakeshores and bogs; Boreal forest. **Very poisonous.**

*Cicuta maculata* L. var. *angustifolia* Hook. (Fig. 161)     WATER-HEMLOCK

A stout-stemmed plant from a swollen, bulbous rootstock, which is divided horizontally into chambers. Plants 50–200 cm high, often much-branched. Leaves twice-pinnate with lanceolate to linear-lanceolate leaflets 5–8 cm long, 1–2 cm wide, sharply toothed. Base of leaf stalk swollen and sheathing the stem. Flowers small and white, in compound umbels 3–10 cm across. Seeds oval, slightly less than 3 mm long, yellow, with dark brown ribs, and slightly grooved between the two carpels. Common; in wet and marshy

A.C. Budd

Fig. 161.  Water-hemlock, *Cicuta maculata* L. var. *angustifolia* Hook.

*Dicots*

places, stream banks, and lakeshores; throughout Prairies and Parklands. **Very poisonous** to all stock and humans. Syn.: *C. occidentalis* Greene; *C. douglasii* (DC.) Coult. & Rose. In var. *maculata*, the leaflets are 2–3 cm wide, and the fruit somewhat larger. Marshes and wet places; southeastern Parklands. **Very poisonous.**

### *Conium*    hemlock

*Conium maculatum* L.    POISON HEMLOCK

An erect branching annual or biennial 60–300 cm high; usually glabrous, nauseous smelling when cut or bruised. Leaves large, divided into numerous leaflets. Umbels terminal, with 10–15 rays. Introduced; occasionally established. **Very poisonous.**

### *Cryptotaenia*    honewort

*Cryptotaenia canadensis* (L.) DC.    HONEWORT

An erect branching plant 30–80 cm high, with lower leaves long-stalked, of 3 thin, incised leaflets 3–10 cm long. Small white flowers on unequal stalks in compound umbels, with neither bracts nor bractlets. Fruit narrowly oblong, 5–6 mm long. Scarce; but has been found in woodlands; southeastern Parklands and Boreal forest.

### *Cymopterus*    cymopterus

*Cymopterus acaulis* (Pursh) Raf. (Fig. 162)    PLAINS CYMOPTERUS

A very low plant from a deep, thick root, with a short stem rarely more than 5 cm high. Leaves bright green, usually ascending, from pinnate to twice pinnate, 7–20 cm long. Flowers white, in an umbel up to 3 cm across. Fruit oval, about 6 mm long, with winged ribs on side and back. Fairly common; on dry prairie and hillsides; throughout Prairies.

### *Daucus*    carrot

*Daucus carota* L.    WILD CARROT

Erect annual or biennial with a taproot, 30–90 cm high. Leaves much divided, with the ultimate segments lanceolate or linear. Umbels terminal, with numerous rays; bracts about as long as the rays, with lobes linear. Introduced; occasional weed in disturbed areas and shores.

### *Eryngium*    eryngo

*Eryngium planum* L.    CROSS-THISTLE

Perennial plants, with stems 25–100 cm high. Basal leaves persistent, 5–10 cm long, 3–6 cm wide, ovate-oblong, cordate at the base, the margins serrate, spinulose. Inflorescence usually bluish with several ovoid-globose heads, 1–2 cm long, 10–15 mm wide; bracts 15–25 mm long, linear-lanceolate, with up to 4 pairs of spiny teeth. Introduced ornamental; occasionally escaped.

Fig. 162.   Plains cymopterus, *Cymopterus acaulis* (Pursh) Raf.

***Heracleum*** cow-parsnip

*Heracleum lanatum* Michx. (Fig. 163)   COW-PARSNIP

A tall, coarse plant, with somewhat woolly or hairy stem, 1–2.5 m high. Leaves stalked, 10–30 cm wide, divided into 3 broad segments, very hairy below. Flowers white, in large flat umbels 15–30 cm across. Fruit oval, 8–12 mm long, with very fine hairs, pale tawny when mature. Common; in shady woodland and moist places; throughout the Prairie Provinces. Easily distinguished from *Cicuta* or *Sium* by its broad leaves and unpleasant odor. Known to cause dermatitis in humans. Syn.: *H. maximum* Bartr.

***Levisticum*** lovage

*Levisticum officinale* Koch   LAVAS

Stout perennials, with stems 1–2 m high. Leaves to 70 cm long, 60 cm wide, triangular in outline; leaflets long-cuneate, entire below, dentate in the upper part. Umbels 3–10 cm across, with 12–20 rays; bracts numerous, deflexed; fruit 5–7 mm long. Introduced; cultivated as an aromatic herb; the root useful in veterinary medicine; occasionally escaped.

***Lomatium*** prairie parsley

Very short-stemmed perennial growing from a thickened tuberous root, with ternately or pinnately dissected somewhat hairy leaves, and yellow, purple, or white flowers. No bracts below head of flowers but usually small bractlets below the separate small umbels making up the head. Plants covered with fine gray hairs.

1. Leaf divided into few elongate, linear
   leaflets. ............................................................ *L. triternatum*
   Leaf divided into numerous short, narrow
   ultimate segments. ........................................................ 2
2. Ovary and fruit densely puberulent. ........................................ 3
   Ovary and fruit glabrous. ................................................... 4
3. Bractlets fused at the base; plants villous
   throughout. ...................................................... *L. villosum*
   Bractlets free from the base; plants spar-
   ingly pubescent. ............................................... *L. sandbergii*
4. Bractlets oblanceolate; root bulbous. ........................... *L. cous*
   Bractlets lanceolate, widest at base; root
   not bulbous. ............................................................... 5
5. Stem glabrous; foliage puberulent. ................. *L. dissectum* var. *multifidum*
   Stem and foliage more or less densely
   pubescent. ................................................................ 6
6. Stems usually with at least one pair of
   leaves near the base. ....................................... *L. macrocarpum*
   Stems leafless or with a single leaf at the
   base. ............................................................. *L. orientale*

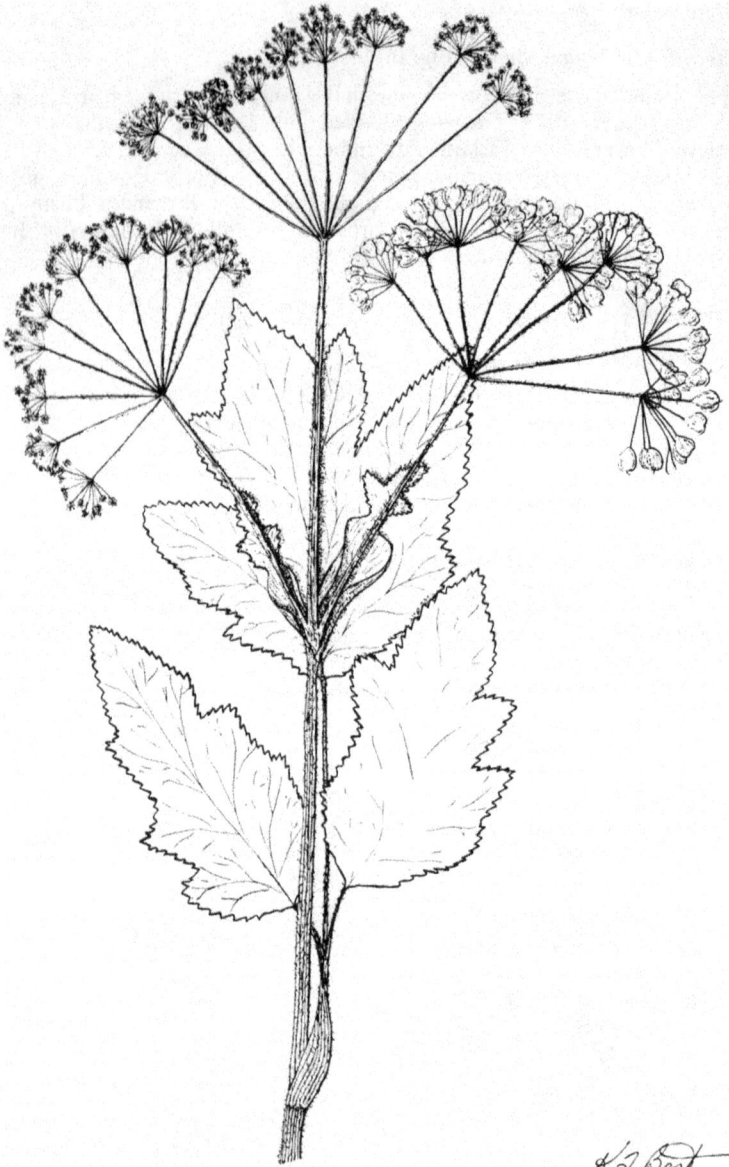

Fig. 163.  Cow-parsnip, *Heracleum lanatum* Michx.

*Lomatium cous* (Wats.) Coult. & Rose                    BISCUIT-ROOT

Plants stemless or with a leaf at the base, 20–25 cm high, from a globose or somewhat elongate tuberous root. Leaves ternate, with divisions 2- or 3-pinnate; the ultimate segments 1–5 mm long, usually glabrous. Umbel with 10–20 unequal rays; flowers yellow; fruit 7–10 mm long, with the wings narrower than the body. Very rare; grassland slopes; Cypress Hills.

*Lomatium dissectum* (Nutt.) Mathias & Constance
var. *multifidum* (Nutt.) Mathias & Constance        MOUNTAIN WILD PARSNIP

A plant with a thick, knobby spindle-shaped root, 30–90 cm high. Leaves in 2 or 3 sections of several times divided leaflets, the ultimate divisions being linear. Flowers yellow, in a large umbel, followed by flattened fruits 8–12 mm long, with thickened corky side wings. Found occasionally; Foothills region. Syn.: *Leptotaenia multifida* Nutt.

*Lomatium macrocarpum* (Hook. & Arn.) Coult. & Rose LONG-FRUITED PARSLEY

A semidecumbent plant growing from a thick, deep taproot, usually with very short stems. Leaves and leaf stalks covered with fine grayish hairs. Leaves 3 or 4 times pinnately divided. Flowers white, in compound umbels. Fruit flattened, about 9 mm or more long, with broad creamy white wings almost as wide as body of fruit. Found occasionally; on dry and rocky hillsides; Prairies. Syn.: *Cogswellia macrocarpa* (Nutt.) M. E. Jones.

*Lomatium orientale* Coult. & Rose              WHITE-FLOWERED PARSLEY

Resembling the preceding species but having white or pinkish flowers, translucent-margined bractlets, and fruit not longer than 6 mm. Rare; but has been reported from dry hillsides; southeastern Parklands. Syn.: *Cogswellia orientalis* (Coult. & Rose) M. E. Jones.

*Lomatium sandbergii* Coult. & Rose             SANDBERG'S WILD PARSLEY

Plants with a stout taproot; the stems 10–25 cm high, more or less clearly leafy at base. Umbels 2–5 cm across; the bractlets narrowly elongate, sometimes lobed at the apex; flowers yellow; fruit 5–7 mm long, densely puberulent. Rare; open alpine areas; southern Rocky Mountains.

*Lomatium triternatum* (Pursh) Coult. & Rose       WESTERN WILD PARSLEY

A species 30–60 cm high, with doubly ternate leaves; the leaflets narrow, 5–10 cm long. Yellow flowers, in a compound umbel with no bracts, but usually a few bractlets. Fruit winged and hairless or finely puberulent. Grassland, openings, and slopes; southern Rocky Mountains. Includes *L. simplex* (Nutt.) Macbr.

*Lomatium villosum* Raf. (Fig. 164)              HAIRY-FRUITED PARSLEY

A low, usually prostrate finely hairy plant, from an enlarged taproot. Leaf stalks usually short; leaves finely dissected, and covered with fine grayish hairs. Flowers yellow, in compound umbels. Very early flowering. Fruit flattened, about 6 mm long, and finely hairy. Common; particularly on heavier soils; Prairies. Syn.: *L. foeniculaceum* (Nutt.) Coult. & Rose.

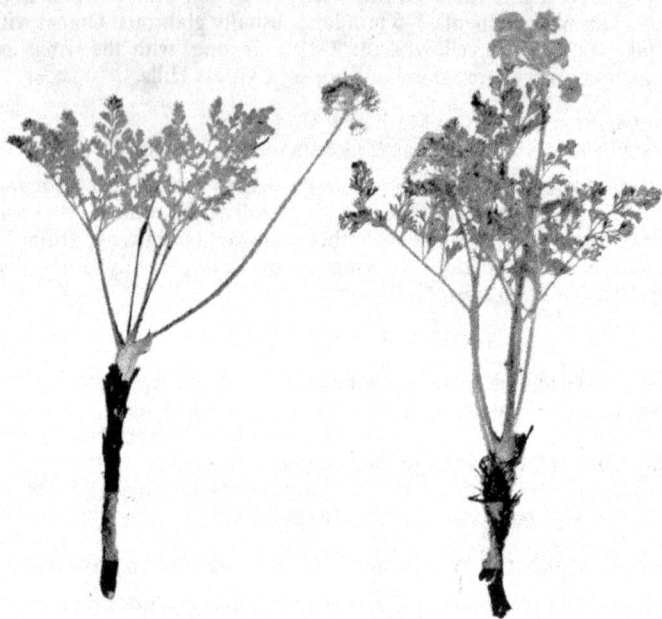

Fig. 164.   Hairy-fruited parsley, *Lomatium villosum* Raf.

***Musineon***     musineon

Low-growing plants with long thickened roots and much-dissected bright green leaves. Flowers yellow, in compound umbels; fruit not flattened as in *Lomatium*. Differing also from *Lomatium* by having no hairs on stem or leaves.

*Musineon divaricatum* (Pursh) Nutt.          LEAFY MUSINEON

A low-growing plant 10–20 cm high, with doubly pinnate bright green leaves and yellow flowers in compound umbels 2–6 cm across. Fruit smooth, 2–3 mm long. Fairly common; on dry ground; throughout Prairies and southern Parklands. Includes *M. trachyspermum* Nutt.

***Osmorhiza***     sweet cicely

Perennial herbs from thick sweet-scented roots. Stems tall and leafy with white or yellow flowers and narrow, linear, bristly-ribbed, long-stalked fruit.

1. Flowers yellow or greenish; fruit
   glabrous. ................................................................................ *O. occidentalis*
   Flowers white; fruit pubescent. ................................................................. 2
2. Involucral bracts present. ........................................................ *O. aristata*
   Involucral bracts absent. ........................................................... *O. chilensis*

*Osmorhiza aristata* (Thunb.) Mak. & Yabe     SMOOTH SWEET CICELY

A plant 30–90 cm high, from thick, rough, fleshy roots. Stems smooth; leaf stalks twice or three times divided into 3's, ending in lanceolate or ovate leaflets 2–7 cm long, pointed at apex, and coarsely toothed. Flowers small, white, in compound umbels, with green leaf-like bracts at base of each umbellet or terminal small umbel. Fruit 15–25 mm long, pointed at apex, on a long stalk. Fairly common; in shady moist woodlands. Two varieties occur. In var. *brevistylis* (DC.) Boiv. the foliage pubescent and the style 0.5–2.0 mm long; found in southeastern Parklands and Boreal forest. In var. *longistylis* (Torr.) Boiv. (Fig. 165) the stem glabrous, the foliage glabrous or nearly so, and the style 2.0–3.5 mm long; found throughout Prairies, Parklands, and Boreal forest.

*Osmorhiza chilensis* Hook. & Arn.     BLUNT-FRUITED SWEET CICELY

A plant 30–90 cm high with much-branched stems and fleshy roots. Leaves thin; stalks twice divided into 3's. The compound umbels bearing small white flowers. Fruit bearing a conical beak slightly longer than 1.5 mm. Fairly plentiful in woods. Syn.: *O. divaricata* (Britt.) Suksd. Three varieties are distinguished in this species:

1. Fruits 20–25 mm long, longer than the
   pedicels. ................................................................................ var. *chilensis*
   Fruits less than 20 mm long. ................................................................. 2
2. Fruits about 15 mm long, mostly shorter
   than the pedicels. ............................................... var. *cupressimontana* Boiv.
   Fruits about 10 mm long, shorter than the
   pedicels, blunt. ............................................... var. *purpurea* (Coult. & Rose) Boiv.

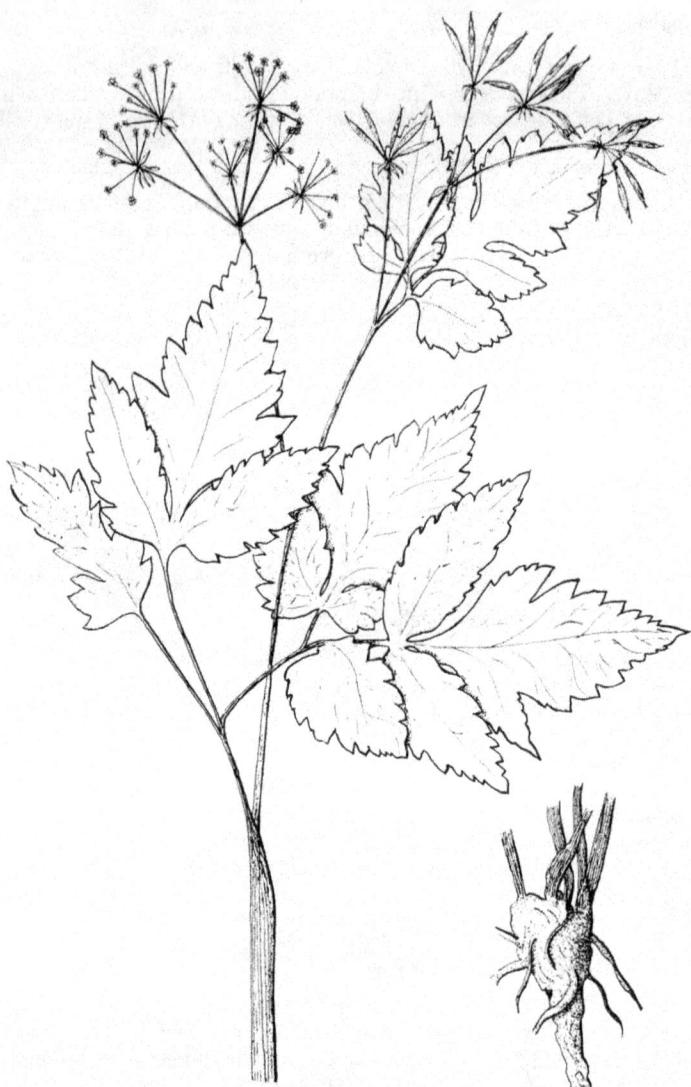

A.C. Budd

Fig. 165.  Smooth sweet cicely, *Osmorhiza aristata* (Thunb.) Mak. & Yabe
var. *longistylis* (Torr.) Boiv.

Of these, var. *chilensis* occurs in Cypress Hills and Rocky Mountains; var. *cupressimontana* (*O. obtusa* (Coult. & Rose) Fern.) throughout Parklands; and var. *purpurea* (*O. purpurea* (Coult. & Rose) Suksd.) in southern Rocky Mountains.

*Osmorhiza occidentalis* (Nutt.) Torr.                    WESTERN SWEET CICELY

Plants 30–100 cm high, villous at the nodes; leaves 10–20 cm long, ovate or oblong, 1- to 3-ternate, with the ultimate divisions 2–10 cm long, lanceolate to ovate, serrate to lobed. Umbels 3–7 cm across, with 5–12 rays, stiffly ascending to spreading; fruit linear-fusiform 12–20 mm long; styles 1 mm long or less, glabrous or rarely bristly at base. Open woods and meadows; southern Rocky Mountains.

**Pastinaca**     parsnip

*Pastinaca sativa* L.                                       WILD PARSNIP

Plants biennial, to 1.5 m high. Leaflets 5–15, usually ovate or oblong, 5–10 cm long, serrate or lobed. Umbel 10–20 cm across, with 15–25 rays; fruit 5–7 mm long. Introduced; occasionally escaped from cultivation.

**Perideridia**     squawroot

*Perideridia gairdneri* (Hook. & Arn.) Mathias             SQUAWROOT

A narrow, erect plant 30–80 cm high. Fleshy tuberous roots, often with a fascicle or cluster of tubers. Leaves pinnate, of very narrow leaflets 2–15 cm long. Flowers very small, white, in compound umbels 3–8 cm across. Fruit somewhat flattened, brown, slightly longer than 1.5 mm. The roots were used by Indians for food. Found in meadows, woodlands, and ravines; in southern Rocky Mountains and Cypress Hills. Syn.: *Atenia gairdneri* Hook. & Arn.

**Sanicula**     snakeroot

*Sanicula marilandica* L. (Fig. 166)                       SNAKEROOT

An erect plant 30–100 cm high, with long-stalked basal leaves and stalkless upper leaves. Leaves of 5 or 7 palmately arranged leaflets, oblanceolate, sharply toothed, 4–20 cm long. Flowers greenish white, in compound umbels of several almost globular umbellets, each 6–15 mm in diam. Fruit about 6 mm long, densely covered with fine hooked bristles. Common; in moist, rich woodlands; throughout the Prairie Provinces.

**Scandix**     shepherd's-needle

*Scandix pecten-veneris* L.                                VENUS'-COMB

An annual plant 20–40 cm high, branching from the base, and hispidulous throughout. Leaves pinnately decompound, with the ultimate segments 2–5 mm long, linear. Umbels 2–4 cm across, with 1–3 primary rays; fruit 5–12 mm long, with a flat straight beak, usually about 4 cm long. An introduced weed, rare.

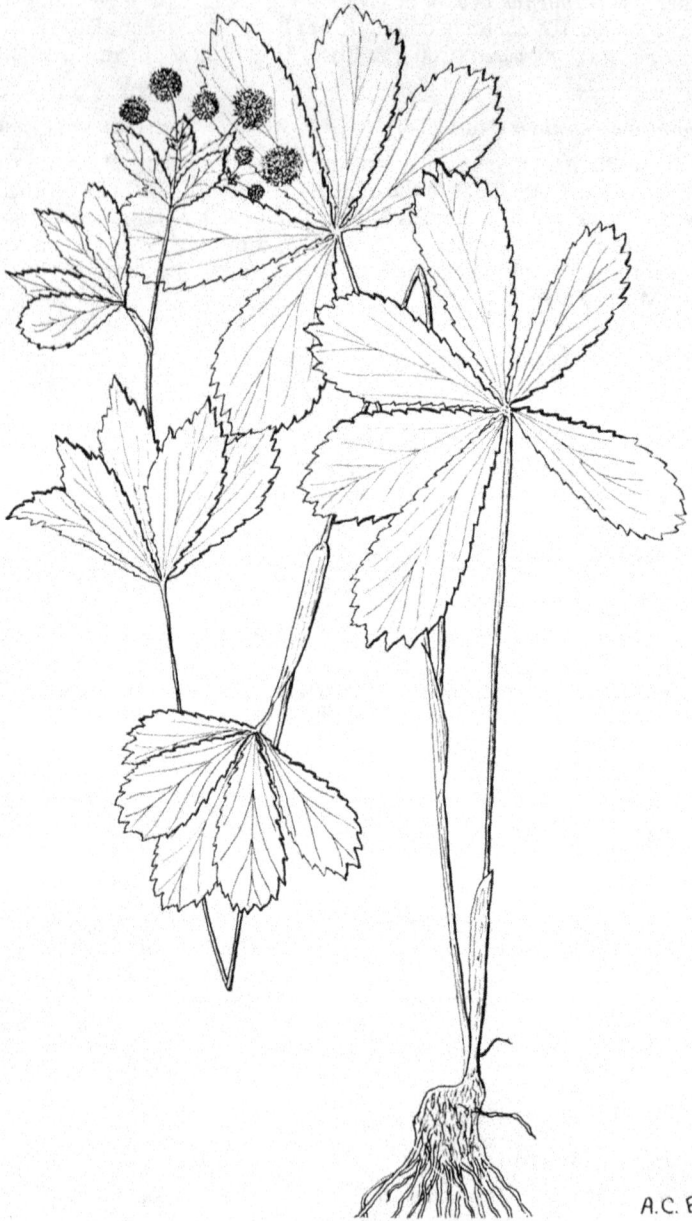

Fig. 166. Snakeroot, *Sanicula marilandica* L.

***Sium***   water-parsnip

*Sium suave* Walt. (Fig. 167)

Tall, erect marsh plants from stout rootstocks, 60–200 cm high, with smooth, hollow stems. Two kinds of leaves: early underwater leaves twice or three times pinnate with thread-like leaflets; later growing abovewater leaves singly pinnate with linear or narrowly lanceolate sharply toothed leaflets 2–10 cm long. Flowers white, in compound umbels 5–8 cm across. Fruit about 3 mm long, ovate and somewhat compressed, bearing prominent ribs. Very common; in sloughs and wet places; throughout the Prairie Provinces. Thought to have slightly **poisonous** properties, but not so dangerous as the water-hemlock, which it closely resembles. Eaten readily by all classes of livestock without bad effect. Water-parsnip has singly pinnate leaves, with narrow leaflets and many small bracts at base of compound flower umbel; water-hemlock has compound pinnate leaves, with lanceolate leaflets and usually no bracts at the base of the umbel. Both species have bracts at the base of each separate umbellet or single portion of the compound umbel. Syn.: *S. cicutaefolium* Schrank.

***Zizia***   alexanders

Smooth, shiny plants growing to 60 cm high. Leaves simple or divided into broad serrated leaflets. Flowers yellow, with bracts only at base of umbellets.

Basal leaves simple and cordate. ......................................................................... *Z. aptera*
All leaves divided into 3 leaflets. ......................................................................... *Z. aurea*

*Zizia aptera* (Gray) Fern.                    HEART-LEAVED ALEXANDERS

An erect plant 30–60 cm high, with long-stalked cordate basal leaves and thrice-divided stem leaves having ovate leaflets. Flowers bright yellow, in compound umbels, early flowering. Very common; in moist places; throughout the Prairie Provinces. Syn.: *Z. cordata* (Walt.) Koch.

*Zizia aurea* (L.) Koch                    GOLDEN ALEXANDERS

Very similar to *Z. aptera,* but all leaves twice or three times divided into leaflets. Fairly common; in meadows and woodlands; eastern Prairies and Parkland.

## CORNACEAE—dogwood family

***Cornus***   dogwood

Herbs, shrubs, or small trees with simple leaves and perfect flowers; the floral parts in 4's. Fruit a small drupe with a 2-seeded stone.

Fig. 167.    Water-parsnip, *Sium suave* Walt.

A.C. Budd

1. Low herbs, with 4 large petal-like bracts
   surrounding minute heads of flowers. ............................................ *C. canadensis*
   Shrubs or small trees, with cymes of
   flowers not surrounded by bracts. ........................................................................ 2
2. Leaves alternate; fruit bluish. .............................................. *C. alternifolia*
   Leaves opposite; fruit white or blue. ................................................................ 3
3. Twigs pale green, mottled with purple;
   fruit blue. ............................................................................................ *C. rugosa*
   Twigs not purplish-mottled; fruit white. ........................................................ 4
4. Leaves mostly lanceolate; branches and
   inflorescence practically hairless; pith
   of young branches tawny; bark gray-
   ish. ....................................................................................... *C. racemosa*
   Leaves mostly ovate; branches and
   inflorescence with appressed hairs and
   often woolly; pith of young branches
   white; bark reddish or reddish brown. ......................................... *C. alba*

*Cornus alba* L.                                                    RED-OSIER DOGWOOD

A shrub 1–2 m high, with bright reddish-colored twigs and opposite
leaves. Leaves generally ovate, with rounded base and pointed apex, 2–8 cm
long, paler beneath, and with a few short appressed hairs. Small white flowers
borne in flat-topped clusters 2–5 cm across. Flowering in early June, producing globular white fruit about 6 mm in diam. Very conspicuous in winter
because of its reddish branches. Common; in woodlands and coulees;
throughout the Prairie Provinces. Syn.: *C. stolonifera* Michx.; *Svida stolonifera*
(Michx.) Rydb. Besides the typical form, two varieties are recognized. In var.
*interior* (Rydb.) Boiv. the lower leaf surface, young twigs, and inflorescence are
densely pubescent; found mainly in shrubbery in Prairies. The var. *baileyi*
(Coult. & Evans) Boiv. differs from the species by undersides of leaves becoming woolly at maturity; common in southeastern Parklands.

*Cornus alternifolia* L. f.                                   ALTERNATE-LEAVED DOGWOOD

A shrub or small tree 2–6 m high, with green branches striped with white.
Leaves borne alternately on the stems, ovate, long, 5–12 cm long, pale
beneath. Flowers white, followed by bluish fruit. Very rare; but has been
found in southeastern Parklands and Boreal forest.

*Cornus canadensis* L. (Fig. 168)                                      BUNCHBERRY

Low-growing herbs from a horizontal slender rootstock, with woody-
based stems 5–15 cm high. One pair of small leaves about the middle of stem
and an apparent whorl of ovate leaves 2–8 cm long near the head of the stem.
Four petal-like white involucral bracts 10–25 mm long, at head of stem, surrounding a cluster of tiny greenish flowers. Fruit a bright red drupe about 6
mm across, borne in a cluster. Very common; in shady woodlands; throughout
the Prairie Provinces. Syn.: *Chamaepericlymenum canadense* (L.) Aschers. &
Graebn.

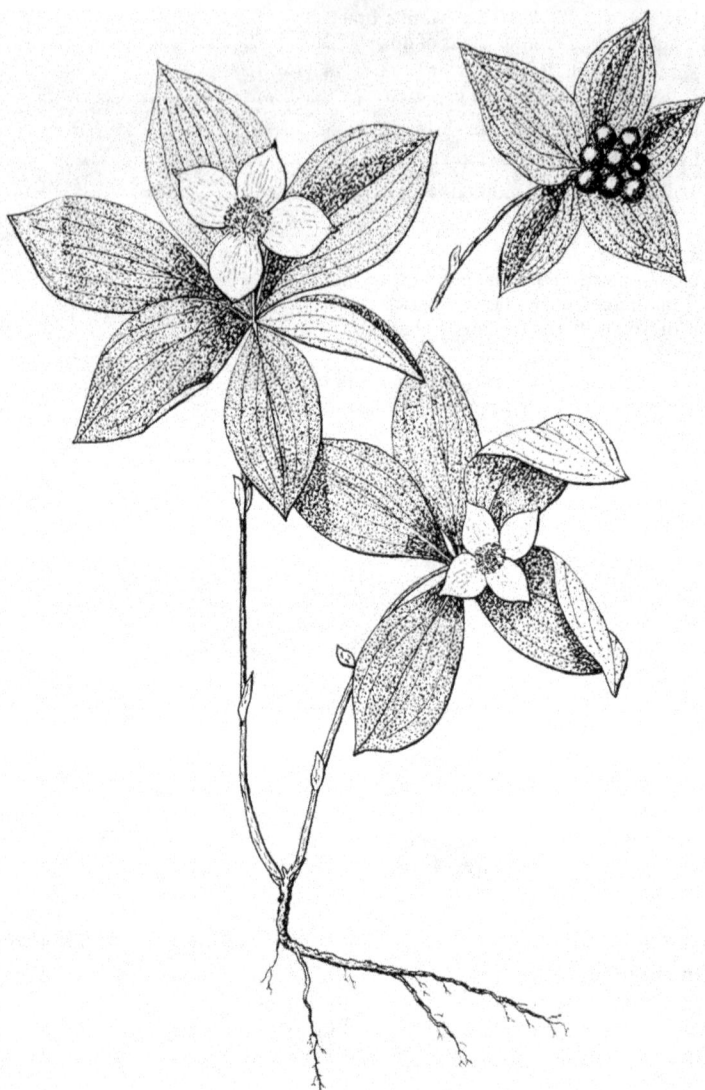

A.C. Budd

Fig. 168.  Bunchberry, *Cornus canadensis* L.

*Cornus racemosa* Lam.                                    PANICLED DOGWOOD

A shrub 2–3 m high, with smooth gray bark. Leaves usually lanceolate, opposite, 4–10 cm long. Flowers and fruit white, stone of fruit slightly grooved. Not common; but has been found in woodlands and along streams; southeastern Boreal forest.

*Cornus rugosa* Lam.                                      SPOTTED DOGWOOD

A shrub 1–3 m high, with the young branches yellow green, mostly mottled with purple or red. Leaves ovate to rotund, 7–12 cm long, softly white pubescent below. Inflorescence flat-topped or slightly convex, 3–6 cm across; fruit about 6 mm across, light blue. Shady woods; southeastern Parklands and Boreal forest.


# PYROLACEAE—wintergreen family

Perennial, usually evergreen herbs with long rootstocks. Leaves thick and leathery. Flowers perfect, in racemes or corymbs, with 4 or 5 sepals, 4 or 5 petals, and 8 or 10 stamens. Fruit a capsule.

1. Plants with leafy stems and corymbose
   inflorescence. ............................................................................ *Chimaphila*
   Plants with leaves in a basal rosette;
   flowers singly or in a raceme. ................................................................ 2
2. Flowers borne singly. ........................................................................ *Moneses*
   Flowers in a raceme. ............................................................................ *Pyrola*


**Chimaphila**        prince's-pine

*Chimaphila umbellata* (L.) Bart. var. *occidentalis* (Rydb.) Blake        PIPSISSEWA

A low perennial herb with decumbent stems 10–20 cm high, with whorled oblanceolate leaves 2–8 cm long, dark green and shiny above, paler beneath. Flowers pinkish white, borne 4–7 in a cyme. Not common; in forested areas on dry soil; Parklands, Boreal forest, and Cypress Hills.


**Moneses**        one-flowered wintergreen

*Moneses uniflora* (L.) Gray (Fig. 169*A*)        ONE-FLOWERED WINTERGREEN

A small herb 5–15 cm high, from a slender rootstock. Leaves round to ovate, 10–25 mm long, and borne in 1 or 2 pairs or in whorls near base of stem. Flowers (Fig. 7) solitary and nodding at head of short stem, each flower usually with 5 waxy white petals, fragrant, 10–15 mm across. Two stamens for each petal and a straight style with a knobbed summit. Found in cool, moist woodlands; Parklands, Boreal forest, and Cypress Hills. Syn.: *Pyrola uniflora* L.

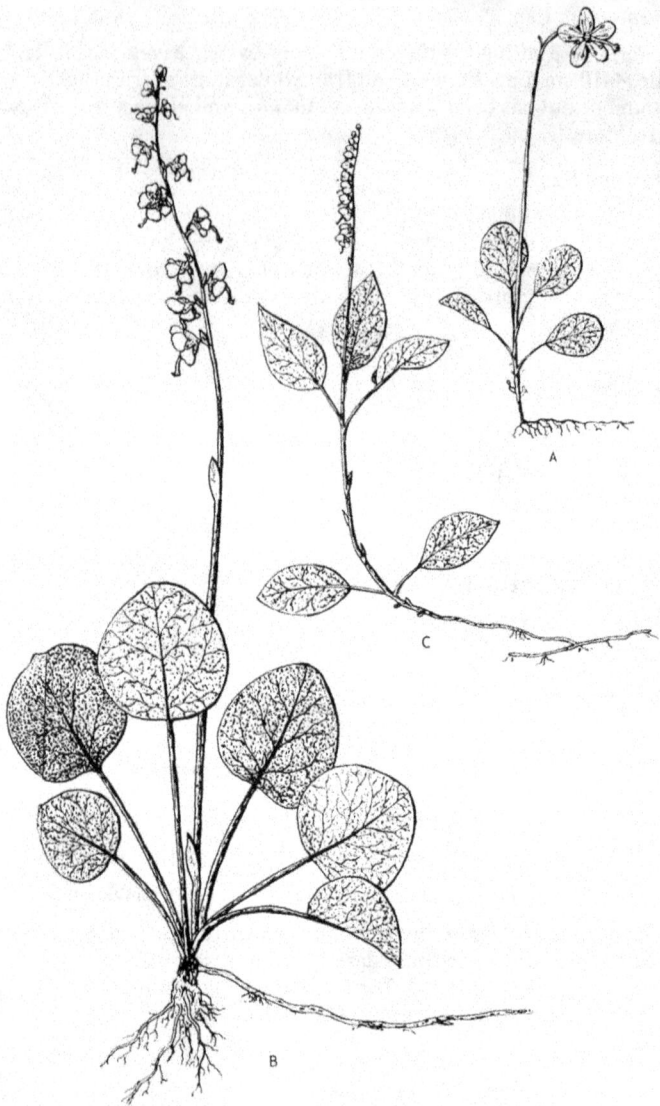

A.C. Budd

Fig. 169.   Wintergreens: *A*, one-flowered wintergreen, *Moneses uniflora* (L.) Gray; *B*, pink wintergreen, *Pyrola asarifolia* Michx. var. *incarnata* Fern.; *C*, one-sided wintergreen, *Pyrola secunda* L.

***Pyrola***     wintergreen

Low-growing evergreen herbs from slender creeping rootstocks, with round or oval stalked basal leaves. Inflorescence a long narrow raceme of perfect, regular, nodding flowers, usually with a protruding style. Fruit a small round capsule.

1.  Style straight, not protruding conspicuously from flower. ...................................................................... *P. minor*
    Style conspicuously protruding from flower. ............................................................................................ 2

2.  Flowers crowded on one side of the stem (secund); style straight. ........................................... *P. secunda*
    Flowers not all on one side of stem; style curved. ................................................................................. 3

3.  Petals pink or purplish. ................................................................................................................................ 4
    Petals white or greenish. ............................................................................................................................. 5

4.  Leaves subentire; the apex rounded, dull above, thin. .................................................................. *P. asarifolia*
    Leaves denticulate-margined; the apex acute, shiny above, firm. ............................................. *P. bracteata*

5.  Leaves ovate-lanceolate, acute; the main veins whitish. ........................................................................ *P. picta*
    Leaves rotund to elliptic; the veins not whitish. ...................................................................................... 6

6.  Sepals oblong to elliptic, much longer than wide; leaves subrotund. ........................................ *P. rotundifolia*
    Sepals more or less triangular, as wide as long; leaves elliptic or rotund. ............................................. 7

7.  Leaves broadly elliptic to oblong, 3–7 cm long, mostly longer than the petioles. ....................... *P. elliptica*
    Leaves obovate to subrotund, 1–3 cm long, often shorter than the petioles. ............................... *P. virens*

*Pyrola asarifolia* Michx.                               PINK WINTERGREEN

A plant 10–30 cm high with leathery, shiny basal leaves, cordate at base, 2–5 cm wide. Flowers 7–15, pinkish, 8–12 mm across when fully opened, in an open raceme and usually nodding; with 5 sepals, 5 petals, and a protruding style. Fairly common; in moist woods; throughout the Prairie Provinces. The var. *incarnata* Fern. (Fig. 169*B*) differs from the typical form by having leaf blades rounded or tapering at base instead of being cordate. Found in similar locations as the typical form.

*Pyrola bracteata* Hook.                                 LARGE WINTERGREEN

Plants with stems 20–30 cm high, with 1 or 2 scarious bracts on the stem below the inflorescence. Leaves with petioles about the same length as the blades, 3–8 cm long, ovate to subrotund, dark green above, pale or reddish brown below. Flowers rose purple or dull red; petals 6–8 mm long; floral bracts conspicuous, to twice as long as the pedicel. Rare; damp coniferous woods; southern Rocky Mountains.

*Pyrola elliptica* Nutt.                                    COMMON SHINLEAF

Leaves broadly elliptic or oblong, with the petioles often very short. Stems 10–30 cm high; flowers 8–12 mm across, white, often greenish- or pinkish-veined. Not common; damp coniferous woods; Boreal forest.

*Pyrola minor* L.                                    LESSER WINTERGREEN

A small species with thin dark green oval or rounded leaves 1–3 cm long, on fairly long basal stalks. Flowers small, about 6 mm across, white or faintly pinkish, in a rather crowded raceme on a stem 5–20 cm high. Found occasionally; in woodlands; Boreal forest, Riding Mountains, and Cypress Hills. Syn.: *Braxilia minor* (L.) House.

*Pyrola picta* Sm.                                    WHITE-VEINED SHINLEAF

Stems 10–20 cm high; leaves 2–7 cm long, entire or denticulate, with the petioles about the same length as, or shorter than, the blade, white-mottled along the principal veins. Flowers yellowish white; petals 7–8 mm long; floral bracts shorter than the pedicel. Rare; coniferous forests; southern Rocky Mountains.

*Pyrola rotundifolia* L.                                    COMMON WINTERGREEN

Stems 10–30 cm high, with 1 or 2 bracts; leaves 3–8 cm long, subrotund or somewhat ovate or obovate, entire to crenate. Flowers white; petals 6–10 mm long; floral bracts the same length as, to longer than, the pedicel. Coniferous woods; Boreal forest.

*Pyrola secunda* L. (Fig. 169*C*)                                    ONE-SIDED WINTERGREEN

A rather small species, usually growing in colonies from a much-branched rootstock. Leaf blades thin, oval to lanceolate, pointed at both ends, 2–6 cm long. Flowers small, about 6 mm across, and crowded on one side of the short stem, 5–20 cm high. Fairly common; in woodlands and bluffs; throughout the Prairie Provinces. Syn.: *Orthilia secunda* (L.) House.

*Pyrola virens* Schweigg.                                    GREENISH-FLOWERED WINTERGREEN

Basal leaves round or broadly oval, rounded at the apex. Leaves thick and dull, on rather long stalks; blades 1–3 cm wide. Flowers greenish white, 8–12 mm across when opened, and borne racemosely on a stem 10–20 cm high; 3–10 flowers in an inflorescence. Found in moist coniferous forest; throughout the Prairie Provinces. Syn.: *P. chlorantha* Swartz.

## MONOTROPACEAE—Indian-pipe family

Plants parasitic on the roots of other plants or feeding on dead organic matter (saprophytic). No green coloring in leaves, which are reduced to scales. Flowers perfect, usually drooping, with 6–12 stamens. Fruit a single-celled capsule with numerous seeds.

Petals united; stem sticky with an enlarged
bulb-like base. ................................................................................................. *Pterospora*
Petals distinct; stem not enlarged and bulb-
like at base, not sticky. ................................................................................................. *Monotropa*

***Monotropa***     pinesap

Flowers solitary at summit of stem. ................................................................... *U. uniflora*
Flowers several in a raceme. ............................................... *M. hypopithys* var. *latisquama*

*Monotropa hypopithys* L. var. *latisquama* (Rydb.) Kearn. & Peebles     PINESAP

A pinkish- or yellowish-stemmed plant 15–30 cm high, with drooping flowers 10–15 mm long. No leaves, but merely stalkless scales on stem. Very rare; found in rich soil in forests; southern Rocky Mountains and Cypress Hills. Syn.: *Hypopithys latisquama* Rydb.

*Monotropa uniflora* L.     INDIAN-PIPE

Stems 10–30 cm high, scaly, white, with a single drooping flower. Plants turning black during drying. Rare; found in damp woodlands; Boreal forest.

***Pterospora***     pinedrops

*Pterospora andromedea* Nutt.     PINEDROPS

Tall sticky-stemmed plants with enlarged bases, growing from a rounded mass of roots. Stems 15–90 cm high, often more than 2 cm in diam, usually pinkish or purplish. Leaves reduced to numerous narrow scales. Flowers borne in an open raceme, whitish, 6–10 mm across. Fruit a capsule containing numerous small seeds. Rare; found in rich coniferous woods; southern Rocky Mountains and Cypress Hills.


## ERICACEAE—heath family

Perennial plants, usually shrubby, with perfect flowers. Fruit usually a capsule, but sometimes a berry or drupe, with the ovary superior.

1. Stems erect. ................................................................................................................ 2
   Stems prostrate to decumbent or creeping. ..................................................................................................... 7
2. Leaves scaly resinous or rusty tomentose below. ................................................................................... *Ledum*
   Leaves not scaly resinous or rusty tomentose below. ................................................................................ 3
3. Leaves white glaucous below. ............................................................................................. 4
   Leaves not white glaucous below. ........................................................................................ 5
4. Leaves opposite; corolla rotate, 6–12 mm across. ........................................................................... *Kalmia*
   Leaves alternate; corolla urn-shaped, 5–7 mm long. ................................................................ *Andromeda*
5. Leaves brownish scurfy; inflorescence a leafy secund raceme. ................................................... *Chamaedaphne*
   Leaves green, not scurfy; inflorescence not as above. ........................................................................ 6

6. Flowers large, about 2 cm across; leaves
   narrowly oblanceolate. .................................................................. *Rhododendron*

   Flowers smaller, about 1 cm long; leaves
   obovate. ................................................................................................ *Menziesia*

7. Leaves 2–6 cm long, oval or
   spatulate-obovate. .................................................................................. 8

   Leaves smaller, usually less than 2 cm
   long. ...................................................................................................... 10

8. Leaves and petioles pilose; leaf base
   rounded or cordate. .......................................................................... *Epigaea*

   Leaves and petioles glabrous; leaf base
   tapered. .................................................................................................. 9

9. Leaves crenate-serrate, conspicuously
   reticulate-veined below. ...................................................... *Arctostaphylos*

   Leaves very shallowly crenate, with teeth
   bristle-tipped, not reticulate-veined
   below. ................................................................................................ *Gaultheria*

10. Leaves and branches densely brown scaly
    and scurfy; flowers 1–2 cm across. .................................... *Rhododendron*

    Leaves and branches not scaly or scurfy;
    flowers smaller. ...................................................................................... 11

11. Leaves leathery, elliptic or obovate, 6–20
    mm long. .................................................................................. *Arctostaphylos*

    Leaves not as above. .............................................................................. 12

12. Leaves linear, closely crowded on the
    stem. ...................................................................................................... 13

    Leaves not as above. .............................................................................. 14

13. Leaves thick, blunt, 4-ranked; flowers
    solitary. .................................................................................................. *Cassiope*

    Leaves flat, grooved below, alternate;
    flowers in terminal clusters. ............................................................ *Phyllodoce*

14. Plants shrubby, with decumbent woody
    stems, diffusely branched. .................................................... *Loiseleuria*

    Plants not shrubby, with stems creeping,
    very slender. ........................................................................................ *Gaultheria*

## *Andromeda*    bog-rosemary

*Andromeda polifolia* L.    BOG-ROSEMARY

A shrub 10–40 cm high; linear-oblong leaves, with margins usually rolled toward underside, dark green above and white below, 2–5 cm long. Inflorescence a few-flowered umbel of pinkish white urn-shaped flowers, each about 6 mm long. Found in swamps and muskeg; Boreal forest.

## *Arctostaphylos*    bearberry

Leaves 2–5 cm long, strongly reticulate below,
serrulate, not evergreen. .................................................................... *A. alpina*

Leaves 1–3 cm long, not strongly reticulate
below, entire, evergreen. ................................................................................. *A. uva-ursi*

*Arctostaphylos alpina* (L.) Spreng.                    ALPINE BEARBERRY

Stems tufted or prostrate, with branches 10–20 cm long. Leaves thin, oblanceolate to spatulate-obovate. Flowers 4–5 mm long; berry black, or red in var. *rubra* (Rehder & Wilson) Bean, 6–10 mm long. Swamps, bogs, and muskeg; Boreal forest.

*Arctostaphylos uva-ursi* (L.) Spreng.                    BEARBERRY

A prostrate, trailing shrub forming large mats on ground. Leaves spatulate, entire-margined, evergreen and shiny dark green, 1–3 cm long, usually with a number of brown or reddish leaves intermixed. Flowers pinkish white, urn-shaped, about 5 mm long, in short few-flowered racemes. Fruit a bright red berry 6–10 mm in diam. Common; in woodlands on sandy hills and eroded slopes; throughout the Prairie Provinces.

***Cassiope***      mountain-heather

Leaves distinctly grooved on the back. ........................................................... *C. tetragona*
Leaves not grooved, but keeled on the back. ............................................. *C. mertensiana*

*Cassiope mertensiana* (Bong.) G. Don      WESTERN MOUNTAIN-HEATHER

Low, creeping alpine shrub, with branches 10–30 cm high, ascending. Leaves 3–6 mm long, rounded or keeled on the back. Flowers solitary, 4–6 mm long, white, on pedicels 5–20 mm long, puberulent. Rocky alpine slopes and forests; southern Rocky Mountains.

*Cassiope tetragona* (L.) D. Don      WHITE MOUNTAIN-HEATHER

Very similar to *C. mertensiana,* but the leaves with a conspicuous groove down the back, densely ciliate margins, and the pedicels to 3 cm long and finely glandular. Alpine and subalpine slopes and forests. In the southern Rocky Mountains var. *saximontana* (Small) C. L. Hitchc. occurs, with shorter pedicels and smaller flowers than var. *tetragona,* occurring in the northern Rocky Mountains.

***Chamaedaphne***      leatherleaf

*Chamaedaphne calyculata* (L.) Moench. (Fig. 170)      LEATHERLEAF

Branching shrubs 20–100 cm high, with oblong or obovate, slightly toothed scurfy leaves 1–5 cm long. Flowers white, somewhat urn-shaped, about 6 mm long, and borne in 1-sided racemes. Fruit an angular round capsule about 5 mm across. Found in swamps, bogs, and muskeg; Boreal forest.

***Epigaea***      mayflower

*Epigaea repens* L.                    MAYFLOWER

Trailing plants with stems 20–40 cm long, branched, hirsute. Leaves ovate to oblong, 2–10 cm long, obtuse to acute, rounded to cordate at the base, more or less pilose. Inflorescence 2–5 cm long; flowers 10–15 mm long. Rare; coniferous forest; southeastern Boreal forest.

Fig. 170.   Leatherleaf, *Chamaedaphne calyculata* (L.) Moench.

## Gaultheria      wintergreen

1. Leafy stems prostrate; branches less than
   10 cm high. ............................................................................................... 2
   Leafy stems upright; branches 10–20 cm
   high. .................................................................................... *G. procumbens*
2. Leaves 5–10 mm long, entire; fruit white. ................................. *G. hispidula*
   Leaves 10–15 mm long, serrulate; fruit
   scarlet. .................................................................................... *G. humifusa*

### *Gaultheria hispidula* (L.) Muhl.      CREEPING SNOWBERRY

Plants with prostrate stems 20–40 cm long, leafy, bristly, especially when young. Leaves short-petioled, broadly elliptic to rotund, bristly beneath. Flowers few, on pedicels about 1 mm long; fruit white, 5–10 mm long. Damp woods, muskeg, and bogs; Boreal forest. Syn.: *Chiogenes hispidula* (L.) Torr. & Gray.

### *Gaultheria humifusa* (Graham) Rydb.      ALPINE WINTERGREEN

Creeping shrub with branches usually less than 10 cm high, slender, glabrous or puberulent. Leaves oval to round-oval, rarely over 15 mm long. Flowers solitary on short-bracted peduncles, axillary; fruit 5–7 mm long, scarlet. Mountain slopes; southern Rocky Mountains.

### *Gaultheria procumbens* L.      CHECKERBERRY

Plants with leafy stems from a creeping rootstock, erect, 10–20 cm high; leaves 2–5 cm long, crowded at summit of stem. Flowers 8–10 mm long, on nodding pedicels 5–10 mm long; fruit 7–10 mm long, bright red. Coniferous forests; southeastern Boreal forest.

## *Kalmia*      sheep-laurel

### *Kalmia polifolia* Wang.      PALE LAUREL

A shrub growing 30–60 cm high, with sharply 2-edged twigs. Leaves opposite, almost stalkless, 1–3 cm long, linear-lanceolate, green above but white and finely hairy below, edges often rolled. Flowers in clusters from upper leaf axils, 10–15 mm across, deep pink to red. Fruits ovoid to spherical capsules about 6 mm long. Found in bogs and swampy places; Boreal forest.

## *Ledum*      Labrador-tea

Branching evergreen shrubs with leaves alternate, entire-margined, rolled edges, and covered on underside with rusty-colored woolly hairiness or glandular. Flowers with 5 white petals, borne in umbels or short corymbs. Fruit an oblong or ovate many-seeded capsule.

Leaves densely woolly or felty tomentose
below. ...................................................................................... *L. palustre*
Leaves glaucous, glandular-dotted below. ................................. *L. glandulosum*

*Ledum glandulosum* Nutt.                                          TRAPPER'S-TEA

Stout erect shrub to 2 m high, with twigs puberulent and glandular. Leaves pale green, rugose above, with margins little revolute, oblong to broadly elliptic, glaucous, and glandular-dotted below. Flowers white; petals 5–8 mm long. Moist woods; southern Rocky Mountains.

*Ledum palustre* L.                                               LABRADOR-TEA

Erect shrubs to 1.5 m high, with twigs tomentose. Leaves with strongly rolled margins, densely white or rusty red tomentose below, rugose, dark green above. Flowers white, 5–8 mm long. Muskeg, bogs, and wet woods; in alpine and boreal to arctic regions. Two varieties occur:

Leaves lanceolate, length-to-width ratio about
  6:1 ....................................................................................... var. *latifolium* (Jacq.) Michx.
Leaves linear, length-to-width ratio about
  12:1. ....................................................................................... var. *decumbens* Ait.


The var. *latifolium (L. groenlandicum* Oeder) is common in northern fringes of Parklands, Boreal forest, and Rocky Mountains; var. *decumbens* occurs in the northern parts of Boreal forest.


**Loiseleuria**      alpine azalea

*Loiseleuria procumbens* (L.) Desv.                              ALPINE AZALEA

Stems woody; branches to 30 cm long, about 5 or 6 cm high. Leaves opposite, narrowly elliptic, 5–8 mm long, evergreen, strongly revolute. Flowers pink or white; petals 3–4 mm long. Rocky tundra; Boreal forest.


**Menziesia**      menziesia

*Menziesia ferruginea* Sm. var. *glabella* (Gray) Peck      WESTERN MINNIEBUSH

An erect or straggling shrub 1–2 m high, with shredding bark; young twigs finely glandular pubescent. Leaves oblong to obovate, 3–6 cm long, somewhat serrate and ciliate, appressed pubescent above, pubescent on the veins below. Flowers 7–8 mm long, whitish or yellowish to purplish pink, on pedicels 15–20 mm long, glandular pilose. Moist woods and slopes; Rocky Mountains.


**Phyllodoce**      mountain-heather

1. Flowers yellow; peduncles and calyx
    densely glandular. ............................................................ *P. glanduliflora*
   Flowers pink to purple. ............................................................................ 2

2. Peduncles, calyx, and twigs densely
    glandular. ............................................................................ *P. caerulea*
   Peduncles glandular; calyx and twigs
    glabrous. ............................................................................ *P. empetriformis*

*Phyllodoce caerulea* (L.) Bab.　　　　　　　PURPLE MOUNTAIN-HEATHER

Shrubby plants with a stout, woody root; stems decumbent, 10–20 cm high. Flowers terminal, 1–4 together on peduncles 1–2 cm long, elongating in fruit. Tundra; northeastern Boreal forest.

*Phyllodoce empetriformis* (Sm.) D. Don　　　BLUE MOUNTAIN-HEATHER

Similar to *P. caerulea,* but flowers few to many together, calyx glabrous, and plants usually less glandular. Mountain meadows, forest openings; Rocky Mountains.

*Phyllodoce glanduliflora* (Hook.) Cov.　　　YELLOW MOUNTAIN-HEATHER

Similar to *P. empetriformis,* but flowers yellow, peduncles conspicuously long glandular pubescent, hairs as long as the diameter of the peduncle, and plants usually very glandular. Mountain meadows, ridges, and forest openings; Rocky Mountains.

**Rhododendron**　　　rose-bay

Flowers borne in 1- to 3-flowered lateral clusters, axillary. ................................................................. *R. albiflorum*
Flowers borne in a terminal umbelliform cluster. ....................................................................... *R. lapponicum*

*Rhododendron albiflorum* Hook.　　　　　　WHITE ROSE-BAY

Erect shrubs to 2 m high, with slender branches and shredding bark. Leaves 2–7 cm long, thin, deciduous. Flowers creamy white, about 2 cm across. Mountain forests; southern Rocky Mountains.

*Rhododendron lapponicum* (L.) Wahl.　　　　LAPLAND ROSE-BAY

Dwarf shrubs 10–30 cm high, freely branched. Leaves leathery, evergreen, 10–15 mm long. Flowers 1–5 in a cluster, with the corolla about 15 mm wide, bright purple. Arctic and alpine tundra; southeastern Boreal forest, southern Rocky Mountains.

## VACCINIACEAE—huckleberry family

Mostly shrubs or shrubby plants with alternate leaves. The members of this family are sometimes considered genera within Ericaceae, but have the ovary wholly inferior.

Petals appearing separate, turned backward; dwarf plants with very slender creeping stems. ............................................................................... *Oxycoccus*
Petals united into an urn-shaped or campanulate tube. ................................................................ *Vaccinium*

*Oxycoccus*   cranberry

*Oxycoccus palustris* Pers.   SWAMP CRANBERRY

A dwarf creeping plant with slender trailing stem 10–40 cm long, bearing small ovate leaves 3–8 mm long, green above and whitish below. Flowers in terminal umbels, pink, about 8 mm across, and with petals soon turning backward (reflexed). Fruit a small berry, 6–10 mm in diam, red, and often dark-spotted when young. Fairly common; in cold bogs, swamps, and damp woods; Boreal forest. Syn.: *Vaccinium oxycoccus* L.

Often divided into two species (or two varieties of *V. oxycoccus*):

Leaves 5–8 mm long, mostly elliptical; berry
8–10 mm long. ............................................................................ *O. quadripetalus* Gilib.
Leaves 3–5 mm long, mostly ovate; berry 6–8
mm long. ............................................................................ *O. microcarpus* Turcz.

However, these characters and others are variable. *O. quadripetalus* (*V. oxycoccus* L. var. *oxycoccus*) is described as having peduncles 20–35 mm long, with small bracts at the base; in *O. microcarpus* (*V. oxycoccus* L. var. *microphyllum* (Lange) R. & R.) the peduncles are 10–20 mm long, and the bractlets are halfway up the peduncle. In many collections characters are mixed: long peduncles with small or very small leaves, short peduncles with larger leaves, and bractlets often missing.

*Vaccinium*   blueberry

Usually shrubs or shrubby plants with alternate leaves, evergreen or deciduous. Fruit a blue or red many-seeded berry. Freely hybridizing between species.

1. Stems mostly creeping; leaves leathery,
   evergreen. ................................................................ *V. vitis-idaea*
   Stems mostly upright; leaves thin,
   deciduous. ................................................................ 2

2. Flowers in terminal racemes. ................................ *V. angustifolium*
   Flowers axillary, solitary or few together. ................ 3

3. Flowers in clusters of 1–4, from axils of
   bud scales; leaves blunt, entire. ........................ *V. uliginosum*
   Flowers solitary in leaf axils; leaves
   toothed. ................................................................ 4

4. Branches round or nearly so, puberulent. ............ *V. caespitosum*
   Branches more or less sharply angled. ................ 5

5. Pedicels 3 mm long, about as long as the
   flowers; berries red. ............................................ *V. scoparium*
   Pedicels longer than 3 mm, longer than
   the flowers; berries blue or black. ........................ 6

6. Shrubs to 40 cm high; leaves to 25 mm
   long. ................................................................ *V. myrtillus*
   Shrubs to 1 m or higher; leaves to 40 mm
   long. ................................................................ *V. membranaceum*

*Vaccinium angustifolium* Ait.  BLUEBERRY

Small shrubs to 30 or 40 cm high; young branches and twigs round or nearly so. Leaves 1–3 cm long, elliptic to lanceolate. Flowers white or pinkish, 3–5 mm long; pedicel recurved, about 6 mm long; berry blue, 4–7 mm in diam. Two varieties can be distinguished:

Leaves serrate, pubescent below on the veins,
  glabrous above; twigs and branches
  glabrous. ........................................................................................ var. *angustifolium*
Leaves entire or nearly so, pubescent on both
  sides; twigs and branches pubescent. .......................... var. *myrtilloides* (Michx.) House

Both varieties form large colonies in coniferous forest, particularly on light soils. The var. *angustifolium* is limited to southeastern Boreal forest; var. *myrtilloides* occurs throughout Boreal forest and Rocky Mountains. Syn.: *V. myrtilloides* Michx.

*Vaccinium caespitosum* Michx.  DWARF BILBERRY

A very low shrub 5–30 cm high, bearing thin obovate to oblanceolate leaves 10–25 mm long, green and shiny on both sides. Flowers pink or white, ovoid, about 5 mm long, and borne either singly or in groups of 3 or 4. Fruit an edible blue berry with a bloom, about 6 mm in diam. Found in pine woods; Boreal forest and Cypress Hills.

*Vaccinium membranaceum* Dougl.  BILBERRY

Shrubs to 1 m or higher, glabrous throughout; twigs and young branches slightly but clearly angled. Leaves 2–6 cm long, thin, ovate to oval; petioles 1–2 cm long. Flowers yellowish, 5–6 mm long; pedicel 10–15 mm long; berry black, 8–10 mm in diam. Coniferous forests; southern Rocky Mountains.

*Vaccinium myrtillus* L.  WHORTLEBERRY

Shrubs 15–30 cm high, with divergent branches; young branches and twigs sharply angular. Leaves 1–2 cm long, ovate to elliptic, rounded to sub-cordate at base. Flowers greenish white, 3–6 mm long, almost globular; pedicel the same length as or longer than flower; berry 6–10 mm in diam, at first red, later blue to purplish black, glaucous. Coniferous forest; southern Rocky Mountains.

*Vaccinium scoparium* Leib.  RED WHORTLEBERRY

Low shrubs 10–20 cm high, glabrous throughout; young branches and twigs conspicuously angled. Leaves 5–15 mm long, oval to broadly elliptic, serrulate. Flowers about 3 mm long; pedicels equaling or exceeding the flowers; fruit 3–5 mm in diam, red. Coniferous forests; southern Rocky Mountains.

*Vaccinium uliginosum* L.  BOG WHORTLEBERRY

Shrubs 10–50 cm high, much-branched; branches round and glabrous. Leaves 10–25 mm long, obovate to oval, firm, rounded or obtuse at the apex, conspicuously reticulate veined below. Flowers mostly in clusters of 2–4, pink, 5–7 mm long; berry 6–7 mm in diam, dark bluish black. Marshes and bogs; Boreal forest and Rocky Mountains.

*Vaccinium vitis-idaea* L. var. *minus* Lodd.     DRY-GROUND CRANBERRY

A low shrub 10–20 cm high, with erect branches growing from a trailing stem. Leaves obovate with rolled edges, dark green and shiny above, paler with black dots beneath, 1–2 cm long. Bell-shaped flowers pink or white, 5–8 mm long, and borne in small terminal clusters. Fruit a dark red, acid berry about 6 mm in diam. Common; in swamps, muskegs, and sandy woodlands; throughout Boreal forest, Rocky Mountains. Syn.: *Vitis-idaea punctata* Moench.

## DIAPENSIACEAE—diapensia family

**Diapensia**     diapensia

*Diapensia lapponica* L.     NORTHERN DIAPENSIA

Dwarf shrubs with matted stems 5–10 cm long, ascending to 3–5 cm high. Leaves crowded, overlapping, 6–15 mm long, spatulate. Flowers solitary, white, on peduncles 1–4 cm long; capsule about 5 mm long. Dry tundra; northeastern Boreal forest.

## PRIMULACEAE—primrose family

Herbs with simple leaves. Flowers perfect and regular, usually with 5 sepals, 5 petals, and 5 stamens that are borne opposite the petals.

1. Leaves all basal (but leafy bracts below
   umbels of flowers in *Androsace*). ............................................................................... 2
   Leaves not all basal. ............................................................................................................. 5

2. Flowers solitary or 1–3 on a peduncle, 1–3
   cm high. ...................................................................................................... *Douglasia*
   Flowers in umbels of few to many flowers. ................................................................. 3

3. Sepals and petals reflexed (turning
   backward). .................................................................................................. *Dodecatheon*
   Sepals and petals not reflexed. ..................................................................................... 4

4. Tube of corolla as long as or longer than
   calyx; flowers pink or lilac; perennials. ...................................................... *Primula*
   Tube of corolla shorter than calyx;
   flowers small and white; annuals. ............................................................ *Androsace*

5. Leaves mostly alternate, the lower ones
   sometimes opposite. ............................................................................... *Centunculus*
   Leaves opposite or in whorls. ....................................................................................... 6

6. Petals absent, flower pink; low-growing
   plant of saline areas. ................................................................................... *Glaux*
   Petals present. ....................................................................................................................... 7

7. Plants with scale-like lower stem leaves;
   upper leaves in a whorl below flowers;
   usually with sepals, petals, and stamens
   in 7's. ......................................................................................................... *Trientalis*

Plants with normal opposite stem leaves. ................................................................. 8

8. Flowers scarlet; plants annual,
    procumbent. .......................................................................................... *Anagallis*
    Flowers yellow; plants perennial, erect. ..................................... *Lysimachia*

## *Anagallis*    pimpernel

### *Anagallis arvensis* L.                                          SCARLET PIMPERNEL

Annual plants with decumbent 4-sided stems 10–30 cm long. Leaves sessile or clasping, oval to ovate, 5–20 mm long. Flowers 4–7 mm across, rotate; capsule 3–4 mm in diam. Introduced; a rare garden weed.

## *Androsace*    pygmyflower

1. Plants perennial; flowers large, crowded. ......................................... *A. chamaejasme*
   Plants annuals; flowers small; inflor-
   escence diffuse. ......................................................................................... 2

2. Bracts below inflorescence elliptic, ovate,
   and more or less oblong. .................................................................. *A. occidentalis*
   Bracts below inflorescence lanceolate to
   awl-shaped. ........................................................................................ *A. septentrionalis*

### *Androsace chamaejasme* Host                              PYGMY SHIELDWORT

Plants with dense rosettes; leaves lanceolate, 5–10 mm long, ciliate. Scapes 2–10 cm high, villous. Flowers 8–10 mm across, white with yellow center. Rocky slopes, riverbanks, open forests, and disturbed areas; Rocky Mountains.

### *Androsace occidentalis* Pursh                         WESTERN PYGMYFLOWER

A very small species, usually with few flowering stems. Uncommon; found on dry, sandy soil; throughout southern part of the Prairie Provinces.

### *Androsace septentrionalis* L. (Fig. 171*A*)              PYGMYFLOWER

Flowering stems few, usually only one well-developed. The var. *diffusa* (Small) Knuth. has many flowering stems, usually of almost equal lengths; calyx almost free of hairs. Occasionally found in eastern parts and in Riding Mountains. The var. *puberulenta* (Rydb.) Knuth. differs from var. *diffusa* by having a hairy calyx. A common species found on eroded and dry soils, and very plentiful on stubble fields and cultivated land in early spring. A common plant, but so small that it is often not noticed.

## *Centunculus*    chaffweed

### *Centunculus minimus* L.                                       CHAFFWEED

A small depressed annual 2–10 cm high, with obovate to spatulate leaves, almost stalkless, 3–8 mm long. Tiny pink flowers borne in axils of leaves, about 3 mm across. Rare; but has been found in wet places and lake edges; in Prairies.

*Dicots*                                                   PRIMULACEAE — 581

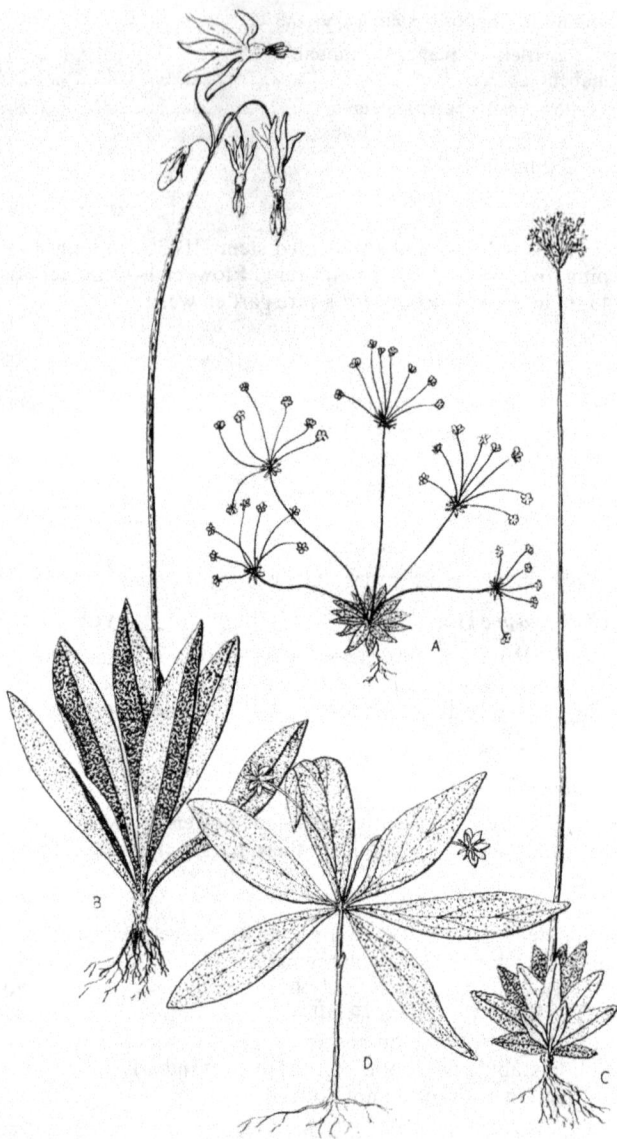

A.C. Budd

Fig. 171.   Primroses: *A*, pygmyflower, *Androsace septentrionalis* L.; *B*, mountain shootingstar, *Dodecatheon conjugens* Greene var. *beamishii* Boiv.; *C*, mealy primrose, *Primula incana* M. E. Jones; *D*, northern starflower, *Trientalis borealis* Raf.

**Dodecatheon**       shootingstar

Perennial herbs with basal leaves. Nodding flowers very showy, with reflexed petal lobes in small umbel at head of leafless stem. Five stamens very prominent, united at base, and bearing long purple anthers. Fruit a cylindrical capsule containing many minute seeds.

Leaves glandular pubescent. ........................................................................ *D. conjugens*
Leaves glabrous. ........................................................................................ *D. pulchellum*

*Dodecatheon conjugens* Greene var. *beamishii* Boiv. (Fig.171*B*)
MOUNTAIN SHOOTINGSTAR

An early flowering, rather low-growing species with oblanceolate or spatulate leaves 4–10 cm long, pale green, and hairless. Flowers 1–5, borne in an umbel on a long stem from base, varying in color from purple to white, but usually pink; petals reflexed showing prominent purple stamens borne on orange filaments. Petals often having a yellowish inner base with a zigzag purple line, making flowers very handsome when closely examined. In the white phase, which is not unusual, anthers and petals are white with no contrasting colors. Common; in moist areas in grassland; Prairies, Cypress Hills, and Rocky Mountains. Syn.: *D. cylindrocarpum* Rydb.

*Dodecatheon pulchellum* (Raf.) Merr.       SALINE SHOOTINGSTAR

A midseason-flowering species, with long, spatulate, hairless basal leaves 3–10 cm long. Flowers pink or lilac with a yellowish throat, often with a purple wavy line, 1–12 in an umbel on a stem 10–30 cm high. Fairly common; in wet areas and around saline sloughs, often associated with *Primula incana* M. E. Jones; Prairies, Parklands, and Rocky Mountains.

**Douglasia**       douglasia

*Douglasia montana* Gray       MOUNTAIN DWARF-PRIMULA

Small cushion plants with the leaves in rosettes. Leaves 5–15 mm long, thick, ciliate. Flowers on a peduncle 1–3 cm high, usually solitary or 1–3 in an umbel. Rare; alpine areas; southern Rocky Mountains.

**Glaux**       sea-milkwort

*Glaux maritima* L.       SEA-MILKWORT

A low-growing rather succulent leafy perennial from a creeping rootstock. Leaves opposite, stalkless, oval to linear-oblong, 3–12 mm long. Small pinkish flowers, in the axils of the leaves, about 3 mm long. If closely examined, they will be seen to have no petals, merely one floral ring. Fairly common in moist saline locations; throughout the Prairie Provinces.

**Lysimachia**       loosestrife

1. Plants not flowering but with bulblets in
   the upper leaf axils. ........................................................................ *L. terrestris*
   Plants flowering with or without bulblets. .......................................................... 2

2. Flowers in short crowded spikes in axils of leaves, or terminal. ............................................................................. 5

   Flowers borne on separate stalks in axils of leaves. ......................................................................................... 3

3. Leaves usually lanceolate, rounded or almost cordate at base; flowers 2–3 cm across. ........................................................................................... 4

   Leaves linear, tapering to base, sessile, in whorls; flowers 10–15 mm across. ...................................... *L. quadrifolia*

4. Lower leaves with petioles 6–10 mm long, densely ciliate. ................................................................ *L. ciliata*

   Lower leaves with petioles 10–30 mm long, not ciliate. ................................................................ *L. hybrida*

5. Inflorescence terminal, loose; petals elliptic. ...................................................................... *L. terrestris*

   Inflorescence axillary, dense; petals linear-lanceolate. ...................................................... *L. thyrsiflora*

*Lysimachia ciliata* L. (Fig. 172)          FRINGED LOOSESTRIFE

   An erect plant 30–100 cm high, with opposite leaves 5–10 cm long, pale green, pointed at apex and rounded at base, borne on stalks 6–10 mm long with a row of hairs on one side of stalk. Flowers 15–25 mm across, with 5 yellow petals somewhat unevenly pointed at tips, and borne in twos or threes in upper leaf axils. There are 5 stamens and 5 infertile stamens (staminodia). Fruit a many-seeded ovoid capsule. Fairly common; in woodlands and in moist spots; throughout the Prairie Provinces. Syn.: *Steironema ciliatum* (L.) Raf.

*Lysimachia hybrida* Michx.          LANCE-LEAVED LOOSESTRIFE

   Very similar in most respects to previous species, but leaves short-stalked and narrower, tapering to base. Found occasionally in moist meadows, but not nearly so common as *L. ciliata.* Syn.: *Steironema hybridum* (Michx.) Raf.

*Lysimachia quadrifolia* L.          WHORL-LEAVED LOOSESTRIFE

   Stems erect, 30–70 cm high, glabrous to sparsely pubescent. Leaves usually in whorls of 4 (3–6). Flowers 10–15 mm across; petals yellow with dark lines. Rare; open woods; southeastern Boreal forest.

*Lysimachia terrestris* (L.) BSP.          SWAMP-CANDLES

   Stems erect, 40–80 cm high, simple or branched. Leaves 5–10 cm long, narrowly lanceolate. Racemes erect, 10–30 cm long, many-flowered, loose, with pedicels to 20 mm long; flowers 10–15 mm across, yellow. Late in the season small elongate purplish bulblets appear in the upper leaf axils. Uncommon; swampy areas and lakeshores; southeastern Boreal forest.

Fig. 172.  Fringed loosestrife, *Lysimachia ciliata* L.

*Lysimachia thyrsiflora* L.

An erect plant 20–50 cm high with lanceolate to linear-lanceolate stalkless leaves 3–10 cm long. Small flowers, about 6 mm or less in diam, borne in dense spike-like racemes in axils of leaves about halfway up stem. Common in swamps and moist places in Parklands and Boreal forest, but scarce in other areas. Syn.: *Naumburgia thyrsiflora* (L.) Duby.

**Primula**      primrose

1. Leaves with sulfur yellow mealiness on
   undersides. ................................................................................. *P. incana*
   Leaves green on both sides. ...................................................................... 2

2. Leaves entire; flowers less than 1 cm
   across. ............................................................................... *P. egaliksensis*
   Leaves dentate or crenate; flowers larger. ................................................ 3

3. Pedicels 1–3 cm long, many times longer
   than the bracts. ............................................................... *P. mistassinica*
   Pedicels less than 1 cm long, at most twice
   as long as the bracts. .............................................................. *P. stricta*

*Primula egaliksensis* Wormsk.

Slender plants with stems to 20 cm high; leaves thin, ovate to oblong or spatulate, to 5 cm long. Flowers 5–9 mm across, greenish white to violet; calyx lobes glandular ciliate. Meadows and calcareous shores; Boreal forest.

*Primula incana* M. E. Jones (Fig. 171*C*)            MEALY PRIMROSE

A low plant with a basal rosette of spatulate to oval leaves 2–10 cm long, tapering at base to a stalk and blunt at apex. Upper sides green and underside covered with a sulfur yellow mealiness. Flowers pale lilac with a yellow center, 6–10 mm across, borne in an umbel-like cluster at head of a leafless stem 10–30 cm high. Found occasionally; in grass in saline meadows and moist spots; throughout the Prairies and Parklands.

*Primula mistassinica* Michx.                          DWARF PRIMROSE

A small plant with a very slender stem 5–20 cm high. Leaves 1–5 cm long, oblanceolate to cuneate-obovate, dentate, green on both sides, or slightly mealy below. Flowers 10–20 mm across, pink, lilac, or white with yellow center. Marshy areas, shores, and bogs; Boreal forest, rare elsewhere.

*Primula stricta* Horn.                                ERECT PRIMROSE

Plants with upright stems 10–30 cm high. Leaves 2–6 cm long, green on both sides, crenate, ovate to lanceolate. Flowers 8–12 mm across. Wet areas; northeastern Boreal forest.

**Trientalis**      starflower

Leaves almost entirely in a whorl at tip of
stem, acute or acuminate; pedicels shorter
than the leaves. ....................................................................... *T. borealis*

Leaves not restricted to tip of stem, obtuse or
  rounded; pedicels as long as the leaves. ........................................................ *T. europaea*

*Trientalis borealis* Raf. (Fig. 171*D*)                    NORTHERN STARFLOWER

A perennial from horizontal creeping roots, which send up simple stems
8–30 cm high, bearing a whorl of 5–10 lanceolate leaves 3–10 cm long and
tapering at both ends. The few flowers are 10–12 mm across, white, with 7
pointed petals, and borne on slender stalks from the center of a leafy whorl.
Found occasionally in damp woodlands; throughout Boreal forest. Syn.: *T.
americana* Pursh.

*Trientalis europaea* L.                                                STARFLOWER

Similar to *T. borealis,* but with the leaves more or less spread out along the
upper part of the stem; pedicels much longer, and also axillary from lower
leaves. Muskeg and bogs; northwestern Boreal forest, Peace River.


## OLEACEAE—olive family

*Fraxinus*        ash

Hardwood trees with pinnate leaves. Inconspicuous flowers, usually unisexual, appearing about the same time as the leaves. Fruit, a samara, borne in
pendulous clusters and consisting of a seed with a long membranous yellowish
green wing.

Body or seed of samara flattened; wing
  extending to bottom of seed. ............................................................ *F. nigra*
Body or seed of samara almost circular in
  cross section; wing extending halfway down
  body of seed. .................................................... *F. pennsylvanica* var. *austinii*

*Fraxinus nigra* Marsh.                                                BLACK ASH

A tall tree with pinnate leaves of 7–11 leaflets; the difference between the
wing and the body of the samara rather indistinct. Southeastern Parklands and
Boreal forest.

*Fraxinus pennsylvanica* Marsh. var. *austinii* Fern. (Fig. 173)      GREEN ASH

A tree to 10 m high with 5–7 leaflets. Branchlets densely short pubescent;
the leaf rachis and underside of leaflets sparingly pubescent. Shores, thickets,
and along rivers; southeastern Parklands and Boreal forest. Trees with glabrous branches and leaflets are var. *subintegerrima* (Vahl.) Fern., occurring
along rivers and in ravines in the eastern part of the Prairies. Syn.: *F.
campestris* Britt.

(Fig. 173 overleaf)

*Dicots*                                                OLEACEAE − 587

Fig. 173. Green ash, *Fraxinus pennsylvanica* Marsh. var. *austinii* Fern.

# GENTIANACEAE—gentian family

Mostly low herbs with a bitter sap and, with one exception, opposite, simple, and stalkless leaves. Flowers regular and perfect, with a tubular corolla, usually with 4 or 5 lobes at mouth, with as many stamens as corolla lobes. Fruit a capsule opening by valves and containing many seeds.

1. Perennial bog plants with trifoliolate leaves (3 leaflets). ............................................. *Menyanthes*
   Plants with simple opposite leaves. ............................................. 2

2. Flowers with 4 hollow spurs at base. ............................................. *Halenia*
   Flowers not spurred at base. ............................................. 3

3. Corolla campanulate or bell-shaped. ............................................. *Gentiana*
   Corolla rotate. ............................................. *Lomatogonium*

## *Gentiana*    gentian

1. Flowers with plaits or folds between lobes of corolla. ............................................. 2
   Flowers without plaits or folds between lobes of corolla. ............................................. 8

2. Leaves with white margins; flowers solitary and terminal. ............................................. *G. aquatica*
   Leaves without white margins; flowers several, in leaf axils. ............................................. 3

3. Mouth of corolla almost or quite closed; corolla lobes absent or very minute. ............................................. *G. andrewsii*
   Mouth of corolla open; corolla lobes distinct. ............................................. 4

4. Leaves linear-oblong; flowers terminal and axillary. ............................................. *G. linearis*
   Leaves ovate to lanceolate. ............................................. 5

5. Plants with basal rosettes present at flowering. ............................................. *G. glauca*
   Plants without basal rosettes at flowering. ............................................. 6

6. Calyx lobes broadly ovate or oval. ............................................. *G. calycosa*
   Calyx lobes linear to linear-lanceolate. ............................................. 7

7. Flowers 25–30 mm long; leaves glabrous. ............................................. *G. affinis*
   Flowers 35–45 mm long; leaves puberulent. ............................................. *G. puberulenta*

8. Lobes of corolla fringed or toothed; flowers long-peduncled. ............................................. 9
   Lobes of corolla not fringed or toothed; flowers short-peduncled. ............................................. 10

9. Calyx finely but distinctly glandular puberulent on the keels. ............................................. *G. crinita*
   Calyx glabrous. ............................................. *G. detonsa*

10. Throat of corolla with a fringe of hairs. ..................................................... *G. amarella*
    Throat of corolla hairless. ........................................................................ *G. propinqua*

## *Gentiana affinis* Griseb.                                    OBLONG-LEAVED GENTIAN

A leafy-stemmed perennial 15–30 cm high, often prostrate, usually with several stems from a deep taproot. Leaves oblong to lanceolate, 10–35 mm long. Flowers dark blue or purple, 25–30 mm long, in raceme-like dense clusters on upper end of stems. Sometimes merely one or a few flowers to a stem. Fairly common; in sandy areas and moist, even saline, meadows; throughout Prairies and Parklands. Syn.: *Dasystephana affinis* (Griseb.) Rydb.; *G. interrupta* Greene.

## *Gentiana amarella* L.                                          NORTHERN GENTIAN

An annual 15–50 cm high. Upper leaves lanceolate and rather sharply pointed; lower leaves usually spatulate or ovate and blunt at apex. Flowers varying from purplish and blue to greenish yellow or white, 10–20 mm, borne in clusters in upper leaf axils. Found fairly often; throughout the Prairie Provinces. Because of wide variations, authorities have made several separate species of various forms of this plant, such as *G. acuta, G. plebeia, G. scopulorum,* and *G. strictiflora,* but they are probably all local and variations of the same species.

## *Gentiana andrewsii* Griseb.                                    CLOSED GENTIAN

An upright rather sturdy perennial herb 30–80 cm high. Leaves ovate to lanceolate, 5–10 cm long, with 3–7 veins. Flowers blue with whitish folds, or sometimes all white, 3–4 cm long, usually closed at the mouth and borne in clusters at the end of the stem or in axils of upper leaves. In wet meadows and among bushes; in eastern Parklands. Syn.: *Dasystephana andrewsii* (Griseb.) Small. Plants with white flowers are distinguished as f. *albiflora* Britt. Syn.: *G. flavida* Gray; *Dasystephana flavida* (Gray) Britt.

## *Gentiana aquatica* L.                                            MOSS GENTIAN

A small annual or biennial, usually branching from base, 3–10 cm high. Leaves many, less than 6 mm long; basal leaves obovate; upper ones linear-lanceolate. Small purplish green flowers solitary at summit of stems, 5–8 mm long, and followed by capsule containing tiny seeds. Capsule opening into two valves and spreading outward at maturity. Found very rarely; around sloughs and marshy places; in Prairies and Parklands. Syn.: *Chondrophylla fremontii* (Torr.) A. Nels.; *G. fremontii* Torr.; *G. prostrata* Haenke.

## *Gentiana calycosa* Griseb.                                    MOUNTAIN GENTIAN

Plants with simple or branched root crown, and several to many erect or ascending stems 15–40 cm high. Leaves 2–4 cm long, round-oval to ovate. Flowers solitary, or 1–3 at the tip of the stem, sometimes also 1 or 2 flowers in the upper axils. Corolla blue, 25–35 mm long; calyx tube 6–8 mm long, the oblong lobes equaling or exceeding the tube; capsule 12–16 mm long. Rare; alpine meadows; southern Rocky Mountains.

*Gentiana crinita* Froel.

An annual species 15–50 cm high, with somewhat clasping lanceolate leaves 2–5 cm long. Flowers large, 2–6 cm long, at end of stems, sky blue, with a conspicuous fringe around the lobes. Fairly common; in moist woods or low areas; in southeastern Parklands and Boreal forest. Syn.: *Anthopogon crinitus* (Froel.) Raf. Three varieties can be distinguished:

1. Leaves lanceolate, 5–20 mm wide. ................................................................ var. *crinita*
   Leaves linear, 3–5 mm wide. ................................................................................ 2

2. Flowers large, mostly 4–6 cm long,
   corolla lobes strongly fringed. ................................... var. *browniana* (Hook.) Boiv.
   Syn.: *G. procera* Holm

   Flowers small, mostly 2–4 cm long,
   corolla lobes little fringed. ............................................... var. *tonsa* (Lunell) Vict.
   Syn.: *G. macounii* Holm

*Gentiana detonsa* Rottb. var. *raupii* (Pors.) Boiv.

Similar to *G. crinita,* but with the calyx glabrous. Stems 20–40 cm high, leafy below. Leaves lanceolate to linear-lanceolate, 2–5 mm wide. Corolla 3–5 cm long, with lobes short-fimbriate. Shores and marshes; Boreal forest.

*Gentiana glauca* Pall.

Low plants with rootstocks and basal rosettes; stems 2–10 cm high. Basal leaves 5–15 mm long, rather thick; stem leaves 2 or 3 pairs, about 1 cm long. Corolla 12–18 mm long, blue. Alpine meadows; southern Rocky Mountains.

*Gentiana linearis* Froel.

Rather stout plants with stems 30–70 cm high. Leaves 4–9 cm long, linear-oblong, to 1 cm wide. Flowers usually 2–4 in a terminal cluster, and solitary in the upper axils. Corolla 3–4 cm long, blue to white. Rare; marshy areas; southeastern Parklands and Boreal forest.

*Gentiana propinqua* Rich.

Annual plants, often branching from the base, 5–30 cm high. Basal leaves spatulate or oblanceolate; stem leaves lanceolate, 1–2 cm long. Flowers 1–3 in axils; corolla blue, 10–15 mm long; calyx lobes unequal: the outer two 5–7 mm long, ovate and the others 3–5 mm long, linear. Arctic and alpine grassland; northeastern Boreal forest, Rocky Mountains.

*Gentiana puberulenta* Pringle

Perennial, usually with a single stem 20–50 cm high, often covered with minute hairs. Leaves lanceolate, usually rough or minutely hairy on edges and midrib. Flowers blue, 3–4 cm long, in upper leaf axils. Very rare; grassland; southeastern Parklands. Syn.: *Dasystephana puberula* (Michx.) Small; *G. puberula* Michx.

*Halenia*      spurred-gentian

*Halenia deflexa* (Smith) Griseb.          SPURRED-GENTIAN

An annual with slender upright stems 15–50 cm high. Basal leaves spatulate or obovate; stem leaves oblong to ovate, 2–5 cm long. Flowers in clusters at head of stems and in axils of upper leaves, about 6 mm long, purplish to greenish or yellowish white. Distinguished from gentians by 4 hollow spurs projecting downward from the base of flowers. Fairly common in moist woodlands; Parklands and Boreal forest.

*Lomatogonium*      marsh felwort

*Lomatogonium rotatum* (L.) Fries         MARSH FELWORT

An erect slender annual 10–45 cm high, with spatulate basal leaves and linear to lanceolate stem leaves 1–5 cm long. White or bluish flowers 1–2 cm across, borne singly or in clusters in axils of leaves; corolla deeply cleft into 4 or 5 segments. Rare; occasionally reported from marshy land; throughout the Prairie Provinces. Syn.: *Pleurogyne rotata* (L.) Griseb.

*Menyanthes*      buck-bean

*Menyanthes trifoliata* L.         BUCK-BEAN

A perennial bog plant arising from a thick scaly rootstock. Leaves trifoliolate, of 3 elliptic leaflets 5–10 cm long, on basally sheathed stems 5–20 cm long. Flowers clustered in a raceme at the head of a separate stalk, whitish or pinkish purple, 10–15 mm long. Fruit an ovoid capsule containing a few shiny seeds. Common; in bogs and wet swampy places; in Boreal forest.

## APOCYNACEAE—dogbane family

Perennial herbs with acrid milky sap, and entire opposite leaves. Flowers regular, perfect, and bell-shaped, with 5 sepals and 5 partly united petals. Fruits long, narrow follicles, in pairs, and containing many seeds; each seed bearing a tuft of hairs.

*Apocynum*      dogbane

1. Petals fully twice as long as sepals; stem
   leaves drooping or spreading. ............................................................... 2

   Petals less than twice as long as sepals;
   stem leaves ascending. ........................................ *A. cannabinum*

2. Petals three times as long as sepals; flower
   clusters at end of stems and in axils of
   leaves. ........................................ *A. androsaemifolium*

   Petals twice as long as sepals; flower clusters at end of stems only. ........................................ *A. medium*

*Apocynum androsaemifolium* L. var. *incanum* DC. (Fig. 174)

A somewhat bushy perennial from a horizontal rootstock, 30–150 cm high. Plant much branched, and stems when broken exuding a milky sap. Leaves opposite, ovate or oval, somewhat paler and often slightly hairy on lower side, 2–7 cm long. Flowers in clusters at ends of branches and in axils of leaves, pink, 6–8 mm long; lobes of petals spreading and often curved downward. Fruits in pairs of long narrow follicles or pods, to 10 cm long, tubular, containing many hairy-tipped seeds. Common; in woodland and on light sandy soil; throughout the Prairie Provinces.

*Apocynum cannabinum* L. var. *hypericifolium* Gray          INDIAN-HEMP

A deep-rooted perennial 30–150 cm high, with fairly erect branches. Leaves lanceolate-oblong or ovate-oblong, pale green above and often somewhat whitened beneath, 3–10 cm long, narrowed at either end. Flowers greenish white, about 3–5 mm long, in clusters at ends of branches and in leaf axils. Fruit similar to that of *A. androsaemifolium.* Found occasionally; in thickets; throughout the Prairie Provinces. Includes *A. sibiricum* Jacq.

*Apocynum medium* Greene                    INTERMEDIATE DOGBANE

Similar to spreading dogbane, for which it is often mistaken, but having a shorter flower, about 6 mm or less in length, usually greenish or white with a pink tinge. Flower clusters only borne at ends of branches, not in leaf axils. A hybrid of the foregoing two species, sparingly found; throughout the Prairie Provinces where the parents are growing.

## ASCLEPIADACEAE—milkweed family

Perennial herbs, usually with milky juice, and flowers borne in umbels. Flowers rather complicated, but having 5 corolla lobes with a 5-lobed crown joining the stamens and the corolla lobes. Corolla lobes usually reflexed or turned downward. Fruit a large and conspicuous follicle, or large pod, which opens down one side to release numerous seeds, each of which bears a tuft of silky hairs. Pollen grains in this family united into masses (pollinia), which are pear-shaped and attached in pairs. The peculiar structure of the flower causes these pairs of pollinia to adhere to visiting insects and to be transferred to other flowers, thereby ensuring cross-fertilization.

*Asclepias*          milkweed

Coarse perennial herbs with deep taproots and white milky juice. Leaves opposite or whorled. Petals and sepals reflexed (turned downward). Fruit a large pod, or follicle, containing many seeds, each with a tuft of white silky hairs.

(Fig. 174 overleaf)

Fig. 174.  Spreading dogbane, *Apocynum androsaemifolium* L. var. *incanum* DC.

*Dicots*

1. Hoods of the crown of flower without an
   incurved horn within. ...................................................................................... 2

   Hoods of the crown of flower with an
   incurved horn within. ...................................................................................... 3

2. Umbels solitary, terminal. ................................................ *A. lanuginosa*
   Umbels several, lateral. ..................................................... *A. viridiflora*

3. Leaves linear and in whorls. ............................................... *A. verticillata*
   Leaves broader and opposite. ................................................................. 4

4. Flowers rose or red; plant almost hairless. ........................... *A. incarnata*
   Flowers purplish or greenish; plants hairy
   or downy. ...................................................................................................... 5

5. Leaves ovate or lanceolate, tapering to
   base; pods without tubercles. ............................................. *A. ovalifolia*

   Leaves blunt or almost cordate at the
   base, oblong or oval; pods with soft
   tubercles. ...................................................................................................... 6

6. Hoods of corolla long and lanceolate,
   three times longer than stamens. .......................................... *A. speciosa*

   Hoods of corolla short and blunt, not
   much longer than stamens. ................................................... *A. syriaca*

*Asclepias incarnata* L.                                    SWAMP MILKWEED

A tall slender-stemmed perennial, 50–150 cm high, almost devoid of hairs, with lanceolate opposite leaves 3–10 cm long. Small flowers usually rosy red but sometimes paler, in numerous many-flowered umbels. Pods of seeds borne upright, usually in pairs, 5–8 cm long, and almost smooth. Fairly common; in swamps, wet spots, and roadside ditches; throughout southeastern Parklands.

*Asclepias lanuginosa* Nutt.                                 HAIRY MILKWEED

Tall perennial, with stems 1–2 m high, and villous. Leaves linear-oblong or lanceolate, 4–7 cm long, pubescent on both sides. Umbel erect, short-peduncled, 4–5 cm across. Rare; open grasslands in sandhills; southeastern Parklands.

*Asclepias ovalifolia* Dcne.                                 DWARF MILKWEED

A low species 20–50 cm high, with ovate to lanceolate leaves narrowing or tapering to base, 3–7 cm long. Flowers greenish white, on long stalks in umbels. Found occasionally; on moist prairie; throughout Parklands and southern margin of Boreal forest.

*Asclepias speciosa* Torr. (Fig. 175)                        SHOWY MILKWEED

A stout erect perennial 30–100 cm high, usually found in large colonies. Leaves broad and oval, 7–15 cm long, rounded or somewhat heart-shaped at base, and with a whitish downiness. Flowers flesh-colored or pinkish purple, 8–12 mm across, in very dense almost globular umbels 5–7 cm in diam. Inflorescence having a strong sweet smell, which may make a person sleepy. The nectar also appearing to have a stupefying effect on insects, which may often be found in a drowsy condition below the plants. Pods 6–10 cm long,

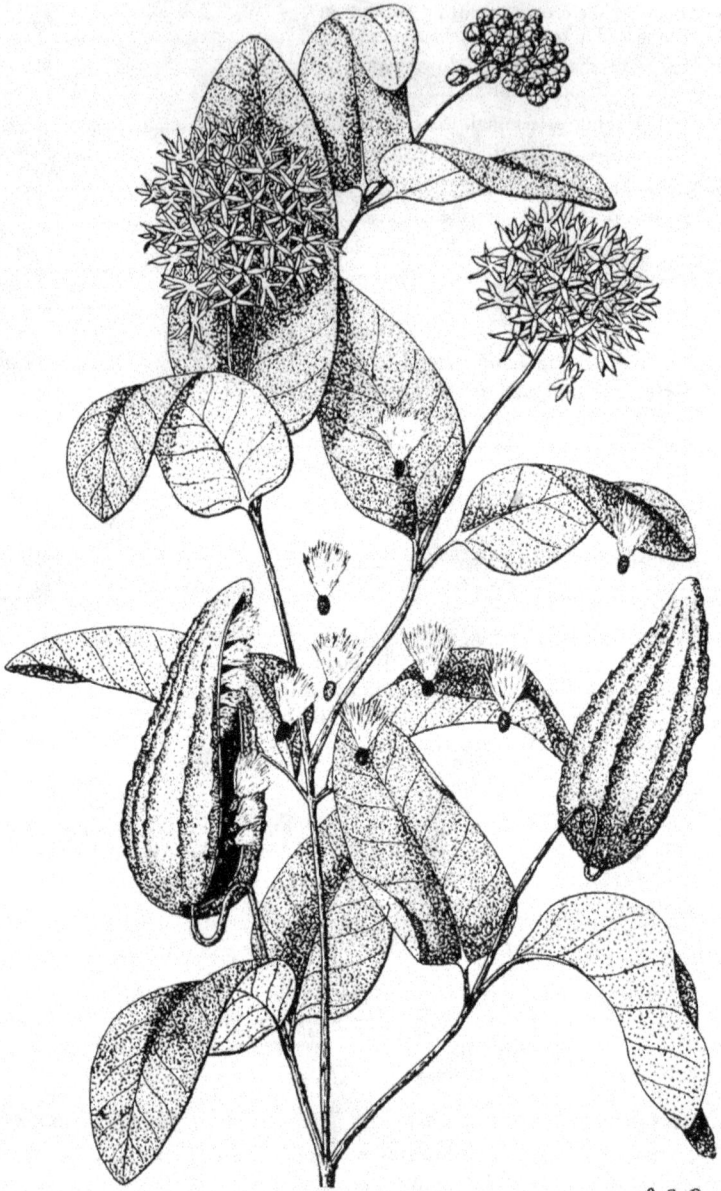

Fig. 175.   Showy milkweed, *Asclepias speciosa* Torr.

densely white woolly, and covered with soft tubercles, borne on recurved stalks. The commonest milkweed; in moist places; throughout Prairies and Parklands.

*Asclepias syriaca* L.                                        SILKY MILKWEED

Very similar to *A. speciosa,* but with elliptic leaves, rounded and not heart-shaped at base. Flowers differ in having a short blunt hood on corolla. Fairly common; on moist sandy soil and riverbanks; southeastern Prairies and Parkland.

*Asclepias verticillata* L.                                WHORLED MILKWEED

A slender-stemmed plant 20–50 cm high with very narrow linear leaves 3–7 cm long, borne in whorls of twos or fours up the stem. Small flowers greenish white, in small umbels. Found occasionally; on dry soil; extreme southeastern Parkland.

*Asclepias viridiflora* Raf.                                 GREEN MILKWEED

Perennial herb with stems sometimes reclining at base, 20–120 cm high. Leaves usually ovate-lanceolate, 2–8 cm long. Flowers borne in umbels at head of stem and also in leaf axils, greenish yellow to dull purple with purple hood. The var. *linearis* (A. Gray) Fern. has linear leaves. Found occasionally; in dry or sandy soil; eastern and south central parts of Prairies and Parkland. Syn.: *Acerates viridiflora* (Raf.) Eat.


## CONVOLVULACEAE—convolvulus family

Leafy twining plants with large funnel-like
flowers. ................................................................................................. *Convolvulus*
Parasitic twining plants with leaves reduced to
scales. ............................................................................................................ *Cuscuta*


***Convolvulus***        morning-glory

Twining or climbing perennials from creeping roots. Leaves alternate, usually arrow- or spear-shaped. Flowers usually large, funnel-form or bell-shaped, with 5 stamens. Fruit a several-seeded capsule.

Calyx not enclosed in two large bracts. ............................................. *C. arvensis*
Calyx enclosed in two large bracts. ................................................... *C. sepium*

*Convolvulus arvensis* L. (Fig. 176*A*)                    FIELD BINDWEED

A deep-rooted perennial with a very extensive system of white roots, and slender twining stems. Leaves 2–4 cm long, somewhat bluntly triangular with hastate bases. Flowers varying from pink to white, 20–25 mm across, borne either singly or 2–4 in a bunch. Fruit a round 2-celled capsule containing large, dark brown, angular seeds. An introduced weed; often found in gardens and roadsides, and very difficult to eradicate.

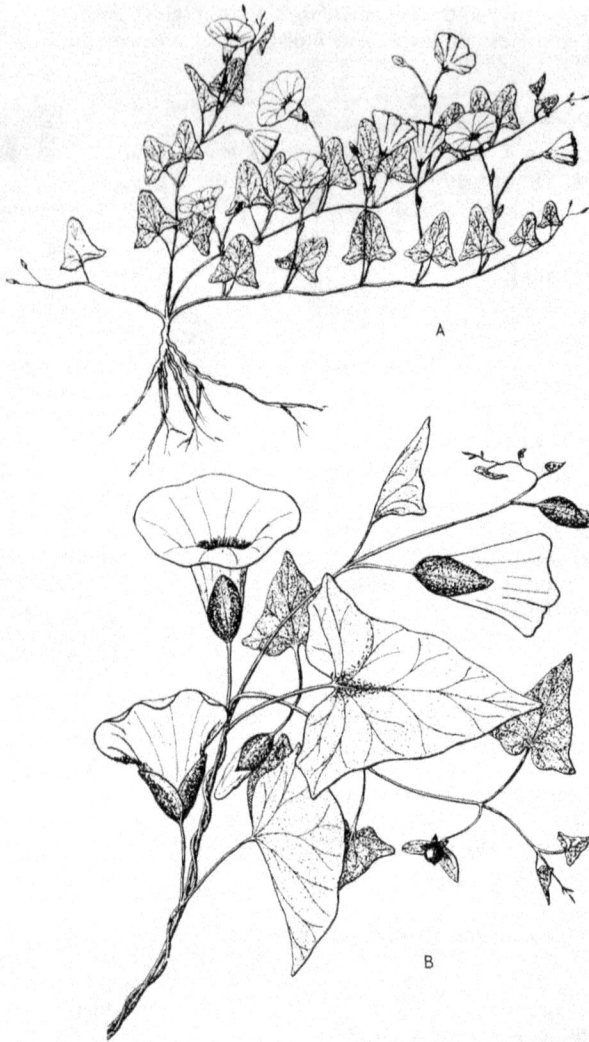

A.C. Budd

Fig. 176. *A*, Field bindweed, *Convolvulus arvensis* L.; *B*, wild morning-glory, *Convolvulus sepium* L. var. *americanus* Sims.

*Dicots*

*Convolvulus sepium* L.

A perennial plant, naturally twining on bushes and shrubs, but in culti-vated places often becoming a creeping field weed. Rhizomes white and creep-ing, sending up many shoots. Leaves varying in size and shape but roughly tri-angular with hastate or sagittate bases. Two large green bracts enclosing sepals and lower portion of flower. The large funnel-shaped flowers varying from pink to pure white. Fruit a capsule containing large angular brown or black seeds. The species has lobes of leaves directed downward, leaves 5–12 cm long. Flowers 4–5 cm long and up to 6 cm across. Probably introduced from Europe or Asia; found around bushes and waste places; especially in eastern Prairies and Parkland. The following key separates the two main varieties of this species.

Leaves with basal lobes spreading sideways;
    hairless or only slightly hairy. ................................................................. var. *americanus*
Leaves with rounded or arrow-shaped, down-
    ward-pointing basal lobes; stems and leaves
    covered with dense fine hairs. ................................................................. var. *pubescens*

The var. *americanus* Sims (Fig. 176*B*), wild morning-glory, is a native twining perennial with broadly hastate leaves and either pink or white funnel-like flowers. When the area on which they grow is brought under cultivation, the roots are spread by tillage throughout the land, and the plant spreads rap-idly by both root portions and seed, often taking full control of large areas and choking out crops. When growing in large masses, leaves and flowers often become much smaller, almost to suggest a different species. The seeds have a long dormant period and may remain viable in soil for many years, even under good germination conditions. Common; found climbing on bushes along water courses and in moist areas; throughout the Prairie Provinces. Syn.: *C. americanus* (Sims) Greene.

The var. *pubescens* (Gray) Fernald, inland bindweed, is a low-growing, twining or crawling perennial with leaves 25–35 mm long, rounded at angles, arrow-shaped or heart-shaped at base, and covered with fine downy hairs beneath. Flowers white, 3–4 cm long. Found occasionally; on sandy soils; throughout southern prairies. Syn.: *C. interior* House.

### *Cuscuta*        dodder

Parasitic annual plants with leaves absent or reduced to scales, found twining on host plants. Dodders grow from seed, and after they become fas-tened to their host plants by aerial roots and suckers, the root and basal por-tion of the stem decays and sustenance is entirely derived from the host plant. Flowers pinkish or whitish, in clusters, with a bell-shaped or almost globular corolla of 4 or 5 overlapping lobes, and with small scales inside the throat, alternating with lobes. Fruits are somewhat globular capsules borne in clus-ters, each capsule containing 2–4 seeds. A rather difficult genus to identify, because the small flowers need to be closely examined for differing characteris-tics.

*Cuscuta gronovii* Willd. COMMON DODDER

An orange- or yellow-stemmed species found twining on various coarse herbs and shrubs. Flowers 2–4 mm long, sessile or subsessile in dense clusters. Corolla lobes broadly ovate, shorter than the tube; calyx short, with lobes round-ovate to subrotund; capsules about 3 mm in diam. Rare; on various shrubs along shores and in moist areas; Prairies and Parklands. Includes *C. curta* (Engelm.) Rydb.; *C. pentagona* Engelm.; *C. planiflora* Tenor.

## POLEMONIACEAE—phlox family

Usually low-growing annual or perennial herbs. Flowers perfect, generally regular, with 5 partly united sepals and 5 united petals. Flowers usually funnel-like or salverform (with a long tube, abruptly flattened at the end). Fruit a 3-celled capsule containing the seeds.

1. Leaves compound with many small
   leaflets. ............................................................................ *Polemonium*
   Leaves simple. ............................................................................ 2
2. Leaves not cleft or divided. .................................................... 3
   Leaves divided or pinnately cleft. ........................................... 4
3. Leaves opposite. ......................................................................... *Phlox*
   Leaves alternate. ........................................................................ *Collomia*
4. Leaves opposite. ........................................................................ *Linanthus*
   Leaves alternate. ........................................................................ 5
5. Calyx spine-tipped, and as long as corolla
   tube. .......................................................................................... *Navarretia*
   Calyx not as long as corolla tube. ......................................... *Gilia*

**Collomia**    collomia

*Collomia linearis* Nutt. (Fig. 177*A*) NARROW-LEAVED COLLOMIA

An erect annual herb 10–40 cm high, somewhat sticky hairy. Leaves alternate, lanceolate or linear-lanceolate, entire, 2–6 cm long; lower leaves usually shorter than upper ones. Flowers very small, pink or pale purple, in a dense head-like leafy cluster at top of stem. Common; on dry soils and sandy places; throughout the Prairie Provinces.

**Gilia**    gilia

Flowers usually red, about 2 cm long, in a
   panicle-like cluster. ................................................................ *G. aggregata*
Flowers white, less than 1 cm long, in head-
   like clusters. ............................................................................. *G. congesta*

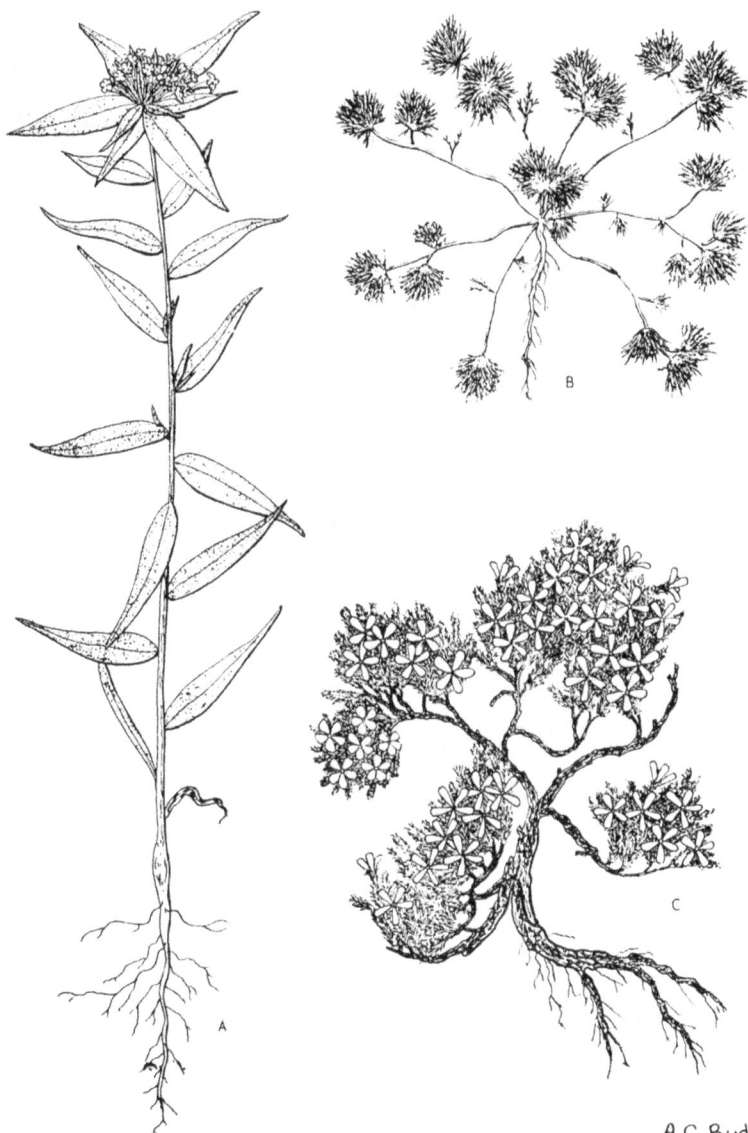

A.C. Budd

Fig. 177.   Phloxes: *A*, narrow-leaved collomia, *Collomia linearis* Nutt.; *B*, small navarretia, *Navarretia minima* Nutt.; *C*, moss phlox, *Phlox hoodii* Richardson.

*Gilia aggregata* (Pursh) Spreng.                                    SCARLET GILIA

An erect perennial 30–50 cm high, usually few-branched. Leaves pinnately divided into narrow segments, 2–7 cm long. Red or scarlet flowers very
conspicuous, in a large, loose, narrow cluster, usually 2–4 cm, with petal lobes
about 10 mm long. Very rare; southern Rocky Mountains.

*Gilia congesta* Hook.                                          CLUSTERED GILIA

A basally-branched perennial 10–15 cm high, usually covered with cob-
webby hairs. Leaves pinnately cleft. White flowers in head-like clusters.
Very rare, but may possibly be found; on dry, sandy soil; southern Rocky
Mountains.

**Linanthus**        linanthus

*Linanthus septentrionalis* H. L. Mason                              LINANTHUS

A much-branched, very fine stemmed annual 5–30 cm high. Leaves
divided to the base into very narrow thread-like segments, about 5–10 mm
long, making them appear clustered. Flowers white, about 3 mm long, on long
fine stalks. Very rare, but has been found; on sandy roadsides; Prairies. Syn.:
*L. harknessii* (Curran) Greene var. *septentrionalis* (Mason) Jepson & Bailey.

**Navarretia**       navarretia

*Navarretia minima* Nutt. (Fig. 177*B*)                      SMALL NAVARRETIA

A low-growing depressed annual with much-branched stems 3–10 cm
high. Leaves 20–35 mm long, deeply divided into needle-like segments. Flowers about 6 mm long, white, and almost hidden in round clusters of leaves and
spiny sepals. Locally abundant; on bottom lands, sandy places, and slough
margins; throughout southwestern Prairies.

**Phlox**        phlox

1. Stem erect; leaves linear to lanceolate. ........................................................................ 2
   Stem decumbent or tufted; leaves short
   and awl-shaped, usually not more than
   10–12 mm long. ........................................................................................................ 3
2. Plants annual; stems 10–20 cm high;
   upper stem leaves alternate. ............................................................ *P. gracilis*
   Plants perennial; stems to 60 cm high;
   upper stem leaves opposite. .............................................................. *P. pilosa*
3. Leaves with cobwebby hairs; tube of
   corolla less than 12 mm long, slightly
   longer than calyx. ............................................................................. *P. hoodii*
   Leaves without cobwebby hairs; tube of
   corolla more than 12 mm long, much
   longer than calyx. ......................................................................... *P. alyssifolia*

*Phlox alyssifolia* Greene                                          BLUE PHLOX

A stout-stemmed prostrate plant with oblong or linear leaves 6–12 mm
long, with sharp-pointed tips. Flowers few, purplish or bluish, with a tube

about 15 mm long. Rare; found occasionally on dry benchland; Prairies and southern Rocky Mountains.

*Phlox gracilis* (Hook.) Greene                                    SLENDER PHLOX

Stems usually erect, simple below and sparsely to freely branching above; glandular pubescent above, sparsely tomentose below. Inflorescence branched, cymose, glandular; corolla 9–12 mm long, with lobes pinkish or purplish; calyx 5–6 mm long. Rare; moist, grassy slopes; southern Rocky Mountains.

*Phlox hoodii* Richardson (Fig. 177*C*)                             MOSS PHLOX

A low tufted mat-forming plant with coarse woody roots. Leaves awl-shaped, sharp-pointed, and somewhat imbricated or overlapping, 3–8 mm long. Flowers white or occasionally pale blue or purple, about 1 cm across, with 5 petal lobes, borne very freely in early spring. One of the most conspicuous of early spring prairie flowers, forming large masses of white on the plains and hillsides. After flowering, plants becoming rather inconspicuous but forming a large proportion of the ground cover, especially on eroded areas and in shallow soil. Common on open prairie throughout Prairies, less common in Parklands.

*Phlox pilosa* L.                                                   DOWNY PHLOX

An erect species 30–60 cm high, usually with soft downy hairs. Leaves linear or lanceolate, 2–10 cm long and stalkless. Flowers purplish pink or white, with a tube about 12 mm long, in a cymose cluster at summit of stem. Found occasionally; in sandy places; southeastern Parklands.

**Polemonium**        Jacob's-ladder

1. Corolla lobes shorter than the tube;
     flowers in dense, capitate cymes. ......................................................... *P. viscosum*
   Corolla lobes longer than the tube;
     flowers in loose, open cymes.  .................................................................................... 2

2. Plants usually less than 30 cm high, several to many stems. ....................................................... *P. pulcherrimum*
   Plants usually more than 30 cm high, solitary. ...................................................................... *P. occidentale*

*Polemonium occidentale* Greene                                    WESTERN PHLOX

A leafy-stemmed erect perennial 30–80 cm high, somewhat glandular hairy above. Leaves of 15–27 ovate to lanceolate leaflets 15–35 mm long. Flowers bell-shaped, violet or blue, 8–12 mm long, in a narrow cluster. May be found occasionally, but not common; in open woodlands and valleys; Rocky Mountains.

*Polemonium pulcherrimum* Hook. (Fig. 178)            SHOWY JACOB'S-LADDER

A glabrous many-stemmed perennial; leaves with 5–11 pairs of ovate to rotund leaflets 2–8 mm long. Inflorescence a cyme; flowers blue, 10–15 mm across; calyx glandular pubescent. A very showy species, found on slopes and in open areas; Rocky Mountains.

Fig. 178.   Showy Jacob's-ladder, *Polemonium pulcherrimum* Hook.

*Polemonium viscosum* Nutt.                                    STICKY JACOB'S-LADDER

Perennials with rootstock; stems 10–40 cm high. Leaves mostly basal; leaflets 3- to 5-lobed, with lobes 1–3 mm long, glandular pubescent. Inflorescence subcapitate; bracts pinnatifid, glandular pubescent; flowers blue, 20–25 mm across, with the calyx glandular pubescent. Rare; rocky slopes; southern Rocky Mountains.


# HYDROPHYLLACEAE—waterleaf family

Mostly hairy annual herbs with a watery sap. Leaves lobed or pinnatifid and flowers perfect with 5 more or less united sepals and a 5-lobed bell-shaped corolla. Fruit a capsule.

1. Flowers solitary in leaf axils. ................................................................................ 2
   Flowers several to many in cymes. ..................................................................... 3
2. Leaves pinnately divided with 7–13 den-
     tate segments. ................................................................................. *Ellisia*
   Leaves with 3–5 obovate, entire-margined
     segments. ..................................................................................... *Nemophila*
3. Leaves palmately lobed; lobes deltoid. ................................... *Romanzoffia*
   Leaves pinnately lobed or parted; lobes
     elongate. ......................................................................................................... 4
4. Flowers in dense capitate cymes. ........................................... *Hydrophyllum*
   Flowers in simple or branched scorpioid
     cymes. .......................................................................................... *Phacelia*


*Ellisia*        ellisia

*Ellisia nyctelea* L.                                           WATERPOD

A low-growing plant 10–30 cm high, with scattered hairs. Leaves opposite or alternate, pinnatifid with toothed segments, 2–10 cm long. Flowers bluish white, 6–12 mm across, on stalks opposite leaf axils. Calyx lobes (sepals) enlarging as fruiting capsule forms. Fruit a globular capsule about 6 mm in diam. Not common; found on river flats, shady spots, lake margins; throughout southern part of the Prairie Provinces.


*Hydrophyllum*        waterleaf

*Hydrophyllum capitatum* Dougl.                                WOOLLEN-BREECHES

Short-stemmed plants with a short rhizome, bearing finger-like tuberous roots. Leaves 5–12 cm long, 3–12 cm wide, with 5–7 primary lobes. Cymes one to several, globose; peduncles 1–5 cm long; flowers purplish blue to white, 6–10 mm across; calyx densely pubescent; capsule 4 mm long; seeds usually 2, 2–3 mm across. Rare; exposed areas; southern Rocky Mountains.

*Nemophila*     baby-blue-eyes

*Nemophila breviflora* Gray                    SMALL BABY-BLUE-EYES

   Weak-stemmed plants 5–20 cm high, with stems sharply angled, bearing minute reflexed prickles. Leaves divided into 3–7 oblong-lanceolate lobes. Flowers short-pediceled; corolla 1.5–3 mm across, purplish blue to white; capsule 3–5 mm in diam, usually containing a single brick red regularly and deeply pitted seed. Rare; open disturbed soils; southern Rocky Mountains.


*Phacelia*     scorpionweed

1. Stamens included in the corolla. ................................................................ *P. thermalis*
   Stamens exserted. ................................................................................................. 2
2. Leaves entire or shallowly dentate, subor-
      bicular to lanceolate. ....................................................................................... 3
   Leaves deeply lobed to divided. ...................................................................... 5
3. Leaves suborbicular, shallowly dentate or
      crenate. ....................................................................................... *P. campanularia*
   Leaves linear to lanceolate or elliptic. ............................................................ 4
4. Plants annual; leaves linear, the lower
      ones often with 3–5 linear divisions. ....................................................... *P. linearis*
   Plants perennial; leaves oblanceolate to
      elliptic. ....................................................................................................... *P. hastata*
5. Leaves pinnately divided, the divisions
      entire to pinnatisect. ....................................................................................... 6
   Leaves more or less deeply lobed, but not
      divided. ........................................................................................................... 7
6. Leaves divided into 7–15 acute lobes; the
      lobes shallowly incised to entire. ....................................................... *P. franklinii*
   Leaves divided into 5–9 oblong lobes; the
      lobes deeply incised to pinnatisect. ................................................ *P. tanacetifolia*
7. Leaves divided halfway to the midrib;
      stamens twice as long as the corolla. ...................................................... *P. lyallii*
   Leaves divided almost to the midrib; sta-
      mens three times as long as the corolla. .................................................. *P. sericea*

*Phacelia campanularia* Gray                    DESERT BLUEBELLS

   Annual plants 10–50 cm high, glandular hispid throughout. Basal leaves oblong-ovate to suborbicular, 15–75 mm long, 10–50 mm wide, truncate or cordate at base. Inflorescence a lax, open cyme, simple or few-branched; flowers 15–40 mm long, blue or rarely white; capsule 8–12 mm long. Introduced ornamental; occasionally reseeding itself, and locally established.

*Phacelia franklinii* (R. Br.) Gray                    FRANKLIN'S SCORPIONWEED

   A hairy erect annual 15–40 cm high. Leaves 3–7 cm long, pinnately divided into linear-oblong toothed segments. Flowers in a dense coiled or scorpioid raceme, blue or bluish white, about 8 mm long. Fairly common; on dry sandy soil; in the Boreal forest and in Riding Mountains.

*Phacelia hastata* Dougl. SILVER-LEAVED SCORPIONWEED

Perennial plants with a branched caudex, ascending stems 15–30 cm long, canescent and appressed hirsute throughout. Leaves linear- or ovate-lanceolate, 3–10 cm long, mostly entire. Cymes 2–10 cm long, numerous, many-flowered; corolla white to light blue; capsule about 3 mm long. Sandy or rocky soil and eroded slopes; southern Rocky Mountains.

*Phacelia linearis* (Pursh) Holz. LINEAR-LEAVED SCORPIONWEED

An annual 10–40 cm high with leaves 2–5 cm long, entire or divided into linear segments. Flowers bright blue, about 1 cm long, in a scorpioid panicle. Found occasionally; in valleys, on hillsides, and on light soil; in southern Rocky Mountains.

*Phacelia lyallii* (Gray) Rydb. LYALL'S SCORPIONWEED

Perennial with sparingly hirsute stems 10–20 cm high; leaves oblanceolate in outline, 5–10 cm long, green, sparingly pubescent. Inflorescence short, dense; corolla dark blue, 7–10 mm long. Rare; alpine slopes; southern Rocky Mountains.

*Phacelia sericea* (Graham) Gray (Fig. 179) SILKY SCORPIONWEED

Perennial with a woody caudex, densely silvery and silky pubescent, 10–40 cm high. Inflorescence an elongate panicle of short cymes, with numerous flowers; corolla dark bluish purple, rarely white; capsule 4–6 mm long. Mountain slopes and disturbed areas; Rocky Mountains.

*Phacelia tanacetifolia* Benth. TANSY SCORPIONWEED

An annual plant 20–80 cm high, sparsely pubescent throughout. Basal leaves oblong-oval to ovate, 6–20 cm long, pinnately divided. Inflorescence densely short-pubescent and hispid, of several corymbosely branched cymes; flowers numerous; corolla 7–10 mm long, bluish purple; capsule 3–4 mm long. Introduced; occasional weed of gardens and roadsides.

*Phacelia thermalis* Greene GLANDULAR SCORPIONWEED

An annual with stem branched from base, 5–30 cm high. Whole plant covered with whitish glandular hairs. Leaves pinnatifid with oblong lobes, 1–3 cm long. Flowers light blue with a paler tube; corolla about as long as calyx, approx 6 mm. A plant of the dry intermountain plains of Washington and Idaho that has been found in the southern Prairies.

***Romanzoffia*** romanzoffia

*Romanzoffia sitchensis* Bong. SITKA ROMANZOFFIA

A slender, few-branched perennial 5–25 cm high, slightly villous; leaf sheaths widened, ciliate. Leaves round-reniform, 10–25 mm across; petioles 1–6 cm long. Inflorescence of a few cymes; corolla 6–9 mm long, white, on pedicels 1–3 cm long; capsule 4–7 mm long. Rare; wet rocks and cliffs; southern Rocky Mountains.

Fig. 179.   Silky scorpionweed, *Phacelia sericea* (Graham) Gray.

# BORAGINACEAE—borage family

Rough-hairy entire-leaved herbs with perfect, usually regular flowers, with 5-lobed calyx and corolla. Stamens attached at their base to inside of corolla tube; style simple. Fruit usually of 4 nutlets.

1. Ovary of flower merely grooved, and style
   at summit of the ovary. ........................................................... *Heliotropium*
   Ovary deeply divided into 4 lobes, and
   style arising from the center. ............................................................... 2

2. Nutlets with hooked prickles. ............................................................. 3
   Nutlets with no prickles or, if prickly, the
   prickles not hooked. ............................................................................ 4

3. Nutlets prickly all over and spreading
   horizontally. ................................................................. *Cynoglossum*
   Nutlets usually prickly on edges and
   borne vertically. ................................................................... *Lappula*

4. Flowers axillary or in leafy cymes. .................................................... 5
   Flowers in cymes, with small bracts or
   bractless. ............................................................................................. 8

5. Upper leaves opposite or in threes. ................................... *Asperugo*
   Upper leaves alternate. ...................................................................... 6

6. Flowers white, small, 3–5 mm across. ......................... *Plagiobothrys*
   Flowers light or dark yellow, larger, 8–15
   mm across. ........................................................................................... 7

7. Leaves ovate-lanceolate; styles conspicu-
   ously long exserted. ...................................................... *Onosmodium*
   Leaves oblong-lanceolate; styles
   included. ..................................................................... *Lithospermum*

8. Inflorescence of cymes bracteolate to the
   tip. ........................................................................................................ 9
   Inflorescence of bractless cymes, or only
   the lower flowers bracted. ................................................................ 13

9. Flowers drooping; the pedicels recurved,
   longer than the calyx. ........................................................... *Borago*
   Flowers not drooping; the pedicels erect
   or ascending. ...................................................................................... 10

10. Flowers white, usually less than 1 cm
    across. ....................................................................... *Cryptantha*
    Flowers bluish or purplish, mostly more
    than 1 cm across. ............................................................................... 11

11. Stamens and style long exserted;
    inflorescence racemose. ........................................................ *Echium*
    Stamens and style included; inflorescence
    not racemose. ..................................................................................... 12

12. Corolla tube curved in the middle, barely
    exceeding the calyx. ........................................................................... *Lycopsis*
    Corolla tube straight, distinctly longer
    than the calyx. ...................................................................................... *Nonea*
13. Flowers yellow, small, about 7 mm long. ................................... *Amsinckia*
    Flowers white, blue, or reddish purple. ........................................................ 14
14. Plants to 1 m high; flowers 12–18 mm
    long, purplish. ................................................................................. *Symphytum*
    Plants smaller; flowers smaller, white or
    blue. ................................................................................................................... 15
15. Flowers blue. ........................................................................................................ 16
    Flowers white. ........................................................................................ *Cryptantha*
16. Corolla tubular; inflorescence more or
    less congested. .......................................................................................... *Mertensia*
    Corolla not tubular; inflorescence elon-
    gate; flowers distant. .................................................................................. *Myosotis*

### *Amsinckia*      fiddle-neck

*Amsinckia menziesii* Nels. & Macbr.                    FIDDLE-NECK

Plants 20–50 cm high, simple or branched. Leaves ovate-lanceolate to linear-lanceolate, 2–7 cm long. Inflorescence elongate, 5–10 cm long, often
strongly curled at the tip; flowers 7–12 mm long, pale yellow; nutlets wrinkled
or papillate. Rare; disturbed areas, railways; Prairies and Parklands.

### *Asperugo*      madwort

*Asperugo procumbens* L.                              MADWORT

A weak-stemmed annual, with stems 20–70 cm long, spreading or procumbent. Leaves oblanceolate, 3–6 cm long. Flowers solitary in the leaf axils,
or sometimes 2–4 together; corolla blue, about 3 mm wide. Introduced; found
occasionally as a weed.

### *Borago*      borage

*Borago officinalis* L.                               OXTONGUE

An erect annual herb 20–60 cm high. Leaves elliptic to oblong or ovate,
3–10 cm long. Inflorescence in terminal many-flowered cymes; corolla bright
blue to white, 15–25 mm across. Introduced; occasionally planted as ornamental, and escaped.

### *Cryptantha*      cryptanthe

1. Plants perennial; cymes in a raceme. ....................................... *C. nubigena*
   Plants annual; cymes not in a raceme. ........................................................ 2
2. Cymes bracteolate to the tip. ......................................................... *C. minima*
   Cymes with only the lower flowers
   bracteolate. .................................................................................... *C. fendleri*

*Cryptantha fendleri* (A. Gray) Greene          FENDLER'S CRYPTANTHE

A very hairy gray or whitish annual with linear leaves 2–6 cm long. Plant sometimes growing to 50 cm high, but usually staying crowded and short. Flowers small and white, borne at ends of branches in scorpioid clusters. Fruits 4 smooth shiny brown nutlets. Common on sand dunes; in western Prairies, sometimes forming ground cover of large areas.

*Cryptantha minima* Rydb.          SMALL CRYPTANTHE

A small several-stemmed annual, blooming when only 1 or 2 cm high. Leaves 5–15 mm long, spatulate; corolla 2.5–3 mm long. Rare or commonly overlooked; eroded soils; Prairies.

*Cryptantha nubigena* (Greene) Payson          CLUSTERED OREOCARYA

A low erect species 10–30 cm high, with lower leaves spatulate and upper ones linear. Basal leaves 2–5 cm long; upper ones often shorter. Whole plant covered with white bristly short hairs, often giving a grayish appearance. Flowers white, about 6 mm across, in compact spike-like clusters at ends of branches. Common on dry hillsides and prairies; throughout Prairies and southern edge of Parklands. Syn.: *C. bradburiana* Payson; *Oreocarya glomerata* (Pursh) Greene.

**Cynoglossum**          hound's-tongue

Inflorescence leafy; flowers reddish to white. .................................................... *C. officinale*
Inflorescence leafless; flowers blue. .................................................... *C. boreale*

*Cynoglossum boreale* Fern.          WILD COMFREY

An erect unbranched perennial with stems 40–60 cm high. Basal leaves elliptic to lanceolate, 10–20 cm long, tapering to the petiole; stem leaves sessile, narrowed at the base or somewhat clasping. Racemes 1–3, terminating an elongate leafless peduncle; corolla blue, 6–8 mm across. Rare; in moist woods; southeastern Boreal forest.

*Cynoglossum officinale* L.          HOUND'S-TONGUE

A soft hairy biennial plant with erect leafy stems 40–80 cm high. Lower leaves oblong-lanceolate, 15–25 cm long, with slender stalks; upper leaves stalkless or clasping, and lanceolate. Flowers in scorpioid racemes at ends of branches, reddish purple, about 5–10 mm across. Fruit a pyramid of 4 nutlets 10–15 mm across. An introduced weed; found occasionally in pastures and waste places.

**Echium**          viper's bugloss

*Echium plantagineum* L.          PURPLE BUGLOSS

Cymes elongate from the upper 1 or 2 leaf axils; flowers 20–25 mm, purple, with a large calyx. Reported as a very rare weed.

*Echium vulgare* L.          VIPER'S BUGLOSS

A coarse erect biennial, with stems 30–70 cm high and hairs somewhat prickly. Basal leaves to 15 cm long; stem leaves progressively smaller. Cymes numerous, in the axils of upper stem leaves; corolla 12–20 mm long, blue. Introduced; weedy in several areas, but rare.

## *Heliotropium*   heliotrope

*Heliotropium curassavicum* L. var. *obovatum* DC.

SPATULATE-LEAVED HELIOTROPE

A perennial from thick fleshy white running roots, 5–30 cm high. Plants smooth with a slight bloom, giving them a somewhat waxy appearance. Leaves fleshy and spatulate, 2–5 cm long, with nerves very indistinct. Flowers white or faintly bluish-tinged, about 6 mm across, in several scorpioid spikes, 1–10 cm long, at ends of branches. Not common but very plentiful where found; in saline slough margins; Prairies and Parklands. Syn.: *H. spathulatum* Rydb.

## *Lappula*   bluebur

1. Plants grayish pubescent; leaves 2–6 cm
   long. ............................................................................................ *L. echinata*
   Plants green, pubescent; leaves 5–15 cm
   long. ............................................................................................................. 2
2. Flowers 8–12 mm across. ..................................... *L. deflexa* var. *americana*
   Flowers less than 8 mm across. ................................................................ 3
3. Branches of inflorescence strictly ascend-
   ing, densely flowered. ............................................................. *L. floribunda*
   Branches of inflorescence widely spread-
   ing, loosely flowered. ............................................................... *L. jessicae*

*Lappula deflexa* (Wahl.) Garcke var. *americana* (Gray) Greene

NODDING STICKSEED

A biennial, with downward-pointing hairs, 30–90 cm high. Leaves oblong-lanceolate; lower ones stalked and upper ones sessile (stalkless), 5–15 cm long. Inflorescence usually not leafy; pale blue flowers on reflexed stalks in a slender raceme. Found occasionally; in moist, shady woodlands; throughout Prairies and Parkland. Syn.: *L. americana* (Gray) Rydb.; *Hackelia americana* (Gray) Fern.

*Lappula echinata* Gilib.

BLUEBUR

A hairy annual or winter annual weed 15–60 cm high, much branching. Leaves 2–7 cm long, with only the lower ones stalked. Flowers pale blue, about 3 mm across, on erect leafy-bracted racemes at ends of branches. Fruit containing 4 nutlets with 2 rows of hooked prickles around margins. Whole plant has a strong smell resembling a mouse-infested building. Introduced from Europe, but now very common and widespread; on waste places in cultivated fields and in overgrazed pastures; throughout Prairies and Parkland.

The var. *occidentalis* (Wats.) Boiv. with nutlets having a single marginal row of hooked prickles. A fairly common native plant; on sandhills, light dry soils, and railway grades; throughout Prairies. Syn.: *L. occidentalis* (S. Wats.) Greene.

*Lappula floribunda* (Lehm.) Greene      <small>LARGE-FLOWERED STICKSEED</small>

A rough hairy biennial or perennial 40–100 cm high. Oblong-lanceolate or linear-lanceolate leaves 5–10 cm long; lower ones stalked and upper ones stalkless. Pale blue flowers in numerous erect racemes. Fairly common; in moist woodlands; throughout Western Parklands and Rocky Mountains. Syn.: *Hackelia floribunda* (Lehm.) Johnston.

*Lappula jessicae* McGregor      <small>JESSICA'S STICKSEED</small>

Stems erect or ascending, 30–60 cm high; basal leaves 8–15 cm long, 15–20 mm wide. Inflorescence an open panicle, with several racemes; flowers pale blue; corolla 3–5 mm across. Margins of woods and openings; southern Rocky Mountains.

**Lithospermum**      puccoon

Herbs with narrow hairy alternate leaves. Flowers perfect, regular, usually bright yellow, funnelform or salverform with rounded spreading lobes. Fruit usually 4 very hard nutlets.

1. Annual plants; nutlets brown, wrinkled,
   and pitted. ................................................................................................ *L. arvense*
   Perennial plants; nutlets white, smooth. ................................................................. 2

2. Corolla tube not longer than calyx;
   flowers white or pale yellow. ................................................................. *L. officinale*
   Corolla tube longer than calyx. ................................................................................ 3

3. Lobes of corolla either fringed or toothed. ................................................. *L. incisum*
   Lobes of corolla entire, neither fringed
   nor toothed. ............................................................................................................ 4

4. Leaves lanceolate; plant very leafy;
   flowers crowded, dull greenish yellow. ................................................. *L. ruderale*
   Leaves linear to oblong; plant not very
   leafy; flowers bright yellow or orange. ............................................... *L. canescens*

*Lithospermum arvense* L.      <small>CORN GROMWELL</small>

An erect usually branched plant, with stems 10–30 cm high. Leaves narrowly lanceolate to almost linear, 20–25 mm long. Flowers white; corolla 6–8 mm across. Introduced; found occasionally as a weed.

*Lithospermum canescens* (Michx.) Lehm.      <small>HOARY PUCCOON</small>

A softly hairy, somewhat hoary perennial growing erect to 15–50 cm high. Leaves linear-oblong, 1–4 cm long, without stalks. Flowers orange yellow, with tube 10–12 mm long, without stalks, in a rather compact cluster at summit of plant. Smooth white nutlets about 3 mm high. Fairly common; in bluffy country grasslands; eastern Parklands.

*Lithospermum incisum* Lehm. (Fig. 180)      <small>NARROW-LEAVED PUCCOON</small>

A deep taprooted perennial, often decumbent. Leaves linear, 1–5 cm long, covered with short stiff hairs. Early flowers stalked, bright yellow, with a tube about 2 cm long, and fringed spreading lobes. Later flowers smaller, self-fertilized while in the bud, and very fertile. Nutlets about 3 mm high, white, shiny,

Fig. 180.   Narrow-leaved puccoon, *Lithospermum incisum* Lehm.

with minute pits, very hard. Common; on dry prairie land; throughout Prairies and Parklands. Syn.: *L. angustifolium* Michx.

*Lithospermum officinale* L.                                              GROMWELL

A finely haired perennial 15–60 cm high. Leaves ovate to ovate-lanceolate, 1–7 cm long. Flowers pale yellow or yellowish white, few, solitary in leaf axils, tube almost as long as calyx. Shiny nutlets about half as long as the sepals. An eastern weed, introduced from Europe, but has been found in southeastern Parklands and Boreal forest.

*Lithospermum ruderale* Lehm.                                      WOOLLY GROMWELL

A coarse hairy perennial from thick roots, with several erect stems; densely leafy. Leaves lanceolate with a prominent midrib, 5–10 cm long, stalkless, and crowded on stem. Numerous dull greenish yellow flowers crowded at top of stem in a leafy cluster. White, shining, ovoid nutlets about 3 mm high. Fairly common; in grasslands; Rocky Mountains and Cypress Hills.

## *Lycopsis*          bugloss

*Lycopsis arvensis* L.                                                SMALL BUGLOSS

An annual rough hairy plant 30–50 cm high. Lanceolate alternate leaves 2–5 cm long. Flowers crowded into somewhat scorpioid terminal racemes, bluish, up to 6 mm across. Tube of corolla slightly curved or bent, and throat closed with stiff fine hairs. The 4 nutlets shorter than calyx. An introduced weed; very uncommon; has been found in southeastern Parklands and in Prairies.

## *Mertensia*          lungwort

Perennial plants with alternate leaves and fairly large blue or purple funnelform flowers with lobes slightly spreading. Fruit erect wrinkled nutlets.

1. Plants entirely glabrous. ........................................................... *M. maritima*
   Plants pubescent. ...................................................................................... 2
2. Plants 30–70 cm high; stem leaves 5–15
   cm long, ovate to ovate-lanceolate. ..................................... *M. paniculata*
   Plants 10–30 cm high; stem leaves 2–10
   cm long, oblong to lanceolate. ................................................................ 3
3. Plants with tuberous roots; flowers 15–25
   mm long. ............................................................................... *M. longiflora*
   Plants with taproot; flowers 10–15 mm
   long. ..................................................................................... *M. lanceolata*

*Mertensia lanceolata* (Pursh) DC.                          LANCE-LEAVED LUNGWORT

A plant 15–30 cm high with linear to lanceolate leaves 5–10 cm long. Leaves usually hairless, but may occasionally be somewhat short hairy on upper side. Blue flowers 10–15 mm long, in few-flowered panicles at ends of branches. Form with no hairiness on upper surface sometimes called *M. linearis* Greene. Both forms fairly common; on open prairie and hillsides; Prairies.

*Mertensia longiflora* Greene                    LARGE-FLOWERED LUNGWORT

Plants usually with a solitary stem, or rarely 2 or 3 stems, 10–30 cm high, from a shallow tuberous root. Basal leaves 2–5 cm long, mostly confined to sterile stems; stem leaves 2–7 cm long. Inflorescence often dense; flowers deep blue, usually about 2 cm long. Grasslands and slopes; southern Rocky Mountains.

*Mertensia maritima* (L.) S. F. Gray              SEASIDE LUNGWORT

Plants with spreading or decumbent stems to 60 cm long. Leaves 2–6 cm long, ovate to spatulate-ovate, fleshy. Cymes often numerous, lax, leafy-bracted; corolla 8–10 mm long, pinkish blue to white. Beaches; northeastern Boreal forest.

*Mertensia paniculata* (Ait.) G. Don (Fig. 181)              TALL LUNGWORT

An erect species 30–70 cm high. Leaves lanceolate, 5–15 cm long, somewhat hairy on both sides. Flowers in few-flowered clusters at ends of stems, purplish blue, 10–15 mm long. Found in the woodlands and shady stream banks; throughout Parklands, Boreal forest, and Rocky Mountains.

## *Myosotis*          forget-me-not

1. Pubescence of the calyx of appressed
   straight hairs. ............................................................................................... 2
   Pubescence of the calyx of spreading
   hooked hairs. .............................................................................................. 3

2. Calyx lobes half as long as the tube;
   corolla lobes much larger than calyx
   lobes. ................................................................................ *M. scorpioides*
   Calyx lobes equaling the tube; corolla
   lobes equaling calyx lobes. ........................................................... *M. laxa*

3. Flowers 5–8 mm across; plants perennial. ................... *M. sylvatica* var. *alpestris*
   Flowers 2–3 mm across; plants annual or
   biennial. ......................................................................................... *M. arvensis*

*Myosotis arvensis* (L.) Hill                    FIELD FORGET-ME-NOT

Plants with erect or ascending branching stems 15–40 cm high. Basal leaves 1–2 cm long; stem leaves to 3 cm long. Racemes loosely flowered; flowers blue or white, 2–3 mm across. Introduced; rare weed; in disturbed areas.

*Myosotis laxa* Lehm.                    SMALL FORGET-ME-NOT

Perennial plants with decumbent or spreading stems, rooting at the nodes, 15–30 cm long. Leaves to 3 cm long, oblong or oblong-lanceolate. Racemes loosely flowered; corolla about 4 mm across. Introduced; a rare weed; in marshy areas.

*Myosotis scorpioides* L.                    MARSH FORGET-ME-NOT

Perennial plants with slender rootstocks; stems decumbent or ascending, 15–40 cm long, rooting at the lower nodes. Leaves 2–8 cm long, oblong-lanceolate to oblanceolate. Racemes loosely flowered; corollas 6–8 mm across. Introduced; a rare weed; in cultivated and disturbed moist areas.

Fig. 181.   Tall lungwort, *Mertensia paniculata* (Ait.) G. Don.

*Myosotis sylvatica* Hoffm. var. *alpestris* (F. W. Schmidt) Koch.  FORGET-ME-NOT

Perennial plants with a short rootstock; stems usually upright, 10–20 cm high. Leaves 2–5 cm long, oblong-lanceolate to spatulate. Racemes mostly densely flowered; corolla 5–8 mm across. Alpine meadows and slopes; Rocky Mountains.

### *Nonea*        monk's-wort

*Nonea vesicaria* (L.) Reich.                                   RED MONK'S-WORT

Perennial plants with stout roots; stems 10–40 cm high. Leaves 3–10 cm long, oblong-lanceolate. Racemes more or less densely flowered; corollas 10–15 mm across, dark red to brown red. Introduced; a very rare weed.

### *Onosmodium*        false gromwell

*Onosmodium molle* Michx. var. *hispidissimum* (Mack.) Cronq.
WESTERN FALSE GROMWELL

An erect coarse perennial 50–100 cm high, with rough hairy stems. Leaves lanceolate to ovate-lanceolate, 5–8 cm long, coarsely appressed hairy, with venation very prominent on undersurfaces. Flowers yellowish white or greenish, 10–20 mm long, in leafy terminal scorpioid spikes. Nutlets 3–4 mm long, distinctly constricted at the base. Margins of woods, openings, and grassland; southeastern Parklands and Boreal forest. Smaller plants, 30–60 cm high, with nutlets 2.5–3.5 mm long, not constricted at the base, are var. *occidentale* (Mack.) Cronq. Margins of woods and shrubbery; southeastern Parklands, southern Rocky Mountains.

### *Plagiobothrys*        allocarya

*Plagiobothrys scouleri* (H. & A.) Johnston var. *penicillatus* (Greene) Cronq.
SCOULER'S ALLOCARYA

A low, spreading much-branched annual herb, with linear leaves 1–5 cm long; the lower ones sometimes opposite but the upper ones alternate. Flowers white, very small, in a small scorpioid inflorescence, and also in axils of leaves. Fruits consisting of 4 nutlets, rough on back. Fairly common locally; in slough margins, sandy places, and moist bottomlands; throughout Prairies. Syn.: *Allocarya californica* (F. & M.) Greene.

### *Symphytum*        comfrey

Leaves all petiolate, not decurrent. ...................................................................... *S. asperum*
Leaves decurrent, the upper ones sessile. .......................................................... *S. officinale*

*Symphytum asperum* Lepechin                              ROUGH COMFREY

A perennial herb with erect branched stems 50–100 cm high. Leaves 10–20 cm long, lanceolate to ovate, petioled; the petiole not decurrent on the stem. Flowers blue, 12–18 mm long, numerous in scorpioid cymes. Introduced; a rare weed in disturbed areas.

*Symphytum officinale* L.

A large coarse hairy perennial 60–100 cm high. Leaves lanceolate to ovate-lanceolate, 8–25 cm long; lower ones with stalks decurrent along the stem; upper ones sessile or stalkless. Flowers purplish or yellowish, 10–15 mm long, in dense terminal scorpioid clusters. Fruit consists of shiny brown somewhat wrinkled nutlets. An introduced European plant; appears to be established in a few places.


# VERBENACEAE—vervain family

***Verbena***        vervain

Annual or perennial hairy herbs with 4-sided stems and opposite leaves. Flowers purplish or blue, tubular with 5 somewhat irregular lobes. There are 4 stamens, one pair longer than the other. Fruit of 4 linear nutlets.

1. Plants with decumbent stems; spikes conspicuously bracteose. .......................................................... *V. bracteata*

    Plants with erect stems; spikes not conspicuously bracteose. ........................................................... 2

2. Spikes densely flowered, thick; flowers and fruit overlapping. .......................................................... *V. hastata*

    Spikes loosely flowered, thin; flowers and fruit distant. ............................................................... *V. urticifolia*


*Verbena bracteata* Lag. & Rodr.                                BRACTED VERVAIN

A prostrate or decumbent annual or short-lived perennial, much branched, with a 4-sided stem. Plants 30–80 cm across and forming mats on ground. Leaves 2–5 cm long, roughly spatulate, pinnately incised. Purplish blue flowers about 2 mm across, on dense spikes 5–10 cm long; conspicuous bracts about 6 mm long. Not common, but locally abundant; on prairie and waste places on lighter soils; throughout Prairies. Syn.: *V. bracteosa* Michx.

*Verbena hastata* L.                                            BLUE VERVAIN

A tall erect perennial with a 4-sided stem 60–150 cm high. Leaves opposite, lanceolate, sharply toothed, 5–10 cm long; lower leaves sometimes hastately lobed at base. Bluish white flowers on numerous terminal spikes 5–10 cm long; fruit densely overlapping on spikes. Fairly common; in woodlands and river valleys; eastern Parklands and Boreal forest.

*Verbena urticifolia* L.                                  NETTLE-LEAVED VERVAIN

Similar to *V. hastata*, but the leaves broadly lanceolate to oblong-ovate, rounded at base, and decurrent into the petiole; the inflorescence more open and more sparsely flowered; the corolla white. Thickets, moist fields, and marshes; southeastern Boreal forest.

# LABIATAE—mint family

Usually scented plants with square stems. Leaves simple, opposite, or whorled, usually with small glandular pits. Flowers perfect, usually irregular in shape, with 2 lips to corolla. Stamens usually 4, one pair longer than the other, and sometimes only one pair bearing anthers. Fruit of 4 nutlets.

1. Inflorescences terminal. ........................................................................ 2
   Inflorescences axillary. ......................................................................... 13

2. Inflorescence a more or less globose head. ................................. *Monarda*
   Inflorescence a more or less elongated raceme. ............................................................................. 3

3. Calyx distinctly 2-lipped; the upper lip 3-toothed, the lower one 2-lobed. ............................................... 4
   Calyx indistinctly or not at all 2-lipped; one lobe may be larger than the others. ...................................... 5

4. Plants with decumbent or creeping stems; spikes dense, erect. ................................................ *Prunella*
   Plants with erect stems; spikes loose and lax. ............................................................................. *Salvia*

5. Inflorescence a raceme of oppositely placed flowers. ......................................................... *Physostegia*
   Inflorescence a raceme of oppositely placed clusters of flowers. ............................................................ 6

6. Bracts much shorter than and differing from the stem leaves. .............................................................. 7
   Bracts grading into the stem leaves. ............................................................................ 9

7. Plants annual; spikes strongly one-sided. ................................... *Elsholtzia*
   Plants perennial; spikes symmetrical. ......................................................... 8

8. Calyx longer on upper than lower side, somewhat oblique at throat. ..................................... *Agastache*
   Calyx regular, clearly 5-lobed. ................................................... *Mentha*

9. Leaves entire, linear to linear-lanceolate. ................................... *Hyssopus*
   Leaves crenate to dentate, broader. ......................................................... 10

10. Upper lip of the corolla inconspicuous, reduced to two small lobes. ........................................ *Teucrium*
    Upper lip of corolla distinctly visible. ......................................................... 11

11. Upper lip of calyx much wider than the other 4. ........................................... *Dracocephalum*
    Upper lip of calyx equal to at least the adjacent ones. ......................................................... 12

12. Flowers white; leaves deltoid or deltoid-ovate, cordate to subcordate at base; calyx oblique. ............................ *Nepeta*
    Flowers purplish; leaves oblong to ovate, obtuse to rounded at base; calyx regular. ............................................... *Stachys*

### *Agastache*       giant-hyssop

*Agastache foeniculum* (Pursh) Ktze. (Fig. 182)                GIANT-HYSSOP

An erect, branched perennial, with smooth or minutely hairy stems 30–80
cm high. Leaves ovate or triangular-ovate, green above and pale below, 2–7
cm long, with short stalks. Flowers blue, 6–12 mm long, in a dense spike,
sometimes interrupted, 2–10 cm long and 12–18 mm thick at ends of stems.
Plant has a pleasant anise-like odor. Common; in open woodlands and semi-
open prairies; throughout Parklands. Syn.: *A. anethiodora* (Nutt.) Britton.

### *Dracocephalum*       false dragonhead

Fig. 182.   Giant-hyssop, *Agastache foeniculum* (Pursh) Ktze.

*Dracocephalum parviflorum* Nutt. AMERICAN DRAGONHEAD

An erect, usually branched, annual or biennial herb, with a finely hairy stem 30–50 cm high. Leaves oblong to lanceolate, 2–5 cm long, stalked, and with rather large pointed teeth. Flower clusters dense, at ends of branches, usually 2–5 cm long and 2–3 cm wide. Calyx membranous, stiff and spiny, giving heads a prickly stiffness. Light blue corollas scarcely longer than calyx. Locally abundant but not generally common; found in openings in woodlands and old pastures; throughout Parklands; rare in Prairies. Syn.: *Moldavica parviflora* (Nutt.) Britton.

*Dracocephalum thymiflorum* L. DRAGONHEAD

Similar to *D. parviflorum*, but the leaves with more rounded teeth, and the inflorescence elongate. Introduced; a rare weed. Syn.: *Moldavica thymiflora* (L.) Rydb.

### *Elsholtzia*

*Elsholtzia ciliata* (Thunb.) Hyl.

An annual herb with erect or ascending branched stems 30–50 cm high. Leaves 3–7 cm long, ovate or ovate-lanceolate, petioled. Inflorescence of terminal and axillary spikes, with the flowers crowded on one side of the axis. Introduced; a rare weed; in wet and shady places.

### *Galeopsis*     hemp-nettle

Corolla tube 2–3 times as long as the calyx,
2–3 cm long, yellow. ...................................................................................... *G. speciosa*
Corolla tube scarcely longer than the calyx,
1.5–2 cm long, purple. ...................................................................................... *G. tetrahit*

*Galeopsis speciosa* Mill. YELLOW HEMP-NETTLE

An annual herb with stems 30–100 cm high. Leaves coarsely toothed, 3–8 cm long. Flowers large, yellow; the lower lip often with a purple spot. Introduced; a rare weed.

*Galeopsis tetrahit* L. HEMP-NETTLE

A coarse rough hairy weedy annual 30–100 cm high. Stems usually swollen below nodes where leaves and axillary flower clusters join. Leaves ovate, coarsely toothed, 5–10 cm long. Flowers in both terminal and axillary clusters, with sharp needle-pointed calyx teeth. Corolla 15–20 mm, purple or pink variegated with white. Seeds grayish brown, egg-shaped, about 3 mm long, somewhat similar to common hemp seed. Introduced from Europe; has become a field weed; in various parts of Parklands and Boreal forest.

### *Glechoma*     ground-ivy

*Glechoma hederacea* L. GROUND-IVY

A creeping perennial, rooting at nodes, with thin dull greenish purple stems 30–40 cm long. Leaves ovate-rounded, cordate at base, 2–4 cm across, often with a purplish tinge. Flowers 15–20 mm long, light blue, in clusters of 2 or 3 in leaf axils. An introduced plant; has become common in waste places in southeastern Parklands and often found as a garden weed elsewhere.

***Hedeoma***        mock pennyroyal

*Hedeoma hispida* Pursh                                      ROUGH PENNYROYAL

A low-growing annual 10–20 cm high, with erect branched stems. Leaves narrowly linear, entire, 10–25 mm long. Numerous flower clusters in leaf axils. Flowers bluish purple, about 6 mm long. Found occasionally; on sandy soil, eroded slopes, and abandoned fields; Prairies.

***Hyssopus***        hyssop

*Hyssopus officinalis* L.                                              HYSSOP

A perennial with a stout woody rootstock, erect stems 30–60 cm high, and entire, lanceolate to oblanceolate leaves 1–3 cm long. Flowers 3–7 in a cluster, with the upper ones on each stem or branch forming a spike-like terminal inflorescence; corolla blue, about 1 cm long. Introduced; occasionally cultivated for medicinal purposes, and rarely weedy.

***Lamium***        dead-nettle

Leaves all stalked, ovate; flowers white;
   perennial. ................................................................................................. *L. album*
Upper leaves not stalked, roughly circular;
   flowers red or purplish; annual or biennial. ........................................... *L. amplexicaule*

*Lamium album* L.                                            WHITE DEAD-NETTLE

A perennial from creeping rootstocks, 30–50 cm high. Leaves ovate, sometimes somewhat cordate at the base, slightly hairy, 2–7 cm long. Pure white flowers 20–30 mm long, in clusters of 7 or 8 in each leaf axil. An introduced plant, has been found in the Prairie Provinces, but is, as yet, rare.

*Lamium amplexicaule* L.                                              HENBIT

A sparingly hairy annual or biennial with slender stems, branched from base, often somewhat decumbent, 15–50 cm high. Leaves almost round, with rounded teeth; the lower ones with slender stalks; the upper ones usually stalk-less and somewhat clasping the stem, 1–5 cm across. Flowers red or purplish with spots on middle lower lobe, 10–20 mm long, in few-flowered axillary and terminal clusters. An introduced plant; uncommon; but found throughout the Prairie Provinces.

***Leonurus***        motherwort

Calyx strongly 5-angled and 5-ribbed; bracts
   small. ................................................................................................. *L. cardiaca*
Calyx not angled, 10-ribbed; bracts large. ........................................... *L. sibiricus*

*Leonurus cardiaca* L.                                            MOTHERWORT

An introduced perennial 50–150 cm high, with slightly hairy or downy stems. Leaves stalked, 2–10 cm long, decreasing in size toward upper part of stem, deeply and irregularly cleft into 3–7 lobes. Pink flowers about 1 cm long, in dense whorls around axils of upper leaves. Upper lip of corolla densely white hairy. Occasionally persists as an intruder in shelterbelts.

*Leonurus sibiricus* L.                                SIBERIAN MOTHERWORT

Biennial plants, with softly retrorse-pubescent stems to 1 m high. Leaves ovate or rotund in outline, deeply 3-parted, with the divisions toothed. Flowers purple, about 1 cm long; upper lip of the corolla pubescent. Introduced; a rare weed.

## *Lycopus*        water-horehound

Perennial plants of swamps and wet places, with flowers in dense axillary clusters.

1. Leaves more or less deeply incised or
   pinnatifid. ................................................................................ *L. americanus*
   Leaves toothed but not deeply incised. ................................................................. 2
2. Calyx lobes narrow and acute, longer
   than nutlets. ................................................................................ *L. asper*
   Calyx lobes ovate and blunt, not as long
   as nutlets. ................................................................................ *L. virginicus*

*Lycopus americanus* Muhl.                            WATER-HOREHOUND

An erect perennial 30–80 cm high. Leaves at definite intervals on stem, usually almost horizontally, giving plant an open and regular appearance. Leaves 2–8 cm long, lanceolate, short-stalked, and except for the upper ones, deeply cut or incised. Bluish white flowers about 3 mm across, in dense clusters around stem at axils of leaves. Fairly common; in moist places, stream banks, and swamps; throughout the Prairie Provinces.

*Lycopus asper* Greene                          WESTERN WATER-HOREHOUND

A perennial 30–50 cm high; narrowly lanceolate leaves 2–7 cm long, with small-toothed margins. Leaves either stalkless or with a very short stalk. Tiny flowers in close clusters in leaf axils. Very common; in wet places and swamps in the western Prairies and Parklands, but comparatively scarce toward the eastern part.

*Lycopus virginicus* L. var. *pauciflorus* Benth.     NORTHERN WATER-HOREHOUND

An erect perennial 15–70 cm high, with somewhat toothed margined lanceolate leaves 2–7 cm long. Flowers about 3 mm long, in axils of leaves. Rare; found in wet areas; northwestern Boreal forest. Syn.: *L. uniflorus* Michx.

## *Marrubium*        horehound

*Marrubium vulgare* L.                              COMMON HOREHOUND

A strongly aromatic perennial, with stout, erect, white pubescent stems 40–60 cm high. Leaves ovate, 3–5 cm long, with the lower ones petioled, white pubescent. Flowers about 6 mm long, densely crowded in upper axils; calyx teeth awl-shaped, smooth and hooked at the apex, catching on fur and clothing. Introduced; occasionally cultivated and escaped.

## *Melissa*    balm

### *Melissa officinalis* L.    HONEY BALM

Perennial plants with erect stems 40–80 cm high. Leaves 4–7 cm long, long-petioled, ovate to deltoid-ovate, coarsely crenate. Inflorescence of few-flowered clusters; corolla blue to white, 10–15 mm long. Introduced; occasionally cultivated for honey, flavoring, and medicinal purposes; rarely spreading.

## *Mentha*    mint

Flowers in axillary clusters. ............................................................................ *M. arvensis*
Flowers in terminal spikes. ............................................................................ *M. spicata*

### *Mentha arvensis* L. var. *villosa* (Benth.) S. R. Stewart    FIELD MINT

Erect perennial herbs with a strong but pleasant mint odor, growing to 10–50 cm high. Square stems with a line of hairs running down each angle. Leaves almost hairless, but with minute glandular dots on both surfaces. Ovate to lanceolate in shape, 1–5 cm long, sometimes long-stalked and sometimes short-stalked. Flowers pink, about 3 mm long, in crowded whorls around stems at leaf axils. Common; in sloughs and wet places, often growing in water; found throughout the Prairie Provinces. Various authorities have split the mints into several species using hairiness of leaves or length of leaf stalks in relation to the size of flower clusters as distinguishing characteristics. It would appear, however, that the mints of the Prairie Provinces should at present be considered as a variety of one species.

### *Mentha spicata* L.    SPEARMINT

Erect perennial herbs to 50 cm high, glabrous or nearly so. Leaves 2–6 cm long, sessile or subsessile, oblong-lanceolate, sharply serrate. Spikes several, terminating stem and short branches from the upper axils, 3–12 cm long, continuous or somewhat interrupted; flowers about 5 mm long. Introduced; cultivated for flavoring, and occasionally escaped.

## *Monarda*    wild bergamot

### *Monarda fistulosa* L. (Fig. 183)    WILD BERGAMOT

Erect perennial plants with a strong but pleasant odor, 60–100 cm high. Leaves lanceolate or ovate-lanceolate, rounded or somewhat cordate based, 2–10 cm long. In the type species, leaves are only slightly hairy and have stalks 10–25 mm long. Inflorescence terminal and head-like, in clusters 3–6 cm across. Calyxes narrow green tubes with small purplish teeth. Corollas hairy, 20–25 mm long, protruding far above calyxes, pink or lilac, very conspicuous. Fairly common; on hillsides, thickets, and in shady places; eastern Parklands and Boreal forest, being replaced farther west by the variety.

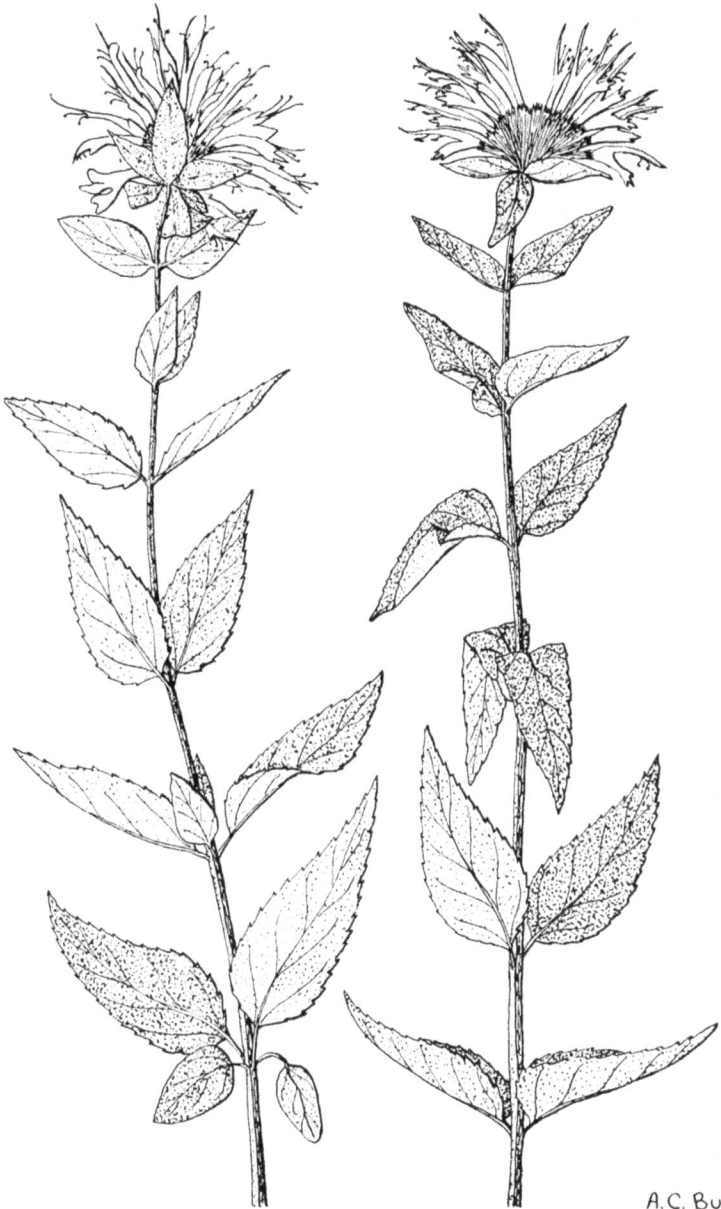

Fig. 183.  Wild bergamot, *Monarda fistulosa* L.

The var. *menthaefolia* (Graham) Fern., western wild bergamot, which is much more common than the species, differs by having very short leaf stalks, less than 6 mm long, and leaves often downy-hairy. Occasionally white-flowered specimens are found. Common; in edges of woods, along shady creek banks and coulees, and in shelter of shrubby patches on prairies; throughout almost all of the Prairie Provinces. Syn.: *M. menthaefolia* Graham.

## Nepeta          catnip

*Nepeta cataria* L.                                                                                    CATNIP

A very hairy branching perennial 30–75 cm high, with ovate, stalked leaves, cordate at base, 2–7 cm long. Flowers white or pale purple with dark spots, hairy, about 12 mm long, in fairly dense terminal heads, somewhat similar to American dragonhead. An introduced plant; found rarely in disturbed areas.

## Physostegia          false dragonhead

*Physostegia virginiana* L. var. *formosior* (Lunell) Boiv.          FALSE DRAGONHEAD

An erect, rarely branched perennial 30–100 cm high. Leaves usually rhomboid-lanceolate, 6–10 cm long, and stalkless, with short sharp teeth. Flowers in terminal spikes, purple, 20–25 mm long. Fairly common locally; along stream banks and moist places; southeastern Parklands and Boreal forest. Plants with narrower leaves, oblong-lanceolate to lanceolate, are var. *ledinghamii* Boiv.; Boreal forest. Plants with linear-lanceolate leaves, and flowers 10–15 mm long are var. *parviflora* (Nutt.) Boiv. Syn.: *P. parviflora* Nutt.; *Dracocephalum nuttallii* Britton.

## Prunella          selfheal

*Prunella vulgaris* L.                                                                               SELFHEAL

An introduced perennial plant growing from running rootstocks, 5–30 cm high. Leaves ovate-lanceolate, rather blunt at apex, 2–5 cm long. Flowers in short dense terminal spikes, violet purple, 8–12 mm long. Found occasionally; in woodlands and moist places in grassland; Parklands and Rocky Mountains.

## Salvia          sage

Flowers 2–4 at each node of the raceme. ............................................................ *S. reflexa*
Flowers numerous in whorls at the nodes. ...................................................... *S. nemorosa*

*Salvia nemorosa* L.                                                                              WOOD SAGE

Coarse densely downy perennial plants 30–80 cm high. Leaves oblong to ovate-lanceolate, 2–8 cm long; lower leaves with stalks; upper ones stalkless. Flowers in long, terminal, narrow spikes 5–15 cm long and 1–2 cm thick. Corollas deep violet blue, 8–12 mm long. Introduced in alfalfa seed; has been found on two or three occasions in widely scattered spots throughout the Prairie Provinces.

*Salvia reflexa* Hornem.                                          LANCE-LEAVED SAGE

Erect, much-branched annuals 30–60 cm high; stems finely pubescent with recurved hairs. Leaves 3–5 cm long, linear-lanceolate to lanceolate; petioles 1–2 cm long. Racemes 5–10 cm long, erect; flowers usually 2, rarely 3 or 4 at a node, about 1 cm long, blue; the corolla scarcely longer than the calyx. Found occasionally as a weed; in southeastern Parklands.

**Scutellaria**     skullcap

Perennial herbs with few flowers, usually borne in leaf axils. Corolla has arched upper lip and is much longer than calyx, which has a crest-like protuberance on upper lip.

1. Flowers borne singly or in pairs in axils of
   leaves. ............................................................................................................ 2
   Flowers borne in several-flowered axillary
   or terminal racemes. ...................................................... *S. lateriflora*

2. Flowers 15–25 mm long. ......................................................... *S. galericulata*
   Flowers 7–10 mm long. ....................................... *S. parvula* var. *leonardii*

*Scutellaria galericulata* L.                                   MARSH SKULLCAP

An erect-stemmed perennial from creeping roots, 30–60 cm high. Leaves oblong to oblong-lanceolate, wavy-margined, 2–6 cm long; lower ones short-stalked; upper ones stalkless. Flowers borne either singly or in pairs at axils of leaves, blue, 15–25 mm long. Fairly common; in wet places and along stream banks; throughout the Prairie Provinces. Syn.: *S. epilobiifolia* Hamilton.

*Scutellaria lateriflora* L.                                     BLUE SKULLCAP

Very similar to *S. galericulata*, with ovate leaves. Flowers only about 6–10 mm long, in loose several-flowered racemes at leaf axils and occasionally at the end of the stem. Fairly common; along stream banks, in swamps, and wet places; southeastern Parklands.

*Scutellaria parvula* Michx. var. *leonardii* (Epling) Fern.      SMALL SKULLCAP

Small perennials, with rootstocks deeply constricted between bead-like segments; stems 10–20 cm high. Main stem leaves sessile, 10–16 mm long, ovate-lanceolate to somewhat deltoid, glabrous or slightly scabrous. Flowers axillary, blue; calyx glabrous or somewhat scabrous. Very rare; in peaty soil over rock outcrops; southeastern Boreal forest.

**Stachys**     hedge-nettle

*Stachys palustris* L. var. *pilosa* (Nutt.) Fern. (Fig. 184)     MARSH HEDGE-NETTLE

A hairy, branched perennial, usually erect, but occasionally decumbent, 30–80 cm high. Leaves somewhat coarse, hairy, lanceolate to oblong-lanceolate, 2–10 cm long, generally rounded at the base, sometimes with very short stalks but usually without. Flowers pale purplish with darker spots, 10–15 mm long, in axillary clusters in the top portion of the plant, sometimes appearing as a leafy spike. Common; in moist places and along stream banks; throughout the Prairie Provinces.

A.C. Budd

Fig. 184. Marsh hedge-nettle, *Stachys palustris* L. var. *pilosa* (Nutt.) Fern.

***Teucrium***      germander

*Teucrium canadense* L. var. *occidentale* (Gray) McCl. & Epl.

A very hairy, branching perennial 30–75 cm high. Leaves narrowly ovate to oblong-lanceolate, short-stalked, 2–7 cm long, white hairy beneath. Purplish flowers 6–10 mm long, in a spike-like terminal raceme. Very scarce, but has been reported from east central Saskatchewan.

## SOLANACEAE—potato family

Herbs or vines with alternate leaves without stipules. Flowers perfect, and, with one exception, regular. Corolla varies from funnel-like and bell-like to rotate (wheel-shaped), with 5 stamens. Fruit either a berry or a capsule. Many native members of this family have either narcotic or poisonous properties, although some parts are edible.

1. Plants shrubby, often spiny. ........................................................................... *Lycium*
   Plants herbaceous. ........................................................................................... 2
2. Corolla rotate (wheel-shaped). ....................................................................... *Solanum*
   Corolla bell-shaped or funnel-shaped. ............................................................ 3
3. Fruit a berry, often enclosed in an
   inflated bladder-like calyx. ............................................................................ *Physalis*
   Fruit a capsule. .............................................................................................. 4
4. Capsule prickly. ............................................................................................. *Datura*
   Capsule not prickly, top falling off when
   mature. ......................................................................................................... *Hyoscyamus*

***Datura***      stramonium

*Datura stramonium* L.                                          JIMSONWEED

An annual, **very poisonous**, weedy plant with stout smooth green to purplish stems, 50–150 cm high. Leaves simple, ovate, irregularly toothed, 7–20 cm long. Flowers funnel-shaped, white or purplish, 5–10 cm long and about 5 cm across, with slender-tipped lobes, borne in the axils of leaves. Fruit a large, prickly capsule 3–6 cm long, breaking into four segments to release the many seeds. A weed, probably escaped from gardens; has been found in two or three places; throughout the Prairie Provinces.

***Hyoscyamus***      henbane

*Hyoscyamus niger* L. (Fig. 185)                                BLACK HENBANE

Sticky, hairy, evil-smelling biennial plants 30–100 cm high from a spindle-shaped root. Leaves roughly ovate or oblong, irregularly toothed or lobed; upper ones clasping the stem, 7–20 cm long. Flowers funnelform, about 2 cm long and 25–35 mm across, greenish yellow with purplish veins, with purple anthers. Borne on one side of stem, crowded at head of stem. Fruit a capsule

Fig. 185.   Black henbane, *Hyoscyamus niger* L.

almost 12 mm long, enclosed in a swollen calyx. An introduced **poisonous** plant, fortunately still rare; found in waste places and around gardens; in several locations throughout Prairie Provinces.

*Lycium*        matrimony vine

*Lycium halimifolium* Mill.

Shrubs with branches to 3 m long, glabrous or nearly so, arched or recurved, sometimes climbing; spines, if present, at the nodes. Leaves 2–5 cm long, lanceolate to spatulate, grayish green. Flowers pinkish violet, 10–12 mm wide, in clusters of 1–4; berries scarlet, ellipsoid, 1–2 cm long. Occasionally cultivated, and escaped.

*Physalis*        ground-cherry

1. Corolla rotate, white; flowers 2–4 at a
   node. .................................................................................. *P. grandiflora*
   Corolla funnelform, white or yellowish;
   flowers solitary. ................................................................................. 2

2. Filaments slender, uniform; corolla usually less than 15 mm long; plants
   annual. .................................................................................................. 3
   Filaments flat, expanded; corolla usually more than 15 mm long; plants
   perennial. ............................................................................................. 4

3. Pubescence villous, with the hairs long
   and spreading. ................................................................ *P. pubescens*
   Pubescence, if present, of minute hairs. ...................... *P. ixocarpa*

4. Corolla white, distinctly 5-lobed; fruiting
   calyx bright red or orange. ........................................ *P. alkekengi*
   Corolla yellow or greenish, not distinctly
   lobed; fruiting calyx brownish or
   green. .................................................................................................. 5

5. Pubescence of stem villous, of soft spreading hairs. ................................................................ *P. heterophylla*
   Pubescence of stem of short, stiff, deflexed or recurved hairs. ........................................ *P. virginiana*

*Physalis alkekengi* L.

Perennial with creeping rootstocks. Stems soft, short pubescent, usually several together, erect, unbranched, 40–60 cm high. Leaves petioled, ovate, to 15 cm long, to 8 cm broad, glabrous except on the veins. Fruiting calyx to 5 cm long, bright red or orange, and finely reticulately veined at maturity. Berry 10–15 mm in diam, reddish or orange. Introduced as an ornamental; occasionally escaped.

*Physalis grandiflora* Hook.                                  LARGE WHITE GROUND-CHERRY

   A tall erect annual plant with hairy and somewhat sticky stems 40–60 cm high. Large ovate to lanceolate leaves entire and stalked, 6–12 cm long, somewhat hairy and sticky. Rotate or shallowly bell-shaped flowers, white with pale yellow centers, 25–35 mm across. Fruit a berry partly enclosed in the persistent calyx, about 12 mm long. In disturbed areas, open woodlands, especially on sandy soils; northeastern Boreal forest. Sometimes called *Chamaesaracha grandiflora* (Hook.) Fern.; *Leucophysalis grandiflora* (Hook.) Rydb.

*Physalis heterophylla* Nees                              YELLOW GROUND-CHERRY

   Plants with erect or spreading stems, usually with many branches; pubescence distinctly villous, the hairs spreading. Leaves mostly ovate to rhombic, 3–8 cm long, shallowly sinuate-dentate. Corolla 15–20 mm long; the pedicels about 1 cm in flower, elongating to 3 cm in fruit; fruiting calyx 3–4 cm long. Introduced; rarely found as a weed.

*Physalis ixocarpa* Brot.                                            TOMATILLO

   Stems erect or spreading, branched; nearly glabrous, except on young shoots. Leaves ovate to rhombic, 2–6 cm long, glabrous below. Corolla 10–15 mm wide, yellowish, with a dark center; pedicels about 5 mm long, scarcely elongating in fruit; berry purplish. A rare garden weed.

*Physalis pubescens* L.                               SMALL YELLOW GROUND-CHERRY

   Stems branching near the base, spreading or prostrate. Upper part of stem and leaves densely pubescent. Leaves 4–8 cm long, sinuate-dentate, rounded or cordate at base. Corolla 6–10 mm long, yellow; pedicels 3–6 mm in flower, elongating to 1 cm in fruit; berry yellow. A rare garden weed.

*Physalis virginiana* Mill.                              PRAIRIE GROUND-CHERRY

   A perennial from creeping rootstocks, about 45 cm high, somewhat hairy. Leaves oblanceolate to spatulate, entire-margined and stalked, 25–50 mm long. Bell-shaped flowers, dull yellow with a brownish center, about 15 mm across. Fruit a yellow or greenish berry, entirely enclosed in the inflated bladder-like ovoid calyx. Rare; but found on sandy prairies and plains; in the eastern part of the Prairie Provinces.


***Solanum***       nightshade

1. Climbing perennial vines with deep purple flowers. ............................................................ *S. dulcamara*
   Nonclimbing annual herbs with white or yellow flowers. ............................................................ 2

2. Plants prickly; flowers yellow. ................................................................. *S. rostratum*
   Plants not prickly; flowers white. ................................................................. 3

3. Leaves entire or merely wavy-margined; erect plants. ................................................................. *S. nigrum*
   Leaves pinnatifid or incised; much branched and decumbently spreading. ................................................................. *S. triflorum*

*Solanum dulcamara* L. (Fig. 186)                                    BITTERSWEET

A woody-based perennial climbing or twining vine, with branches several
feet long. Leaves 5–10 cm long; upper leaves ovate, cordate at base, and
pointed at apex; lower ones often 3-lobed, with a large ovate middle lobe and
two small basal side lobes. Flowers in panicles or on compound cymes opposite a leaf stalk, somewhat rotate or wheel-shaped, with separate corolla lobes.
Corolla purple with yellow anthers erect and pyramidal around female organ
in center of flowers. Flowers 8–12 mm across, succeeded by oval or globose
berries, which are red and about 1 cm long. This introduced plant is **very
poisonous,** and as yet the only records of it in the area appear to be at Morden,
Man., Eastend, Sask., and Edmonton, Alta.

*Solanum nigrum* L.                                            BLACK NIGHTSHADE

An erect annual weed 10–30 cm high, with entire-margined ovate leaves
2–5 cm long. White flowers rotate, about 1 cm across, in clusters of 3–10
flowers. Fruits green berries, about 5 mm across, turning black when ripe. An
introduced species; now a fairly common weed; in gardens and waste places;
across the Prairie Provinces. There appears to be considerable variation in the
hairiness of this species, some specimens, especially from the west, being quite
hairy, others almost hairless. Pubescent forms with black berries have been
called *S. interius* Rydb.; densely pubescent plants with yellow or orange berries, *S. sarachoides* Sendt. Some cultivated forms of this species have edible
fruit and are known as wonderberry or garden huckleberry. There is always a
danger, however, that the fruit may be injurious to some people.

*Solanum rostratum* Dunal                                          BUFFALOBUR

An annual yellowish hairy prickly species 10–50 cm high. Leaves lobed
and pinnatifid, 5–8 cm long, and yellowish hairy. Bell-shaped flowers yellow,
about 2.5 cm across. Fruit a berry enclosed in a prickly calyx. A straggler from
farther south, found occasionally; in Prairies. Syn.: *Androcera rostrata* (Dunal)
Rydb.

*Solanum triflorum* Nutt. (Fig. 187)                               WILD TOMATO

A low, spreading annual, forming mats 15–60 cm across. Leaves deeply
lobed, oblong or ovate, 2–6 cm long, with scattered hairs. Flowers white,
rotate, 6–10 mm across, usually in clusters of three. Fruit a smooth green berry
about 12 mm in diam. A native plant found on disturbed areas such as gopher
and badger mounds; in southern Prairies. With cultivation it persists as a very
bad garden weed. If pulled and turned upside down, it will develop rootlets
along the stems and continue to grow. Although it is enjoyed as preserves by
some people, it causes violent sickness in others.

(Figs. 186 and 187 overleaf)

Fig. 186.  Bittersweet, *Solanum dulcamara* L.

Fig. 187.  Wild tomato, *Solanum triflorum* Nutt.

# SCROPHULARIACEAE—figwort family

Herbs with opposite, alternate, or whorled leaves with no stipules. Flowers perfect, but generally irregular in shape, with petals partly united into a tube and mostly 2-lipped. Perfect stamens, often only 2, usually 4, rarely 5. Fruit a many-seeded capsule.

1. Five anther-bearing stamens; flowers rotate, yellow, almost regular, in dense spike-like racemes. ................................................................ *Verbascum*
   Only 2 or 4 stamens anther-bearing, others sterile or absent. ................................................................ 2

2. Corolla spurred at the base. ................................................................ 3
   Corolla not spurred at the base. ................................................................ 4

3. Flowers axillary, blue; low slender annuals. ................................................................ *Chaenorrhinum*
   Flowers in terminal racemes, yellow; perennials. ................................................................ *Linaria*

4. Four anther-bearing stamens and 1 long sterile stamen. ................................................................ *Penstemon*
   Two or four anther-bearing stamens and no sterile stamen. ................................................................ 5

5. Anther-bearing stamens 2, any others sterile. ................................................................ 6
   Anther-bearing stamens 4. ................................................................ 9

6. Sepals and petal lobes 5. ................................................................ *Gratiola*
   Sepals and petal lobes 4. ................................................................ 7

7. Petals absent; inflorescence very hairy. ................................................................ *Besseya*
   Petals present. ................................................................ 8

8. Corolla rotate; stamens not protruding. ................................................................ *Veronica*
   Corolla short-tubular; stamens protruding; inflorescence a very long, narrow, spike-like raceme. ................................................................ *Veronicastrum*

9. Floral leaves or bracts brightly colored, usually red or yellow, or shades of these colors. ................................................................ *Castilleja*
   Floral leaves or bracts usually green, but may be brownish in *Pedicularis*. ................................................................ 10

10. Leaves pinnately lobed or cleft. ................................................................ *Pedicularis*
    Leaves simple, neither pinnately lobed nor cleft. ................................................................ 11

11. Leaves alternate. ................................................................ *Orthocarpus*
    Leaves opposite, whorled, or basal. ................................................................ 12

12. Leaves in a basal rosette; stem leafless; flower solitary; annuals of mud or water. ................................................................ *Limosella*
    Stems leafy; flowers not solitary. ................................................................ 13

## *Agalinis*     agalinis

*Agalinis aspera* (Dougl.) Britt.          ROUGH AGALINIS

An annual herb 10–30 cm high, with leaves narrowly linear, very scabrous above, often with axillary clusters of reduced leaves. Flowers on pedicels to 3 cm long; corolla 20–25 mm long, pinkish, with the tube pubescent. Rare; moist grasslands; southeastern Parklands.

*Agalinis purpurea* (L.) Pennell var. *parviflora* (Benth.) Boiv.    PURPLE AGALINIS

An annual herb 10–50 cm high, simple or branched; leaves to 4 mm wide. Flowers on short pedicels; corolla 15–25 mm long. Very rare; moist grassland; southeastern Parklands.

*Agalinis tenuifolia* (Vahl) Raf. var. *parviflora* (Nutt.) Pennell SLENDER AGALINIS

An annual herb 10–50 cm high, with stems very slender and usually much branched. Leaves linear, 1–6 mm wide. Pedicels filiform; corolla 10–15 mm long, pinkish. Rare; moist areas; southeastern Boreal forest.

**Bartsia**    bartsia

*Bartsia alpina* L.    VELVET BELLS

Plants with a somewhat woody rootstock; stems erect, 5–15 cm high, vis-cid-villous. Leaves sessile, ovate, 10–25 mm long, dentate, clasping the stem. Inflorescence a dense raceme; flowers to 2 cm long, glandular. Plants blackening on drying. Arctic meadows; northeastern Boreal forest.

**Besseya**    kittentails

*Besseya wyomingensis* (A. Nels.) Rydb.    KITTENTAILS

Softly hairy perennial plants 10–30 cm high, often with reddish-tinged leaves. Basal leaves stalked, ovate or oblong; stem leaves without stalks, alternate, and smaller. Inflorescence a dense terminal spike, 2–5 cm long in flower, and lengthening to 5–15 cm in fruit, very hairy, and usually with a purplish tinge. Flowers bearing 2 protruding stamens but no petals. Fruits many-seeded capsules. Found on open hillsides in the Foothills region and also on bench-land in the Cypress Hills. Syn.: *B. cinerea* (Raf.) Pennell.

**Castilleja**    Indian paintbrush

Annual, biennial, or perennial plants, usually partly parasitic on roots of other plants. Stem generally simple or with a few branches above. Leaves stalkless, alternate, lobed or entire. The bracts (leaves of the inflorescence) usually red or yellow. Flowers in terminal spikes with a 2-lipped corolla; the upper lip arched and called the galea, the lower one with 3 lobes.

1. Plants annual or biennial; inflorescence
   red or yellow. ............................................................................................ *C. coccinea*

   Plants perennial, with more or less woody
   rootstock. ............................................................................................................ 2

2. Flowers yellowish, 40–55 mm long,
   strongly curved. ....................................................................... *C. sessiliflora*

   Flowers smaller, usually straight and
   erect. ................................................................................................................... 3

3. Inflorescence pinkish purple to deep
   purple. ................................................................................................. *C. raupii*

   Inflorescence whitish to scarlet. ........................................................................ 4

4. Upper stem leaves 3- to 5-lobed, with the
   lateral lobes linear. ........................................................................ 5

   Upper stem leaves entire; floral leaves
   coarsely 3-lobed. ........................................................................... 7

5. Inflorescence bright red to scarlet. ........................................ *C. hispida*
   Inflorescence yellow or whitish. ............................................................. 6

6. Calyx 20–25 mm long, with the lobes
   rounded; corolla 18–22 mm long. ........................................ *C. cusickii*

   Calyx 15–20 mm long, with the lobes
   acute; corolla 15–20 mm long. .............................................. *C. lutescens*

7. Inflorescence reddish to scarlet. .......................................... *C. miniata*
   Inflorescence whitish to yellow. ............................................................... 8

8. Flowers 20–30 mm long. ...................................................... *C. occidentalis*
   Flowers 15–20 mm long. ................................................................................ 9

9. Stem stout, densely pubescent; leaves
   lanceolate. ................................................................................... *C. lutescens*

   Stem slender, finely pubescent to glab-
   rous; leaves narrowly linear-lanceolate. ................................. *C. pallida*

*Castilleja coccinea* (L.) Spreng.         SCARLET PAINTBRUSH

An annual or biennial species with hairy slender stems 30–50 cm high.
Basal rosette leaves entire but stem leaves deeply divided into 3–5 linear divisions. Bracts crimson-tipped, 3- to 5-lobed, and usually about as long as
flowers. Corolla greenish yellow, with the tube shorter than the calyx, and the
upper lip much longer than the lower lip. Fairly common; in meadows and
open woods; southeastern Prairies and Parklands.

*Castilleja cusickii* Greenm.         YELLOW PAINTBRUSH

A hairy simple-stemmed perennial species 20–30 cm high and very leafy.
Leaves 2–4 cm long, hairy, and 3- to 5-ribbed; lower leaves lanceolate and
entire; upper ones broader and 3- to 5-cleft at tip. The broad pale sulfur yellow
bracts about as long as flowers. Calyx 15–20 mm long; corolla rarely protruding beyond calyx. Found occasionally; in extreme southwestern Alberta. Syn.:
*C. lutea* Heller.

*Castilleja hispida* Benth.         HISPID PAINTBRUSH

Stems several, 20–40 cm high, hirsute, not glandular. Leaves lanceolate to
ovate, with 2 or 3 pairs of ascending lobes. Inflorescence villose, scarlet to red;
calyx cleft into oblong, rounded lobes. Montane grasslands and open woods;
southern Rocky Mountains.

*Castilleja lutescens* (Greenm.) Rydb.         STIFF YELLOW PAINTBRUSH

Stems several, 30–50 cm high, scabrous to short-pubescent. Leaves linear-lanceolate to lanceolate, entire or the upper ones shallowly lobed.
Inflorescence pale yellow, somewhat hirsute; calyx cleft into ovate acute-to-
rounded lobes. Montane grasslands; southern Rocky Mountains.

*Castilleja miniata* Dougl. (Fig. 188)                    RED INDIAN PAINTBRUSH

A perennial growing to 40-60 cm high; stem usually without hairiness below inflorescence, sometimes branched above. Linear pointed leaves 2-5 cm long. Bracts broader than leaves, scarlet or bright red; the lower ones usually having 2 or 5 teeth near summit. Flowers longer than bracts, green with red margins, lip a little shorter than tube. Found in open pine woods; in the Foot-hills region and also in the Cypress Hills.

*Castilleja occidentalis* Torr.                    LANCE-LEAVED PAINTBRUSH

A smooth-stemmed perennial 15-50 cm high. Leaves 5-10 cm long; lower leaves narrower than upper ones, and having 3-5 nerves. Yellowish, greenish white, or purple bracts oblong, oval or obtuse, sometimes with small teeth, and about as long as flower. Corolla 12-18 mm long; the upper lip 2-4 times as long as the lower. Fairly abundant; in open woodlands; southeastern Park-lands. Syn.: *C. acuminata* (Pursh) Spreng.; *C. sulphurea* Rydb.

*Castilleja pallida* (L.) Spreng. var. *septentrionalis* (Lindl.) Gray
                                                   LABRADOR PAINTBRUSH

Plants finely pubescent to glabrous; stems several, 15-30 cm high, simple. Leaves linear-lanceolate, entire, with upper ones sometimes shallowly lobed. Inflorescence pale yellow; calyx cleft into lance-ovate, acute lobes; corolla puberulent above. Marshy areas; northeastern Boreal forest.

*Castilleja raupii* Pennell                    PURPLE PAINTBRUSH

Stems usually several, 15-30 cm high, thinly pubescent. Leaves linear or linear-lanceolate, 2-5 mm wide. Inflorescence narrow to fairly dense, violet to deep violet purple; corollas shorter than bracts. Open areas, open forest, and forest margins; Boreal forest, Peace River district.

*Castilleja sessiliflora* Pursh                    DOWNY PAINTBRUSH

A perennial pale ashy gray downy-haired species 10-30 cm high. Leaves 2-4 cm long; lower leaves generally linear and entire; upper ones cleft into narrow, spreading segments. Bracts green, shorter than flowers. Corolla 40-55 mm long, yellowish white; the upper lip about twice as long as the 3-lobed lower lip. Found occasionally; on dry hills and prairies; throughout Prairies, southeastern Parklands, and in Cypress Hills.

**Chaenorrhinum**        small-snapdragon

*Chaenorrhinum minus* (L.) Lange                    SMALL-SNAPDRAGON

A low glandular hairy annual 15-30 cm high, usually branched. Leaves alternate, linear to linear-spatulate, 10-25 mm long. Flowers blue, 6-8 mm long, with a short spur, borne on short stalks in axils of leaves. An introduced plant; not common, but found occasionally; on or in the vicinity of railroad grades; throughout the Prairies and Parklands.

Fig. 188.   Red Indian paintbrush, *Castilleja miniata* Dougl.

*Chelone*        turtlehead

*Chelone glabra* L. var. *linifolia* Colem.                    TURTLEHEAD

   Perennial plants 50–80 cm high, with stems simple or branched above. Leaves linear or nearly so, serrate, short-petioled to subsessile. Inflorescence 3–8 cm long; flowers greenish or yellowish white, 20–25 mm long. Rare; wet areas and shrubbery; southeastern Boreal forest.


*Collinsia*        bluelips

*Collinsia parviflora* Dougl.                    BLUE-EYED MARY

   A low much-branched spreading annual with slender purplish stems 10–30 cm long. Leaves linear to lanceolate, 10–25 mm long, opposite or with upper leaves in whorls of 3 or 5. Flowers blue, about 6 mm long, borne either singly or in clusters on stalks in axils of upper leaves. Not common, but plentiful locally; found in shady woods or openings in woodlands; throughout southern Parklands, Cypress Hills, and southern Rocky Mountains.


*Euphrasia*        eyebright

*Euphrasia arctica* Lange                    NORTHERN EYEBRIGHT

   Annual plants with slender stems 10–20 cm high. Leaves in 3–10 pairs, ovate, 5–15 mm long, with 3–5 teeth on each margin. Flowers white with purple lines, 4–7 mm long. In more or less disturbed soils; northern Boreal forest. In var. *dolosa* Boiv. the flowers only 3–4 mm long, and lacking the purple lines. Northern Boreal forest and Rocky Mountains.


*Gratiola*        hedge-hyssop

*Gratiola neglecta* Torr.                    CLAMMY HEDGE-HYSSOP

   Annual or perennial herbs 10–20 cm high, somewhat sticky hairy. Leaves opposite, linear to oblong-lanceolate, 1–5 cm long, without stalks. Flowers pale yellow to whitish, 6–10 mm long, borne singly on long stalks in axils of leaves. Found occasionally; in mud or shallow water; particularly throughout the Prairies and southern Parklands.


*Limosella*        mudwort

*Limosella aquatica* L.                    MUDWORT

   A low annual with stems that root at nodes, 7–10 cm high. Leaves linear or spatulate, 5–30 mm long, on long stalks from plant crown. White or purplish flowers solitary on short stalks from base of plants, about 3 mm long. Found occasionally; rooted in mud or floating in shallow water around lakes and in streams; throughout most of the Prairie Provinces.


*Linaria*        toad-flax

1. Plants perennial, with extensive root-
   stocks; flowers yellow and orange. ........................................................................ 2

Plants annual; flowers white, pink, or purple, not yellow. .................................................................. 3

2. Stem leaves ovate to lanceolate, clasping. ............................................... *L. dalmatica*
   Stem leaves linear to linear-lanceolate, not clasping. ...................................................................... *L. vulgaris*

3. Lower lip of corolla violet; spur 6–10 mm long; pedicels 2–4 mm long. .............................................................. *L. canadensis*
   Lower lip of corolla with an orange spot; spur 8–15 mm long; pedicels 5–8 mm long. ......................................................................................................... *L. maroccana*

*Linaria canadensis* (L.) Dum. var. *texana* (Scheele) Pennell     FIELD TOAD-FLAX

Stems erect, slender, 20–60 cm high, glabrous, with several sterile shoots spreading from the base. Leaves 1–3 cm long. Racemes at first dense, later elongating; corolla 10–14 mm long, with the spur filiform; seeds distinctly rugose. Rare; an introduced weed.

*Linaria dalmatica* (L.) Miller     BROAD-LEAVED TOAD-FLAX

An introduced perennial with a coarse branching growth 50–80 cm high. Leaves numerous, 1–5 cm long, ovate to lanceolate, often clasping the stem at the base. Flowers borne on a long spike-like raceme, 2–5 cm long, with long spurs, bright yellow, sometimes with orange-colored throat entrance. Grown as an ornamental in gardens, but has escaped in many places and is becoming a persistent weed.

*Linaria maroccana* Hook. f.     MAROCCAN TOAD-FLAX

Plants to 50 cm high, with remote linear leaves. Raceme becoming lax; corolla violet; the lower lip with a large orange spot; the spur conical. Rare; an introduced weed.

*Linaria vulgaris* Miller     BUTTER-AND-EGGS, YELLOW TOAD-FLAX

A perennial from creeping rootstocks, 20–60 cm high. Leaves linear, stalkless, alternate, 2–7 cm long. Flowers bright yellow with orange throat entrance and a long spur at base, mouth of flowers closed, 20–35 mm long, in dense terminal raceme. Introduced as a garden plant from Europe, and found as a persistent weed of fields and waste places; throughout the Prairies and Parklands.

*Melampyrum*     cow-wheat

*Melampyrum lineare* Desr. (Fig. 189)     COW-WHEAT

Annual plants 15–50 cm high with lanceolate to linear-lanceolate leaves 2–6 cm long. Lower leaves entire-margined; upper floral leaves more ovate and often bearing bristle-pointed teeth. Flowers 8–12 mm long, generally whitish with a yellowish lower lip, borne either solitary in upper leaf axils or in a terminal leafy spike. Fairly common; on dry, sandy soil in woodlands; Boreal forest.

Fig. 189.   Cow-wheat, *Melampyrum lineare* Desr.

*Mimulus*     monkeyflower

Perennial herbs of streams and very wet places, with large showy flowers. Calyx angled, and corolla open-throated with upper lip spreading or reflexed.

1. Flowers blue or crimson. ............................................................................ 2
   Flowers yellow. ...................................................................................... 3
2. Flowers blue to violet; plants glabrous. ................................................. *M. ringens*
   Flowers crimson to reddish; plants glandular pubescent. ............................................................. *M. lewisii*
3. Calyx regular or nearly so; corolla 5–12 mm long. ................................................. *M. floribundus*
   Calyx distinctly 2-lipped; corolla 12–25 mm long. ...................................................................... 4
4. Stems mostly creeping; calyx 5–12 mm long; corolla to 25 mm long. ................................................. *M. glabratus*
   Stems mostly erect; calyx 12–17 mm long; corolla 25–45 mm long. ................................................. *M. guttatus*

*Mimulus floribundus* Dougl.           MONKEYFLOWER

An annual herb, with stems weak, villosely glandular pubescent, much branched, 5–35 cm long. Leaves ovate, dentate, rounded or cordate at base, the larger ones 1–3 cm long. Flowers yellow; the corolla 5–12 mm long, on pedicels 5–25 mm long. Wet places; southern Rocky Mountains.

*Mimulus glabratus* HBK.           SMOOTH MONKEYFLOWER

Stems weak, glabrous or minutely pubescent. Leaves 1–4 cm long, ovate to rotund, rounded at base. Flowers on pedicels 1–2 cm long, bright yellow, pubescent in the throat. Rare; wet places, springs; southeastern Parklands.

*Mimulus guttatus* DC.           YELLOW MONKEYFLOWER

Perennial plants from basal branches that root at nodes. Stem 10–50 cm high, bearing opposite ovate or rounded leaves 1–5 cm long; lower leaves stalked; upper ones usually clasping the stem. Conspicuous flowers 25–45 mm long, bright yellow with an open mouth, somewhat hairy inside the lower lip; calyx inflating somewhat at maturity to contain the many-seeded capsule. Not common, but plentiful locally; found in running streams; in Prairies, southern Rocky Mountains, and in Cypress Hills. Syn.: *M. langsdorfii* Donn.

*Mimulus lewisii* Pursh           LEWIS MONKEYFLOWER

A somewhat sticky hairy perennial 30–60 cm high. Leaves lanceolate to oblong, 3–7 cm long. Crimson or reddish flowers 3–5 cm long and somewhat hairy inside open throat. Found along stream banks; in southern Rocky Mountains.

*Mimulus ringens* L.           BLUE MONKEYFLOWER

A hairless perennial from rootstocks 30–100 cm high. Leaves lanceolate to oblong, opposite, 5–10 cm long, without stalks. Flowers blue or violet, 20–35 mm long, with a narrow throat. Fairly common; in swamps, along streams and lakes; southeastern Parklands and Boreal forest.

***Odontites***     eyebright

*Odontites serotina* (Lam.) Dum.                    LATE-FLOWERING EYEBRIGHT

An annual herb with stems 10–40 cm high, finely retrorse-pubescent, usually branched. Leaves lanceolate, 1–3 cm long, roughly pubescent. Flowers
nearly sessile, about 1 cm long, pubescent, light red. Introduced; a rare weed;
along roads; in Boreal forest.


***Orthocarpus***     owl's-clover

*Orthocarpus luteus* Nutt.                                      OWL'S-CLOVER

A short erect annual 10–30 cm high, sometimes with erect growing
branches. Leaves 10–25 mm long, linear or narrowly lanceolate, crowded and
ascending. Yellow flowers 10–15 mm long, on very leafy narrow terminal
spikes, and followed by numerous many-seeded capsules. Common; on open
dry prairie; throughout the Prairies and Parklands.


***Pedicularis***     lousewort

Annual or perennial herbs with pinnately cleft or lobed leaves and flowers
in dense terminal spikes or racemes.

1. Galea prolonged into a tubular recurved
    beak. ................................................................................................. 2
    Galea not prolonged, the apex broad. ............................................... 5
2. Flowers purplish red or pink. ............................................ *P. groenlandica*
    Flowers white or yellow. .................................................................... 3
3. Leaves serrulate to crenate. ................................................... *P. racemosa*
    Leaves deeply lobed or divided. ......................................................... 4
4. Beak of galea bent into a half circle. .......................................... *P. contorta*
    Beak of galea slightly bent, at most one-
        quarter of a circle. ..................................................... *P. lapponica*
5. Flowers in an elongating inflorescence,
        becoming axillary. ....................................................... *P. parviflora*
    Flowers in well-defined, more or less
        crowded racemes. ..................................................................... 6
6. Stem leaves opposite or subopposite. ................................ *P. lanceolata*
    Stem leaves alternate. ....................................................................... 7
7. Flowers 3–5 in a short raceme; corolla
        30–35 mm long. .......................................................... *P. capitata*
    Flowers more numerous, usually less than
        25 mm long. ............................................................................. 8
8. Inflorescence glabrous. ........................................................ *P. flammea*
    Inflorescence pubescent to glandular
        pubescent. ................................................................................. 9

*Pedicularis bracteosa* Benth.　　　　　　　　　WESTERN LOUSEWORT

A perennial species with an erect stem 30–90 cm high. Leaves alternate, 10–30 cm long, divided so deeply into toothed leaflets that they appear pinnate. Flowers pale yellow, about 2 cm long, in a dense terminal spike 10–30 cm long. Bracts of inflorescence almost as long as flowers, but a shorter-bracted variety has been named *P. montanensis* Rydb. Fairly plentiful; among bushes and in moist places; Rocky Mountains and Cypress Hills.

*Pedicularis canadensis* L.　　　　　　　　　　COMMON LOUSEWORT

A somewhat hairy, erect, perennial herb 15–50 cm high. Oblong leaves deeply incised or lobed, 7–15 cm long. Flowers 20–25 mm long, yellow or reddish, in a dense spike at head of stem. Fairly common; in woodlands; southeastern Parklands and Boreal forest.

*Pedicularis capitata* Adans.　　　　　　　LARGE-FLOWERED LOUSEWORT

Small plants, with the stems usually solitary, 10–20 cm high, leafless, glabrous to somewhat pubescent. Inflorescence a terminal raceme with 2–5 flowers, yellowish white to pinkish-tinged, 25–40 mm long. Alpine tundra; southern Rocky Mountains.

*Pedicularis contorta* Benth.　　　　　　　COILED-BEAK LOUSEWORT

Plants with stems 30–60 cm high, glabrous throughout. Leaves to 15 cm long, 2–3 cm wide. Inflorescence dense; the bracts the same length as or shorter than the flowers; flowers 15 mm long, white or pale yellow, with the hood of the galea recurved in a semicircle, finely purple-spotted. Dry montane and alpine slopes; southern Rocky Mountains.

*Pedicularis flammea* L.                                          FLAME-COLORED LOUSEWORT

Small plants with spindle-shaped roots; stems 5–20 cm high, with 1 or 2 stem leaves. Inflorescence few-flowered, 2–5 cm long; flowers 15–20 mm long, with the lower lip yellow, the galea purple to crimson. Arctic and alpine meadows; northeastern Boreal forest, northern Rocky Mountains.

*Pedicularis groenlandica* Retz. (Fig. 190)                      ELEPHANT'S-HEAD

An erect hairless perennial 20–60 cm high, from rootstocks. Leaves lanceolate, very deeply incised; lower ones stalked, 5–15 cm long. Flowers borne on a terminal spike 5–15 cm long, usually purple or deep red, 10–15 mm long. Upper lip of corolla long, curved downward and then upward, making flowers resemble small red elephant heads. Found in swampy places and stream banks; in Boreal forest and Rocky Mountains.

*Pedicularis labradorica* Wirsing                                LABRADOR LOUSEWORT

Low very branchy plants to 30 cm high; stems and foliage puberulent to retrorsely pilose. Inflorescence about 5 cm long; flowers about 15 mm long, at first yellow, and then fading purplish. Bogs and tundra; northeastern Boreal forest.

*Pedicularis lanata* C. & S.                                     WOOLLY LOUSEWORT

Small plants, with densely woolly stems; leaves 2–3 cm long, glabrous. Inflorescence 3–5 cm long, with flowers 20–25 mm long, reddish purple to rose. The lower lip the same length as the galea. Mountain meadows; southern Rocky Mountains.

*Pedicularis lanceolata* Michx.                                  SWAMP LOUSEWORT

A practically hairless perennial, with stout erect stems 30–90 cm high. Leaves both alternate and opposite, 5–12 cm long, with short marginal lobes. Yellow flowers 20–25 mm long with slight difference between lengths of upper and lower lips. Borne in a short dense terminal spike. Not common; but has been found in swamps; eastern Parklands and Boreal forest.

*Pedicularis langsdorffii* Fisch.                                ARCTIC RATTLE

Similar to *P. lanata*, but not so woolly, the flowers deep pink, and the lower lip only half as long as the galea. Alpine slopes; southern Rocky Mountains.

*Pedicularis lapponica* L.                                       LAPLAND RATTLE

Plants 10–20 cm high, with stems and inflorescence densely retrorse-pubescent, otherwise glabrous. Inflorescence few-flowered, with the flowers 15–20 mm long, yellow. Tundra; northeastern Boreal forest.

*Pedicularis oederi* Vahl. var. *albertae* (Hulten) Boiv.        OEDER'S LOUSEWORT

Low plants 5–15 cm high, with glabrous stems and leaves; leaves 3–6 cm long. Inflorescence 3–10 cm long, woolly, and somewhat glandular; densely flowered; flowers 20–25 mm long, yellow, with the galea purple-tinged. Alpine tundra; southern Rocky Mountains.

Fig. 190. Elephant's-head, *Pedicularis groenlandica* Retz.

*Pedicularis parviflora* Smith                          PURPLE LOUSEWORT

An annual or biennial species 15–50 cm high, much branched. Leaves 2–4 cm long, deeply divided. Flowers 10–15 mm long, purple, with galea rounded at apex. Uncommon; but has been found in marshlands and bogs; Boreal forest.

*Pedicularis racemosa* Dougl.                           LEAFY LOUSEWORT

Plants with stems 30–50 cm high; leaves all on the stem, 4–7 cm long, 1–2 cm wide, glabrous. Inflorescence loose, leafy-bracted; the bracts as long as or longer than the flowers; flowers 10–15 mm long, pale violet to purple, with the galea strongly recurved. Moist open areas in montane woods; southern Rocky Mountains.

*Pedicularis sudetica* Willd.                           PURPLE RATTLE

Plants 10–20 cm high, with stems and leaves glabrous or somewhat short-pubescent. Leaves mostly basal, typically with a solitary stem leaf. Inflorescence 3–5 cm long, densely woolly; flowers 15–20 mm long, violet purple, with the lower lip pale pink, purple-dotted. Wet tundra; northeastern Boreal forest.

**Penstemon**      beardtongue

Perennial herbs with opposite leaves and irregular entire flowers. There are 4 fertile stamens and 1 sterile stamen that is usually more or less bearded. Fruit an ovoid many-seeded capsule. Some authorities spell the generic name *Pentstemon* or *Pentastemon*.

1. Plants shrubby; stems woody, decumbent
    to ascending. ............................................................................................................ 2
    Plants herbaceous; stems erect or nearly
    so. ......................................................................................................................................... 3

2. Stems 20–50 cm high; leaves lanceolate. ............................... *P. fruticosus*
    Stems matted, to 20 cm high; leaves ellip-
    tic to orbicular. ........................................................................ *P. ellipticus*

3. Flowers 8–12 mm long. ........................................................................................... 4
    Flowers 15–40 mm long. ........................................................................................ 5

4. Flowers dark bluish purple. ...................................................... *P. procerus*
    Flowers white to yellowish. ...................................................... *P. confertus*

5. Style yellowish, long-pilose, exserted. ............................... *P. eriantherus*
    Style not yellow-pilose, included. ................................................................. 6

6. Flowers 35–40 mm long. ........................................................... *P. lyallii*
    Flowers 15–25 mm long. ....................................................................................... 7

7. Flowers white, finely purplish lined or
    spotted. ...................................................................................... *P. albidus*
    Flowers blue or purplish. ..................................................................................... 8

8. Flowers blue; foliage glaucous; lower
    bracts suborbicular. ................................................................... *P. nitidus*

*Penstemon albertinus* Greene                    BLUE BEARDTONGUE

Stems 10–30 cm high, mostly decumbent at the base, glabrous to the inflorescence. Leaves 2–4 cm long; lower ones petioled, glabrous; upper stem leaves denticulate. Inflorescence interrupted; flowers deep blue to purplish, 15–20 mm long; calyx 2.5–4.0 mm long. Slopes, openings in woods, and disturbed areas; southern Rocky Mountains.

*Penstemon albidus* Nutt.                    WHITE BEARDTONGUE

A rather short, stout, clustered, and erect species 15–30 cm high, with fine hairs on stem. Lower leaves oblong to spatulate and stalked; upper ones stalkless and lanceolate, 2–7 cm long. Flowers about 2 cm long, white; whole flower somewhat sticky hairy. Inflorescence a narrow terminal raceme. Flowering in late May and early June, usually a little later than *P. nitidus*. Common; on dry prairies and hillsides; throughout the Prairies and southern parts of Parklands.

*Penstemon confertus* Dougl.                    YELLOW BEARDTONGUE

A perennial species, usually with several slender stems 10–50 cm high, not hairy. Leaves 5–10 cm long, lanceolate to oblanceolate or linear, and entiremargined. Flowers yellow, 10–15 mm long, in dense clusters on an interrupted terminal spike. Common; on hillsides and dry areas; southern Rocky Mountains; introduced at Swift Current, Sask.

*Penstemon ellipticus* C. & F.                    CREEPING BEARDTONGUE

A mat-forming species, woody at the base; flowering stems 10–20 cm high, puberulent. Leaves thick, elliptic to orbicular, 5–15 mm long. Inflorescence few-flowered, glandular pubescent; flowers 20–35 mm long, purple violet; calyx 7–10 mm long. Alpine slopes and ridges; southern Rocky Mountains.

*Penstemon eriantherus* Pursh                    CRESTED BEARDTONGUE

A somewhat hairy-stemmed erect species 15–40 cm high. Upper portion of stem sticky hairy and glandular. Lower leaves oblong to spatulate and stalked; upper ones often somewhat clasping, acute, 2–5 cm long. Flowers red or purple, 20–25 mm long, with sterile stamens densely hairy; inflorescence a fairly dense, leafy, terminal raceme. Found occasionally; in grassland; southern Rocky Mountains.

*Penstemon fruticosus* (Pursh) Greene                    SHRUBBY BEARDTONGUE

Plants forming dense clumps; stems 10–50 cm high, glabrous below the inflorescence. Leaves thick, leathery, narrowly lanceolate, 1–5 cm long, serrulate to denticulate. Inflorescence somewhat glandular pubescent, somewhat one-sided; flowers 25–40 mm long, purplish blue to violet. Slopes and ridges; southern Rocky Mountains.

*Penstemon gracilis* Nutt. (Fig. 191*A*)     LILAC-FLOWERED BEARDTONGUE

A slender erect species 15–40 cm high. Leaves linear-oblong to linear-lanceolate, slightly toothed at margins, 2–7 cm long. Usually flowering in mid-June; flowers borne in clusters of 2 or 3 in the axils of upper leaves, forming an open panicle, pale purple or lilac, 20–25 mm long. Fairly common; on moist prairie, slough margins; throughout Prairies and Parklands.

*Penstemon lyallii* Gray     LARGE-FLOWERED BEARDTONGUE

Plants with a woody caudex; stems 20–30 cm high, grayish pubescent. Leaves lanceolate to linear-lanceolate, 2–7 cm long, entire to finely denticulate. Inflorescence somewhat glandular pubescent; flowers 35–40 mm long, deep blue to purple; calyx 10–12 mm long. Mountain slopes; southern Rocky Mountains.

*Penstemon nitidus* Dougl. (Fig. 191*B*)     SMOOTH BLUE BEARDTONGUE

A stout-stemmed, often branching, hairless species 15–30 cm high. Lower leaves lanceolate to oblanceolate, 2–5 cm long; upper leaves smaller and ovate. Inflorescence a fairly dense, somewhat leafy raceme; flowers 15–20 mm long, usually deep blue, but also varying from purple through all shades of red and pink to pure white; earliest of penstemons, flowering in May. Very common; on dry hills, and eroded areas and banks; throughout Prairies, Parklands, and Rocky Mountains.

*Penstemon procerus* Dougl. (Fig. 191*C*)     SLENDER BEARDTONGUE

A slender-stemmed, low-growing perennial 15–30 cm high; stems often decumbent at base. Basal leaves stalked and usually oblanceolate; stem leaves not stalked, oblong to lanceolate, 2–7 cm long. Inflorescence a dense, terminal spike, often interrupted; usually flowering from end of May until early August; flowers crowded on the stem; corolla dark blue, 10–15 mm long. Common; usually found in large colonies around slough margins, in shelter of shrubs, and in openings in woodlands; Prairies, Parklands, and Rocky Mountains.

**Rhinanthus**     yellowrattle

*Rhinanthus crista-galli* L.     YELLOWRATTLE

Somewhat branching annual plants 30–60 cm high. Leaves opposite, linear to lanceolate, 2–5 cm long, with finely toothed margins. Flowers pale yellow, 6–10 mm long, borne either in upper leaf axils or in a one-sided leafy terminal spike; calyx hairy, greenish yellow, compressed, membranous, and at maturity inflated, enclosing a capsule. Not common; has been found at widely scattered locations in moist places and open woodlands; throughout Parklands, Boreal forest, and Rocky Mountains.

**Scrophularia**     figwort

*Scrophularia lanceolata* Pursh     HARE FIGWORT

An erect plant 50–200 cm high, with stems and inflorescence somewhat glandular. Leaves ovate to lanceolate, coarsely toothed, 2–15 cm long, on short stalks borne oppositely on stem. Flowers greenish, 8–12 mm long, on stalks in a tall terminal inflorescence. Rare; has been found on moist ground; Prairies.

Fig. 191. Beardtongues: *A*, lilac-flowered beardtongue, *Penstemon gracilis* Nutt.; *B*, smooth blue beardtongue, *Penstemon nitidus* Dougl.; *C*, slender beardtongue, *Penstemon procerus* Dougl.

***Verbascum***     mullein

1. Lower stem leaves long-petioled, ovate-cordate. ......................................................... *V. nigrum*

    Lower stem leaves sessile, ovate-lanceolate. ....................................................................... 2

2. Leaves long-decurrent; filaments of upper stamens 3–4 times as long as the anthers. ......................................................... *V. thapsus*

    Leaves short-decurrent, ovate; filaments of upper stamens 1.5–2 times as long as the anthers. ......................................... *V. phlomoides*

***Verbascum nigrum*** L.     BLACK TORCH

Biennial plants 50–150 cm high, pubescent. Leaves stellate-pubescent below; lower leaves long-petioled, mostly cordate at the base; upper stem leaves sessile. Inflorescence a raceme; flowers 2–5 together, 15–20 mm across, light yellow; filaments purplish pubescent. Introduced; a rare roadside weed.

***Verbascum phlomoides*** L.     WOOLLY MULLEIN

Biennial plants with stems 50–200 cm high, densely yellow woolly pubescent with branched hairs. Inflorescence spike-like; flowers 30–35 mm across, yellow; stigma club-shaped. Introduced; a rare garden and roadside weed.

***Verbascum thapsus*** L.     COMMON MULLEIN

Biennial plants, tall, erect, woolly, 30–200 cm high. Stems very stout and straight, bearing large, densely woolly spatulate to elliptic leaves 10–30 cm long. Flowers yellow, almost regular, 5-lobed, 20–25 mm across, on a dense woolly spike 10–50 cm high; lower flowers usually opening first. Rare; has been found on railway grades and waste places; at widely separated locations throughout Prairies and Parklands. Its scarcity is rather surprising, because in British Columbia and in Eastern Canada this species is a common roadside weed, appearing to thrive best in dry, dusty, and exposed locations.

***Veronica***     speedwell

Annual or perennial herbs usually associated with wet places. Leaves opposite. Flowers perfect, 4-lobed, rotate, slightly irregular, with 2 stamens. Fruit a several-seeded capsule.

1. Calyx 5-lobed, with the upper 2 long and the lower 3 short. ............................................ *V. teucrium*

    Calyx 4-lobed. ............................................................................. 2

2. Flowers solitary in axils of leaves; annuals. ....................................................................................... 3

    Flowers in axillary or terminal racemes; perennials. ............................................................................. 6

3. Flower stalks shorter than or equaling leaves. ..................................................................................... 4

    Flower stalks as long as or longer than leaves. ..................................................................................... 5

4. Leaves narrowly oblong to oblong-lanceolate; capsule glabrous, veinless. .................... *V. peregrina* var. *xalapensis*

Leaves oblong-ovate to almost rotund; capsule reticulate-veined. ........................... *V. agrestis*

5. Flowers 10–15 mm across; capsule reticulate-veined. .................................................. *V. persica*

Flowers 4–6 mm across; capsule glandular pubescent. ............................................. *V. polita*

6. Flowers in terminal racemes. .............................................. 7

Flowers in axillary racemes. .............................................. 9

7. Racemes well-defined, dense; leaves 5–15 cm long, often in whorls. ........................... *V. longifolia*

Racemes not well-defined, loose; leaves mostly less than 5 cm long. ............................. 8

8. Leaves all sessile; capsule longer than wide, barely notched. ........................ *V. alpina* var. *unalaschensis*

Lower leaves petioled; capsule as long as wide, deeply notched. ................... *V. serpyllifolia* var. *humifusa*

9. Leaves ovate to oblong, all with short stalks. .............................................. 10

Leaves linear to lanceolate, those of flowering shoots without stalks. ........................... 11

10. Leaves and stems glabrous; semiaquatic plants. .............................................. *V. americana*

Leaves glabrous or pubescent, and stem with 2 rows of hairs; a plant of gardens and waste areas. ........................................ *V. chamaedrys*

11. Leaves broadly lanceolate; fruit not much flattened, with its stem less than twice as long as fruit. ........................ *V. comosa* var. *glaberrima*

Leaves linear to linear-lanceolate; fruit much flattened, with its stem several times as long as fruit. ........................... *V. scutellata*

*Veronica agrestis* L.                          PROSTRATE SPEEDWELL

Annual plants with prostrate or ascending stems 10–30 cm long. Leaves ovate to rotund, crenately serrate, rounded or truncate at the base, 1–2 cm long. Flowers 6–8 mm across, blue or white, with blue veins; pedicels 6–10 mm long, elongating in fruit. Introduced; a rare garden weed.

*Veronica alpina* L. var. *unalaschensis* C. & S.           ALPINE SPEEDWELL

Stems erect or decumbent at the base, 10–20 cm high, more or less villous throughout. Leaves all sessile, elliptic to oblong, 15–35 mm long, pilose on both sides. Inflorescence densely villous, elongating during and after flowering; flowers about 5 mm across, blue; capsules 5–8 mm long, about 3–5 mm wide, barely notched. Alpine meadows and moist areas; Rocky Mountains.

*Veronica americana* (Raf.) Schwein.                                       AMERICAN SPEEDWELL

An aquatic or semiaquatic, rather weak-stemmed plant 10–50 cm long. Leaves all stalked, oblong-lanceolate to ovate, 2–6 cm long. Flowers blue or white, 3–5 mm across, in long loose racemes in axils of leaves. Not common; found in streams and around springs; Boreal forest, Rocky Mountains, and Cypress Hills.

*Veronica chamaedrys* L.                                                   GERMANDER SPEEDWELL

A sparsely pubescent perennial; stems prostrate, 20–40 cm long. Leaves sessile or nearly so, ovate to deltoid-ovate, 2–3 cm long, serrate. Racemes few, erect from the upper axils, loosely 10- to 20-flowered, 8–15 cm long; flowers about 1 cm across, blue, with darker lines; capsules rarely produced, 4–5 mm wide, shallowly notched. Introduced; a rare garden weed.

*Veronica comosa* Richt. var. *glaberrima* (Pennell) Boiv.       WATER SPEEDWELL

A branched species 20–70 cm long, with stalkless, often clasping, lanceolate leaves 2–7 cm long. Flowers blue, 8–10 mm across, in loose axillary racemes. Uncommon; has been found in wet places and along rivers; Boreal forest and Cypress Hills. Syn.: *V. catenata* Pennell; *V. connata* Raf. ssp. *glaberrima* Pennell.

*Veronica longifolia* L.                                                  SPIKED SPEEDWELL

Perennial plants with erect stems 40–100 cm high. Leaves opposite or in whorls of 3, lanceolate or ovate-lanceolate. Racemes erect, spike-like; axis pubescent; flowers 8–10 mm across, blue. Introduced ornamental; occasionally escaped and established.

*Veronica peregrina* L. var. *xalapensis* (HBK.) St. John & Warren

                                                                          HAIRY SPEEDWELL

A glandular hairy erect species 10–30 cm high. Leaves spatulate to linear, 10–25 mm long. Flowers whitish, about 3 mm across, borne singly in axils of leaves. Not common but locally abundant; on moist sandy soils and in low areas; across the Prairie Provinces. Syn.: *V. xalapensis* HBK.

*Veronica persica* Poir.                                                  BIRD'S-EYE

A low-spreading, very branching, finely hairy annual. Leaves oval, short-stemmed, 10–25 mm long, coarsely toothed. Flowers blue, about 8–15 mm across, borne singly on long stalks from axils of leaves. Introduced; becoming increasingly abundant as a garden weed.

*Veronica polita* Fr.                                                     SMOOTH SPEEDWELL

A species similar to *V. persica*, but with flowers at most 6 mm across and usually smaller. Introduced; a rare garden weed.

*Veronica scutellata* L.                                                  MARSH SPEEDWELL

A species with decumbent or ascending stems 20–50 cm high. Leaves linear to linear-lanceolate, sharp-pointed, 2–7 cm long. Flowers about 6 mm across, blue, in long loose racemes in axils of leaves. Fairly common; in moist meadows and swamps and around springs; throughout the Prairie Provinces.

*Veronica serpyllifolia* L. var. *humifusa* (Dicks.) Vahl. THYME-LEAVED SPEEDWELL

A slender-stemmed perennial 5–20 cm high, much-branched from base. Leaves ovate to oblong or almost orbicular, 5–15 mm long. Flowers white or pale blue, 5–8 mm across, in short narrow leafy terminal racemes. An eastern species, very rare; has been found in low swampy ground; Cypress Hills.

*Veronica teucrium* L. BROAD-LEAVED SPEEDWELL

An erect finely pubescent perennial 30–60 cm high. Leaves sessile or nearly so, ovate to lanceolate, 2–4 cm long, serrate. Racemes 2–4, from the upper axils, 5–10 cm long, on pedicels to 10 cm long, rather densely flowered; flowers about 12 mm across, blue. Introduced ornamental; occasionally escaped and established.

### *Veronicastrum* Culver's-root

*Veronicastrum virginicum* (L.) Farwell CULVER'S-ROOT

A coarse erect leafy-stemmed perennial 30–150 cm high, with opposite or whorled leaves. Leaves oblong to lanceolate, 7–15 cm long, varying from 2 to 7 at a node. Flowers white, with prominent protruding stamens 4–6 mm long, crowded on long narrow terminal racemes 6–20 cm long. An eastern species; has been found on roadsides and meadows; southeastern Parklands and Boreal forest.


## LENTIBULARIACEAE—bladderwort family

Small plants found growing in water or wet locations, carnivorous or insectivorous, catching small insects and aquatic life by means of sticky leaves or submersed bladders. Leaves basal. Flowers irregular, perfect, with 2 stamens.

Calyx with 5 sepals; land plants; leaves entire, basal; flowers solitary. ................................................................... *Pinguicula*

Calyx with 2 sepals; plants floating on or submersed in water; leaves dissected, bearing small bladders; flowers several in a raceme above water. ................................................................... *Utricularia*


### *Pinguicula* butterwort

1. Stem villous below; flower 6–8 mm long. ....................................... *P. villosa*

   Stem glandular pubescent; flower more than 17 mm long. ...................................................................... 2

2. Lower corolla lip over 12 mm long; flower dark purple with a straight spur. ......................................... *P. macroceras*

   Lower corolla lip less than 12 mm long; flower pale purple with slightly recurved spur. ...................................................................... *P. vulgaris*

*Pinguicula macroceras* Willd.                     WESTERN BUTTERWORT

A low bog plant with a basal rosette of sticky leaves. Flower single, 2–3 cm long, dark purple, on a stalk 5–10 cm high. Not common; found occasionally in bogs and swamps; southern Rocky Mountains.

*Pinguicula villosa* L.                            SMALL BUTTERWORT

A low bog plant, similar to *P. macroceras*, but even smaller, 2–5 cm high; stem densely villous below, glandular puberulent above. Tundra and subarctic bogs; northern parts of Boreal forest.

*Pinguicula vulgaris* L.                           BUTTERWORT

A low plant with leaves in a basal rosette. Leaves oval or elliptic, 2–3 cm long, usually somewhat rolled at edge, and having a sticky secretion. Flower single, 12–18 mm long, pale purple, on a stalk 3–10 cm high. Very rare; has been found in bogs and cold swampy lands; Boreal forest and Rocky Mountains.

**Utricularia**      bladderwort

1. Bladders minute; leaves thread-like, simple, not readily seen. ................................................................... *U. cornuta*
   Bladders readily seen; leaves finely dissected. .................................................................................... 2

2. Stems either submersed or floating; flowers 15–25 mm long. ................................... *U. vulgaris* var. *americana*
   Stems creeping on bottom in shallow water; flowers 3–10 mm long. .................................................... 3

3. Flower stalks ascending; flowers 10–15 mm long. ............................................................................ *U. intermedia*
   Flower stalks recurved; flowers 3–8 mm long. ................................................................................ *U. minor*

*Utricularia cornuta* Michx.                       HORNED BLADDERWORT

A delicate terrestrial plant; stems creeping underground. Leaves linear-filiform, with bladders in the margins. Scapes erect, 3–15 cm high, bearing 1–3 yellow flowers 15–25 mm long; spur 7–12 mm long. Rare; peaty or muddy shores and bogs; Boreal forest.

*Utricularia intermedia* Hayne                     FLAT-LEAVED BLADDERWORT

An aquatic plant creeping on mud in shallow water, with somewhat floating branches 6–15 cm long. Leaves scattered and very finely dissected into thread-like segments; some branches leafless, bearing several bladders 3–10 mm long. Flowers yellow, with a large lower lip, about 10 mm long, in a 1- to 4-flowered raceme on a leafless upright stem. Found occasionally; in bogs and shallow water; Parkland, Boreal forest, and Cypress Hills.

*Utricularia minor* L.                             LESSER BLADDERWORT

A small submersed plant with alternate leaves. Leaves having a few thread-like divisions, very minute, less than 6 mm long, bearing a few bladders slightly longer than 1.5 mm. Flowers pale yellow, less than 8 mm long, on curved stalks on an upright flowering stem. Very rare; has been found in bogs and shallow water; Boreal forest.

*Utricularia vulgaris* L. var. *americana* Gray (Fig. 192)   GREATER BLADDERWORT

A species often floating on the surface of the water, with leaves submersed. Stems 30–80 cm long, branched. Many much-divided leaves 2–5 cm long, bearing numerous bladders 3–5 mm long. Flowers yellow, 15–25 mm long, in a raceme of 6–15 flowers on a long, flowering stem 10–30 cm high. Fairly common; in lakes and sloughs; throughout the Prairie Provinces. Syn.: *U. macrorhiza* Le Conte.

## OROBANCHACEAE—broom-rape family

Low pinkish herbs without green foliage, parasitic on roots of other plants (particularly *Artemisia*). Leaves scale-like. Flowers perfect and irregular; calyx 5-lobed; corolla 2-lipped; stamens 4. Fruit a many-seeded capsule.

Calyx deeply cleft on the lower side only;
  glabrous throughout. ................................................................................ *Conopholis*
Calyx nearly regular or cleft on both upper
  and lower sides; glandular puberulent. ........................................................ *Orobanche*

**Conopholis**      squawroot

*Conopholis americana* (L.) Wallr.                                 CANCERROOT

Stems stout, erect, 5–20 cm high, pale brown or yellowish throughout, covered by ovate fleshy leaf scales to 2 cm long. Spike about half the height of the stem, 15–20 cm thick; corolla 10–15 mm long. Very rare; parasitic on shrubs and trees; southeastern Parklands and Boreal forest.

**Orobanche**      broom-rape

1. Plants with a solitary flower; scale leaves
   blunt. ................................................................................................ *O. uniflora*
   Plants with more than one flower; scale
   leaves acuminate. ........................................................................................ 2

2. Flowers borne singly on a naked stalk,
   with no bracts beneath each flower. ................................................ *O. fasciculata*
   Flowers borne in a racemose cluster, with
   1 or 2 long bracts beneath each flower. ........................................ *O. ludoviciana*

*Orobanche fasciculata* Nutt.                         CLUSTERED BROOM-RAPE

A low pinkish-stemmed plant 3–10 cm high, parasitic on roots of *Artemisia* and other Compositae. Leaves scale-like. Flowers 1–10, purplish to yellow, about 25 mm long, each on a naked unbracted stalk. Common; wherever *Artemisia* is found; throughout Prairies. Syn.: *Anoplanthus fasciculatus* (Nutt.) Walp.

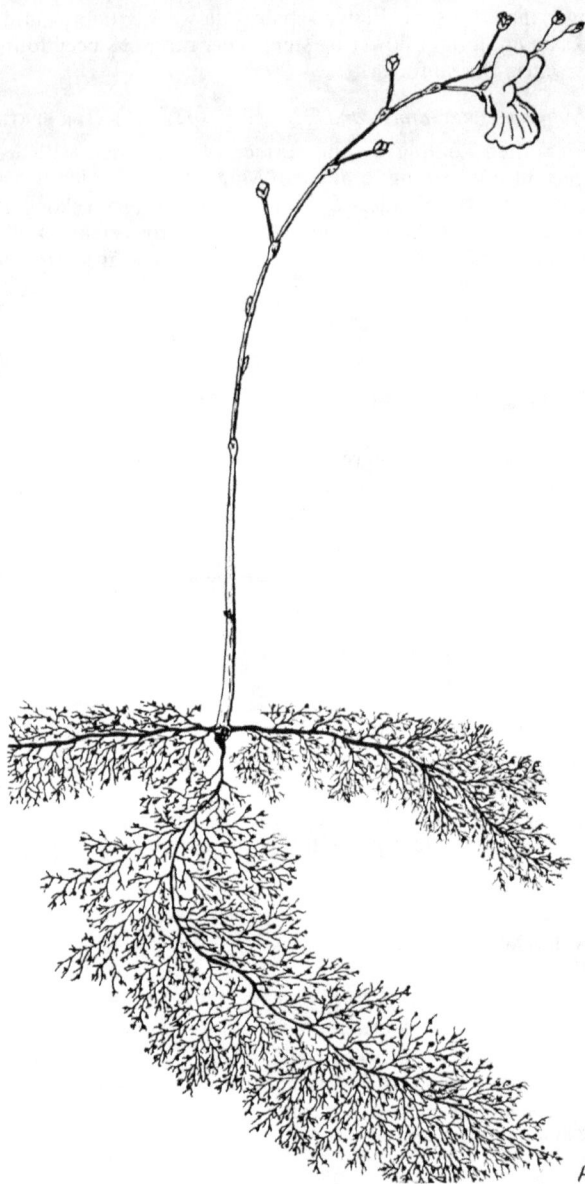

Fig. 192.   Greater bladderwort, *Utricularia vulgaris* L. var. *americana* Gray.

*Orobanche ludoviciana* Nutt.                                    LOUISIANA BROOM-RAPE

A stout, somewhat sticky, glandular hairy, pink-stemmed plant 5–25 cm high, parasitic on roots of *Artemisia* and other Compositae. Leaves scale-like. Flowers purplish, 1–2 cm long, in a dense terminal spike, with 1 or 2 long bracts beneath each flower. Not so common as *O. fasciculata*; found on sandy soil; throughout Prairies. Syn.: *Myzorrhiza ludoviciana* (Nutt.) Rydb.

*Orobanche uniflora* L.                                          SMALL BROOM-RAPE

A small thin-stemmed plant; stem almost entirely underground, only 1–3 cm showing, bearing a few overlapping, glabrous, oblong-obovate scale leaves. Pedicels 1–4, usually 2, erect, 6–20 cm long, finely glandular pubescent, each bearing a single flower, about 2 cm long, white to violet. Rare; margins of woods and wooded slopes; southern Rocky Mountains and Cypress Hills.

## PHRYMACEAE—lopseed family

***Phryma***        lopseed

*Phryma leptostachya* L.                                         LOPSEED

Perennial plants with stems 50–100 cm high, erect, simple or with a few branches. Petioles to 5 cm long in lower leaves, shorter in higher leaves, and absent or nearly absent in uppermost leaves; leaves ovate, 6–15 cm long, coarsely toothed. Inflorescence a spike-like raceme; flowers 6–8 mm long, pale purple to white, opposite, subtended by 3 small bracts; fruiting calyx to 1 cm long, reflexed-appressed. Rare; moist woods; southeastern Boreal forest.

## PLANTAGINACEAE—plantain family

***Plantago***        plantain

Plants with leaves usually having several prominent longitudinal ribs. Flowers borne in long narrow spikes, having a 4-lobed calyx, a 4-lobed corolla, and 2 or 4 stamens. Fruit a pyxis (capsule), with a conical top falling off at maturity and releasing the seeds.

1. Plants with stem leaves opposite or
   whorled. ......................................................................... *P. psyllium*
   Plants with all leaves basal. ........................................................... 2

2. Leaves coarsely toothed to pinnately
   lobed. ......................................................................... *P. coronopus*
   Leaves all entire, or distantly and shal-
   lowly toothed. ........................................................................... 3

3. Leaves lanceolate to ovate. ........................................................... 4
   Leaves linear to filiform (thread-like). ........................................... 8

4. Leaves lanceolate, gradually tapering to
   stalks. ........................................................................................ 5
   Leaves ovate, abruptly joining stalk. ................................................. 7

5. Spikes of inflorescence long and narrow;
   plants of saline areas. ................................................. *P. eriopoda*
   Spikes of inflorescence short, dense, and
   oblong; stamens protruding and very
   conspicuous. ................................................................................. 6

6. Stem and leaves densely pubescent;
   flower bracts smaller than buds, acute. ...................... *P. canescens* var. *cylindrica*
   Stems and leaves not densely pubescent;
   flower bracts larger than buds, blunt. ............................... *P. lanceolata*

7. Leaves glabrous to sparsely pubescent,
   long-petioled; peduncle slightly longer
   than    leaves;    filaments    white,
   inconspicuous. ..................................................................... *P. major*
   Leaves short-pubescent, short-petioled;
   peduncle many times longer than
   leaves; filaments lilac, conspicuous. ........................... *P. media*

8. Plants densely white silky-woolly. ........................... *P. patagonica*
   Plants not densely woolly. ................................................................. 9

9. Flower spike densely long-pubescent. ....................... *P. aristata*
   Flower spike not long-pubescent. ...................................................... 10

10. Leaves thick, fleshy; flowers pubescent. ..................... *P. maritima*
    Leaves thread-like, not fleshy; flowers
    glabrous. ............................................................................ *P. elongata*


*Plantago aristata* Michx.                                        BUCKHORN

An annual plant with scapes 15–25 cm high, more or less villous. Leaves
linear or very narrowly lanceolate, 10–15 cm long, more or less pubescent.
Spikes cylindric, 3–6 cm long; bracts linear, with the lowest ones to 2 cm long
and the upper ones shorter, hirsute to long-villous. Introduced; a rare weed in
disturbed soils; Prairies.

*Plantago canescens* Adams var. *cylindrica* (J. M. Macoun) Boiv.

WESTERN RIBGRASS

A perennial species with a woolly-pubescent scape about 15 cm high.
Leaves narrowly lanceolate, 10–15 cm long, woolly-pubescent, especially
below. Spike about 5 cm long, rather loose in fruit. Dry grasslands; southern
Rocky Mountains.

*Plantago coronopus* L.                                BUCKHORN PLANTAIN

A biennial or short-lived perennial; scapes 5–20 cm high. Leaves 3–15 cm
long, with several pairs of linear lobes 5–20 mm long, pubescent. Spike 3–8 cm
long, dense; flowers puberulent. Introduced; a rare weed; southeastern Parklands.

*Plantago elongata* Pursh                                   LINEAR-LEAVED PLANTAIN

An annual plant 6–15 cm high. Leaves very narrow, thread-like, 3–12 cm long, about 2 mm wide, single-ribbed. Inflorescence a loosely flowered very narrow spike 5–10 cm long. Not common; found occasionally in wet places and low flats; throughout Prairies.

*Plantago eriopoda* Torr.                                   SALINE PLANTAIN

A somewhat fleshy perennial species from a long coarse rootstock. Leaves oblanceolate, 5–20 cm long, with several longitudinal ribs borne on long stalks usually arising from a mass of long brown hairs at crown of root. Inflorescence a somewhat dense narrow spike 2–10 cm long, on a long stem. Common; on saline or alkaline soils, river flats, and slough margins; throughout Prairies and Parklands.

*Plantago lanceolata* L.                                    RIBGRASS

An introduced biennial or perennial species with a short rootstock. Leaves numerous, narrowly lanceolate, 3- to 5-ribbed, 5–30 cm long, tapering to very short stalks, usually some tufts of brownish hair at crown of root. Flowering stems much longer than leaves, 30–50 cm; inflorescence a short thick dense spike 2–8 cm long; stamens much protruding, often forming a conspicuous yellow or white ring around flower head. Found occasionally where imported lawn grass has been sown and as a weed in newly seeded lawns.

*Plantago major* L. (Fig. 193)                             COMMON PLANTAIN

A perennial species from a short thick rootstock. Leaves numerous, very dark green, oval or ovate, 3–25 cm long, with many conspicuous longitudinal ribs contracting abruptly at the base into a long stalk. Flowering stems fairly long; inflorescence a dense narrow spike 7–30 cm long. Common; a weed in lawns, waste places, and yards; throughout the Prairie Provinces. Possibly introduced, but, if so, having spread across the country.

The var. *asiatica* (L.) Dcne., with tapering leaf bases and more upright leaves, has been found in Boreal forest.

*Plantago maritima* L.                                     SEASIDE PLANTAIN

A short-lived perennial with scapes 5–20 cm high. Leaves thick and fleshy, almost 3-angled in cross section, entire or somewhat distantly toothed. Spikes 2–10 cm long, often somewhat interrupted below, with the axis pubescent; sepals and corolla tube pubescent. Shores and saline areas; Boreal forest, rare in southern edge of Boreal forest and Parklands.

*Plantago media* L.                                        HOARY PLANTAIN

Perennial plants. Leaves elliptic to obovate, 10–20 cm long, narrowed at the base to a petiole. Scapes 20–40 cm high, somewhat pubescent; spikes at first narrowly conic, elongating to short-cylindric, 3–10 cm long. Introduced; a rare weed of disturbed soils and waste areas.

Fig. 193.  Common plantain, *Plantago major* L.

*Plantago patagonica* Jacq. PURSH'S PLANTAIN

An annual plant, very pale green and whitish silky-woolly all over. Leaves 2–6 cm long, linear, with 1–3 nerves. Inflorescence a dense spike 2–12 cm long, on a stem 7–10 cm long, and very woolly. Found in sandy soil, on river flats, and dry soils; throughout southwestern part of the Prairie Provinces. An indicator of an overgrazed condition. Syn.: *P. purshii* R. & S.; *P. spinulosa* Dcne.

*Plantago psyllium* L. SAND PLANTAIN

Annual plants with erect stems, 10–50 cm high, pubescent throughout. Leaves linear, 3–8 cm long. Heads 10–15 mm long, on peduncles 3–5 cm long arising from leaf axils; bracts broadly scarious-margined. Introduced; a rare weed of disturbed soils.

# RUBIACEAE—madder family

Plants with opposite or whorled leaves, stipules, and perfect regular flowers. As many stamens as corolla lobes. Fruit a capsule, berry, or drupe.

1. Leaves opposite, in pairs; low plants. ........................................................... *Houstonia*
   Leaves apparently whorled; the stipules
   resembling the leaves; stems square. ........................................................... 2

2. Inflorescence of open cymes. ........................................................... *Galium*
   Inflorescence a head, subtended by
   bracts. ........................................................... *Asperula*

**Asperula**      woodruff

*Asperula arvensis* L. FIELD WOODRUFF

Annual plants, with stems 20–30 cm high, erect; the angles retrorsely scabrous. Leaves in whorls of 4–8, lanceolate or the lower ones obovate. Flowers subsessile in a terminal head, blue to purplish; bracts pubescent. Introduced; a very rare weed of disturbed areas.

**Galium**      bedstraw

1. Ovary and fruit bristly or hairy. ........................................................... 2
   Ovary and fruit glabrous. ........................................................... 4

2. Principal stem leaves in whorls of 4. ........................................................... *G. boreale*
   Principal stem leaves in whorls of 6 or 8. ........................................................... 3

3. Leaves mostly in whorls of 8; plants
   annual. ........................................................... *G. aparine*
   Leaves mostly in whorls of 6; plants
   perennial. ........................................................... *G. triflorum*

4. Flowers yellow; leaves strongly revolute,
   in whorls of 6–12. ........................................................... *G. verum*
   Flowers white or greenish; leaves in
   whorls of 2–6. ........................................................... 5

5. Flowers in many-flowered cymes, short-pedicellate. ......................................................................... *G. palustre*

Flowers solitary or in few-flowered cymes, long-pedicellate. ...................................................................... *G. trifidum*

### *Galium aparine* L.                                CLEAVERS

A trailing or decumbent annual with a square stem 30–100 cm long and covered with retrorse (backward-pointing) bristly hairs. Leaves in whorls of 6 or 8, oblong-linear to oblanceolate, sharp-tipped, rough bristly on margins and midrib, and 2–5 cm long. Long-stalked cream-colored flowers in axillary clusters of 4–9. Fruits in pairs, bearing hooked hairs. Found often; in moist woodlands and along riverbanks; a troublesome weed in Parklands. Syn.: *G. aparine* L. var. *vaillantii* (DC.) Koch; *G. vaillantii* DC.

### *Galium boreale* L. (Fig. 194)                     NORTHERN BEDSTRAW

An erect perennial 20–50 cm high, from thin brown rootstocks, with a slender square stem. Leaves in whorls of 4 around stem, linear to lanceolate, 3-ribbed, 2–6 cm long. Inflorescence a terminal leafy panicle, fairly dense, many-flowered; flowers white, about 3 mm across, with 4 corolla lobes, and faintly fragrant. Fruits in pairs, about 1.5 mm long, covered with short whitish hairs. Common; in openings in woodlands, along roadsides and moister places on prairies; almost throughout the Prairie Provinces. In some years, almost the dominant roadside flower. Most of the plants of this species found in the prairies are var. *intermedia* DC., with short curved hairs on fruit. Syn.: *G. septentrionale* R. & S.

### *Galium palustre* L.                             MARSH BEDSTRAW

A slender perennial with simple or branched stems 20–60 cm long, sparsely retrorse-scabrous on the angles. Leaves 5–15 mm long, linear to narrowly oblanceolate, commonly in whorls of 4 or 6. Inflorescences many-flowered, often forked; pedicels widely spreading; corolla white, about 4 mm wide. Rare; wet places, bogs, and marshes; Boreal forest.

### *Galium trifidum* L.                              SMALL BEDSTRAW

A slender-stemmed perennial with decumbent or erect weak stems 20–40 cm long. Leaves borne in whorls of 4, linear to spatulate, 6–12 mm long. Flowers terminal or axillary, on long stalks, greenish white, in 2's or 3's, very small, usually with 3 blunt corolla lobes. Fruits in pairs, smooth, without any hairiness. Found occasionally; in moist places; Parkland, Boreal forest, and Cypress Hills.

### *Galium triflorum* Michx.                     SWEET-SCENTED BEDSTRAW

A spreading, branched perennial 30–100 cm long, often trailing or decumbent, with an almost hairless stem. Leaves usually in whorls of 6, narrowly oval or oblanceolate, with a sharp, pointed tip, 2–6 cm long. Flowers long-stalked, greenish white, in clusters of 3, with 4 corolla lobes. Fruits in pairs, covered with long hooked hairs. Fairly common; in moist places and damp woodlands; throughout the Prairie Provinces.

Fig. 194.   Northern bedstraw, *Galium boreale* L.

*Galium verum* L.

A perennial with a woody rootstock; stems erect, 30–80 cm high, finely pubescent. Leaves linear, 1–3 cm long, mostly in whorls of 8, often deflexed at maturity. Inflorescences several from the upper axils, many-flowered. Introduced; a very rare weed; mostly along roadsides; southeastern Parkland.

***Houstonia***     bluets

*Houstonia longifolia* Gaertn.     LONG-LEAVED BLUETS

Low, tufted perennials 6–25 cm high, usually with a purplish square stem. Leaves opposite, linear to linear-oblong, 10–25 mm long, with small whitish or purplish stipules at base. Flowers pinkish or purple, 3–5 mm long, in very leafy terminal clusters. Fruit an ovoid capsule. Found occasionally; on grasslands and sandy soils; Parklands and Boreal forest.

# CAPRIFOLIACEAE—honeysuckle family

Trees, shrubs, vines, or perennial herbs with opposite leaves without stipules. Flowers perfect; corolla varying from rotate or campanulate to urn-shaped; usually having 5 stamens, except for one genus, *Linnaea*, with only 4. Fruit a berry, drupe, or capsule.

1. Leaves pinnately compound. ......................................................................... *Sambucus*
   Leaves simple, at most deeply lobed. ....................................................... 2

2. Low, trailing evergreen shrub; flowers
   with 4 stamens. ...................................................................................... *Linnaea*
   Shrubs or small trees; flowers with 5
   stamens. ................................................................................................. 3

3. Corolla rotate; inflorescence all terminal
   in compound cymes; fruit a drupe con-
   taining a flattened stone. ..................................................................... *Viburnum*
   Corolla tubular or bell-shaped;
   inflorescence axillary as well as terminal; fruit a capsule or berry with 2 or
   more seeds. .......................................................................................... 4

4. Leaves with finely toothed margins; calyx
   with 5 linear lobes; fruit a capsule. .................................................... *Diervilla*
   Leaves with entire or wavy margins; calyx
   with short lobes; fruit a berry. ........................................................... 5

5. Flowers regular, bell-shaped, in clusters;
   fruit a white berry. ............................................................................. *Symphoricarpos*
   Flowers often slightly irregular, tubular,
   usually in pairs; fruit a red or black
   berry. ................................................................................................... *Lonicera*

***Diervilla***   bush-honeysuckle

*Diervilla lonicera* Mill.   BUSH-HONEYSUCKLE

A shrub 50–100 cm high, with opposite simple leaves. Leaves short-stalked, ovate to oval, finely toothed on margins, 5–12 cm long. Flowers narrow, funnel-shaped, about 2 cm long, yellow, in axillary and terminal clusters of 1–5. Fruit a slender capsule with linear stamens remaining on the end. Not common; found in rocky woodlands; along the eastern Boreal forest.

***Linnaea***   twinflower

*Linnaea borealis* L. var. *americana* (Forbes) Rehder (Fig. 195A)   TWINFLOWER

A very low, creeping or trailing evergreen plant 15–75 cm long. Leaves opposite, short-stalked, oval or orbicular, 10–15 mm across, with a somewhat wavy margin. Flowers pink, fragrant, funnelform, about 1 cm long, with only 4 stamens, in pairs, pendent or hanging downward from the top of a stem 3–10 cm high. Common; in cool woodlands and often found on decayed, fallen tree trunks; throughout the entire wooded parts of the Prairie Provinces. Syn.: *L. americana* Forbes.

***Lonicera***   honeysuckle

1. Somewhat twining shrub; inflorescence a
   terminal cluster. ........................................................................ 2
   Erect shrubs; leaves not connate-perfoli-
   ate; inflorescence axillary and flowers
   in pairs. ................................................................................... 3

2. Uppermost connate leaves glaucous,
   obtuse or acute. ......................................... *L. dioica* var. *glaucescens*
   Uppermost connate leaves green, pointed
   at tip. ............................................................ *L. hirsuta* var. *schindleri*

3. Bracts below flowers large, broad, and
   leaf-like; berries of the 2 flowers
   separate. ...................................................................... *L. involucrata*
   Bracts below flowers thin or minute; ber-
   ries of the 2 flowers more or less united. ................................ 4

4. Ovaries completely fused, ripening into a
   single bluish black berry. ......................................... *L. caerulea* var. *villosa*
   Ovaries only partly united, with berries
   distinguishable. ....................................................................... 5

5. Pith of branches brown, not filling the
   entire core. ................................................................. *L. tatarica*
   Pith of branches white, filling the entire
   core. ......................................................................................... 6

6. Leaves densely puberulent, at least below. ..................... *L. oblongifolia*
   Leaves subglabrous to lightly pilose
   below. ............................................................................ *L. utahensis*

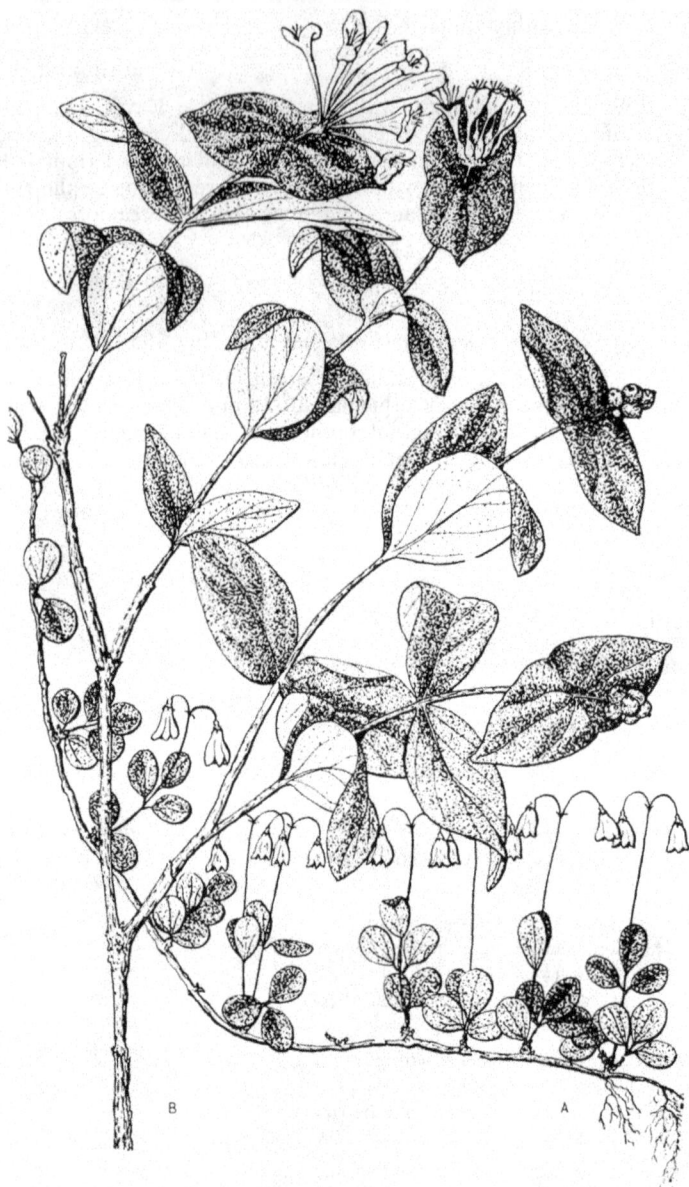

Fig. 195. *A*, Twinflower, *Linnaea borealis* L. var. *americana* (Forbes) Rehder
*B*, twining honeysuckle, *Lonicera dioica* L. var. *glaucescens* (Rydb.) Butt.

*Dicots*

*Lonicera caerulea* L. var. *villosa* (Michx.) T. & G.    BLUE FLY HONEYSUCKLE

Low erect shrubs 30–100 cm high; twigs sometimes finely hairy. Leaves oval to obovate, slightly downy when young, very short-stalked, 2–4 cm long. Flowers in pairs in axils of leaves, yellow, about 12 mm long. Fruit an oblong or globose bluish black berry, covered with a bloom, edible. Fairly common in moist or swampy places in woodlands; Boreal forest and Rocky Mountains. Syn.: *L. villosa* (Michx.) R. & S. var. *solonis* (Eat.) Fern.

*Lonicera dioica* L. var. *glaucescens* (Rydb.) Butt. (Fig. 195*B*)

TWINING HONEYSUCKLE

A more or less twining shrub with light-colored shreddy bark. Leaves obovate or oval, all opposite, often connate (joined at bases) and perfoliate, especially the upper ones, 5–8 cm long, pale, and often hairy below, particularly on veins. Flowers in a terminal cluster, yellow, later turning reddish, 20–25 mm long. Fruits in clusters of red berries. Fairly common; in woodlands; throughout the Prairie Provinces. Syn.: *L. glaucescens* Rydb.

*Lonicera hirsuta* Eat. var. *schindleri* Boiv.    HAIRY HONEYSUCKLE

Similar to *L. dioica*, but with leaves dull green, and, except for the upper ones, short-petioled; young stems glandular pubescent. Rare; moist woods; southeastern Boreal forest.

*Lonicera involucrata* (Richards.) Banks    INVOLUCRATE HONEYSUCKLE

A shrub 1–3 m high, sometimes with downy stems. Leaves oblong to oval, 5–10 cm long, somewhat glandular-dotted and downy below, with a prominent midrib. Flowers in pairs, yellow, with large leaf-like bracts at bases. Fruits 2 large berries, very dark purple to black, about 8 mm across, surrounded by persistent bracts. Fairly common; in woodlands; Boreal forest and Rocky Mountains. Syn.: *Distegia involucrata* (Richards.) Cockerell.

*Lonicera oblongifolia* (Goldie) Hook.    SWAMP FLY HONEYSUCKLE

A shrub 50–150 cm high, with ascending branches. Leaves oblong, downy below when young, 2–5 cm long. Flowers yellow with a purplish tinge inside, 10–15 mm long, 2-lipped, in pairs on a long stem. Fruits in pairs of more or less united purplish red berries. Found occasionally; in swamps and marshy lands; eastern Boreal forest. Syn.: *Xylosteon oblongifolium* Goldie.

*Lonicera tatarica* L.    TARTARIAN HONEYSUCKLE

A shrub 1–3 m high, with thin, ovate, somewhat cordate-based leaves 2–8 cm long. Flowers pink or white, about 15 mm long, in pairs, very numerous. Fruit an orange or yellow berry about 6 mm in diam. An introduced species, commonly used for ornamental planting and hedges. Has become established in woodlands and moist spots; in several locations in the Prairie Provinces. Syn.: *Xylosteon tataricum* (L.) Medic.

*Lonicera utahensis* Wats.    UTAH HONEYSUCKLE

A low shrub 60–150 cm high; leaves oblong to oblong-ovate, 2–5 cm long, entire, thin glabrous or sparsely pubescent below, obtuse to blunt at the base. Flowers in pairs; peduncle slender, 10–15 mm long; bracts 2, very narrow, 1–3 mm long; corolla 15–20 mm long. Berries red. Moist, coniferous woods; southern Rocky Mountains.

## *Sambucus*     elder

*Sambucus racemosa* L.     RED ELDERBERRY

Shrubs 1–6 m high, with 3–7 leaflets and yellowish brown pith. Inflorescence ovate in outline, 5–8 cm long, 3–5 cm wide; flowers somewhat greenish white to yellowish.

In var. *pubens* (Michx.) Wats., twigs and leaflets pubescent, inflorescence rather oblong-ovate, berries small, and stems rarely more than 3 m high. Moist open woods; eastern Boreal forest, rarely in eastern Parklands.

In var. *arborescens* (T. & G.) Gray, leaves and twigs subglabrous, inflorescence more compact and broadly ovate, berries large and deep red, and stems to 6 m high. Moist woods; southern Rocky Mountains.

## *Symphoricarpos*     snowberry

Low shrubs with strongly creeping roots. Leaves opposite, simple, short-stalked. Flowers perfect, small, pink and white, campanulate (bell-shaped). Fruit a small, round, 2-seeded, white berry.

Stamens and styles not protruding from
  corolla. ................................................................................................................ *S. albus*
Stamens and styles protruding from corolla. ................................................. *S. occidentalis*

*Symphoricarpos albus* (L.) Blake     SNOWBERRY

An erect shrub 50–150 cm high. Leaves thin, oval 1–5 cm long, sometimes slightly toothed. Flowers in several-flowered clusters. Berry waxy, white, 6–10 mm across. Found occasionally; eastern Parklands and Boreal forest. The var. *pauciflorus* (Robbins) Blake, few-flowered snowberry, a low-growing spreading shrub 10–50 cm high. Leaves oval or round, 1–3 cm long, softly hairy beneath. Flowers borne singly in upper leaf axils or in terminal clusters of 2 or 3, about 6 mm long. Berries white, waxy, about 6 mm in diam. This variety is more common than the species; in rocky and sandy woodlands; throughout bush areas of the Prairie Provinces and in Cypress Hills. Syn.: *S. pauciflorus* (Robbins) Britt.

*Symphoricarpos occidentalis* Hook.     WESTERN SNOWBERRY. BUCKBRUSH

A shrub 50–100 cm high, growing from creeping roots. Leaves oval, ovate, or almost round, somewhat softly hairy beneath, 2–6 cm long. Flowers pink and white, with styles and stamens conspicuously projecting from corolla, in rather dense terminal and axillary spikes. Fruits snow white waxy berries, often in large numbers. One of the most widespread and commonest shrubs; in dense stands on open prairie, in ravines, coulees, and woodlands; throughout the Prairie Provinces.

*Viburnum*        bush-cranberry

Shrubs or small trees with opposite leaves and flowers all in terminal clusters. Corolla rotate or wheel-shaped to short bell-shaped, regular, 5-lobed; flowers perfect, with 5 stamens and a short 3-cleft style. Fruits single-seeded edible drupes; stone somewhat compressed or flattened.

1. Leaves usually 3-lobed, with veins radiating from base. ........................................................................... 2
   Leaves not lobed, with veins pinnate from midrib. ............................................................................. 3

2. Outer flowers of cluster large and neutral (sexless), with large lobes. ............................. *V. opulus* var. *americanum*
   None of flowers large and neutral. ................................................. *V. edule*

3. Leaves densely hairy beneath, with stalks not more than 6 mm long. .......................................... *V. rafinesquianum*
   Leaves not hairy beneath, with slender stalks more than 6 mm long. ................................................... *V. lentago*

### *Viburnum edule* (Michx.) Raf. (Fig. 196)            LOW BUSH-CRANBERRY

A rather straggly shrub 50–200 cm high. Leaves usually shallowly 3-lobed, 4–10 cm across; leaf bases usually flat or cordate. Inflorescence few-flowered, 1–3 cm across; flowers all perfect. Fruit a red drupe about 3 mm long. Fairly common; in rich moist woodlands; throughout the heavily wooded areas of the Prairie Provinces. Syn.: *V. eradiatum* (Oakes) House.

### *Viburnum lentago* L.            NANNYBERRY

A tall shrub or small tree up to 6 m high. Leaves 4–10 cm long, ovate, with sharp, small, marginal teeth tapering to a point at apex. Flowers in large clusters 5–12 cm across. Fruit an edible bluish black berry 8–12 mm long, covered with a bloom. Common; in woodlands; southeastern Boreal forest.

### *Viburnum opulus* L. var. *americanum* (Mill.) Ait.            HIGH BUSH-CRANBERRY

A shrub 1–4 m high, with smooth branches. Leaves palmately veined, broad, 3-lobed, 5–10 cm across. Outer flowers in clusters 10–15 mm across, with 5 large petals, but neutral (without fully formed styles or stamens); inner flowers smaller, creamy white, and perfect. Fruit a red berry, about 1 cm in diam, very acid. Fairly common; in woodlands throughout heavily wooded areas. Syn.: *V. trilobum* Marsh.

### *Viburnum rafinesquianum* Schultes            DOWNY ARROWWOOD

A shrub 50–150 cm high, with many slender gray branches. Leaves coarsely toothed, oval or ovate, slightly cordate at base, 2–7 cm long; undersurface densely velvet-hairy. Inflorescence rounded, 2–7 cm across; flowers all perfect, white. Fruits in large clusters, almost black, about 6 mm long; stone grooved slightly on both sides. Fairly common; in woodlands; eastern Boreal forest. Syn.: *V. affine* Bush var. *hypomalacum* Blake.

Fig. 196.   Low bush-cranberry, *Viburnum edule* (Michx.) Raf.

*Dicots*

# ADOXACEAE—moschatel family

***Adoxa***     moschatel

*Adoxa moschatellina* L.                                        MOSCHATEL

A dwarf perennial plant from a scaly rootstock, 10–20 cm high, with several long-stalked basal leaves divided into 3 long-stalked, somewhat 3-cleft leaflets. Flowers very small, in a cluster about 6 mm across, greenish or yellowish, at head of slender stem bearing 2 leaves. Found occasionally; in cool mossy woodlands; along margins of Boreal forest.

# CUCURBITACEAE—gourd family

***Echinocystis***     balsam-apple

*Echinocystis lobata* (Michx.) T. & G.                          WILD CUCUMBER

An annual, succulent, trailing vine 3–6 m long, climbing by means of tendrils over bushes and shrubbery. Stem somewhat angled, bearing long, spirally twisted tendrils. Leaves thin, pale green, rough on both sides, palmately veined, with 3–7 large lobes 5–12 cm across. Flowers greenish white, unisexual; male flowers in panicles or racemes; female flowers 1 or 2, in axils of leaves. Fruit a pepo (a large, ovoid, fleshy berry with a thick skin), pale green, covered with weak spines. Fruits 4–10 cm long, containing several large, flat, roughened, almost black seeds. Fairly common; in moist locations among bushes; southeastern Parklands and Boreal forest. Syn.: *Micrampelis lobata* (Michx.) Greene.

# CAMPANULACEAE—bluebell family

***Campanula***     bellflower

Perennial plants with rootstocks and alternate simple leaves. Flowers perfect, usually blue, campanulate or funnelform, in panicles; stamens 5; stigmas 3–5. Fruit an ovoid capsule containing many seeds.

1. Stem simple, bearing a solitary terminal
   flower. ................................................................................................ 2
   Stem usually bearing many flowers. ................................................. 3
2. Leaves sharply dentate; corolla 15–30
   mm long. ............................................................... *C. lasiocarpa*
   Leaves entire or glandular denticulate;
   corolla 10 mm long. ............................................................... *C. uniflora*
3. Flowers in part crowded into a head. ....................................... *C. glomerata*
   Flowers not crowded. ................................................................................ 4
4. Stem leaves broad, ovate to ovate-
   lanceolate. ............................................................... *C. rapunculoides*
   Stem leaves narrow, linear to filiform. ........................................................ 5

5.  Basal leaves ovate or cordate, soon with-
    ering; flowers more than 12 mm long. ............................................. *C. rotundifolia*
    Basal leaves linear or narrowly lanceo-
    late; flowers not over 12 mm long. .................................................. *C. aparinoides*


## Campanula aparinoides Pursh

MARSH BELLFLOWER

A very thin slender-stemmed plant 20–50 cm high. Stem 3-angled, somewhat roughened, with downward-pointing stiff short hairs. Leaves lanceolate
to linear-lanceolate, 2–5 cm long. Flowers at ends of branches, usually white
or with a faint bluish tinge, 10–15 mm long. Found often; in wet meadows and
marshes; eastern Parklands and Boreal forest. Includes *C. uliginosa* Rydb.


## Campanula glomerata L.

CLUSTERED BLUEBELL

An erect perennial with a short rootstock and glabrous to softly pubescent
stems 30–70 cm high. Leaves 2–5 cm long, oblong-lanceolate; lower leaves
petiolate; upper ones sessile to more or less clasping. Flowers 2–3 cm long, violet blue, sessile in a terminal head subtended by leafy bracts. Introduced ornamental; locally escaped and established.


## Campanula lasiocarpa Cham.

ALASKA BLUEBELL

Small perennial plants 4–15 cm high, from a creeping rootstock. Leaves
2–7 cm long, spatulate to narrowly oblanceolate; lower leaves petioled; upper
ones sessile, with margin laciniate-denticulate. Flowers solitary and terminal;
calyx long villous; lobes lanciniate. Mountain meadows; southern Rocky
Mountains.


## Campanula rapunculoides L.

CREEPING BLUEBELL

Erect perennial plants with a creeping rootstock; stems 40–100 cm high,
usually unbranched. Leaves coarse, serrate, usually sparsely pubescent below.
lower leaves petioled; upper ones sessile or subsessile, 3–7 cm long.
Inflorescence one-sided; flowers 2–3 cm long, blue. Introduced ornamental;
locally escaped and established.


## Campanula rotundifolia L. (Fig. 197)

HAREBELL

A perennial growing from rootstocks, to 10–50 cm high, often with many
stems. Basal leaves soon disappearing, ovate or deeply cordate-based, 10–25
mm long, on long stalks; stem leaves linear or linear-oblong, 1–5 cm long.
Flowers blue, campanulate, 15–25 mm long, occasionally solitary, but more
often in a raceme of 3 or 4 flowers. Common; throughout the Prairie Provinces.


## Campanula uniflora L.

ALPINE BLUEBELL

A dwarf perennial 10–20 cm high, with a single flower terminating the
stem. Leaves and calyx lobes entire or finely glandular denticulate. Arctic and
alpine meadows; northeastern Boreal forest and Rocky Mountains.

A.C. Budd

Fig. 197. Harebell, *Campanula rotundifolia* L.

# LOBELIACEAE—lobelia family

Annual or perennial herbs often with a bitter, milky sap. Leaves alternate or basal. Flowers perfect, irregular, with a 2-lipped corolla split to the base on one side in one genus; stamens 5. Fruit a many-seeded capsule.

Annual plants; corolla not split; fruit 1-celled. ................................................... *Downingia*
Perennial plants; corolla split; fruit 2-celled. .......................................................... *Lobelia*

## *Downingia*  downingia

### *Downingia laeta* Greene  DOWNINGIA

Low-growing annual herbs 5–15 cm high. Leaves alternate, oblong, to 15 mm long. Flowers in axils of upper leaves or bracts, about 1 cm long, light blue with darker blue veins and with a whitish throat spotted with yellow. Very scarce; has been found in moist and saline places; southwestern Prairies. Syn.: *Bolelia laeta* Greene.

## *Lobelia*  lobelia

1. Plants aquatic, with leaves all submerged. ............................................. *L. dortmanna*
   Plants terrestrial. ............................................................................................... 2
2. Stem leaves linear; plants of moist mar-
   shy places; racemes few-flowered. ............................................. *L. kalmii*
   Stem leaves lanceolate to ovate; plants of
   drier soils; racemes many-flowered. .................................... *L. spicata* var. *hirtella*

### *Lobelia dortmanna* L.  WATER LOBELIA

A perennial plant with a submerged rosette of hollow basal leaves 2–9 cm long and fleshy, linear. Stem 40–70 cm high, hollow, with minute filiform stem leaves. Raceme emersed, few-flowered; flowers usually about 15 mm long, pale blue to white. Shallow lakes; Boreal forest.

### *Lobelia kalmii* L.  KALM'S LOBELIA

A leafy-stemmed biennial or perennial branching plant 10–30 cm high. Lower leaves spatulate, 10–25 mm long; upper leaves linear. Flowers light blue, about 1 cm long, in loose racemes; each flower stalk usually bearing 2 small bracts or 2 glands. Fairly common; in bogs and wet meadows; throughout Boreal forest, rare in Parklands. Syn.: *L. strictiflora* (Rydb.) Lunell.

### *Lobelia spicata* Lam. var. *hirtella* A. Gray  SPIKED LOBELIA

A perennial with a simple erect stem 30–100 cm high. Lower leaves spatulate, 2–10 cm long; stem leaves lanceolate to spatulate. Flowers on a long, spike-like terminal inflorescence, pale blue, about 1 cm long. In the type, sepals and bracts slightly or not at all hairy. Found occasionally; in dry sandy soil; throughout eastern Parklands and Boreal forest.

# VALERIANACEAE—valerian family

***Valeriana***    valerian

Perennial herbs with scented roots. Leaves opposite, without stipules, varying; lower leaves often entire and spatulate; upper ones pinnately lobed with a large terminal lobe. Flowers usually rose or white, 5-lobed, small, in many-flowered clusters, lengthening and becoming paniculate as the plant matures. Seed an achene.

Stem leaves with 9–15 segments; basal leaves
   oblong to spatulate, petioled. ........................................................ *V. dioica* var. *sylvatica*
Stem leaves with 3–7 segments; basal leaves, if
   present, pinnate. ............................................................................................ *V. sitchensis*

*Valeriana dioica* L. var. *sylvatica* (Sol.) Gray (Fig. 198)    NORTHERN VALERIAN

An erect rather weak-stemmed plant 20–60 cm high. Basal leaves entire, usually spatulate; stem leaves pinnate, with a large terminal lobe pinnately veined. Flowers white, about 3 mm across, in dense terminal clusters, later lengthening into a short cymose panicle. Fairly common; in wet places; throughout Boreal forest. Syn.: *V. septentrionalis* Rydb.

*Valeriana sitchensis* Bong.    MOUNTAIN HELIOTROPE

A robust species with a stout rootstock; stems 40–100 cm high, glabrous or nearly so. Stem leaves 10–15 cm long, 3- to 5-pinnate; leaflets 3–5 cm long, with those of the lower leaves ovate and those of the upper ones lanceolate and coarsely sinuate-dentate; basal leaves similar to stem leaves or simple, usually absent at flowering time, sparingly pilose to glabrous. Inflorescence in dense terminal clusters; corolla 6–8 mm long, white or pinkish. In var. *scouleri* (Rydb.) M. E. Jones, basal leaves larger than stem leaves, leaflets less deeply toothed, and corolla 5–6 mm long. Both varieties in mountain meadows and moist woods; southern Rocky Mountains.

# DIPSACACEAE—teasel family

***Knautia***    bluebuttons

*Knautia arvensis* (L.) Duby    FIELD SCABIOUS

An erect hairy perennial 30–100 cm high. Lower leaves stalked, oblong-lanceolate, 9–25 cm long; upper stem leaves opposite, pinnatifid or lobed, without stalks. Flowers pale lilac or blue, in heads 2–4 cm broad, with involucral bracts resembling heads of Compositae. Introduced from Europe; found occasionally; western Alberta and near Winnipeg. Syn.: *Scabiosa arvensis* L.

(Fig. 198 overleaf)

Fig. 198.   Northern valerian, *Valeriana dioica* L. var. *sylvatica* (Sol.) Gray.

# COMPOSITAE—composite family

Compositae is a very large and diverse family (Fig. 199). The flower head is composed of many florets or small flowers borne on a common receptacle, which is surrounded by an involucre composed of one or more rows of bracts. There are 2 main types of floret: tubular (or disk) and ray (or ligulate). Tubular florets are regular and tube-shaped; a flower head entirely composed of this type of floret is called discoid. Ray florets are irregular in shape and have a single strap-like petal; a head entirely composed of this type of floret is called ligulate. When both forms of floret are present, the flower head is called radiate. In such cases, the tubular florets occupy the center (or disk) and the ray florets form the margins. In ligulate and discoid flowers the florets are all perfect (having both pistil and stamens). In radiate flowers the tubular florets are usually perfect and the ray florets are either female (pistil only) or neutral (without either pistil or stamens). Often there are bracts or scales, which are called the chaff, among the florets on the top of the receptacle. If there are no bracts, the receptacle is said to be naked. There are 5 stamens, usually with their anthers united to form a ring around the pistil. In some genera the flowers are unisexual, the male and female florets being borne in separate flowers. The fruit is an achene, sometimes enclosed in a bur-like closed involucre, sometimes provided with a pappus (or tuft of hairs) to aid in dissemination, and sometimes having awns or scales. In one group all the species contain latex, a sticky, often milky sap. In this publication, for convenience in making identifications, this large family has been divided into three groups.

1. Stamens not united to form a tube around
   the pistil. ..................................................................................... Group 1, p. 683
   Stamens united to form a tube around the
   pistil. ................................................................................................................. 2
2. Some or all of the florets tubular. ........................................ Group 3, p. 703
   None of the florets tubular, heads ligu-
   late; plants with a milky or sticky sap. ......................................... Group 2, p. 689

## GROUP 1

The plants in this group are annual or perennial herbs, usually with alternate leaves. There are no perfect florets, and the flowers are inconspicuous. In one genus, both male and female florets are borne on the same head, but in the others they are borne on separate heads in different parts of the plant. Heads composed entirely of male florets and those with both male and female florets hang downward; this trait is helpful in recognizing this group. When the male and female florets are borne in separate heads, the fruit is often enclosed in a bur-like or nut-like involucre. Pollen from flowers of this group is often responsible for causing hay fever.

1. Both male and female florets on one
   head; achenes not contained in bur-
   like involucre. .............................................................................................. *Iva*
   Male and female florets on separate
   heads; involucre of female flowers bur-
   like or nut-like. ............................................................................................... 2

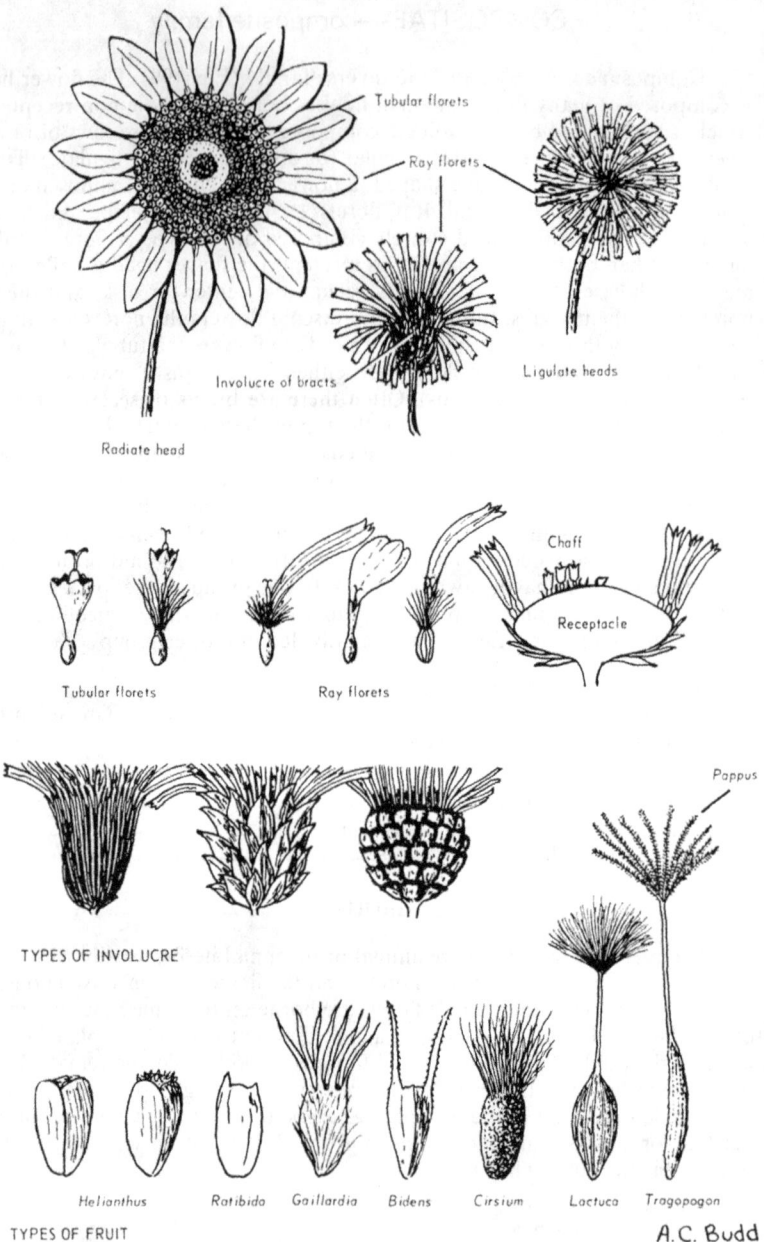

Fig. 199.   Characteristics of composite flowers and fruit.

2. Leaves not lobed or dissected; female
   heads with 2 florets; involucre forming
   an oblong bur with hooked prickles. ..................................................... *Xanthium*

   Leaves lobed or divided; female heads
   usually with one floret; prickles on
   involucre not hooked. ................................................................................. 3

3. Involucre of female heads with a single
   series of prickles or tubercles, and with
   spines short. ............................................................................................... *Ambrosia*

   Involucre of female heads with several
   rows of spines, and with spines long
   and prominent. ........................................................................................... *Franseria*

***Ambrosia***       ragweed

1. Leaves all opposite, some entire, some
   with 3–5 lobes. ......................................................................................... *A. trifida*

   Leaves opposite or alternate, all once- or
   twice-divided. ............................................................................................ 2

2. Annuals; leaves with stalks; fruit with
   sharp spines or tubercles. ..................................... *A. artemisiifolia* var. *elatior*

   Perennials with running rootstocks;
   leaves without stalks; fruit without
   sharp spines. ....................................... *A. psilostachya* var. *coronopifolia*

*Ambrosia artemisiifolia* L. var. *elatior* (L.) Descourtils       COMMON RAGWEED

   An erect annual 30–90 cm high. Leaves much-divided, stalked, somewhat
grayish hairy below. Many male heads borne in long narrow racemes; fewer
female heads borne in small clusters. Common; a weed of roadsides and waste
places; in eastern Parklands, but scarcer farther west. Syn.: *A. elatior* L.

*Ambrosia psilostachya* DC. var. *coronopifolia* (T. & G.) Farwell
                                                                      PERENNIAL RAGWEED

   A grayish hairy erect perennial 30–90 cm high, from running rootstocks.
Leaves once- or twice-divided, without stalks. Inflorescence similar to *A.
artemisiifolia*. Fruit almost devoid of spines. Fairly common; along roadsides
and waste places; eastern Parklands, but scarcer farther west. Syn.: *A.
coronopifolia* T. & G.

*Ambrosia trifida* L.                                                 GREAT RAGWEED

   A rather stout-stemmed erect annual 60–150 cm high, with a rough hairy
stem. Leaves all opposite, 5–25 cm across, stalked, mostly 3- to 5-lobed, with 3
main palmate veins. Male flowers in long terminal racemes; female flowers in
clusters in axils of upper bract-like leaves. Common; along roadsides and in
waste places and fields; eastern Parklands, but scarce elsewhere.

***Franseria***       bur-ragweed

*Franseria acanthicarpa* (Hook.) Coville                             BUR-RAGWEED

   An annual, usually rather decumbent and spreading plant 15–60 cm high,
much-branched. Leaves doubly divided, 5–10 cm long. Male flowers in long

terminal racemes; female flowers clustered in leaf axils; involucre with long straight spines making the plant very prickly. May be plentiful locally; only found in sand dunes; Prairies.

## *Iva*      marsh-elder

Low-growing perennials with creeping woody
  roots; flowers solitary on stalks in leaf axils. .................................................. *I. axillaris*
Tall-growing coarse annuals with fibrous
  roots; flowers in terminal and axillary spike-
  like panicles. ..................................................................................... *I. xanthifolia*

### *Iva axillaris* Pursh                                        POVERTYWEED

An erect herbaceous perennial 10–50 cm high, with woody running roots. Leaves stalkless, entire, 3-nerved, obovate to linear-oblong, 1–4 cm long; lower leaves opposite; upper ones alternate. Flowers small, yellow, reflexed in leaf axils. A serious weed in many farming areas where soil conditions favor its growth. Very plentiful locally; in heavy somewhat saline soils; Prairies.

### *Iva xanthifolia* Nutt. (Fig. 200)                          FALSE RAGWEED

A tall, erect, branching annual with a rough downy stem 0.5–2 m high. Leaves broadly ovate, rough above and downy below, long-stalked, 5–15 cm long. Heads small, inconspicuous, crowded on terminal and axillary panicles. Plant looks like a sunflower except for its inflorescence. Very common; in vacant lots and waste places in towns and settlements; throughout the Prairie Provinces, a common weed along roadsides and fields in Prairies and Parklands. Syn.: *Cyclachaena xanthifolia* (Nutt.) Fresn.

## *Xanthium*      cocklebur

Annual herbs with alternate leaves. Male flowers in terminal spikes or racemes; female flowers in axils of leaves. Fruit a bur with 2 stout beaks at one end, covered with hooked prickles, and containing 2 long flat seeds; one seed germinating rapidly, but the other having a long-delayed germinating period.

### *Xanthium strumarium* L. (Fig. 201)                          COCKLEBUR

An annual herb, rather decumbent, coarse-stemmed, much-branched, 15–60 cm high. Leaves roughly ovate, with wavy or slightly lobed margins, 2–8 cm long. Male flowers clustered at ends of branches; female flowers in clusters below; involucre of female flower closed, ending in 2 beaks; involucre forming a bur over fruit. Fruit with 2 terminal beaks, and covered with stout hairy hooked prickles. Common; around slough margins and low places, especially somewhat saline areas; throughout entire southern portion of the Prairie Provinces. Cocklebur seedlings considered **poisonous** to swine. Authorities disagree on the identification of species of *Xanthium*: *X. echinatum* Murr., *X. commune* Britt., *X. italicum* Moretti, *X. macounii* Britt., and *X. glanduliferum* Greene have been reported for Western Canada, but the identification is doubtful.

A.C. Budd

Fig. 200.   False ragweed, *Iva xanthifolia* Nutt.

Fig. 201.    Cocklebur, *Xanthium strumarium* L.

## GROUP 2

The plants in this group have a ligulate (strap-like) irregular corolla in each of the florets. All the florets are perfect (with both pistils and stamens), and the anthers of the stamens are united in a ring around the pistil. The leaves are either basal or alternate, and all the plants contain a milky sap. The primary leaves, those appearing after the cotyledons or seed leaves of seedlings, are a single leaf instead of a pair of equal leaves.

1. Plants with leaves all basal or with a few
   reduced leaves on the stem. ................................................................... 2
   Plants with normal stem leaves. ........................................................... 7

2. Leaves tomentose ciliate; achene about 8
   mm long. ............................................................................ *Microseris*
   Leaves glabrous, or if pubescent, not
   tomentose. ................................................................................... 3

3. Flower heads solitary on a long stalk. ................................................. 4
   Heads several on a flowering stalk. ..................................................... 5

4. Bracts of the involucre imbricated (over-
   lapping like shingles). ..................................................... *Agoseris*
   Main bracts of the involucre equal, with
   shorter spreading bracts at their base. ........................... *Taraxacum*

5. Pappus hairs plumose (with fine hairs on
   either side like a feather); bracts of
   involucre few. ................................................................. *Hypochoeris*
   Pappus hairs simple, not plumose. ..................................................... 6

6. Bracts in a single series, all equal with
   their midribs thickened. ............................................................ *Crepis*
   Bracts in series of 1–3, with their midribs
   not thickened. ................................................................... *Hieracium*

7. Pappus not hairy, merely covered with
   scales. ........................................................................................ 8
   Pappus hairy. ................................................................................ 9

8. Flowers blue. ......................................................................... *Cichorium*
   Flowers yellow. ......................................................................... *Lapsana*

9. Pappus bristles plumose (branched). ............................................... 10
   Pappus bristles simple. ................................................................... 13

10. All leaves long, narrow, grass-like; heads
    solitary on the stem. ...................................................... *Tragopogon*
    Not all leaves grass-like; more than one
    head on the stem. ........................................................................ 11

11. Ligules pink. ...................................................... *Stephanomeria*
    Ligules yellow to red. ................................................................... 12

12. Stem leaves few, small, mostly basal. .......................... *Hypochoeris*
    Stem leaves many, reaching to the
    inflorescence. ................................................................... *Picris*

13. Pappus double, the outer ring consisting
    of 5 small scale-like appendages, the
    inner ring bristles. .......................................................................... *Krigia*

    Pappus simple, consisting of bristles only. ........................................ 14

14. Leaves narrowly linear or scale-like;
    flowers pink. ................................................................................ *Lygodesmia*

    Leaves broader and leaf-like; flowers yel-
    low, white, or purplish. ..................................................................... 15

15. Seeds flattened; pappus usually white. ........................................... 16

    Seeds not flattened; pappus tawny or
    white. ...................................................................................................... 17

16. Seeds with a long beak at the apex. ............................................ *Lactuca*

    Seeds without a long beak at the apex. .......................................... *Sonchus*

17. Stems leafy at the base only; leaves
    mostly basal. ............................................................................. *Microseris*

    Stems leafy throughout or at least to mid-
    way up the stem. ................................................................................. 18

18. Pappus white; involucre of one series of
    equal bracts. ...................................................................................... *Crepis*

    Pappus brownish or tawny; involucre
    usually of unequal bracts. ................................................................ 19

19. Heads borne erect; leaf margins entire or
    merely toothed; bracts of involucre
    narrow and green. ....................................................................... *Hieracium*

    Heads usually nodding; leaves divided or
    lobed; bracts of involucre broad and
    colored. .............................................................................................. *Prenanthes*

## *Agoseris*      false dandelion

Perennial herbs from a taproot, with leaves all basal and usually long and narrow. Large yellow flower heads, about 2 cm across, borne singly on a long stalk or scape. Pappus composed of white hairs.

1. Achene beaked, with the beak 2–4 times
   as long as the body; outer bracts much
   broader than the inner ones. ................................................. *A. grandiflora*

   Achenes with the beak less than twice as
   long as the body. ................................................................................. 2

2. Ligules yellow; beak half as long as the
   body or less. ................................................................................ *A. glauca*

   Ligules orange; beak longer than the
   body. ........................................................................................ *A. aurantiaca*

### *Agoseris aurantiaca* (Hook.) Greene      ORANGE FALSE DANDELION

A species with narrowly linear leaves 5–35 cm long, 1–30 mm wide, and entire or with a few teeth or a short lobe. Involucral bracts narrow, often purplish-dotted; ligules deep orange, usually drying to purple or pinkish. Achenes 5–9 mm long; beak thin, from shorter to longer than the body. Alpine meadows and slopes; southern Rocky Mountains.

*Agoseris glauca* (Pursh) Raf.                                    FALSE DANDELION

A hairless species with linear-oblanceolate sometimes toothed leaves up to 25 cm long. Flowers 2–5 cm across, light yellow when young but turning pinkish at maturity, on a stem 15–45 cm high. Common; on moist prairie; throughout most of the Prairie Provinces, but occasionally replaced by var. *dasycephala* (T. & G.) Jepson, with leaves oblanceolate to linear-oblanceolate, 10–30 cm long, usually entire-margined, and pubescent. Involucre of large flower head 2–3 cm across; bracts hairy, at least along the margins. Common; on hillsides and prairies; Cypress Hills, Rocky Mountains, Peace River.

*Agoseris grandiflora* (Nutt.) Greene        LARGE-FLOWERED FALSE DANDELION

Stems 25–60 cm high, from a branched caudex. Leaves linear to spatulate, entire to irregularly lobed, glabrous to canescent, usually pubescent along the midribs. Scapes tomentose below; flower heads 3–6 cm across; involucral bracts in 4 or 5 series; outer bracts conspicuously ciliate; ligules yellow, shorter than the bracts. Beak of achene filiform. Moist meadows; western Parklands, Peace River.

### *Cichorium*        chicory

Upper ·stem leaves large, triangular or
  rounded, entire. .................................................................................... *C. endivia*
Upper stem leaves small, lanceolate, dentate. .................................................... *C. intybus*

*Cichorium endivia* L.                                                ENDIVE

Annual or biennial plants, with stems 15–60 cm high. Leaves mostly basal, 10–20 cm long, deeply toothed to pinnatifid; upper stem leaves 3–5 cm long, clasping. Flowers 3–5 cm across, bright blue, in clusters of 3 or 4 along upper part of stem. Introduced, cultivated for salad; occasionally escaped.

*Cichorium intybus* L.                                              CHICORY

A perennial plant from a long thick deep root, with branching hairy stems 60–100 cm high. Leaves mostly near base of stem, somewhat spatulate, but deeply toothed or pinnatifid with backward-pointing lobes, 7–15 cm long; upper stem leaves small, lanceolate, entire or lobed, stalkless. Flowers bright blue, almost 5 cm across, in clusters of 3 or 4 without stalks, at intervals along upper leafless part of stems. Seed without hairy pappus, but having a series of short blunt scales. An introduced weedy plant; has become fairly common in some localities in Parklands and at scattered locations in Prairies.

### *Crepis*        hawk's-beard

1. Annuals; stem with many stalkless leaves. ............................................... 2
   Perennials; few and smaller stem leaves. .................................................... 3
2. Basal rosettes persistent; stems several to
     many. ................................................................................................ *C. capillaris*
   Basal rosettes absent when flowering;
     stems solitary. ................................................................................ *C. tectorum*
3. Plants dwarfed, with stems 5–10 cm high. ........................................ *C. nana*
   Plants with well-developed stems. ...................................................... 4

*Dicots*                              COMPOSITAE (ASTERACEAE) — 691

4. Stems and foliage glabrous and glaucous,
   with stems much-branched. ........................................................ *C. elegans*

   Stems or foliage or both pubescent. ........................................................ 5

5. Plants glabrous except for the rosette
   leaves. ........................................................ *C. runcinata*

   Plants lightly to densely pubescent
   throughout. ........................................................ *C. occidentalis*

*Crepis capillaris* (L.) Wallr.                    GREEN HAWK'S-BEARD

An annual or biennial plant with several stems from a taproot. Stems 2–60 cm high, slender, somewhat pubescent at the base. Basal leaves 3–20 cm long, lanceolate, runcinate-pinnatifid, glabrous or somewhat pubescent with yellow hairs. Inflorescence corymbiform, with several to many heads 10–15 mm across, light yellow. Achenes at first gray, later black. Introduced; a rare weed of open areas; southern Rocky Mountains.

*Crepis elegans* Hook.                    YOUNGIA

Many-stemmed plants from a deep taproot; stems 10–25 cm high. Basal leaves petiolate, 1–3 cm long, lanceolate to spatulate, entire to crenate; stem leaves numerous, linear-lanceolate to linear. Flower heads numerous, 6–8 mm high, about 1 cm across. Gravel flats along creeks and rivers; Rocky Mountains.

*Crepis nana* Rich.                    DWARF HAWK'S-BEARD

Stems 5–10 cm high, leafless, as long as or shorter than the leaves. Basal leaves to 7 cm long, long-petioled, ovate to suborbicular, glaucous, often purplish, clustered on the long thick taproot. Heads 10–13 mm high; ligules short, yellow tinged with purple. Rare; loose gravel and slides; southern Rocky Mountains.

*Crepis occidentalis* Nutt.                    SMALL-FLOWERED HAWK'S-BEARD

A gray hairy or scurfy plant 20–60 cm high. Basal leaves deeply lobed or divided, 10–15 cm long; stem leaves smaller and less divided. Several yellow heads, 12–20 mm across, borne near top of flowering stem. Uncommon; may be found on hillsides; Prairies and southern Rocky Mountains. Syn.: *C. intermedia* A. Gray.

*Crepis runcinata* (James) T. & G.                    SCAPOSE HAWK'S-BEARD

A species with a stem 20–80 cm high. Basal leaves oblanceolate to spatulate, 5–15 cm long, usually entire, but occasionally somewhat toothed. Flower heads 20–25 mm across, on stalks near top of flowering stem. Found in meadows, wooded areas, and low places; throughout the Prairie Provinces.

The var. *glauca* (Nutt.) Boiv. (Fig. 202), smooth hawk's-beard, with involucral bracts neither hairy nor glandular. Found in low often alkaline areas; Prairies. The var. *hispidulosa* Howell, larger than the typical form, with leaves usually glandular along the midrib. Grasslands; southern Rocky Mountains and Cypress Hills.

A.C. Budd

Fig. 202.  Smooth hawk's-beard, *Crepis runcinata* (James) T. & G. var. *glauca* (Nutt.) Boiv.

*Crepis tectorum* L.

An introduced annual with a slender, branched, leafy stem 10–50 cm high. Stem leaves linear and stalkless; basal leaves 10–15 cm long, usually with backward-pointing teeth. Many small yellow flowers, 10–15 mm across. Plentiful as a weed; on light soils and roadsides; in Boreal forest and Parklands, and spreading farther southward.

## *Hieracium*        hawkweed

Perennial herbs with alternate or basal leaves. Bracts of involucre in series of 1–3 with a few smaller bracts at their base. Pappus composed of brownish or tawny hairs.

1. Flowers white or cream. ........................................................... *H. albiflorum*
   Flowers yellow or orange. ...................................................................... 2

2. Rootstock long, slender, with offshoots;
   flowers orange. ................................................................ *H. aurantiacum*
   Rootstock short, stout, without offshoots;
   flowers yellow. ...................................................................................... 3

3. Bracts of involucre imbricate (overlap-
   ping like shingles). ............................................................ *H. umbellatum*
   Bracts of involucre almost in a single row. ................................................ 4

4. Involucres and flower stalks with dense
   long yellow hairs. ........................................................ *H. cynoglossoides*
   Involucres blackish or glandular, with
   very few yellow hairs. ............................................ *H. triste* var. *gracile*

*Hieracium albiflorum* Hook.                                   WHITE HAWKWEED

A species growing to 30–60 cm high. Basal leaves stalked, hairy, spatulate, 5–12 cm long; stem leaves smaller and without stalks. Flowers white or pale cream, about 12 mm across, in an open-branched cluster at the head of the stem. Found fairly often; in woodlands; southern Rocky Mountains and Cypress Hills.

*Hieracium aurantiacum* L.                                    ORANGE HAWKWEED

A perennial with slender running roots. Basal leaves hairy, spatulate, 5–15 cm long. Stem glandular hairy, 15–50 cm high, and bearing a cluster of orange red flowers 20–25 mm across and sometimes 1 or 2 small leaves. Rare; has been found in scattered locations in the Prairie Provinces. A common weed in Eastern Canada.

*Hieracium cynoglossoides* Arv.-Touv.                         WOOLLY HAWKWEED

A species 30–50 cm high, with long white or yellowish hairs. Leaves 5–15 cm long; lower leaves having winged stalks; upper ones oblong to lanceolate, stalkless, hairy. Not common; found occasionally on hillsides and in woodlands; southern Rocky Mountains. Syn.: *H. albertinum* Farr.

*Hieracium triste* Willd. var. *gracile* (Hook.) Gray ALPINE HAWKWEED

A plant very similar to *H. cynoglossoides*, but smaller, to 30 cm high, with the foliage glabrous or nearly so, and with the involucre densely black pubescent. Mountain meadows; Rocky Mountains.

*Hieracium umbellatum* L. CANADA HAWKWEED

An erect plant 30–100 cm high, with a leafy stem. Leaves ovate to lanceolate, 3–8 cm long. Numerous yellow heads 20–25 mm across. Fairly common; in dry woodlands; Boreal forest and Rocky Mountains. Syn.: *H. canadense* Michx.

**Hypochoeris**   cat's-ear

*Hypochoeris radicata* L. CAT'S-EAR

A perennial with stems 20–50 cm high and with a basal rosette of lobed, somewhat hairy leaves. Leaves 5–15 cm long, oblanceolate, with backward-pointing lobes or teeth; the few stem leaves scale-like. Flowers yellow, stalked, 3 or 4 at summit of stems; flower heads 20–25 mm across. Achene (seed) with a long beak and a pappus of plumose white hairs. Introduced; a rare weed; Prairie Provinces.

*Krigia*   dwarf dandelion

*Krigia biflora* (Walt.) Blake DWARF DANDELION

Perennial plants with fibrous roots; stems 20–80 cm high. Leaves mostly basal, 3–25 cm long, oblanceolate to broadly elliptic, entire or toothed or occasionally lobed; stem leaves few, sessile, clasping, usually small. Heads several; involucre 7–14 mm high; ligules orange yellow, about 15 mm long. Rare; woods, openings, and fields; southeastern Parklands and Boreal forest.

*Lactuca*   lettuce

Tall leafy plants, annual, biennial, or perennial, with yellow, blue, or white flowers and alternate leaves. Involucres in 1 or 2 rows of almost equal inner main bracts, and with smaller bracts at base. Seeds with white or brown pappus hairs.

1. Pappus brown; achenes without a beak. ..................................................... *L. biennis*
   Pappus white; achenes with a distinct
   beak. .......................................................................................................... 2

2. Flowers blue; achenes with a short beak;
   plant smooth, with a bluish bloom, or
   blue. ......................................................................................... *L. pulchella*
   Flowers yellow or blue; achenes with a
   long slender beak; annual or biennial. ................................................. 3

3. Heads with 6–12 florets. ........................................................... *L. serriola*
   Heads with 12–20 florets. ........................................................................ 4

4. Leaves with hairs or bristles on underside
   of midrib. ..................................................................................... *L. ludoviciana*
   Leaves smooth, without hairs or bristles. ........................................... 5

5. Pappus on a long beak; flowers yellow;
   inflorescence crowded. ........................................................................ *L. canadensis*
   Pappus sessile; flowers blue; inflorescence
   open. ............................................................................................................ *L. floridana*

### *Lactuca biennis* (Moench) Fern.                    TALL BLUE LETTUCE

An annual or biennial plant 1–2.5 m high, with a leafy stem. Leaves large, 10–25 cm long, deeply lobed or pinnatifid, with sharp-toothed margins. Flowers very numerous, about 5 mm across, blue or creamy white, in a large dense panicle. Found occasionally; in swampy or moist places; throughout Boreal forest. Syn.: *L. spicata* (Lam.) Hitchc.

### *Lactuca canadensis* L.                    TALL YELLOW LETTUCE

An annual or biennial plant, smooth, hairless, 1–3 m high, very leafy. Leaves with wavy or lobed margins, lanceolate, 5–20 cm long. Numerous flowers pale yellow, about 6 mm across, in an open, long, terminal panicle. Rare; has been reported from eastern Boreal forest. Native to Eastern Canada.

### *Lactuca floridana* (L.) Gaertn.                    ANNUAL BLUE LETTUCE

An annual or biennial plant with stems 50–150 cm high. Leaves mostly petiolate, often pubescent on the underside of the main veins, toothed and often also pinnatifid, 8–30 cm long. Heads numerous in a large panicle; involucre 10–15 mm high; ligules bluish. Margins of woods; southeastern Boreal forest.

### *Lactuca ludoviciana* (Nutt.) Riddell                    WESTERN LETTUCE

A biennial species, hairless, 50–150 cm high, with a leafy stem. Leaves ovate-oblong, 5–10 cm long, wavy-lobed; lobes often spiny-tipped. Flowers numerous, yellow or pale lilac, 5–10 mm across, in an open panicle. Not common; found occasionally along riverbanks; southeastern Parklands.

### *Lactuca pulchella* (Pursh) DC.                    BLUE LETTUCE

A pale bluish green, smooth, glaucous perennial 30–100 cm high, with white running rootstocks. Leaves usually linear-lanceolate, often with backward-pointing lobes. Flowers bright blue, nearly 25 mm across, in a few-flowered panicle. Seed with a short thick beak and a pappus of white hairs. Common; a weed in cultivated lands and along roadsides; throughout the Prairie Provinces.

### *Lactuca serriola* L.                    LOBED PRICKLY LETTUCE

An annual, winter annual, or biennial plant, erect, 30–150 cm high. Leaves 5–20 cm long, deeply lobed; upper leaves clasping stem and often eared at base; underside of midrib bearing a row of short stiff prickles. Flowers 6–8 mm across, yellow, with 6–12 florets; many florets on a large open panicle. Seeds long-beaked, bearing a pappus of white hairs. Introduced, but very common; a weed along roadsides, slough margins, waste places, and on cultivated land; throughout the Prairie Provinces.

The var. *integrata* Gren. & Godr., dentate prickly lettuce, is similar to the species in every way except the leaves are neither lobed nor pinnatifid, except for those at the base. As plentiful in most locations as the type; both often found growing together. Syn.: *L. virosa* L.

### *Lapsana*     nipplewort

*Lapsana communis* L.                              COMMON NIPPLEWORT

An annual plant with hirsute to subglabrous stems 15–150 cm high. Leaves 3–10 cm long, thin, ovate to subrotund, toothed or occasionally lyrate. Inflorescence corymbiform or paniculate, with several to many heads; involucre 5–8 mm high; ligules yellow. Very rare; a weed of shaded places; southeastern Boreal forest.

### *Lygodesmia*     skeletonweed

Stiff-stemmed branching plants with leaves very narrow or reduced almost to scales. Flower heads pink to rose, with 3–10 florets. Pappus with white or very light brown hairs.

Perennials from deep tough rootstocks; leaves
  small or scale-like. ................................................................................... *L. juncea*
Annuals from a tough taproot; leaves long and
  very narrow; flowers racemose among
  branches. ................................................................................... *L. rostrata*

*Lygodesmia juncea* (Pursh) D. Don (Fig. 203)          SKELETONWEED

A much-branched skeleton-like perennial herb 10–40 cm high, from deep, tough, sticky rootstocks. Lower leaves linear-lanceolate, sometimes 36 mm long; upper ones smaller or reduced to scales. Flower heads with 3–5 florets, pink, 12–15 mm across, borne singly at ends of branches. Seeds short, without a beak, and bearing a pale brownish pappus. Common; on light sandy soil and in sandhills; throughout the southern and central portions of the Prairie Provinces.

*Lygodesmia rostrata* A. Gray                ANNUAL SKELETONWEED

A branching thin-stemmed annual 10–60 cm high, from a tough thin root. Leaves linear, 3-nerved, sharp-tipped, 7–16 cm long; upper leaves much smaller than lower ones. Flowers numerous, with 6–10 florets, pink, about 12 mm across, borne in racemes toward ends of branches. Seeds almost 12 mm long, with a tapering beak and a pappus of white hairs. Somewhat rare; has been found in sandhills and similar locations; in southwestern part of the Prairie Provinces.

### *Microseris*     scorzonella

A plant very similar to *Agoseris*, but with the achenes beakless and the pappus subsessile.

Leaves all basal; scape with a solitary flower. ................................................. *M. cuspidata*
Leaves not all basal, usually at least one stem
  leaf. ................................................................................... *M. nutans*

Fig. 203.   Skeletonweed, *Lygodesmia juncea* (Pursh) D. Don.

*Microseris cuspidata* (Pursh) Schultz-Bip.  PRAIRIE FALSE DANDELION

Leaves linear, 10–20 cm long, somewhat hairy, very narrow, tapering to a long narrow point; margins somewhat crinkled. Flowering stem usually no longer than leaves, and bearing a single head with involucres about 20 mm high. Rare, has been found occasionally; eastern Parklands. Syn.: *Agoseris cuspidata* (Pursh) Raf.

*Microseris nutans* (Hook.) Schultz-Bip.  NODDING SCORZONELLA

A perennial with fleshy taproot; stems few, erect or curved at base, 10–50 cm high, usually branched and leafy above. Leaves 10–30 cm long, linear or linear-lanceolate, entire to toothed or pinnatifid. Inflorescence with 1 to several heads, nodding before flowering; involucres 8–22 mm high; bracts glabrous to scurvy puberulent; ligules yellow, 3–8 mm long. Mountain slopes; southern Rocky Mountains.

**Picris**      bitterweed

*Picris echioides* L.  BRISTLY OXTONGUE

Coarse spiny-hispid annual plants, with stems 30–80 cm high. Leaves 10–20 cm long and 7 cm wide, toothed or entire; lower leaves oblanceolate, petioled; upper ones oblong to lanceolate, sessile, often clasping. Involucre 1–2 cm high; bracts in two series; inner bracts narrow, with spiny tip; ligules about 1 cm long, yellow. Introduced; a rare weed of waste places; southern fringe of Boreal forest, Peace River.

**Prenanthes**      rattlesnakeroot

Perennial plants with alternate, usually stalked leaves. Heads clustered, usually nodding or pendent. Bracts of involucre usually colored, not green. Pappus hairs brownish or tawny.

1. Bracts of involucre hairy; basal leaves
   tapering to winged stalks. ................................................................ *P. racemosa*
   Bracts of involucre not hairy; basal leaves
   cordate- or hastate-based, abruptly
   joining stalk. ................................................................................................ 2

2. Flower heads nodding; leaves lobed or
   cleft. ........................................................................................................... *P. alba*
   Flower heads erect; leaves dentate. ............................................ *P. sagittata*

*Prenanthes alba* L.  WHITE LETTUCE

A smooth hairless plant with an erect stem 50–150 cm high, sometimes purplish. Stalkless lower leaves hastate, cordate, or triangular, sometimes lobed, 2–15 cm long; upper leaves lanceolate and without stalks. Heads in a long terminal panicle, drooping, greenish or yellowish white with purplish bracts, slightly scented. Pappus cinnamon brown, very conspicuous. Found often; in open woodlands; throughout the Prairie Provinces. Syn.: *Nabalus albus* (L.) Hook.

*Prenanthes racemosa* Michx. GLAUCOUS WHITE LETTUCE

A very erect stout perennial 50–150 cm high; stem usually covered with a white bloom. Lower leaves oblanceolate to spatulate, 10–20 cm long, tapering to a winged stalk; upper leaves smaller and somewhat clasping. Heads in a long spike of crowded clusters, purplish, about 12 mm long and 6 mm broad, usually not as pendent as in *P. alba*. Pappus straw-colored. Fairly abundant; in wooded areas; across the Prairie Provinces. Syn.: *Nabalus racemosus* (Michx.) DC.

*Prenanthes sagittata* (Gray) Nels. PURPLE RATTLESNAKEROOT

An erect glabrous perennial with stems 30–70 cm high. Lower leaves petioled, sagittate or hastate, dentate, glabrous, with blades 10–15 cm long; upper leaves sessile, lanceolate. Heads in a narrow panicle, at first erect, later spreading; involucres about 12 mm high; ligules purplish-tinged. Mountain woods; northwestern Boreal forest, Rocky Mountains.

**Sonchus**    sow-thistle

1. Flowering heads 3–5 cm across; perenni-
   als from creeping rootstocks. ...................................................................... *S. arvensis*

   Flowering heads not over 2.5 cm across;
   annuals growing from deep taproots. ................................................................ 2

2. Ears at base of clasping leaves rounded. ........................................................ *S. asper*

   Ears at base of clasping leaves acutely
   pointed. ................................................................................................... *S. oleraceus*

*Sonchus arvensis* L. (Fig. 204) PERENNIAL SOW-THISTLE

A weedy plant 50–150 cm high, with vigorous creeping rootstocks. Stems usually hollow and slightly branched. Lower leaves runcinate-pinnatifid or with backward-pointing lobes, 10–25 cm long, and narrowed to a short stalk; upper leaves less lobed and without stalks; teeth of leaves spiny-pointed. Flowers numerous, showy, bright yellow, on bristly stalks in a corymbose panicle; involucres glandular hairy. Very common; in moister districts and wet places; throughout the Prairie Provinces.

The var. *glabrescens* Guenth., Grab. & Wimm., smooth perennial sow-thistle, differing from the species in having a smooth, hairless, not glandular involucre. Common; in many locations; throughout the Prairie Provinces. Syn.: *S. uliginosus* Bieb.

*Sonchus asper* (L.) Hill PRICKLY SOW-THISTLE

An annual weedy plant 50–150 cm high. Leaves clasping stem at base, slightly lobed or divided; leaf lobes usually having spine-tipped teeth; basal lobes rounded. Flowers 12–25 mm across, pale yellow. Fairly common; a weed of gardens and roadsides; throughout the Prairie Provinces.

*Sonchus oleraceus* L. ANNUAL SOW-THISTLE

An annual plant 50–200 cm high, slightly branched. Leaves deeply lobed, with rather soft prickles; lower leaves stalked; upper ones clasping the stem.

A.C. Budd

Fig. 204. Perennial sow-thistle, *Sonchus arvensis* L.

Heads pale yellow, 12–25 mm across. Fairly common; in gardens and waste places. Leaves undivided in f. *integrifolius* (Wallr.) G. Beck.

## *Stephanomeria*    skeletonweed

*Stephanomeria runcinata* Nutt.    RUSH-PINK

Plants similar to *Lygodesmia*, with creeping rootstocks. Leaves toothed or pinnatifid, well-developed. Heads terminal; ligules 10–15 mm long, pink. Pappus plumose. Rare; badlands, loose shale beds; Prairies.

## *Taraxacum*    dandelion

1. Involucral bracts all appressed or
   ascending. ............................................................................ *T. ceratophorum*
   Involucral bracts reflexed. ........................................................................ 2

2. Leaves deeply divided, almost to midrib,
   into narrow segments, with terminal
   lobe small; seeds reddish. .................................................... *T. laevigatum*
   Leaves shallowly divided, with terminal
   lobe large; seeds greenish or brownish
   yellow. ............................................................................... *T. officinale*

*Taraxacum ceratophorum* (Ledeb.) DC.    DANDELION

Plants similar to *T. officinale*, but usually less robust; leaves less lobed, commonly glabrous or nearly so. Flower heads 15–20 mm across; involucre about 15 mm high; outer bracts appressed to spreading, often with horned tip. Open ground; throughout the Prairie Provinces.

*Taraxacum laevigatum* (Willd.) DC.    RED-SEEDED DANDELION

A stemless perennial plant, with all leaf and flower stalks arising from a fleshy root crown, and growing from a deep taproot. Leaves 10–20 cm long, deeply divided into narrow segments. Flowers 25–35 mm across, with many florets, bright yellow, borne singly on stems or scapes 10–30 cm high. Seeds bright red, with a white hairy pappus on a long thin beak. Fairly common; in waste places and roadsides; throughout the Prairie Provinces. Introduced from Europe. Syn.: *T. erythrospermum* Andrz.

*Taraxacum officinale* Weber    DANDELION

An introduced stemless perennial with deep fleshy taproots. Leaves coarsely incised, with triangular lobes and a large terminal lobe; leaf stalks and flowering stems arising from a root crown. Flower heads yellow, 35–50 mm across, with very many florets. Seed greenish buff, with a long beak and a pappus of white hairs. Often the earliest plant blooming in spring. Very common; on lawns, along roadsides, and in waste places; throughout the Prairie Provinces.

## *Tragopogon*    goat's-beard

A fairly tall biennial or perennial plant, with deep fleshy taproots and grass-like leaves. Flowering heads large. Seeds and plumose seed heads very large; seeds narrow and long, with 5–10 ribs, ending in a long beak terminating in a pappus of plumose hairs.

1. Flowers purple; involucral bracts much
   longer than the florets. ........................................................... *T. porrifolius*
   Flowers yellow; involucral bracts varying
   in length. ................................................................................................ 2

2. Involucral bracts longer than florets; stem
   usually thickened below flower head. ...................................... *T. dubius*
   Involucral bracts not longer than florets;
   stem slightly thickened below flower
   head. .................................................................................... *T. pratensis*

*Tragopogon dubius* Scop.                    YELLOW GOAT'S-BEARD

A coarse biennial plant from a deep fleshy taproot 30–60 cm high. Leaves narrow, erect, grass-like, 10–30 cm long, stalkless, clasping at base. Flowers sulfur yellow, 3–5 cm across; involucral bracts 10–14, longer than yellow florets; heads borne singly on summit of an erect scape or stem that is decidedly thickened just below head. Seed heads plumose, very conspicuous, usually 7–10 cm in diam; seeds long, tapering to a long beak, having ribs of minute tubercles and a pappus of plumose white hairs; body of seed about 12 mm long, with beak slightly longer. An introduced plant, apparently coming from southwestern United States through Colorado. Common; along roadsides and in waste places; almost throughout the Prairie Provinces.

*Tragopogon porrifolius* L.                    SALSIFY

A perennial or biennial with an edible fleshy taproot 30–80 cm high. Leaves grass-like. Flower heads 5–10 cm across; florets purple; bracts much exceeding the florets. A garden plant sometimes found as a weed; in southeastern Parklands and southern Rocky Mountains.

*Tragopogon pratensis* L.                    GOAT'S-BEARD

A biennial very similar to *T. dubius*, but somewhat smaller. Florets chrome yellow; bracts of involucre usually 8 or 9, not longer than florets. Seed somewhat shorter than that of *T. dubius*, but similar in appearance. Not as aggressive and rapid-spreading as *T. dubius*, which came into the Prairies from the west. A fairly common weed; in eastern Parklands.

## GROUP 3

This very large group consists of plants in which some or all florets are tubular. The tubular (central) florets are usually perfect or bisexual, whereas the marginal or ligulate (strap-shaped) florets are either female or neutral. Plants are without milky sap. The primary leaves (those following the cotyledons or seed leaves) are usually in pairs.

1. Flower heads with all florets tubular (dis-
   coid heads). ........................................................................................ 2
   Flower heads with florets both tubular
   and ray (radiate heads). ................................................................ 33

2. Bracts of involucre dry, parchmenty or
   membranous, not green. ........................................................... 3
   Bracts of involucre not dry or parch-
   menty, usually green. ............................................................... 7

3. Some florets bisexual, with both male and
   female florets on same plant. .................................................. 4
   No fertile bisexual florets, with male and
   female florets on separate plants. .......................................... 6

4. Leaves alternate; flowers all perfect. ..................... *Gnaphalium*
   Leaves opposite. ...................................................................... 5

5. Pappus consisting of terminal awns; inner
   bracts about twice as long as outer
   ones. .............................................................................. *Thelesperma*
   Pappus absent; heads subtended by foli-
   age leaves. ................................................................... *Psilocarphus*

6. Tall plants; stems leafy. ........................................... *Anaphalis*
   Low plants; leaves mostly basal, with
   stem leaves reduced. ................................................. *Antennaria*

7. Pappus none or inconspicuous. ............................................. 8
   Pappus present, distinct. .......................................................11

8. Leaves large, deltoid-ovate. ................................. *Adenocaulon*
   Leaves elliptic, lobed or
   divided. ................................................................................... 9

9. Inflorescence glandular sticky; leaves
   linear-elliptic. ............................................................... *Madia*
   Inflorescence not glandular; leaves lobed
   or divided. .............................................................................. 10

10. Heads small, numerous, in open or spike-
    like panicles. ............................................................. *Artemisia*
    Heads larger, 4–7 mm across. .............................................. 11

11. Leaves 2–3 times pinnatifid, or
    pinnately divided into toothed
    segments, inflorescence has no
    ray florets. .................................................................. *Tanacetum*
    Leaves entire or toothed, not
    pinnately divided. Inflorescence
    usually with white ray florets. ......................... *Chrysanthemum*

12. Involucral bracts spiny-tipped or with
    hooked bristles. ..................................................................... 13
    Involucral bracts not spiny-tipped and
    without hooked bristles. ....................................................... 16

13. Involucral bracts with hooked bristles;
    leaves large, cordate, and not prickly. ..................... *Arctium*
    Some bracts spiny-tipped. .................................................... 14

14. Pappus hairs not longer than achene. .................... *Centaurea*
    Pappus hairs much longer than achene. .............................. 15

58. Ray florets 50 or more; bracts in 1 or 2
    series. ........................................................... *Erigeron*

    Ray florets 10–50; bracts in several series. ........................... *Aster*

59. Receptacle with chaffy scales. ....................................... 60

    Receptacle naked, without chaffy scales. ............................. 63

60. Leaves opposite. ...................................................... *Galinsoga*

    Leaves alternate. ...................................................... 61

61. Flowers large and purple; leaves
    undivided. ............................................................ *Echinacea*

    Ray florets white; leaves divided. ..................................... 62

62. Heads small, numerous, in dense almost
    flat-topped clusters; plant not strongly
    scented. .............................................................. *Achillea*

    Heads fewer, about 25 mm across; plant
    unpleasantly scented. ................................................. *Anthemis*

63. Pappus composed of bristly scales or with
    2–4 slender bristles. .................................................. *Boltonia*

    Pappus absent, or composed of merely a
    border. ............................................................... 64

64. Leaves entire, not divided; heads solitary
    on stem, 25–50 mm across; plants not
    scented. .............................................................. *Chrysanthemum*

    Leaves much divided into narrow seg-
    ments; heads several to a stem, less
    than 25 mm across; plants usually
    somewhat scented. ..................................................... *Matricaria*

*Achillea*      yarrow

1. Leaves pinnately dissected into linear-lan-
   ceolate segments. ..................................................... *A. millefolium*

   Leaves subentire to incised, not pinnately
   dissected. ............................................................ 2

2. Leaves subentire to serrate; ligules 3–5
   mm long. ............................................................. *A. ptarmica*

   Leaves incised; ligules 1–2 mm long. ................................. *A. sibirica*

*Achillea millefolium* L.                         YARROW. MILFOIL

   A perennial from rootstocks, with stems 20–100 cm high, and somewhat
aromatic. Leaves pinnately dissected, with the ultimate segments 1–2 mm
wide; blades 3–15 cm long and to 25 mm wide. Heads numerous, in a flat-
topped inflorescence, white to pinkish. Moist meadows, woods, and openings;
Parklands, Boreal forest, and Rocky Mountains. In f. *purpurea* (Gouan) Schinz
& Thell., ligules purple.

   The var. *occidentalis* DC., woolly yarrow, a perennial from shallow rootstocks, 10–30 cm high, usually covered with silky hairs. Leaves 3–10 cm long,
finely divided into segments. Flowers in a compact round-topped cluster at
head of stem, white, rarely pink, about 6 mm across; involucral bracts with

straw-colored margins. Common; on prairies and along roadsides; throughout the Prairie Provinces. Syn.: *A. lanulosa* Nutt.

In var. *nigrescens* E. May., tips and margins of bracts dark brown to blackish. This form more common than the others in the northern part of Boreal forest. Syn.: *A. borealis* Bong.

*Achillea ptarmica* L. (Fig. 205)                                    SNEEZEWEED

A perennial with creeping rootstocks; stems 30–60 cm high, villous-tomentose above, subglabrous below. Leaves glabrous or nearly so, 8–10 cm long, 2–6 mm wide, closely serrate to subentire. Heads several to numerous in a corymbiform inflorescence; involucre 4–5 mm high; ligules white, 3–5 mm long. Introduced; commonly the "double-flowered" form, occasionally escaped and established; Boreal forest.

*Achillea sibirica* Ledeb.                                    MANY-FLOWERED YARROW

A perennial 30–60 cm high. Leaves linear, deeply toothed, 3–7 cm long. Flowers very similar to those of *A. ptarmica*, but bracts having dark brown margins. Found occasionally; in wooded areas and disturbed places; Boreal forest. Syn.: *A. multiflora* Hook.

**Adenocaulon**

*Adenocaulon bicolor* Hook.                                    TRAILPLANT

A slender fibrous-rooted perennial 30–70 cm high. Leaves mostly basal, long-petioled, 3–15 cm wide, deltoid-ovate to cordate, subglabrous above, white lanate below. Inflorescence an open panicle; heads discoid; involucral bracts about 2 mm long, reflexed in fruit. Moist shady woods; southern Rocky Mountains.

**Anaphalis**          pearly everlasting

*Anaphalis margaritacea* (L.) C. B. Clarke                  PEARLY EVERLASTING

A woolly-stemmed perennial growing in a cluster from numerous running rootstocks; stem usually erect, 30–80 cm high, often branched near top. Leaves stalkless, alternate, linear-lanceolate, white woolly beneath, grayish above, 5–12 cm long. Flowers numerous, in a fairly dense terminal cluster up to 15 cm across; each flower head discoid, 6–8 mm across, white, with pearly white papery bracts. Found in open woodlands; in eastern Parklands.

The var. *subalpina* Gray, with very congested inflorescence, found in southern Rocky Mountains and in Cypress Hills.

**Antennaria**          everlasting

Perennial herbs, usually mat-forming, with leaves mostly basal and usu-ally woolly or hairy. Flowers unisexual, sometimes with both sexes on the same plant, but usually on separate plants; female heads sometimes producing fertile seed without pollination; involucral bracts papery. Achenes bearing a white, hairy pappus.

Fig. 205.   Sneezeweed, *Achillea ptarmica* L.

11. Rosette leaves 5–10 mm wide; tips of bracts golden brown. ................................ *A. russellii*

Rosette leaves 1–5 mm wide; tips of bracts dark brown. ................................ *A. alpina*

12. Involucre 8–10 mm high; leaves narrowly oblanceolate. ................................ 13

Involucre 4–7 mm high; leaves oblanceolate. ................................ 14

13. Leaves green, glabrous or nearly so above. ................................ *A. glabrata*

Leaves grayish tomentose above, especially those of rosettes. ................................ *A. angustata*

14. Flower heads typically solitary; bracts squarrose. ................................ *A. monocephala*

Flower heads typically 3–5 on a stem: bracts appressed. ................................ *A. alpina*

15. Terminal part of involucral bracts roseate to deep pink, even at maturity. ................................ *A. rosea*

Terminal part of involucral bracts whitish or light pink when young. ................................ 16

16. Involucral bracts with a large dark spot at the base. ................................ *A. corymbosa*

Involucral bracts green or whitish at the base. ................................ 17

17. Rosette leaves to 50 mm long, 20 mm wide. ................................ *A. neodioica*

Rosette leaves to 25 mm long, usually less than 10 mm wide. ................................ 18

18. Involucre 5–7 mm high; stems with 8–12 leaves. ................................ *A. parvifolia*

Involucre 8–13 mm high; stems with 5–7 leaves. ................................ *A. aprica*

19. Rosette leaves clearly 3- to 5-veined, often to 50 mm long by 30–40 mm wide. ................................ 20

Rosette leaves with only the midrib clearly visible, and usually smaller than 50 mm long by 40 mm wide. ................................ 21

20. Rosette leaves glabrous; involucre 5–9 mm high. ................................ *A. plantaginifolia*

Rosette leaves at first somewhat pubescent above; involucre 7–10 mm high. ................................ *A. howellii*

21. Rosette leaves glabrous or glabrate, 20–40 mm long; involucre 7–10 mm high. ................................ *A. parlinii*

Rosette leaves usually thinly pubescent above to glabrate, 10–50 mm long; involucre 6–9 mm high. ................................ *A. neodioica*

*Antennaria alpina* (L.) Gaertn.                    ALPINE EVERLASTING

A mat-forming perennial with more or less leafy stolons; stems 3–15 cm high. Basal leaves 5–20 mm long, 1–5 mm wide, oblanceolate, loosely pubescent on both sides. Heads 3–5 in a small cyme; involucres mostly 4–7 mm high; bracts pointed with terminal part dark brown or blackish green. Rare; alpine meadows, openings, and slopes; southern Rocky Mountains.

The common form is var. *canescens* Lange, with leaves persistently pubescent on both sides.

*Antennaria anaphaloides* Rydb.                    TALL EVERLASTING

A tall species, not mat-forming, 20–50 cm high, with oblanceolate basal leaves 5–10 cm long and long narrow stem leaves. Many-flowering heads in an open corymb; bracts with white tips, and with or without very small dark spots. Found often; in southern Rocky Mountains and locally in Cypress Hills.

*Antennaria angustata* Greene                    PUSSY-TOES

A mat-forming perennial without stolons. Rosette leaves linear-oblanceolate, tomentose, often glabrescent above. Heads usually solitary; involucres 8–10 mm high; tips of bracts greenish black. Alpine slopes; southern Rocky Mountains.

*Antennaria aprica* Greene                    LOW EVERLASTING

A low mat-forming perennial usually less than 15 cm high. Spatulate- or wedge-shaped rosette leaves 1–2 cm long, densely whitish woolly on both sides; stem leaves 5–7, linear, about 1 cm long. White flower heads occasionally with a faint pinkish tinge, in compact, short-stalked clusters on a stem 5–15 cm high. Common; on dry prairies; throughout the Prairie Provinces.

*Antennaria corymbosa* E. Nels.                    CORYMBOSE EVERLASTING

A medium-tall species with stems 20–30 cm high. Rosette leaves oblanceolate, 1–3 cm long, grayish, finely woolly; stem leaves linear and sharp-tipped. Flower heads stalked, in a terminal corymb; bracts having white upper portions. Found in mountain meadows; in southern Rocky Mountains and in open woodlands in Cypress Hills.

*Antennaria dimorpha* Nutt.                    CUSHION EVERLASTING

A densely cespitose grayish tomentose perennial with a many-branched caudex forming small mats or cushions. Stems 1–4 cm high, leafy, terminated by a solitary flower head; involucre 10–15 mm high. Rare, but probably often overlooked; more or less open grassland; Prairies and southern Rocky Mountains.

*Antennaria glabrata* (Vahl) Greene                    PUSSY-TOES

Possibly only a much less pubescent, often almost glabrous, form or variety of *A. angustata*. Rare; moist slopes; southern Rocky Mountains.

*Antennaria howellii* Greene (Fig. 206)     HOWELL'S EVERLASTING

A tall species 20–35 cm high. Basal leaves ovate, wedge-shaped, 2–5 cm long, closely silky-woolly beneath, but bright green above; stem leaves small and narrow. Heads in a corymb. Three varieties are distinguished in this species:

1. Rosette leaves 1–2 cm wide. ........................................................ var. *howellii*
   Rosette leaves 5–10 mm wide. ................................................................... 2
2. Stems to 20 cm high when in flower, elon-
   gating to 35 cm at maturity. .............................. var. *athabascensis* (Greene) Boiv.
   Stems 15–20 cm high at maturity. ............................ var. *campestris* (Rydb.) Boiv.

The var. *howellii* is commonly found in pine woods and open coniferous forests in Boreal forest, Riding Mountain, and Cypress Hills; var. *athabascensis*, in Parklands and dry open areas in Boreal forest; var. *campestris*, in moist grasslands in Prairies and on slopes in Parkland.

*Antennaria lanata* (Hook.) Greene     WOOLLY EVERLASTING

A perennial with a rather thick caudex or rootstock. Stems densely grayish tomentose, stout, erect, leafy, 10–20 cm high. Basal leaves in tufts, linear-oblanceolate to oblanceolate, 2–6 cm long, 3–15 mm wide, obscurely to distinctly 3-nerved. Heads 6–12 in a dense cyme; involucre densely tomentose, 5–8 mm high; outer bracts with a dark brown tip. Alpine meadows; southern Rocky Mountains.

*Antennaria luzuloides* T. & G.     SILVERY EVERLASTING

A perennial with a somewhat woody caudex and erect or ascending leafy stems 10–50 cm high. Basal leaves erect in tufts, 3–10 cm long, 1–5 mm wide, thinly tomentose. Heads often numerous, in a dense cyme or cymose panicle; involucres 4–5 mm high, glabrous or nearly so; bracts pale green or brownish with white or somewhat pinkish tips. Alpine meadows and slopes; southern Rocky Mountains.

*Antennaria monocephala* DC.

Plants with dense basal rosettes and stems 5–10 cm high. Rosette leaves 8–15 mm long, silky tomentose below; stem leaves linear, 5–10 mm long. Heads usually solitary; involucre about 5 mm high; bracts with greenish brown tips, squarrose. Alpine meadows and slopes; southern Rocky Mountains.

*Antennaria neodioica* Greene     COMMON PUSSY-TOES

A fairly tall species with rosette leaves indistinctly 3-ribbed, oblanceolate to obovate, woolly, 2–3 cm long, yellowish green, more or less tomentose above. Flower heads with rather long stalks, on a leafy stem 15–25 cm high, strongly stoloniferous. Syn.: *A. obovata* E. Nels. Plants with the leaves glabrous above distinguished as var. *randii* (Fern.) Boiv. Syn.: *A. canadensis* Greene. Both forms found in dry open woods, shrubbery, and moist grassland; throughout Parklands and southern fringes of Boreal forest.

Fig. 206. Howell's everlasting, *Antennaria howellii* Greene.

*Antennaria parlinii* Fern.                     EASTERN EVERLASTING

A mat-forming stoloniferous plant with stems 20–40 cm high. Rosette leaves 2–6 cm long, 1–4 cm wide, bright green and glabrous above, obovate-spatulate. Heads 4–8 in a corymb; involucres 7–10 mm high. Open woods; southeastern Boreal forest.

*Antennaria parvifolia* Nutt.                 SMALL-LEAVED EVERLASTING

A mat-forming perennial 15–25 cm high. Rosette leaves very small, barely exceeding 15 mm long, white, densely woolly on both sides, somewhat angular but roughly spatulate. Stem leaves 7–12, about 10 mm long, linear. Flower heads nodding or pendent when young. Very plentiful; on dry prairie and in saline meadows; throughout Prairies.

*Antennaria plantaginifolia* (L.) Hook.        PLANTAIN-LEAVED EVERLASTING

A tall species; basal leaves oval to spatulate, 3–7 cm long, woolly above and below. Flower heads on an erect leafy stem 20–50 cm high. Found occasionally; southeastern Boreal forest.

*Antennaria pulcherrima* (Hook.) Greene        SHOWY EVERLASTING

Perennials, not mat-forming plants, 30–50 cm high. Basal leaves oblanceolate, 5–10 cm long, 3-ribbed, woolly; stem leaves narrow and smaller. Heads in a cluster at the top of the stem; bracts usually brownish with gray white tips. Found occasionally; in moist soils; Boreal forest and Rocky Mountains.

*Antennaria racemosa* Hook.                RACEMOSE EVERLASTING

Plants with elongate stolons and tufts of basal leaves; stems 10–60 cm high, erect, leafy. Basal and stolon leaves elliptic to elliptic-obovate, 2–8 cm long, 1–5 cm wide, 1- to 3-nerved, glabrous or nearly so above, more or less tomentose below. Heads in open racemes; lower ones on peduncles 2–5 cm long; involucres 6–8 mm high. Coniferous woods; southern Rocky Mountains.

*Antennaria rosea* Greene                    ROSY EVERLASTING

A somewhat mat-forming white woolly perennial 15–50 cm high. Basal rosette leaves 10–25 mm long, oblanceolate to spatulate, stalked, white woolly, pointed at apex. Flower heads in a close terminal cluster, having conspicuous pinkish-tipped involucral bracts. Found in meadows and moist hillsides in favored localities; throughout the Prairie Provinces, but more particularly in Cypress Hills and Rocky Mountains.

*Antennaria russellii* Boiv.                RUSSELL'S EVERLASTING

A species very similar to *A. neodioica*, but with the tips of the involucral bracts golden brown or straw-colored, and with the involucre smaller (only 6–7 mm high). Open woods; Cypress Hills.

*Antennaria umbrinella* Rydb.     BROWN-BRACTED MOUNTAIN EVERLASTING

A low mat-forming species 2–6 cm high, with small, whitish, silky rosette leaves and linear-lanceolate stem leaves. Bracts of the few small flower heads dark brown. Very local, but found on dry hillsides; southern Rocky Mountains and Cypress Hills. Syn.: *A. aizoides* Greene.

*Anthemis*     mayweed

*Anthemis cotula* L.                                    STINKING MAYWEED

A hairless annual weed, much-branched, 30–70 cm high, having a fetid and unpleasant odor. Leaves deeply dissected into very narrow lobes, 2–5 cm long. Flowers about 2 cm across, with numerous yellow disks; ray florets 10–18, white, borne on heads of stems, forming a large flat cluster. An introduced weed, plentiful in Eastern Canada, rare in southeastern and northwestern Parklands. Most plants recently reported in the Prairie Provinces as stinking mayweed now have been identified as *Matricaria maritima* var. *maritima* (*M. inodora*). Syn.: *Maruta cotula* (L.) DC. Another species, *A. tinctoria* L., with yellow flowers and somewhat broader leaf divisions, also reported, but very rare.

*Arctium*     burdock

Large, coarse, biennial plants with broad, oval, or cordate leaves, which are long-stalked and paler beneath. Flowers discoid; involucres globose and much-imbricated; bracts stiff, hook-tipped.

1. Bracts of involucre densely cottony. ..................................................... *A. tomentosum*
   Bracts of involucre not cottony, sometimes slightly woolly. ............................................................................. 2

2. Involucre 2 cm or more across; inner bracts as long as, or longer than, flower head. ..................................................................................................... *A. lappa*
   Involucre 10–25 mm across; inner bracts not longer than flower head. ................................................... *A. minus*

*Arctium lappa* L.                                    COMMON BURDOCK

A coarse, branching plant 1–3 m high. Leaves broadly ovate, stalked, pale beneath, often cordate-based, up to 45 cm long. Flowers purple, discoid, with a globose involucre 25–40 mm across; bracts tipped with hooked bristles. An introduced weed becoming increasingly plentiful; in waste places; eastern Parklands and southeastern Boreal forest.

*Arctium minus* (Hill) Bernh.                          LESSER BURDOCK

A coarse, tall, branching biennial 1–2 m high, from deep thick taproots. Leaves large, cordate, pale and downy beneath, up to 30 cm long. Flower heads numerous, usually in a leafy one-sided raceme, discoid with purple florets, 15–25 mm across. Involucre somewhat globose, green to purplish, bracts with hooked bristles. Burs (seed-bearing heads) very prickly, and can be carried on clothing and the coats of animals. An introduced and widely distributed weed, very plentiful locally; as soon as it becomes established, becoming very common.

*Arctium tomentosum* Mill.                             COTTON BURDOCK

A species similar to *A. minus*, but with involucres slightly larger and covered with a cottony web. Involucral bracts having hooked tips. Introduced; found in a few locations, widely separated, but still quite rare.

***Arnica*** arnica

Perennial plants from rootstocks, with opposite leaves and large radiate yellow or orange heads. Pappus a single series of rough bristles.

1. Leaves cordate to oblong. ........................................................................ 2
   Leaves oblong-lanceolate to linear. ........................................................ 4

2. Stem leaves long-petioled; the large ones
   cordate or ovate. ....................................................... *A. cordifolia*
   Stem leaves with the petiole shorter than
   the blade, oblong. ................................................................................ 3

3. Pappus pale brown; lower stem leaves
   deltoid, dentate. ........................................................ *A. diversifolia*
   Pappus white; lower stem leaves
   elliptic-lanceolate. ....................................................... *A. latifolia*

4. Stem leaves numerous, in 4–8 pairs. ...................................................... 5
   Stem leaves less numerous, in 1–3 pairs. ............................................... 8

5. Lower stem leaves ovate or deltoid,
   sharply dentate. ....................................................... *A. diversifolia*
   Lower stem leaves lanceolate, entire to
   remotely dentate. ................................................................................ 6

6. Heads solitary or few. ............................................................. *A. mollis*
   Heads several to many. ........................................................................ 7

7. Bracts with a distinct tuft of hairs at the
   tip. ....................................................................... *A. chamissonis*
   Bracts not with a tuft of hairs, and with
   the tip acute. .............................................................. *A. longifolia*

8. Heads discoid; pappus tawny to
   brownish. ...................................................................... *A. parryi*
   Heads radiate; pappus white to brownish. ............................................... 9

9. Pappus straw-colored or brownish;
   flowering stems without tufts of basal
   leaves. ................................................................................................ 10
   Pappus white; basal leaves mostly
   present. .............................................................................................. 11

10. Heads usually several, narrow,
    top-shaped. .............................................................. *A. diversifolia*
    Heads usually solitary or few, nearly
    half-round. ................................................................... *A. mollis*

11. Basal leaves usually petiolate, ovate to
    cordate. ...................................................................... *A. latifolia*
    Basal leaves linear to lanceolate or
    oblanceolate. .................................................................................... 12

12. Lower leaves mostly more or less
    long-petioled. ................................................................................... 13
    Lower leaves sessile to short-petioled. ................................................. 15

13. Leaves dentate or denticulate; heads 1–5,
    top-shaped. ............................................................. *A. lonchophylla*

Leaves entire; heads mostly solitary, half-round. ........................................................................... 14

14. Base of stem with a tuft of brown hairs; leaves lanceolate. .............................................. *A. fulgens*

    Base of stem without brown hairs; leaves linear-lanceolate. ............................................... *A. sororia*

15. Involucres 9–11 mm high; stem leaves remotely dentate. ............................................. *A. rydbergii*

    Involucres 10–15 mm high; stem leaves entire. ................................................................................. 16

16. Achenes densely pubescent; involucral bracts villose-glandular. ....................... *A. alpina* var. *ungavensis*

    Achenes glabrous or nearly so; involucral bracts glabrous, at least above the middle. ........................................................... *A. louiseana*

*Arnica alpina* (L.) Olin var. *ungavensis* Boiv.                    ALPINE ARNICA

Plants with stems 10–30 cm high. Stem leaves usually 2 or 3 pairs, linear-lanceolate, acuminate, entire. Heads usually solitary; involucre villous and glandular; ligules distinctly toothed at apex. Pappus white. Alpine and arctic tundra; northeastern Boreal forest, southern Rocky Mountains.

The var. *vestita* Hulten, densely soft woolly pubescent, especially on the involucral bracts. Alpine tundra; southern Rocky Mountains.

*Arnica chamissonis* Less.                                         LEAFY ARNICA

A species 30–60 cm high, with a very leafy softly hairy stem. Leaves oblong-lanceolate; lower leaves tapering to a winged stalk clasping stem at base, 6–15 cm long; upper leaves stalkless and opposite, usually in several pairs. Flowers lemon yellow, 2–5 cm across, in cluster at the head of the stem. Widespread, but nowhere very common; in moist places; throughout Boreal forest and Peace River district.

*Arnica cordifolia* Hook. (Fig. 207*A*)                    HEART-LEAVED ARNICA

A species 20–50 cm high, with cordate (heart-shaped) leaves. Basal leaves long-stalked, 2–8 cm long; stem leaves smaller, usually without stalks. Flowers lemon yellow, 3–7 cm across, on the top of the stem. Plentiful; in wooded areas; in Rocky Mountains and Cypress Hills, rarer in Boreal forest.

*Arnica diversifolia* Greene                                      LAWLESS ARNICA

A perennial with freely branching, creeping rootstocks; stems solitary or in small tufts, 15–40 cm high, glandular pubescent to subglabrous. Stem leaves 4–8 cm long, ovate or deltoid to elliptic; upper leaves sessile; lower ones wing-petioled. Heads several; ligules 15–20 mm long; bracts glandular. Slopes and creek banks; southern Rocky Mountains.

*Arnica fulgens* Pursh (Fig. 207*B*)                              SHINING ARNICA

A plant 20–40 cm high. Basal leaves stalked, oblanceolate, 5–8 cm long, usually 3-ribbed, and entire-margined. Stem leaves linear-lanceolate, smaller,

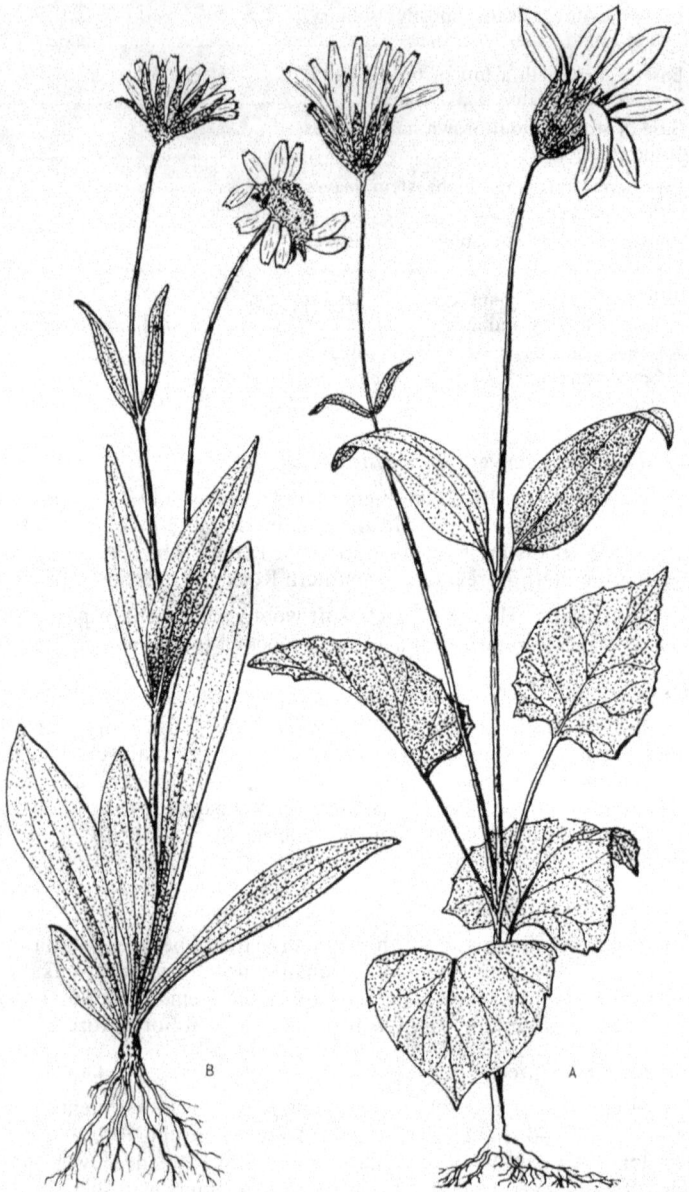

Fig. 207.   *A*, Heart-leaved arnica, *Arnica cordifolia* Hook.; *B*, shining arnica, *Arnica fulgens* Pursh.

opposite, in 2 or 3 pairs. Flower heads orange yellow, 2–5 cm across, usually solitary on stem, but occasionally 2 or 3. Fairly abundant; in meadows and slightly moister spots on prairie; throughout Prairies and Parklands. Plentiful some years and scarce in others.

*Arnica latifolia* Bong.                                    MOUNTAIN ARNICA

Plants with long creeping rhizomes; stems solitary or few together, 10–60 cm high. Basal leaves long-petioled, often on separate sterile shoots; stem leaves 3–10 cm long, ovate or elliptic, more or less dentate, the upper ones sessile, the lower ones petiolate. Heads usually 1–3, top-shaped; ligules 10–15 mm long; bracts 10–15 mm high, more or less glandular pubescent. Moist to wet montane forests; southern Rocky Mountains.

*Arnica lonchophylla* Greene                              SPEAR-LEAVED ARNICA

Plants with dark, scaly, branching rootstocks; stems usually solitary, 15–40 cm high, finely stipitate-glandular, and more or less pubescent. Leaves 5–15 cm long, lance-elliptic to lanceolate, petiolate, mostly finely stipitate-glandular, denticulate to dentate. Heads 1 to several; ligules 1–2 cm long; bracts 8–12 mm high, glandular pubescent. Lakeshores and riverbanks, particularly on calcareous soils; Boreal forest.

*Arnica longifolia* Eat.                                   LONG-LEAVED ARNICA

Plants with a short rootstock or branching caudex; stems clustered, usually 20–40 cm high, with many short sterile shoots at the base. Leaves 5–12 cm long, narrowly lanceolate or lance-elliptic, the lower ones connate or nearly so, the others shortly petiolate to connate. Heads several to numerous; ligules 1–2 cm long; bracts 7–10 mm high, glandular puberulent. Along montane and alpine creeks; southern Rocky Mountains.

*Arnica louiseana* Farr                                        ROCK ARNICA

Perennial plants with long, creeping, densely scaly rootstocks; stems solitary or several together, 10–25 cm high, softly villous, and usually more or less glandular. Leaves 2–6 cm long, oblanceolate to elliptic, entire or distantly denticulate, mostly subglabrous to sparsely villous, and more or less glandular. Heads solitary; ligules 1–2 cm long; bracts about 1 cm high, glabrous above the middle, somewhat pubescent toward the base. On rock cliffs and shale slides at high altitudes; southern Rocky Mountains.

*Arnica mollis* Hook.                                     CORDILLERAN ARNICA

Plants with branching rootstocks; stems 20–40 cm, puberulent to long pubescent, and glandular. Stem leaves 3–6 cm long, ovate to elliptic, irregularly dentate to entire, sessile or the lower ones short petiolate. Heads solitary or few; ligules 15–25 mm long; bracts 10–15 mm high, pubescent below, glandular toward the tips. Wet, boggy areas; southern Rocky Mountains. Plants with very leafy stems and 5–12 pairs of stem leaves, as compared with 3–5 pairs in the typical variety, are distinguished as var. *aspera* (Greene) Boiv.

*Arnica parryi* Gray                                         NODDING ARNICA

Plants with creeping rhizomes; stems usually solitary, 20–40 cm high, often woolly-pubescent toward the base, somewhat glandular above. Stem leaves 5–15 cm long, lanceolate to lance-ovate, the lowermost petiolate. Heads usually several, nodding in bud; discoid bracts 10–15 mm high, usually glandular to glandular pubescent. Mountain meadows; southern Rocky Mountains.

*Arnica rydbergii* Greene                    NARROW-LEAVED ARNICA

Plants with short, branched, scaly rootstocks; stems 10–30 cm high, glandular pubescent to subglabrous. Basal leaves petiolate, oblanceolate or spatulate, to 7 cm long; stem leaves 3–10 cm long, lanceolate to spatulate, entire or nearly so. Heads solitary or few; rays 1–2 cm long; bracts 10–15 mm high, glandular, and sparsely pubescent to subglabrous. High slopes and shale slides; southern Rocky Mountains.

*Arnica sororia* Greene                              TWIN ARNICA

Plants with slender, freely rooting rootstocks; stems 10–40 cm high, glandular pubescent. Basal leaves lanceolate to narrowly lanceolate, glandular pubescent. Heads solitary, rarely 2 or 3; ligules 15–25 mm long; bracts 10–15 mm high, sparingly pubescent and glandular. Montane grasslands and forest openings; southern Rocky Mountains.

***Artemisia***      wormwood

Biennial or perennial shrubs or herbs, usually with a conspicuous odor. Leaves alternate, varying from entire to much dissected, green and hairless to dense white woolly. Flowers small, usually in spike-like panicles, discoid with only tubular florets. Seeds bearing no pappus. The wind-borne pollen of some species often causing hay fever.

1. Shrubs with woody bases and branches. ....................................................... 2
   Herbs or subshrubs with bases not woody
   and branches usually herbaceous. ........................................................ 3
2. Leaves entire, lanceolate to oblanceolate. ........................................... *A. cana*
   Leaves 3-lobed at the tip, cuneate. .............................................. *A. tridentata*
3. Plants glabrous or nearly so. ................................................................ 4
   Plants hairy, silky, or woolly. ............................................................ 6
4. Leaves mostly entire and undivided,
   rarely 3-cleft. ................................................................ *A. dracunculus*
   Leaves pinnatifid and divided. ............................................................ 5
5. Leaves cleft into very narrow linear
   divisions. .......................................................................... *A. campestris*
   Leaves cleft into lanceolate toothed
   divisions. ............................................................................. *A. biennis*
6. Leaves hairy or silky hairy, but not
   woolly. ................................................................................... 7
   Leaves white woolly, at least on
   underside. ................................................................................ 8
7. Leaves 2 or 3 times divided. ........................................................ *A. frigida*
   Leaves only once or twice divided. ............................................. *A. absinthium*

8. Leaves entire or at most coarsely lobed. ............................................................... 9
   Leaves much divided; lobes narrow, mostly linear. ................................................................. 11

9. Leaves white woolly on both sides. ...................................................... *A. ludoviciana*
   Leaves white woolly only on lower side. ................................................................. 10

10. Leaves entire or sharply toothed. ........................................ *A. tilesii* ssp. *unalaschensis*
    From coarse, woody roots; leaves 5–10 cm long and usually with rolled margins. ............................................................................... *A. longifolia*

11. Ultimate leaf segments usually less than 1 mm wide, entire or subentire. ................................................................. 12
    Ultimate leaf segments usually more than 1 mm wide, usually toothed or lobed. ................................................................. 13

12. Plants usually sterile; herbaceous; ultimate leaf segments often 1- or 2-toothed. ................................................................... *A. pontica*
    Plants usually fertile; subshrubs; ultimate leaf segments entire. ........................................................................... *A. abrotanum*

13. Pubescence of leaves loosely woolly or absent; most or all of the leaves basal. .......................... *A. norvegica* var. *saxatilis*
    Pubescence densely woolly; most of the leaves cauline. ........................................................................... 14

14. Leaves deeply dissected to near midrib into linear divisions with rolled margins. ........................................................................... *A. michauxiana*
    Leaf segments doubly divided with broad ultimate divisions. ........................................................................... *A. vulgaris*

*Artemisia abrotanum* L. LEMONWOOD

A perennial plant, somewhat shrubby at the base, 50–200 cm high. Leaves 3–6 cm long, lemon-scented when bruised, thinly tomentose below, green and subglabrous above. Inflorescence ample; involucres 2–3.5 mm high. Introduced; occasionally cultivated and locally escaped.

*Artemisia absinthium* L. (Fig. 208) ABSINTHE

A somewhat shrubby plant 50–100 cm high. Stems with many branches, finely hairy. Leaves 5–10 cm long, several times divided into ovate to oblong segments, finely grayish hairy. Inflorescence a large, many flowered, somewhat spike-like panicle; heads stalked, drooping, about 5 mm across. Found in waste places where it has escaped from gardens.

*Artemisia biennis* Willd. BIENNIAL WORMWOOD

An annual or biennial plant 30–100 cm high, with hairless coarse stems, usually reddish for about half their length. Early leaves twice or three times dissected into toothed segments, usually forming a rosette on the ground. Stem leaves and upper leaves 2–7 cm long, once or twice divided into narrow segments, and hairless. Flowers in short compact spikes in axils of upper leaves, forming a dense leafy spike-like panicle. Common; in moist places, slough margins, roadsides, and cultivated fields; throughout the Prairie Provinces.

Fig. 208.   Absinthe, *Artemisia absinthium* L.

*Dicots*

*Artemisia campestris* L.

A very variable glabrous to pubescent perennial or biennial, usually 10–80 cm high. Most of the leaves in a basal rosette, 3–10 cm long, 2 or 3 times divided into narrow segments. Stem leaves, when present, sessile and smaller than basal leaves. Inflorescence a leafy panicle with numerous heads; involucres 2–4.5 mm high. Three varieties are distinguished:

1. Stems usually 10–30 cm high; foliage
   glabrous or nearly so; most leaves
   basal. ............................................................ var. *wormskjoldii* (Bess.) Cronq.
   Stems higher; foliage more pubescent;
   stems leafier. ............................................................................................. 2

2. Stems 30–80 cm high; leaves green,
   sparsely pubescent, mostly cauline;
   basal leaves withering. ........................................ var. *scouleriana* (Bess.) Cronq.
   Stems 20–40 cm high; leaves grayish
   pubescent. ................................................ var. *douglasiana* (Bess.) Boiv.

The var. *wormskjoldii* occurs on subarctic shores and slopes in northern fringe of Boreal forest. Syn.: *A. borealis* Pall. The var. *scouleriana*, usually biennial, found in sandy woods, shores, and openings in Boreal forest and Parklands. Syn.: *A. canadensis* Michx.; *A. caudata* Michx. The var. *douglasiana* is the common, usually short-lived perennial of grasslands and disturbed areas in Prairies and Parklands. Syn.: *A. camporum* Rydb.; *A. bourgeauiana* Rydb.

*Artemisia cana* Pursh                                       HOARY SAGEBUSH

A shrub with somewhat gnarled and twisted shreddy-barked woody stems 30–150 cm high. Leaves 1–3 cm long, silvery hairy on both sides, linear to linear-lanceolate, usually entire, or rarely with toothed points. Yellow flowers crowded into a leafy panicle. Very common; on lighter soils; Prairies, rare in Parklands.

*Artemisia dracunculus* L.                                   LINEAR-LEAVED WORMWOOD

An entirely hairless perennial 50–100 cm high. Leaves 1–6 cm long, narrowly linear, usually entire, but basal leaves occasionally cleft or divided. Inflorescence in a leafy compound panicle. Common; on dry prairie; throughout Prairies.

*Artemisia frigida* Willd. (Fig. 209)                        PASTURE SAGE

A densely silky hairy silvery gray perennial with a somewhat woody base, 15–50 cm high. Leaves 1–3 cm long, 2 or 3 times divided into linear segments. Plants having a distinct odor when handled. Numerous yellowish heads borne in terminal somewhat leafy racemes. Common; on unforested land and in overgrazed pastures; throughout the Prairies and Parklands. Abundant on prairies, its unpalatability enabling it to increase with heavy grazing at the expense of more palatable plants; therefore a useful indicator of overgrazing.

*Artemisia longifolia* Nutt.                                 LONG-LEAVED SAGE

A perennial from a coarse woody much-branched root crown, with densely white woolly stems 40–80 cm high. Leaves white on both sides, 5–10

A.C. Budd

Fig. 209.   Pasture sage, *Artemisia frigida* Willd.

cm long, linear or linear-lanceolate, and often with rolled margins. Upper sides of leaves sometimes losing the tomentum or woolliness with age. Inflorescence a narrow leafy spike-like panicle. Found occasionally; in saline areas, badlands, and shaly outcrops; throughout Prairies.

*Artemisia ludoviciana* Nutt. var. *ludoviciana*                    PRAIRIE SAGE

A white woolly often much-branched perennial 15–60 cm high, very variable in form and size. Leaves 1–7 cm long, white woolly on both sides, but usually slightly less so on upper side. Lower leaves oblanceolate, usually entire, but occasionally with a few lobes; upper leaves lance-linear and entire. Inflorescence in dense axillary spikes, making a leafy panicle; flower heads brownish. Very common; on prairie, especially where conditions are a little moist; throughout the Prairie Provinces. The var. *gnaphalodes* (Nutt.) T. & G., slender sage, is a low perennial from creeping rootstocks, 15–40 cm high, but often growing prostrate on the ground. Stems white woolly and slender. Leaves white woolly or more often pale yellowish woolly on both sides, 2–5 cm long, narrowly linear-lanceolate, usually entire, but the lower ones sometimes toothed and usually conduplicate (folded lengthwise). Flower heads small, in narrow terminal spike-like inflorescences. Common; along slough margins and moister prairie; throughout Prairies and Parklands.

*Artemisia michauxiana* Besser                    MICHAUX'S SAGE

A perennial from a much-branched root crown, 20–50 cm high. Leaves 1–5 cm long, green above, white woolly beneath, and cleft into narrow linear sometimes toothed segments, with margins often rolled. Rather large flower heads in a sparingly leaved terminal spike-like panicle. Found on hillsides and along rivers; southern Rocky Mountains.

*Artemisia norvegica* Fries var. *saxatilis* (Besser) Jepson                    MOUNTAIN SAGE

Plants with a branching caudex; stems 15–50 cm high, loosely villous to subglabrous. Leaves 2–10 cm long, pinnately dissected, green and glabrous above, green or more or less tomentose below. Inflorescence spike-like; heads 4–6 mm high. Alpine slopes; southern Rocky Mountains.

*Artemisia pontica* L.                    ROMAN WORMWOOD

Perennial with a creeping rootstock, often somewhat shrubby, 40–100 cm high. Leaves 1–3 cm long, fragrant, white tomentose on both sides, or glabrate above, 2- or 3-pinnate, with the ultimate segments usually less than 1 mm wide. Inflorescence rarely present; if so, rather narrow, elongate, with involucres 2–3 mm high, more or less tomentose. Introduced and cultivated; occasionally escaped.

*Artemisia tilesii* Ledeb. ssp. *unalaschensis* (Besser) Hulten                    HERRIOT'S SAGE

An erect perennial from a coarse woody base, 30–90 cm high, with a simple fine-hairy stem. Leaves usually linear, 5–15 cm long, sometimes with a few sharp linear teeth, densely white woolly beneath, and smooth and green above. Numerous heads in a dense spike-like panicle. Quite unusual; but found in badlands and river breaks; southern Rocky Mountains. Syn.: *A. herriotii* Rydb.

*Artemisia tridentata* Nutt. SAGEBRUSH

An evergreen shrub 30–150 cm high, strong-scented, with a single short trunk. Leaves 1–4 cm long, wedge-shaped; the apex usually with 3 blunt teeth, but sometimes 4- to 9-toothed. Inflorescence equaling or exceeding the sterile branches, panicle-like, narrow to fairly broad; heads 3–4 mm high. Grasslands; southern Rocky Mountains; very rare, possibly introduced in Prairie.

*Artemisia vulgaris* L. COMMON WORMWOOD

A coarse weedy perennial 30–100 cm high, with a much-branched stem. Leaves 2–8 cm long, dark green above, and densely white woolly below, several times divided; segments somewhat broad and oblanceolate. Numerous heads in erect leafy panicles. An introduced plant; escaped from gardens.

*Aster*        aster

Perennial or biennial herbs with alternate leaves and usually showy blue or white radiate heads. Ray florets and disk florets perfect (bisexual). Ray florets usually 10–50 with bracts in several series. Pappus of capillary bristles or hairs. Asters usually flower in late summer and early fall.

1. Plants annual, with fibrous roots. ............................................................................ 2
   Plants perennial, with rootstocks or root crowns. .................................................... 3
2. Ray florets lacking or inconspicuous. ..................................................... *A. brachyactis*
   Ray florets purple; involucral bracts glandular sticky. ...................................... *A. canescens*
3. Involucres and usually branches of the inflorescence glandular. .............................. 4
   Involucres and branches not glandular. .................................................................... 9
4. Leaves to 20 cm long and 10 cm wide, coarsely toothed or closely serrate. ........... 5
   Leaves smaller, entire or distantly serrate. .............................................................. 6
5. Lower leaves petioled, cordate or subcordate at the base, serrate. ................. *A. macrophyllus*
   Lower leaves sessile, somewhat clasping, coarsely toothed to subentire. ........ *A. conspicuus*
6. Leaves 3–8 cm long, 3–8 mm wide; ray florets 5–8 mm long. ................................ 7
   Leaves 5–15 cm long, 8–20 mm wide; ray florets 10–20 mm long. ........................ 8
7. Plants glabrous except for the glandular pubescence; leaves rather thick, grass-like. ........................................................ *A. pauciflorus*
   Plants glandular pubescent more or less throughout; leaves not thick or grass-like. ................................................. *A. campestris*
8. Ray florets numerous, usually 45–100, 10–20 mm long; leaves entire. ............. *A. novae-angliae*

      Ray florets fewer, usually 20–40, 10–15
      mm long; leaves serrate. ........................................................... *A. modestus*

9.  Heads solitary or 1–3 on a stem. ............................................... 10
     Heads few to many on a stem. ................................................. 11

10.  Stem leaves greatly reduced; stem and
      leaves densely pubescent; plants with a
      root crown. ........................................................................ *A. alpinus*
     Stem leaves not greatly reduced; stem
      and leaves sparsely pubescent; plants
      strongly rhizomatous. ......................................................... *A. sibiricus*

11.  Basal and lower stem leaves petioled or
      narrowed to a petiolar base. ................................................ 12
     Basal and lower stem leaves sessile or
      subsessile. ............................................................................ 16

12.  Lower leaves cordate or subcordate to
      rounded at the base, usually ovate. .................................... 13
     Lower leaves not cordate or rounded at
      the base, usually lanceolate to
      linear-lanceolate. ................................................................ 14

13.  Lower leaves narrowly ovate; stem pubes-
      cent in lines above middle. .............................................. *A. ciliolatus*
     Lower leaves lanceolate; stem glabrous
      except in the inflorescence. .............................................. *A. maccallae*

14.  Upper stem leaves clearly serrate, 10–25
      mm wide; heads 25–30 mm across;
      some foliaceous involucral bracts often
      present. ............................................................................... *A. subspicatus*
     Upper stem leaves entire or obscurely
      toothed, 5–10 mm wide; heads 15–25
      mm across; foliaceous involucral
      bracts absent. .................................................................... 15

15.  Stem leaves linear, the upper ones not
      reduced in size; branches of the
      inflorescence strongly ascending. ..................................... *A. adscendens*
     Stem leaves linear-lanceolate, the upper
      ones reduced in size; branches of the
      inflorescence spreading. .................................................... *A. occidentalis*

16.  Plants glabrous, except sometimes in the
      inflorescence; leaves clasping,
      glaucous. ............................................................................ *A. laevis*
     Plants more or less pubescent. .......................................... 17

17.  Leaves silvery silky on both sides, entire. ......................... *A. sericeus*
     Leaves not silvery silky. ................................................... 18

18.  Pappus double, with the inner series of
      firm long bristles, the outer series of
      bristles about 1 mm long. ................................................. *A. umbellatus*
     Pappus single. .................................................................... 19

19. Involucre usually with some foliaceous
    outer bracts. ........................................................................ *A. eatoni*
    Involucre without foliaceous bracts. ......................................... 20
20. Stem leaves auriculate clasping; plant
    rough pubescent. .......................................................... *A. puniceus*
    Stem leaves sessile or somewhat clasping;
    plant not rough pubescent. ......................................................... 21
21. Leaves narrowly linear, seldom more than
    6–7 mm wide. ............................................................................. 22
    Leaves broadly linear to linear-lanceolate
    or lanceolate-ovate, usually more than
    10 mm wide. ............................................................................... 24
22. Plants very slender; stems arising from
    slender rhizomes; leaves glabrous;
    stem pubescent in lines below the leaf
    bases. ............................................................... *A. junciformis*
    Plants coarse; leaves and stems more or
    less densely pubescent. ......................................................... 23
23. Heads solitary or few at the end of
    branches, 20–25 mm across. ................................... *A. falcatus*
    Heads numerous, more or less secund
    along the branches, 10–15 mm across. .................... *A. ericoides*
24. Ray florets 10–15 in a head. ........................................................ 25
    Ray florets 20–40 in a head. ........................................................ 26
25. Heads 20–40 mm across; leaves in the
    inflorescence ovate-lanceolate. ............................ *A. engelmannii*
    Heads 12–18 mm across; leaves in the
    inflorescence narrowly linear. ................................. *A. lateriflorus*
26. Leaves serrate, 10–35 mm wide; involucre
    4–5.5 mm high; ray florets white. ............................... *A. simplex*
    Leaves entire to subentire, seldom more
    than 18 mm wide; involucre 5–8 mm
    high; ray florets blue. ............................................... *A. hesperius*

*Aster adscendens* Lindl.                                    WESTERN ASTER

A slender erect species 15–50 cm high, with the stem sometimes decumbent at base. Leaves 2–7 cm long, with the lower ones short-stalked and spatulate, the upper ones stalkless and more or less clasping and linear. Heads
about 25 mm across, not very numerous, in a panicle. Bracts oblanceolate,
blunt-pointed, in series of 3–5. Found occasionally; in valleys and favored
localities; Prairies and southern Rocky Mountains.

*Aster alpinus* L.                                          ALPINE ASTER

Stems 5–20 cm high, pilose, and somewhat glandular throughout. Basal
leaves numerous, 2–5 cm long, oblong to lanceolate, rounded at the apex.
Heads solitary, 3–5 cm across; ligules violet to whitish; involucres pilose.
Mountain slopes and forests; southern Rocky Mountains.

*Aster brachyactis* Blake                              RAYLESS ASTER

A slender purplish-stemmed somewhat branching annual 15–60 cm high. Leaves linear, 2–7 cm long, without hairs except for some on margin. Flower heads numerous, often almost hiding the foliage and appearing to be mainly composed of white hairy pappus, 8–12 mm broad in a raceme; the few florets purple and occasionally with a few rudimentary ligulate florets. Involucral bracts very narrow, often purplish. Common; in saline soil and moist places; throughout most of the Prairies and Parklands. Syn.: *Brachyactis angusta* (Lindl.) Britton; *A. angustus* (Lindl.) T. & G.

*Aster campestris* Nutt.                               MEADOW ASTER

Plants with slender creeping rootstocks; stems 10–50 cm high, simple or branched, leafy, densely stipitate glandular. Stem leaves linear to linear spatulate, 2–5 cm long, almost uniform in size, sessile and often slightly clasping, entire, glabrous to somewhat hispid. Heads 1 to many, 15–20 mm wide, in a narrow cyme or panicle; the ligules 5–8 mm long, violet to purple; involucral bracts densely glandular. Montane grasslands; southern Rocky Mountains.

*Aster canescens* Pursh                                CANESCENT ASTER

A low, branching biennial, from a taproot, 10–30 cm high; stems covered with short fine hairs. Leaves 10–35 mm long, with the lower ones spatulate or oblanceolate, the upper ones often linear, varying from entire to slightly toothed. Teeth and apex of leaves bearing a short bristle-like tooth. Flowers 15–25 mm across, bluish purple, usually very numerous. Common; especially along roadsides; Prairies. Syn.: *Machaeranthera canescens* (Pursh) A. Gray; *M. pulverulenta* (Pursh) A. Gray.

*Aster ciliolatus* Lindl. (Fig. 210)                   LINDLEY'S ASTER

A stout erect-stemmed species 30–75 cm high. Leaves thick, sometimes slightly hairy on veins. Basal leaves ovate or cordate, 5–10 cm long, with long stalks; the lower stem leaves with wing-margined stalks, the upper ones often stalkless. Flower heads usually few, blue to violet, 15–30 mm across. Common; in woodlands; throughout Parklands and Boreal forest. Syn.: *A. lindleyanus* T. & G.

*Aster conspicuus* Lindl. (Fig. 211)                   SHOWY ASTER

An erect coarse species 40–100 cm high, with rough and hairy stems. Leaves ovate to obovate, coarsely toothed, rough on upper surface and finely hairy below, 10–15 cm long. Inflorescence a large corymb of violet to blue flowers, each about 4 cm across; bracts and flower stems glandular. Very palatable to livestock. Common; in woodlands; Parklands and Boreal forest, except in the eastern part, and also found abundantly in Cypress Hills.

*Aster eatoni* (A. Gray) Howell                        EATON'S ASTER

A branching species 30–80 cm high. Leaves linear to lance-linear, 5–10 cm long, stalkless, and entire, but not clasping. Outer bracts large and leaf-like. Heads 20–25 mm across, with pinkish or white ray florets. Not common; has been found in river valleys; southern Rocky Mountains. Syn.: *A. mearnsii* Rydb.

Fig. 210.   Lindley's aster, *Aster ciliolatus* Lindl.

*Dicots*

A.C. Budd

Fig. 211. Showy aster, *Aster conspicuus* Lindl.

*Aster engelmannii* (Eat.) Gray                                    ELEGANT ASTER

Plants with woody root crowns; stems tufted, 30–100 cm high, branched in the inflorescence. Leaves 4–10 cm long, elliptic to oblong, rounded or narrowed at the base, sessile, entire or nearly so, glabrous above except along the midrib, sparsely pilose on the veins below. Heads few to many, mostly terminal on short branches, 2–4 cm across; involucre 7–10 mm high, glabrous or somewhat pubescent. Mountain meadows; southern Rocky Mountains.

*Aster ericoides* L.                                    MANY-FLOWERED ASTER

A branching perennial from a thick tufted rootstock, 20–60 cm high. Stems finely hairy. Leaves linear to narrowly linear-lanceolate, 1–5 cm long, entire-margined. Heads numerous, 8–12 mm across, white, usually on one side of the recurved branches. Common; on open prairie, roadsides; throughout Prairies and Parklands. Syn.: *A. pansus* (Blake) Cronquist; *A. multiflorus* Ait.; *A. adsurgens* Greene.

*Aster falcatus* Lindl.                                    WHITE PRAIRIE ASTER

A much-branched perennial, from running rootstocks, 30–80 cm high. Stem rough hairy. Leaves linear to linear-oblong, 1–4 cm long, stalkless, and entire-margined. Inflorescence of a few or single heads at ends of branches; flowers white, 20–25 mm across. Bristle-tipped bracts almost equal in length, with the outer ones at least as high as the involucre. Fairly common; throughout Prairies and Parklands. Syn.: *A. commutatus* (T. & G.) A. Gray.

*Aster hesperius* A. Gray                                    WILLOW ASTER

A species 40–80 cm high, with a much-branching stem. Leaves narrow, 5–15 cm long, sometimes entire-margined and sometimes somewhat toothed. Flower heads numerous, in a branching panicle, varying from white to violet or pink. A very variable species, known under many names. Common; along stream banks, ditches, and moist spots; throughout the Prairie Provinces. Syn.: *A. osterhoutii* Rydb.

*Aster junciformis* Rydb.                                    RUSH ASTER

A slender-stemmed, erect, little-branched species 20–60 cm high. Leaves 2–8 cm long, narrow, and linear, usually with entire margins, stalkless, and having clasping stems. Inflorescence an open panicle; flower heads white, 15–20 cm across. Fairly common; in swamps and bogs; throughout Parklands, Boreal forest, and in the Cypress Hills.

*Aster laevis* L. (Fig. 212)                                    SMOOTH ASTER

A stout-stemmed hairless species 30–100 cm high. Leaves thick, 2–10 cm long, often toothed, ovate or lanceolate, and hairless, with the basal leaves on wing-margined stalks but the upper ones stalkless and often clasping. Numerous flower heads 2–3 cm across, with blue ray florets; pappus somewhat tawny-colored. Common; on moist prairie, around bluffs or clumps of shrubbery, and in open woodlands; throughout the Prairie Provinces.

*Aster lateriflorus* (L.) Britt.                                    WOOD ASTER

A perennial with short stout rhizomes or a branching caudex; stems several, 30–90 cm high, curly villous to subglabrous. Leaves scabrous to subglab-

Fig. 212. Smooth aster, *Aster laevis* L.

rous above, glabrous below, except on the midrib; the principal leaves 5–12 cm long, to 3 cm wide, broadly linear or linear-lanceolate. Inflorescence mostly branched, or rarely simple; heads 10–15 mm across, with rays white or purple-tinged; involucre 4–5.5 mm high. Open woodlands, openings, or beaches; southeastern Parklands and Boreal forest.

*Aster maccallae* Rydb.                                    MACCALLA'S ASTER

Plants with creeping rootstocks; stems 30–60 cm high, often purplish; glabrous below the inflorescence, pubescent in lines above. Lower leaves 8–15 cm long; petioles winged; blades lanceolate, usually serrate, glabrate or nearly so. Inflorescence with few heads; ligules about 15 mm long, blue or purplish; involucres 8–10 mm high, glabrous. Open areas, along mountain streams, and in grassland; southern Rocky Mountains.

*Aster macrophyllus* L.                                    WHITE WOOD ASTER

A tufted perennial with a zigzag twisted brittle stem 40–100 cm high. Leaves thin, slender-stalked, coarsely toothed, with the lower ones cordate-based, 2–10 cm long. Flower heads usually white, 2–3 cm across, usually in a flattish wide corymb. Found occasionally; in extreme southeastern Boreal forest.

*Aster modestus* Lindl.                                    LARGE NORTHERN ASTER

Plants with a long creeping rootstock; stems usually solitary, 30–80 cm high, branching in the inflorescence, densely glandular above or throughout. Leaves rather uniformly 6–15 cm long, lanceolate, entire to sharply serrate, with the base more or less clasping. Heads 1 to many, 2–3 cm wide; ligules about 1 cm long, purple or violet; involucre about 7 mm high, densely stipitate-glandular. Wet areas and bogs; western Parklands and Boreal forest.

*Aster novae-angliae* L.                                    NEW ENGLAND ASTER

A tall erect stout-stemmed species 60–200 cm high. Leaves lanceolate to oblong, entire, 3–12 cm long, stalkless, and clasping stem by a cordate or an eared base. Flower heads numerous, 2–5 cm across, with reddish purple to violet purple ray florets in compact clusters at ends of branches. Fairly common; in moist woodlands and low ground; in eastern Parklands and Boreal forest.

*Aster occidentalis* (Nutt.) T. & G.                       WESTERN MOUNTAIN ASTER

Plants with creeping rootstocks; stems mostly less than 50 cm high, glabrous below, pubescent in lines above. Lower leaves 4–12 cm long, oblanceolate, tapering to a petiolate ciliate base. Heads 1 to few, about 25 mm across; ligules 6–10 mm long; involucre about 6 mm high. Mountain meadows; Rocky Mountains.

*Aster pauciflorus* Nutt.                                   FEW-FLOWERED ASTER

A hairless species, growing from creeping rootstocks, much-branched, 15–50 cm high. Leaves entire-margined, somewhat fleshy, 2–8 cm long, with the upper ones linear and stalkless, the lower ones linear-lanceolate and stalked. Flower heads few, 15–25 mm across, with blue or white ray florets. Rare; found in saline soil; Prairies and Parklands.

*Aster puniceus* L. PURPLE-STEMMED ASTER

A stout purplish-stemmed branching species from a thick rootstock, 60–200 cm high. Leaves 7–15 cm long, lanceolate to oblong-lanceolate, often hairy on midrib below, usually sharp-toothed. Flower heads numerous, 25–35 mm across, with light violet to pale purple ray florets. Fairly common; in swamps and marshlands; throughout Boreal forest.

*Aster sericeus* Vent. WESTERN SILVERY ASTER

A slender-stemmed branching species 30–60 cm high. Leaves 10–35 mm long, covered with dense silvery white silky hairs on both sides; the lower leaves with short winged stalks and the upper ones stalkless, oblanceolate to oblong. Numerous heads about 35 mm across, with reddish violet to violet blue ray florets and tawny pappus. Rare; southeastern Parklands.

*Aster sibiricus* L. ARCTIC ASTER

Plants with slender creeping rootstocks; stems 3–30 cm high, simple or with few branches, usually purple, sparingly pubescent. Leaves 1–6 cm long, obovate or elliptic, entire to serrate above the middle, with the lower ones short petiolate, the upper ones sessile and narrowly clasping. Heads solitary to few in a leafy cymose panicle, with the lateral heads on axillary peduncles 2–3 cm across; ligules 7–12 mm long, purple to violet or sometimes whitish; involucre 5–10 mm high. Open slopes, creeks, and riverbanks; Rocky Mountains, Peace River.

*Aster simplex* Willd. SMALL BLUE ASTER

Plants with stout, long creeping rootstocks; stems stout, 60–150 cm high, glabrous below, pubescent in lines above. Leaves 8–15 cm long, lanceolate to linear, glabrous on both sides, serrate, rarely entire, sessile, narrowly clasping. Heads more or less numerous in a leafy inflorescence; ligules 5–10 mm long; involucres 3–6 mm high, glabrous, with ciliolate margins. Moist areas and shores; eastern Parklands and Boreal forest.

*Aster subspicatus* Nees LEAFY-BRACTED ASTER

An erect species 30–100 cm high. Lower leaves 10–15 cm long, oblanceolate, entire-margined, and with wing-margined stalks; upper leaves shorter with clasping bases. Flowers violet, 15–25 mm across. A mountain species, which may be found in Cypress Hills. Syn.: *A. frondeus* (A. Gray) Rydb. Smaller plants, usually less than 25 cm high, with leaves narrowly oblanceolate and bracts purple-tipped are var. *apricus* (Gray) Boiv. Syn.: *A. foliaceus* Lindl. var. *apricus* Gray.

*Aster umbellatus* Mill. FLAT-TOPPED WHITE ASTER

A tall erect species, from a woody rootstock, 60–200 cm high. Leaves 6–15 cm long, narrowly elliptic to lanceolate, tapering to apex and at base to a short stalk. Very numerous flower heads 12–20 mm across, with white ray florets in a large flat-topped terminal cluster. Pappus has an outer row of short stiff bristles, the inner bristles being long and hair-like. Common; in moist woodlands; throughout Manitoba. Syn.: *Doellingeria umbellatus* (Mill.) Nees. The var. *pubens* Gray differs from the species by the hairy undersides of leaves. Found occasionally in woodlands; eastern Parklands and Boreal forest. Syn.: *Doellingeria pubens* (A. Gray) Rydb.

## Bahia   picradeniopsis

*Bahia oppositifolia* (Nutt.) DC.                                    PICRADENIOPSIS

A much-branched perennial, from creeping rootstocks, 10–25 cm high, with a somewhat woody base. Stem very finely hairy and very leafy. Leaves 1–3 cm long, gray green, very finely hairy, and often several times divided into narrow linear segments. Yellow flower heads, about 12 mm across, at ends of branches. Many tubular florets, but only a few short ray florets. Quite unusual, but has been found on a few occasions in Prairies as a rather persistent weed in cultivated land. It occurs on saline flats and dry plains in the USA. Syn.: *Picradeniopsis oppositifolia* (Nutt.) Rydb.

## Balsamorhiza   balsamroot

*Balsamorhiza sagittata* (Pursh) Nutt.                              BALSAMROOT

A low perennial from a thick edible spindle-shaped root (often more than 5 cm thick) that exudes a balsam or sticky substance with a turpentine-like odor. Long-stalked leaves mostly basal, 10–25 cm long, varying in shape from cordate to hastate or sagittate (arrow-shaped). White woolly on both sides, densely below, sparsely above. Flowers bright yellow, usually borne singly on stems about 30 cm long arising from root crown. Bracts lanceolate and densely white woolly; seeds without pappus. Fairly plentiful; on hillsides and prairie; southern Rocky Mountains.

## Bidens   beggarticks

Annual herbs, with leaves usually opposite. Heads discoid or radiate, yellow or orange. Achenes (seeds) bearing 2 or 4 retrorsely (downward-pointing) barbed awns, by which the seed attaches to coats of animals or clothing and is disseminated.

1. Plants aquatic; the submerged leaves
   filiform-dissected. ............................................................................................ *B. beckii*
   Plants mud or land plants, not aquatic. ...................................................................... 2

2. Ray florets conspicuous; leaves stalkless
   and clasping. ..................................................................................................... *B. cernua*
   Ray florets absent or inconspicuous;
   leaves usually with short stalks, not
   clasping. ............................................................................................................... 3

3. Leaves simple, deeply incised or 3-lobed. ............................................... *B. tripartita*
   Leaves 1- to 3-pinnatifid, with the termi-
   nal leaflet usually petioled. ..................................................................... *B. frondosa*

*Bidens beckii* Torr.                                               WATER-MARIGOLD

A perennial, with the submerged leaves finely filiform-dissected and the emergent leaves simple, lanceolate, sessile, and serrate. Heads terminal, solitary, about 1 cm wide; ligules 10–15 mm long; outer involucral bracts several; pappus of 3–6 awns, longer than the achene. In shallow waters; eastern Parklands and Boreal forest. Syn.: *Megalodonta beckii* (Torr.) Greene.

*Bidens cernua* L.                                   SMOOTH BEGGARTICKS

An erect hairless annual 30–80 cm high. Leaves opposite, 5–15 cm long, toothed, linear-lanceolate, stalkless, clasping the stem, somewhat paler on underside. Flower heads usually nodding, 20–35 mm across; ray florets conspicuous. Long outer bracts usually as long as or longer than ray florets and reflexed. Common; in water and very wet soil; throughout almost the entire Prairie Provinces. Syn.: *B. glaucescens* Greene.

*Bidens frondosa* L.                                   COMMON BEGGARTICKS

An erect often branching species 30–100 cm high, usually with a purplish stem. Leaves slender-stalked, 5–10 cm long, usually divided into 3 or 5 lanceolate segments, slightly hairy below, and toothed. Flowers orange, 12–20 mm across, with 4–8 large outer involucral bracts. Fairly common; in wet places and along stream banks; southeastern Parklands and Boreal forest. The var. *puberula* Wieg. is coarser and has 10–16 outer bracts.

*Bidens tripartita* L.                                   TALL BEGGARTICKS

Annual plants with stems 30–200 cm high, glabrous or nearly so. Leaves simple, usually deeply 3-lobed, serrate, 3–15 cm long. Heads erect; disk about 10–25 mm wide; pappus of 2–4 awns. Introduced; very rare; southwestern Prairies.

**Boltonia**      boltonia

*Boltonia asteroides* (L.) L'Hér.                                   BOLTONIA

A rather stout-stemmed perennial 60–200 cm high. Leaves linear to lanceolate, 5–8 mm long, without stalks. Flower heads numerous, radiate, 20–35 mm across; ray florets varying from white to pink. Achenes bearing a pappus of short scales, often with 2–4 slender bristles, differing from asters, which have a hairy pappus. Rare; has been found in moist soil; eastern Parklands and Boreal forest.

**Brickellia**      brickellia

*Brickellia grandiflora* (Hook.) Nutt.                                   LARGE-FLOWERED BRICKELLIA

Perennial plants with long spindle-shaped rootstocks; stems 30–70 cm high, minutely hirsute throughout. Leaves 3–12 cm long, deltoid to cordate, toothed. Heads several to many, discoid; flowers creamy or greenish white; disks 8–12 mm across and high. Wet places and rocky areas; southern Rocky Mountains.

**Carduus**      plumeless thistle

*Carduus nutans* L.                                   NODDING THISTLE

A branching biennial 60–100 cm high. Leaves deeply divided, lanceolate, 7–15 cm long, very prickly. Flower heads borne singly on long stems, nodding or drooping, 35–65 mm across, purple, rarely white. Involucral bracts in many series, each with a prominent midrib prolonged into a spine. Pappus hairs about 25 mm long, white, roughed but not feathery, thus distinguishing this genus from *Cirsium*, which has plumose or feathery pappus hairs. An introduced weed found, though rarely, at various widely separated locations.

### *Centaurea*    cornflower

Plants somewhat resembling thistles, with a globular involucre, but pappus hairs either very short or absent.

1. Bracts of involucre definitely spine-tipped; flowers cream or yellow. .................................................................. 2

   Bracts of involucre not definitely spine-tipped; flowers pink or purple. ..................................................... 3

2. Stems winged by extensions of leaf bases (decurrent); spines of involucre about 12 mm long. ............................................................ *C. solstitialis*

   Stems not winged; spines of involucre not 12 mm long. ............................................................ *C. diffusa*

3. Most stem leaves deeply cleft into long narrow segments; bracts of involucre fringed at end. ............................................................ *C. maculosa*

   Most stem leaves entire or dentate; bracts of involucre not fringed. ............................................ *C. repens*

### *Centaurea diffusa* Lam.                               DIFFUSE KNAPWEED

A much-branched introduced annual weed 15–60 cm high, sometimes with a fine web-like covering on stems. Leaves once or twice pinnately divided into very narrow segments, some of the uppermost ones entire. Numerous flowering heads in terminal panicles, usually pale yellowish or cream, about 1 cm high. Bracts of involucre spiny-margined, with a terminal spine about 3 mm long. This weed, though not reported in the area, should be looked for because it is a noxious weed in south central British Columbia.

### *Centaurea maculosa* Lam.                             SPOTTED KNAPWEED

An introduced biennial species 30–100 cm high; not as much branched as the preceding species. Leaves pinnately divided into narrow lobes, except those of the inflorescence, which are smaller and entire. Flower heads on long stems; flowers purplish, occasionally white. Bracts of involucre tipped with a short dark fringe, not stiff and spiny. A common weed of roadsides in southern British Columbia; it is expected to invade prairie regions.

### *Centaurea repens* L.                                 RUSSIAN KNAPWEED

An erect perennial 30–100 cm high, growing from coarse, woody, running roots. Stems grooved and ridged; young stems covered with whitish woolly hairs. Leaves 12–75 mm long, pale green, sometimes woolly when young, linear to lanceolate. Lower leaves deeply lobed, upper ones entire-margined. Flower heads numerous at ends of branches, with a hard globular involucre 8–12 mm high and wide, with broad pale green to almost white bracts, with membranous tips. Florets all tubular, pale pink or sometimes purplish. Seeds white. An extremely persistent introduced weed; has been found in shelterbelts and fields; various localities throughout Prairies. Syn.: *C. picris* Pall.

*Centaurea solstitialis* L. YELLOW STAR-THISTLE

An annual plant, with branching stems bearing cottony hairs and growing 30–60 cm high. Basal leaves deeply lobed, up to 12 cm long; upper leaves entire, lanceolate to linear, 10–25 mm long. Involucre has yellowish bracts, many of them tipped by a yellow spine 12–20 mm long. Corollas yellow and all florets tubular. A rare, introduced weed; has been found in gardens; throughout Prairie Provinces.

**Chrysanthemum**     ox-eye daisy

1. Heads with tubular florets only. .................................................................. *C. balsamita*
   Heads with ray florets. ................................................................................................ 2

2. Leaves long, wedge-shaped, tapering into
   the petiole. ...................................................................................................... *C. arcticum*
   Leaves petiolate below, with the middle
   and upper stem leaves sessile. ...................................................... *C. leucanthemum*

*Chrysanthemum arcticum* L. var. *polaris* (Hulten) Boiv.     ARCTIC DAISY

Plants with creeping rootstocks; stems 10–20 cm high, glabrous or nearly so. Leaves fleshy and mostly basal, 4–8 cm long. Heads 2–3 cm across; bracts with conspicuous black tips. Arctic coasts and gravel banks; northeastern Boreal forest.

*Chrysanthemum balsamita* L.     COSTMARY

A fragrant perennial with stems 50–100 cm high, strigose above, glabrous below. Leaves 10–25 cm long, silvery strigose when young, becoming glabrate. Heads in a corymbiform inflorescence; disk 4–7 mm wide; rays mostly lacking, but if present 5–7 mm long, white; bracts with a conspicuous hyaline tip. The rayless form is *tanacetoides* (Boiss.) Boiv. Occasionally spreading from cultivation.

*Chrysanthemum leucanthemum* L.     OX-EYE DAISY

An erect perennial with few branches, 30–60 cm high. Lower leaves somewhat stalked, obovate to spatulate, toothed or incised, 2–8 cm long; upper leaves not stalked, clasping, oblong, and toothed near base. Heads usually borne singly at summit of stem, 2–5 cm across, radiate, with yellow disk florets and white ray florets. Seeds lacking pappus. Plants found in the Prairie Provinces are var. *pinnatifidum* Lecoq & Lamotte. An introduced plant, escaped from gardens and found in meadows and moist roadsides adjacent to forested areas; Parklands and Boreal forest. Syn.: *Leucanthemum vulgare* Lam.

**Chrysopsis**     golden-aster

Perennial much-branched plants with a tufted root crown, usually low-growing and decumbent. Stems very leafy, with stalkless, alternate, and usually entire-margined leaves. Flowers radiate, medium in size, with yellow ray florets. Pappus of the hairy achenes double, the inner ones consisting of rough hairs and the outer ones of small scales or minute bristles.

*Chrysopsis villosa* (Pursh) Nutt.                                    HAIRY GOLDEN-ASTER

A much-branched species 15–60 cm high, from a woody, branching taproot. Stems covered with coarse stiff hairs. Leaves numerous, grayish green, oblong or oblanceolate, alternate, 2–5 cm long, covered with short stiff appressed hairs; the lower leaves occasionally with a short stalk; the upper ones usually stalkless. Flower heads not numerous, radiate, with bright yellow ray florets, 25 mm or more across. This species has been divided into several species or varieties by various authorities, a linear-leaved variety being sometimes found in southern Alberta. Common; on dry sandy prairies and hillsides; throughout Prairies and less common in Parklands. Including *C. hispida* (Hook.) DC.

## *Chrysothamnus*       rabbitbrush

*Chrysothamnus nauseosus* (Pall.) Britt.                             RABBITBRUSH

A low shrubby much-branched plant with a very coarse thick woody root often protruding some distance above the surface of the soil, and appearing very large for the size of the plant. Growing 20–60 cm high with white woolly upright branches bearing very narrowly linear pale grayish green leaves 1–5 cm long, usually erect. Inflorescence dense, in terminal panicles; flower heads discoid, with no ray florets; flowers pale yellow, about 10–15 mm high, with bracts in 2 or 3 series. Copious pappi consisting of dull white hairs. A very local plant; abundant on badlands, eroded hillsides, and occasionally on saline clay flats; Prairies. Syn.: *C. frigidus* Greene.

## *Cirsium*       thistle

Stout erect biennial or perennial herbs with alternate lobed or dentate very prickly spiny leaves. Involucres ovoid or globose with imbricated bracts, which are usually spine-tipped. Flower heads discoid, with all florets tubular; achenes bearing a pappus of plumose (feathery) hairs.

1. Bracts of involucre covered with cob-
    webby hairs, and all bracts spine-tip-
    ped; upper leaf surface with short stiff
    bristles. ................................................................................. *C. vulgare*
   Bracts of involucre only slightly, if at all,
    cobwebby, and if so the inner bracts
    twisted and not spine-tipped. ............................................................. 2
2. Perennial plants from deep creeping root-
    stocks; involucres rarely over 12 mm
    wide; male and female flowers on sepa-
    rate plants. ......................................................................... *C. arvense*
   Plants without creeping rootstocks; invo-
    lucres usually more than 12 mm wide;
    male and female florets on the same
    plant or flower head. ......................................................................... 3

*Cirsium altissimum* (L.) Spreng. var. *discolor* (Muhl.) Fern.     FIELD THISTLE

A tall perennial 90–150 cm high, with a grooved stem. Leaves deeply cleft into linear-lanceolate lobes, stalkless, and large, the lower ones sometimes 30 cm long, prickly, with rolled margins, deep green above, and white woolly beneath. Flower heads 35–50 mm across, with florets pale purple or pink, occasionally white. Rare; has been found in rich soil; southeastern Boreal forest.

*Cirsium arvense* (L.) Scop. (Fig. 213)     CANADA THISTLE

A persistent perennial from deep running rootstocks, 30–100 cm high, usually in large patches. Leaves stalkless, often somewhat clasping, curled and wavy-surfaced, 5–15 cm long, roughly lanceolate, but deeply incised with toothed prickly segments; basal leaves sometimes stalked. Numerous flower heads in large loose corymbs at tops of stems and bearing purple or occasionally white florets. Plants bearing florets of only one sex; some all-male florets with heads often 25 mm across and others about 12 mm across; female florets bearing large quantities of seed. Achene bearing a pappus of white plumose hairs. A common introduced weed, found in great quantities; in waste places,

Fig. 213.  Canada thistle, *Cirsium arvense* (L.) Scop.

fields, and roadsides; across the Prairie Provinces. Caterpillars of the painted lady (or thistle butterfly) and other closely related butterflies feed on thistles and will occasionally almost eradicate it by continual defoliation. The var. *integrifolium* Wimm. & Grab., entire-leaved Canada thistle, differs from the species by having flat almost entire leaves, with spiny margins, the lower leaves being slightly lobed. Found occasionally; throughout the Prairie Provinces.

*Cirsium drummondii* T. & G.                                   SHORT-STEMMED THISTLE

A low-growing species 10–30 cm high, with a hairy, slightly cobwebby stem. Leaves oblanceolate, green on both sides, with triangular lobes and weak spines, occasionally somewhat cobwebby when young, but not white woolly. Heads purple or rose purple, 35–50 mm high, with twisted inner bracts and spiny outer ones. Found occasionally; on somewhat open prairie in wooded areas; Rocky Mountains, Parklands, and Boreal forest.

*Cirsium flodmanii* (Rydb.) Arthur                              FLODMAN'S THISTLE

A slender-stemmed perennial 40–100 cm high, from a deep root. Stem usually branched and with loose cottony hairs. Leaves deeply cleft into lanceolate spiny lobes, white cottony or woolly beneath, and somewhat cottony above, 5–15 cm long. Flower heads rose to rose purple, 3–4 cm across. Not common; found on moist prairie and valleys; Parklands and Prairies. An albino form with cream-colored flowers can be found occasionally. This species can be distinguished from *C. undulatum*, which it resembles, by its perennial habit and by the lower and newer basal leaves often being entire-margined, and its numerous new shoots around the base of the older plant.

*Cirsium foliosum* (Hook.) DC.                                  DWARF THISTLE

Plants biennial or short-lived perennials with a thick taproot, stemless, or stems to 1 m high, very leafy; the uppermost leaves much overtopping the inflorescence. Rosette leaves to 50 cm long, 5 cm wide, entire or shallowly lobed, with margins spinulose-ciliate. Heads solitary or in clusters at the top of the stem, 3–5 cm high; florets whitish to pink or rose. Mountain meadows; Rocky Mountains.

*Cirsium hookerianum* Nutt.                                    HOOKER'S THISTLE

A short-lived perennial with deep taproot; almost stemless or to 1.5 m high; the uppermost leaves overtopping the inflorescence. Foliage thinly tomentose; basal leaves to 30 cm long, to 15 cm wide, pinnately lobed or divided; lobes with stiff marginal spines and spinulose-ciliate. Inflorescence racemose or subspicate, or densely clustered; heads 3–4 cm high, with flowers whitish or pale yellowish. Grasslands and slopes; southern Rocky Mountains.

*Cirsium muticum* Michx.                                        SWAMP THISTLE

A biennial 60–150 cm high, with a branched leafy stem. Leaves 10–30 cm long, deeply cleft into oblong or lanceolate segments with slender spines, white woolly beneath when young, but becoming hairless when more mature. Basal leaves stalked, the upper ones smaller and stalkless. Flower heads few, about 35 mm across, purple; bracts of involucre not spiny, but somewhat sticky and hairy or cobwebby. Found in wet marshy lands; eastern Parklands and Boreal forest.

*Cirsium undulatum* (Nutt.) Spreng. (Fig. 214)          WAVY-LEAVED THISTLE

A white woolly biennial species 30–100 cm high, with a stout, branched, leafy stem. Leaves oblong or lanceolate, with triangular lobes, very prickly, with the lower leaves stalked and the upper ones stalkless, often continuing for some distance down the stem, usually densely white woolly on both sides, although with age the upper surface often becoming bare. Flower heads solitary at ends of the branches, purple or pink, and 3–8 cm across. Very common; on dry prairies and roadsides; throughout Prairies and Parklands. A variety with much larger flowers and the involucre over 4 cm high has been named var. *megacephalum* (Gray) Fernald. Found occasionally; Parklands.

*Cirsium vulgare* (Savi) Tenore          BULL THISTLE

A stout-stemmed biennial, more or less woolly, 50–150 cm high, branched, and leafy up to the heads. Leaves 7–15 cm long, dark green, hairy on both sides, deeply cleft, and very prickly. Leaves usually continuing down the stem, forming prickly lobed wings. Numerous flower heads at ends of branches, purple, 3–5 cm broad and high. Involucral bracts cobwebby, all spine-tipped. An introduced species, found occasionally; on waste land and field borders; throughout Prairies and Parklands. Syn.: *C. lanceolatum* Scop.

## *Coreopsis*          tickseed

*Coreopsis tinctoria* Nutt.          COMMON TICKSEED

A much-branched annual plant with a slender hairless stem 30–70 cm high. Leaves once or twice divided into linear segments, with the very uppermost ones entire, the lower ones sometimes stalked. Numerous flower heads on slender stalks, about 25 mm across, with brownish disk florets and 6–10 broad yellow ray florets having brownish bases. Seeds without a pappus and somewhat resembling small insects, from which the common name of the plant was derived. Very plentiful locally; in moist places, slough margins, low clay flats, irrigation ditches; throughout Prairies. The plant may be very abundant for one season in a particular location and then apparently disappear for several seasons. Often used as a garden flower. Most of the western tickseeds are *C. atkinsoniana* Dougl., distinguishable by a narrow wing around the seed.

## *Echinacea*          purple coneflower

*Echinacea angustifolia* DC.          PURPLE CONEFLOWER

An erect perennial plant 30–60 cm high, with a stiff hairy stem. Lower leaves lanceolate, 2–20 cm long, pointed at apex, and narrowed at base to a slender stalk; upper leaves stalkless or short-stalked, and all leaves either stiffly short hairy or at least very rough to touch. Flowers borne singly at the head of the stems; the conic disk bearing awned, stiff, purplish chaff that almost hides disk florets; ray florets purple, 20–25 mm long, often somewhat reflexed. Found fairly often; on dry benchland; southeastern Prairies and Parklands.

## *Echinops*          globe-thistle

Stems white tomentose, densely so above. ........................................................ *E. exaltatus*
Stem glandular tomentose, with the hairs long
   and colored. ................................................................................ *E. sphaerocephalus*

Fig. 214.  Wavy-leaved thistle, *Cirsium undulatum* (Nutt.) Spreng.

*Echinops exaltatus* Schrad.

A perennial herb with stem 60–150 cm high, simple or with few branches above. Leaves to 40 cm long, 20 cm wide, tomentose and pale below. Inflorescence 3–4 cm in diam; heads 15–20 mm long, lavender blue. Cultivated, and rarely escaped.

*Echinops sphaerocephalus* L.                                   GLOBE-THISTLE

A coarse, branching perennial to 2.5 m high, with the stem spreading hairy. Leaves to 35 cm long, 20 cm wide, white tomentose below. Inflorescence 3–6 cm in diam; heads 15–20 mm long. Cultivated, and occasionally escaped.

**Erigeron**      fleabane

Herbs with basal or alternate leaves. Flower heads usually radiate with many ray florets, usually more than 50. Involucral bracts usually in 1 or 2 series and imbricated (overlapping). Disk florets yellow and ray florets white to pink or purple, although in one rare species they are yellowish. A very large and widespread genus.

1. Heads discoid or with short inconspicuous ray florets. ............................................................... 2
   Heads with well-developed ray florets. .................................................................................. 8
2. Leaves deeply dissected into at least 2 or 3 lobes. ........................................................... *E. compositus*
   Leaves entire. ........................................................................................................................ 3
3. Heads solitary; plants usually less than 10 cm high. ........................................................... 4
   Heads 2 or more; plants usually taller. ................................................................................ 5
4. Involucre 6–8 mm high, densely blue lanate. ....................................... *E. uniflorus* var. *unalaschkensis*
   Involucre 4–6 mm high, pubescence of hyaline hairs. ...................................................... *E. scotteri*
5. Involucre glabrous, 2.5–5 mm high. ................................................................. *E. canadensis*
   Involucre pubescent to glandular pubescent, higher. ........................................................... 6
6. Involucre finely glandular, not or sparsely pubescent. ........................................ *E. acris* var. *asteroides*
   Involucre pubescent, not glandular. ..................................................................................... 7
7. Leaves linear-oblong to oblong, shorter than peduncles of lower flower heads. ........................ *E. elatus*
   Leaves narrowly long linear, exceeding the lower heads in the inflorescence. ................... *E. lonchophyllus*
8. Plants usually less than 20 cm high; heads solitary. .............................................................. 9
   Plants usually more than 20 cm high; heads 2 to many, or both. ..................................... 19
9. Stem leaves 0–3, mainly basal. ........................................................................................... 10
   Stem leaves numerous. ......................................................................................................... 19

10. Leaves deeply dissected into at least 2 or 3 lobes. ............ *E. compositus*

Leaves entire or merely 3-toothed at apex. ............ 11

11. Ray florets bright to deep yellow. ............ *E. aureus*

Ray florets white to pink or bluish. ............ 12

12. Involucres usually more than 10 mm high, densely lanate. ............ 13

Involucres usually less than 8 mm high, not lanate, pubescence short or pilose. ............ 14

13. Ray florets white; some or all leaves 3-toothed at apex. ............ *E. lanatus*

Ray florets reddish purple to whitish; leaves entire. ............ *E. grandiflorus*

14. Leaves long linear; ray florets broad. ............ 15

Leaves narrowly obovate to oblanceolate or spatulate. ............ 16

15. Leaves usually 1–3 cm long; stems usually leafless. ............ *E. radicatus*

Leaves usually 3–7 cm long; stems with 1–3 leaves. ............ *E. ochroleucus* var. *scribneri*

16. Leaves in part with 3- to 5-toothed apex. ............ *E. pallens*

Leaves all entire. ............ 17

17. Plants long stoloniferous; leaves strigose. ............ *E. flagellaris*

Plants not stoloniferous; leaves pilose with spreading hairs. ............ 18

18. Ray florets inconspicuous, about 3 mm long, very narrow; plants usually less than 5 cm high. ............ *E. scotteri*

Ray florets about 10 mm long; plants usually about 10 cm high. ............ *E. arthurii*

19. Ray florets white. ............ 20

Ray florets colored. ............ 25

20. Stem leaves at middle of stem longer than those below or above. ............ *E. hyssopifolius*

Stem leaves gradually reduced upward. ............ 21

21. Ray florets with a minute pappus. ............ *E. annuus*

Ray florets with a normal pappus. ............ 22

22. Leaves 1–3 mm wide, pilose; stem leaves all about 2 mm wide. ............ *E. pumilus*

Leaves more than 3 mm wide; stem leaves reduced in width and length. ............ 23

23. Stem leaves usually more than 10; plants with thick taproot. ............ *E. caespitosus*

Stem leaves usually 5–7; plants with caudex, no taproot. ............ 24

24. Leaves scabrous or pilose on both sides. ........................................................ *E. asper*
    Leaves glabrous below, lightly pubescent
    above. ...................................................................................................... *E. glabellus*
25. Stem leaves at middle of stem longer than
    those above and below. .............................................................. *E. hyssopifolius*
    Stem leaves gradually reduced upward. .......................................................... 26
26. Leaves all or mostly clasping and
    auriculate. ................................................................................. *E. philadelphicus*
    Leaves not clasping or auriculate. ................................................................... 27
27. Ligules filiform, less than 1 mm wide. ............................................................ 28
    Ligules wider than 1 mm. ................................................................................ 29
28. Stem leaves usually 5–7; involucre
    pubescent. ................................................................................................ *E. glabellus*
    Stem leaves numerous; involucre
    glandular. ............................................................................................... *E. speciosus*
29. Stem leaves usually 3–5, sparsely
    pubescent. ........................................................................................... *E. grandiflorus*
    Stem leaves numerous, with the margins
    ciliate, only the midrib pubescent. ........................................ *E. peregrinus*
    ssp. *callianthemus* var. *scaposus*

*Erigeron acris* L. var. *asteroides* (Andrz.) DC.   NORTHERN DAISY FLEABANE
    Biennial or short-lived perennial plants, with a simple or branched cau-
dex; stems 30–80 cm high, hirsute to subglabrous, glandular in the
inflorescence. Basal leaves 5–15 cm long, lanceolate; stem leaves narrower,
sessile. Inflorescence more or less corymbiform, with several to numerous
heads; ligules inconspicuous; involucre glandular to glandular hirsute. Open
areas and moist woods; Boreal forest.

*Erigeron annuus* (L.) Pers.   WHITETOP
    Annual or sometimes biennial plants; stems 30–80 cm high, sparsely to
copiously long pubescent below the inflorescence. Leaves 4–10 cm long, elliptic to broadly ovate, coarsely toothed. Inflorescence leafy, with more or less
numerous heads; ligules 5–10 mm long, white or bluish; involucre finely glandular. Open woods and waste places; Boreal forest.

*Erigeron arthurii* Boiv.   CLIFF DAISY
    A cespitose perennial plant, with stems about 10 cm high. Rosette leaves
2–10 cm long, spatulate to oblanceolate; stem leaves 1–4, smaller than basal
leaves, more or less densely glandular pubescent. Heads solitary; ligules about
1 cm long, white or pinkish; involucre 6–7 mm high, with tips purple, finely
glandular. Alpine gravel slopes; southern Rocky Mountains.

*Erigeron asper* Nutt.   ROUGH FLEABANE
    A woody-rooted perennial 20–30 cm high; stems erect, covered by short
stiff hairs. Basal leaves linear-oblanceolate, 2–10 cm long, with short stiff hairs.
Upper leaves linear to linear-lanceolate. Heads 1–4, 20–25 mm across, with

100–150 very narrow white ray florets. Common on dry prairies and hillsides in Prairies and Parklands. By some authorities considered merely a form of *E. glabellus.*

*Erigeron aureus* Greene                                              YELLOW DAISY

Perennial plants with fibrous roots and branched caudex; stems 2–15 cm high; leaves 2–5 cm long, elliptic to obovate, finely pubescent, with hairs appressed or somewhat loose; stem leaves few and reduced. Heads solitary; ligules 6–10 mm long, yellow; involucre 5–8 mm high, sometimes purplish, sparsely to densely villous; hairs often with purple cross walls. Alpine slopes and meadows; southern Rocky Mountains.

*Erigeron caespitosus* Nutt. (Fig. 215)                      TUFTED FLEABANE

A deep-rooted tufted perennial 10–20 cm high. Leaves finely hairy, with the lower ones spatulate, stalked, 2–8 cm long, the upper ones smaller, oblong, and stalkless. Flower heads borne singly or 3 or 4 to a stem, 2–3 cm across, with many narrow white ray florets. Involucre has 3 or 4 series of unequal bracts, thickened on back. Plentiful; on dry hillsides and prairie; throughout Prairies and Parklands.

*Erigeron canadensis* L.                                        CANADA FLEABANE

A slender bristly hairy-stemmed annual 10–100 cm high, usually with many branches toward the top. All leaves usually somewhat hairy, the lower ones spatulate, short-stalked, 2–10 cm long, and slightly toothed, the upper ones linear, entire, stalkless, and smaller. Flower heads very numerous, in a large open panicle, small, not over 5 mm across; florets, though numerous and white, usually hidden in the pappus. A common native weed; on dry soils, slough margins, and fields; throughout the Prairie Provinces. Syn.: *Leptilon canadense* (L.) Britt.

*Erigeron compositus* Pursh                                  COMPOUND FLEABANE

A low, tufted perennial from a woody root crown, 2–15 cm high, but usually very low. Leaves mostly basal, crowded, and usually twice divided into 3 linear or spatulate divisions. Flower heads 10–15 mm across, with white or rarely violet ray florets, borne singly on short stems. Found occasionally; on eroded hillsides, badlands, and dry or gravelly ridges; Prairies and southern Rocky Mountains.

*Erigeron elatus* (Hook.) Greene                                   TALL FLEABANE

Plants perennial, with slender stems 5–20 cm high, sparingly hirsute. Rosette leaves 2–7 cm long, oblanceolate; stem leaves 3–8, linear-oblong or oblanceolate. Heads 1–8, on flexuous or arching hirsute peduncles; ligules about 3 mm long; involucre 7–8 mm high, hirsute. On light soils; Prairies and Parklands.

*Erigeron flagellaris* Gray                                    CREEPING FLEABANE

Perennial plants, with stems 10–20 cm high, stoloniferous, strigose. Basal leaves 2–4 cm long, spatulate to oblanceolate, strigose; stem leaves 1–3, small, linear. Heads on peduncles 4–10 cm long; ligules about 5 mm long, white to purplish; involucre about 4 mm high, hirsute. Mountain meadows; southern Rocky Mountains.

Fig. 215.   Tufted fleabane, *Erigeron caespitosus* Nutt.

*Erigeron glabellus* Nutt. SMOOTH FLEABANE

A perennial, from a somewhat tufted rootstock, 15–40 cm high. Stems either hairless or sparingly hairy, and usually somewhat decumbent or horizontal at base. Basal leaves 5–10 cm long, oblanceolate and hairless; the upper leaves much smaller and linear-lanceolate. Leaves have only one prominent nerve. Flower heads 1–3, 1–2 cm across, with numerous narrow purple ray florets. Fairly common; in moist woods; Parklands and Boreal forest.

*Erigeron grandiflorus* Hook. LARGE-FLOWERED DAISY

Perennials with stems to 10 cm high, white villous. Basal leaves 2–4 cm long; stem leaves 1–2 cm long, oblanceolate, long pilose. Heads solitary, 3–4 cm across; ligules 8–10 mm long, reddish purple to white; involucre 1 cm high, densely white lanate. Alpine prairies; southern Rocky Mountains.

*Erigeron hyssopifolius* Michx. WILD DAISY

Perennial with extensive fibrous roots; stems solitary to several, 15–35 cm high, subglabrous to occasionally densely spreading villous. Leaves thin, lax; the lowermost leaves linear, reduced, and scale-like, the higher ones to 3 cm long, often with axillary leafy shoots. Heads solitary, rarely 2–5, borne on long peduncles; ligules 4–8 mm long, white to rose purple; involucre 4–6 mm high. Open woods and grassy openings; Boreal forest, rare in Parklands.

*Erigeron lanatus* Hook. WOOLLY DAISY

Low perennials with black scaly rootstocks and branching caudex; stems 3–10 cm high, densely villous. Leaves 1–2 cm long, with the apex rounded; some leaves 3-toothed, densely woolly. Heads solitary; ligules about 1 cm long, white or purplish; involucres 10–12 mm high, densely woolly. Alpine meadows and slopes; southern Rocky Mountains.

*Erigeron lonchophyllus* Hook. HIRSUTE FLEABANE

A somewhat hairy-stemmed biennial with one to several bunched stems 15–60 cm high. Lower stem leaves stalked and narrow oblanceolate, 5–12 cm long; upper stem leaves linear and shorter. All leaves usually smooth and hairless, but often with lower margins hairy. Few heads, about 15 mm across, with short white ray florets about 3 mm long borne erect on head. Not common; found occasionally in wet places; throughout the Prairie Provinces.

*Erigeron ochroleucus* Nutt. var. *scribneri* (Canby) Cronq. YELLOW ALPINE DAISY

Perennial plants with a cespitose caudex; stems about 10 cm high, grayish strigose. Leaves 3–7 cm long, narrowly linear, strigose, and ciliate. Heads usually solitary; ligules about 1 cm long, light yellow to white or lavender; involucres about 5 mm high, with tips purple, squarrose, somewhat lanate. Alpine meadows; southern Rocky Mountains.

*Erigeron pallens* Cronq. PALE DAISY

Low perennial plants, with a cespitose caudex; stems about 10 cm high, long villous, and somewhat glandular throughout. Leaves spatulate, 2–4 cm long, mostly 3-toothed at apex. Heads solitary; ligules about 1 cm long, white; involucre 4–8 mm high, yellowish lanate. Alpine shale slides; southern Rocky Mountains.

*Erigeron peregrinus* (Pursh) Greene ssp. *callianthemus* (Greene) Cronq.
var. *scaposus* (T. & G.) Cronq.                                    WANDERING DAISY

Perennial plants with short rhizomes or caudex; stems 20–40 cm high, glabrous. Leaves linear-oblanceolate to spatulate, tapering to the petiole; stem leaves much reduced; stem often subscapose. Heads solitary; ligules 10–15 mm long; involucres 7–10 mm high, densely glandular. Openings and open woods; southern Rocky Mountains.

*Erigeron philadelphicus* L.                                    PHILADELPHIA FLEABANE

A slender-stemmed upright perennial 30–60 cm high, with a stem either smooth or downy. Lower and basal leaves spatulate, blunt-rounded, toothed, 2–8 cm long, tapering to a short stalk, with upper leaves lanceolate, stalkless, partly clasping, and shorter. Flower heads in a terminal corymb 12–25 mm across, with many narrow pinkish or white ray florets. Fairly common; in moist places, open woodlands; throughout the Prairie Provinces.

*Erigeron pumilus* Nutt.                                              HAIRY DAISY

Usually a short-lived perennial with a taproot; stems 10–20 cm high, simple; spreading-pubescent leaves 3–7 cm long, linear-oblanceolate, copiously villous; cauline leaves somewhat smaller. Heads mostly solitary or few on axillary peduncles; ligules 6–15 mm long, white or occasionally rose purple; involucre 4–7 mm high, densely villous. Dry grassland, especially on light soils; Prairies and Parklands.

*Erigeron radicatus* Hook.                                       DWARF FLEABANE

A perennial with erect stem 3–10 cm high, somewhat stiff hairy. Basal leaves 10–35 mm long, linear to oblanceolate, with stem leaves few and linear. Flower heads few or solitary, 1–2 cm across, with white ray florets. Rare; slopes and crests of eroded hills; Prairies and southern Rocky Mountains.

*Erigeron scotteri* Boiv.                                            DWARF DAISY

Perennial with a short taproot; stems 2–10 cm high, solitary or few together. Rosette leaves petiolate, entire, ciliate, more or less pubescent. Heads solitary; ligules about 3 mm long, pinkish; involucre 5–6 mm high, densely glandular. Alpine slopes; Rocky Mountains.

*Erigeron speciosus* (Lindl.) DC.                                 SHOWY FLEABANE

An erect-growing species, with smooth or slightly hairy leafy stems 30–40 cm high. Basal leaves 5–10 cm long, linear-oblanceolate, tapering at base to a winged stalk, each with 3 ribs and a fringe of marginal hairs; stem leaves linear-lanceolate, clasping at broad base, with a marginal fringe of hairs. Flower heads large, 25–35 mm across, with numerous, narrow, blue or violet ray florets. Southern Rocky Mountains.

*Erigeron uniflorus* L. var. *unalaschkensis* (DC.) Boiv.            PURPLE DAISY

Low perennial plants with short rootstocks; stems 3–10 cm high, villous with blackish hairs. Leaves mostly basal, 1–4 cm long, spatulate, hirsute. Heads solitary; ligules 2–3 mm long, purplish or white, involucre black purple, densely pubescent with long hairs with blackish purple cross walls. Arctic and alpine tundra; northeastern Boreal forest and southern Rocky Mountains.

***Eupatorium***    thoroughwort

Tall erect perennials with opposite or whorled leaves. Inflorescence a large terminal corymb-like cluster of discoid flower heads (without ray florets), either white or purplish, with several series of overlapping involucral bracts, unequal in length, outer ones quite small.

Leaves stalked and usually in whorls of 3–6
   around the stem; flowers purple or
   purplish-tinged. ................................................................................ *E. maculatum*
Leaves connate-perfoliate (leaves opposite
   with their bases united around the stem);
   flowers white. ................................................................................ *E. perfoliatum*

*Eupatorium maculatum* L.                                    SPOTTED JOE-PYE WEED

An erect-stemmed somewhat branching plant 50–200 cm high; stem somewhat purplish or purple-spotted, with soft short hairs near the top. Leaves ovate to ovate-lanceolate, sharp-pointed at the apex, coarsely toothed, somewhat hairy beneath, 10–15 cm long, short-stalked, in whorls of 3–6 at internodes of stems. Inflorescence consisting of many flower heads forming rounded to flat-topped clusters up to 15–20 cm across, very conspicuous. Flower heads pinkish or purplish and consisting of pinkish lilac tubular florets; each head about 1 cm high and about 6 mm across. Pappus of white hairs. Pinkish involucral bracts unequal in size. Frequent in moist ground and in low moist woodland openings; southeastern Parklands and Boreal forest. The var. *bruneri* (Gray) Breitung, bruner's trumpetweed, very similar to the species, but with the underside of somewhat narrower leaves densely covered with a fine, soft, velvety hairiness. Stem hairy in its entire length. Found in similar locations, and of same distribution as the species, possibly advancing slightly farther west. Syn.: *E. bruneri* Gray.

*Eupatorium perfoliatum* L.                                                    BONESET

An erect stout hairy-stemmed perennial 50–150 cm high. Lanceolate leaves 10–20 cm long, opposite, each pair joined at base and encircling stem. Leaves wrinkled, hairy on undersides. Inflorescence a large terminal cluster of white flowers. Flower heads about 6 mm high, with all tubular florets and bracts of unequal length. Fairly common; in wet places; southeastern Parklands and Boreal forest.

***Gaillardia***    gaillardia

*Gaillardia aristata* Pursh (Fig. 216)          GREAT-FLOWERED GAILLARDIA

An erect-stemmed perennial 20–60 cm high, with a somewhat hairy stem. Lower leaves oblong to spatulate, sometimes lobed or pinnatifid, grayish hairy, 5–12 cm long, tapering to a stalk; upper leaves stalkless, smaller, and usually entire or slightly lobed. Flower heads terminal, 3–7 cm across, radiate, with a rounded purple disk. Ray florets 10–18, wedge-shaped, with 3 short triangular lobes at apex, yellow, but often with a purplish tinge at base. After flowers fall, receptacle somewhat globose and bearing achenes or seeds, which have a pappus of lanceolate papery bristled scales. Common; on dry prairie; a roadside plant throughout entire unforested portion of the Prairies and Parklands.

A.C. Budd

Fig. 216.   Great-flowered gaillardia, *Gaillardia aristata* Pursh.

*Dicots*

*Galinsoga*      galinsoga

*Galinsoga ciliata* (Raf.) Blake                                   GALINSOGA

An erect or spreading annual 30–60 cm high, with many branches; stems often rooting at nodes when decumbent. Leaves opposite, stalked, ovate with rounded teeth, 15–50 mm long. Numerous terminal or axillary flower heads, 3–6 mm across, with 4 or 5 very small ray florets. Introduced from South and Central America, a weed that is invading the southeastern part of the Prairies and Parklands as a garden weed. Another species, *G. parviflora* Cav., yellow-weed, subglabrous to strigose, has been found once in the southeastern Parklands.

*Gnaphalium*      cudweed

Annual or biennial woolly and sometimes glandular herbs, with narrow entire alternate leaves. Flower heads discoid, with no ray florets, borne in panicles of crowded clusters on the stem. Bracts of involucres usually dry and membranous, and in several overlapping series. Achenes bearing a pappus of white hairs.

1. Tall plants without leafy bracts below
   flower heads; bracts of involucre well
   overlapping,    white,    dry,    and
   membranous. ............................................................................................... 2

   Low plants with leafy bracts under flower
   heads; bracts of involucre slightly over-
   lapping, yellowish or white. ............................................................................................... 3

2. Stem and leaves glandular; leaves continuing down the stem and forming a
   wing (decurrent). .................................................................... *G. viscosum*

   Stem and leaves not glandular; leaves not
   decurrent. .................................................................... *G. microcephalum*

3. Leaves broad, oblanceolate to spatulate;
   plant loosely woolly tufted. .................................................................... *G. palustre*

   Most of leaves narrow, linear, or oblanceolate;  plant  covered  with  short
   appressed hairs. .................................................................... *G. uliginosum*

*Gnaphalium microcephalum* Nutt.                      COMMON CUDWEED

A perennial white tomentose plant, with an erect stem 30–60 cm high. Leaves 3–8 cm long, stalkless, linear-lanceolate, densely white woolly beneath. Flower heads about 6 mm high, in numerous clusters at head of stem. Involucral bracts membranous and dry, white tinged with brown. In open places in moist wood; southern Rocky Mountains.

*Gnaphalium palustre* Nutt.                      WESTERN MARSH CUDWEED

A low-growing much-branched loosely woolly annual plant 5–20 cm high. Leaves 1–2 cm long, spatulate to oblong, stalkless, and loosely white woolly. Heads small, about 3 mm high, with woolly white or yellowish involucres, in leafy-bracted small clusters. Found occasionally; in sloughs and wet places; Prairies and Parklands.

*Gnaphalium uliginosum* L. LOW CUDWEED

Somewhat similar to *G. palustre*, but with narrow linear leaves; the woolliness or hairiness appressed and not loose and spreading. Rare in the Prairies and Parklands, but found in wet places in Boreal forest.

*Gnaphalium viscosum* HBK. CLAMMY CUDWEED

A biennial, from a taproot, 30–70 cm high, with a glandular hairy stem. Leaves linear to lanceolate, 5–10 cm long, bright green but glandular hairy above and white woolly beneath, stalkless and decurrent, continuing as a narrow wing for a distance down the stem. Flower heads discoid, about 6 mm high, white, with dry membranous white or straw-colored involucral bracts. Heads borne in several globose clusters. Rare; has been found in southern Rocky Mountains. Syn.: *G. macounii* Greene.

### *Grindelia*    gumweed

*Grindelia squarrosa* (Pursh) Dunal GUMWEED

Biennial or perennial branching smooth-stemmed plants 20–60 cm high. Leaves oblanceolate, hairless, stalkless, finely and closely toothed, 1–4 cm long, alternate. Plants may be found occasionally with long narrowly spatulate basal leaves up to 7 cm long. Flower heads 2–3 cm across, with an involucre of many series of very sticky gummy bracts, in large numbers at heads of stems. Ray florets bright yellow. Achenes or seeds bearing 2 or 3 awns. Common; on dry prairie, roadsides, and especially on somewhat saline flats and slough margins; throughout entire unwooded part of the Prairie Provinces. Syn.: *G. perennis* A. Nels.

### *Gutierrezia*    broomweed

Erect many-stemmed perennials, with many branches from a deep woody taproot, to 50 cm long. Leaves narrow; flowers small, but very numerous. Native plants, unpalatable to livestock, increasing with overgrazing.

*Gutierrezia sarothrae* (Pursh) Britt. & Rusby COMMON BROOMWEED

A low erect many-stemmed perennial, from a branching crowned woody taproot, 10–30 cm high. The numerous leaves stalkless and entire, narrowly linear, 1–4 cm long. The very numerous small flowers about 3 mm high, in close clusters at the ends of the branches. Unpalatable to livestock and therefore tending to increase in abundance when native pastures are heavily grazed. Extremely drought-tolerant, with deep roots and narrow leaves. Common; on dry prairie lands; throughout Prairies and Parklands. Syn.: *G. diversifolia* Greene.

### *Haplopappus*    ironplant

1. Flowers having only tubular florets. ............................................................ *H. nuttallii*
   Flowers with both tubular and ray florets. .................................................................. 2
2. Leaves pinnately dissected. ........................................................ *H. spinulosus*
   Leaves not pinnately dissected. ................................................................................. 3

3. Leaves entire-margined; plants low, tufted with woody base; pappus white. .................................................................. 4

Leaves usually somewhat toothed; plants with stem dying down to base annually; pappus reddish brown or yellow. ............................................... *H. lanceolatus*

4. Stem leaves few, reduced; plants not glandular. ................................................................................. *H. armerioides*

Stem leaves several, not reduced; plants glandular throughout. ............................................................................... *H. lyallii*

*Haplopappus armerioides* (Nutt.) A. Gray          NARROW-LEAVED STENOTUS

A tufted hairless perennial from a large branching woody root crown, 10–15 cm high. Leaves mostly basal, linear or very narrowly spatulate, 2–7 cm long, pale greenish. Flowering stems almost leafless or with 2 or 3 narrow reduced leaves, 4–8 cm high; flower head about 25 mm across, yellow, with 8–10 ray florets. Not common, but found occasionally on dry eroded hills and badlands; Prairies. Syn.: *Stenotus armerioides* Nutt.

*Haplopappus lanceolatus* (Hook.) T. & G.          LANCE-LEAVED PYRROCOMA

A herbaceous leafy-stemmed plant 10–40 cm high, with stems practically hairless. Basal leaves lanceolate, 2–7 cm long, long-stalked, with a few spine-pointed teeth; stem leaves shorter and stalkless. Flowers in a raceme or panicle of 3–15 heads, 20–25 mm across, yellow. Involucral bracts in 2 or 3 series of unequal length, dry, white at base, but with a greenish tip, sharp-pointed. Not uncommon; in meadows and moist saline areas; throughout Prairies. Syn.: *Pyrrocoma lanceolata* (Hook.) Greene.

*Haplopappus lyallii* Gray          STICKY STENOTUS

A perennial plant with a large underground branching root crown, with few to several erect leafy stems 5–15 cm high. Basal leaves tufted, oblanceolate to obovate, 1–5 cm long, entire; cauline leaves several. Heads solitary; ligules short; involucre 8–12 mm high, like the stem and leaves densely stipitate glandular. Rocky ridges and slopes; southern Rocky Mountains.

*Haplopappus nuttallii* T. & G.          TOOTHED IRONPLANT

A low tufted perennial from a deep woody root, 10–30 cm high. Leaves either oblong or lanceolate to spatulate, 1–3 cm long, grayish green, with short spiny teeth. Flowers borne either singly or in groups of 2 or 3 at ends of branches, about 12 mm across, and without ray florets. Bracts of involucres yellowish with a faint green tip. Fairly common; on dry eroded hillsides and plains; Prairies. Syn.: *Sideranthus grindelioides* (Nutt.) Britt.

*Haplopappus spinulosus* (Pursh) DC. (Fig. 217)          SPINY IRONPLANT

A much-branched perennial from a thick woody root, 15–40 cm high. Leaves bluish green, often finely hairy, 1–4 cm long, and very deeply dissected into narrow segments, which have bristle-pointed teeth. Heads yellow, 6–15 mm across, often very numerous, and with narrow ray florets. Pappus soft, faintly tawny white. Plentiful on dry plains and hillsides; throughout Prairies and Parklands. Syn.: *Sideranthus spinulosus* (Pursh) Sweet.

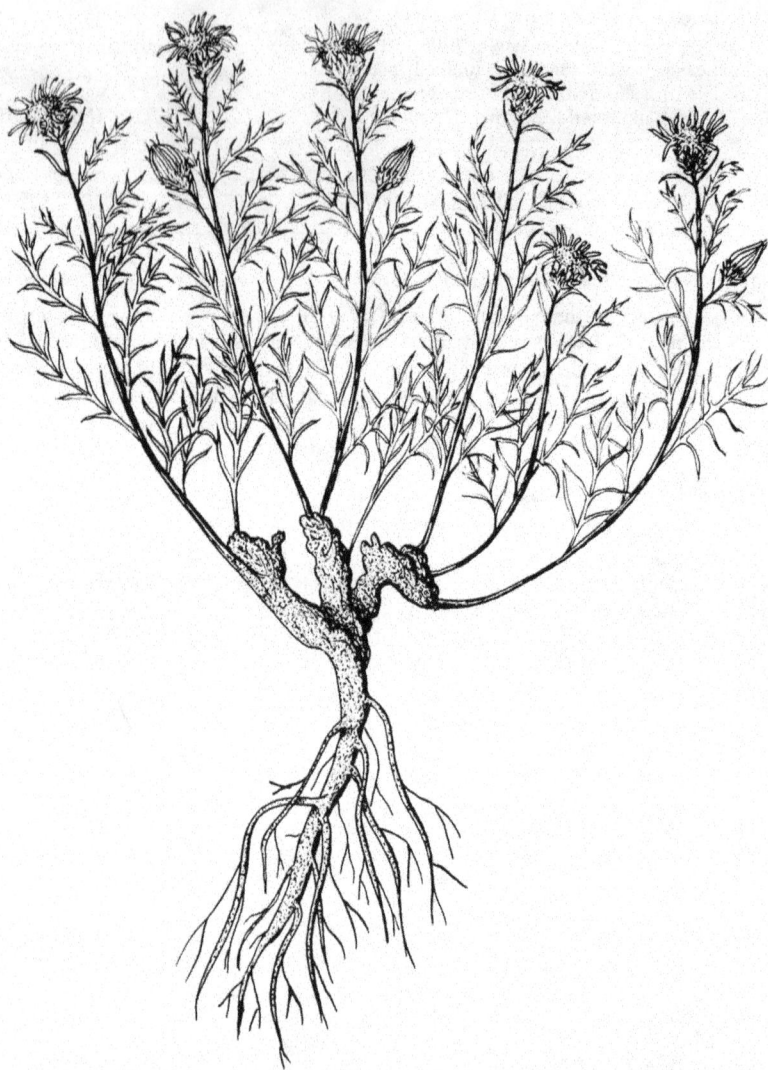

Fig. 217.  Spiny ironplant, *Haplopappus spinulosus* (Pursh) DC.

*Helenium*       sneezeweed

*Helenium autumnale* L. var. *montanum* (Nutt.) Fern.   MOUNTAIN SNEEZEWEED
  A perennial 20–70 cm high, with a stout erect stem. Leaves 4–10 cm long, lanceolate, sometimes slightly toothed, stalkless, and continuing down the sides of the stem. Numerous flower heads at ends of branches, 2–3 cm across, with a high rounded yellow disk, and yellow ray florets up to 12 mm long. Bracts of involucre short, reflexed; ray florets usually somewhat reflexed. Achenes, which remain on almost globose receptacle after maturity, bearing a pappus of several sharp-pointed long scales. Common in low meadows, beside water courses, and in low places; Prairies, less common in Parklands. Thought to be somewhat **poisonous** to livestock and producing a taint in milk. Syn.: *H. montanum* Nutt.

*Helianthus*       sunflower

  Tall, coarse, erect annual or perennial herbs with opposite or alternate undivided leaves. Flowers large, with yellow ray florets. Receptacle or disk flat or rounded, broad, and chaffy. Disk florets yellow, brown, or purple. Achenes with a pappus of small scales or awns soon falling off. A difficult genus to classify.

1. Annual plants. ......................................................................................... 2
   Perennial plants. ..................................................................................... 3
2. Lower leaves usually cordate-based and toothed; bracts of involucre conspicuously hairy on margins; scales between disk florets not very hairy. ........................ *H. annuus*
   Leaves not cordate-based, entire; bracts of involucre not conspicuously hairy on margins; scales between disk florets each with tuft of hairs. ...................... *H. couplandii*
3. Disk of flower heads dark brown or purple. ....................... *H. laetiflorus* var. *subrhomboideus*
   Disk of flower heads yellow or light brown. ............................................................... 4
4. Leaf blades ovate; rootstocks bearing tubers. ............................................. *H. tuberosus*
   Leaf blades lanceolate or linear-lanceolate; rootstocks not tuber-bearing. ...................... 5
5. Leaves rough on both sides and usually somewhat folded lengthwise. ...................... *H. maximilianii*
   Leaves rough above but somewhat hairy beneath, not folded lengthwise. ...................... *H. nuttallii*

*Helianthus annuus* L. var. *giganteus* Hort.       SUNFLOWER
  The sunflower with large flower head 15–30 cm across, cultivated for its seed, and occasionally escaped. More common is *H. annuus* f. *lenticularis* (Dougl.) Boiv., showy sunflower. A tall stout-stemmed erect annual 50–200 cm high. Stem rough, sometimes with short bristly hairs. Leaves coarsely toothed,

*Dicots*                                    COMPOSITAE (ASTERACEAE) — 761 ovate, and often cordate-based, usually alternate, 10–20 cm long. Heads 7–15 cm across, with a dark brown or purple disk, usually about 5 cm across. Common; on clay and heavier soils and plentiful along roadsides; Prairies. Syn.: *H. annuus* L. ssp. *lenticularis* (Dougl.) Cockerell.

*Helianthus couplandii* Boiv.                                    PRAIRIE SUNFLOWER

An annual species 30–100 cm high, with a somewhat hairy stem. Leaves 2–7 cm long, with fairly long stalks, entire-margined, ovate to ovate-lanceolate, rough on both sides, usually with somewhat wedge-shaped bases, never cordate-based. Flower heads 3–8 cm across, usually with a rather small brown raised disk. Scales between disk florets each bearing a small tuft of hairs, which aid in distinguishing this annual species from *H. annuus*. Some authorities refer to this species as *H. aridus* Rydb. This species is found only on sandy soils. Plentiful; along roadsides in sandy areas and on sand dunes. Often where sandy soils merge into heavier ones, the demarcation line can be seen by the change from this species to the showy sunflower. Syn.: *H. petiolaris* AA.

*Helianthus laetiflorus* Pers. var. *subrhomboideus* (Rydb.) Fern.
                                                          BEAUTIFUL SUNFLOWER

An erect perennial from a rootstock, 30–100 cm high, with a sparingly hairy stem often tinged with red. Leaves opposite, 3-veined, 5–10 cm long, somewhat rhombic (obliquely 4-sided); the upper leaves rhombic-ovate or rhombic-lanceolate, the lower ones somewhat spatulate. Flower heads 2–7 cm across, with ray florets 15–35 mm long, disk purplish. Found occasionally; on moist grassland; Prairies and Parklands.

*Helianthus maximilianii* Schrad. (Fig. 218)    NARROW-LEAVED SUNFLOWER

A plant spreading by underground stems. A short-stemmed perennial 50–200 cm high. Stem rough, usually with coarse stiff hairs on upper part. Leaves stalkless or very short-stalked, 7–15 cm long, narrowly lanceolate, usually folded lengthwise along midrib, rough on both sides; the lower leaves opposite, the upper ones alternate. Yellow-disked flower heads 5–8 cm broad, on short stalks. Common; along roadsides, prairies, and valleys; throughout eastern Parklands and Boreal forest.

*Helianthus nuttallii* T. & G.              TUBEROUS-ROOTED SUNFLOWER

A perennial with creeping rootstocks, often with spindle-shaped fleshy roots, 1–3 m high. Leaves lanceolate, stalkless or very short-stalked, 5–15 cm long, somewhat narrow, and very rough on both sides. Flower heads on long stalks, 35–65 mm across, with a yellowish disk. Common; in moist and saline soils; throughout almost entire Prairie Provinces, especially in the southern portion. The var. *subtuberosus* (Britt.) Boiv. is distinguishable by having leaves lanceolate to linear-lanceolate, 6–15 cm long; the lower leaves opposite but the upper ones sometimes alternate. Flowers with yellowish brown disk, 3–9 cm across. A plant of moist meadows, slough margins, and wet roadsides; sometimes found in Prairies.

Fig. 218.   Narrow-leaved sunflower, *Helianthus maximilianii* Schrad.

*Helianthus tuberosus* L.                                    JERUSALEM ARTICHOKE

A perennial with fleshy edible tubers on roots, 1–3 m high. Stems hairy and with many branches. Ovate or oblong leaves 10–20 cm long, rough above and finely hairy below, tapering at the base to a stalk. Numerous heads 5–8 cm across, with a yellow disk and 12–20 yellow ray florets. Often cultivated for edible tubers. Fairly common; in moist soil, river flats, and other alluvial soil; southeastern Parklands and Boreal forest.

**Heliopsis**        ox-eye

*Heliopsis helianthoides* (L.) Sweet var. *scabra* (Dunal) Fern.

ROUGH FALSE SUNFLOWER

A perennial from a cluster of fibrous roots, 40–100 cm high. Leaves ovate to ovate-lanceolate, opposite, 3-ribbed, 5–12 cm long, rough on both sides, abruptly narrowed at base to a short stalk, with large sharp teeth on margins. The few flower heads usually borne singly on long stalks, 5–6 cm across, with yellow ray florets 15–25 mm long. Found occasionally; on dry soil, banks, and sandhills; southeastern Parklands and Boreal forest.

**Hymenopappus**        hymenopappus

*Hymenopappus filifolius* Hook.                    TUFTED HYMENOPAPPUS

A perennial from a deep woody root, 15–30 cm high. Numerous stems from root crown, usually tufted, white woolly when young, becoming smooth and purple with age. Leaves 2–7 cm long, white woolly when young, turning smooth and pale green when older. Lower leaves stalked, each divided several times into very narrow almost thread-like segments. Flower heads usually few, borne terminally, 12–20 mm across, discoid, yellow. Densely woolly bracts in 1 or 2 series. A very drought-tolerant plant found on badlands, eroded slopes, and gravelly hills; Prairies and southern Rocky Mountains. Mostly considered to be var. *polycephalus* (Osterh.) B. L. Turn.

**Hymenoxys**        rubberweed

Heads solitary on a long scape or stem; leaves
  all basal and entire. ...................................................................................... *H. acaulis*
Heads several to many; leaves alternate and
  divided into linear segments. ........................................................ *H. richardsonii*

*Hymenoxys acaulis* (Pursh) Parker            STEMLESS RUBBERWEED

A low grayish perennial from a coarse taproot, 10–15 cm high. Linear-oblanceolate leaves, all basal and clustered, 1–5 cm long, silky hairy, entire-margined. Yellow flowers 2–3 cm across, with 10–15 rather broad ray florets, borne singly on leafless stems 10–15 cm long. Not common; found on eroded hillsides; western Prairies. Syn.: *Tetraneuris acaulis* (Pursh) Greene.

*Hymenoxys richardsonii* (Hook.) Cockerell (Fig. 219)   COLORADO RUBBERWEED

A perennial, 10–20 cm high, from a coarse woody taproot, often with a divided woolly crown. Stems almost hairless. Leaves alternate, mostly basal, 5–10 cm long, divided into very narrow linear lobes, their divisions 2–3 cm

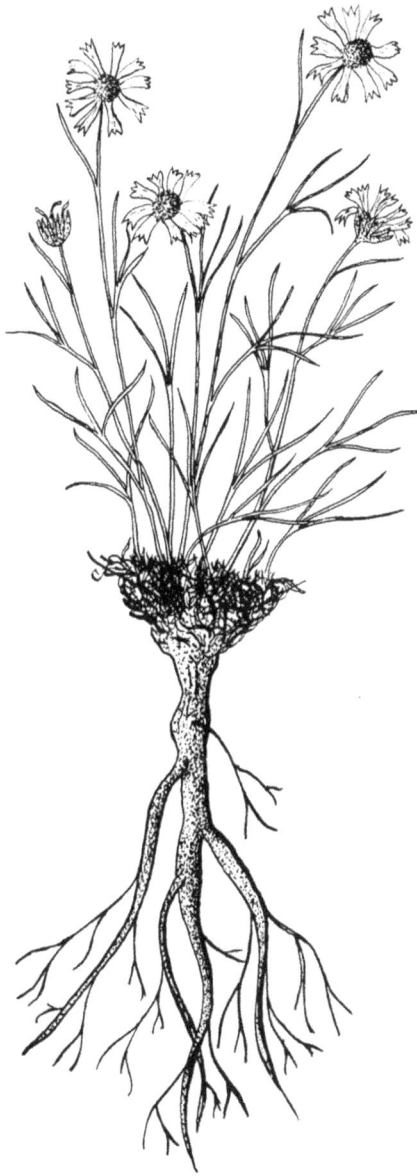

Fig. 219.    Colorado rubberweed, *Hymenoxys richardsonii* (Hook.) Cockerell.

long. Flowers usually in a flat-topped cluster at ends of branches, yellow, about 2 cm across. Unpalatable to livestock, therefore tending to increase with heavy grazing. Common; on open prairie; throughout Prairies. Syn.: *Actinea richardsonii* (Hook.) Kuntze.

***Liatris***    blazingstar

Perennial plants from a globular corm or tuber-like root, with narrow undivided leaves. Flowers purple, in racemes or spikes, discoid, very showy; all florets tubular; bracts in several overlapping series; pappus of white, purplish, or tawny hairs.

Flower heads with 4–6 florets; pappus plumose or feathery. ............................................................................ *L. punctata*

Flower heads with more than 15 florets; pappus barbed but not plumose. ........................................................ *L. ligulistylis*

*Liatris ligulistylis* (A. Nels.) K. Schum.    MEADOW BLAZINGSTAR

An erect plant 30–60 cm high, from a corm or tuber. Leaves 3–10 cm long, linear-oblanceolate, bright green with a rather conspicuous whitish midrib. Heads 2–3 cm across, reddish purple, on short stalks in a long raceme. Bracts of involucre in many series, green with purple tips; the rounded tops irregularly jagged, as if torn. Fairly common; in moist places, slough margins, forest openings, and sandhills; throughout Parklands, rare in Prairies.

*Liatris punctata* Hook. (Fig. 220)    DOTTED BLAZINGSTAR

A perennial from a stout, often corm-like rootstock, 10–30 cm high, often decumbent. Stiff linear leaves 5–15 cm long, densely covered with minute dots or depressions, smooth, but with a marginal fringe of short white hairs. Heads in a dense crowded spike, usually about 15 mm wide, with 4–6 pinkish purple tubular florets, usually showing many white plumose pappus hairs. A common species; on dry hillsides and grasslands; Prairies and Parklands.

***Madia***    tarweed

*Madia glomerata* Hook.    TARWEED

A very sticky, strong, and peculiarly scented annual 20–50 cm high, with glandular hairy stem and leaves. Leaves alternate, linear, entire, 2–7 cm long. Heads about 6 mm high and 3 mm wide, in close congested terminal and axillary clusters, with up to 5 small yellow ray florets, often none, usually hidden by the sticky glandular hairy involucral bracts. Found occasionally; in moist open spots in woodlands or slough margins; Prairies, especially in areas adjacent to Rocky Mountains and Cypress Hills.

***Matricaria***    chamomile

1. Ray florets absent; leaf segments linear; plants with a pleasant pineapple-like odor. ............................................................................ *M. matricarioides*

   Ray florets present; leaf segments thread-like. ........................................................................................ 2

Fig. 220.   Dotted blazingstar, *Liatris punctata* Hook.

2. Plant with distinct odor; disk conical ................................................. *M. chamomilla*
   Plant odorless; disk merely rounded. ........................................................ *M. maritima*

### *Matricaria chamomilla* L.                         WILD CHAMOMILE

A branching annual plant 15–60 cm high, with finely divided leaves. Flower heads about 2 cm across, with white ray florets and a conical disk of yellow disk florets. Whole plant has a distinctive odor, faintly resembling pineapple, but well known to older people, who were dosed with chamomile tea in their childhood. An introduced weed, which has been reported from several locations in the Prairie Provinces.

### *Matricaria maritima* L. (Fig. 221)               SCENTLESS CHAMOMILE

An annual with a much-branched hairless stem 20–70 cm high. Numerous leaves stalkless and several times divided into narrow thread-like segments. Flower heads numerous, at the ends of the branches, 10–25 mm across, radiate, with yellow disk and white ray florets. An introduced plant found occasionally; in waste places and roadsides; across the Prairie Provinces, and becoming more common in Parklands. Syn.: *M. inodora* L.

### *Matricaria matricarioides* (Less.) Porter         PINEAPPLEWEED

A hairless annual plant 5–40 cm high, with leaves several times divided into linear segments. Leaves usually compact and copious, much more so than in *M. maritima*. Numerous flower heads, discoid, with tubular florets only, at the ends of the branches, about 6 mm across, conical, with yellow florets and greenish yellow involucral bracts with membranous margins. Plants, when squeezed, have a strong pleasant odor of pineapple. An introduced plant, which has become very persistent and plentiful locally; in waste places, and especially around farm or ranch yards and driveways; rapidly spreading throughout the Prairie Provinces. Syn.: *Chamomilla suaveolens* (Pursh) Rydb.

### *Petasites*      colt's-foot

Perennial, usually woodland plants, with thick creeping rootstocks and scaly-bracted flowering stems, which usually appear before the leaves. Leaves long-stalked and all basal, usually white woolly on underside. Flower heads in terminal clusters on stem, radiate, often unisexual, with male and female florets on separate plants. Achenes bearing a pappus of soft white hairs.

1. Leaves cordate to triangular, toothed but
    not deeply cleft. ................................................................................. *P. sagittatus*
   Leaves round, kidney-shaped, or triangu-
    lar and deeply cleft. ...................................................................................... 2
2. Leaves almost round in outline, deeply
    cleft almost to base. ............................................................................. *P. palmatus*
   Leaves somewhat triangular, cleft only
    half or one-third of the way to midrib. ......................................... *P. vitifolius*

Fig. 221.  Scentless chamomile, *Matricaria maritima* L.

*Petasites palmatus* (Ait.) A. Gray (Fig. 222*A*)   PALMATE-LEAVED COLT'S-FOOT

A species with leaves almost circular, 7–30 cm across, deeply cleft almost to the base into several divisions, green and smooth above, somewhat white woolly below when young, but losing the woolliness with age. Flowering stem appearing before the leaves, stout and scaly-bracted, 15–50 cm high, with heads in a corymbose cluster on long stalks. Flower heads white, 8–12 mm across, with male and female flowers on separate plants. Common; in moist woodlands; Parklands and Boreal forest.

*Petasites sagittatus* (Pursh) A. Gray (Fig. 222*B*)   ARROW-LEAVED COLT'S-FOOT

This species has triangular-ovate to cordate leaves, with rounded marginal teeth, but not cleft or divided. Leaves 10–30 cm long, dull green above, densely white woolly below, on long stalks from root crown. Flowers in a dense terminal cluster at head of stem 20–50 cm high, which is scaly and somewhat bracted, appearing before the leaves. Fairly plentiful; in wet places, slough margins; Parklands and Boreal forest.

*Petasites vitifolius* Greene (Fig. 222*C*)   VINE-LEAVED COLT'S-FOOT

A species with leaves somewhat triangular, usually with hastate basal lobes, and cleft as much as halfway to midrib, 7–15 cm long, green on both sides or sometimes white woolly below. Flowers nearly white, in a corymbose cluster at head of a scaly-bracted stem 20–30 cm high. Found occasionally; in wet places in woods; Boreal forest.

**Psilocarphus**   woolly-heads

*Psilocarphus elatior* Gray   TALL WOOLLY-HEADS

An annual species with gray woolly stems 5–15 cm high, erect, sometimes branched; depauperate plants sometimes no more than 1 cm high. Leaves 1–3 cm long, 2–5 mm wide, linear or linear-oblong. Heads solitary or clustered at the tip of stems and axillary branches 4–8 mm thick; involucre densely woolly or tomentose. Rare; damp areas, dried-up sloughs; Prairies.

**Ratibida**   coneflower

*Ratibida columnifera* (Nutt.) Woot. & Standl. (Fig. 223)

LONG-HEADED CONEFLOWER

A perennial from a taproot, 30–70 cm high, usually branched from near base. Stems somewhat stiff hairy, with longitudinal grooves and angles. Leaves 5–10 cm long, very deeply pinnately divided into narrow segments. Flower heads borne at ends of long stalks, conspicuous by the cylindrical disk or receptacle, gray to purple, about 6 mm wide, 10–35 mm high. Ray florets yellow, 15–25 mm long, usually reflexed. Common; on dry prairie and roadsides; throughout Prairies, less common in Parklands. Syn.: *Lepachys columnifera* (Nutt.) Rydb. The f. *pulcherrima* (DC.) Fern., brown coneflower, similar to the species, but with ray florets partly or entirely brownish purple. This form has been found occasionally, but is very rare.

Fig. 222.  *A*, Palmate-leaved colt's-foot, *Petasites palmatus* (Ait.) A. Gray; *B*,
arrow-leaved colt's-foot, *Petasites sagittatus* (Pursh) A. Gray; *C*, vine-leaved
colt's-foot, *Petasites vitifolius* Greene.

A.C. Budd

Fig. 223.    Long-headed coneflower, *Ratibida columnifera* (Nutt.) Woot. &
Standl.

***Rudbeckia***     rudbeckia

Leaves neither lobed nor divided; disk or
  flower heads dark brown. ................................................................................ *R. hirta*

Leaves, at least the lower ones, deeply lobed
  and divided; disk of flower heads greenish
  yellow. ................................................................................ *R. laciniata*


*Rudbeckia hirta* L.                                           BLACK-EYED SUSAN

A rather coarse hairy biennial 30–60 cm high, with erect sometimes tufted
stems. Leaves lanceolate to oblanceolate, 5–15 cm long, 3-ribbed and hairy,
with the lower ones stalked, but the upper ones stalkless and smaller. Flowers
borne singly on long stalks, 5–8 cm across, with a dark brown hemispheric disk
and 10–20 orange yellow ray florets. Fairly common; on prairies and woodlands; Parklands and Boreal forest. Some botanists consider the western form
with oblanceolate to linear-lanceolate leaves as a separate species, *Rudbeckia
serotina* Nutt., and reserve the name *R. hirta* L. for the eastern plants with
ovate leaves.


*Rudbeckia laciniata* L.                                          TALL CONEFLOWER

A branching perennial 1–2 m high, with a smooth stem. Leaves, except
the uppermost, deeply divided into 3–7 segments and stalked. Leaves varying
in size up to about 20–25 cm long, with the upper ones progressively smaller.
Stems usually bearing several long-stalked heads 5–10 cm across, with a greenish yellow disk, and 6–10 bright yellow ray florets. Frequent in open woodlands and forest edges; southeastern Parklands and Boreal forest.


***Saussurea***     sawwort

Plants usually less than 20 cm high; involucral
  bracts equal, acuminate at tip. ................................................................ *S. nuda* var. *densa*

Plants to 40 cm high; involucral bracts un-
  equal, with the tip dilated. ................................................................ *S. glomerata*


*Saussurea glomerata* Poir.                                      TALL SAWWORT

Plants with rootstocks; stems 20–40 cm high, pubescent. Leaves 5–10 cm
long, lanceolate to linear-lanceolate, densely brown glandular punctate below.
Heads 10–15 mm across, with the outer bracts much shorter than the inner
ones, which have a pink petaloid segment 1–2 mm wide. Rare; introduced;
Peace River.


*Saussurea nuda* Led. var. *densa* (Hook.) Hulten              DWARF SAWWORT

Plants with short thick leafy stems 5–20 cm high, more or less densely
woolly or tomentose. Leaves 5–10 cm long; the lower ones with winged
petioles; the upper ones sessile, lanceolate, loosely pubescent when young,
somewhat dentate. Heads crowded; inflorescence 4–7 cm across; involucres
about 1 cm high and wide; inner and outer bracts equal. Alpine meadows and
slopes; Rocky Mountains.

*Senecio*      groundsel

Annual or perennial plants with alternate leaves. Flowers yellow, usually radiate, although one introduced species is discoid. Bracts is one series or a few basal bracts forming an outside row. Pappus consisting of many soft white hairs.

1. Flower heads solitary. ........................................................................................................ 2
   Flower heads few to many. .......................................................................................... 3

2. Heads 15–20 mm across; leaves entire to
   denticulate. ................................................................................... *S. megacephalus*
   Heads 10–15 mm across; leaves toothed
   to pinnatifid. ................................................................................ *S. resedifolius*

3. Annual plants with taproots. ............................................................................... 4
   Perennial plants, usually with rootstocks. ................................................... 6

4. Stems and leaves densely glandular
   pubescent. ......................................................................................... *S. viscosus*
   Stems and leaves not glandular. .......................................................................... 5

5. Plants branching; flower heads discoid. ........................................ *S. vulgaris*
   Plants with few branches, tall; flower
   heads radiate. ................................................................................. *S. congestus*

6. Leaves all or mostly subentire to dentate. ................................................. 7
   Leaves all or mostly coarsely lobed to
   pinnately divided. ................................................................................................ 13

7. Flower heads solitary or with a small
   head at base. ................................................................................ *S. megacephalus*
   Flower heads 2 to many. ....................................................................................... 8

8. Stem leaves and basal leaves equal in size
   and shape. ....................................................................................... *S. triangularis*
   Stem leaves different from basal leaves. ...................................................... 9

9. Middle and upper stem leaves much
   reduced in size. ................................................................................................... 10
   Middle and upper stem leaves as large as
   or slightly smaller than lower leaves. ........................................................ 11

10. Inflorescence mostly terminal; axillary
    pedicels short. ............................................................................. *S. integerrimus*
    Inflorescence mostly axillary; pedicels
    long. ...................................................................................................... *S. foetidus*

11. Leaves all cauline; basal leaves entirely
    lacking. ............................................................................................. *S. fremontii*
    Basal leaves present. ......................................................................................... 12

12. Leaves and stems somewhat tomentose,
    at least till flowering. ............................................................ *S. tridenticulatus*
    Leaves and stems glabrous; leaves rather
    succulent. ................................................................... *S. streptanthifolius*

13. Basal and stem leaves all deeply pinnately
    lobed or divided; plants often 1 m high. ......................... *S. eremophilus*

*Senecio aureus* L.                                              GOLDEN RAGWORT

A plant 30–80 cm high, with long-stalked cordate-based almost circular leaves 3–10 cm long. Stem leaves pinnatifid, and reduced in size. Flowers deep yellow, 6–8 mm high, in a terminal cluster. Found in meadows and moist places; southeastern Parklands and Boreal forest, Cypress Hills, and southern Rocky Mountains.

*Senecio canus* Hook. (Fig. 224)                         SILVERY GROUNDSEL

White woolly perennial from horizontal rootstocks, 10–30 cm high. Basal leaves entire-margined, oval to spatulate, 3–7 cm long, stalked, and somewhat white woolly; stem leaves much smaller, stalkless, oblong, and often somewhat pinnatifid or lobed. Heads in a terminal cluster 10–25 mm across. Not common; on dry hills; across the Prairies and Parklands.

*Senecio congestus* (R. Br.) DC.                          MARSH RAGWORT

A coarse hollow-stemmed annual 15–60 cm high, with fleshy stems somewhat cobwebby when young but hairless when mature. Lower leaves lanceolate to spatulate, 5–15 cm long, with wavy margins, and winged stalks; upper leaves smaller, stalkless, somewhat lobed or dentate, linear-lanceolate, and clasping the stem. Flower heads in a very crowded, dense, terminal cluster, pale yellow, 1–2 cm across. Fairly common; around sloughs, stream banks, and lakes, often forming a solid belt around a small lake; throughout the Prairie Provinces. Syn.: *S. palustris* (L.) Hook.

Fig. 224. Silvery groundsel, *Senecio canus* Hook.

*Dicots*

*Senecio eremophilus* Richards. (Fig. 225)          CUT-LEAVED RAGWORT

A leafy-stemmed species with a stout often purplish stem 30–100 cm high. Leaves lobed or pinnatifid, 5–15 cm long; the lower leaves stalked; the upper ones without stalks, slightly smaller than the lower ones. Stems leafy up to inflorescence. Numerous yellow flower heads 10–25 mm across, with black-tipped bracts, in compact terminal clusters. Found occasionally; on moist soil; Parklands, more common in Boreal forest.

*Senecio foetidus* Howell          MARSH BUTTERWEED

A fibrous-rooted perennial with solitary or sometimes clustered stems 30–100 cm high, glabrous throughout. Basal leaves 7–20 cm long, to 7 cm wide, rather sharply dentate; stem leaves progressively reduced, sessile. Inflorescence dense, with the heads several to many, more or less corymbosely arranged; ligules about 8 mm long; involucre 6–9 mm high, with the bracts black-tipped. Moist woods; southern Rocky Mountains.

*Senecio fremontii* T. & G.          MOUNTAIN BUTTERWEED

A glabrous perennial with taproot and branching caudex, freely branching; stems decumbent, 10–20 cm high. Leaves thick, 1–4 cm long, obovate to spatulate, rather sharply dentate. Heads terminating short branches; ligules 6–10 mm long; involucre 7–12 mm high. Slopes and open areas in mountain woods; southern Rocky Mountains.

*Senecio indecorus* Greene          RAYLESS RAGWORT

A fibrous-rooted perennial with a simple or occasionally branched caudex and glabrous stems 30–80 cm high. Leaves rather thin, lightly floccose when young; basal leaves elliptic or ovate, tapering to truncate at base, serrate or somewhat incised, petiolate; stem leaves incised to pinnatifid, with toothed lobes. Heads several to many; involucre 7–10 mm high; bracts often purple-tipped. Moist meadows; Boreal forest and occasionally in Parklands.

*Senecio integerrimus* Nutt.          ENTIRE-LEAVED GROUNDSEL

A stout-stemmed perennial from coarse, fleshy, fibrous roots, 20–60 cm high. When young, the stem is somewhat hairy, but later the plant is entirely smooth and hairless. Leaves entire, thick and fleshy, 5–20 cm long; basal leaves oblong to lanceolate with long stalks; upper leaves stalkless, lanceolate to linear, and reduced in size. Flower heads often 12–20 mm across, in dense terminal clusters.

The species is represented by three varieties:

1. Involucral bracts green to tip. .......................................................... var. *integerrimus*
   Involucral bracts with a black tip. .......................................................................... 2
2. Black tip about 1 mm long. ......................................... var. *exaltatus* (Nutt.) Cronq.
   Black tip about 3 mm long, triangular. ......................... var. *lugens* (Richards.) Boiv.

The var. *integerrimus* is found throughout Prairies and Parklands, var. *exaltatus* in Cypress Hills and southern Rocky Mountains, and var. *lugens* in Rocky Mountains, northwestern Parklands, and Peace River.

Fig. 225.   Cut-leaved ragwort, *Senecio eremophilus* Richards.

*Senecio megacephalus* Nutt.                                      LARGE-FLOWERED RAGWORT

Plants with a thick rootstock, and simple stout stems 10–20 cm high, more or less floccose pubescent. Leaves linear-oblanceolate to lanceolate, entire or denticulate. Heads solitary or seldom 2 or 3, 20–25 mm high; ligules orange, 15–20 mm long; involucre with the outer bracts often surpassing the inner ones. Alpine slopes or ridges; southern Rocky Mountains.

*Senecio pauciflorus* Pursh                                       FEW-FLOWERED RAGWORT

A fibrous-rooted perennial with a simple caudex; stems 10–30 cm high, glabrous, or somewhat floccose-tomentose when young. Basal leaves 2–4 cm long, elliptic-ovate to subrotund, abruptly contracted or truncate at base, long-petioled; stem leaves reduced, with the upper ones sessile, toothed to pinnatifid. Heads usually 2–6, discoid or rarely with a few short rays; involucre usually 6–8 mm high; bracts more or less purplish. Arctic and alpine meadows; Boreal forest and Rocky Mountains.

*Senecio pauperculus* Michx.                                      BALSAM GROUNDSEL

A slender-stemmed perennial with a woolly-based stem 20–40 cm high. Basal leaves stalked, oblong or oval, with wavy margins, 2–8 cm long; stem leaves smaller, not stalked, and lobed or toothed. Leaf blades hairless and thin. Few flower heads, in a loose cluster on thin stalks at the head of the stem, 10–15 mm across.

Three varieties of this species can be found:

1. Involucre 4–5 mm high; basal leaves
   about 1 cm wide. ........................................................................ var. *pauperculus*
   Involucre 5–9 mm high; basal leaves 1–2
   cm wide. ........................................................................................................ 2
2. Involucre 5–7 mm high; basal leaves
   10–15 mm wide. ............................................... var. *firmifolius* Greenman
   Involucre 6–9 mm high; basal leaves
   15–20 mm wide. ........................................ var. *thomsoniensis* (Greenman) Boiv.

The var. *pauperculus* is found in wet, often somewhat saline meadows throughout Prairies and Parklands; the other varieties are usually found in wet meadows, especially on sandy soils, in Parklands and Boreal forest.

*Senecio resedifolius* Less.                                      ARCTIC BUTTERWEED

Plants with a rather thick rhizome and branched caudex; stems 5–30 cm high, essentially glabrous. Leaves thick, 1–2.5 cm long, ovate or reniform to suborbicular, toothed or entire; stem leaves reduced, with the upper ones sessile. Heads solitary or rarely 2; ligules to 15 mm long or sometimes missing; involucre 6–8 mm high, purplish. Boggy mountain meadows and slopes; southern Rocky Mountains.

*Senecio streptanthifolius* Greene                                NORTHERN RAGWORT

A fibrous-rooted perennial with a short rootstock; stems 10–50 cm high, glabrous or lightly pubescent. Lower leaves 2–7 cm long, oval, long-stalked, and usually toothed or lobed; upper leaves stalkless, reduced, linear-lanceo-late, and deeply toothed; most of the leaves somewhat fleshy. Orange yellow flower heads 1–2 cm across, in a rather loose terminal cluster on long stalks. Found occasionally; on moist soils; in valleys in Parklands and lightly wooded parts of northwestern Boreal forest.

*Senecio triangularis* Hook.                                   BROOK RAGWORT

A leafy-stemmed often tufted perennial, from fleshy fibrous roots, 30–60 cm high. Stem hairless and leafy to the inflorescence. Leaves almost all stalked, 2–8 cm long, with conspicuous teeth, triangular to lanceolate. Flower heads 1–2 cm across, in a terminal cluster. Along streams and in wet places; southern Rocky Mountains.

*Senecio tridenticulatus* Rydb.                          COMPACT GROUNDSEL

A low slender plant from a short rootstock, 15–30 cm high, slightly white hairy at the base of the stem and at the axils of the leaves. Lower leaves 2–7 cm long, short-stalked, oblanceolate, and often somewhat toothed; upper leaves linear, stalkless, much smaller, and lobed. Flower heads few in number, 1–2 cm across, with rather pale yellow ray florets, in a fairly compact terminal cluster. Found occasionally; on moist prairie; southeastern Parklands.

*Senecio viscosus* L.                                      STINKING GROUNDSEL

An annual weed with taproot; stems 10–60 cm high, usually branched. Herbage densely glandular pubescent, strong-scented. Leaves pinnatifid, 3–12 cm long, 1–5 cm wide. Heads several to numerous, long pedicellate; ligules inconspicuous; involucre 6–8 mm. Introduced, rare; southeastern Parklands.

*Senecio vulgaris* L.                                       COMMON GROUNDSEL

A low hollow-stemmed much-branched annual, 15–40 cm high. Leaves lobed, 5–15 cm long; lower leaves stalked and upper ones clasping at base. Flower heads discoid (without any ray florets), about 6 mm across, often with some black-tipped bracts, in clusters at ends of branches. An introduced weed, found often in gardens and around settlements; across the Prairie Provinces, but nowhere very common.

*Solidago*        goldenrod

Perennial herbs from rootstocks, with alternate entire leaves. Inflorescence of terminal panicles, in racemes or corymbose clusters, with many radiate yellow flower heads. Two species are white or cream. Pappus of rough bristly hairs in one or two series. Most goldenrods bloom in mid- or late summer and are good honey-producing plants.

1. Inflorescence corymbose, with the branches slanting upward, forming a rounded or flat-topped inflorescence. .................................................................. 2

   Inflorescence a panicle or spike. ................................................................. 6

2. Ray florets white or pale yellow, large, 5–8 mm long. ................................................................. *S. ptarmicoides*

Ray florets yellow, usually less than 5 mm
  long. ................................................................................ 3

3. Leaves firm, very rough, gray pubescent;
    stems rigid and gray puberulent. ........................... *S. rigida* var. *humilis*
  Leaves not rough or gray pubescent. ........................................... 4

4. Stem leaves, at least the lower ones,
    petioled, the upper ones much reduced. ........................... *S. multiradiata*
  Stem leaves sessile or subsessile, all about
    the same size. ................................................................... 5

5. Leaves finely glandular pubescent, flat;
    inflorescence with 10–50 heads. ........................... *S. graminifolia*
  Leaves not glandular, commonly folded
    lengthwise; inflorescence mostly with
    more than 50 heads. ........................................... *S. riddellii*

6. Inflorescence racemose, with branches
    ascending along the stem; inflorescence
    more or less cylindrical or narrowly
    oval. ................................................................................ 7
  Inflorescence a panicle with more or less
    spreading or arching branches; heads
    often secund on the branches. ........................................... 11

7. Flowers white or cream. ........................................... *S. bicolor*
  Flowers yellow. ................................................................ 8

8. Leaves and stems densely hairy; lower
    leaves ovate, with the petiole as long as
    the blade. ........................................... *S. bicolor* var. *concolor*
  Leaves and stems not densely hairy;
    lower leaves not ovate, or if so, not
    long-petioled. ................................................................ 9

9. Leaves obovate, subsessile, 2.5–3 times as
    long as wide, grayish green. ........................... *S. mollis*
  Leaves lanceolate, at least the lower ones
    petioled, more than 4 times as long as
    wide. ................................................................................ 10

10. Branches of the inflorescence shorter than
     or as long as the subtending stem
     leaves; inflorescence cylindrical, more
     or less interrupted. ........................... *S. spathulata* var. *spathulata*
   Branches of the inflorescence much
     longer than the subtending stem
     leaves; inflorescence more or less oval,
     not interrupted. ........................................... *S. uliginosa*

11. Basal leaves much longer than stem
     leaves, often forming a rosette. ........................................... 12
   Basal leaves not much longer than stem
     leaves, seldom forming a rosette. ........................................... 14

12. Stem and leaves minutely gray pubescent. ........................... *S. nemoralis*
   Stem and leaves not gray pubescent. ........................................... 13

13. Lower and middle stem leaves distinctly
   3-nerved, to 10 cm long, 15 mm wide,
   entire or obscurely dentate. ............................................................... *S. missouriensis*

   Lower and middle stem leaves not distinctly nerved, usually 20 cm or more
   long, 20–75 mm wide, clearly serrate. ...................................................... *S. juncea*

14. Leaves obovate, subsessile, 2.5–3 times as
   long as wide, grayish green. ......................................................................... *S. mollis*

   Leaves lanceolate to oval, more than 4
   times as long as wide, green. ...................................................................... 13

15. Stem below the inflorescence glabrous or
   very sparsely pubescent. ...................................................... *S. gigantea* var. *serotina*

   Stem below the inflorescence densely
   pubescent. ...................................................................................... *S. canadensis*

*Solidago bicolor* L.                                           PALE GOLDENROD

   A stout plant with white hairy stems 20–60 cm high. Lower leaves long-
stalked, obovate to oblong, 5–10 cm long; upper leaves stalkless, smaller and
narrow; all leaves hairy on both sides. Inflorescence narrow, 5–15 cm high,
with short branches bearing flower heads, which have white or pale cream ray
florets. Found occasionally; on dry soil; southeastern Boreal forest and Parklands. The var. *concolor* T. & G. differs only in having yellow ray florets.
Found in dry to moist woods; throughout Parklands and Boreal forest. Syn.:
*S. hispida* Muhl. var. *lanata* (Hook.) Fern.

*Solidago canadensis* L. var. *canadensis* (Fig. 226)     GRACEFUL GOLDENROD

   A slender-stemmed leafy plant from a horizontal rootstock, 30–80 cm
high. Leaves narrowly lanceolate, 3-veined, 5–10 cm long, with fine teeth,
slightly hairy along the veins on the underside, green. Inflorescence a broad,
pyramid-like panicle; flower heads small, borne on one side of spreading
branches. Found commonly; in woodlands and forests; Parklands and Boreal
forest, also in Cypress Hills. The var. *gilvocanescens* Rydb., canescent goldenrod, is a slender-stemmed yellowish gray species from a horizontal rootstock,
25–60 cm high, with a finely downy-haired stem. Leaves 2–5 cm long, 3-veined,
narrowly lanceolate, entire or somewhat toothed, with the lower ones falling
off early. Inflorescence pyramid-shaped with somewhat recurved branches.
Fairly common; on disturbed moist areas; Prairies. Syn.: *S. pruinosa* Greene.

*Solidago gigantea* Ait. var. *serotina* (Ait.) Cronq.          LATE GOLDENROD

   A stout erect species from stout rootstocks, 60–150 cm high, with a
smooth leafy stem. Leaves thin, lanceolate, 2–12 cm long, without stalks, 3
nerves, usually sharply toothed, smooth on both sides. Inflorescence large,
pyramidal; heads crowded on spreading recurved branches. Common; around
bluffs, woodlands, and also in coulees and low areas; throughout Prairie Provinces. Syn.: *S. serotina* Ait.

*Solidago graminifolia* (L.) Salisb. var. *graminifolia*     FLAT-TOPPED GOLDENROD

   A much-branched plant from a long rootstock, 30–60 cm high. Leaves
numerous and somewhat crowded, 2–10 cm long, linear-lanceolate, 15–20

Fig. 226.   Graceful goldenrod, *Solidago canadensis* L. var. *canadensis*.

times as long as wide, and pointed at both ends. Small yellow flowers, in very short-stemmed terminal clusters, forming a somewhat flat-topped inflorescence. Fairly common; along stream banks and moist places; eastern Parklands and Boreal forest. The var. *major* (Michx.) Fern. has wider leaves and the length-to-width ratio is usually 8–10:1; plants smaller; inflorescence with fewer heads. The variety is more common in the western part of the distribution area.

*Solidago juncea* Ait.                                      SHARP-TOOTHED GOLDENROD

A stout smooth-stemmed species from a horizontal rootstock, 40–80 cm high. Lower leaves single-ribbed, broadly oblanceolate, stalked, hairless, 15–30 cm long; upper leaves stalkless, smaller, and narrower. Inflorescence a somewhat flat-topped panicle with recurved spreading branches; flower heads borne on one side of branches. Rarely found; a species of woodlands and banks; southeastern Boreal forest.

*Solidago missouriensis* Nutt.                                      LOW GOLDENROD

A rather low smooth-stemmed species from horizonal rootstocks, 15–50 cm high. Stems hairless, often tufted, usually somewhat reddish. Leaves 3-ribbed, linear-lanceolate, 2–10 cm long, often reddish and hairless except for sparse short marginal hairs. Inflorescence a compact terminal panicle with erect branches. Earlier flowering than other goldenrods. Common on dry prairies and hillsides; throughout Prairies and southern fringe of Parkland. The var. *fasciculata* Holzinger, low goldenrod, differs from the species by having branches of inflorescence recurved and spreading instead of erect. Probably the eastern form of the species. Common; on dry prairie; in south central parts of the distribution area. Syn.: *S. glaberrima* Martens.

*Solidago mollis* Bartl.                                      VELVETY GOLDENROD

A stout, low, erect plant from a horizontal rootstock, 20–50 cm high. Whole plant covered with very fine, short, velvety hairs. Leaves pale green, almost entire-margined, obovate to oval (or the upper leaves elliptic), 3-nerved, 2–7 cm long; lower leaves short-stalked; upper leaves stalkless and crowded on stem. Inflorescence a pyramid-shaped dense panicle. Common; on dry prairie land and roadsides; probably the most common species in Prairies.

*Solidago multiradiata* Ait.                                      ALPINE GOLDENROD

A rather small species with a short rootstock or branching caudex; stems 5–40 cm high, pilose above and in the inflorescence. Heads rather large; involucre 4–8 mm high, few, seldom more than 10–20. Boreal and montane or alpine meadows; Boreal forest and Rocky Mountains.

*Solidago nemoralis* Ait.                                      SHOWY GOLDENROD

A species growing in clumps of several stems from a thick rootstock, 30–50 cm high. Stems often somewhat decumbent, reddish, covered with fine downy hairs. Basal leaves narrowly oblanceolate, entire, 5–10 cm long; upper leaves oblong or linear; all leaves somewhat ashy gray, with very minute hairs. Inflorescence narrow, usually bent over or somewhat nodding at top. Plentiful; on sandy soil and in sandhills; throughout Prairies and Parklands.

*Solidago ptarmicoides* (Nees) Boiv.                    UPLAND WHITE GOLDENROD

A somewhat tufted plant from creeping rootstocks, 20–60 cm high. Leaves linear-lanceolate, 2–15 cm long, sometimes slightly toothed, firm, and shiny. Flower heads numerous, 10–25 mm across, white, in a somewhat flat-topped terminal cluster. Rare, but found occasionally; in dry, saline, or gravelly soil; eastern Parklands and Boreal forest. Syn.: *Unamia alba* (Nutt.) Rydb.; *Aster ptarmicoides* (Nees) T. & G. Plants with somewhat smaller heads and pale yellow ligules are considered to be a hybrid between *S. ptarmicoides* and *S. rigida* var. *humilis*, and have been named × *S. lutescens* (Lindl.) Boiv. Syn.: *Aster ptarmicoides* (Nees) T. & G. var. *lutescens* (Lindl.) A. Gray.

*Solidago riddellii* Frank                              RIDDELL'S GOLDENROD

A tall species with a more or less developed caudex and sometimes short rootstocks; stems 40–100 cm high, glabrous below the inflorescence. Leaves 5–15 cm long, glabrous, with margins scabrous, entire, tending to be 3-nerved; stem leaves numerous, somewhat clasping or sheathing at the base. Inflorescence corymbiform; branches puberulent; heads numerous, rarely less than 50, often several hundred; involucre 5–6 mm high. Rare; swamps and wet meadows; southeastern Parklands and Boreal forest.

*Solidago rigida* L. var. *humilis* Porter (Fig. 227)    STIFF GOLDENROD

An erect stout-stemmed species, with a densely fine-hairy rough stem, from a thick woody rootstock, 10–40 cm high. Basal leaves long-stalked, oval, 4–10 cm long, thick, and densely fine-hairy on both sides; stem leaves oval, stalkless, and smaller. Inflorescence a dense corymbose cluster, somewhat flat-topped, with an involucre 6–8 mm high. Common; on prairies and openings in woodlands; throughout the Prairie Provinces.

*Solidago spathulata* DC. var. *spathulata* (Fig. 228)   MOUNTAIN GOLDENROD

Rather low plants 20–50 cm high. Stems decumbent at base, often reddish-tinged. Lower leaves spatulate, 2–10 cm long, usually blunt-tipped, often with rounded teeth; stem leaves smaller and entire; basal leaves usually somewhat crowded at root crown. Inflorescence a narrow erect panicle; heads fairly large, usually about 8 mm high. Common; in grasslands; throughout Prairies, Parklands, and southern Boreal forest; often mistaken for *S. missouriensis*, with smaller flowers and triple-veined leaves. Syn.: *S. decumbens* Greene; *S. oreophila* Rydb.

*Solidago uliginosa* Nutt.                              MARSH GOLDENROD

A large species with a long much-branched rootstock or caudex; stems 60–150 cm high, glabrous below the inflorescence. Basal and lower stem leaves 6–35 cm long, lanceolate to narrowly elliptic, subentire to dentate, tapering to a long winged petiole; upper stem leaves gradually reduced. Inflorescence puberulent, elongate; branches straight or recurved-secund; involucre 3–5 mm high. Bogs and marshes; southeastern Boreal forest.

Fig. 227.   Stiff goldenrod, *Solidago rigida* L. var. *humilis* Porter.

Fig. 228.   Mountain goldenrod, *Solidago spathulata* DC. var. *spathulata*.

## *Tanacetum*    tansy

Leaf segments very fine, about 1 mm wide. ................................................... *T. huronense*

Leaf segments much coarser, 3–10 mm wide. ..................................................... *T. vulgare*

### *Tanacetum huronense* Nutt.    INDIAN TANSY

A perennial with long rootstocks; stems 10–60 cm high, more or less villous throughout. Leaves 5–20 cm long, 2–8 cm wide, 2- or 3-pinnatifid. Heads 1–15, with the disk 10–18 mm across; ligules inconspicuous, or rarely to 4 mm long. Rare; sandy lakeshores; Boreal forest.

### *Tanacetum vulgare* L. (Fig. 229)    TANSY

A stout erect-stemmed perennial 30–100 cm high. Leaves 5–25 cm long, pinnately divided into narrow toothed segments, very aromatic when bruised. Flower heads discoid (without ray florets), 6–8 mm across, in a somewhat flat-topped cluster. An introduced garden plant, which has become established and often forms extensive colonies in roadsides, waste areas, and ditches; in many locations throughout the Prairie Provinces.

## *Thelesperma*    tickseed

### *Thelesperma marginatum* Rydb.    TICKSEED

A perennial with rootstock or woody caudex; stems 10–20 cm high, leafy at the base. Leaves once or twice pinnately divided, with the divisions linear, 1–2 mm wide. Peduncles mostly 7–10 cm long; heads solitary or few, about 1 cm wide; involucral bracts fused; lobes with a distinct white margin. Rare; eroded slopes; Prairies.

## *Townsendia*    townsendia

Tufted plants with clustered basal alternate entire leaves and large aster-like radiate heads. The pappus consisting of bristly hairs.

Heads on a stalk from the root crown. ................................................... *T. parryi*

Heads stalkless, growing directly on the root
  crown. ........................................................................................... *T. exscapa*

### *Townsendia exscapa* (Rich.) Porter    LOW TOWNSENDIA

An almost stemless plant from a deep woody branching root. Leaves narrowly spatulate to linear, 2–5 cm long. Flower heads stalkless, borne amongst rosettes of leaves, 20–35 mm across, and with bluish or white ray florets. Found occasionally; on eroded prairies and dry stony hillsides; throughout Prairies and southern fringe of Parklands. Syn.: *T. sericea* Hook. A similar species, *T. hookeri* Beaman, is smaller, with narrower leaves, and involucral bracts very densely ciliate throughout and tipped with a tuft of hairs. However, the pubescence and size of plants in any location can vary greatly, and there seems little reason to assume the presence of two species of low townsendia.

Fig. 229. Tansy, *Tanacetum vulgare* L.

*Townsendia parryi* D. C. Eaton (Fig. 230)          PARRY'S TOWNSENDIA

A biennial plant, usually cushion-like, with crowded basal leaves. Leaves spatulate and thick, 2–5 cm long; stem leaves few, very small. Flower head on a short stem 2–15 cm high, about 5–7 cm across, with many narrow violet or purplish blue ray florets and a wide disk about 25–30 mm across. Found occasionally; on open benchland; southern Rocky Mountains.

## *Vernonia*          ironweed

*Vernonia fasciculata* Michx. var. *corymbosa* (Schwein.) Schub.

WESTERN IRONWEED

A coarse erect perennial, usually with a red smooth stem 40–100 cm high. Leaves stalkless, lanceolate to ovate-lanceolate, 7–15 cm long, smooth, and sharply toothed. Inflorescence a loose terminal cluster. Flower heads discoid (with no ray florets), about 6 mm across, dark purple. Rare; has been found beside sloughs and in river valleys; along southeastern Parklands and Boreal forest.

Fig. 230.   Parry's townsendia, *Townsendia parryi* D. C. Eaton.

# Tabular classification

The heading "varieties" includes subspecies, forms, and small or included species in addition to varieties. Plants represented only by a subspecies or variety are included under species.

|                  | Families | Genera | Species | Varieties |
|------------------|----------|--------|---------|-----------|
| Pteridophyta     | 7        | 21     | 58      | 2         |
| Spermatophyta    |          |        |         |           |
|   Gymnospermae | 3 | 8 | 22 | 3 |
|   Angiospermae |   |   |    |   |
|     Monocotyledoneae | 16 | 125 | 429 | 66 |
|     Dicotyledoneae | 92 | 436 | 1465 | 149 |
| Totals           | 118      | 590    | 1974    | 220       |

# Glossary

**acaulescent**   Stemless, or having a very short stem.

**achene**   A 1-celled, 1-seeded, dry hard fruit that does not open when ripe (Fig. 9, p. 19).

**acrid**   Pungent, bitter.

**acuminate**   Gradually tapering to a point.

**acute**   Somewhat abruptly tapering to a point.

**adnate**   Of a plant part, united with or attached to another part, usually of a different kind.

**alternate**   Distributed, as leaves, at different positions on the stem, not opposite each other (Fig. 5, p. 16).

**ament**   A scaly spike of flowers of one sex only; also called catkin.

**androgynous**   Of spikes in *Carex*, having both staminate and pistillate flowers, the latter at the base.

**angiosperm**   A plant bearing seed in a closed ovary.

**annual**   A plant germinating, flowering, and ripening seed in 1 year.

**annular**   Arranged in rings.

**anther**   The pollen container of a stamen or male organ (Fig. 7, p. 17).

**anthesis**   In full flower, usually referring to the flowering period.

**apetalous**   Without petals.

**apex**   The summit or point (Fig. 5, p. 16).

**appressed**   Lying flat and close to some part of a plant, usually referring to hairs.

**approximate**   Close together, but not overlapping.

**aquatic**   Living in water.

**articulation**   Natural separation, joint.

**ascending**   Growing upward or turned up.

**astringent**   Binding, contracting.

**attenuate**   Becoming very narrow.

**auricle**   An ear-shaped appendage, or the ear at the base of a leaf (Fig. 5, p. 16).

**auriculate**   Having auricles.

**awl-shaped**   Broad at base, tapering to a sharp point.

**awn**   A bristle, often found on grass flowers.

**axil**   The upper angle formed where a leaf stalk or a branch joins a stem (Fig. 5, p. 16).

**axillary**   In an axil.

**axis**   The central line of an organ.

**barb**   A short, stiff point, or short bristle, often bent backward.

**basifixed**   Attached at or near the base, at one end only.

**beak**   A tip or point, somewhat resembling the beak of a bird.

**bearded**   Hairy, often used of a stamen or of the throat of a flower.

**berry**   A pulpy fruit with several seeds, as that of currant and grape (Fig. 5, p. 16).

**biennial**   Of 2 years' duration.

**bifid**   Two-cleft.

**bilabiate**   Two-lipped.

**bisexual**   Having both stamens and pistils.

**blade**   The expanded part of a leaf (Fig. 5, p. 16).

**bloom**   A whitish, powdery covering.

**bract**   A small leaf or scale, often borne below a flower or flower cluster.

**bracteose**   Having numerous or conspicuous bracts.

**bractlet**   A secondary bract, as one borne on the pedicel of a flower.

**bud**   The rudimentary state of a stem or branch; an unexpanded flower.

**bulb**   A thick underground organ composed of successive fleshy layers.

**bulblet**   A small bulb, borne on the stem or inflorescence.


**callus**   A small, hard protuberance.

**calyx**   The outer floral ring, or sepals, usually green, but sometimes brightly colored (Fig. 7, p. 17).

**campanulate**   Bell-shaped.

**canescent**   Densely fine pubescent, giving a gray appearance.

**capitate**   Gathered into a head, as a cluster at the end of a stem.

**capsule**   A dry fruit, as that of the gentian, consisting of more than one chamber, and opening at maturity.

**carpel**   An ovule-bearing chamber at the base of the pistil or female organ of a flower.

**caryopsis**   A grain, as in the grasses.

**catkin**   As used in this book, a scaly spike of flowers of one sex.

**caudex**   The woody base of a plant from which the stems arise.

**caulescent**   Having a well-defined stem above the ground.

**cauline**   Belonging or attached to the stem.

**cell**   A chamber of the ovary or anther; an individual unit of plant structure.

**cespitose**   Growing in tufts; matted or turf-forming; also spelled caespitose.

**chaff**   A small, thin, dry, and membranous scale or bract.

**chartaceous**   Having the texture of paper, papery.

**chlorophyll**   The green coloring matter within the cells of a plant.

**choripetalous**   Having petals separated from each other.

**ciliate**   Having marginal hairs (Fig. 5, p. 16).

**circumscissile**   Splitting all the way around, as the lid of a fruit, such as that of purslane.

**clasping**   Partly or entirely surrounding the stem; used of leaf bases (Fig. 5, p. 16).

**cleft**   Deeply lobed.

**cleistogamous**   Of flowers, small, inconspicuous, permanently closed, and hence self-fertilizing.

**compound**   Of a leaf, composed of 2 or more leaflets; of a branch, composed of 2 or more parts, forming a common whole.

**compressed**   Flattened, especially laterally.

**cone**   A dense, usually elongated collection of flowers or fruits borne beneath scales; or a collection of spore-bearing leaves on an axis, the whole mass forming a fruit-like body.

**coniferous**   Cone-bearing.

**connate**   Of leaves, united at the base (Fig. 5, p. 16); joined together.

**convex**   Curved outward, as the surface of a sphere.

**convolute**   Rolled together, coiled.

**cordate**   Of a leaf, heart-shaped, with the point away from the base (Fig. 3, p. 15).
**corm**   A thick enlarged base of a stem, as found in crocus and gladiolus.
**corolla**   The petals or inner floral ring (Fig. 7, p. 17).
**corymb**   A cluster of flowers in which the flower stalks arise from different points on the stem; the cluster has a flat or rounded top (Fig. 6, p. 17).
**cosmopolite**   Occurring in all or most parts of the world.
**cotyledon**   The first leaf from the seed, sometimes called the seed leaf.
**crenate**   Having the margin cut into rounded scallops.
**crown**   The place where stem and root meet.
**culm**   The stem of a grass or sedge.
**cuneate**   Of a leaf, broadly rounded at the apex and tapering rather abruptly toward the point of attachment (Fig. 3, p. 15).
**cylindric**   In the form of a cylinder.
**cyme**   A cluster of flowers in which the central flowers open first (Fig. 6, p. 17).
**cymose**   Of the form or nature of a cyme.

**deciduous**   Having leaves that fall off in autumn; not evergreen.
**decompound**   More than once compound or divided.
**decumbent**   Of a stem, the base lying on the ground, but the tip growing upright.
**decurrent**   Of a leaf, the blade extending down the stem (Fig. 5, p. 16).
**deflexed**   Turned abruptly downward.
**deltoid**   Of a leaf, triangular (Fig. 3, p. 15).
**dentate**   Toothed, with teeth pointed and directed outward.
**diadelphous**   Of stamens, united in 2, often unequal, bundles.
**dichotomous**   Regularly and repeatedly branching in twos.
**dicotyledon**   A plant bearing 2 cotyledons, or seed leaves.
**digitate**   Of a leaf, with divisions somewhat like fingers (Fig. 4, p. 15).
**dilate**   Spreading out in all directions.
**dioecious**   Having male and female flowers on separate plants.
**discoid**   Having only disk flowers, without ray flowers.
**disk** or **disc**   A more or less fleshy or elevated development of the receptacle about the pistil; the receptacle in the head of Compositae (Fig. 153, p. 524); a flattened extremity, as on tendrils of Virginia creeper.
**dissected**   Divided into many segments.
**divaricate**   Spreading very far apart; extremely divergent.
**divergent**   Spreading away from each other.
**divided**   Of a leaf, cleft to the midrib or base (Fig. 4, p. 15).
**dorsal**   The backside of an organ, facing away from the axis.
**drupe**   A pulpy or fleshy fruit containing a single seed enclosed in a hard shell or stone, as that of the plum.
**drupelet**   One part of a fruit composed of aggregate drupes, as in the raspberry.

**edentate**   Lacking teeth.
**ellipsoid**   Solid but with an elliptical outline.
**elliptical**   Of a leaf, oval or oblong with the ends rounded and widest in the middle (Fig. 3, p. 15).

**elongate**  Stretched out, lengthened.

**entire**  Of a leaf or leaflet, having the margin not toothed or cleft.

**equitant**  Of leaves, enfolding each other or borne astride, as in the lily and the iris.

**erose**  Appearing as though gnawed at the margin.

**excurrent**  Projecting beyond the margin or tip, as an awn.

**exserted**  Projecting outward, as stamens from a corolla.

**falcate**  Curved and flat; sickle-shaped.

**farinose**  Covered with a whitish, mealy powder.

**fascicle**  A dense cluster; used of roots, leaves, or flowers.

**fertile**  Referring to stamens that bear pollen, and fruits that contain seeds.

**filament**  The stalk of a stamen below the anther (Fig. 7, p. 17).

**filiform**  Thread-like, long and very slender.

**flaccid**  Limp, floppy.

**flexuous**  Curved or bending in alternate directions, zigzag.

**floret**  A single flower, usually used of a composite head or cluster.

**-foliate**  Of a leaf, composed of 2 or more parts.

**follicle**  A fruit with a single chamber that opens along one side, as in milkweed (Fig. 9, p. 19).

**frond**  The expanded leaf-like portion of a fern.

**fruit**  The seed-bearing product of a plant.

**funnelform**  Of a flower, shaped like a funnel, cone-shaped.

**galea**  A hooded or helmet-shaped portion of a perianth, especially the upper part of some 2-lipped corollas, such as in elephant's-head.

**geniculate**  Bent, like a knee.

**glabrate**  Without pubescence.

**glabrescent**  Becoming glabrous with age or at maturity.

**glabrous**  Smooth, without hairs.

**gland**  An organ that secretes sticky or resinous matter.

**glandular**  Bearing glands.

**glaucous**  Covered with a bloom; bluish white or bluish gray.

**globose**  Spherical or nearly so.

**globular**  Globe-like, spherical.

**glomerule**  A small compact cluster of flowers, as in Compositae.

**glume**  A scaly bract on the floral parts of grasses and sedges.

**glutinous**  Sticky.

**grain**  A fruit resembling an achene but having the seed coat and thin pericarp fused into one body, particularly the fruit of grasses.

**granular**  Covered with very small grains.

**gynaecandrous**  Of spikes in *Carex*, having both staminate and pistillate flowers, the latter at the apex.

**hastate**  Resembling an arrowhead; of a leaf, with basal lobes protruding sideways.

**head**  A dense cluster of flowers or fruits on a very short axis or receptacle.

**herb**  A plant without a woody stem above the ground.

**hip**  The berry-like enlarged calyx tube containing many achenes, found in roses.

**hirsute**  Having coarse spreading hairs.

**hispid**  Bearing stiff hairs or bristles.

**hispidulous**  Minutely hispid.

**hoary**  Grayish white.

**hyaline**  Transparent or translucent, glassy.

**hypanthium**  The place on a flower head where the sepals and petals are attached.

**imbricate**  Overlapping like shingles.

**incised**  Having a margin that is cut or slashed irregularly; between toothed and lobed.

**indehiscent**  Not opening by valves; remaining closed.

**indurate**  Hardened.

**indusium**  In ferns, a membranous cover of a spore cluster or sorus; plural, indusia.

**inferior**  Of an ovary, situated below the calyx and corolla (Fig. 7, p. 17).

**inflorescence**  Arrangement of flowers in a cluster (Fig. 6, p. 17).

**internode**  The part of an axis between 2 nodes.

**interrupted**  Not continuous.

**introduced**  Imported from another region for ornamental or cultivation purposes.

**involucre**  The whorl of bracts below a flower cluster, or around a flower of Compositae (Fig. 199, p. 684).

**involute**  Rolled inward.

**irregular**  Of a flower, petals and sepals differing in shape or size (Fig. 8, p. 18).

**keel**  The 2 lower united petals of a leguminous flower (Fig. 7, p. 17); a sharp ridge.

**lacerate**  Irregularly cleft or cut, as if torn.

**lanate**  Covered with soft, intertwined hairs; woolly.

**lanceolate**  Of a leaf, much longer than wide, broadest near the base and tapering toward the tip (Fig. 3, p. 15).

**lax**  Loose, the opposite of stiff or congested.

**leaflet**  A division of a compound leaf (Fig. 4, p. 15).

**legume**  A dry pod-like fruit, splitting down one or both sides at maturity (Fig. 9, p. 19).

**lemma**  The lower of the 2 bracts enclosing a grass flower (Fig. 16, p. 53).

**lenticular**  Lens-shaped, biconvex.

**ligulate**  Having a ligule.

**ligule**  A strap-shaped organ, as in ray florets of Compositae (Fig. 199, p. 684); also, a collar of a grass blade (Fig. 16, p. 53).

**linear**  Of a leaf, long and narrow with parallel margins (Fig. 3, p. 15).

**lip**  The main lobe of a 2-lobed corolla or calyx; the odd and peculiar petal of Orchidaceae.

**lobe**  A rounded projection of a leaf (Fig. 4, p. 15), or a leaf-like part of a plant.

**locule**  The cavity of an ovary or anther.

**loment**  A legume or pod, constricted between the seeds, the joints separating at maturity (Fig. 9, p. 19).

**lyrate**  Of a leaf, having a large terminal lobe and smaller lobes toward the base.

**malpighian**  Of hairs, attached in the middle.
**membranous**  Rather soft, thin, and somewhat translucent.
**mericarp**  One of the two parts of the fruit of certain families, especially Umbelliferae.
**-merous**  Having 2 or more parts; -parted.
**midrib**  The central vein of a leaf or other organ.
**monocotyledon**  A plant bearing only 1 cotyledon or seed leaf.
**monoecious**  Having male and female flowers on the same plant.
**mucro**  A stiff, sharp point; plural, mucrones.
**mucronate**  Terminated by a mucro.

**nectar**  A sweet liquid secreted by the nectaries of plants.
**nerve**  A simple or unbranched vein or slender rib.
**neutral**  Of a flower, without stamens or pistils.
**node**  The place on a stem where leaves grow or normally arise; the solid part of a culm.
**nut**  A single-seeded fruit with a woody, hard outer coat (Fig. 9, p. 19).

**obcordate**  Of a leaf, heart-shaped with the attachment at the pointed end.
**oblanceolate**  Of a leaf, much longer than wide, broadest near the tip, and tapering toward the place of attachment (Fig. 3, p. 15).
**oblong**  Longer than broad, having the sides nearly parallel for most of their length (Fig. 3, p. 15).
**obovate**  Of a leaf, egg-shaped having the wide part near the tip (Fig. 3, p. 15).
**obovoid**  Of a plant part, such as a fruit, that is wider near the tip.
**obpyramidal**  Inversely pyramidal.
**obtuse**  Blunt or rounded at the end.
**ocrea**  A loose sheath, composed of 1 or 2 membranous stipules at the base of a leaf stalk.
**opposite**  Borne 2 at a node, on opposing sides of an axis (Fig. 5, p. 16).
**orbicular**  Somewhat circular in outline (Fig. 3, p. 15).
**oval**  Of a leaf, egg-shaped (Fig. 3, p. 15).
**ovary**  The part of a pistil or female organ of a flower containing the cells that become seeds after fertilization.
**ovate**  Of a leaf, egg-shaped with the broad part toward the base (Fig. 3, p. 15).
**ovoid**  Egg-shaped, having the wide part near the point of attachment.
**ovule**  The seed-containing unit of the ovary.

**palea**  The inner of 2 bracts enclosing a grass flower (Fig, 16, p. 53).
**palmate**  Of a leaf, having the shape of a hand with the fingers spread (Fig. 4, p. 15).
**paludose**  Of, or growing in, marshes.
**panicle**  A branched cluster of flowers, each stalked, the lower branches longest and opening first (Fig. 6, p. 17).
**paniculate**  Resembling a panicle.
**papilla**  A minute, nipple-shaped projection.

**papillate**   Bearing papillae; also called papillose.

**pappus**   The bristly or scale-like appendage on fruits of Compositae (Fig. 199, p. 684).

**paraphysis**   A slender sterile filament among the spores of some ferns; plural, paraphyses.

**parasitic**   Growing on, and deriving nourishment from, another living plant.

**pectinate**   Comb-like.

**pedicel**   The stalk of a single flower in a cluster.

**pedicellate**   Having or attached by a pedicel.

**peduncle**   Stem of a solitary flower or of a flower cluster.

**pedunculate**   Having a peduncle.

**peltate**   Shield-shaped; of a leaf, attached to its stalk inside the margin (Fig. 3, p. 15).

**pendulous**   Hanging down.

**pepo**   A berry-like fruit of Cucurbitaceae, with a hard rind and pulpy interior filled with seeds.

**perennial**   A plant that persists for more than 2 years.

**perfect**   Of a flower, complete, having both stamens and pistil.

**perfoliate**   A stalkless leaf with basal portions encircling the stem.

**perianth**   Petals and sepals referred to together.

**perigynium**   The papery sheath that envelopes the fruit in *Carex*; plural, perigynia.

**persistent**   Remaining attached.

**petal**   A separate part of a corolla or inner floral ring, usually brightly colored (Fig. 7, p. 17).

**petiole**   A stalk of a leaf.

**petiolate**   Stalked, having a stalk.

**phyllodium**   A leaf-like petiole with no leaf; plural, phyllodia.

**pilose**   Sparsely pubescent with long straight hairs.

**pinna**   A leaflet or primary division of a pinnate leaf or frond; plural, pinnae.

**pinnate**   Of a compound leaf, or frond, with leaflets (pinnae) arranged on each side of a common axis.

**pinnatifid**   Cleft or parted in a pinnate way.

**pinnatisect**   Cleft pinnately to or almost to the midrib.

**pinnule**   A secondary pinna.

**pistil**   The female part of a flower, composed of style and stigma.

**pistillate**   Having pistils, or female organs; generally used when no male parts are present.

**pith**   The spongy center of the stems of most angiosperms.

**plicate**   Folded lengthwise, as in a fan.

**plumose**   Having fine hairs, resembling a feather.

**pod**   A dry fruit, opening when mature.

**pollinium**   A mass of waxy pollen, as in Orchidaceae; plural, pollinia.

**polygamous**   Having both perfect and unisexual flowers on the same plant.

**pome**   A fleshy fruit, as the apple.

**procumbent**   Trailing along the ground without rooting at the nodes.

**protuberance**   A swelling or bulge.

**puberulent**   Minutely pubescent, downy.

**pubescent**   Covered with hairs.

**punctate**   Dotted, with glandular depressions or colored dots.

**puncticulate**   Dotted with very small dots.

**pyxis**   A capsule, the upper part of which falls off as a lid, as in the fruit of purslane (Fig. 9, p. 19).

**raceme**   A flower cluster with each flower borne on a short stalk from a common stem (Fig. 6, p. 17).

**racemose**   Composed of or resembling racemes.

**rachilla**   Axis of a spikelet (Fig. 17, p. 54).

**rachis**   The axis of a spike or compound leaf (Fig. 17, p. 54).

**radiate**   Having ray florets (Fig. 199, p. 684).

**ray**   In some Compositae, a modified marginal floret with a strap-like extension of the corolla.

**receptacle**   The part of a flower stalk bearing the floral organs; the part of a capitate flower cluster bearing the florets (Fig. 199, p. 684).

**recurved**   Curved backward.

**reflexed**   Bent sharply backward, or downward.

**regular**   Of a flower in which all respective parts are the same size and shape (Fig. 8, p. 18).

**reniform**   Of a leaf, somewhat kidney-shaped.

**reticulate**   Having the appearance of a net.

**retrorse**   Bent or curved over, backward or downward.

**retuse**   Having a shallow notch at an otherwise rounded apex.

**revolute**   Rolled backward or downward.

**rhizome**   An underground, root-like stem; rootstock.

**rhombic**   Having the outline of an equal-sided oblique diamond shape.

**rootstock**   A rhizome.

**rosette**   A dense cluster of leaves on a very short stem or axis (Fig. 5, p. 16).

**rotate**   Of a flower, wheel-shaped (Fig. 8, p. 18).

**rudimentary**   Imperfectly developed and nonfunctional.

**rugose**   Wrinkled, corrugated.

**runcinate**   Coarsely toothed or cut, the pointed teeth turned toward the base of the leaf.

**sac**   A pouch, especially the cavity of an anther.

**sagittate**   Resembling the head of an arrow; of a leaf, having the basal lobes pointing toward the place of attachment.

**salverform**   Of a flower, having a tube with wheel-shaped expansion on top, as in flax and collomia (Fig. 8, p. 18).

**samara**   A winged fruit that does not split open at maturity (Fig. 9, p. 19).

**saprophyte**   A plant that lives on dead organic matter.

**scaberulous**   Somewhat roughened.

**scabrous**   Rough to the touch.

**scale**   A dry and appressed modified or reduced leaf or bract.

**scape**   A flowering stem growing from the root crown and not bearing proper leaves, as in the tulip.

**scapose**   Bearing flowers on a scape.

**scarious**   Thin and dry, not green, as in margins of sheaths or bracts.

**scorpioid**   Of an inflorescence, uncoiling as the flowers develop.

**secund**   One-sided; of flowers that appear to be borne on one side.

**seed**   The ripened ovule, consisting of the embryo and its proper coats.

**sepal**   One of the separate parts of a calyx, usually green and leaf-like.

**sericeous**   Silky hairy.

**serrate**   Having sharp teeth pointing forward.

**serrulate**   Minutely and finely serrate.

**sessile**   Without a stalk (Fig. 5, p. 16).

**sheath**   A long tubular structure surrounding some part of a plant.

**shrub**   A woody plant that remains low and produces shoots or trunks from the base or caudex.

**silicle**   A short silique.

**silique**   A capsule with 2 valves separating from a thin longitudinal partition (Fig. 9, p. 19).

**simple**   Of a leaf, having a single blade not divided into leaflets.

**sinuate**   Wavy-margined.

**sorus**   A cluster of spore cases, as in the ferns; plural, sori.

**spadix**   A dense or fleshy spike of flowers, as in *Calla*.

**spathe**   A large leaf-like bract enclosing a flower cluster.

**spatulate**   Spoon-shaped; of a leaf, having a broad rounded tip, gradually narrowing to the point of attachment (Fig. 3, p. 15).

**spike**   A flower cluster, the individual flowers of which are stalkless, borne on a common stalk (Fig. 6, p. 17).

**spikelet**   A secondary spike, especially in grasses and sedges.

**spinose**   Bearing many spines.

**spinulose**   Bearing small spines.

**sporangium**   A spore case; plural, sporangia.

**spore**   A reproductive body, usually of a single detached cell without embryo, as in ferns.

**sporocarp**   A receptacle containing spores.

**spur**   A hollow projection, usually at the base of a flower, as in the snapdragon.

**squarrose**   Sharply recurved at the tips, spreading at right angles.

**stalk**   The stem of an organ, such as the petiole, peduncle, filament, or stipe.

**stamen**   The male organ of a flower (Fig. 7, p. 17).

**staminodium**   A false stamen.

**staminate**   Having stamens; the term usually used when no female organs are present.

**standard**   The large upper petal (or banner) of a leguminous flower (Fig. 7, p. 17).

**stellate**   Star-shaped.

**steppe**   Native grassland, prairie.

**stigma**   The summit of the style; the part that receives the pollen (Fig. 7, p. 17).

**stipe**   Any short stalk, especially that of the pistil; the petiole of a fern leaf.

**stipitate**   Borne on a stipe or short stalk.

**stipule**   An appendage at the base of a leaf (Fig. 5, p. 16).

**stolon**   Basal branch that roots at nodes, often underground.

**striate**   Marked with fine longitudinal lines or ridges.

**strigose**   Pubescent with appressed straight hairs.

**strobilus**   A cone-like aggregation of sporophylls; plural, strobili.

**style**   The part of the pistil between the stigma and ovary (Fig. 7, p. 17).
**submersed**   Being or growing under water.
**subulate**   Awl-shaped, tapering along its entire length.
**succulent**   Juicy, fleshy; soft and thickened in texture.
**superior**   Of an ovary, above the calyx or hypanthium (Fig. 7, p. 17).
**sympetalous**   Having the petals wholly or partly united.

**tendril**   A slender outgrowth by which some plants attach themselves to objects (Fig. 5, p. 16).
**terete**   Circular in cross section.
**terminal**   At the end of a stem or branch.
**ternate**   In threes.
**tomentose**   Densely hairy with matted wool.
**translucent**   Partly transparent.
**trifid**   Cleft part way into 3.
**trifoliate**   Having 3 leaves (Fig. 4, p. 15).
**trifoliolate**   Having a leaf or leaves of 3 leaflets, as most clovers.
**tuber**   A thick underground branch bearing buds.
**tubercle**   A small tuber; a rounded protruding body.
**tubular**   Hollow and of an elongated or pipe-like form.
**tussock**   A dense tuft or bunch, mostly used of grasses and sedges.

**umbel**   A flower cluster in which all flower stalks arise from a common point (Fig. 6, p. 17).
**umbellet**   A secondary umbel.
**unarmed**   Without spines, prickles, or other sharp appendages.
**unisexual**   Having only male or female organs.
**utricle**   A small bladdery one-seeded fruit, as in amaranth.

**valve**   The units or pieces of a capsule or pod; the enlarged inner sepals in *Rumex*.
**venation**   The arrangement or disposition of veins.
**ventral**   The inside of an organ, facing the axis.
**villous**   Bearing long straight hairs.
**verticil**   A whorl of similar organs, as flowers or leaves, implanted at a node.
**verticillate**   Arranged in verticils.
**viscid**   Sticky.

**whorl**   A group of 3 or more leaves arising from the same node (Fig. 5, p. 16).
**wing**   A thin, dry, or membranous expansion or flat extension or appendage of an organ (see silicle, Fig. 9, p. 19); also the lateral petals of a leguminous flower (Fig. 7, p. 17).
**winter annual**   A plant that germinates in the fall and produces seed and dies in the following spring or summer.
**woolly**   Covered with entangled soft hairs.

**xerophyte**   A plant of very dry habitat.

# The spelling of common names

1.  One word
    (a)  When the modified word is *plant* or a type of plant, as in lead*plant*, skunk*bush*, pea*tree*, pea*vine*, jewel*weed*, lung*wort*; except when the modifier is more than one word, as in Kentucky coffee *tree*, or a hyphened word, as in silk-tassel *bush*, or a proper noun, as in Virginia *creeper*.
    (b)  When the modified word is a part of a plant, as in june*berry*, tway*blade*, buffalo*bur*, cone*flower*, leather*leaf*, bladder*pod*, balsam*root*, bug*seed*, twisted*stalk*, blue*stem*, buck*thorn*, worm*wood*.
    (c)  When the modified word is a part of animal anatomy, as in arrow*head*, blue*lips*, cat*tail*, beard*tongue*; except when the modifier is in the possessive case, as in bird's-*eye*, crane's-*bill*.
    (d)  When the words are figurative or suggestive, as in *beggarticks*, *fairybells*, *meadowsweet*, *paintbrush*; except when the modifier is in the possessive case, as in *baby's-breath*, or when it is a proper noun or the adjectival form of a proper noun, as in *Venus-slipper*, *Indian-pipe*, or when letters demand separation for ease in reading or pronunciation, as in *morning-glory*.

2.  Two words
    (a)  When the modified word is taxonomically correct, as in red *clover* (genus *Trifolium*), alkali *grass* (family Gramineae), fringed *milkwort* (genus *Polygala*), common *plantain* (genus *Plantago*).
    (b)  When the modifier is the word *common*, *false*, *mock*, *wild*, as in *common* camas, *false* flax, *mock* pennyroyal, *wild* chives.
    (c)  When the modifier is a proper noun, as in *Douglas* hawthorn, *Mackenzie's* hedysarum.
    (d)  Exceptions in 1(a).

3.  Hyphened
    (a)  When the modified word is not taxonomically correct, as in sweet-*clover* (genus *Melilotus*, not genus *Trifolium*), whitlow-*grass* (genus *Draba*, not family Gramineae), sea-*milkwort* (genus *Glaux*, not genus *Polygala*), water-*plantain* (genus *Alisma*, not genus *Plantago*), except in a few instances of spelling of long standing, as *burdock*, *buckwheat*.
    (b)  When the modifier is a compound, as in *round-leaved* hawthorn, and whether or not the modified word is taxonomically correct, as in *salt-meadow* grass (family Gramineae), *blue-eyed* grass (genus *Sisyrinchium*, not family Gramineae).
    (c)  In certain three- or four-word groups, as *balm-of-Gilead*, *butter-and-eggs*, *grass-of-Parnassus*, *lily-of-the-valley*, *touch-me-not*.
    (d)  Exceptions in 1(c), 1(d).

# Additions and Corrections to the 1979 Edition

p. 106   *Bromus tectorum*: add to last sentence ". . . and cultivated fields"

p. 123   second and third lines, change "floret" to "spikelet"

p. 256   second line on lead 3 in Liliaceae should read
Leaves variously shaped, not more than 5–6 times as long as wide. ............................................ 12

p. 271   add to lead 11 in Orchidaceae as an extra choice
Leaves 2–5, ovate-lanceolate; flowers greenish. ............................................ *Malaxis*

p. 304   *Parietaria*: first line, change "opposite" to "alternate"

p. 313   *Polygonum*: fourth line in paragraph, change "seeds" to "fruits"

p. 314   *Polygonum*: lead 2, change "fruit" to "outer calyx"

p. 315   Lead 6 should read
6. Achene 1.5–2.0 mm wide; flowers pale pink to white. ...................................... *P. lapathifolium*
Achene 2.0–3.0 mm wide; flowers greenish. ............................................ *P. scabrum*

p. 324   Chenopodiaceae: in last line of lead paragraph, change "one" to "two" (to read "but two species are poisonous")

p. 407   lead 2 should read:
2. Petals longer than 10 mm. ............................................ *E. asperum*
Petals shorter than 10 mm. ............................................ 3

p. 587   bottom full line: change "eastern" to "western"

p. 612   *Lappula*: lead 2 should read
2. Flowers 1.5–3 mm across. ............................................ *L. deflexa* var. *americana*
Flowers 5–8 mm across. ............................................ 3

p. 702   *Taraxacum*: lead 1 in second half should read
Outer involucral bracts reflexed ............................................ 2

p. 734   *Aster junciformis* Rydb.: in fourth line of paragraph, change "cm" to "mm" (to read "15–20 mm")

p. 738   *Bidens*: lead 2 should read
2. Ray florets usually conspicuous; leaves stalkless and clasping. ...................................... *B. cernua*

NOTE: The European–Russian classification of perennial Triticeae (mainly *Agropyron* and *Elymus* species) has come into more general acceptance in North America recently. Because in the European–Russian classification many of the *Agropyron* genus are in the *Elymus* genus, and because the new names are increasingly used in scientific and semi-popular articles, cross-references are given in the following table.

| Traditional North American name | European–Russian name |
|---|---|
| *Agropyron albicans* (Scribn. & Sm.) | *Elymus lanceolatus* subsp. *albicans* (Scribn. & Sm.) D.R. Dewey |
| *Agropyron bakeri* E. Nels. | *Elymus bakeri* (E. Nels.) Löve |
| *Agropyron cristatum* (L.) Gaertn. | *Agropyron cristatum* (L.) Gaertn. |
| *Agropyron desertorum* (Fisch. ex Link) Schult. | *Agropyron desertorum* (Fisch. ex Link) Schult. |
| *Agropyron dasystachyum* (Hook.) Scribn. | *Elymus lanceolatus* (Scribn. & Sm.) Gould |
| *Agropyron elongatum* (Host) Beauvois | *Elytrigia pontica* (Podp.) Holub |
| *Agropyron intermedium* (Host) Beauvois | *Elytrigia intermedia* (Host) Nevskii |
| *Agropyron latiglume* (Scribn. & Sm.) Rydb. | *Elymus alaskanus* (Scribn. & Merrill) Löve subsp. *scandicus* (Nevskii) Melderis |
| *Agropyron repens* (L.) Beauvois | *Elytrigia repens* (L.) Nevskii |
| *Agropyron riparium* Scribn. & Sm. | *Elymus lanceolatus* (Scribn. & Sm.) Gould |
| *Agropyron scribneri* Vasey | *Elymus scribneri* (Vasey) M.E. Jones |
| *Agropyron smithii* Rydb. | *Pascopyrum smithii* (Rydb.) Löve |
| *Agropyron spicatum* (Pursh) Scribn. & Sm. | *Pseudoroegneria spicata* (Pursh) Löve |
| *Agropyron subsecundum* (Link) Hitchc. | *Elymus subsecundus* (Link) A & D Löve |
| *Agropyron trachycaulum* (Link) Malte | *Elymus trachycaulus* (Link) Gould ex Shinn. |
| *Agropyron trichophorum* (Link) Richt. | *Elytrigia intermedia* subsp. *barbulato* (Schur) Löve |
| *Agropyron triticeum* Gaertn. | *Eremopyrum triticeum* (Gaertn.) Nevskii |
| *Elymus angustus* Trin. | *Leymus angustus* (Trin.) Pilg. |
| *Elymus arenarius* L. | *Leymus arenarius* (L.) Hochst. |
| *Elymus canadensis* L. | *Elymus canadensis* L. |
| *Elymus cinereus* Scribn. & Merrill | *Leymus cinereus* (Scribn. & Merrill) Löve |
| *Elymus glaucus* Buckl. | *Elymus glaucus* Buckl. |
| *Elymus hirtiflorus* Hitchc. | considered a sterile hybrid |
| *Elymus innovatus* Beal | *Leymus innovatus* (Beal) Pilg. |
| *Elymus interruptus* Buckl. | *Elymus interruptus* Buckl. |
| *Elymus junceus* Fisch. | *Psathyrostachys juncea* (Fisch.) Nevskii in Komarov |
| *Elymus virginicus* L. | *Elymus virginicus* L. |

# Index to common names

(Page numbers of illustrations are in boldface.)

*Common Names*

*Common Names*

*Common Names*

rye 140
  Italian 140
  perennial 140
salt 119
salt-meadow 166
  Nuttall's 166, **167**
  slender 166
sand 112, **113**
scratch 144
scurvy- 401
sea lime 124
slough 100, **102**
spear 177, **178**
star- 268
  yellow 268
sweet 135, **136**
switch 152, **153**
tumble 166
wedge 174
  prairie 174
  slender 174
whitlow- 403
  alpine 404
  creeping 406
  few-seeded 406
  golden 404
  hairy 405
  thick-leaved 404
  yellow **390**, 405
witch 151
wood 112
  slender 112
wool- 243
grass-of-Parnassus 425
  fringed 425
  glaucous 425
  northern 425, **426**
  small 425
grass-pink 271
  purple 271
greasewood 339, **340**
gromwell 615
  corn 613
  false 618
  western false 618
  woolly 615
ground-cherry 633
  large white 634
  prairie 634
  small yellow 634
  yellow 634
ground-fir 52
ground-ivy 623
ground-pine 52, **54**
ground-plum 19, 469

groundsel 774
  balsam 779
  common 780
  compact 780
  entire-leaved 777
  silvery 775, **776**
  stinking 780
gumweed 758

hackberry 303
hair grass 117
  annual 117
  mountain 117
  rough 92, **93**
  slender 119
  tufted 117, **118**
harebell 18, 678, **679**
hare's-ear mustard 19, **390**, 401
hawk's-beard 691
  dwarf 692
  green 692
  narrow-leaved 694
  scapose 692
  small-flowered 692
  smooth 692, **693**
hawkweed 694
  alpine 695
  Canada 695
  orange 694
  white 694
  woolly 694
hawthorn 440
  Douglas 440
  long-spined 440
  round-leaved 440, **441**
hazel, American 19
  beaked 19
hazelnut 301
  American 301, **302**
  beaked 301, **302**
heads, fiddle 42
heath 571
heather, false 526
  mountain- 573, 576
    blue 577
    purple 577
    western 573
    white 573
    yellow 577
  sand- 526
hedge-hyssop 644
  clammy 644
hedge-nettle 629
  marsh 18, 629, **630**

*Common Names*

*Common Names*

morning-glory 18, 597
  wild **598**, 599
moschatel 677
moss, club- 51
  little club- 55
  mountain 419
  spike- 55
  stiff club- 52
  trailing club- 52, **53**
motherwort 624
  Siberian 625
mountain ash, western 460
mountain-avens 442
  white 442
  yellow 442
mountain-heather 573, 576
  blue 577
  purple 577
  western 573
  white 573
  yellow 577
mountain-marigold 368
mountain moss 419
mountain parsnip 550
mousetail 374
  least 374
mud-purslane 525
mudwort 644
muhly 143
  bog 144
  foxtail 144, **145**
  marsh 146
  mat 146, **147**
  prairie 144
  wood 146
mullein 656
  common 656
  woolly 656
musineon 559
  leafy 559
mustard 18, 389, **390**, 396
  ball **390**, 412
  dog 407
  gray tansy 402
  hare's-ear 19, **390**, 401
  hedge 416
  Indian 397
  narrow-leaved 416
  perennial 414
  short-fruited tansy 402
  tall hedge 416
  tansy 401
  tower **390**, 395
  treacle 407

  tumbling 416
  white 397
  wild **390**, 397
  wormseed 407

naiad 66
  slender 66
nannyberry 675
navarretia 602
  small **601**, 602
nettle 304
  dead- 624
  English 305
  hedge- 629
  hemp- 623
  marsh hedge- 18, 629, **630**
  stinging 305, **306**
  white dead- 624
  wood 304
  yellow hemp- 623
nettletree 303
New Jersey tea 518
nightshade 634
  black 635
  enchanter's- 536
  large enchanter's- 536
  small enchanter's- 536
ninebark 444
  mallow-leaved 444
nipplewort 697
  common 697
nut-grass 234
  awned 235
  Houghton's 235
  sand 235
  straw-colored 235

oak 19, 303
  bur 303
oakfern 41
oat 99, 100
  wild 100, **101**
oat grass 135
  California 115
  Hooker's 135
  one-spike 115, 117
  Parry 115
  poverty 115
  purple 168
  timber 115, **116**
oleaster 533
olive 587
  Russian 533

*Common Names*

*Common Names*

*Common Names*

# Index to scientific names

(Page numbers of illustrations are in boldface; synonyms are in italic type.)

corymbosa 712, 713
dimorpha 711, 713
glabrata 712, 713
howellii 712, 714, **715**
  var. athabascensis 714
  var. campestris 714
  var. howellii 714
lanata 711, 714
luzuloides 711, 714
monocephala 712, 714
neodioica 712, 714, 716
  var. randii 714
*obovata* 714
parlinii 712, 716
parvifolia 712, 716
plantaginifolia 712, 716
pulcherrima 711, 716
racemosa 711, 716
rosea 712, 716
russellii 712, 716
umbrinella 711, 716
Anthemis 708, 717
  cotula 717
  tinctoria 717
*Anthopogon crinitus* 591
*Anticlea elegans* 268
Apocynaceae 31, 592
Apocynum 592
  androsaemifolium 592
    var. incanum 593, **594**
  cannabinum 592
    var. hypericifolium 593
  medium 592, 593
  *sibiricum* 593
Aquilegia 363, 366
  brevistyla 366, 368
  canadensis 368
  flavescens 366, 368
  formosa 368
  jonesii 366, 368
Arabis 392–394, 409
  arenicola 394
    var. arenicola 394
    var. pubescens 394
  divaricarpa 394, 395
    var. dacotica 395
  drummondii 394, 395
  glabra **390**, 394, 395
  hirsuta 394, 395
    var. glabrata 395
  lemmonii 394, 395
    var. drepanoloba 395
  lyallii 394, 395
  lyrata var. kamchatica 394, 395

nuttallii 394, 396
retrofracta 394, 396
  var. collinsii 396
  var. multicaulis 396
Araceae 22, 246
Aralia 544
  hispida 544, 545, **546**
  nudicaulis 544, 545, **547**
  racemosa 544, 545
Araliaceae 27, 544
Arceuthobium 308
  americanum 308
  pusillum 308
Arctagrostis 82, 99
  latifolia 99
Arctium 704, 717
  lappa 717
  minus 717
  tomentosum 717
Arctophila 79, 99
  fulva 99
Arctostaphylos 572
  alpina 572, 573
    var. rubra 573
  uva-ursi 573
Arenaria 348
  capillaris var. americana 348, 350
  congesta var. lithophila 348, 350
  humifusa 348, 350
  laricifolia var. occulta 348, 350
  *lithophila* 350
  lateriflora 348, 350
  macrophylla 348, 350
  nuttallii 348, 350
  *obtusiloba* 350
  peploides var. diffusa 348, 351
  rossii var. columbiana 350, 351
  serpyllifolia 348, 351
  stricta ssp. dawsonensis 350, 351
  verna 350, 351
Arenariae 186
Arethusa 270, 271
  bulbosa 271
Aristida 82, 99
  longiseta 99
Aristolochiaceae 24, 308
Armoracia 391, 396
  rusticana 396
Arnica 705, 706, 718
  alpina var. ungavensis 719
    var. vestita 719
  chamissonis 718, 719
  cordifolia 718, 719, **720**
  diversifolia 718, 719

*Scientific Names*

*Scientific Names*

Hypericaceae 26, 522
Hypericum 522
  canadense 522
  formosum var. nortoniae 522, 525
  majus 522, 525
  virginicum var. fraseri 522, 525
Hypochoeris 689, 695
  radicata 695
*Hypopithys latisquama* 571
Hypoxis 268
  hirsuta 268
Hyssopus 620, 624
  officinalis 624

Iliamna 520
  rivularis 520
Impatiens 508
  biflora 508, 510
  *capensis* 508
  noli-tangere 508, 510
  *occidentalis* 510
Intermediae 187
Iridaceae 22, 268
Iris 268, 269
  pseudacorus 269
  versicolor 269
Isoetaceae 20, 50
Isoetes 50
  bolanderi 50, 51
  echinospora 51
    var. braunii 51
Iva 683, 686
  axillaris 686
  xanthifolia 686, **687**

Juncaceae 22, 74, 246
Juncaginaceae 22, 68
Juncus 247
  albescens 248, 249
  alpinus 249
  arcticus 247, 249
  articulatus 249, 250
  balticus 247, 250
    var. littoralis 250
    var. montanus 250, **251**
  biglumis 248, 250
  brevicaudatus 249, 250
  bufonius 247, 250
  canadensis 249, 250
  castaneus 248, 252
  compressus 248, 252
  confusus 248, 252
  drummondii 247, 252
  dudleyi 248, 252
  ensifolius 248, 252

filiformis 247, 252
longistylis 248, 253
mertensianus 248, 253
nodosus 248, 253
parryi 247, 253
saximontanus 248, 253
tenuis 248, 253
  var. dudleyi 252
torreyi 249, 254
tracyi 248, 254
vaseyi 248, 254
Juniperus 60
  communis 60
    var. depressa 60
    horizontalis 60, 61, **62**
    scopulorum 60, 61
    *sibirica* 60

Kalmia 28, 571, 575
  polifolia 575
*Kentrophyta montana* 472
Knautia 681
  arvensis 681
Kobresia 182, 241
  myosuroides 241
  simpliciuscula 241
Kochia 326, 338
  scoparia 338
Koeleria 81, 137
  gracilis 137, **139**
    ssp. eugracilis 140
    ssp. nitida 140
    var. glabra 140
    var. typica 140
Krigia 690, 695
  biflora 695

Labiatae 30, 31, 620
Lactuca **684**, 690, 695
  biennis 695, 696
  canadensis 696
  floridana 696
  ludoviciana 695, 696
  pulchella 695, 696
  serriola 695, 696
    var. integrata 697
  *spicata* 696
  *virosa* 697
Lamium 621, 624
  album 624
  amplexicaule 624
Laportea 304
  canadensis 304
Lappula 609, 612, 804
  *americana* 612

deflexa var. americana 612, 804
    echinata 612
        var. occidentalis 612
    floribunda 612, 613
    jessicae 612, 613
    *occidentalis* 612
Lapsana 689, 697
    communis 697
Larix 56, 57
    laricina 57
    lyallii 57
    occidentalis 57
Lathyrus 462, 479
    japonicus 479, 480
        ssp. maritimus 480
    ochroleucus 479, 480, **481**
    odoratus 479
    palustris 479, 480
    sativus 479, 480
    tuberosus 479, 480
    venosus 479, 480
        var. intonsus 480
Lavatera thuringiaca 520
*Lavauxia flava* 541
Laxiflorae 190
Lechea 525, 526
    intermedia 526
    stricta 526
Ledum 571, 575
    glandulosum 575, 576
    *groenlandicum* 576
    palustre 575, 576
        var. decumbens 576
        var. latifolium 576
Leersia 83, 140
    oryzoides 140
Leguminosae 27, 28, 462
Lemna 245
    minor 245
    trisulca 245
Lemnaceae 21, 245
Lentibulariaceae 31, 659
Leonurus 621, 624
    cardiaca 624
    sibiricus 624, 625
*Lepachys columnifera* 770
Lepidium 389, 410
    *apetalum* 411
    campestre 410
    densiflorum 410, 411
        var. bourgeauanum 411
    latifolium 410, 411
    perfoliatum 410, 411
    ramosissimum 410, 411
    ruderale 410, 411

sativum 410, 411
Leptarrhena 421, 422
    pyrifolia 422
*Leptilon canadense* 751
*Leptotaenia multifida* 557
Lesquerella 391, 412
    alpina var. spathulata 412
    arctica 412
        var. purshii 412
    ludoviciana var. arenosa **390**, 412, **413**
*Leucanthemum vulgare* 741
*Leucophysalis grandiflora* 634
Levisticum 548, 555
    officinale 555
Lewisia 345, 346
    pygmaea 346
Liatris 705, 766
    ligulistylis 766
    punctata 766, **767**
Lilaea 65, 66
    scillioides 66
    *subulata* 66
Liliaceae 22, 256, 804
Lilium 256, 262
    philadelphicum 262
        var. andinum 262, **264**
*Limnobotrya lacustris* 428
*Limnorchis dilatata* 276
    *viridiflora* 277
Limosae 191, 198
Limosella 638, 644
    aquatica 644
Linaceae 28, 507
Linanthus 600, 602
    *harknessii* var. *septentrionalis* 602
    septentrionalis 602
Linaria 638, 644
    canadensis 645
        var. texana 645
    dalmatica 645
    maroccana 645
    vulgaris 645
Linnaea 670, 671
    *americana* 671
    borealis var. americana 671, **672**
Linum 507
    *compactum* 508
    lewisii 507
    pratense 507, 508
    rigidum 507, 508, **509**
    sulcatum 507, 508
Liparis 271, 277
    loeselii 277
Listera 270, 277
    borealis **272**, 278

empetriformis 576, 577
  glanduliflora 576, 577
Phyllostachyae 186, 189
Physalis 631, 633
  alkekengi 633
  grandiflora 633, 634
  heterophylla 633, 634
  ixocarpa 633, 634
  pubescens 633, 634
  virginiana 633, 634
Physaria 389, 414
  didymocarpa 414
Physocarpus 436, 444
  malvaceus 444
Physostegia 620, 628
  *parviflora* 628
  virginiana var. formosior 628
    var. ledinghamii 628
    var. parviflora 628
Picea 56, 57
  glauca 58
    var. albertiana 58
  mariana 58
*Picradeniopsis oppositifolia* 738
Picris 689, 699
  echioides 699
Pinaceae 21, 56
Pinguicula 659
  macroceras 659, 660
  villosa 659, 660
  vulgaris 659, 660
Pinus 56, 58
  albicaulis 59
  banksiana 58, 59
  contorta 58
    var. latifolia 59
  *divaricata* 59
    var. *latifolia* 59
  flexilis 59
  monticola 59
  *murrayana* 59
  ponderosa 58, 59
  resinosa 58, 60
  strobus 59, 60
*Pisophaca flexuosa* 470
Plagiobothrys 609, 618
  scouleri var. penicillatus 618
Plantaginaceae 30, 663
Plantago 663
  aristata 664
  canescens var. cylindrica 664
  coronopus 663, 664
  elongata 664, 665
  eriopoda 664, 665

lanceolata 664, 665
  major 664, 665, **666**
    var. asiatica 665
  maritima 664, 665
  media 664, 665
  patagonica 664, 667
  psyllium 663, 667
  *purshii* 667
  *spinulosa* 667
*Pleurogyne rotata* 592
Poa 80, 158
  *agassizensis* 165
  alpina 160
  ampla 158, 160
  annua 159, 160
  arctica 158, 159, 161
  arida 158, 161
  canbyi 159, 161, **162**
  compressa 158, 161
  cusickii 159, 161
  epilis 159, 161
  fendleriana 158, 163
  glauca 159, 163
  glaucifolia 159, 163
  gracillima 160, 163
  *interior* 164
  juncifolia 160, 163
  leptocoma 159, 163
  nemoralis 159, 164
    var. firmula 164
    var. interior 164
  nervosa 158, 164
  palustris 160, 164
  pattersonii 159, 164
  pratensis 158, 164
    var. arenaria 165
    var. humilis 165
  rupicola 159, 165
  secunda 159, 165
  stenantha 160, 165
  trivialis 160, 165
Polanisia 386, 389
  dodecandra 389
    var. trachysperma 389
  *graveolens* 389
Polemoniaceae 30, 31, 600
Polemonium 600, 603
  occidentale 603
  pulcherrimum 603, **604**
  viscosum 603, 605
Politrichoideae 186
Polygala 510
  alba 510
  paucifolia 510

uncinatus 375, 381
*verecundus* 380
Raphanus 391, 414
raphanistrum **390**, 414
Rapistrum 391, 414
perenne 414
Ratibida **684**, 707, 770
columnifera 770, **772**
f. pulcherrima 770
Reseda 417
alba 417
lutea 417
Resedaceae 27, 417
Rhamnaceae 29, 518
Rhamnus 518
alnifolius 518
catharticus 518
frangula 518, 519
Rhinanthus 639, 654
crista-galli 654
Rhododendron 572, 577
albiflorum 577
lapponicum 577
Rhus 514
aromatica var. trilobata 515
glabra 514, 515
radicans var. rydbergii 515, **516**
Rhynchospora 182, 241
alba 241
capillacea 241
Ribes 29, 420, 427
americanum 428
aureum 427, 428
diacanthum 428
*floridum* 428
glandulosum 427, 428
*hirtellum* 429
hudsonianum 428
lacustre 427, 428
laxiflorum 427, 429
oxyacanthoides var. oxyacanthoides
427, 429, **430**
var. saxosum 429
rubrum var. propinquum 429
*setosum* 429
*triste* 429
viscosissimum 427, 429
Romanzoffia 605, 607
sitchensis 607
Rorippa 391–393, 414
austriaca 414
*hispida* 415
islandica 415
var. fernaldiana 415
var. hispida 415

*palustris* 415
sinuata 415
sylvestris 415
tenerrima 415
Rosa 437, 455
acicularis 457
arkansana 457
blanda 457
woodsii 457
Rosaceae 27, 29, 436
Rubiaceae 31, 667
Rubus 436, 437, 458
acaulis 458
chamaemorus 458
idaeus var. aculeatissimus 458, **459**
*melanolasius* 458
parviflorus 458, 460
pedatus 458, 460
pubescens 458, 460
*strigosus* 458
Rudbeckia 707, 773
hirta 773
laciniata 773
serotina 773
Rumex 309, 320
acetosa 320, 321
acetosella 320, 321, **322**
confertus 321
crispus 321
dentatus 320, 323
*domesticus* var. *pseudonatronatus* 323
fennicus 321, 323
longifolius 321, 323
maritimus var. fueginus 320, 323
*mexicanus* 324
occidentalis 321, 323
orbiculatus 321, 323
paucifolius 320, 323
salicifolius 321, 324
stenophyllus 321, 324
*triangulivalvis* 324
venosus 320, 324, **325**
Rupestres 186, 190, 196
Ruppia 65, 67
maritima 67
occidentalis 67

*Sabina horizontalis* 61
Sagina 348, 354
caespitosa 355
decumbens 354, 355
nodosa 355
saginoides 355

Schedonnardus 83, 166
  paniculatus 166
Scheuchzeria 68
  palustris 68
Schizachne 79, 168
  purpurascens 168, **169**
Scirpinae 185
Scirpus 182, 242
  acutus 242, 243
  americanus 242, 243
  atrovirens 243
  caespitosus 242
    var. callosus 243
  clintonii 242, 243
  cyperinus 243
  fluviatilis 242, 244
  hudsonianus 242, 244
  microcarpus 243, 244
  nevadensis 242, 244
  paludosus 243, 244
  pumilus 242, 244
  rufus 242, 244
  torreyi 242, 244
  validus 242, 244
Scolochloa 79, 168
  festucacea 168, **170**
Scrophularia 639, 654
  lanceolata 654
Scrophulariaceae 30, 31, 638
Scutellaria 621, 629
  *epilobiifolia* 629
  galericulata 629
  lateriflora 629
  parvula var. leonardii 629
Secale 80, 168
  cereale 168
Sedum 418
  acre 419
  aizoon 419
  hybridum 419
  lanceolatum 419
  rosea 419
    var. integrifolium 420
  stenopetalum 419, 420
  telephium 419, 420
Selaginella 55
  densa 55
  rupestris 55
  selaginoides 55
  wallacei 55
Selaginellaceae 20, 55
Senecio 705, 706, 774
  aureus 775
  canus 775, **776**
  congestus 774, 775

eremophilus 774, 777, **778**
  foetidus 774, 777
  fremontii 774, 777
  indecorus 775, 777
  integerrimus 774, 777
    var. exaltatus 777
    var. integerrimus 777
    var. lugens 777
  megacephalus 774, 779
  *palustris* 775
  pauciflorus 775, 779
  pauperculus 775, 779
    var. firmifolius 779
    var. pauperculus 779
    var. thomsoniensis 779
  resedifolius 774, 779
  streptanthifolius 774, 775, 779
  triangularis 774, 780
  tridenticulatus 774, 775, 780
  viscosus 774, 780
  vulgaris 774, 780
Setaria 83, 168
  glauca 168
  viridis 168, 171, **172**
Shepherdia 534
  argentea 534, **535**
  canadensis 534
*Sibbaldiopsis tridentata* 454
*Sideranthus grindelioides* 759
  *spinulosus* 759
*Sieversia triflora* 444
Silene 347, 355
  acaulis var. exscapa 357
  anthirrina 357
  cserei 357
  cucubalus 357
  menziesii 357, 358
  noctiflora 357, 358
  parryi 357, 358
  sibirica 357, 358
  *vulgaris* 357
Sisymbrium 392, 415
  altissimum 415, 416
  linifolium 415, 416
  loeselii 416
  officinale 415, 416
    var. leiocarpum 416
Sisyrinchium 268, 270
  *angustifolium* 270
  montanum 270
Sitanion 81, 171
  hystrix 171
Sium 548, 555, 563
  *cicutaefolium* 563
  suave 563, **564**

ciliatisepala 359, 360
crassifolia 359, 360
crispa 359, 360
humifusa 359, 360
laeta 359, 360
longifolia 359, 360
longipes 359, 360
media 359, 361
obtusa 359, 361
umbellata 359, 361
Stellulatae 188
Stenanthium 257, 266
occidentale 266
*Stenotus armerioides* 759
Stephanomeria 689, 702
runcinata 702
Stipa 82, 175
columbiana 177
comata 175, 177, **178**
richardsonii 177
spartea 175, 177
var. curtiseta 175, 177
viridula 177, 179, **180**
Streptopus 257, 266
amplexifolius **261**, 266
var. americanus 266
var. denticulatus 266
roseus var. perspectus 266
Suaeda 326, 339
depressa 339
var. erecta 339
*maritima* var. *americana* 339
Subularia 389, 416
aquatica 416
Suckleya 326, 339
suckleyana 339, **341**
Suksdorfia 421, 435
violacea 435
*Svida stolonifera* 565
*Swainsona salsula* 472
Sylvaticae 189, 190, 197
Symphoricarpos 670, 674
albus 674
var. pauciflorus 674
occidentalis 674
*pauciflorus* 674
Symphytum 610, 618
asperum 618
officinale 618, 619

Tanacetum 704, 788
huronense 788
vulgare 788, **789**
Taraxacum 689, 702, 804

ceratophorum 702
*erythrospermum* 702
laevigatum 702
officinale 702
*Taraxia breviflora* 541
Taxaceae 21, 56
Taxus 56
brevifolia 56
canadensis 56
Telesonix 421, 435
jamesii var. heucheriformis 435
*Tetraneuris acaulis* 764
Teucrium 620, 631
canadense var. occidentale 631
Thalictrum 363, 381
dasycarpum 381, 382, **383**
occidentale var. palousense 381, 382
sparsiflorum var. richardsonii 381, 382
venulosum 381, 382
var. lunellii 382
var. turneri 382
var. venulosum 382
Thelesperma 704, 788
marginatum 788
Thellungiella 393, 417
salsuginea 417
Thermopsis 463, 497
rhombifolia 497, **500**
Thlaspi 389, 417
arvense **390**, 417
Thuja 61
occidentalis 61
plicata 61
Tiarella 421, 435
trifoliata 435
unifoliata 435
Tilia 520
americana 520
Tiliaceae 29, 520
*Tium drummondii* 470
*racemosum* 474
Tofieldia 257, 266
glutinosa 266
pusilla 266
Townsendia 707, 788
exscapa 788
hookeri 788
parryi 788, 790, **791**
*sericea* 788
*Toxicodendron rydbergii* 515
*Toxicoscordion gramineum* 268
Tradescantia 246
occidentalis 246

*Scientific Names*

www.ingramcontent.com/pod-product-compliance
Lightning Source LLC
Chambersburg PA
CBHW050212270326
41914CB00003BA/379